COLD ROLLING
OF STEEL

MANUFACTURING ENGINEERING AND MATERIALS PROCESSING

A Series of Reference Books and Textbooks

SERIES EDITORS

Geoffrey Boothroyd

Department of Mechanical Engineering
University of Massachusetts
Amherst, Massachusetts

George E. Dieter

Dean, College of Engineering
University of Maryland
College Park, Maryland

1. Computers in Manufacturing, *U. Rembold, M. Seth, and J.S. Weinstein*
2. Cold Rolling of Steel, *William L. Roberts*

OTHER VOLUMES IN PREPARATION.

COLD ROLLING OF STEEL

William L. Roberts

MARCEL DEKKER, INC.   New York and Basel

Library of Congress Cataloging in Publication Data

Roberts, William L. [Date]
    Cold rolling of steel.

    (Manufacturing engineering and materials processing ; 2)
    Includes indexes.
    1. Rolling (Metal-work) I. Title. II. Series.
TS340.R54   672'.3'2   78-15340
ISBN 0-8247-6780-2

COPYRIGHT © 1978 by MARCEL DEKKER, INC.   ALL RIGHTS RESERVED

Neither this book nor any part may be reproduced or transmitted in any form or by any means, electronic or mechanical, including photocopying, microfilming, and recording, or by any information storage and retrieval system, without permission in writing from the publisher.

MARCEL DEKKER, INC.
270 Madison Avenue, New York, New York 10016

Current printing (last digit):
10 9 8 7 6

PRINTED IN THE UNITED STATES OF AMERICA

Preface

Although greater tonnages of metals are processed by rolling than by any other metalworking technique, relatively few books have been written on this method of deformation. Most of the treatises that have appeared on the subject are out of print, are difficult to obtain, and, of course, do not reflect the contemporary state of the art in rolling technology. Moreover, authors have generally felt impelled to discuss the subject in the broadest terms, thereby treating both the hot and cold rolling of ferrous and nonferrous metals. Naturally this extensive scope limits the depth to which the subject may be examined and, as a consequence, it is unusual to find any references to such topics as roll cooling, rolling lubrication, and shape of the rolled product.

Accordingly, it was my belief that a current book on rolling should be written but that its contents should be examined in as great detail as possible. This book was therefore written with the intent of providing the metalworking student, the mill builder, the rolling mill operator, the user of cold-rolled products, and, indeed, anyone who has an interest in the subject, with as much information as possible on the technology.

As will be noted, the book is basically a synopsis of information published in the technical literature, organized in such a manner as to acquaint the reader with the history of cold rolling, the equipment currently in use, the behavior of the rolling lubricant, the thermal and metallurgical aspects of the subject, mathematical models relating to rolling force and power requirements, the subject of strip shape, and the further processing of cold-rolled steel. For those readers wishing to pursue the subject in even greater detail, copious references are conveniently given as footnotes on the relevant pages.

In manuscript form, this book has been used in teaching courses in cold rolling and rolling lubrication in both the United States and Mexico. In its published form, it is hoped that it will find much more extensive use, not only as a textbook, but also as a reference book and bibliography on cold-rolling technology.

William L. Roberts

Contents

Preface		iii
Chapter 1	The History of Rolling	1
Chapter 2	Various Types of Cold Mills	23
Chapter 3	The Components of Cold Rolling Mills	64
Chapter 4	Mill Rolls and Their Bearings	109
Chapter 5	The Instrumentation and Automatic Control of Cold Rolling Mills	187
Chapter 6	Cold Rolling Lubrication	243
Chapter 7	Thermal Aspects of the Cold Rolling Process	332
Chapter 8	The Alloys of Iron, Their Physical Nature, and Behavior During Deformation	398
Chapter 9	Mathematical Models Relating to Rolling Force	478
Chapter 10	Torque Equations and Tandem Mill Control Models	568
Chapter 11	Strip Shape: Its Measurement and Control	655
Chapter 12	The Rolled Strip—Its Properties and Further Processing	717
Name Index		777
Subject Index		786

Chapter I   The History of Rolling

1-1 The Early History of Rolling

In its earliest beginnings, the rolling of flat materials was undoubtedly limited to those metals of sufficient ductility to be worked cold, and it is probable that it was first performed by goldsmiths or those manufacturing jewelry or works of art.* Yet, as is the case with many other important processes, metal rolling cannot be traced to a single inventor.† □

During the fourteenth century, small hand-driven rolls about half an inch in diameter were used to flatten gold and silver and perhaps lead. However, the first true rolling mills of which any record exists were designed by Leonardo da Vinci in 1480. (See Figures 1-1 and 1-2.) Sketches in his notebook show two mills, driven by worm gears, for rolling lead sheets and also a machine for producing tapered lead bars by means of a die and a spiral roll. Yet, there is no evidence that these mills were ever built, and there is a fair degree of certainty that metal rolling was not of any importance before the middle of the sixteenth century.

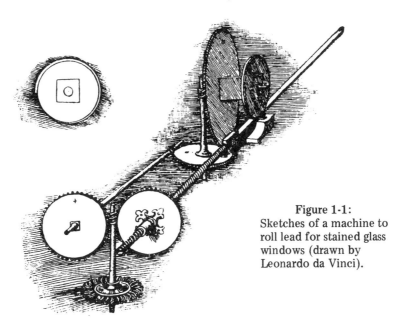

Figure 1-1:
Sketches of a machine to roll lead for stained glass windows (drawn by Leonardo da Vinci).

Before the end of the sixteenth century, however, at least two mills embodying the basic ideas of rolling are known to have been in operation. A Frenchman named Brulier in 1553 rolled sheets of gold and silver to obtain uniform thickness for making coins and mills for rolling mint flats were in use in 1581 at the Pope's mint, in 1587 in Spain, and in 1599 in Florence.□ In 1578, Bevis Bulmer received a patent for the operation of a slitting mill which consisted of a series of discs mounted on two spindles, one above the other, in such a manner that the flat bar passing between the revolving discs was cut into strip. A mill of this type was set up at Dartford, in Kent, in 1590, by Godefroi de Bochs, a native of Liege, Belgium.

During the same period, lead was also beginning to find increasing use for roofing, for flashing, for the fabrication of gutters, and for other purposes. Salomon de Caus of France, in 1615,

---

* D. Eppelsheimer, "The Development of Continuous Strip Mills", Journal of the Iron and Steel Institute, 1938, No. II, pp. 185-303.
† "Rolling Mills, Rolls and Roll Making", Mackintosh-Hemphill Company, Pittsburgh, Pennsylvania, 1953.
□ J. R. Adams, "Hardened and Ground Rolls", Yearbook of the AISI, 1924, pp. 115-147, AISI, New York, 1925.

built a hand-operated mill for rolling sheets of lead and tin used in making organ pipes, the rolls being turned by a "strong-armed cross" attached to the lower axle as illustrated in Figure 1-3.*

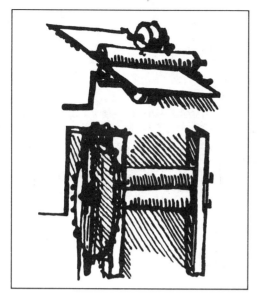

Figure 1-2:
Sketch of a Rolling Mill by Leonardo da Vinci about A.D. 1495.

Figure 1-3:
Mill Built by Salamon of Caus to Roll Sheets of Lead.

With the exception of the Bulmer slitting mill, mentioned above, all of these early developments pertain to the rolling of the softer metals, presumably at ambient temperatures. Johannsen, in "Geschichte des Eisens" says, "The use of rolls in an iron works was a German development of the 16th century. Belgium and England both started to use rolls about the same time, and they are both sometimes cited as the birthplace of rolling." All three nations probably reached this development at about the same time, but there is little evidence of anything other than slitting mills in the 16th century, and still less evidence to give any nation a clear claim to priority. Such information as is available indicates that, in the rolling of iron, Great Britain led the way. No record of developments during the first half of the 17th century exists, but we know that in 1665 a rolling mill was in operation in the Parish of Bitton, near Bristol, and it is stated that, from 1666 on, iron was rolled into thin flats for slitting.†

The rolling of bars was foreshadowed during this period but was not brought to fruition. In 1679 a patent was issued covering the finishing of bolts by rolling, and in 1680 bars were being passed through plane-surfaced rolls to flatten out irregularities. A pamphlet on the "British Iron Trade" published in 1725 states, however, that even at that date all bars were hammered.†

However, by 1682, large rolling mills for the hot rolling of ferrous materials were in operation at Swalwell and Winlaton, near Newcastle, England. Using these mills, bars were rolled into sheets and the sheets cut into rods at the slitting mills. Soon after this date, at Pontypool in Wales, John Hanbury began using at his ironworks a rolling mill as an independent machine for the production of thin sheet iron. Edward Llwyd, in a letter dated June 15, 1697 wrote, "One Major Hanbury of this Pontypool shew'd us an excellent invention of his own, for driving hot iron (by the help of a rolling machine mov'd by water) into as thin plates as tin . . . They cut their common iron bars into pieces of about 2 ft. long, and heating them glowing hot, place them betwixt these iron

---

* D. Eppelsheimer, "The Development of Continuous Strip Mills", Journal of the Iron and Steel Institute, 1938, No. II, pp. 185-303.

† "Rolling Mills, Rolls and Roll Making", Mackintosh-Hemphill Company, Pittsburgh, Pennsylvania, 1953.

THE HISTORY OF ROLLING

rollers, not across, but their ends lying the same way as the ends of the rollers. The rollers, moved with water, drive out these bars to such thin plates, that their breadth, which was about 4 in., becomes their length, being extended to about 4 foot, and what was before the length of the bars is now the breadth of the plate." *

Although Major Hanbury designed the rolling mill described in the letter and illustrated in Figure 1-4, there is no evidence that he originated the idea of hot-rolling bars into thin sheets since it is believed that the practice was general throughout Europe by 1660 being known in Germany very early in the century. At any rate, Germany monopolized the growing English market for tinplate from shortly after 1620 until Major Hanbury began tinplate manufacture in this same Pontypool works sometime before 1720.■ After that, for more than 150 years, Wales was the major source of tinplate and terne plate.

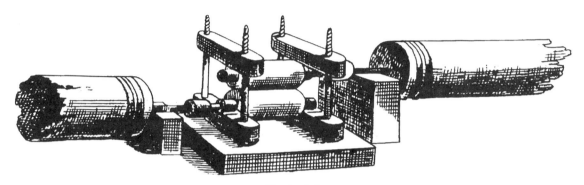

Figure 1-4:
John Hanbury's Mill for Rolling Iron Sheets for the Manufacture of Tinplate (Sketch by Angerstein in 1755).

During the early part of the eighteenth century, there is no doubt that rolling mills were in common use both in England and on the continent. Christopher Polhem (1720 – 1746), Sweden's great mechanical genius, wrote of rolling mills at about this time and his writing indicates that he assumed his readers were familiar with them. Polhem himself designed a mill very similar to the modern Lauth mill, except that his mill utilized four rolls, with the backup rolls driven.

A "machine for rolling sheets of lead", that really foretold the shape of subsequent rolling mills, was brought to France from England in 1728. This mill, shown in Figure 1-5, used rolls five feet long and twelve inches in diameter and was equipped with a roller table 24 feet long at front and back. It was a radical departure from the accepted mill design of the day, being a reversing mill, controlled by a clutch and gearing system. The plain rolls could be replaced with other, grooved rolls sixteen inches in diameter, with grooves ranging from four to two inches in diameter, with which hollow cast lead ingots were rolled into pipe over a mandrel.

In 1728, a patent for a mill to roll hammered bars "into such shapes and forms as shall be required" was issued to John Payne in England. However, Payne's concepts (shown in Figure 1-6) do not appear to have been reduced to practice, but the rolling of iron bars and shapes was of interest to steelmakers and appears to have been practiced. For example, in 1747, the Academie des Sciences appointed a commission to visit a new mill at Essonne, France, which rolled iron bars.† By this time, the practice of polishing iron plates for tinning by cold rolling was also in vogue.□ In 1759, a patent was granted to Thomas Blockley of England for "polishing and rolling metals" — a

---

* See also, Thomas Turner's, "The Metallurgy of Iron", published by Charles Griffin and Company of London in 1920.
■ F. H. Fanning, "Wide Strip Mills -- Evolution or Revolution", Yearbook of the AISI, 1952, pp. 194-221.
† "Rolling Mills, Rolls and Roll Making", Mackintosh-Hemphill Company, Pittsburgh, Pennsylvania, 1953.
□ J. R. Adams, "Hardened and Ground Rolls", Yearbook of AISI, 1924, pp. 115-147.

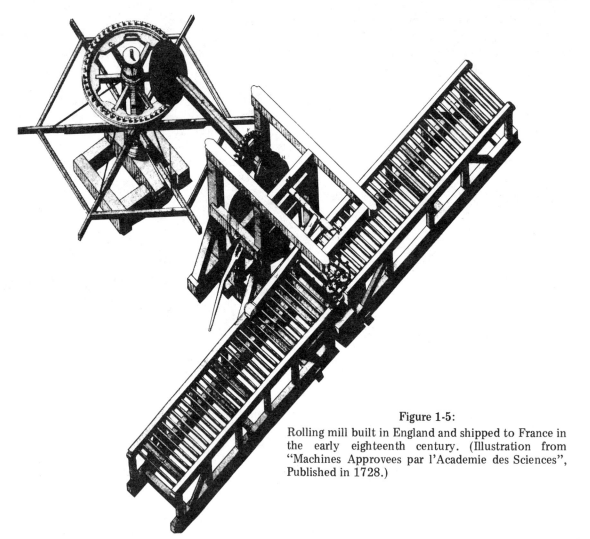

Figure 1-5:
Rolling mill built in England and shipped to France in the early eighteenth century. (Illustration from "Machines Approvees par l'Academie des Sciences", Published in 1728.)

Figure 1-6:
Rolls designed by John Payne in 1728 to produce round bars. (Illustration from "Machines Approvees par l'Academie des Sciences". Edition published in Paris in 1728.)

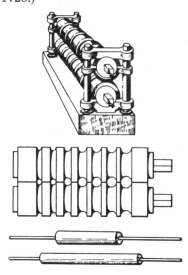

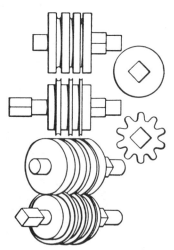

Figure 1-7:
English patent issued to John Purnell in 1766 for grooved mills driven in unison by coupling boxes and nut pinions. (Staffordshire Iron and Steel Institute Proceedings.)

THE HISTORY OF ROLLING

broad description which really boiled down to rolls which the user could groove to suit his requirements; and in 1766, another Englishman, John Purnell received a patent for grooved rolls with coupling boxes and nut pinions for turning the rolls in unison. (See Figure 1-7.) Until this time, rolls were individually driven, and the unequal rates of revolution caused excessive roll wear, as well as making it necessary to install guides on both sides of each roll.

In this same period, the general appearance of these hot mills was beginning to change to the modern form. For example, the cast housing and the single screw each side of the mill were featured in English patent specifications filed by William Playfield in 1783.† (See Figure 1-8.)

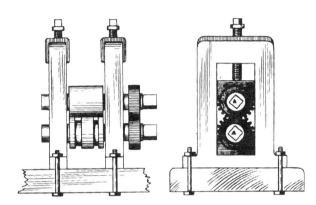

Figure 1-8:
English patent specifications filed by William Playfield in 1783. (Staffordshire Iron and Steel Institute of Proceedings.)

In the manufacture of hand-forged plates for tinplating, it had been the practice to hammer several layers of metal at one time. Accordingly, when the rolling mill came into use for producing sheets for tinning, it became the practice, about 1756, to double the sheet after some elongation and then proceed with the further rolling of the two thicknesses of metal. Sometimes in this "pack rolling" or "ply rolling" the doubling was repeated so that four thicknesses of steel were rolled simultaneously.

The aforementioned hot-rolling mills consisted essentially of a single-stand of two rolls, one above the other. As described by Shannon,* "the operating principle of this type of hot mill consists briefly of the following steps: heat the piece, pass the piece through the rolls, push the piece back over the top roll with hand tongs, pass the piece through the rolls again, and so on until the piece being rolled either is of the required thickness or else has cooled down to the point where it must be reheated before the rolling can continue. This is only a bare outline of the elements of the operation, which is very much further complicated by matching (putting two or more sheets together), doubling the pack, adjustments of the rolls, and various other factors, according to the nature of the sheets being rolled. Owing to the fact that the piece is fed to the rolls by the roller, passes through the rolls, is caught on the other side and handed back to the roller for passing through the rolls again, and since there are only two rolls, one above the other in the set, this conventional sheet hot mill or sheet mill is described as a "two-high, pull-over mill". More than one of these mills may be employed in the rolling unit, each performing a separate stage of the work, but the principle of each remains the same."

The eighteenth century also saw the advent of the tandem mill in which the metal is rolled in successive stands. The first true tandem mill of which we have record was patented in England by Richard Ford, in 1766, for the hot-rolling of wire rods. James Cockshutt and Richard Crawshay, about 1790, erected a four-high tandem mill near Sheffield, England. This mill was about five feet in length and less than two feet high, with a capacity of, probably, one or two tons per day. A later patent, issued in 1798, refers to a tandem mill for the rolling of plates and sheets. In the same year, John Hazeldine added mechanical guides to a rod mill, as shown in Figure 1-9.

---

† "Rolling Mills, Rolls and Roll Making", Mackintosh-Hemphill Company, Pittsburgh, Pennsylvania, 1953.
* R. W. Shannon, "Sheet Steel and Tinplate", The Chemical Catalog Company, Inc., New York, New York, 1930.

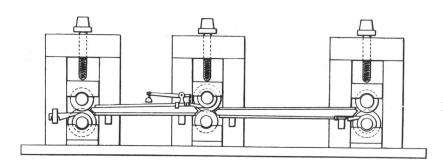

Figure 1-9:
Early English Tandem Mill Using Guides.

1-2 Later Developments in the Hot Rolling of Steel in Europe

The advent of modern rolling practice may, however, be said to date from 1783 when a patent was granted to Henry Cort of Fontley Iron Mills, near Fareham, England, for utilizing grooved rolls for rolling iron bars. A mill with rolls of this design could produce at least 15 times the output per day obtainable with a tilt hammer. However, the claim to innovation put forward by Cort and his successors was strenuously contested at a later date. He was not the first to use grooved rolls, but he was the first to combine the use of all the best features of the various steelmaking and shaping processes known at that time. This fact alone justifies the term "father of modern rolling", which has been applied to him by modern writers.

In the beginning of the nineteenth century, the industrial revolution in England was gathering momentum, creating an unprecendented demand for iron and steel. Accordingly, rolling mill developments were numerous and important. John Birkenshaw started the first rail rolling mill in 1820 producing fish-bellied wrought iron rails in lengths of 15 to 18 feet. In 1831 the first T-rail (of the same basic design as that used today) was rolled in England and the first I-beams were rolled by Zores in Paris in 1849.

Both the sizes of the mills and the sizes of rolled product grew rapidly. At the British Great Exposition of 1851, a plate 20 feet long, 3-1/2 feet wide, and 7/16 of an inch thick was exhibited by the Consett Iron Company. This plate weighed 1,125 pounds and was the largest plate rolled up to that time.

Three-high mills were also introduced about the middle of the century. A British patent for such a mill designed for rolling heavy sections was granted in 1853 to R. B. Roden of the Abersychen Iron Works. In this mill, the middle roll was driven and fixed in the housing while the upper and lower rolls were adjustable in position. On the same mill, a steam-operated lifting table raised and lowered the material to be rolled. This design was improved on a few years later by Bernard Lauth who used a middle roll of smaller diameter than the upper and lower rolls, as illustrated in Figure 1-10. This modification to the mill provided it with a higher productivity with less power utilization.

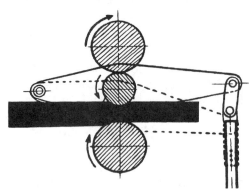

Figure 1-10:
Lauth's Mill with Smaller Middle Roll, (From Staffordshire Iron and Steel Institute Proceedings).

THE HISTORY OF ROLLING

In mid century, the first reversing plate mill was put into operation at the Parkgate Works in England, and in 1854 it was used to roll the plates for the "Great Eastern" steamship. In 1848, the universal mill was invented by R. M. Daelen of Lendersdorf, Germany, who built the first mill of this type about seven years later. Although patents for "continuous" hot mills were issued to Sir Henry Bessemer in 1857 and to Dr. R. V. Leach in 1859, "the first mill constructed on the continuous principle of rolling iron or steel" was the subject of a patent issued to Charles While of Pontypridd, Wales.

However, an apparently more successful continuous mill was patented in 1862 by George Bedson of the Bradford Iron Works at Manchester, England, in which he claims the employment of a series of rolls placed at varying angles, whereby the necessity of turning the metal is avoided.◆ This was a rod rolling mill in which a 100-pound billet of 1-1/16-inch square cross section was drawn through 16 pairs of rolls in line, 8 horizontal and 8 vertical as illustrated in figure 1-11. Its production rate was such that 20 tons of No. 5 iron wire rods could be rolled in 10 hours.

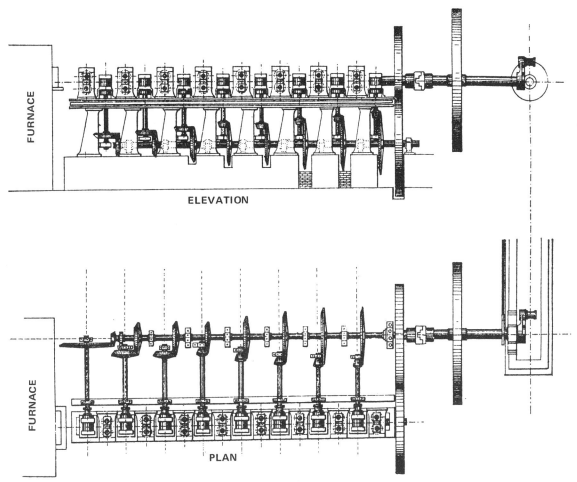

Figure 1-11: The First Continuous Wire Rolling Mill Developed by George Bedson, 1862.

A British patent issued in 1862 to J. T. Newton of Ystalyfeva, Wales, described a predecessor to the modern cluster mill, in as much as it used small work rolls backed up by others of larger diameter. The work rolls were driven but the pressure was applied by the large backup rolls, a principle utilized in both the hot and cold mills of today.

◆ J. P. Bedson, "Continuous Rolling Mills: Their Growth and Development", Journal of the Iron and Steel Institute, 1924, No. I, Volume CIX, pp. 43-66.

The four-high mill with its rolls in the same vertical plane was introduced by Bleckley of Warrington, England, in 1872, to finish wrought-iron piles from which rails were rolled. Mills to produce Z-bars were put in use in Germany in 1863, and in 1867, beams 8 to 12 inches deep were rolled on a mill designed by Menelaus, of the Dowlais Works in Wales. This mill contained two pairs of rolls, one pair placed in a vertical plane, somewhat higher than the other pair of rolls. Petin, Gaudet et Cie, Rive-de-Giev, France, was rolling beams on a universal mill in 1872, and four years later, Joseph de Buigne, of France, rolled the first H-beams produced on a continuous mill, utilizing the diagonal method of rolling.

Tandem rolling of hot steel took an upsurge around 1890, and in 1892, a semi-continuous hot strip mill, with a mechanically geared two-high tandem finishing train, was built at Teplitz, Bohemia.+ It was reported to have rolled sheets up to 50 inches in width, in thicknesses from 0.080 inch to 0.120 inch and in lengths up to 60 feet. The mill utilized a roughing train of two three-high stands and a finishing train consisting of five stands of 24-5/8 inch by 59-inch rolls spaced on 9-foot centers. Each train was powered by a 1,000-HP engine as illustrated in Figure 1-12. Since the works at Teplitz were abandoned in 1907, it is to be assumed that the mill was not a commercial success.□

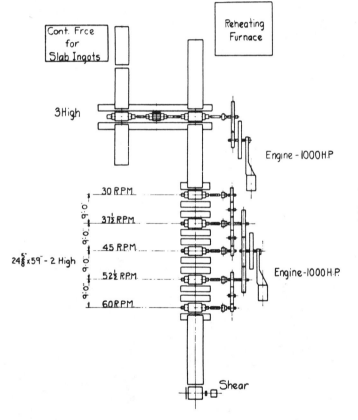

Figure 1-12: Continuous Sheet Mill Built at Teplitz, Bavaria.

1-3 The Early Rolling of Steel in the United States

To all intents and purposes, the history of metalworking in the United States began with the arrival of colonists from Europe. Since skilled metalworkers were present in every colony, the colonists supplied a great part of their own needs for metals. However, in 1750, the British Parliament decreed that "no mill or other engine for slitting or rolling iron, or any plating forge to

---

+ S. Badlam, "The Evolution of the Wide Strip Mill", Yearbook of the AISI, 1927, pp. 382-444.
□ M. D. Stone, "Reversing and Tandem Cold Mills", Iron and Steel Engineer Year Book, 1947, pp. 265-272.

work with a trip hammer, or any furnace for making steel" should be built in the American Colonies. This law was generally disregarded so that by 1775, the colonies were producing 30,000 tons of iron per year, only one third of which was exported to England as pig iron and bar iron.

The first American rolling mill was built in 1751 for Peter Oliver, one of the Crown Judges in the province of Middlesboro, Massachusetts.‡ It was used to roll down 3-inch wide hammered bars made at the charcoal forges from a thickness of about 3/4-inch to 1/4-inch suitable for slitting into nail rods in 4 passes. The rolls were each driven by an undershot waterwheel 18 feet in diameter with a face length of 10 feet through cogwheels and the speed of the rolls could be matched by adjustment of the gates in the mill streams. The chilled iron rolls were 36 inches long by 15 inches in diameter and were designed with roll necks 9 inches in diameter.

At the outbreak of the American Revolution, the colonies possessed a flourishing iron industry from which all restrictions were ended with the establishment of independence. But it did force competition from a European iron industry with its century and a half of experience. Undaunted and unhampered by tradition, the Americans succeeded, however, within the next two centuries to develop the largest national steel industry in the world.

Some of the more notable steps in the development of the new steel industry were as follows. Isaac Pennoch established a slitting mill on Bush's Run (near Coatesville, Pa.) in 1793, and by 1810 this plant was rolling plates with mills using rolls 16 to 18 inches in diameter and 3 to 4 feet long, driven by an over-shot waterwheel. (See Figure 1-13.) In 1820, Dr. Charles Lukens, Pennoch's son-in-law, rolled boiler plates for the first time at this plant which eventually developed into the present Lukens Steel Corporation.

Figure 1-13:
The First Boiler Plate Rolling Mill in America, Built by Lukens Iron & Steel Co., Coatesville, Pa. ("Fifty Years of Iron and Steel", Butler.)

However, at the beginning of the nineteenth century, it was apparent that the Pittsburgh area was becoming the focal point of the industry. Christopher Cowan built the first rolling mill in Western Pennsylvania, and, incidentally, the first one known to have been powered by steam. It utilized a 70-HP steam engine which also supplied power for a slitting mill and a tilt hammer. At Plumsock (about midway between Connellsville and Brownsville, Pa.) Isaac Meason, in 1816, built the first American mill for puddling iron and rolling flat bars. Two years later, the Pittsburgh Steam Engine Company built a sheet rolling mill, and in 1819, the first angle iron rolled in the United States was produced at the Union Rolling Mill in Pittsburgh. By 1825 five rolling mills were in operation in Pittsburgh and a sixth was under construction.

By the middle of the nineteenth century, iron production in the USA had risen to 350,000 tons per year and the increasing availability of metal fostered considerable inventiveness designed to further process and fabricate metal parts. The rolling of corrugated plates was patented

---

‡ Journal of the Iron and Steel Institute, Volume XVIII, No. I, 1881, pp. 271-272.

in 1850, and in the same year an Ohioan patented an improved machine for rolling irregular forms of metal such as strip hinges, plane irons, elliptic springs, socket chisels, axle trees, shovels, axes, hammers, and spades. (See Figure 1-14.)

The first three-high beam mill in the U. S. was used by the Trenton Iron Works, at Trenton, N. J. Built in 1852, it employed three vertical rolls and the iron was reduced in each pass, in each direction. Another beam mill departing radically from previous designs was supposed to have been built in 1853 by Charles Hewitt of Cooper, Hewitt and Co., Trenton, N. J. However, the first thoroughly satisfactory three-high mill is generally attributed to John Fritz, who built such a mill for the Cambria Iron Works, Johnstown, Pa. in 1857 to roll rails between 18-inch diameter rolls. (See Figure 1-15.) This mill is of interest because it established the practice of mounting the mill housings on heavy, cast guide rails. However, Jones and Lauth of Pittsburgh, Pa. introduced the Lauth mill to the U. S. in 1859 buying all U. S. rights from the British patent owner.

Figure 1-14:
Machine for rolling irregular forms. (Steel Engraving from Scientific American, 1850.)

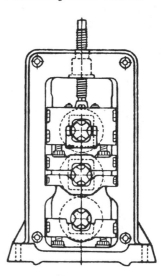

Figure 1-15:
Three-high mill, built by John Fritz. (Drawing in Staffordshire Iron and Steel Institute Proceedings.)

After the Civil War ended in 1865, the rapid expansion of the railroads proved to be a tremendous stimulus to the American iron and steel industry. The first steel rails were rolled in 1865 by the North Chicago Rolling Mill Company. In 1867, George Fritz (brother to John Fritz) started the first successful blooming mill operation in the U. S. and, in the same year, the first beams were produced in Pittsburgh on a 20-inch structural mill. Also in the same year, Andrew Kloman, with the help of John Zimmer, built the first universal mill in Pittsburgh capable of rolling plates from 7 to 24 inches wide and from 3/10 inch to 2 inches thick. In 1877, Mackintosh-Hemphill designed and installed a 30-inch reversing blooming mill for Schoenberger and Company (later the Schoenberger Works of the American Steel and Wire Company). This was the first mill of its kind in the Pittsburgh area and possibly the first in the United States. In 1881, the firm of Mackintosh-Hemphill, Inc. built the first rolling mill of wholly American construction, this being a two-high reversing blooming mill at the Pittsburgh Bessemer Works (the forerunner of the present Homestead Works of the U. S. Steel Corporation).

The rolling of sheet steel in the United States began around 1880 using mills similar to that shown in Figure 1-16, and it is recorded that a three-high mill with 84-inch long rolls at the Brandywine Rolling Mill (later part of Lukens Steel Corp.) was used for this purpose. Sometime later, the same company installed a three-high roughing mill and a three-high finishing mill, both using chilled iron rolls 34 inches in diameter and 120 inches long.

THE HISTORY OF ROLLING

Figure 1-16:
Typical Sheet Mill of the 1860's.

1-4 Energy Sources for Rolling Mills

The earliest mills, as has been seen in Section 1-1, were operated by handpower, usually by turning a strong-armed cross or an adjustable crank attached to either or both of the rolls. With such limited power available, the only materials that could be rolled were the softer metals such as gold, silver, tin, and lead.

Waterwheels were next used to turn mill rolls, and this was a convenient development since such wheels were widely used in iron works for the operation of bellows. The first mill likely to have been powered by a waterwheel was the slitting mill built at Dartford in Kent, England, in 1590 by Godefroi de Bochs under a patent granted in 1588 to Bevis Bulmer.[†] In the four-high mill erected near Sheffield, England, about 1790, by James Cockshutt and Richard Crawshay, the top and bottom rolls were each driven by separate waterwheels, which were weighed by a heavy stone rim bolted together in segments to act as a flywheel. Figure 1-17 illustrates the use of an undershot waterwheel to drive a plate mill in 1734.

Figure 1-17:
Undershot Waterwheel Driving a Plate Mill.

---

† "Rolling Mills, Rolls and Roll Making", Mackintosh-Hemphill Company, Pittsburgh, Pennsylvania, 1953.

The first commercially successful steam engine was invented in 1698 by Thomas Savery of England, who, incidentally, was the first to evaluate an engine in terms of horsepower. However, his steam engine was not first used to drive the mill rolls but to pump back into the reservoir the water that had already passed over the waterwheel. However, a Boulton and Watt steam engine was used to power a rolling and slitting mill at John Wilkinson's Bradley Works and a steam engine was used to power a tinplate rolling mill in 1798. (See Figure 1-18.)

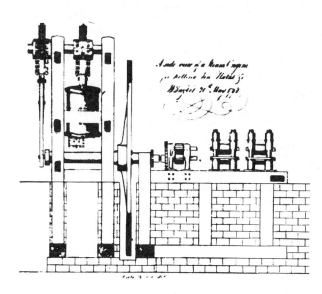

Figure 1-18:
Earliest drawing of a steam engine to operate a tinplate rolling mill at Lower Redbrook, 1798.

Improvements to steam engines occurred rapidly at the beginning of the nineteenth century and they were soon commonly used for driving mills, the power being transmitted to the rolls by direct mechanical connections through shafts, couplings, and gears. In the latter half of the nineteenth century, there was a constant demand for larger and larger engines so that by 1875, engines were being built capable of delivering in excess of 1,000-HP, some being as large as 3,000 to 4,000-HP units (Figure 1-19).

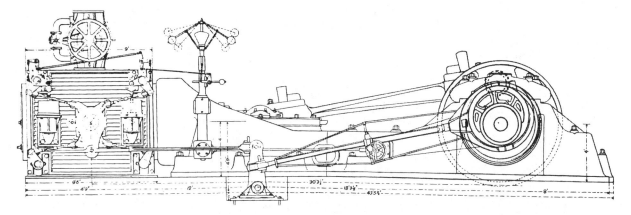

Figure 1-19:
A valve-side view of the engine designed for Carnegie, Phipps & Company's Homestead Steel Works in 1890. Largest mill engine in America, it was 43-1/2 feet long, had a 27-foot flywheel weighing 90 tons, and developed 3,500 Horsepower. (Iron Age, December, 1890.)

Initially flywheels found extensive use in powering of mills, particularly after the development of 3-high mills (such as the Lauth mill) which required no reversals. It became apparent, however, towards the end of the nineteenth century, that two-high mills with drives

THE HISTORY OF ROLLING

capable of rapid reversal were to be preferred, especially in view of the fact that they could be quickly stopped if necessary.⊕ ‡

Yet steam power, at its best, was costly and inefficient and it was fortunate that the latter part of the nineteenth century saw the development of electric generators and motors. With the generators driven remotely from the mills, electric power could be conveniently transmitted over wires to motors directly attached to the mills. Some of the generators were driven by internal combustion engines as, for example, at the Gary Steel Works. This plant was designed in 1908 to be the first sizable steel mill built for the use of electric power and it had 15 gas-driven generators each capable of outputs up to 2,000 kilowatts. The size was increased a few years later to 3,000 kilowatts equalling the largest reciprocating steam-driven generator at that time.

Even earlier, direct current motors had been installed to operate some smaller mills. In 1903, two 1500-HP motors powered a light rail mill at the Edgar Thomson Works at Braddock, Pa., and the first reversing d-c main drive motor was installed the same year on a 36-inch universal plate mill at South Works in Chicago.

Both gas and steam engines operated at relatively slow speeds which put a physical limitation on generators. For example, 5,000 kw generators so powered had diameters in excess of 30 feet. On the other hand, steam turbines operating at high rotational speeds enabled generators driven by them to be much more compact.

Other improvements in the generation and distribution of electrical power led to a steady conversion to electric motors in steel mills not only in the USA but throughout the world. The use of variable speed d-c motors on main drives began in the early 1940's and has been gaining popularity ever since. At the same time, the power utilized by mills has gradually increased so that some of the more recent hot mill stands are driven by multi-armature motors providing 12,000 horsepower usually enclosed in dust free motor rooms§ (see Figure 1-20). In the case of modern

Figure 1-20: Motor Room of Modern Hot-Strip Mill Showing Two Triple-Armature Motors In the Foreground.

⊕ Journal of the Iron and Steel Industry, Volume XVII, No. II, 1880, pp. 652-653.
‡ Journal of the Iron and Steel Institute, Volume LV, No. I, 1899, pp. 395-399.
§ V. E. Verheyden, "New Concepts in Power-Control Rooms For Steel Mill Drive Systems", Iron and Steel Engineer Year Book, 1961, pp. 711-717.

cold reduction facilities, the stands of wide sheet mills are typically powered by motors of 8,000-HP, which are generally located on the open mill floor rather than in specially-built motor rooms.

1-5 The Historical Development of Cold Rolling

In spite of the fact that the first rolling of metals was, in effect, cold rolling, the flat, cold reduction of iron and steel does not seem to have been successfully carried out until the end of the eighteenth century, although the flat, hot rolling of steel had been undertaken on 2-high mills since about 1660. However, it should be noted that cold rolling, in the form of a planishing operation, was practiced on tin plate in England as early as 1747 and, in 1783, in the same country, John Westwood proposed the cold reduction of steel bands for watch springs. From 1825 to 1860, due mainly to improvement in the manufacture of rolls, considerable amounts of high-carbon flat wire, corset stays, etc., were produced by cold rolling.□

When the first cold-rolling operations were undertaken in the U.S.A. is uncertain. It appears, however, that the flattening of wire carried out by the Washburn and Moen Company, in Worcester, Massachusetts (later the Worcester Works of the U. S. Steel Corporation) constituted the first commercial operations of this type.▫

The development of the flat, cold-rolling of steel as a production process, however, gained real impetus only after the evolution of the Lauth 3-high cold mill with its smaller diameter middle (work) roll. Yet the advantages of smaller diameter work rolls had been recognized much earlier for Christopher Polhem described a 4-high mill for flat, hot rolling using "slender" wrought-iron work rolls, supported by large cast-iron backing rolls, because "small rolls possessed much more power of stretching (elongating) material than did large ones". The commercialization of the Lauth cold mill, however, as discussed in Section 1-4, was carried out principally in America by the old American Iron & Steel Company of Pittsburgh, later acquired by Jones & Laughlin.

As the superior properties of cold rolled strip became more and more appreciated, cold rolling spread even more widely both in this country and abroad, being practiced primarily on 2-high and double 2-high mills, although 4-roll and 6-roll cluster mills of the Wilmot and Mann types were later used in this country. The first 4-high mill for the cold rolling of steel was first used on an experimental basis as recently as 1923 by the Allegheny Ludlum Steel Corporation.

Improvements to roll neck bearings also contributed to the increasing use of cold reduction mills. Roller bearings were first used on 2-high cold mills as early as 1890, on the backup rolls of cluster-type cold mills in 1909 and on 4-high mills in 1926.

Reversing cold mills of the 2-high type were first used in Germany in the 1920's (having been disclosed in the patent literature as early as 1917) and of the 4-high type in 1932. The first such cold mill in this country was installed at Gary in 1933.

The first record of tandem cold rolling of steel strip goes back to about 1904, when the West Leechburg Steel Company installed and operated a 2-high 4-stand tandem mill, each stand being driven by a separate, adjustable speed d-c motor. Real tandem mill operation, with tension between stands, and a tension reel, was developed around 1915, on mills installed by Superior Steel Company and the Morris & Bailey Steel Company of Pittsburgh. And in 1926, the first 4-high, 4-stand tandem cold mill was put into operation by the American Rolling Mill Company at their Butler plant.

□ M. D. Stone, "Reversing and Tandem Cold Mills", Iron and Steel Engineer Year Book, 1947, pp. 265-272.
▫ J. R. Adams, "Hardened and Ground Rolls", Yearbook of the AISI, 1924, AISI, New York, 1925, New York, pp. 115-147.

THE HISTORY OF ROLLING

In the operation of both reversing and tandem mills, as the strip being handled got longer, and was rolled at higher and higher speeds, the matter of handling the material necessarily demanded attention. This occurred first in the hot rolling of wire rods and the first reels on record were used around 1860, being manually rotated by a boy, who visually synchronized the reel with the mill. These reels were however, subsequently mechanized. The cold strip reel seems to have preceded the hot strip reel or coiler by some ten years, the first cold reels being built in Germany by August Schmitz Company, around 1893. The reels were well designed units, having wedge-type collapsing segments, and driven from the mills by slipping belts for tension control. The first high-tension reel, as such, was patented in 1905 by W. F. Conklin, of Pittsburgh, using a slipping friction clutch. Around 1920, the separate, electrically-driven reel, maintaining constant tension by current control, was developed jointly by the Superior Steel Company and the Westinghouse Electric Corporation.

The concurrent development of reels and the electric drive resulted in the development of strip tension control, first between mill and reel, and later between adjacent stands of the tandem mill. Although the first tandem cold mill was operated with slack strip between stands, the art was later advanced to the using of slack or loop take-ups, and finally to the adoption of high interstand tension around 1920.

As late as 1930, Shannon* stated, "It should be remembered that ordinarily sheet steel receives its shaping entirely in hot rolling, the cold rolling of sheets being merely a surfacing, flattening, or stiffening operation.... In this connection, another recent development requires mention. Within the past few years, cold rolls have been designed which will actually thin down and draw out steel sheets of considerable width while cold; in other words, these powerful rolls will effect cold reduction and consequently are capable of shaping cold steel into sheets of predetermined thickness and length, which cannot be done by the ordinary sheet cold mills. These newly developed mills are not widely used yet, however, so the statement that sheets receive their shaping from hot rolling still holds good for the sheet industry as a whole. Nevertheless, before many years, the cold rolling of sheets may undergo radical changes and developments."

The last words of the quotation indeed proved to be prophetic because of the extensive use of tandem and reversing cold mills for the production of sheet and tinplate products. However, it must be realized that hot rolling must still be used in reducing an ingot to a strip with a thickness on the order of a quarter of an inch. Attempts to cold roll strip of greater thickness would necessitate excessive rolling forces and energy requirements.

1-6 Modern Primary Cold-Reduction Facilities

In the 1930's, the cold reduction of hot-rolled steel strip evolved from a rather specialized, small-scale process to a position of prime importance in the production of cold-rolled bars, sheet and strip. These products differ from each other principally in dimensions as defined in Table I.

The maximum available widths of cold-rolled strip increased rapidly from 1925 onwards as illustrated in Figure 1-21, and the minimum thickness for a given width decreased.♦ By 1937, the thickness-width limits for both hot- and cold-rolled strip had reached values shown in Figure 1-22,• and today flat-rolled products are available in even greater width-to-thickness ratios than those corresponding to Figure 1-22.

---

* R. W. Shannon, "Sheet Steel and Tinplate", The Chemical Catalog Company, Inc., New York, New York, 1930.
♦ "The Modern Strip Mill", published by the Association of Iron and Steel Engineers, 1941, Edited by T. J. Ess.
• D. Eppelsheimer, "The Development of Continuous Strip Mills", Journal of the Iron and Steel Institute, 1938, No. II, pp. 185-303.

Table 1
Product Classification by Dimensions of Flat, Cold-Rolled Carbon Steel

WIDTH (Inches)	THICKNESS, (Inches)		
	0.2500 and Thicker	0.2499 to 0.0142	0.0141 and Thinner
To 12, inclusive	Bar	Strip[1,2]	Strip[1]
2 to 12, inclusive	Bar	Sheet[3]	Strip
Over 12 to 23.9375, inclusive	Strip[2]	Strip[2]	Strip
Over 12 to 29.9375, inclusive	Sheet[4]	Sheet[4]	Black Plate[4]
Over 23.9375	Sheet	Sheet	Black Plate

1) When the width is greater than the thickness with a maximum width of 0.5 inch and a cross-sectional area not exceeding 0.05 sq. inch and the material has rolled or prepared edges, it is classified as flat wire.
2) When a particular temper as defined in A.S.T.M. specification A109, or a special edge, or a special finish is specified, or when single strand rolling is specified in widths under 24 inches.
3) Cold-rolled sheet coils and cut lengths, slit from wider coils with slit edge (only) and in thicknesses 0.0142 inch to 0.0821 inch inclusive, carbon 0.20 per cent maximum.
4) When no special temper, edge or finish (other than "Dull" or "Luster") is specified, or when single-strand rolling widths under 24 inches is not specified or required.

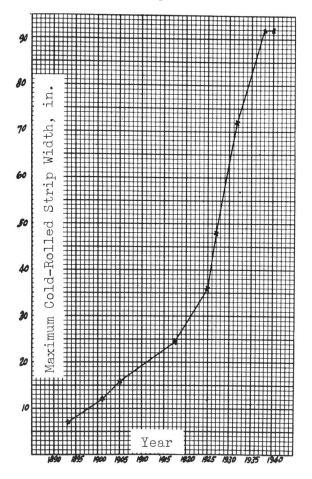

Figure 1-21: Maximum available width of cold-rolled strip.

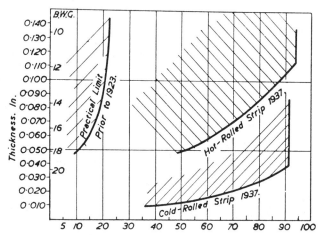

Figure 1-22: Chart Comparing the Limits of Rolling Prior to 1923 and the years following.

THE HISTORY OF ROLLING

Typical of the sheet mills built in the 1930's is the three-stand, 84-inch tandem mill shown in Figure 1-23. This mill utilized 20-1/2 inch diameter work rolls, 56-inch diameter backup rolls, and was driven by motors totalling 6,850-HP at speeds up to 542 feet/minute.

Just before and subsequent to World War II, four-stand sheet mills came into vogue, but in the 1960's five-stand sheet mills were built as discussed in Chapter 2-

For tinplate production (with final thicknesses of rolled strip on the order of 0.010 inch) five-stand tandem mills soon came into use in the 1930's. Such a mill installed at the Irvin Works in 1941 used 21 inch and 53 inch by 48 inch rolls and possessed a total motor drive of 11,100-HP. (See Figure 1-24.) It could deliver strip at speeds up to 3,750 feet/minute. Later mills for tinplate production utilized larger motors and were generally operated at speeds up to 5,000 feet/minute. Six-stand tin mills were introduced in the 1960's with still more installed horsepower and slightly larger workrolls (23 inch as compared with 21 inch diameter). Twin drives are used on such mills and they are operated partially under the control of computers.

Figure 1-23:
View of exit end of 84-inch three-stand tandem mill, used for cold reduction of sheet material.

Although tandem cold mills constructed in recent years have utilized conventional 4-high mill stands, a unique tandem mill facility specifically designed for rolling stainless sheet products up to 50 inches wide was put into service in 1969 by Nisshin Steel Company at the Shunan Works located in Nanyo, Japan.± This mill was the first mill designed for fully continuous operation with the incoming coils fed from pay-off reels through a welder to be joined head-to-tail. It utilizes a train of 6 stands, the first and last stands being 2-high mills and the intermediate stands

---

± T. Ohama, S. Sasaki and M. G. Sendzimir, "World's First Sendzimir Tandem Mill", Iron and Steel Engineer Year Book, 1973, pp. 173-179.

Figure 1-24:
A 48-inch, high-speed, five-stand tandem cold-reduction mill. Product being rolled travels through the mill from right to left in this view.

being Sendzimir mills, one being a type ZR-22N-50 and the other three type ZR-21B-50. A more detailed description of this mill is presented in Section 2-15.

Another fully continuous cold mill for rolling steel strip was put into operation in 1971 at Nippon Kokan's Fukuyama Works in Japan.° This mill also features two pay-off reels, a shear, welder, strip accumulator, five mill stands, a flying shear and two tension reels. Coils to be rolled are welded end to end and, during the welding operation, strip continues to be drawn from the accumulator. The mill utilizes 4-high stands (with hydraulic cylinders in stands 1 and 5 for rapid roll positioning), is as fully automated as present technology permits and is under computer control. Further details of this mill are presented in Section 2-15.

For stainless and silicon steel rolling and for the processing of special alloys with limited markets, single-stand reversing mills (particularly Sendzimir mills) have continued to find popularity. Such mills usually feature powerful main drives (up to 8,000-HP or more) and motors attached to the reels supplying power approximately equal to half that of the main drive motor.

The foregoing mills, discussed in greater detail in Chapter 2, are operated differently from their early slow speed predecessors in a number of ways. High strip tensions are now usually used (whereas in some of the first tandem cold mills, free loops of strip were developed between the stands) and rolling lubricants were found to be necessary. Such lubricants (generally oils such as palm oil or cottonseed oil) were found to greatly facilitate the rolling operation in reducing rolling forces and lessening the rolling energy required. An aqueous recirculated mixture of the rolling lubricant (generally called a rolling solution) was used not only to provide the lubricity in the roll bite between the roll and strip surfaces, but also acted as a coolant for the rolls and the strip being rolled.

Mill speeds have also increased considerably but coincident with this has been a steady improvement in rolled strip quality. Automatic gage control systems have maintained good

---

° T. Okamoto, Y. Kawasoko, S. Fujii and T. Anmura, "Fukuyama's Fully Continuous Tandem Cold Mill", Iron and Steel Engineer Year Book, 1972, pp. 259-263.

THE HISTORY OF ROLLING

uniformity in the thickness of the rolled product and improved lubricants have maintained strip coiling temperatures within permissible limits. Although shape problems still reoccur in rolling operations, roll grinding techniques have improved so that, together with better mill instrumentation, such problems become less frequent.

1-7 Secondary Cold Rolling Mills

Just prior to 1960 double-reduced tin plate was commercialized in the U.S.A. as a packaging material. This light-gage steel strip is rolled on specially designed cold mills variously designated as "secondary cold mills", "DCR (Double Cold Reduction) mills", 2-CR (Secondary Cold Reduction) mills and sometimes as "skinny-tin mills". One-, two- and three-stand mills have been used for this purpose, with the two-stand mills being the most popular (see Figure 1-25).

Figure 1-25: Two-Stand Cold-Reduction Mill Viewed from Operator's Side

The first method of manufacturing double-reduced tin plate involved the following steps; 1) primary cold reduction, 2) electrolytic cleaning, 3) box annealing, 4) tinning, 5) secondary cold rolling (with a reduction close to 50 per cent), 6) cleaning, 7) chemically-treating, and 8) oiling.× It soon became apparent, however, that due to equipment and processing conditions peculiar to each producer, differences in appearance and lustre of the matte-like surface did exist and created problems for the can manufacturer in matching surface appearance (particularly with respect to the lithographing of the tin plate) from the various producers. As a consequence, the product was soon made by reducing box-annealed black plate 30-40 per cent and tinning this light gage steel strip in the conventional manner. Today, it is sold commercially in gages ranging from approximately 0.005 to 0.011 inches in thickness (45-100 lb/base box).

× W. L. Cooper, "Double Reduced Tin Plate", Yearbook of the AISI, 1964, pp. 147-163.

Although the stands of a secondary cold reduction mill generally resemble those of a primary cold-reduction facility, there are some significant differences. The maximum speeds of these mills are generally less than those of primary cold rolling facilities, and for this reason, as well as the fact that the drafts are relatively small, the drive power required is appreciably less in secondary than in primary cold-rolling mills. On the other hand, the strip width-to-thickness ratio is larger so that the lubricity provided by the rolling lubricant becomes more critical. Moreover, because of the light-gage of the rolled product, the mill must be designed so that excessive tensile stresses are not developed in the strip. Lastly, both the surface finish and the shape (see Chapter 11) of the final product must be exceptionally good otherwise surface defects will be visible through a coating (such as a chromium-chromium oxide or "tin-free-steel" coating) and the sheared product could not be satisfactorily handled on can-making lines.

Very recently, the drawn-and-ironed (D & I) can made out of both conventional black plate and tin plate has been developed to a successful stage. In the course of manufacture, the steel, given a primary cold reduction and annealing, is in effect given a "secondary cold reduction" in the drawing and ironing operation. It is conceivable, therefore, that the drawn-and-ironed can could significantly influence the future status of the secondary cold reduction mill.

1-8 Foil Mills

In recent decades, various metals have been rolled to very light gages for such applications as packaging, capacitor manufacture, and printed circuit boards. For virtually all these applications, however, the quantity of rolled product was limited and coil widths of a few inches usually sufficed. Accordingly, small mills (often Sendzimir mills) were used in making such specialized foils in thickness often in the range 0.0001 to 0.001 inch.

Soon after 1960, however, it became apparent that a reasonably large market existed for relatively wide (up to 30 inches or more) steel foils. A foil mill capable of rolling tin coated strips down to a thickness of 0.0015 inch and less in widths up to 30 inches or more was therefore installed at the Gary Works of the U. S. Steel Corporation in 1965 (see Figure 1-26).[ø,▲] This single-stand mill and its rolling lubrication system are described in detail in Section 2-18 and the product characteristics presented in Chapter 12.

Figure 1-26: Foil Mill — Payoff Side.

ø R. C. Haab and M. B. Jacobs, "Steel Foil Production and Applications", Year Book of the AISI, 1967.
▲ "Tin Coated Steel Foil Available in 0.002-Inch Gage from U. S. Steel", Iron and Steel Engineer Year Book, May, 1964, p. 168.

1-9 Temper or Skin Pass Mills

In Section 1-5, the early use of cold mills to planish tin plate was noted, such mills constituting a type of temper or skin-pass mills. Basically such mills give a very light reduction to annealed stock so as to provide a degree of surface hardening, restore temper and prevent stretcher strain or the breaking of the surface in subsequent drawing operations. In addition, temper mills are used to impart a desired finish or luster to the surfaces of the workpiece and are frequently used to impart the desired degree of flatness to the rolled product.

Usually temper mills are operated without a rolling lubricant. This is fortunate in that the rolled surfaces of the strip remain virtually uncontaminated and, therefore, ready for further processing (such as tinning). Moreover, the high friction occurring in the roll bite between the roll and strip surfaces ensures that only a very limited elongation or reduction is given to the workpiece.

Occasionally, however, temper mills are operated "wet" either for the purpose of achieving a larger reduction than would otherwise be obtainable and/or with the intent of leaving a lubricating, corrosion-resistant or other type of film on the surface of the temper rolled material.

Temper mills for the rolling of sheet products have traditionally utilized a single-stand as illustrated in Figure 1-27 while two-stand mills are commonly used to produce the harder tempers required for tinplate products.

Figure 1-27:
Single-Stand, Four-High Sheet Temper Mill, for Temper Rolling Flat-Rolled Steel in Coil Form. Product Travel Through the Mill is from Right to Left.

In recent years, there has been a tendency to utilize one rolling facility for two or more different types of rolling operations. For example, single and two-stand mills have been installed to carry out both cold reduction and temper rolling operations (see Figure 1-28). In such facilities, provision is usually made to accommodate sets of work rolls of different sizes, smaller diameter rolls being used for normal cold rolling operations and larger rolls for temper rolling.

Figure 1-28: Combination Cold Reduction — Temper Mill.

Chapter 2 Various Types of Cold Mills

2-1 Introduction

The many types of rolling mills used are categorized in various ways; single-stand mills are frequently classified on the basis of their roll arrangements; single- and multistand mills either by the number of stands or their commercial use, and unusual types of mills by the manufacturer's name or a generally accepted designation of the mill.

In modern cold-rolling practice, two-high, four-high, and cluster-type mills, including Sendzimir mills, constitute the principal examples of single-stand mills. (See Figure 2-1.) However, the Steckel mill has found considerable commercial use, but in this mill the strip should be thought of as drawn rather than rolled since the deformation energy is supplied by the tensile force exerted on the strip on the exit side of the mill.

Through the years various unusual types of mills have been proposed to provide certain advantages, such as flatter or thinner strip and greater simplicity of construction and/or operation. Although some of these mills have been developed in the laboratory to the extent that they have reduced samples of strip, they have not as yet found acceptance for commercial cold-rolling operations.

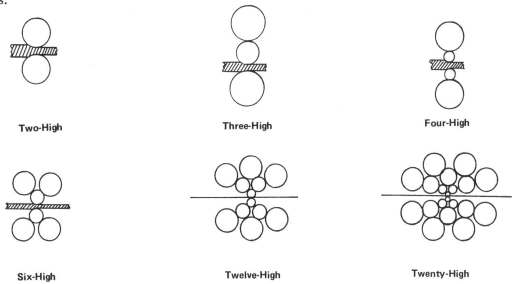

Figure 2-1: Roll Arrangements of Cold Mills.

Generally speaking, mills are usually identified by their function. Primary cold mills are designated "sheet mills" if they reduce hot band to sheet gages or "tin mills" if they roll the same incoming material to tinplate stock. The necessity of reducing tinplate stock to even thinner strip gave rise about 1960 to the introduction of the "secondary-cold reduction" (SCR, DCR or 2CR) mill. Still more recently, the commercialization of steel foil in thicknesses down to 0.0015 inch in widths up to 30 inches has resulted in the installation of a specially-engineered "foil" mill.

Whereas, the foregoing types of mills are designed to provide sizable reductions to the strip, certain mills known as "temper mills" are utilized to impart desired surface finishes, metallurgical properties, and flatness to the rolled product. Where they are used to process annealed sheet product, they are designated "sheet temper mills" and where they process annealed black plate, they are known as "tin-temper mills".

All of the foregoing types of mills are discussed in greater detail in this chapter together with pertinent information relating to their productivity.

2-2 Two-High Mills

Historically speaking, two-high mills are the oldest type used for the cold reduction of steel. However, even as late as the nineteen twenties, such mills were used mainly to flatten the sheet by removing dents, creases, and some types of 'waves', to improve the surface of the sheet and produce either a close and smooth surface (but not necessarily a glossy, polished surface) or a glossy polished surface, and to stiffen the sheet through the hardening effect of cold working.†

In the early days of tinplate, the shaping of the workpiece was generally accomplished by the hot mill and hence the use of the two-high cold mills was essentially for temper rolling. Elongation of the strip during the processing was regarded as a nuisance since it was "not sufficient to be useful and only enough to be troublesome".•

An early example of a two-high mill is illustrated in Figure 2-2. It is similar in design to the hot mills then in use except that the screws were "designed to be moved only by the use of considerable force, as by striking the screw bar with a wooden mallet" since operating experience dictated only infrequent adjustment of cold-mill rolls. Moreover, only the bottom rolls of such mill stands were usually driven• and no rolling lubricants were applied to the mill. The rolls themselves had no camber or crown and possessed hard, chilled surfaces. They were ground and polished prior to use in a manner similar to today's practice, except that marks on the rolls developed during mill use were frequently ground out while the rolls were still in place in the mill stand.

Figure 2-2:
Cold-rolling mill (hot-pack process). The mill comprises 15 stands of rolls in five 3-stand tandems.

Such mills were used only to roll cut lengths of sheet or strip with the rolls constantly rotating in one direction. Wood-faced tables were used to support the workpieces on the entry and exit sides of the mill. In sheet-mill practice, a "cold roller" pushed the sheets into the rolls, one sheet at a time and a "catcher" piled the sheets up on the exit side of the mill, usually on an "annealing box bottom". If a second pass were to be required, the whole pile of partially cold-rolled sheets would be picked up by a crane and set down on the cold roller's side of the mill.

In tin mills, three single stands were usually arranged in tandem with belts to convey the sheets automatically from the first to the second and from the second to the third stands. Highly

---

† R. W. Shannon, "Sheet Steel and Tin Plate", The Chemical Catalog Company, Inc., New York, New York, 1930, p. 156.
• W. E. Hoare and E. S. Hedges, "Tinplate", Edward Arnold & Co., London, 1945, p. 76.

VARIOUS TYPES OF COLD MILLS

polished rolls were used because of the glossy finish desired on the steel strip to be tinned. However, tin mills usually maintained a few single-stand mills (not arranged in tandem) to provide ordinary flattening passes and for nonglossy finishes.

Considerable care was given to the proper lubrication of the bearings in an effort to prevent excessive temperature gradients from occurring in the rolls due to frictional heating at the roll necks. Even so, gas flames were frequently turned against the center of the rolls in an attempt to obtain temperature uniformity within them.

With the advent of the high-speed, continuous tandem mills using 4-high roll stands, the slow-speed, two-high mills fell into general disuse. Narrow two-high mills find occasional use today in research work and commercial rolling operations where the short rolls present no serious bending problems.

Recently, however, two-high mills have staged a come back as skin pass mills, and Usinor-Schloemann have developed a mill of this type with a bending roll positioned below the lower work roll as schematically illustrated in Figure 2-3. The mill shown in Figures 2-4 features hydraulic roll screwdown and polishing of the rolls during mill operation. The rolls of the mill shown are 41 inches in diameter with a face length of 86-1/2 inches and the roll assembly may be rapidly changed by means of a hydraulic changing device.

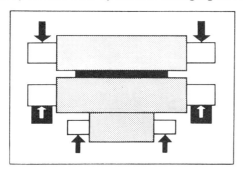

Figure 2-3:
Schloemann two-high mill with bending roll.

Figure 2-4:
Schloemann two-high cold mill with a roll diameter 41 inches (1,040 mm) and a width of 86-1/2 inches (2,200 mm)

2-3 Three-High Mills

A Lauth-type, three-high mill as illustrated in Figure 2-5 now finds application mainly in hot rolling.[†] Here the top roll is usually screwed down electrically, its lifting and lowering mechanism being joined to the screw down gear. The middle roll is an idler roll, i.e. it is set in rotation by the friction developed by the top and bottom rolls. It has no screw down but is raised or lowered by hydraulic pistons according to whether the stock is to pass below or above the middle roll which is slightly crowned. Similarly, the mill tables are raised or lowered depending on the direction the workpiece is to travel through the mill. (See Figure 2-6.)

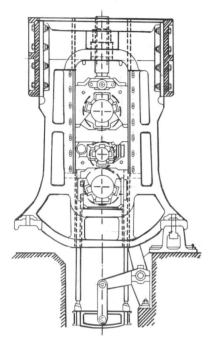

Figure 2-5:
Lauth three-high sheet
or plate-rolling mill.

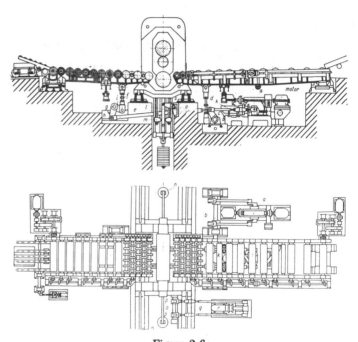

Figure 2-6:
Three-high plate-rolling mill showing mechanisms for roll balancing and positioning the middle roll and the lifting tables.

2-4 Conventional Four-High Mills

By far the most popular type of mill stand for the cold rolling of steel strip is the 4-high stand. In the conventional form, closed housings are always used, the two housings being joined by spacer bars or other means to form a rigid unit. Though it was originally common practice to drive the work rolls, in recent years the trend has been towards backup roll drives. The material of the backup rolls is generally softer than the work rolls, the shore hardness of the former being about 65, whereas it is usually in the range of 95 to 100 for the latter.

To obtain the desired rigidity of a 4-high mill in the rolling of steel strip, the backup rolls should be approximately "square", that is, their diameters should be at least equal to their face lengths. For cold mills constructed during the years 1945 to 1970 in the United States, the relationship of backup roll diameter to face length is illustrated graphically in Figure 2-7. As a margin of safety, and to allow for regrinds, the diameter might be chosen as equal to the face length plus six inches.

Although it might be thought that, with the rigidity provided by the backup roll, there might be no lower limit to the permissible diameter of the work roll, such is not the case. It has

---

[†] A. Geleji, "Forge Equipment, Rolling Mills and Accessories", Akademiai Kiado, Budapest, 1967, pp. 442-446.

VARIOUS TYPES OF COLD MILLS

been pointed out by Greenberger [†] that, if the spalling of steel backup rolls is to be prevented, the maximum stress in the backup roll at the zone of contact between the work and backup rolls must not exceed 300,000 psi. This means that there must not be too great a disparity between the work and the backup roll diameters; the maximum permissible ratio of the two diameters being dependent on the maximum allowable specific rolling force, as illustrated in Figure 2-8.

The choice of the most desirable work roll diameter is not influenced by consideration of spalling alone. Roll cooling at high mill speeds is of major importance, so that such factors as the mill speed, the percentage reduction to be given to the strip, and the roll cooling arrangements must all be taken into account as discussed in Chapter 7.

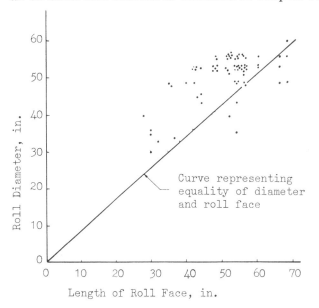

Figure 2-7:
Relationship of diameter to face length for backup rolls of cold mills installed in the USA 1945-1970.

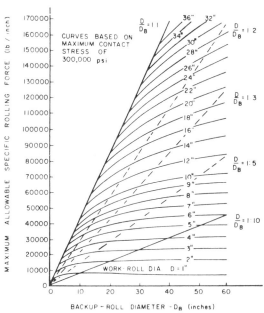

Figure 2-8:
Chart showing relationship between the maximum allowable specific rolling force and the backup roll diameter for various work roll to backup roll diameter ratios.

Where multiple passes are to be taken in the operation of a single-stand mill, a decision must be made as to whether the mill should be unidirectional (one-way) or reversible. In the one-way mill, the workpiece, usually in the form of a coil, must be transferred from the exit side of a mill to the entry side between successive passes. Thus, a materials-handling problem exists which, however, can be solved by making the mill reversible. Unfortunately, the cost of the electric drive for a reversible mill is much higher than for a one-way mill.[•] In spite of this disadvantage, the reversible, single-stand, 4-high mill has retained its popularity, principally because of its versatility and its ability to roll hot band to a number of different gages merely by varying the number of passes given to the workpiece. This type of mill is discussed in greater detail in Section 2-19.

2-5 The Steckel Mill

The Steckel mill, an example of which is shown in Figure 2-9, is a 4-high reversing mill employing working rolls which vary in diameter from 2 to 5 inches with backing-up rolls about 6 to 8 times the working roll diameter and utilizing two separately driven tension reels through which all power is provided.[‡] In this mill, the work rolls act as rotating dies and the reduction process is

---

[†] J. I. Greenberger, "Rolling of Metals", Iron and Steel Engineer Year Book, 1959, pp. 215-223.
[•] C. W. Starling, "The Theory and Practice of Flat Rolling", University of London Press Ltd., London, 1962, p. 29.
[‡] "ABC of Iron & Steel", Sixth Edition, edited by D. Reebel, The Penton Publishing Company, Cleveland, Ohio, 1950, p. 293.

Figure 2-9: Steckel four-high reversing cold mill.

more akin to a drawing rather than a rolling operation. Due to the fact that the working rolls are not driven, it is necessary to raise them apart on the loading (initial feeding) pass in order to insert the coil end into the reel's jaws. The rolls are then run down on the material for the initial reduction and the strip pulled through the mill entirely by tension. The mill is stopped when the tail end is at the proper point for insertion into the jaws of the reel on the entry side of the mill.

2-6 The Frohling Low-Expansion Mill

In the conventional mill, shown diagrammatically on the left in Figure 2-10, the roll separating force is transmitted from the rolls, through the backup roll bearings to the backup chocks, via pressure plates and the screws to the housings. To prevent large elastic strains in these components, they are usually constructed with cross sections as large as possible to provide maximum rigidity.

In the case of the Frohling mill, screws, pressure plates, and chocks are replaced by components of new design.[†] The top and bottom backup roll chocks are each connected by two tie rods which are prestressed to a value exceeding the maximum mill separating force, as shown on the right in Figure 2-10. The deformation of the prestressed mill is therefore restricted to rolls, chocks, and tie rods.

The advantages claimed for this design are as follows:

1) Lesser construction costs for the mill.
2) Greater accessibility to the rolls and the roll bite.
3) More rigid strip guides.
4) Lesser head and tail end wastage of the strip due to its compact construction.

The tie rods connecting the top and bottom backup roll chocks contain threaded ends, which are not only fitted by the nuts but which contain special hydraulic jacks which brace against the chocks and stretch the bolts when actuated by a high pressure hydraulic system. Thus the top and bottom chocks are stressed to a predetermined value. After tightening down the nuts, the hydraulic pressure is released and a residual stress (termed the prestress) is maintained in the system by the tensile stress in the bolts.

[†] "The Low-Expansion Mill: Principle, Construction, and Application", Sheet Metal Industries, May, 1968, pp. 352-355.

VARIOUS TYPES OF COLD MILLS

Top and bottom backup roll chocks are fitted with eccentric bearing sleeves which, when rotated, serve as the means of roll gap adjustment. The bearing sleeves on the top chock are meshed with hydraulically actuated racks whose lateral movement provides the means of rotating the sleeves. (The roll gap is preselected by means of a potentiometer system which, in conjunction with the hydraulic actuation, provides a very rapid response [0.03 second] to control signals, and the adjustment of the gap is displayed remotely by electronic instrumentation.) The eccentric bearings in the bottom chocks may be rotated manually and the pass line maintained at a desired height after roll changes.

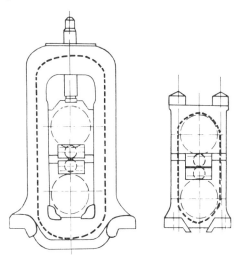

Figure 2-10: Comparison of conventional 4-high mill (left) with the Frohling mill (right)

In the Frohling mill, work rolls with a range of diameters may be used. However, when exchanging one set of work rolls for a set of substantially different diameter, the differently sized work roll chocks are accommodated by placing different size bolsters between the backup roll chocks. This, however, is facilitated by the use of the prestressing cylinders for separating the chocks. Work rolls are mounted in the conventional manner but axial locking devices in their chocks provide high-speed lateral positioning of the work rolls and make rapid roll changing possible.

The Frohling low-expansion mill is currently in use rolling all types of steel, nonferrous and precious metals, and it may be used as a one-way or reversing mill or in tandem arrangements. A reversible mill of this type used for rolling carbon steels and featuring work rolls with diameters of 40, 80, and 120 mm and backup rolls of 390 mm diameter is shown in Figure 2-11.

Figure 2-11: Frohling Mill used for Cold Rolling of Carbon Steels.

2-7 Prestressed Four-High Mills

In a conventional four-high mill, almost any change in one or more of the rolling parameters affects the roll gap. Thus, variations of the gage of the incoming strip or in its yield strength affect the rolling force which in turn changes the bending of the mill rolls, the compression of the backup roll bearings, the chock assemblies and the screw down screws, and the stretch of the mill housing. These elastic changes in the mill components would, therefore, in the absence of any compensating changes wrought manually or automatically, affect the gage and shape of the rolled strip.

The prestressed four-high mill developed by Loewy-Robertson and shown diagrammatically in Figure 2-12 overcomes these problems by a combination of techniques which compensate automatically and immediately for roll deflection or bending, housing stretch and screw and chock compression.† These features are built into the mill in such a way that the overall cost is comparable with that of a conventional unit.

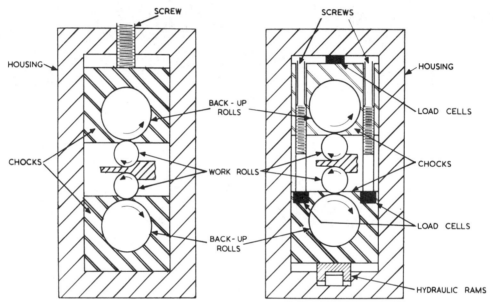

Figure 2-12:
Principle of Loewy-Robertson prestressed mill with constant gap control (right) compared with a conventional mill. Signals from load cells are integrated to control pressure in a hydraulic ram which maintains a constant gap between the work rolls.

Whereas, on the conventional mill, the roll gap is adjusted by screws acting between the housing and the top backup roll chock, in the Loewy-Robertson prestressed mill, the top chock bears against a load cell and the gap is initially set by screws between the top and bottom backup chocks, bearing on the latter through other load cells. Instead of bearing on the housing, the bottom chocks are supported on hydraulic rams.

Compensation for housing stretch and screw and chock compression, to maintain constant chock gap for a given setting of the screws in the face of changes in the roll separating force, is based on the fact that the chock gap can only vary by the extent that the screws are, like a spring, compressed under load. To eliminate this factor, the screws are prestressed in compression and the housing in tension by the hydraulic rams. Pressure in the ram cylinders is automatically adjusted to signals from the load cells under the screws to maintain screw compression, and hence deflection, constant under varying rolling conditions. Since the screws remain at a constant length, the chock gap must remain constant.

---

† "Constant Gap Mill in Service — Highly Successful First Installation at Dolgarrog", Iron and Steel, November, 1966, pp. 520-522.

VARIOUS TYPES OF COLD MILLS

The other part of the problem, roll deflection, is taken care of indirectly by the load cells above the top chocks. These measure the total push-up load, which is the sum of the rolling load and the screw prestressing load. The latter is held constant as seen above, so that the variations in the top load cells are directly related to the rolling load. This provides the source of compensation for the roll deflection factor. In practice, the two sets of signals obtained from the two sets of load cells are integrated and applied to adjust the hydraulic rams, providing constant control of the actual roll gap regardless of the source of its tendency to vary.

Since measurements actually occur in the mill and not several feet after the rolls (as with a flying micrometer), a very fast speed of correction can be obtained by the use of the latest type of fast-response hydraulic control valves developed for aerospace applications. In practice, the total delay between the occurrence of a change in the roll separating force and its full correction is only about 70 milliseconds.

A similar four-high prestressed mill, illustrated by Figure 2-13, has been developed by the Hunter Engineering Co., Riverside, California.[†] Wedge assemblies featuring hardened and ground blocks with antifriction roller and cage assemblies (Figure 2-14) are used between the backup roll chocks. The hydraulic cylinders under the bottom backup roll chocks are maintained at a constant, preset pressure so as to give a force 30 to 40 percent higher than the maximum design separating force. With no metal in the roll bite, the preload force is transmitted through the wedge assemblies to stretch the housings. When a strip is introduced into the roll gap preset for a desired exit gage, the roll separating force merely transfers the load distribution to unload the wedge assemblies without affecting the mill housing prestress.

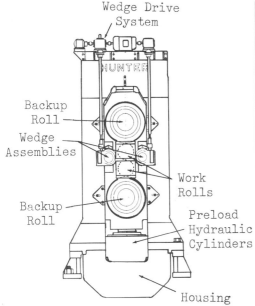

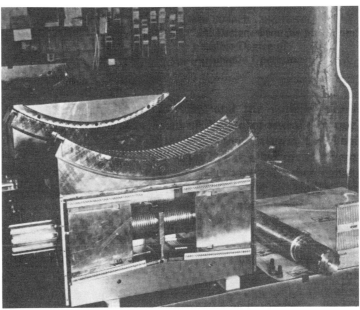

Figure 2-13:
Prestressed Mill Manufactured by
the Hunter Engineering Co.

Figure 2-14:
The Double-Inclined Plank and Adjusting
Screw of a Wedge Assembly.

Comparing a conventional screw down mill with a prestressed mill and designating A, L, and E as the cross-sectional area, effective length and modulus of elasticity respectively, the spring rate k (defined as the change of force with screw setting) can be written for the two systems as follows,

---

[†] S. Sankaran, "Gage Correction with Prestressed Rolling Mills", Iron & Steel Engineer, September, 1967, pp. 207-208.

$$k\begin{pmatrix}\text{conventional}\\ \text{mill}\end{pmatrix} = \frac{1}{\frac{L_S}{A_S E_S} + \frac{L_h}{A_h E_h} + \frac{2L_C}{A_C E_C}} \qquad (1)$$

$$k\begin{pmatrix}\text{prestressed}\\ \text{mill}\end{pmatrix} = \frac{1}{\frac{L_W}{A_W E_W} + \frac{2L_C}{A_C E_C}} \qquad (2)$$

where the subscripts S, h, C, and W denote screw, housing, chocks, and wedge. Since

$$\frac{L_S}{A_S E_S} + \frac{L_h}{A_h E_h} > \frac{L_W}{A_W E_W} \qquad (3)$$

the spring rate for a conventional mill [k (conventional mill)] is smaller than that of the prestressed mill [k (prestressed mill)] and hence the prestressed mill is a "stiffer" mill.

It is interesting to note that prestressed mills of the designs discussed above may be very "soft" mills if the screw or wedge force separating the backup roll chocks is reduced to zero and the pressure in the hydraulic cylinders is held constant.

2-8 The Schloemann Mill

Schloemann Aktiengesellschaft of Dusseldorf, Germany, manufacture a mill of distinctive design which utilizes small diameter work rolls prevented from excessive lateral bending by the use of additional backup roll assemblies located on the exit side of the roll bite. The roll arrangement of a Schloemann MKW mill is illustrated in Figures 2-15 and 16; and except for the work roll bearings and the backup roll drive that is used, the construction of the housing is basically identical with that of the conventional 4-high mill. However, the ratio between the backup roll diameter and the work roll diameter is about 6 to 1 or more.

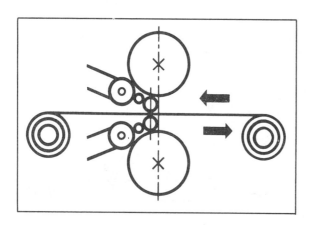

Figure 2-15:
MKW Reversing Strip Mill.

Figure 2-16:
General Arrangement Drawing of MKW Reversing Strip Mill.

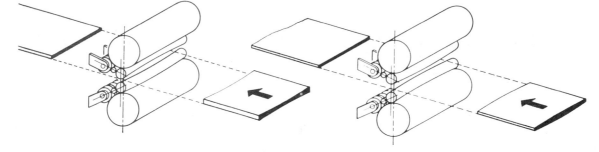

VARIOUS TYPES OF COLD MILLS

The work rolls do not have chocks and, being held in simple gripping jaws, they are easily changed. They are balanced by means of a hydraulic system even when no material is being rolled.

Because of the lesser rolling forces generated by the smaller work rolls, roller bearings are used to support the main backup rolls. Since each backup roll axis remains essentially fixed in this type of bearing, the roll gap remains essentially constant during the acceleration and deceleration of the mill.

Schloemann mills, because of their design, can be operated reversibly; and, because they are backup roll driven, the roll assemblies may be readily changed to provide conventional 2-high or 4-high arrangements as well as the MKW configurations.

A modern MKW cold mill, shown in Figure 2-17, went into operation at the Stahlwerke Bochum Akliengeselleschaft (Bochum) in 1967.† At that time it was believed to be the most powerful reversing cold mill in the world with an overall drive output of 11,000 HP. It was employed for rolling wide, mild steel and silicon steel strip in widths up to 2,150 mm. The final thicknesses for mild steel are 0.10 mm and for silicon steel 0.20 mm, and the mill reaches its maximum rolling speed (600 m per minute) during the second pass.

Figure 2-17:
An MKW Mill Rolling Wide Silicon Steel Strip.

A diagrammatic sketch of the mill is shown in Figure 2-18. The mill is equipped with a Schloemann crown control unit for rolled-strip shape adjustment and the reel turntable gives a rapid coil sequence. Moreover, a hydraulic device mounted on the mill stand enables the work rolls to be changed without the use of a shop crane.

The technical data relating to the mill is illustrated in Table 2-1. In rolling silicon steel strip 1,040 mm wide, three passes are used to reduce the strip thickness from 2.5 mm to 0.50 mm. Mill speeds for this operation are usually in the range 400 to 600 meters per minute and the rolling force lies in the range 750 to 950 tons.

---

† "A Modern Cold-Rolling Mill Installed at Stahlwerke Bochum", Sheet Metal Industries, October, 1967, pp. 693-694.

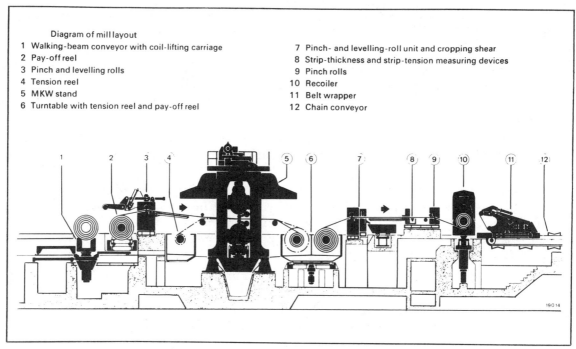

Figure 2-18: Diagram of MKW Mill Layout.

Table 2-1

Data Relating to the MKW Mill

Roll Diameter	1,320/250 mm
Roll Face	1,400 mm
Rolling Speed	300/600 m per minute
Strip Width	1,250 mm maximum
Mild Steel (Initial Thickness)	4.5 mm maximum
Mild Steel (Final Thickness)	0.1 mm minimum
Silicon Steel (Initial Thickness)	2.5 mm maximum
Silicon Steel (Final Thickness)	0.2 mm minimum
Coil Weight	20 Kg/mm

2-9 The Y-Mill

The late 1940's saw the introduction of the reversible Y-mill with a roll arrangement as illustrated in Figure 2-19.[†] Here, the object was to use the smallest practical work rolls with rigid backing both in the vertical and horizontal directions. The fundamental design objective of the cluster part of the mill was to provide means for maintaining an approximate 90-degree angle from the center of the small, friction-driven body or top work roll, through the axes of the intermediate backup or driven rolls and the main top backup rolls. This arrangement was maintained by shims under the top backup roll chocks, the shims being used to compensate for roll grinding and for maintaining the correct alignment of the top roll assembly. The required thickness of shims was read directly from a gage that measured the distance between the top intermediate backup rolls, in relation to their diameter and that of the body roll. The 90-degree angle between the rolls in this part of the mill was found to be essential for its proper operation.

[†] A. B. Montgomery and W. M. McConnell, "The Y-Mill...A New Type of Cold Strip Mill", Iron and Steel Engineer Year Book, 1948, pp. 229-234.

VARIOUS TYPES OF COLD MILLS

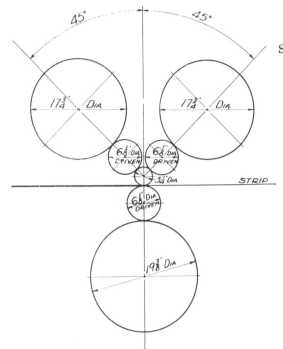

Figure 2-19: Sketch Showing Arrangement of Rolls in a Y-Mill.

Figure 2-20: A 20-Inch Reversing Y-Mill.

The lower half of the mill utilized a work roll of comparable diameter to the intermediate mill work roll supported along its length by a backup roll. These two lower rolls were carried on housing screws and could be positioned in a vertical direction to correspond to the rolled strip thickness.

Each upper backup roll was clamped in place at the top of its housing window with a self-aligning rocker plate. The two upper intermediate rolls and body roll were balanced by springs or hydraulic cylinders, mounted in the lower work roll chocks, to maintain contact between all the rolls when idling the mill. The two upper intermediate rolls and the lower work roll were driven through spindles from the mill pinion stand and all the rolls were adequately lubricated by a flood of filtered soluble oil coolant from a recirculating system.

In one installation of this mill described by Montgomery and McConnell,[†] and illustrated in Figure 2-20, the mill was built to a width of 20 inches and intended to roll strip with a maximum width of 17 inches at speeds up to 1,200 fpm. It was driven by a 600 HP motor connected to the two intermediate and the lower work rolls (each being about 6-1/4 inches in diameter). Up to 600 fpm the motor developed constant torque and constant horsepower from 600 to 1,200 fpm.

The reels were designed to grip the strip for rotation in either direction and were driven by 200-HP motors which could alternately act as motors or generators depending on the direction of operation of the mill. Rotating, constant-current regulating exciters were used to control the generators to maintain approximately constant reel tensions during the unwinding and rewinding of each pass. The control also provided for deceleration to compensate for reel inertia, thereby maintaining the same tension as during normal rolling at constant speed.

2-10 Six-Roll Cluster Mills

To utilize small diameter work rolls of the same size and provide horizontal and vertical rigidity, six-roll cluster mills with a roll arrangement as shown in Figure 2-21 have found limited industrial use. Proposed in Germany as early as 1837, no further mention of such a mill was made

---

[†] A. B. Montgomery and W. M. McConnell, "The Y-Mill... A New Type of Cold Strip Mill, Iron and Steel Engineer Year Book, 1948, pp. 229-234.

until 1910 when Robertson's of Bedford, England,[†] were granted a license to build small units designed by Mathey, a Swiss engineer.[•] Figure 2-22, shows a six-high mill of German construction built about 1930.[‡]

Figure 2-21:
Diagrammatic arrangement of six-high cluster mill.

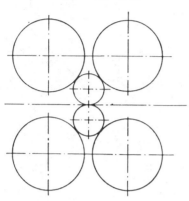
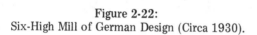

Figure 2-22:
Six-High Mill of German Design (Circa 1930).

Farrington[□] has described a reversing 8-inch wide mill of this type used for rolling stainless steel down to thicknesses as small as 0.006 inch. This particular mill utilized 3-1/4-inch diameter work rolls and 10-inch diameter backup rolls. One top and one bottom backup roll were driven through a pinion stand by a 100-HP 575/1,150 rpm motor in conjunction with a Ward-Leonard control providing a maximum speed of 300 fpm.

Spring pressure was used in the usual manner to separate the top and bottom backup roll assemblies and to take out the slack or looseness that could develop between the screws and the chocks. The work rolls were separated by another set of coil springs to insure contact with the backup rolls when the material was not in the mill. A pair of screws were used in conjunction with each upper roll chock, these screws being driven through gearing by a 3-HP high-torque motor. Misalignment that might occur was corrected by a manually adjusted wedge under one screw. The work roll bearings were anti-friction, whereas the backup roll bearings were oil-flooded. These latter bearings were rated at 130,000 pounds per roll neck equivalent to a maximum specific rolling force of 45,500 pounds per inch.

Each reel on the mill was driven by a 10-HP 350/1,300 rpm motor connected to the same power source as the main drive motor. A coil buildup of over three to one was possible with constant tension throughout the entire rolling of a coil.

2-11 Rohn and Sendzimir Mills

In the six-high mill of the type designed by Mathey, there is a practical limit to which the size of the work rolls can be reduced with respect to the backup rolls. As illustrated in Figure 2-23, this limit is reached when the horizontal tangent of the work roll coincides with the horizontal tangents of the backup rolls.

[†] E. C. Larke, "The Rolling of Strip, Sheet, and Plate", The MacMillan Company, New York, 1957, p. 41.

[•] "New Type of Rolling Mill for Accurately Rolling Steel Strip", Iron and Coal Trades Review, July 15, 1910, 81, p. 87.

[‡] J. Puppe, "Walzwerkswesen Dritterband, Verlag Julius Springer, Berlin, 1939, p. 620.

[□] G. E. Farrington, "The Reversing Cluster Mill", Iron and Steel Engineer Year Book, 1949, pp. 155-157.

VARIOUS TYPES OF COLD MILLS

To overcome the difficulty of reducing the size of the work roll without reducing the relative size of the backup rolls, Rohn, about 1925, proposed mills having ten or eighteen supporting rolls as illustrated by Figure 2-24, showing the twelve-high roll arrangement. A twelve-high mill of the Rohn design with work rolls 0.4 inch in diameter, intermediate rolls of 0.8-inch diameter and peripheral rolls of 1.6-inch diameter was used to reduce coils of nickel from a thickness of 0.020 to 0.0004 inch in six passes.†

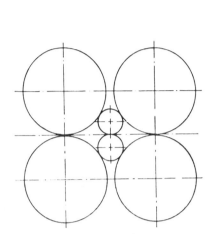

Figure 2-23:
Limiting Reduction in Diameter of Six-High Cluster-Mill Working Rolls.

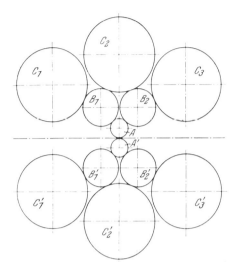

Figure 2-24:
Roll Arrangement in Twelve-High Mill Proposed by Rohn.

The basic concepts of Rohn were successfully developed by Tadeusz Sendzimir, whose mills have gained worldwide acceptance, particularly for the rolling of stainless and electrical steels.

There are two fundamental reasons that brought about the development of the Sendzimir mill. First, was the fact that Sendzimir mills could be built more economically than 4-high mills of comparable size. Second, the small work rolls of the cluster mill promised to roll strip to a much closer gage accuracy than possible on other mills.

The first Sendzimir mill was put into operation in Europe in 1932.* While this mill was a type of cluster mill, there were certain important differences between it and other cluster mills. The fundamental idea of the Sendzimir mill is the backing arrangement, characterized by one single compact and rigid housing, which holds backing rolls and work rolls in place. The roll separating force in the Sendzimir mill passes from the work rolls through the intermediate rolls to the backing shafts. These shafts have concentrically mounted bearings of the roller type which are located eccentrically in the saddle rings, equally spaced between the bearings. These transmit the work roll separating force directly to the rigid housing across every inch of mill width.

The original Sendzimir mill installed in Poland utilized two 4-inch diameter work rolls each of which was supported by two backing shafts which rotated eccentrically in the saddle rings and opened or closed the mill. On subsequent mills it was necessary to give a heavier backing capacity to make heavier reductions on the strip. The use of a smaller diameter work roll was made possible by adding another backing stage. Consequently, the backing shafts were replaced by solid rolls, and each of these rolls was in turn backed up by two backup shafts. In this arrangement the work rolls were driven, one from each side of the mill. Their diameter was about 3 inches, and the backing bearings were 9 inches in diameter.

---

† W. Rohn, "Anwendung Kleinster Walzendurchmesser and Fortbildung von Mehrrollenwalzwerken", Stahl u. Eisen, 1932, 52, p. 821.
* M. G. Sendzimir, "The Sendzimir Cold Strip Mill", Journal of Metals, September, 1956, pp. 1154-1158.

It was recognized that with the reduction in the work roll diameter, the tension on the strip should be somewhat heavier than for comparable reductions with larger diameter work rolls. Furthermore, this tension had to be much more constant and, therefore, necessitated a carefully designed new strip tensioning system.

Tension control was a remarkable feature of the early Sendzimir mills. It consisted chiefly of two large drums on each side of the mill housing, on the outside of which were attached smaller rolls in such a fashion as to form a number of small pinch rolls. This system could apply the necessary front and back tension to the strip. The mill was set up so that a constant elongation was always given to the strip under this tension arrangement and drive, and the roll bite was adjusted accordingly. During the 1940's, the Sendzimir mill acquired the modern tension control of a full reversing mill which has been constantly improved and whose uniformity today permits the Sendzimir mill to roll strip of extremely close gage tolerance in the lengthwise direction.

Most small Sendzimir mills have been equipped with operating tensiometers which control the tension of the winder reels. The larger mills have been equipped with current regulators and some new mills utilize indicating tensiometers.

During World War II and immediately afterwards, the Sendzimir mill underwent a considerable change. First came a 1-2-3 arrangement in which the work rolls were supported by two intermediate rolls that were driven and these two driven rolls were in turn supported by a total of three backing-up shafts on the upper and on the lower side of the mill. (See Figure 2-25.) This arrangement permitted the use of tungsten carbide work rolls and increased the power that the mill could transmit.

In order to get the smallest possible diameter work roll and, at the same time, to give the maximum possible reduction to the strip, a 1-2-3-4 mill was developed in the 1950's.

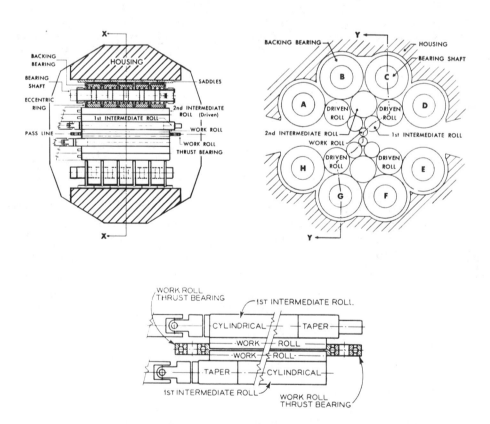

Figure 2-25:
Roll Arrangement and Bearing Details of a 1-2-3-4 Sendzimir Cold-Strip Mill.

VARIOUS TYPES OF COLD MILLS

In the 1-2-3-4 Sendzimir mill, shown in Figure 2-26, there are eight backing shafts, denoted A - H. Shafts B and C are the main screw-down shafts which are equipped with large hydraulic cylinders on the top of the mill. These shafts have roller bearings in the saddle rings and can be easily rotated under the heavy screw-down pressure. All other shafts have plain bearings in the saddle rings and can be rotated only under no-load condition. The other shafts are also self-locking, i.e., in order to open or close the mill the shafts have to be positively moved. Shafts A and H are moved by an electric motor located in the back of the mill and shafts D and E are moved by a similar motor. These shafts are brought closer together or further apart, depending on the size of the rolls in the mill.

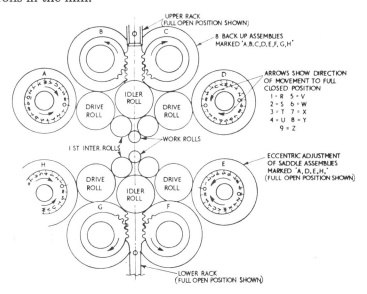

Figure 2-26: Diagram of 1-2-3-4 Sendzimir Mill Showing Roll Positioning Arrangements.

Shafts F and G, the two bottom shafts, are moved by a hydraulic cylinder located in the back of the mill. These shafts are opened or closed in order to change the work rolls in the mill. The movement of these shafts serves two purposes. First, it brings the work rolls to the pass line of the mill and, therefore, provides an even bearing of the work roll end surfaces against the thrust bearings in the front and back door of the mill. Second, and this is more important, the closing of the bottom rolls takes out all the slack between the rolls and enables the full travel of the top screw down of the mill. This permits the operator to reduce thick strip of hot rolled gage down to thinnest gages without changing the work rolls in the mill.

Shaft D is equipped with a crown control. On the right side of the mill are located bolts corresponding to each saddle on shaft D. When there is no load in the mill, these bolts can be rotated, and each individual saddle on shaft D can change its height with respect to the housing; in other words, this shaft has double eccentricity. This adjustment enables the operator to give any shape to the mill he desires by adjusting these bolts.

As a rolling lubricant, mineral oil is used for stainless and nonferrous metals, as well as some low carbon steels, while some Sendzimir mills are lubricated by soluble oil. The lubricant enters from the back of the mill through an annulus around the back door and is then distributed to the centers of each of the backing shafts that are drilled axially for this purpose. The lubricant passes through these shafts, then flows radially through the bearings, and through holes that are drilled in the backing shaft in the radial direction. In this way, the bearings are lubricated and cooled. The oil next flows over the rolls and finally onto the strip and extracts some of the heat from it. The oil finally escapes through two pipes located in the front and back of the mill.

Additional lubrication is provided to the strip and the work roll bite proper, the oil flowing through headers located immediately adjacent to the work roll bite where it is sprayed into the roll bite at a high pressure. The direction of the oil is such that it flows from the center sidewise and thereby washes away any fragments of the metal that may detach themselves from the strip.

This is essential for the edges, which if slit, would tend to have a lot of small detachable particles adhering to them.

A few years ago, the Sendzimir Cartridge mill, illustrated in Figure 2-27, was introduced.[†] The rigid one-piece housing of the mill is provided with just one cylindrical bore instead of the eight individual parallel bores for the background elements. These bores are now provided in the cartridge which is hydraulically inserted, or taken out of the mill, in a matter of one or two minutes.

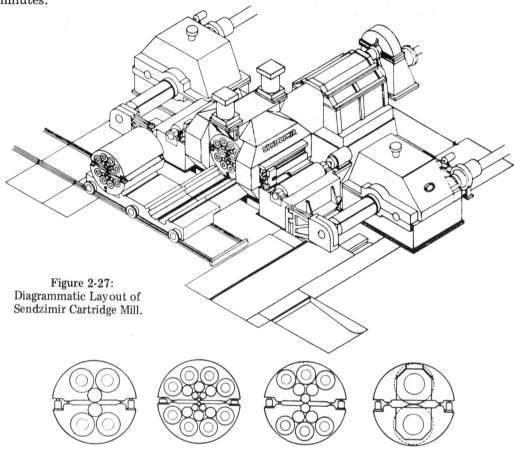

Figure 2-27: Diagrammatic Layout of Sendzimir Cartridge Mill.

Use of the cartridge means that with just one mill a number of roll geometries can be employed. Not only smaller and larger work rolls in the most frequently used 1-2-3-4 roll arrangement (already described as a 20-roll mill) but also other arrangements with 12 rolls, 6 rolls and even 2-high for some polishing skin-pass operations on stainless steel, are possible. Increased work roll utilization can be achieved by a slight variation of the vertical centers of the spare 1-2-3-4 cartridge.

Sendzimir reversing mills cover a wide range of sizes as shown in the Table 2-2, where a ZR.32 employs about 10 HP and the largest, ZR.21, utilizes a total connected horsepower in excess of 17,000.

Most of the modern Sendzimir mills utilize "As-U-Roll" shape control, activated through small hydraulic motors which can be controlled from the pulpit during rolling. This adjustment is provided on shafts B and C acting simultaneously through a very small secondary eccentric gear train. The adjustment can be made under load and, therefore, the operator can change the shape of the strip while the mill is in operation.

[†] H. W. Ward, "The Continuing Development of Sendzimir, Multi-Roll Cold Mills in Worldwide Strip Production", Sheet Metal Industries, May, 1969, pp. 407-411.

VARIOUS TYPES OF COLD MILLS

Table 2-2
Data Relating to Sendzimir Mills

Mill Section	Type	Nominal W.R. Dia.	Backing Brg. Dia.	Torque Capacity HP/100 FPM	Maximum Sep. Force lb./in. W.	Strip Width Min.	Strip Width Max.	Minimum Strip Thickness
ZR.32	1-2-3-4	0.250 in.	1.875 in.	5	4,000	4¼ in.	8¾ in.	0.0001 in. using Tungst. Carbide W.R.
ZR.15	1-2-3	0.468 in.	2.937 in.	10	6,000		8½ in.	
ZR.16	1-2-3	0.800 in.	4.724 in.	30	8,000	8½ in.	18 in.	
ZR.34	1-2-3-4	0.400 in.	3.000 in.	30	8,000	7½ in.	17½ in.	0.0004 in.
ZR.24	1-2-3-4	0.844 in.	4.724 in.	75	12,000	8½ in.	19½ in.	0.0008 in.
ZR.33	1-2-3-4	1.125 in.	6.299 in.	100	15,000	13 in.	48 in.	0.001 in.
ZR.19	1-2-3	1.812 in.	8.858 in.	120	15,000	19 in.	48 in.	
ZR.23		1.578 in.						0.002 in.
ZR.23M	1-2-3-4	2.420 in.	8.858 in.	360	20,000	19 in.	62 in.	0.0025 in.
ZR.22		2.125 in.						0.003 in.
ZR.22B	1-2-3-4	2.500 in.	11.811 in.	800	30,000	26 in.	120 in.	0.0035 in.
ZR.21	1-2-3-4	3.500 in.	16,000 in.	1,200	50,000	33 in.	209 in.	0.0035 in.

2-12 Tandem Mills

For the large volume production of sheet and strip products, tandem mills are now generally employed for primary and secondary cold reduction as well as for temper rolling. It is pertinent, therefore, to examine some of the general characteristics of such mills.

Tandem mills, regardless of the number of stands, are similar in arrangement. The stands are placed as close together as possible and the center line spacing of the stands, which depends upon such factors as the physical size and arrangement of the mill drives, is generally in the range 12-16 feet. The stands are basically identical in construction with housing posts usually exhibiting more than 1,000 square inches in cross-sectional area. Heavy construction is essential as the rolling force, which may be as high as 10 million pounds, should not induce stresses in the housing posts much in excess of 3,000 psi.

The rigidity of construction required for successful cold reduction necessitates the use of the four-high or cluster-type mill, except in the case of some narrow mills used for temper rolling or to provide light reductions. Work roll diameters generally fall within the range 16-23 inches with backup rolls running 42-56 inches in diameter. Roll face, or nominal mill size, runs 35-54 inches for tinplate production and 54-98 inches for sheet production. The ratio of backup roll diameter to work roll diameter is 2.4-2.7.[†]

The smaller work-roll diameter permitted by the use of four-high mill construction offers the further advantage of decreased rolling or separating force for the same reduction.

Generally speaking, backup-roll bearings are of the oil-film type with roller bearings being used on the necks of the work rolls. Moreover, they are usually the largest bearings that can be conveniently fitted to the roll necks and the chocks.

Where conventional screw-down mechanisms are used, steel screws 12-28 inches in diameter work in nuts of high tensile bronze, being operated through a worm gear unit, usually of double reduction. The gearing may give an overall ratio of as much as 2,000:1 from motor to screw, while the screw pitch varies with the kind of mill and the work. The pitch must be fine enough to resist rotation due to the load, yet free for rapid adjustment. Many installations show an overall ratio of about 1,000:1 and a screw pitch of about 0.5 inch, giving a roll movement of about 0.0005 inch per revolution of the motor.

[†] "The Modern Strip Mill", AISE, Pittsburgh, Pennsylvania, 1941, pp. 229-257.

The screw-down drive system must be capable of operation under rolling loads. The design must be rugged as loading torques on the order of 250,000 lb.-ft. may be encountered in addition to frictional effects, which may be considerable, particularly with larger diameter screws. Adequate lubrication of the screw-down system is, therefore, mandatory.

The screw-down system on each stand is usually driven by two motors, with a magnetic clutch joining the two drives so that the screws may be moved independently or in unison. Screw-down motors range in size from 35 to 75 HP and are generally 230 volt d.c., compound wound with reversing, dynamic-braking control. Indicators of roll position may be of mechanical or electrical type, the latter being the more commonly used.

Faster roll positioning is now being attained through the use of hydraulic systems and some of the newest tandem mills have been designed with the so-called "hydraulic stands". One of the first tandem cold mills to be so equipped was that installed by Rasselstein AG, Neuwied/Rhein, W. Germany.[†] This mill, shown in Figure 2-28, operating at 2,500 m/min., utilizes rolls 610 and 1,450 by 1,450 mm with a 3,000-ton maximum rolling load rating. To permit the use of a short stroke in the hydraulic cylinders, the mill stands were equipped with wedge adjustments for top and bottom roll assemblies capable only of driving against the roll balance forces.

Figure 2-28:
Six-Stand Tandem Cold Mill (Rasselstein AG, W. Germany).

In tandem cold mills, it is the general practice to use forged, hardened-steel work rolls with close to 100 Scleroscope hardness and a finish dependent upon the product to be rolled. The work rolls are ground with crowns ranging from zero to 0.010 inch, the crown being generally reduced in successive stands. Thus, it may be as high as 0.007-0.010 inch in the first-stand but only 0.002 inch or less in the last-stand.

Backup rolls are generally of cast steel and may be ground concave to the extent of 0.005 inch in the case of top rolls and of 0.010 inch in the case of bottom rolls. However, the crowns used in practice for both work and back rolls are dependent upon a number of parameters, including the roll-cooling system design and hence these crowns, which provide the desired flatness in the rolled product, are usually established experimentally.

Roll life in cold mills varies considerably. For sheet rolling, work rolls may be used to process up to 3,000 tons or more between dressings at the first-stand, 1,000-1,100 tons at the second-stand, and 600-800 tons on the third-stand. Backup rolls may average 20,000 tons between dressings in the case of bulk sheet or strip production. In tinplate rolling, many mills change the work rolls of the last finishing stand after rolling 150-175 tons.

Modern tandem cold mills are driven by variable speed d-c motors operated under variable speed control. Main drive power to the mill stands has shown a gradual increase through the

---

[†] J. Davies, R. W. Jackson and J. A. Tracy, "Design Criteria, Development, and Test Activities for a Six-Stand Tandem Mill Hydraulic Screw-Down System", Chapter in "Hydraulic Control of Rolling Mills and Forging Plants", Iron and Steel Institute, (Pub. 142), 1972, pp. 55-100.

VARIOUS TYPES OF COLD MILLS

years with the latest cold mills providing as much as 8,000 HP to an individual stand. Mill speeds have also increased until in the early 1950's, they were being designed for maximum speeds of 7,000 fpm or more.[†] However, the problems associated with the adequate cooling of the work rolls have made operating speeds much in excess of 5,000 fpm difficult to attain for prolonged periods.

In tandem-mill operation, the schedule of reductions from pass to pass must be such as to allow the mill stand speeds to fall within their allowable range (or within the "speed cone"). Three-stand mills are designed to provide an overall reduction ratio up to about 4:1, four-stand mills a ratio in the range 1.5:1 to 6:1 and five and six-stand mills a ratio ranging from 2:1 to 10:1.

Total reductions accomplished by tandem cold mills usually range from 20 to 90 per cent, with 40-65 percent reductions being typical for sheet gages and as high as 85-90 per cent for tinplate gages. Individual pass reductions generally average about 30 per cent but may be as high as 45 per cent or more.

In practice, it is found that, without the operation of an automatic gage control system, the final gage of the rolled product becomes heavy if the mill speed is decreased, and vice versa. Thus, the acceleration and deceleration involved in the rolling of each coil in a low mill speed range through which the automatic gage control system is not operative results in heavy head and tail ends. The economic importance of this effect may be lessened by using large coils, by obtaining rapid acceleration and deceleration and by suitable adjustments of the screws and/or the strip tensions.

During the last decade, three fully continuous tandem mills have been placed in operation. The first was commissioned in 1967 at the Trentwood plant of Kaiser Aluminum and Chemical Corporation for the cold rolling of aluminum alloys. The second (and the first with respect to the cold reduction of steel products) was basically a train of Sendzimir mills installed by the Nisshin Steel Company for rolling stainless steels and the third was a 5-stand train of 4-high mills installed by Nippon Kokan for the rolling of low carbon steel strip.

These mills are intended to provide higher productivity by eliminating the threading, acceleration and deceleration associated with the rolling of individual coils. Coils fed to these facilities are welded end-to-end and, during the welding operation, strip for rolling is withdrawn from an accumulator as in the case of other continuous strip processing facilities, such as continuous annealing lines.

It should be pointed out that such mills cannot, in the strictest sense, be fully continuous in that the mill must be periodically stopped to permit the changing of rolls and the clearance of cobbles. Moreover, their design and operation present complex technological problems (such as gage change "on-the-fly") and their successful operation is likely to depend significantly on the product mix rolled. Accordingly, even though a mill similar to Nippon Kokan's was recently placed in operation by National Steel Corp. at Weirton, West Virginia, there has been a reluctance to adopt fully continuous rolling in the steel industry. However, some of the more recent tandem cold mills could be converted to fully continuous operation if necessary.

2-13 Tandem Sheet Mills

Sheet mills are utilized to reduce hot-rolled strip with a thickness in the range 0.080 to 0.250 inch to a finish thickness usually lying in the range 0.0141 to 0.240 inch. Historically, 3 and 4-stand trains of 4-high mills have found widespread use for sheet production. The energy consumption associated with 75-inch wide, 3-stand tandem mills rolling low carbon steels as a function of elongation is illustrated in Figure 2-29.

A similar curve for a 4-stand mill rolling galvanized roofing stock is shown in Figure 2-30.

---

† R. E. Noble, "Trends in Recent Cold Mill Installations", Iron and Steel Engineer Year Book, 1954, pp. 540-551.

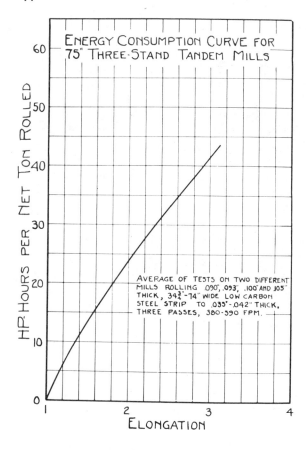

Figure 2-29:
Energy Consumption Curve for 75-Inch Three-Stand Tandem Mills.

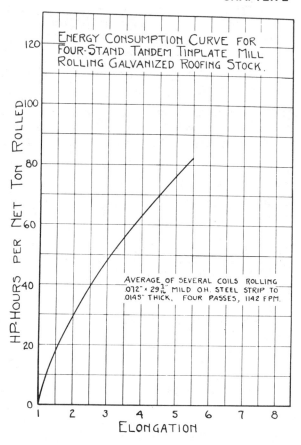

Figure 2-30:
Energy Consumption Curve for Four-Stand Tandem Mill Rolling Galvanized Roofing Stock.

Figure 2-31:
Modern Five-Stand Cold Reduction Mill (Youngstown Sheet and Tube Company).

VARIOUS TYPES OF COLD MILLS

Modern sheet mills generally utilize five stands as exemplified by the 80-inch sheet mill installed by the Youngstown Sheet and Tube Company, at Indiana Harbor, Indiana (Figure 2-31).† Data pertinent to this mill, which features individual drives to all work rolls are listed in Table 2-3.

Table 2-3
Data Pertaining to 80-inch Sheet Mill
(Youngstown Sheet and Tube Company)

No. 1 stand	6,000 HP
No. 2 stand	8,000 HP
No. 3 stand	8,000 HP
No. 4 stand	8,000 HP
No. 5 stand	7,000 HP
Total	37,000 HP
Backup roll size	80 x 60 in.
Work roll size	80 x 26 in.
Mill speed	5,250 fpm
Strip width	18 inches min. to 74 inches max.
Delivery gages	0.009 inch min. to 0.104 inch max.
Maximum coil weight	100,000 lb.
Delivery coil ID	20 inches
Delivery coil OD	90 inches max.

An interesting feature of this sheet mill is the method of changing rolls. The backup rolls, as well as the work rolls, are changed with a hydraulically-operated sled which is an integral part of the mill. Each pair of work rolls is carried out of the mill housings perched on top of the bottom backup roll and is picked up by a motorized carrier attached to the overhead crane and placed in a storage rack. A fresh pair of work rolls is then placed on the bottom backup roll and the assembly retracted into the mill housing.

Rolling lubrication is utilized on all sheet mills but, generally speaking, frictional effects are not as critical in tandem sheet mills as in tin mills and secondary cold reduction mills. The hot rolled strip brought to the sheet mill from the pickle line is usually coated with a "pickle oil" which is basically a mineral oil containing various additives. At the first stand of primary tandem mills, the pickle oil acts as the rolling lubricant and water is applied to the mill rolls merely as a coolant. A conventional rolling solution is usually applied to the intermediate stands but the last stand of a tandem sheet mill may be operated in a number of different ways. Where the last stand must be used to take a significant reduction, the normal rolling solution may be applied at the stand. However, where light reductions are taken, a detergent solution may be applied, either continuously or intermittently, to prevent any buildup from occurring on the rolls and to minimize the quantity of residual oil remaining on the surface of the sheet product. (This is important since oil is not generally removed from sheet products prior to box annealing.)

2-14 Tandem Mills for Tinplate Production

Frequently referred to as "tin mills", such facilities are now generally of the 5- or 6-stand type intended to reduce hot-rolled strip with a thickness in the range 0.070 to 0.100 inch down to a thickness in the range 0.007 to 0.018 inch in widths from 24 to 36 inches. The coiling speeds of these mills are generally 4,000 to 5,000 fpm, although some have been powered for delivery speeds up to 7,000 fpm.•

---

† J. C. Siadak, "Youngstown's 80-Inch Cold Rolled Sheet Mill at Indiana Harbor", Iron and Steel Engineer Year Book, 1966, pp. 959-963.
• G. Perrault, Jr., "High-Speed Mills and Their Application to Ferrous and Non-Ferrous Rolling", Iron and Steel Engineer Year Book, 1953, pp. 494-502.

The high speed of tin mills creates a number of technical problems in their design and operation. Perhaps the most important relates to heat removal from both the rolls and the strip.† Excessive temperatures of the rolls tend to produce poor strip shape and also reduce the effectiveness of the rolling lubricant. At the same time, excessive coiling temperatures result in strip surfaces exhibiting oxide films which adversely affect the quality of tinplate produced from the rolled strip. As a consequence, the use of satisfactory rolling lubricants and the provision of adequate roll cooling is imperative in the operation of tin mills.

Both direct and recirculating systems are used with tandem tin mills. In the former, palm-oil or a palm-oil substitute is applied to the strip and cooling water to the rolls. In the latter system, a recirculated aqueous emulsion of such oils (usually with a concentration of 4-5 per cent) is applied copiously to both the strip and the rolls, being filtered, skimmed and maintained at proper volume and concentration levels during its recycling. Since the emulsion is quasi-stable, it must be kept in rapid motion to avoid separation of the oil and, in addition, since the emulsion temperature generally ranges from 120-170°F, the lubrication system should preferably be thermally insulated and the piping steam-traced. From the recirculating loop, the mixture is often fed to the manifolds at the mill by individual rotary gear-type displacement feeder pumps driven by adjustable speed motors. At the manifolds, remotely-controlled three-way valves allow the pump delivery to be bypassed back to the tank when the mill is standing idle or being threaded. As a result of careful system design and economic operational practices, the usage rate of rolling oils is approximately 3 to 4 lbs. per ton of steel rolled.

In mills using recirculated rolling solutions, the vapor or fog that would otherwise envelop the mill stands necessitates the partial closure of the stands with hoods and baffles that leave openings between the stands just large enough to allow the entry of the operators for strip threading and guide maintenance. The partially enclosed spaces between and around the stands are then connected to an exhaust fan system which draws off the vapors with the air that enters between the stands. The fan system usually discharges through a baffle-type separator into a vertical stack of adequate height and area where vapors escaping the separator have a chance to condense and drop back. Under these circumstances, the discharge at the top of the stack is as free as possible of oil and water vapor.

The production of strip of uniform gage is rendered more difficult by the high speeds of these mills. The variation of the yield strength of the strip with the speed of its deformation and the variation of the effectiveness of the rolling lubricant with mill speed both combine to render gage control more difficult than with slower speed mills. Fortunately, improved gaging instruments and automatic control systems as discussed in Chapter 5 have essentially provided an adequate solution to this problem. However, difficulties in obtaining gage uniformity still occur because of backup roll eccentricity.

The rotational speeds of the work rolls in the high-speed stands of recently built tin mills have outstripped the speeds of the driving motors. Accordingly, driving gears have been used to provide rotational speeds for the work rolls up to twice the speeds of the drive motors. In other cases, the larger diameter backup rolls of the mills have been directly driven by the mill motors.

Modern tandem tin mills must possess good facilities for handling coils weighing from 30,000 to 60,000 lb. Feeding the coil to the mill involves bringing it up to position on a ramp or entry conveyor, lowering it to coil box level, rotating it to present the leading end to the bite of the feeding pinch rolls and releasing it to the mill after the leading end of the new coil has been entered through open entry guides into the bite of the mill rolls. At the delivery end, a belt wrapper is necessary to save time and effort in engaging the strip on the tension-reel block and is used for all except the heaviest gages. An elevating stripper is also generally provided for supporting the coil during its removal from the mandrel and for transferring it to a ramp or conveyor.

---

† R. E. Noble, "Trends in Recent Cold Mill Installations", Iron and Steel Engineer Year Book, 1954, pp. 540-551.

VARIOUS TYPES OF COLD MILLS

A typical example of a five-stand tandem tin mill is that installed in 1951 in the Fairless Works of the U. S. Steel Corporation at Fairless Hills, Pennsylvania.[†] Shown in Figure 2-32, the mill has a capacity in excess of 700,000 tons per annum. The stands are arranged as indicated in Figure 2-33 and utilize rolls of 48 inch face, the work rolls having a nominal diameter of 21 inches and the backup rolls a nominal diameter of 53 inches. The mill has a top speed close to 7,000 fpm and a total rated power of 21,000 HP. The first two stands are driven directly from the motors through conventional pinion stands, whereas stands 3, 4 and 5 have speed-increaser drives with ratios of 1.5 to 1, 1.5 to 1 and 2.1 respectively. On these last three stands, each work roll is driven separately by its own motor with no mechanical connection (other than the workpiece) existing between the rolls. The data relative to the mill motors are as given in Table 2-4.

Figure 2-32:
A High-Speed, 5-Stand Tandem Cold Mill Used for Tinplate Production.

Figure 2-33:
Layout of the 5-Stand Mill Shown in Figure 2-32.

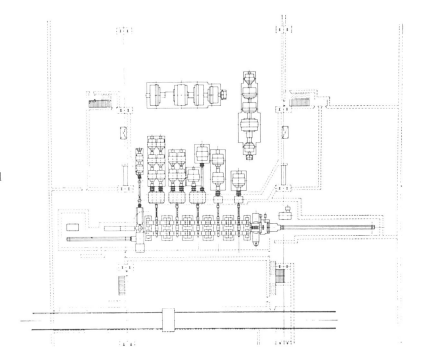

---

† R. E. Noble, "Trends in Recent Cold Mill Installations", Iron and Steel Engineer Year Book, 1954, pp. 541-551.

Table 2-4
Five-Stand Mill Drive Motors

Stand No.	Motor Power (HP)	Type of Motor	Motor Speed (rpm)	Max. Roll Speed (fpm)
1	1,750	Single Armature	110/310	1,710
2	3,500	Double Armature	165/420	2,320
3	4,500	Single Armature Twin Drive	200/415	3,460
4	4,500	Double Armature Twin Drive	300/625	5,210
5	6,000	Triple Armature Twin Drive	345/635	6,985
Tension Reel	1,400		350/1,700	

More recently, six-stand tandem cold mills have been introduced into service for tinplate production. An example of such a mill is the computer-controlled facility shown in Figure 2-34 which was installed at the Fairfield Works of the U. S. Steel Corporation in 1962.[†] Rated at close to 780,000 net tons per annum and designed for coiling speeds up to 7,000 fpm, this 52-inch wide mill features work rolls 21 and 23 inches in diameter and backup rolls 56 inches in diameter. It rolls strip with entry gages in the range 0.070 to 0.106 inch to finish thicknesses of 0.006 to 0.015 inch under manual or computer control. The data relative to the drive motors are as presented in Table 2-5.

Figure 2-34: Six-Stand Tandem Cold Mill for Tinplate Production.

Computer control of the mill was initiated early in 1967, the computer performing the functions of mill setup, control and data-logging.[*] The central processor, console and input-output equipment are located in the computer room (Figure 2-35). The manual data-entry station, card reader, coil ticket typewriter, coil quality display, percent draft display and alarm panel are located on the mill floor. Assorted indicating lights, pushbuttons and decade switches are on the operators' cabinets. The production summary and downtime summary typewriter is located in the production superintendent's office, and the production logging typewriter is located in the Cold Reduction Department's Production Planning Office.

[†] "Cold Strip Mills in U.S.", 33 Magazine, September, 1965.

[*] B. C. Bradley, M. O. Smith, Jr., and H. N. Cox, "U. S. Steel — Fairfield Works' 6-Stand Computer Controlled Cold Mill", Iron and Steel Engineer Year Book, 1970, pp. 257-264.

VARIOUS TYPES OF COLD MILLS

Table 2-5
Six-Stand Mill Drive Motors

Stand	Motor Power (HP)	Type of Motor
1	2,000	Single Armature
2	2 x 2,000	Single Armature
3	2 x 2,500	Single Armature
4	2 x 3,000	Double Armature
5	2 x 3,000	Double Armature
6	2 x 3,500	Double Armature
Tension Reel	2,100	

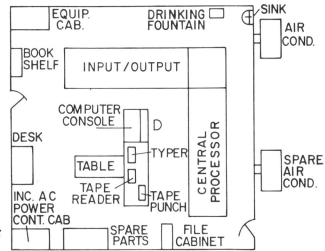

Figure 2-35:
Layout of the Computer Room for the 6-Stand Tandem Cold Mill.

Absolutely essential to the proper performance of the computer is accurate input information relating to the product entering and to be delivered from the mill. Such information is obtained from an order data card, one order card corresponding to each order and a separate coil card for each coil of the order.

Calculation of a mill setup begins with the total elongation which will be given to the strip. Total elongation, steel hardness, strip width, mill delivery speed, and mechanical losses determine the total power required by the drive motors. Occasionally, this theoretical total power will be in excess of the available motor horsepower and the computer program then establishes a lower mill speed compatible to the available drive power.

The drafting criteria distributes the total required horsepower among the six stands taking into account the power associated with the strip tension at the various locations along the pass line of the mill. From an appropriate power-versus-elongation curve, the six-stand elongations are determined from the distributed loads. Mill delivery speed and thickness and the six elongations are used to compute interstand strip thicknesses and speeds. Using appropriate values for the forward slip (1 to 7.5 per cent), the work roll surface and rotational speeds are calculated for each stand and the computed values compared with the rated speeds of the mill motors. If any stand is found to be outside the rated field range, either the mill delivery speed is reduced or the load is redistributed as necessary. Either of these actions requires a recalculation for the new conditions. The rotational speeds of the work rolls, thus calculated, are used to determine the settings of the field rheostats for the corresponding stands.

From data relative to the reduction taken at each stand and the entry and exit strip tensions, the computer predicts the corresponding rolling force. From a curve of force-versus-mill stretch, the stretch is then computed. This anticipated mill stretch, the mill stretch occurring at screw-position zeroing, the loaded roll-gap opening (strip delivery thickness) and a term which takes roll heating into consideration are all used to compute screw positions.

When the mill is operating under computer setup and has reached running speed, the power, force, tension and screw setup models are modified periodically throughout the rolling of a coil to improve the setup for successive coils. This successive updating procedure maintains a total mill model which closely tracks changes in mill roll temperatures and in the rolling solution. The coil-to-coil improvements in calculated setups resulting from the adaptive updating of the mathematical models is one of the significant advantages of calculated setups over setups obtained from stored schedules or mathematical models with no adaptive updating.

In comparing a week's performance of the mill, it was found that under computer control there was 1.14 per cent less strip out of permissible gage tolerance. Figure 2-36 shows a more detailed comparison of manual versus computer operation of the mill with respect to output gage.

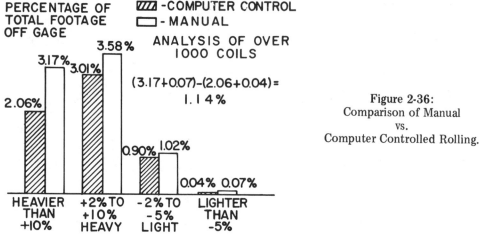

Figure 2-36: Comparison of Manual vs. Computer Controlled Rolling.

2-15 Fully Continuous Cold Mills

The Sendzimir tandem mill installed by the Nisshin Steel Company is illustrated diagrammatically in Figure 2-37. The facility is designed to roll stainless steel coils up to 50 inches wide and weighing a maximum of 24 tons. The hot rolled strips of the 300 and 400 series reach the mill with a thickness in the range 0.063 inch to 0.255 inch and they are reduced to a thickness in the range 0.012 inch to 0.158 inch at a maximum rolling speed of 1,969 fpm.

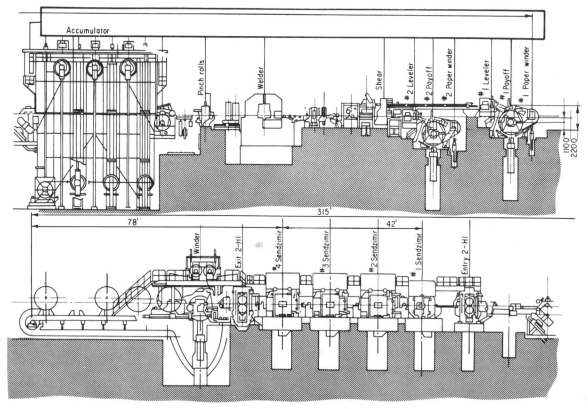

Figure 2-37:
General Arrangement of the Sendzimir Tandem Mill (Entry End at Upper, Right-Hand Side of Illustration).

---

† T. Ohama, S. Sasaki, and M. G. Sendzimir, "World's First Sendzimir Tandem Mill", Iron and Steel Engineer, April, 1973, pp. 25-31.

VARIOUS TYPES OF COLD MILLS

Because the mill is designed to be "fully-continuous", rethreading is required only after a lengthy shutdown or when the strip breaks between stands. When rethreading is required, however, a leader strip is used, this being passed through the entire line from the payoff to the exit winder.

The entry section of the line contains two coil skids, one for each payoff reel. Each reel is designed to accommodate 20-inch ID by 87-inch maximum OD coils up to 50 inches wide and is connected to a 330 HP drag generator geared to supply a maximum back tension of 11,000 lb. up to the maximum payoff speed of 985 fpm. Reels to take up the paper interleaving the wraps of the coils are also provided.

Two levellers flatten the ends of the incoming coils, each machine being driven by a 50 HP a-c motor and consisting of one set of pinch rolls and five levelling rolls.

A hydraulic double-cut shear squares the ends of a coil and the succeeding coil simultaneously and the welder can make either a flash butt weld or a resistance seam weld.

The strip accumulator was designed to provide only adequate storage under conditions of minimum rolling speed. It is capable of storing 220 feet of strip and utilizes 5-foot diameter rolls. Deflector rolls, also 5-feet in diameter, are located at the entry and exit of the accumulator, the exit being unpowered but the roll at the entry, acting as a bridle roll, is powered by a 100 HP d-c motor. An automatic strip centering roll is also incorporated at the exit side of the accumulator to assure proper tracking of the strip through the mill stands.

Data pertaining to the entry 2-high mill, the four Sendzimir stands and the exit 2-high mill are presented in Table 2-6.

Table 2-6
Tandem Mill Characteristics

Equipment	Type	Work roll diameter, in.	Rolling speed, fpm	Motor, hp	Maximum interstand tension, tons
Payoff reel	Overhung mandrel 20-in. diameter	—	0/985	330	5
Entry 2-high	Hydraulic screw down 440 tons	30	0/328/985	2x440	44
No. 1 Sendzimir	ZR22N-50	2.68	0/330/1,080	2x940	66
No. 2 Sendzimir	ZR21B-50	3.38	0/480/1,310	2x1,610	66
No. 3 Sendzimir	ZR21B-50	3.38	0/610/1,640	2x1,610	66
No. 4 Sendzimir	ZR21B-50	3.38	0/740/1,969	2x1,610	66
Exit 2-high	Hydraulic screw down 440 tons	30	0/740/1,969	2x1,000	
Tension reel	Overhung mandrel 20-in. diameter	—	0/1,969	2x1,340	22

Except for modifications associated with the bearings on shafts F and G (see Section 2-11) and the lower screw down (which can be operated to provide constant roll pressure or constant roll gap in addition to the conventional fixed position locked system), the Sendzimir stands are essentially of typical design providing a mill speed cone as shown in Figure 2-38.

Between the exit 2-high mill and the exit-side deflector roll, a hydraulic up-cut shear severs the strip after the required coil buildup has been attained on the single 20-inch diameter collapsible block winder. Supported during operation by an outboard bearing, the mandrel is driven through a gear unit by two 1,350 HP armatures providing a maximum tension of 22 tons up to maximum mill speed. Coiling is initiated by a hydraulically-operated horizontal type two-strand belt wrapper and to protect the material surface from damage, equipment for paper interleaving has been provided.

The mill is equipped with two soluble oil roll coolants and backing bearing spray systems, one servicing the entry 2-high and the first two Sendzimir stands while the other services

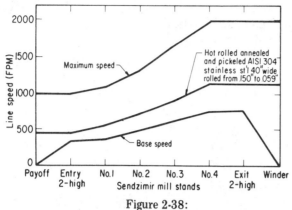

Table 2-7
5-Stand Continuous Cold Mill
Drive Motor Data

Stand No.	Kw	Rpm
1	2 x 1,650	105/315
2	2 x 2,000	225/565
3	2 x 2,000	225/565
4	2 x 2,000	225/565
5	4 x 1,100	250/750
Tension Reel No.		
1	3 x 450	245/1,225
2	3 x 450	245/1,225

Figure 2-38:
Speed Cone of the Sendzimir Tandem Mill.

the remaining two Sendzimir stands. The former uses a 50,000-gallon receiving tank and a 41,000-gallon clean tank, whereas, the latter has 40,000- and 33,000-gallon tanks respectively. For the exit 2-high mill, rolling lubrication is furnished by a relatively small, separate, self-contained system.

The tandem mill is equipped with tension control between each stand and an AGC system (see Section 5-28).

The five-stand fully continuous cold mill installed by Nippon Kokan is illustrated diagrammatically in Figure 2-39. It is intended to roll strip with widths up to 51 inches and thicknesses in the range 0.06 to 0.18 inch down to thicknesses in the range 0.006 to 0.064 inch. Coil ID at entry is 24.5 inches and, on exit, 20.375 inches. The maximum weight of coil that can be handled is 70,000 lb. and the production capacity of the mill is rated at 118,000 net tons per month for strip with an average cross section of 0.0184 inch by 35.25 inches. Maximum rolling speed of the mill is 6,000 fpm and the drive motor data are presented in Table 2-7.

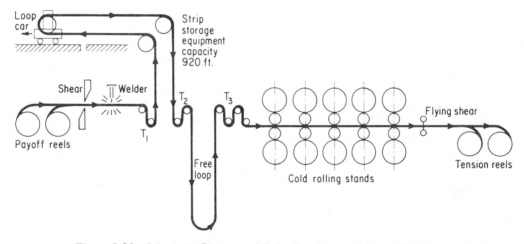

Figure 2-39: Schematic Diagram of Fully Continuous 5-Stand Cold Mill.

The work rolls used are 21.20/24.40 inches in diameter with a 57-inch face length, whereas, the backup rolls are 52.80/58.20-inch in diameter with a face length of 54.80 inches.

The welding operation at the entry end of the mill requires 1.5 minutes and to supply the mill with strip during this period, a strip accumulator is provided with a storage capacity of 920 feet of strip. The mill stands themselves are devoid of strip threading devices but are equipped with mechanisms to change the work rolls even with strip in the mill. Moreover, stands 1 and 5 feature

VARIOUS TYPES OF COLD MILLS

hydraulic as well as electromechanical screw down systems to provide improved shape and gage with respect to the rolled strip.

The mill is under computer control and key locations on the mill are monitored by television. Data concerning the coils before and after rolling are fed to the computer by instruction cards and coil cards and the mill setup is accomplished by the computer through the use of mathematical models of the rolling operation. Important controls unique to this mill include dynamic gage changing (or gage changing on-the-fly to ensure a minimum length of off-gage material) and the coil-tracking control which, by the use of weld detectors, keeps individual coils located throughout the rolling operation.

2-16 Temper or Skin-Pass Mills

Skin-pass or temper rolling is the very light reduction given to annealed stock in order to effect a surface hardening, restore temper and prevent stretcher strains or breaking of the surface in subsequent drawing operations. Reductions may range from approximately 0.5 per cent to perhaps 4 per cent in the case of "dry" temper rolling and 10 per cent in the case of "wet" temper rolling.[†]

Early coil temper mills were single-stand units with back-tension being obtained by means of brakes applied to the payoff reel and with forward tension developed between the stand and the takeup reel. Demands for increased hardness and better finishes necessitated the rerolling of many coils on such mills until, only a few years after the beginning of temper rolling, two-stand temper mills made their appearance. This type of mill, shown in Figure 2-40, designed for single pass coil tempering, was gradually improved mechanically as well as electrically until today these mills roll at speeds up to 6,000 fpm. Quite frequently, because of the very light reductions given to the strip, only a single roll in each stand is driven. Otherwise, with the use of conventional pinion stands, slight mismatches in the diameters of the pairs of work rolls would cause the generation of very high torques in the spindles.

Figure 2-40: Two-Stand, 4-High, Tin-Temper Mill.

---

† "The Modern Strip Mill", AISE, Pittsburgh, Pennsylvania, 1941, p. 254.

In the early 1950's, a review of existing temper mills was made by Sellers et. al.[†] who established the power data for the various mills as presented in Table 2-8. The two values in the "Average" column represent averages of base speed and top speed per unit power for variable speed motor drives.

Table 2-8
Average Installed Horsepower/Inch Mill Width/100 fpm for Temper mills

	Minimum	Maximum	Average
Entry Reel	0.061	0.286	0.148
Entry Tension Bridle	0.145	0.634	0.22 — 0.385
Stand No. 1	0.225	1.625	0.419 — 0.802
Stand No. 2	0.267	1.625	0.500 — 0.914
Delivery Tension Bridle	0.267	1.370	0.442 — 0.842
Reel	0.119	0.610	0.212

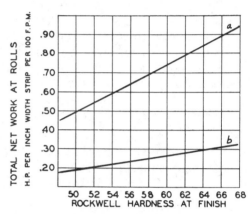

VARIATION OF TOTAL WORK ON STRIP WITH HARDNESS.
a - TWO STAND MILLS WITH 18" DIA. WORK ROLLS.
b - SINGLE STAND MILLS WITH 18" DIA. WORK ROLLS AND TWO STAND MILLS WITH 9" TO 12" DIA. WORK ROLLS.

Figure 2-41

Curves showing the variation of total work with finished strip hardness relating to single- and two-stand mills are shown in Figure 2-41 and the curves showing idling losses for several mills are given in Figure 2-42.

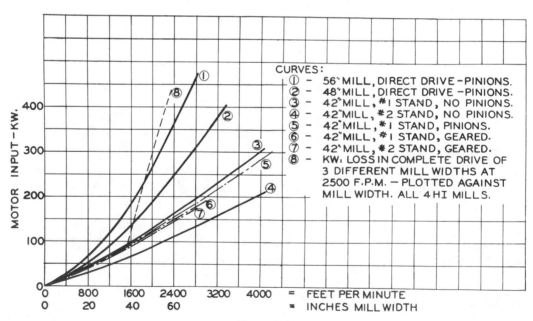

Figure 2-42:
Curves Showing Idle Running Losses for Several Temper Mills.

[†] J. F. Sellers, R. M. Peeples, and A. C. Halter, "Temper Rolling", Iron and Steel Engineer Year Book, 1952, pp. 364-375.

VARIOUS TYPES OF COLD MILLS

A typical 4-high, 2-stand, tandem temper mill used for tinplate production is shown in Figure 2-43. This mill is installed at the Trostre Works of the British Steel Corporation and pertinent data relative to its specifications are given in Table 2-9.

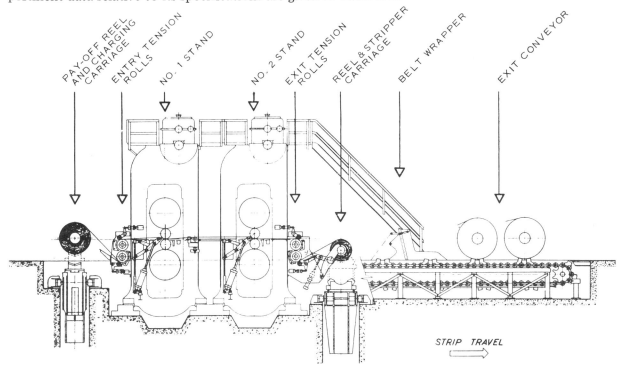

Figure 2-43:
Four-High, Two-Stand, Tandem Temper Mill (Trostre Works, British Steel Corp.).

Table 2-9
Data Pertaining To Two-Stand Temper Mill
(Trostre Works, British Steel Corporation)
Max. Speed — 4,000 fpm

Coil Sizes:	Max. dia.	72"
	Min. dia.	36"
	Max. width	44"
	Min. width	16"
	Max. weight	30,000 lbs.
	Thickness range	.025" to .004".
Roll Drives:	Stands No. 1 & 2, 1,000 HP at 700/950 rpm.	
Entry Conveyor:	Slat type with down tilter, etc., to handle 44" wide coils.	
Backup Rolls:	Max. dia. 53"	Min. dia. 49"
	Body face 47"	Overall length 12'-8¼"
	Weight 45,500 lbs.	
	Bearings — Morgoil taper neck sleeve with SKF roller thrust bearings.	
Work Rolls:	Max. dia. 18"	Min. dia. 16½"
	Body face 48"	Overall length 10'-4"
	Weight 5,150 lbs.	
	Bearings. Timken four-row tapered roller.	
Housings:	Weight 96 tons	
	Post area 870 sq. ins.	
Screw downs:	Two-motor operated through double reduction spur gearing; each drive 50 HP at 550/1,750 rpm	
	Reduction 1,000 to 1.	
	Screwing speed .22 to .66 ins. per minute.	
Hydraulics:	One system for roll balance and roll change rig.	
	One system for auxiliaries.	

2-17 Secondary Cold Reduction Mills

Installed in the early 1960's, these mills were intended to provide reductions in the range 30-40 per cent to black plate with a thickness of 0.010 to 0.018 inch. Continuously annealed strip is usually fed directly to these mills for processing but box-annealed coils are sometimes temper rolled prior to secondary cold reduction.

One, two, three and even five-stand installations are used as secondary cold reduction mills,† • ‡ but as far as the tandem mills are concerned most of the drafting occurs at only one stand. Mill speeds are generally less than those of tandem tin mills but the operating practices are essentially the same. However, because of the large work-roll diameter-to-strip-thickness ratios encountered in secondary cold reduction, only the most effective rolling lubricants must be used. Even so, the attainment of satisfactory flatness in the rolled product is more difficult with secondary than with primary cold reduction mills.

Because of their modern construction, secondary cold reduction mills frequently exhibit some of the more recent trends in mill design. For example, as will be seen later, prestressed stands are sometimes used. Moreover, they often feature the best available instrumentation and automatic gage control systems and, in the not-too-distant future, such mills will probably utilize rolled-strip shape monitoring and controlling systems.

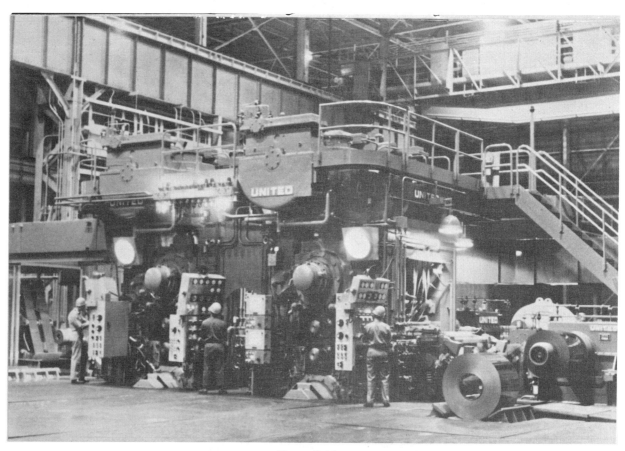

Figure 2-44:
Two-Stand Secondary Cold Reduction Mill.

---

† T. A. Bessent and R. F. Long, "Mechanical Features of the 2-Stand Duo-Reducing Mill at Sparrows Point", Iron and Steel Engineer Year Book, 1966, pp. 189-199.

• O. B. Thomson and V. S. Coleman, "Kaiser Steel's 3-Stand Cold Reducing Mill for Light-Weight Tin Mill Products", Iron and Steel Engineer Year Book, 1967, pp. 543-551.

‡ "French Cold Reduction Mill in Full Production", Sheet Metal Industries, October, 1973, pp. 554-556.

VARIOUS TYPES OF COLD MILLS

A typical two-stand secondary cold rolling mill is shown in Figure 2-44. It is a 21-inch and 53-inch by 48-inch mill designed to provide reductions up to 50 per cent in tinplate or annealed black plate at speeds close to 3,000 fpm. The ratings of the various mill motors are listed in Table 2-10.

Table 2-10
Motor Ratings for Two-Stand Secondary Cold Reduction Mill

Entry Reel	400 hp, 375 v, 870 amp
Entry Tension Bridle	
a) Bottom Roll	100 hp, 240 v, 350 amp
b) Top Roll	150 hp, 240 v, 512 amp
No. 1 Stand	
a) Top Roll	1,000 hp, 600 v, 1,333 amp
b) Bottom Roll	1,000 hp, 600 v, 1,333 amp
No. 2 Stand	1,000 hp, 600 v, 1,333 amp
Exit Tension Bridle	
a) Top Roll	150 hp, 240 v, 512 amp
b) Bottom Roll	100 hp, 240 v, 350 amp
Delivery Reel	400 hp, 375 v, 870 amp

A schematic diagram of the general arrangement of the mill is shown in Figure 2-45. It will be noted that tensiometers have been provided to measure entry and interstand tensions and that X-ray thickness gages are utilized to measure interstand and exit strip thickness. Load cells are located under each bottom backup roll chock to indicate the total and the differential rolling forces at each stand.

The locations of the rolling solution sprays, cooling water sprays and air jets for removing excess moisture are also shown in Figure 2-45. A 9 per cent aqueous solution of a commercial rolling oil is used as the rolling lubricant with the temperature of the solution being maintained at 140°F. Rolling solution and cooling water are applied only at the first-stand, the second-stand being operated "dry". It should be noted, however, that direct-application rolling solution systems are also utilized with mills of this type.

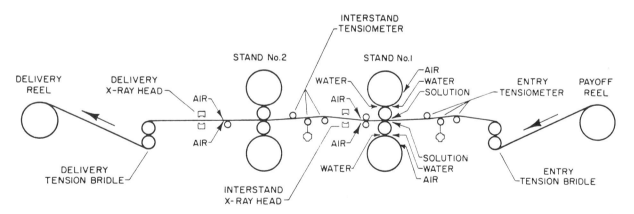

Figure 2-45: Schematic Diagram of 2-Stand Secondary Cold Reduction Mill.

In the operation of this mill, most of the reduction is taken at the first-stand. The energy requirements of the mill in the rolling of annealed black plate are as indicated in Figure 2-46 and the variation of specific rolling at the first-stand with reduction is shown in Figure 2-47.

It is interesting to note that, to prevent the coils of thin strip collapsing on removal from the mandrel, the coils are wound on metal or fiberboard "cores" or "sleeves". These cylinders are placed over the coiling mandrel and are subsequently retrieved from the entry end of the electrotinning or other processing line.

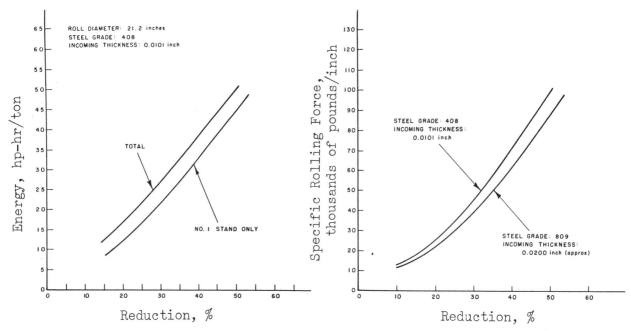

Figure 2-46:
Curves Showing the Variation of Energy With Reduction in the Rolling of Annealed Low-Carbon Steel Strip on 2-Stand SCR Mill.

Figure 2-47:
Curves Pertaining to the Rolling Force at the First Stand of a Two-Stand DCR Mill.

A secondary cold reduction mill of more recent construction installed at the Trostre Works of the British Steel Corporation is shown in Figure 2-48.[†][•] This mill, which is of the prestressed hydraulic variety under thyristor control, is capable of rolling at speeds up to 5,600 fpm and can handle coils weighing up to 40,000 lb. The other pertinent specifications of the mill are as given in Table 2-11.

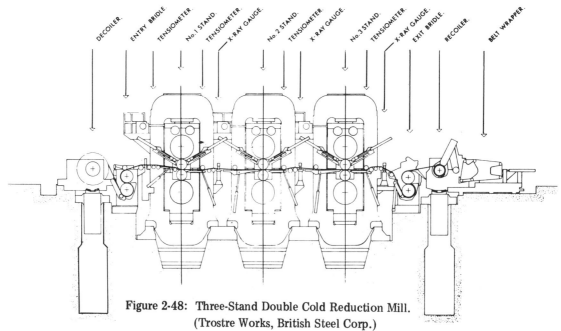

Figure 2-48: Three-Stand Double Cold Reduction Mill.
(Trostre Works, British Steel Corp.)

[†] "Prestressed Mill for Extra-Thin Tinplate", Sheet Metal Industries, September, 1966, pp. 707-709.
[•] D. Veck, P. Thomas, and D. O'Sullivan, "Operating Experience With a Prestressed Hydraulic Rolling Mill", Portion of "Hydraulic Control of Rolling Mills and Forging Plant", Iron and Steel Institute, Pub. 142, 1972, pp. 26-29.

VARIOUS TYPES OF COLD MILLS 59

Table 2-11
Data Pertaining To
Three-Stand Tandem Double Cold Reduction
and Temper Rolling Mill

Coil Sizes:	Max. diameter	72"
	Max. width	44"
	Min. width	24"
	Max. weight	40,000 lbs.
Motor Powers and Strip Speeds:		
	Decoiler Drag Generators:	
	Total 0/750/750 KW	@0/290/1,300 rpm
	Bridle Drag Generators:	
	Total 0/750/750 KW	@0/750/1,000 rpm
	Stands 1, 2 & 3:	
	0/5,000/5,000 HP	@0/2,900/5,800 fpm
	Exit Bridle Motors:	
	Total 0/1,005/1,005 HP	@0/750/1,000 rpm
	Recoiler Motors:	
	Total 0/1,000/1,000 HP	@0/290/1,300 rpm
Backup Rolls:	Max. dia. 53"	Min. dia. 49"
	Body face 49"	Overall length 13'-9-3/4"
	Weight 55,400 lbs.	
	Bearings: Timken tapered roller	
Work Rolls:	Max. dia. 23"	Min. dia. 21-3/8"
	Body face 48"	Overall length 13'-4-1/2"
	Weight 10,250 lbs.	
	Bearings: Timken tapered roller	
Duty:	Double cold reduction of Annealed strip	
	Max. duty .0176" reduced to .011"	
	Min. duty .0066" reduced to .0044"	
	Double cold reduction of tinned strip	
	Duty: .0066" reduced to .002"	
	Double cold reduction of hard rolled strip	
	Max. duty .012" reduced to .0082"	
	Min. duty .008" reduced to .004"	
	Temper Rolling:	
	Max. strip thickness .036"	
	Min. strip thickness .004"	
	Rolling speed 5,600 fpm max.	

2-18 Foil Mills

Although cluster-type mills with small diameter work rolls have been used for many years to cold reduce narrow strip to thicknesses less than 0.001 inch, the high-speed commercial production of low-carbon steel foil in comparable thicknesses yet in widths up to 30 inches or more was not possible until the early 1960's. Using secondary cold reduction mills, it was found that, under carefully controlled conditions, strip as thin as 0.002-inch could be conveniently rolled. It was believed that a greater commercial demand would exist for even thinner strip and, for this reason, the U. S. Steel Corporation installed at its Gary Works the first mill specifically designed to roll foil into coils as large as 40,000 lb.[†] This mill, illustrated in Figure 1-26, is shown schematically in Figure 2-49.

This production facility is a single-stand, hydraulically-loaded mill with 8-inch and 44-inch by 46-inch rolls capable of exerting rolling forces up to 1,000 tons.[•] It is designed to operate at speeds up to 2,000 fpm with the drive arrangements as shown in Figure 2-50. A direct application lubrication system is used with the mill and it should be noted that the effectiveness of the lubricant, for a given set of rolling conditions, exerts a significant influence on the gage of the rolled foil.

Rolling schedules on the foil mill are flexible. For tin-coated foil, reductions of 60-80 per cent are achieved in a single pass using an initial tin-coating weight on the incoming strip of 0.25

[†] R. C. Haab and M. B. Jacobs, "Steel Foil Production and Applications", AISI Yearbook, 1967, AISI, New York, 1968.

[•] "How They'll Roll Steel Foil at U. S. Steel's Gary Mill", 33 Magazine, January, 1965.

lb. per base box. With uncoated foil, lesser reductions are given to the strip. In all cases, the rolled foil is wound onto 16-inch ID steel cores with a wall thickness of 1-inch.

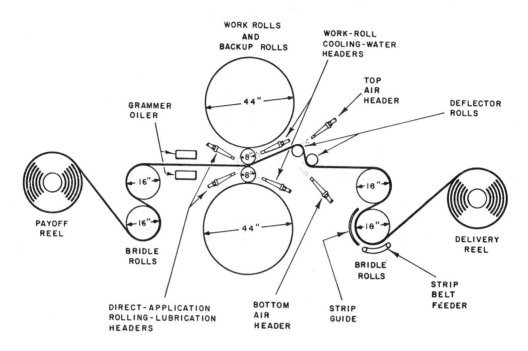

Figure 2-49: Schematic Diagram of Foil Mill.

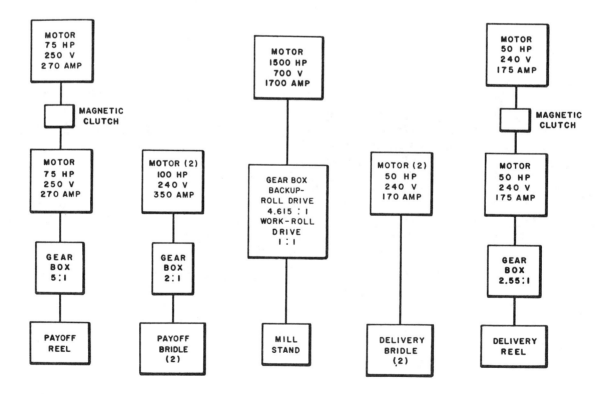

Figure 2-50: Drive Arrangements for Foil Mill.

VARIOUS TYPES OF COLD MILLS

The rolled foil is commercially available either in the as-rolled or the annealed condition. In the annealing operation, the cores are supported horizontally in a frame in the annealing furnace to minimize pressure on the coil wraps.

Sendzimir or similar cluster mills are generally utilized to produce strip to thicknesses less than 0.001 inch. Such mills may have rolls with a diameter of 1-cm or less[†] but all the mill rolls must be ground with very high standards of tolerance and symmetry. The utmost importance is attached to grinding the drive rolls, and the set should be matched to the required tolerance so that the loading on the pinion gears and coupling shafts is evenly distributed and the peripheral speeds of the work rolls matched as closely as possible.[•]

2-19 Reversing Mills

The reversing mill is a single-stand unit, through which the strip is passed back and forth for a number of passes to obtain the required reduction. On one side of the mill, means are provided for uncoiling and feeding the strip through the mill for winding on the tension reel on the opposite side. After the first pass, the tail end of the strip is gripped by the second tension reel on the uncoiler side of the mill. The strip is then rolled in successive passes in alternate directions until the final desired thickness is attained. On the last pass, the tail end of the strip is released from the unwinding tension reel, the coil completely wound on the winding reel and then removed from the mandrel.

During alternate passes on all reversing mills, the reel used during the previous pass as a tension reel is used as a regenerative feed reel for maintaining back tension. It should also be noted that it is customary to take an odd number of passes on a reversing mill in order to speed up the threading operation of the next coil while the cold-reduced coil is being removed from the tension reel on the opposite side of the mill. Also, in the operation of a reversing mill, the pass schedule is usually chosen so as to maintain reasonably constant drive power and rolling force during successive passes.

The tandem mill is generally thought of as a high-production, single-purpose facility, while the reversing mill has been considered applicable to the relatively limited production of one or more types of products. However, a careful consideration of all factors involved should be undertaken before a choice is made in favor of one type of mill or the other.[‡]

One point that must be remembered concerning the operation of the reversing mill is the amount of end scrap produced. Since a certain amount of strip at each end of the coil must be retained on the tension reels, scrap corresponding to about 20 feet of strip per coil at its original gage, must be charged against the economics of reversing cold reduction. Where costly materials are involved, or where short coils are, of necessity, being rolled, in the earlier stages of rolling the mill may frequently be operated as a one-way mill.[‡]

Table 2-12 lists a number of reversing mills installed in the U.S.A. which are used primarily for the rolling of electrical and stainless steels.[*] It is to be noted that the power applied to each reel is about half that associated with the mill drive spindles.

A typical single-stand reversing mill is shown in Figure 2-51. This 12-inch and 26-inch by 24-inch four-high mill is installed at the Fontana, California Plant of the Kaiser Steel Corporation.[□] It is powered by a 1,000 HP reversing motor 400/960 tpm at 600 volts and is driven through a combination gear drive and pinion stand with a gear ratio of 2.64 to 1. The full voltage strip speed with 12-inch working rolls is, therefore, 475 to 1,140 feet/minute. The tension reels on both sides of the mill have 20-inch diameter collapsible blocks and are powered by 300 HP motors through single-reduction gear drives.

[†] F. Wiesner, "The Development of the Rolling of Thin Foils", (In Czechoslovakian), Hutn. Listy, 1968, 23 (11), pp. 805-808.

[•] H. H. Scholefield, J. E. Riley, E. C. Larkman, and D. W. Collins, "Some Experiences in Cold Rolling Thin Alloy Strip", Journal of the Institute of Metals, Vol. 88, 1959-60, pp. 289-295.

[‡] M. D. Stone, "Reversing and Tandem Cold Mills", Iron and Steel Engineer Year Book, 1947, pp. 265-272.

[□] R. E. Noble, "Trends in Recent Cold Mill Installations", Iron and Steel Engineer Year Book, 1954, pp. 540-551.

[*] Dimensions in inches; power in HP or kw and speed in feet/minute.

Table 2-12
A Selection of Reversing Mills Installed
In the U.S.A.

Width	Work Roll Diameter	Backup Roll Diameter	Power to Rolls	Total Power to Reels	Speed	Max. Width of Sheet	Max. Entry Gage	Min. Finish Gage
56	16.5	53	2 x 1,750	100 kw 2 x 500	600/1,200	51	0.14	0.018
30	6.0	35	2 x 750	2 x 350	825/1,650	28	0.060 0.029 Si	0.020
36½	7.0	34	2 x 750	2 x 350	825/1,650	34	0.200	0.003
56	19.0	49.0	2,500	1,000	1,650	50	0.140	0.014
68	19.0	49.0	5,000	2 x 750	1,650	65	0.140	0.014
42	15.5	35.5	1,000	125	282/564	38	0.080	0.009
43	16.5	45.0	1,500	600	500/1,000	48	0.025	0.010
42	16.5	49	2 x 2,750	100 kw D-C Gen	1,040/2,000	40	0.080	0.009
56	17.0	53.5	2,000	100 kw D-C Gen	340/850	52	0.187	0.018
44	10.0	43.5	1,500	150 kw	1,000/2,000	38	0.125	0.010
30	10.5	36						
30	6.5	30						
56	18.0	51.5	3,000	1,250	2,000	56	0.160	0.012
56	19.0	49.0		100 kw	1,750	50	0.250	0.010
44		46	2,500	150 kw	493/1,300	38	0.200	0.010
68	21	56	3,000	2,000	3,000	62	0.187	0.0125
84	21	56						
42	8	53						
54	21¾	53						

(Dimensions in inches; power in HP or kw & speed in ft/min.)

Figure 2-51: A Typical Single-Stand Reversing Mill.

The top work roll and top backup roll of the mill are balanced by springs located within their respective bearing chocks. Pickled coils are fed to the mill for the first pass from a coil feeder through pinch rolls and a three-roll straightener. Successive passes are wound under tension on their respective tension reels. The gage of the strip is read by two roller contact-type electrolimit gages located one on each side of the mill between mill and reels.

A more recent reversing 4-high cold mill of the Algoma Steel Corporation, Ltd. is shown in Figure 2-52.[†] This 80-inch mill, with its backup rolls driven by separate motors, was designed for the maximum feasible degree of automatic operation and features:

1) Speed regulators on both of the main drive motors.

2) Preselection of mill speed, roll force, gage and front and back reel tensions.

3) Coil end detection and automatic reel stop (CEDARS). (After the first pass, the control remembers the coil end and automatically slows down the mill at the predetermined decelerating rate and stops the mill with a minimum of steel left on the mandrel.)

4) Automatic sequencing and mill reversal. (Provided the preselected values have been set and the CEDARS control is in operation, when the mill is stopped, the new

---

† A. J. F. MacQueen, "Finding a Practical Method for Calculating Roll Force in Wide Reversing Cold Mills", Iron and Steel Engineer Year Book, 1967, pp. 425-440.

VARIOUS TYPES OF COLD MILLS

preset parameters are set, the mill is automatically reversed and accelerated to thread speed. If the operator is satisfied that all is well, he pushes the run button and the mill continues to accelerate to the run speed.)

5) Automatic gage control with three modes of operation utilizing an X-ray gage mounted on each side of the mill. (In the first mode, an error signal from the exit side X-ray gage resets the screws. In the second mode, an error signal from the exit X-ray gage changes the real tensions over a predetermined portion of reel capacity. If the tension control range is exceeded, the error signal changes the screw position sufficiently to get back to the tension control range. In the third mode, the mill is operated at constant roll force. An error in the roll force signal changes the reel tensions to maintain the rolling force constant. If the tension control range is exceeded, the error signal changes the screws and the roll force reference is recalibrated.

Figure 2-52:
An 80-inch Cold Reversing Mill Designed for the Maximum Feasible Degree of Automatic Operation.

The mill, capable of operating at speeds up to 2,200 fpm, is also equipped with hydraulic roll-bending cylinders mounted between the 20-inch diameter work rolls and the 60-inch diameter backup roll chocks; the work rolls being bent by the action of these cylinders. It is intended to roll strip ranging in width from 24 to 74 inches and with an incoming thickness in the range of 0.210 to 0.060 inch to a thickness in the range of 0.134 to 0.012 inch. The mill is powered with two 2,200 HP, 750 volt d-c, 55/138 rpm motors, and the reels are also powered by 2,200 HP motors rated at 750 volt d-c, 225/735 rpm, with a gear ratio of 2.094 to 1. The uncoiler is powered by two 125 HP, 375 volt d-c, 400/1,400 rpm motors with a gear ratio of 7.98 to 1.

Yet another example of a recently installed reversing mill is that installed by the U. S. Steel Corporation at its Cuyahoga Plant for the rolling of carbon, alloy and stainless steels.

This mill utilizes rolls 14-inch and 44-inch by 42-inch and is designed for rolling strip with a maximum entry thickness of 0.250-inch and a minimum delivery thickness of 0.008-inch. It will handle coils 20 inches ID to 60 inches OD or 25,000 lb. maximum. The installation includes a payoff stand with coil lift and transfer, automatic edge positioning, combination peeler, pinch roll and flattener, the 4-high mill stand, and two 20-inch diameter collapsible winders with coil lifts and transfers. Two 1,250 HP, 175/700 rpm motors drive the backup rolls of the mill while the reels are each driven by two 900 HP, 350/1,150 rpm motors coupled in tandem.

Chapter 3 The Components of Cold Rolling Mills

3-1 Introduction

Certain basic components are common to most types of mills, these including:

a) the work rolls and their bearings
b) the backup rolls and their bearings
c) the mill housing
d) the mill foundations
e) roll balancing systems
f) roll positioning systems
g) mill roll changing devices
h) mill protection devices
i) roll cooling and lubrication systems
j) spindles and couplings
k) pinions
l) gearing
m) motor couplings
n) drive motors
o) electrical power supply systems
p) idler and bridle rolls
q) the uncoiler and coiler
r) coil handling equipment
s) mill instrumentation, and
t) the operating controls for the mill.

These components are discussed in greater detail in the following sections of this chapter and in portions of certain other chapters. A number of them are identified in Figure 3-1, which depicts a small mill used for experimental rolling studies in a laboratory.

3-2 Work Rolls and Their Bearings

A sketch of a typical work roll for the rolling of strip (a plain roll without its bearings) is shown in Figure 3-2. Made of material harder than it is intended to roll, such as cast iron, forged and heat treated steel or carbides, such rolls are essentially cylindrical with a neck at each end for insertion into a bearing and have ends that are suitably shaped for attachment to the spindle. This shaping of the roll ends may assume various forms. Wobblers are shown in Figure 3-3 and flat roll ends in Figure 3-4. Splines have also been used for coupling rolls to spindles.

The body of the roll (sometimes called the barrel) may be of slightly larger diameter at the center, in which case it is said to possess a "crown". The magnitude of this crown is usually expressed in thousandths of an inch (mils) and is the difference in diameter as measured at the center of the roll and at one end. A roll with a larger center diameter is said to have a positive crown, whereas, a hollow-ground roll is said to have a negative crown. The length of the roll body is generally known as the "face length" and this dimension limits the width of the strip that can be rolled.

The roughness of the surface of the work roll body is important in that it affects both the finish of the rolled strip and the magnitude of the force that must be applied to the rolls to achieve a given reduction. In the rolling of products that must have a bright finish, such as stainless steel strip, the rolls must be ground and polished to a roughness of 1 to 2 microinches. On the other hand, the work rolls of the first stand of tandem cold rolling mills reducing hot rolled and pickled strip may have a surface roughness of 50 microinches or more. However, it should be noted that the rougher the roll, the greater the effective coefficient of friction between the strip and work roll surfaces and, as a consequence, the greater the rolling force required in a given rolling operation.

COMPONENTS OF COLD ROLLING MILLS

Figure 3-1:
Small Single-Stand Cold Rolling Mill Used for Experimental Purposes.

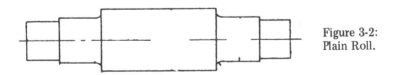

Figure 3-2:
Plain Roll.

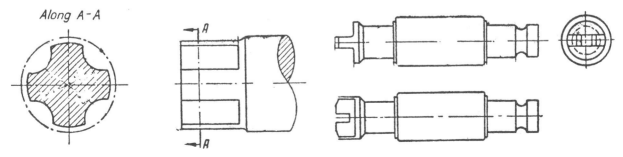

Figure 3-3:
Shape of Wobbler.

Figure 3-4:
Roll with flat end for connecting universal spindle.

Except in the case of two-high mills, the bearings of the work rolls do not have to support the main rolling force, this being carried by the backup roll bearings. Accordingly, the work roll chocks containing the bearings are used primarily to position the roll in the horizontal plane and, therefore, the roll necks are subject primarily to forces in both the rolling direction and a direction parallel to the axis of the roll. However, in shape controlling systems involving work roll bending, vertical forces of a relatively small magnitude are applied to the work roll chocks and, therefore, to the work roll necks.

The work roll chocks are usually fitted into a recess in the backup roll chocks as illustrated in Figure 3-5. In addition, the work rolls as generally offset towards the entry side of the mill from the plane containing the axes of the backup rolls as illustrated in Figure 3.6.*

The manufacture of mill rolls, their heat treatment and finishing as well as their bearings are discussed in detail in Chapter 4.

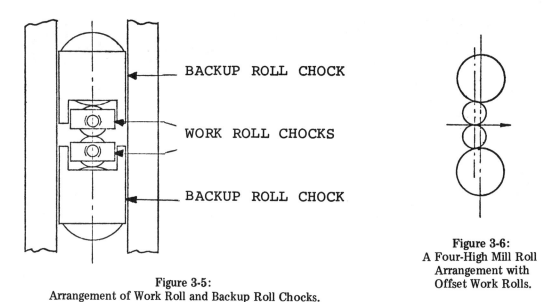

Figure 3-5:
Arrangement of Work Roll and Backup Roll Chocks.

Figure 3-6:
A Four-High Mill Roll Arrangement with Offset Work Rolls.

3-3 Backup Rolls and Their Bearings

Backup rolls (shown in Figure 3-7) are intended to provide the rigid support required by the work rolls to prevent their flexure or bending under the rolling load. Accordingly, they usually have a diameter two or three times the diameter of the work roll and generally of about the same magnitude as the length of the roll face. Under such conditions, they are generally referred to as "square".

Originally, backup rolls were used for support only. However, in recent years, backup rolls are frequently connected to spindles and driven, especially in high-speed stands where the ratio of the work and backup roll diameters may be used to advantage. Moreover, work roll changing is made simpler because the rolls no longer need to be carefully oriented so that the ends of the roll necks mate with the spindle couplings.

Because of their larger diameters, backup rolls on commercial mills, in effect, constitute sizable flywheels. Accordingly, in considering the acceleration of high-speed mills, the spindle torque necessary to provide the rotational acceleration of these rolls must be taken into account.

The surfaces of backup rolls, when used in cold rolling mills, are usually given an intermediate finish in the range of 15 to 20 microinches. A smoother finish of the rolls would gradually be roughened by the work rolls and a rougher finish would become gradually polished.

---

* C. W. Starling, "The Theory and Practice on Flat Rolling", University of London Press Ltd., London, 1962, p. 68.

COMPONENTS OF COLD ROLLING MILLS

The bearings in the backup roll chocks must, in general, support the whole rolling load and are, therefore, ruggedly built. Such bearings fall into two classes, plain bearings and roller bearings. The former are exemplified by plain metal bearings, fabric bearings and oil film lubricated bearings (e.g., Morgoil bearings). A white metal bearing is shown in Figure 3-8. Here the lubrication is provided by grease-filled grooves on the inner surface of the bearing. Fabric bearings (Figure 3-9) have in many cases replaced the plain metal bearing and often represent the most economical bearing for a particular requirement. Composed of a fabric such as cotton impregnated with a resin, they take the form of shell bearings which are fitted into the chocks and retained in position by keep plates. In these bearings, water is used both as a lubricant and a coolant.

Figure 3-7:
Rolls in a Four-High Mill.

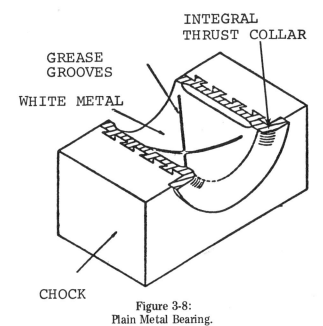

Figure 3-8:
Plain Metal Bearing.

An oil film bearing, illustrated in Figure 3-10, is designed to maintain a thin film of oil between a steel sleeve mounted on the roll neck and the casing or bushing of the bearing. The casings are lined with white metal or fitted with solid aluminum bushes. A separate thrust race is required with this type of bearing and it is customary to fit ball or tapered roller races for this purpose as shown in Figure 3-10. The oil film is maintained by recirculating the lubricating oil at a controlled temperature and pressure.

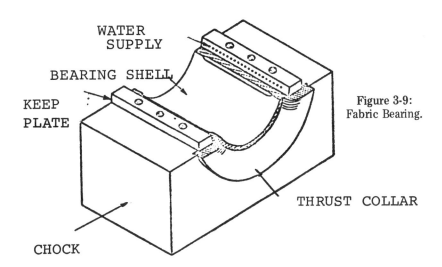

Figure 3-9:
Fabric Bearing.

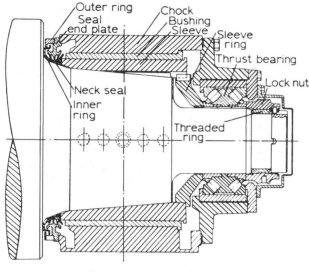

Figure 3-10:
Oil Film Bearing.

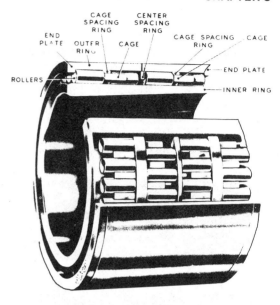

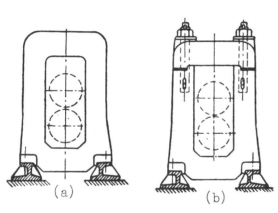

Figure 3-12:
Mill Housings
a — closed-top; b — open-top

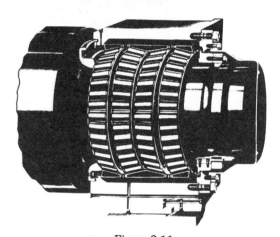

Figure 3-11:
Roller Bearing Arrangements
(a) Cylindrical Roller Bearing.
(b) Tapered Roller Bearing.

Two typical roller bearing arrangements are shown in Figure 3-11. Such bearings exhibit low frictional losses and they find wide use in all types of mills. Tapered rollers can take thrust as well as radial loads but parallel rollers require a separate thrust race. Lubrication may be provided either by packing the bearing with grease or by automatically feeding grease, oil or oil mist to the bearing.

3-4 The Mill Housing

To position and ensure the correct alignment of the rolls, very rigid housings are utilized to accommodate the rolls and their chocks. Generally, for cold rolling operations, closed top housings of the type shown in Figure 3-12 are used since they provide greater rigidity in a more economical manner than the open top type shown in the same illustration. The windows are machined so that roll chocks can conveniently slide within them. Wear plates of hardened materials may be located on the vertical faces of the windows to assist in preserving the proper roll alignment. To prevent lateral movement of the rolls, keeper plates or latches are used to tie the chocks to the mill posts.

COMPONENTS OF COLD ROLLING MILLS

Apart from the rolling force that must be withstood by the mill housing, the housing is also subject to a tilting moment (M). This moment is attributable to the sum of two components M_1 and M_2 where M_1 is the moment acting on the stand from the mill drive as a direct result of rolling, and M_2 is the moment resulting for forces acting on the rolls via the workpiece, for example, the strip tensions.

If the two spindle torques are equal and opposite, M_1 will be zero. On the other hand, M_1 will assume its largest value (M_{max}) if only one roll is driven. The value of M_2 may be computed from the product of the unbalance of the tensile forces in the strip (R) and the elevation of the rolling plane above the housing shoe (c). (See Figure 3-13.) The largest value of R in many cases can be equated to $2M_{max}/D$. Thus, the total tilting moment M may be as large as

$$M_{max}\left(1 + \frac{2c}{D}\right).$$

If the weight of the mill stand and the rolls is G, and the forces on the base each side of the mill housing are F_1 and F_2, then reference to Figure 3-14 shows that, for equilibrium,

$$F_1 = \frac{G}{2} - \frac{M}{b}$$

and

$$F_2 = \frac{G}{2} + \frac{M}{b}$$

where b is the separation of the forces F_1 and F_2.

Calculation of the stresses present in a typical mill housing during rolling operations can be a complicated undertaking.☐ However, the approach to the problem is simplified by considering the posts and crossbeams of the housing to be of equal cross sections (as shown in Figure 3-15) and loaded by a force P. Under these circumstances, the stresses may be calculated using Castigliano's theorem relating to deformation work or strain energy. Posts AD and BC of the housing are under

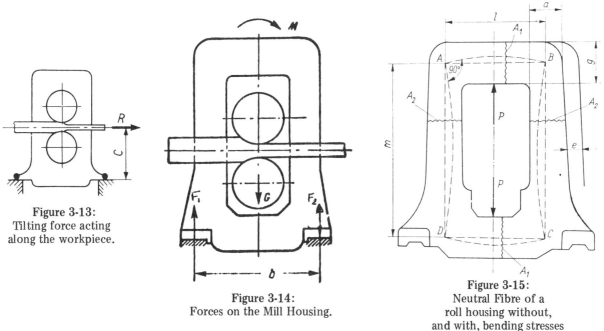

Figure 3-13: Tilting force acting along the workpiece.

Figure 3-14: Forces on the Mill Housing.

Figure 3-15: Neutral Fibre of a roll housing without, and with, bending stresses

---

☐ For a more detailed discussion of mill housing stresses, the reader is referred to "Forge Equipment, Rolling Mills and Accessories" by A. Gelegi, published by Akademiai Kiadó, Budapest, 1967, as well as the paper "Mill Housing Design" by H. Erzurum, Iron and Steel Engineer Year Book, 1969, pp. 472-480.

tensile and bending stresses, whereas, crossbeams AB & DC contain only bending stresses. The stress σ_1 in the posts is given by

$$\sigma_1 = \frac{P}{2A_1} + \frac{M_o}{W_1}$$

and the maximum stress σ_2 in the crossbeams by

$$\sigma_2 = \frac{M_o - \frac{PL}{4}}{W_2}$$

where M_o = statically indeterminate fixed-end moment in the four frame corners
W_1 = section modulus of the post cross section A_1
W_2 = section modulus of the crossbeam of section A_2
L = the horizontal separation of the neutral fibers in the mill posts.

If the thickness of the mill posts as measured in the rolling direction is a, the thickness in the transverse direction is b, and the vertical dimension of each crossbeam is g, then the moments of inertia I_1 and I_2 of the posts and crossbeams respectively about central axes are given by

$$I_1 = \frac{ba^3}{12}$$

and

$$I_2 = \frac{bg^3}{12}$$

Similarly, the corresponding section moduli, W_1 and W_2 may be computed from

$$W_1 = \frac{ba^2}{6}$$

and

$$W_2 = \frac{bg^2}{6}$$

The fixed end moment M_o may then be calculated from the equation

$$M_o = \frac{PL^2}{8\left[L + \frac{I_1}{I_2}m\right]} = \frac{PL^2}{8\left[L + \left(\frac{g}{a}\right)^3 m\right]}$$

where L is the separation of the neutral planes of the crossbeams as illustrated in Figure 3-15.

To prevent damage to rolling facilities, it is not uncommon for mills to carry special safety devices between the screwdowns and the chocks to protect the rolls and housings when the mill is heavily overloaded. Such devices are discussed more fully in Section 3-9.

To secure the mill housing to the bedplate, the former is usually provided with feet capable of being bolted to the latter. Many designs for housing feet exist, one of which is shown in Figure 3-16.

To maintain the proper separation of the mill housings, cross-tie beams or rods are used, the former being shown in Figure 3-17 and the latter in Figure 3-18.

COMPONENTS OF COLD ROLLING MILLS

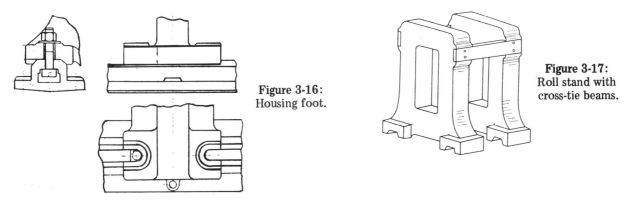

Figure 3-16: Housing foot.

Figure 3-17: Roll stand with cross-tie beams.

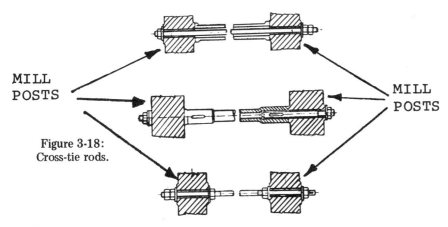

Figure 3-18: Cross-tie rods.

3-5 The Mill Foundations

Because of the considerable weight of most commercial mills and the magnitude of the tilting moments to which they are subjected, such facilities must have firm foundations. As a rule, they are made of reinforced concrete with approximately two tons of concrete for each ton of mill weight.

The weight of the mill is carried by the feet of the housing resting on girders above the foundation, these girders being known as housing shoes or bedplates. Figure 3-19 depicts a shoe with a rectangular bearing surface while Figure 3-20 shows other types of bedplates. These devices may be considered as beams on a resilient base and are fastened with bolts to the foundation. Needless to say, these bolts must be capable of withstanding the same forces as the bolts securing the feet of the housing to the shoes.

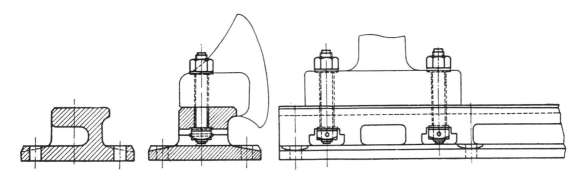

Figure 3-19: Shoe with rectangular bearing surface.

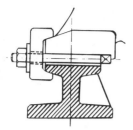 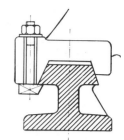

Figure 3-20:
Two methods of securing housing feet to bedplates.

3-6 Roll Balancing Systems

To ensure that the upper work and backup rolls are maintained in a correct position relative to the lower rolls, roll balancing systems are used in mills. Such systems apply forces to the rolls above the pass line to offset their weight and ensure that the upper work and backup rolls remain in contact with each other and the chocks of the latter remain firmly pressed against the screws, wedges or other roll positioning devices. Several roll balancing systems have found acceptance on commercial mills and some of these are discussed below.

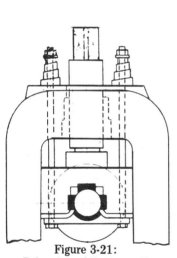 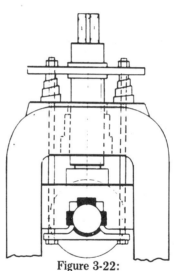 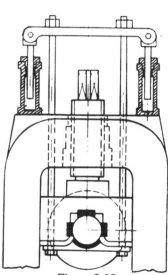

Figure 3-21:
Balancing of the top roll by two draw rods each suspended on buffer springs.

Figure 3-22:
Balancing of the top roll by two buffer springs and a crosspiece.

Figure 3-23:
Balancing of the top roll by draw rods and two pressure cylinders seated on the housing.

One simple method of counterbalancing the upper backup roll is by the use of suspension bolts and buffer springs as illustrated in Figure 3-21. These springs are placed directly on top of the housing, or more rarely on a crosspiece resting on a ring forced into the screw. (See Figure 3-22.)

Balancing can also be accomplished by the use of draw rods and cylinders as shown in Figure 3-23. In the example shown, the pistons of the cylinders are connected to a crosspiece on which connecting rods support the chock. Figure 3-24 shows a modification of this balancing mechanism in which only one cylinder is used.

In four-high mills, hydraulic cylinders incorporated into the roll chocks can perform the roll balancing function. A commonly used arrangement is illustrated in Figure 3-25.

Smaller stands with medium adjustability of the top rolls are usually balanced by counterweights, which may be attached to levers at the top or bottom of the mill housing as shown in Figure 3-26.

COMPONENTS OF COLD ROLLING MILLS

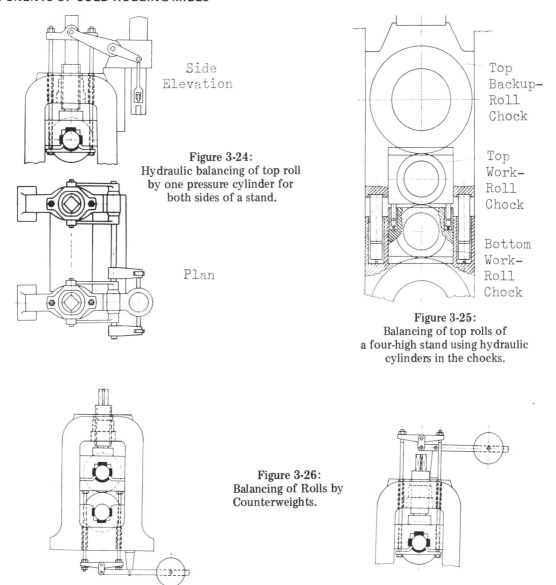

Figure 3-24:
Hydraulic balancing of top roll by one pressure cylinder for both sides of a stand.

Figure 3-25:
Balancing of top rolls of a four-high stand using hydraulic cylinders in the chocks.

Figure 3-26:
Balancing of Rolls by Counterweights.

3-7 Roll Positioning Systems

To control the relative positions of the rolls and the forces that they exert on the workpiece during the rolling operation, a roll positioning system must be employed on the mill stand. Because large screws threaded through the upper crossbeams of the mill housing and bearing upon the upper backup or work roll chocks have been commonly used for this purpose, it is the general practice to refer to the roll positioning system as the "mill screws". However, a number of other types of mechanical systems (some of which do not use screws) have been developed for controlling the rolling force. These systems are also discussed in this section.

Hand operated top roll adjusting mechanisms suitable for small mills are illustrated in Figure 3-27. In the first sketch of the illustration, a wedge is moved horizontally between a flat surface associated with the crossbeam and an inclined surface machined on the top of the upper roll chock. The conventional rotating screwdown is shown in the second illustration, a nut being rigidly fastened in the crossbeam. In the third sketch, the nut in the crossbeam is rotated by a worm gear fixed onto a horizontal shaft and in the last arrangement shown, two rotating screws are used with each roll chock, making for a more precise positioning of the rolls.

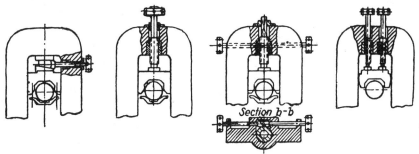

Figure 3-27: Diagrams of manual top roll adjusting mechanisms.
a — with wedge; b — with rotating screw down; c — with rotating nut; d — with 4 screw downs.

In large commercial cold reduction mills, the screws are motor driven with a mechanical arrangement as shown schematically in Figure 3-28 utilizing two pairs of worm gears. In the larger mills, two worm (or two cylindrical) gears may be used as shown in Figure 3-29. It is to be noted that, with the clutch engaged, both screws may be turned together, but all systems provide for individual screw adjustment for levelling the mill and for shape control.

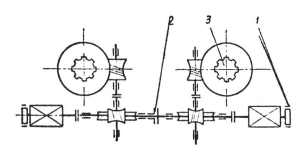

Figure 3-28:
Diagram of screw down drive in four-high reversing cold rolling mill —
1 — brake; 2 — electromagnetic clutch; 3 — screw down.

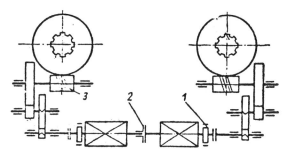

Figure 3-29:
Diagram of screw down system for large four-high cold mill —
1 — brake; 2 — electromagnetic clutch;
3 — single thread worm gear.

In cold strip mills, it is very desirable for reasons of gage and shape control to be able to adjust the screws rapidly under load. Accordingly, the roll positioning system must be able to exert a very high force on the roll chocks, and with reasonably sized drive motors, this implies a relatively slow speed adjustment of roll position. The total moment M_T involved in turning the two screws has been given by Tselikov and Smirnov * as

$$M_T = F\left[\frac{d_1}{3}\mu + d_2 \tan(\rho \pm \alpha)\right]$$

where

    F is the rolling force
    d_1 is the outer diameter of the screw shoulder abutting on the breaking piece or the top of the roll chock
    d_2 is the average diameter of the thread
    ρ is the angle of friction in the thread
    α is the helix angle of the thread, and
    μ is the coefficient of friction between the end of the screw and the surface contacting it.

In lowering the rolls, the tangent term assumes the value $\tan(\rho + \alpha)$, whereas, in raising the rolls, the term becomes $\tan(\rho - \alpha)$.

---

* A. I. Tselikov and V. V. Smirnov, "Rolling Mills" Translated from the Russian by M. H. T. Alford and edited by W. J. McG. Tegart, Pergamon Press, London, 1965.

COMPONENTS OF COLD ROLLING MILLS

To speed up the rate of roll positioning achieved by conventional mill screws, hydraulic systems have frequently been used in combination with mechanical screwdowns. The "Wheeler" nut system is an example of such a hybrid unit as it consists of a motor-operated screw and a hydraulically-actuated nut, the latter providing a rapid, vernier adjustment to roll positions.[‡][▲]

It was, however, inevitable that attempts should be made to completely replace the mechanical screws with hydraulic roll-positioning systems which will hold the mill rolls in the required positions during rolling and in the unloaded condition.[+] A mill stand so equipped, generally referred to as a "hydraulic mill" utilizes pressure cylinders capable of sustaining the maximum anticipated rolling force. It is preferable to locate the cylinders below the bottom backup roll chocks for the following reasons:

i) work roll changes may be executed quickly by lowering the chocks onto fixed rails.

ii) backup rolls may be changed in a similar manner.

iii) a constant pass line can be readily achieved by using hydraulically actuated wedges above the top backup roll chock.

iv) all the hydraulic lines, transducers, etc., are below the pass line and are, therefore, less vulnerable to mechanical damage.

v) no device is required to support the cylinders during backup roll changes.

These advantages are, however, to some degree, offset by the poor environment and the inaccessibility of the lower regions of the mill stand windows.

The window of a reversible, position-controlled, hydraulic mill stand is illustrated in Figure 3-30 and a schematic diagram of a control circuit designed for use with one side of the mill is shown in Figure 3-31. In Figure 3-30 it will be noted that guide rods, carrying wheels, pass through the work roll chocks. These wheels engage with fixed tracks mounted across the housings at floor level to facilitate roll changing. In Figure 3-31, servo valves meter the oil into and out of the cylinder and a linear-position transducer is used to measure the movement of the cylinder during rolling. With respect to the latter instrument, great care should be taken to protect it from damage due to its environment.

The relative responses of electrically operated (mechanical) and hydraulic roll positioning systems has been investigated by Stone.[⊕] He has shown that the latter are from 10 to 30 times faster than the former.

3-8 Work Roll Changing Devices

In the operation of a commercial cold rolling mill, rapid and convenient work roll changing is of considerable economic importance. The conventional method of roll changing involves the use of an overhead crane and a unit designed to attach to the neck of the roll to be removed from or inserted into the mill. A newer method involves the use of automatic roll changers which handle the rolls as a pair.

An example of the former is a C-shaped crossbar shown in Figure 3-32. However, this method requires two hooks on the crane. A simple, yet equally effective method, is illustrated in Figure 3-33. Here a heavy generally-cylindrical member hollowed out at one end and counter-

---

‡ J. I. Greenberger, "Design Concepts in an Hydraulic Plate Mill with AGC and Shape Control", Iron & Steel Engineer Year Book, 1971, pp. 461-468.

▲ J. G. Marshall, G. W. Roos, and J. H. Torrance, "Application of the Wedge Actuator to the Metal Industry Rolling Process", Iron and Steel Engineer Year Book, pp. 688-699.

+ R. Jackman, R. W. Gronbeck, and G. A. Forster, "The Position Controlled Hydraulic Mill" Portion of "Hydraulic Control of Rolling Mills and Forging Plants", Iron and Steel Institute, (Pub. 142), 1971, pp. 30-54.

⊕ M. D. Stone, "Hydraulic vs. Electric AGC Systems — A Comparison of Speed of Response", Iron and Steel Engineer Year Book, 1972, pp. 346-348.

balanced by a large rectangular mass of steel is used in conjunction with a small dolly supported on spring-mounted wheels. Two wire-rope slings are used with a single crane hook, the positioning of the slings being changed as necessary to afford proper balancing of the roll-changing mechanism. Before insertion of the new rolls into the mill, they must be rotated so that the wobblers or flattened portions of the roll necks fit into the coupling units attached to the spindles. Also to prevent scratching of the rolls by the backup rolls, pieces of plywood are often inserted in the mill during roll changing and are then withdrawn when the new rolls are in place.

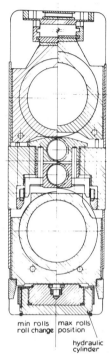
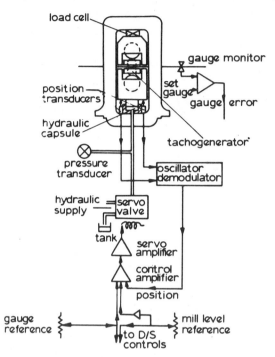
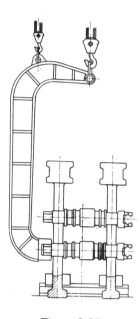

Figure 3-30:
Window of Reversible, position-controlled hydraulic mill stand.

Figure 3-31:
Diagram of Control Circuit for One Side of Hydraulic Mill Stand.

Figure 3-32:
Roll changing using a C-shaped crossbar.

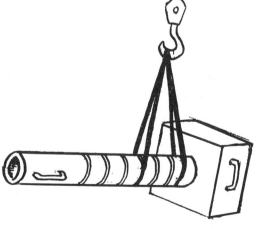

Figure 3-33:
Sketch of Work Roll Changing Equipment.
(When changing the lower work roll, the counterweight is positioned on the lower step of the dolly shown on the right. When changing the upper work roll, the top of the dolly supports the counterweight.)

COMPONENTS OF COLD ROLLING MILLS

Automatic roll changers use some type of carriage as illustrated in Figure 3-34. New pairs of rolls may be conveniently positioned on carriages in front of the mill stands as shown in Figure 3-35, these carriages first receiving the rolls as withdrawn from the housings and then positioning the new rolls ready for insertion into the mill.

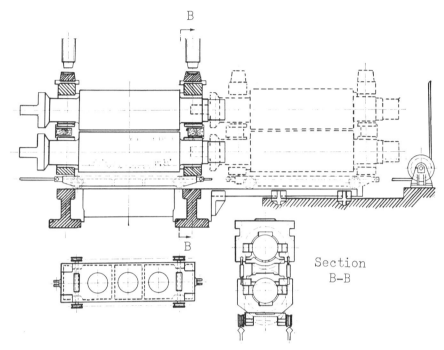

Figure 3-34: System for Changing a Pair of Work Rolls.

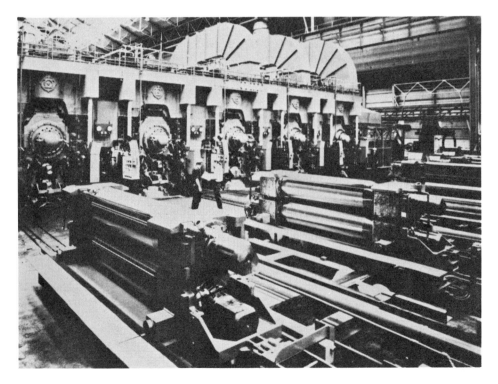

Figure 3-35:
Five-Stand Tandem Cold Mill with Roll Changing Equipment in the Foreground.

3-9 Mill Protection Devices

To ensure that the forces applied to the backup roll chocks are not of such a magnitude as to fracture the roll necks or damage the mill housing, protection devices in the form of breaking pieces are often provided between each screw and the corresponding roll chock. These are designed to disintegrate at a certain rolling force level and are made in the form of sleeves, wedges and plates.

Safety sleeves (see Figure 3-36) are generally made of cast iron and, for the higher forces exerted in plate and sheet mills, of cast steel. According to Tselikov and Smirnov,* the dimensions are usually as follows:

$$h = 0.7\, d_o;\ s = 0.5\, d_o;\ d_1 = 1.2\, d_o \text{ and } d_2 = 0.9\, d_o$$

where d_o is the extended diameter of the thread of the mill screws.

The greatest bending stress σ in the center of the lower surface is given approximately by

$$\sigma = 0.4 \frac{P}{S^2}$$

and the shear stress τ on the periphery of the bottom of the sleeve by the approximate formula

$$\tau = \frac{P}{\pi\, dS}$$

where P is the force on the roll neck
    d is the diameter of the end of the screw where it contacts the breaking piece, and
    S is the thickness of the sleeve bottom.

For safety reasons, the safety sleeve is usually enclosed in a casing which prevents the scattering of splinters when the sleeve breaks.

In wedge-type safety devices (see Figure 3-37), bolts are used as the shearing components. These bolts are specially designed with necks to localize the breakage, the length of the necks being about 0.8 times the bolt diameter.

The force T acting on the bolts of the safety box may be derived from a consideration of the components of the forces acting on each wedge of the safety box as illustrated in Figure 3-38 in both the horizontal and vertical directions.

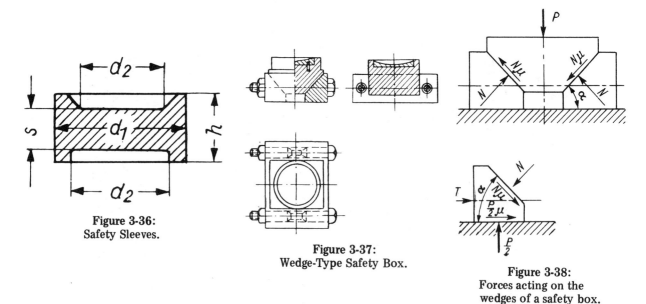

Figure 3-36: Safety Sleeves.

Figure 3-37: Wedge-Type Safety Box.

Figure 3-38: Forces acting on the wedges of a safety box.

---

* A. I. Tselikov and V. V. Smirnov, "Rolling Mills" Translated from the Russian by N. H. T. Alford and edited by W. J. McG. Tegart, Pergamon Press, London, 1965.

COMPONENTS OF COLD ROLLING MILLS

Here it is seen that

$$T - N \sin \alpha + N \mu \cos \alpha + \frac{P}{2} \mu = 0$$

and

$$\frac{P}{2} - N \cos \alpha - N \mu \sin \alpha = 0$$

where N is the force of normal pressure on the sloping surface of the wedge, α is the angle of slope of the wedge and μ is the coefficient of friction. These equations yield

$$T = \frac{P}{2} \left(\frac{\tan \alpha - \mu}{1 + \mu \tan \alpha} - \mu \right)$$

As pointed out by Tselikov and Smirnov,* the disadvantage of this type of safety box is the dependence of the tensile stress in the bolts on the coefficient of friction. To minimize this effect, the angle α is made large (30-45º) and the surfaces in contact are carefully machined and well lubricated.

3-10 Roll Cooling and Lubrication Systems

To facilitate the rolling operation by reducing the rolling forces and to cool the mill rolls and the strip, a rolling "solution" is usually applied in copious quantities to the strip and the rolls at the entry side of the mill. The "solution" is actually a quasi-stable aqueous emulsion of a lubricant which is usually recirculated for reasons of economy. (See Figure 3-39.)♦ The oil from the emulsion "plates out" onto the strip and roll surfaces in the roll bite. Accordingly, the frictional effects at the surfaces and the rolling loads are reduced.

In the rolling operation, the dissipation of frictional energy in the roll bite causes the temperature of the roll surfaces to increase. The application of the solution to the rolls as a coolant removes most of the heat so imparted to the rolls by frictional effects and permits the average roll temperature to assume a value not more than a 100 to 150 degrees F above ambient.

In a so-called direct application system, the lubricant is applied once and then discarded. Under these conditions the oil and water may be pre-mixed in a metering system or applied separately.

In other cases, the lubricant may already exist on the surface of the strip to be rolled as, for example, in the case of pickled and oiled hot band. When such is the case, as in the first stand of a primary cold reduction tandem mill, only cooling water is applied to the mill rolls. In yet other cases, the recirculated rolling lubricant may also act as the lubricant for the roll bearings. This is true for certain types of Sendzimir and other cluster type mills.

3-11 Spindles and Couplings

To transmit power to the mill rolls either directly from the drive motors or through pinion gears, spindles and couplings are used. Spindles are cylindrical steel shafts with approximately the same diameter as the roll necks and because of the design of the couplings, they permit operation of the mill while the positions of the rolls are being changed. Figure 3-40 illustrates in simplified form the side elevation of two slipper-type universal spindles connecting a pinion stand with two mill rolls.

---

* A. I. Tselikov and V. V. Smirnov, "Rolling Mills" Translated from the Russian by N. H. T. Alford and edited by W. J. McG. Tegart, Pergamon Press, London, 1965.
♦ J. P. Wettach, "Recent Advancements in Filtration of Roll Coolants for Cold Rolling Operations", Iron and Steel Engineer Year Book, 1966, pp. 613–619.

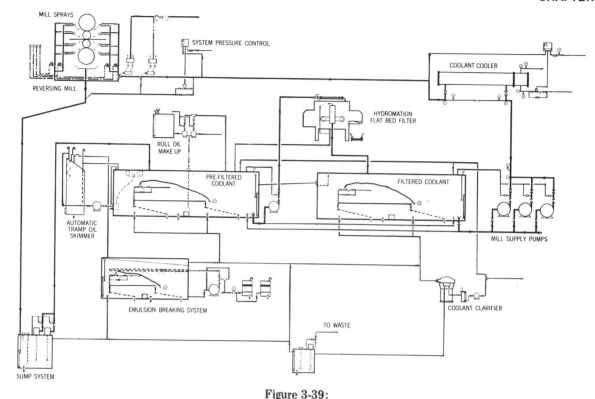

Figure 3-39:
A schematic arrangement of a recirculating lubrication system for a stainless steel reversing mill. The system has automatic roll oil makeup, emulsion disposal facilities and automatic filtration down to the 10-micron range.

Figure 3-40:
Diagram of Arrangement of Universal Spindles.

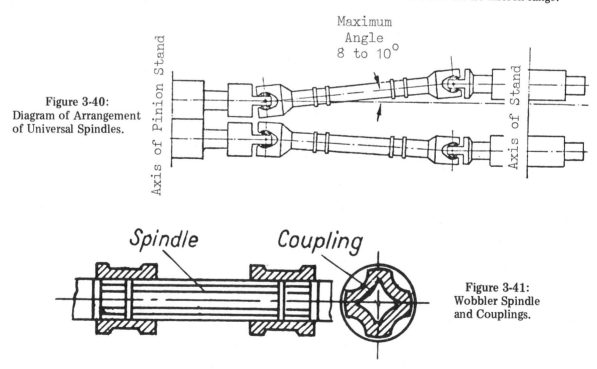

Figure 3-41:
Wobbler Spindle and Couplings.

Two types of coupling arrangements are commonly used. Universal spindles of the type shown in Figure 3-40 allow rotation of the rolls at angles of 8 to 10 degrees between the axis of the spindle and the axes of the rolls or pinions of a pinion stand. Wobbler spindles and couplings, as shown in Figure 3-41, may be used where the angle of inclination does not exceed 1 to 2 degrees.

COMPONENTS OF COLD ROLLING MILLS

(Larger angles would produce relatively large friction losses and a more rapid deterioration of the connection.)

In deriving the stresses associated with universal (or articulated) spindles, the assumption is made that the torque acting on the work roll is the same as that acting on the spindle.

By 1950, it was recognized that wobblers and the slipper-type universal spindles were rapidly becoming obsolete because of their speed and life limitations and their place taken by gear-type couplings.° Basically, the coupling, illustrated in Figure 3-42, consists of a roll-end sleeve with internal gear teeth and a bore to match the roll end, a pinion-end sleeve with internal teeth and a bore to match the pinion shaft, and a spindle with gear hubs mounted on each end. However, the complete spindle assembly also features seal rings, lubrication fittings, key, and retaining bolts. Usage of the gear-type spindle coupling started in rod and bar mills, progressed through temper mills, and ultimately became standard equipment for tandem cold mills and the finishing stands of hot-strip mills.

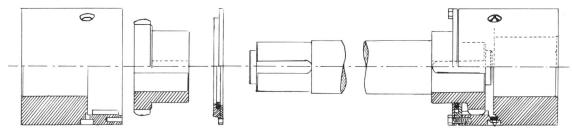

Figure 3-42: Major Components of Gear-Type Spindle Couplings.

The most critical component of the spindle coupling is the gear teeth. To maintain contact stresses to a minimum, the teeth are specially shaped so that, regardless of the operating misalignment angle, the hub-to-sleeve-tooth contact is retained for the full profile height of the teeth. The gear parts are fabricated from chrome-molybdenum or nickel-chrome-molybdenum steels and the teeth are nitrided and hardened.

Proper lubrication is essential for all types of spindle couplings and particularly the slipper and gear types.† The lubricants commonly employed include the lithium-base and the calcium-complex greases which exhibit high film strength, moderate to high thermal stability, and good retention characteristics. To retain the lubricant in the coupling, seals are utilized as shown in Figure 3-43.

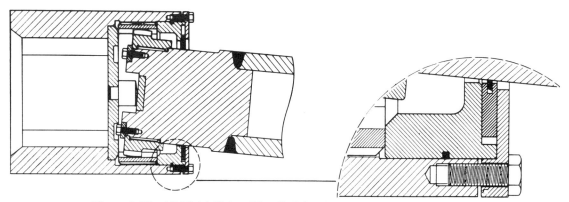

Figure 3-43: All-Metal Piston-Ring Seal for Gear-Type Spindle Coupling.

° J. J. Kimmel and B. W. Deringer, Jr., "The Gear-Type Spindle Coupling", Iron and Steel Engineer Year Book, 1968, pp. 427-439.

† J. J. Winkler, "Lubrication of Gear-Type Flexible Spindles", Iron and Steel Engineer Year Book, 1966, pp. 222-224.

In large commercial mills, both wobbler and universal spindles are sometimes balanced so that the weight of each spindle is not transmitted to the corresponding wobbler couplings or spindle joints. When small movements of the spindles are involved, as is usually the case in cold mills, balancing may be accomplished by springs as illustrated in Figure 3-44. An arrangement such as that shown permits easy removal of the spindles. Where larger movements of the spindles are encountered, hydraulic counterbalancing, as shown in Figure 3-45, is the most convenient method to employ.♦

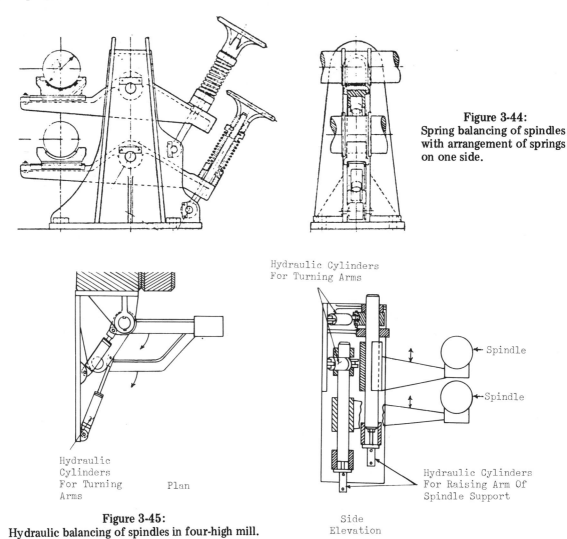

Figure 3-44: Spring balancing of spindles with arrangement of springs on one side.

Figure 3-45: Hydraulic balancing of spindles in four-high mill.

3-12 Pinions

Pinions, such as are illustrated in Figure 3-46, are gears located between the drive motor and the rolls serving to divide the available power between the two spindles and, at the same time, rotating them at exactly the same speed but in opposite directions. (If twin motor drives are used, pinions are not required although gears may be used in association with each drive motor to provide the desired rolling speed.) Early type pinions utilized either spur teeth or a divided face and staggered spur type teeth, but the present practice is to use double helical teeth. The last named gear teeth provide a smoother drive as some parts of the teeth are in contact at all times, making the transmission of power continuous.

♦ T. A. Bessent and R. F. Long, "Mechanical Features of the 2-Stand Duo-Reducing Mill at Sparrows Point", Iron and Steel Engineer Year Book, 1966, pp. 189-199.

COMPONENTS OF COLD ROLLING MILLS

Pinions are made of cast or forged steel and, mounted on babbitted bearings, they are set in housings similar to those used in roll stands. The housings must be of sufficient strength to withstand the tilting effect of the full torque transmitted by the drive motor. They should also be sealed to prevent damage to the gears by contaminants and forced lubrication should be provided for the pinions. Under such conditions, the pinions absorb about 6 percent of the power delivered by the motor.[§]

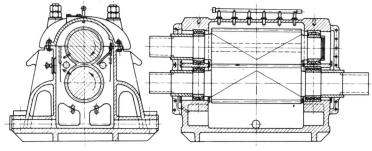

Figure 3-46:
Pinion stand of a sheet mill.

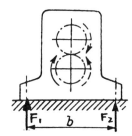

Figure 3-47:
The tilting moments which act on a pinion stand.

As in the case of a mill stand, the forces exerted on the supports of the pinion stand depend upon its mass G and the torque M applied to the stand. Referring to Figure 3-47, the forces F_1 and F_2 are given by

$$F_1 = \frac{G}{2} - \frac{M}{b}$$

$$F_2 = \frac{G}{2} + \frac{M}{b}$$

where b is the separation of the supports.

In normal rolling operations, where the spindle torques are equal and opposite, the torque M is that delivered by the motor. However, if the torque is transmitted only to the lower spindle (due to a breakage of the upper spindle or roll neck), then the value of M is close to zero. However, if only the upper spindle can transmit the torque, the torque M may assume a value twice that delivered by the motor.

The journals of the pinions are subject to bending and torsional stresses. In the section under greatest stress (where the journal is joined to the pinion), the nominal stress in transverse bending (σ) and in torsion (τ) are given by

$$\sigma = \frac{M_b}{0.1d^3} \text{ and } \tau = \frac{M_T}{0.2d^3}$$

where

    d   is the diameter of the journal
    M_b is the bending moment, and
    M_T is the torque transmitted to the pinion

Generally speaking, roller bearings of the tapered or self-aligning spherical type are preferred for use with the shafts in the pinion stand. However, if the diameter is not critical, the use of plain bearings is justifiable where the rotational speed of the shaft is too high for roller bearings. Such plain bearings must be made with considerable precision and are normally babbitted.

Lubrication of the pinions and bearings is usually carried out by a recirculating system (see Figure 3-48).[±] For the pinions, a high viscosity oil is desirable, but for plain bearings operating

---

[§] "The Making, Shaping, and Treating of Steel", Edited by H. E. McGannon, Eighth Edition, Published by United States Steel, Pittsburgh, Pennsylvania, p. 574.

[±] R. Jones and R. Hawley, "Trends in Rolling Mill Drive Design", Iron and Steel Engineer Year Book, 1961, pp. 735-744.

at relatively high speeds, a lower viscosity oil is desirable to prevent overheating.◄ The oil may be applied to the pinions by nozzles placed in proximity to the contact zone and to the bearings through holes bored in the housings and chocks.

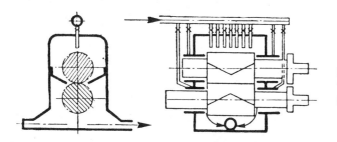

Figure 3-48:
Lubrication diagram of reversible pinion stand with overhead feed to gearing.

3-13 Gearing

To establish the desired rolling speed, gears are frequently used coupling the motor either to the pinion stand or to a spindle. Where low speed rolling is to be accomplished (as for example at the first stand of a five-stand tandem mill), a step down gear ratio would be used, whereas on high speed stands (such as temper mills), step-up speed ratios would be employed. Figure 3-49 illustrates 1, 2 and 3 stage reduction gears, the type used being dependent on the desired rotational speed of the mill rolls.

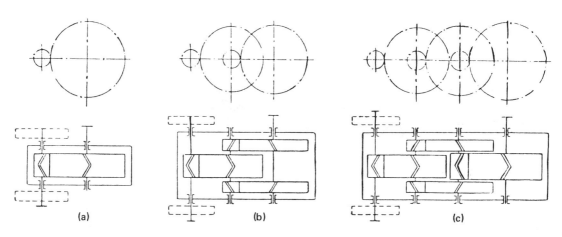

Figure 3-49:
Diagrams of 1, 2 and 3 stage reduction gears.

The gear wheels of large reduction gears on the main drive are usually of the double helical type which operate smoothly with no axial pressure on the bearing. In two and three stage gear boxes, (see Figure 3-49, b and c.) pairs of spiral gears may be used. In this type of reduction gear only one shaft (usually the slow speed one) needs to be fixed in the axial direction. The teeth are usually cut to provide a helix angle of about 30 degrees and the number of teeth is usually high (in the range 25 to 40).

The design of the gear wheels depends primarily on the diameters d and d_s of the gears and shaft respectively. On an approximate basis, where $d < 2d_s$, the gear wheel and the shaft are integral. Where $1.8\ d_s < d < 3.0\ d_s$, the gear wheel is usually attached to the shaft by a heavy key and where $d > 3d_s$, the wheels can either be solid cast or fabricated. To reduce the casting stresses, gear wheels are often made with tripartite hubs, as illustrated in Figure 3-50. Fabricated wheels consist

---

◄ E. S. Reynolds, "Mill Gear Lubrication and Causes of Gear Failure", Iron and Steel Engineer Year Book, 1963, pp. 801-807.

COMPONENTS OF COLD ROLLING MILLS

of a cast iron body and one or two shrink fitted forged steel toothed rims, depending on the width of the wheel (Figure 3-51).

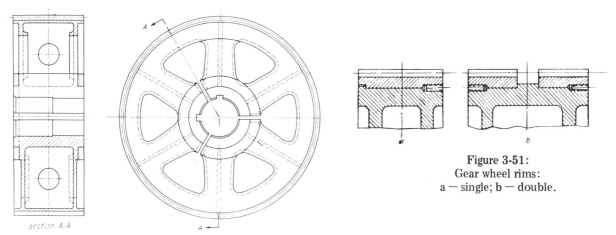

Figure 3-51:
Gear wheel rims:
a — single; b — double.

Figure 3-50:
Gear wheel with divided hub.

Modern mills are often designed with speed reducer and pinion stand units constructed as a single unit, as illustrated in Figure 3-52. Originally, this arrangement was regarded as risky and undesirable because a set of pinions could not be changed without dismantling the entire speed reducer. However, the use of antifriction bearings and replaceable wobbler couplings has made the possibility of pinion failure extremely remote. [±]

Figure 3-52:
Speed Reducer and Pinion Stand Combined in a Single Housing.

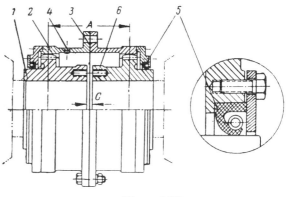

Figure 3-53:
Geared coupling
1 — sleeves; 2 — rims; 3 — paper jointing ring;
4 — plug for oil nipple; 5 — packing;
6 — hole for removing sleeves.

3-14 Motor Couplings

Although other types of couplings have been used to connect the shafts of motors to reduction gears and pinion stands, compensating and flexible couplings are mainly used. The former include gear types, universal joints and universal spindles and they allow for a small amount of misalignment. The latter include couplings with spiral springs (Bibby) and with pins (Bamag) and are intended to protect the drive motor against shock loads and to compensate for slight shaft misalignment.

[±] R. Jones and R. Hawley, "Trends in Rolling Mill Drive Design", Iron and Steel Engineer Year Book, 1961, pp. 735-744.

Gear couplings, such as that illustrated in Figure 3-53, are widely used because they are compact and combine simplicity of manufacture with high precision and low frictional losses. Before being bolted together, the couplings should be filled with oil.

The Schloemann universal joint couplings shown in Figure 3-54 were once widely used as they possess good alignment compensating properties. Each sleeve of the coupling has four trunnions arranged in the shape of a cross and the four trunnions have rectangular bronze blocks which fit into the longitudinal channels of the coupling sleeve. Thus, each sleeve is, in effect, a universal joint, requiring periodic lubrication for satisfactory operation.

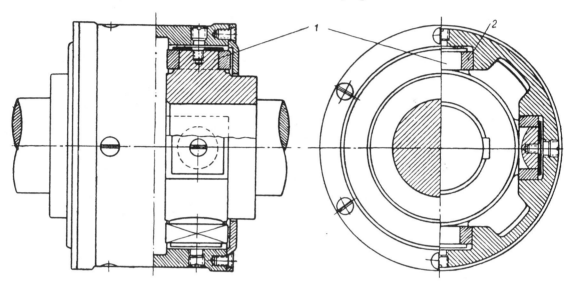

Figure 3-54:
Universal Coupling (Schloemann).

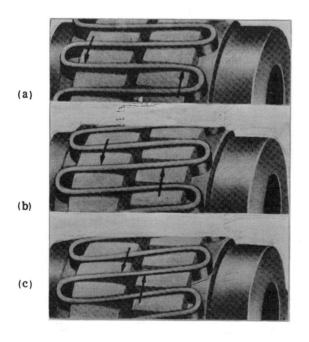

Figure 3-55:
Flexible coupling with spiral springs (Bibby) showing deformation of springs under loads:
　a — nominal;
　b — normal overload;
　c — under very heavy
　　　and impact loads.

The Bibby flexible coupling is shown in Figure 3-55. In this coupling a continuous spring is threaded between the flared teeth in a serpentine manner. Once widely used, these couplings have been largely replaced by gear couplings.

COMPONENTS OF COLD ROLLING MILLS

3-15 D.C. — Drive Motors

Virtually all cold reduction mills are now driven by electric motors ranging in size from a few horsepower for smaller mills rolling narrow foil product to thousands of horsepower for high-speed commercial mills. For mill stands requiring large power inputs, twin drives are commonly used with one or more motors driving each spindle.

Usually D.C. motors are used for such applications.[■] These fall into three categories; shunt, series and compound wound with basic schematic diagrams as illustrated in Figure 3-56. In the case of shunt wound motors, the drop in speed from no load to full load seldom exceeds 5 per cent.[●] The rotational speed is expressed

$$\text{Rpm} = \text{constant} \times (E_a - I_a R_M)/\emptyset$$

where E_a = the emf applied to the armature
 I_a = the armature current
 R_M = the resistance of the armature winding, and
 $\emptyset$ = the total flux per pole.

MOTOR TYPES

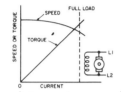

D-c Shunt Motor—main field winding designed for parallel connection to armature—stationary field—rotating armature with commutator—has a no load speed—full speed at full load less than no load speed—torque increases directly with load.

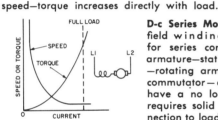

D-c Series Motor—main field winding designed for series connection to armature—stationary field—rotating armature with commutator—does not have a no load speed—requires solid direct connection to load to prevent runaway at no load—speed decreases rapidly with increase in load—torque increases as square of armature current—main motor for crane hoists, excellent starting torque.

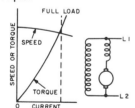

D-c Compound Motor—main field both shunt (parallel) and series—stationary fields—rotating armature with commutator—combination shunt and series fields produce characteristics between straight shunt or series D-c motor—good starting torque—main motor for D-c driven machinery—mills, presses, etc.

Figure 3-56:
Schematics of D-C Motors

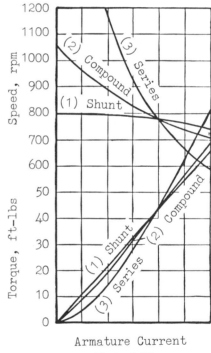

Figure 3-57:
Motor Characteristics

Thus, it can be seen that variation of the field strength gives constant horsepower but variable torque over the speed range, assuming constant armature voltage. Variation of the armature voltage gives constant torque but variable horsepower over the speed range assuming constant armature current. (See Figure 3-57.)

[■] E. P. Smith, J. W. Gough, and B. L. Goss, "Recent Developments of D-C Motors and Generators for Metal Rolling Mill Main Drives", Iron and Steel Engineer Year Book, 1968, pp. 194-204.
[●] A. E. Knowlton (Editor), "Standard Handbook for Electrical Engineers", McGraw Hill, 1957, p. 768.

The above equation applies equally well to series wound motors but in this case the flux $\emptyset$ increases with the armature current I_a. As a consequence, the torque would be proportional to I_a^2 were it not for the fact that the magnetic circuit becomes saturated with increase of current. Since $\emptyset$ therefore increases with load, the speed drops as the load increases as illustrated in Figure 3-57. For a given load (and, therefore, for a given current), the speed of a series motor can be increased by shunting the series winding or by short circuiting some of the series turns, so as to reduce the flux. On the other hand, the speed can be decreased by inserting resistance in series with the armature.

The compound motor is a compromise between the shunt and series types. Because of the series winding which assists the shunt winding, the flux per pole increases with the load, so that the torque increases more rapidly and the speed decreases more rapidly than if the series winding were not connected; but the motor cannot run away on light loads because of the shunt excitation. The speed and torque characteristics for such a motor are illustrated in Figure 3-57, the former being adjusted by armature and field rheostats, just as in the case of a shunt machine.

Indirect compounding is used on some d-c motors. In this case, the heavy strap-wound series field is replaced by a wire-wound field similar to a small shunt field. This field is connected to an unsaturated d-c exciter, usually separately driven at a constant speed. This exciter has its field energized by the line current of the motor for which it supplies the series excitation (see Figure 3-58). The output voltage and current from the exciter are proportional to the main motor current, so a given proportionality exists between the load current of the motor and its wire-wound series field strength. The use of a reversing switch and rheostat in the armature circuit of the series exciter permits variations in strength and even polarity of the series field. This permits an easy method of changing the compounding of the motor to maintain a nearly constant speed regulation over a given speed range. If desired, the series-exciter rheostat can be mechanically connected to the shunt field rheostat to accomplish this automatically.

Numerous types of speed regulators may be used to control the speed of a d-c motor. These may include electronic, magnetic amplifier and the rotating regulator types. A speed sensing signal for the regulator is usually obtained from a small tachometer generator driven by the motor and the regulator controls the armature terminal voltage and/or the strength of the shunt field of the motor.

In order to reverse the direction of rotation of a d-c motor, it is necessary to reverse the current in the field coils or in the armature, but not in both. Almost always, the armature connections are reversed rather than reversing both the shunt and series fields.

D-C mill motor standards for series, shunt and compound wound motors of two enclosure types; (totally enclosed nonventilated (TENV) and totally enclosed force ventilated (TEFV)) have been published by the Association of Iron & Steel Engineers.□

3-16 A-C Drive Motors

A-C motors are occasionally used to power small, low-speed mills such as found in metallurgical laboratories. These are principally of three types; synchronous, squirrel cage and wound rotor. The synchronous motor, if an exciter is not required, usually costs the least and, in certain speed ranges, may be used interchangeably with the squirrel cage design. Wound-rotor motors, on the other hand, cost more than either of the other two types.

The synchronous motor, illustrated by Figure 3-59, has the stator or stationary part wound with core and coils in the same manner as the other two types. However, the rotor has salient poles, north and south, energized by d-c current furnished by an exciter. This, in turn, may be mechanically coupled to the main motor shaft, or driven by a small separate motor. In addition to

---

□ D. C. Mill Motor Standards, AISE Standard No. 1, Revised September, 1968, Published by Association of Iron and Steel Engineers, Pittsburgh, Pa. 15222 — See also Iron and Steel Engineer, September, 1968, pp. 115-118.

COMPONENTS OF COLD ROLLING MILLS

the d-c coils on the pole pieces, another short-circuited winding is buried in the pole faces (the squirrel cage winding) which controls the necessary starting and pull-in torques of the motor.

The speed of a synchronous motor is determined by the number of poles it contains and the power supply frequency. The equation for speed is

$$\text{RPM} = \frac{\text{Frequency} \times 120}{\text{Number of Poles}}$$

The motor will, under steady state conditions, run only at this synchronous speed.

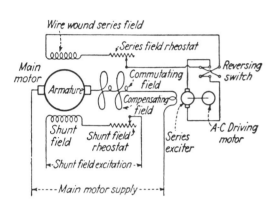

Figure 3-58:
D-C motor with indirect compounding, using a series exciter.

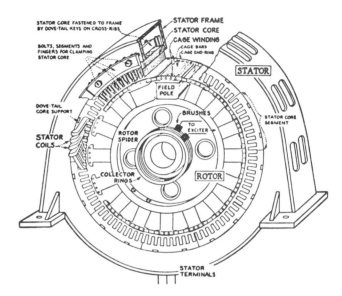

Figure 3-59:
Construction of A Synchronous Motor. (Courtesy of Electric Machinery Manufacturing Company)

The synchronous motor devoid of the cage winding would have no starting torque. With such a winding and with the field circuit de-energized, the motor starts by induction as a squirrel cage motor and will reach a speed slightly below synchronism. At that point, the field may be energized and the motor will then pull into synchronism.

Synchronous motors exhibit higher efficiencies than induction motors and they find application in circumstances requiring constant speed, continuous running conditions. Where the required starting torque is too great for the motor, a magnetic clutch may be installed between the motor and the load, the clutch being energized when the motor has reached synchronous speed. Speed-torque and speed-current curves for a typical, general purpose synchronous motor are illustrated in Figure 3-60.

Squirrel cage induction motors are the most simple with respect to design. They can be built in a wide range of sizes and torques and by varying the design resistance in the rotor winding, the starting torque can be low, normal, or high, as desired. The stator is a laminated framework into which are wound wire coils for connection to the power supply. The rotor is also a laminated structure, having slots into which copper or aluminum bars are fitted, these bars being connected together at each end by metal end rings. Although such motors have been built in large sizes (up to 10,000 HP), they are not normally used on mill drives.

The wound-rotor induction motor (often called the slip ring motor) is like the squirrel-cage motor, but instead of having a series of conducting bars placed in the rotor slots, it has a wire winding in the rotor. If the winding is permanently short-circuited, it becomes just another form of squirrel cage. However, if the ends of the rotor winding are brought out to three continuous slip rings, it may be given certain desirable characteristics. Connections can be made from these rings through external resistance boxes or a slip regulator. (See Figure 3-61.) By this means, the starting power input and accelerating time can be controlled. These motors are available over the complete output and speed ranges required for mill drives, for operation on voltages up to 13,200 volts. As in the case of squirrel cage motors, the efficiency and power factor decrease with the base speeds. The speed-torque and current curves of this type of motor are illustrated in Figure 3-62. Curve (a) represents the motor short circuited, curve (b) relates to the addition of a small resistance to the motor circuit, and (c) and (d) to high values of resistance in the circuit.

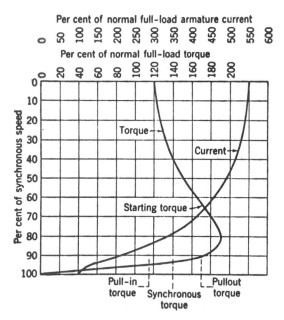

Figure 3-60:
Speed-torque and Speed-current Curves of a Typical General-purpose Synchronous Motor.

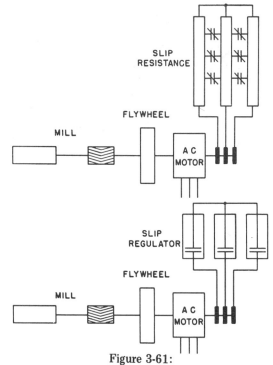

Figure 3-61:
A wound rotor induction motor, with flywheel application and slip regulator, employed to equalize peak loads on a single-speed drive.

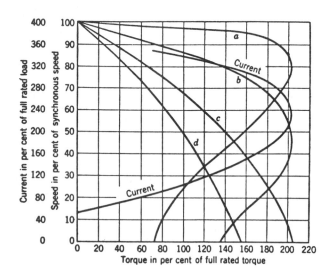

Figure 3-62:
Speed-torque and Current Curves of a Slip-ring Motor. Curve — (a) Rotor short circuited, Curves (b), (c), and (d) — Increasing resistance in Rotor Circuit.

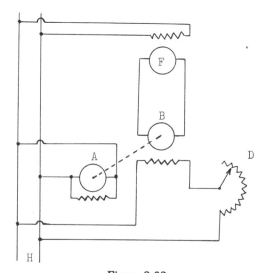

Figure 3-63:
Connections for Ward-Leonard system speed control with a d-c power supply and a non-reversing variable speed d-c motor.

COMPONENTS OF COLD ROLLING MILLS

3-17 Motor Generator Sets for Powering Mill Motors

Where d-c mill drive motors are used, these have traditionally been powered from constant voltage and variable voltage motor-generator sets; the voltages applied to the motors not only controlling their power but also their speed.

The Ward-Leonard system may be utilized to vary the speed of a d-c motor without incurring any rheostatic losses in the main circuit.[※] This is accomplished by interposing a motor-generator set between the variable speed motor and the power supply system. If the power supply is alternating current, it is usual to use a synchronous motor in the set coupled to a d-c generator. The excitation of the latter may be varied over a wide range by means of a rheostat in the field circuit. The generator's armature and the d-c motor's armature constitute a circuit by themselves, which is not interrupted in service except for fault conditions or severe overloads. By simple and efficient manipulations of the field rheostats and switch gear in the exciting circuits of the generator and of the variable speed motor, the latter's speed and direction are under perfect control at all loads. The system is regenerative, returning energy to the supply system when the motor is driven like a generator.

Figure 3-63 shows a diagram of connections for a Ward-Leonard system of speed control with a d-c power supply and nonreversing variable speed d-c motor. A is the d-c motor, B is a d-c generator of the motor-generator set, F is a variable speed d-c motor, H is the d-c power supply, and D is the generator field rheostat used for controlling the speed of motor F.[•]

Figure 3-64 is a similar diagram for a Ward-Leonard system with a 3-phase a-c power supply and a reversing variable-speed d-c motor. A is the a-c motor, B is the d-c generator of the

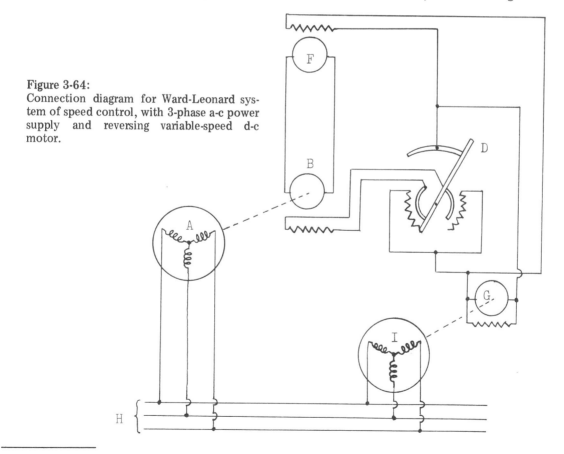

Figure 3-64:
Connection diagram for Ward-Leonard system of speed control, with 3-phase a-c power supply and reversing variable-speed d-c motor.

[※] F. A. Woodbury, "Selecting Electrical Drives for Applications Under 500-HP", Iron and Steel Engineer Year Book, 1961, pp. 315-321.

[•] A. E. Knowlton (Editor), "Standard Handbook for Electrical Engineers", McGraw Hill, 1957, p. 768.

main motor-generator set, G is an exciter generator driven by a separate 3-phase motor I, F is a reversible variable-speed d-c motor and D is a combined rheostat and reversing switch. It should be understood, however, that the reversing and nonreversing features of the systems described above are not dependent upon the generator power source but are contingent on the type of motor and control.

3-18 Static Power Systems for D.C. Motors

For many reasons including high installation costs, noise and maintenance of bearings and commutators, alternate d-c supply sources have long been sought.►↓ Mercury arc rectifiers and magnetic amplifiers in conjunction with dry plate rectifiers found limited application. However, in the late 1950's, a solid-state device known as a silicon controlled rectifier (SCR) and more commonly called a thyristor (Figures 3-65 and 3-66) came into existence and was first used in motor and generator field excitation systems about 1960. Since that time, it has been used with increasing frequency to supply d-c power to larger and larger mill motors including main mill drives.▣◊✹×✦

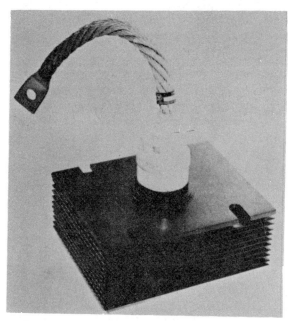

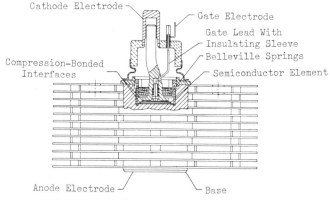

Figure 3-65:
Typical thyristor unit as used in armature power supplies for large motors.

Figure 3-66:
Cross section of high-power thyristor with 470-amp current rating (rms) and capability of blocking 1500-v reverse voltage.

► P. J. Roumanis, "Silicon Controlled Rectifiers on Steel Mill Drives", Iron and Steel Engineer Year Book, 1964, pp. 909-916.

↓ A. M. Curry and J. B. Walker, "Application and Maintenance Experience with Thyristor Powered Rolling Mills", Iron and Steel Engineer Year Book, 1967, pp. 506-514.

▣ L. F. Stringer and E. T. Schonholzer, "Thyristor Drive Systems for Metal Mill Applications", Iron and Steel Engineer Year Book, 1967, pp. 193-208.

◊ H. B. Koehler, "The Selection of Seilicon Rectifiers for Metal Finishing Applications", Iron and Steel Engineers Year Book, 1965, pp. 574-581.

✹ R. G. Wilson, "Static Drives — State of the Art, Drive Configurations", Iron and Steel Engineer Year Book, 1966, pp. 441-444.

× A. F. Kenyon, "Static Power — A New Tool for Steel Mill Control", Iron and Steel Engineer Year Book, 1966, pp. 811-820.

✦ P. J. Tsivitse and A. N. Schiff, "Static Power Sources for D-C Motors", Iron and Steel Engineer Year Book, 1969, pp. 155-164.

COMPONENTS OF COLD ROLLING MILLS

The thyristor has operating characteristics similar to the thyratron mercury vapor controllable rectifier. It can block applied voltages in both the forward and reverse directions and, when gated, can conduct current in the forward direction with a low forward voltage drop. These characteristics are illustrated in Figure 3-67. If a voltage within the rating of the thyristor is applied to the anode and cathode terminals in either direction and the device is in the "off" state, only a small leakage current will flow. However, if the anode is positive, the thyristor can be switched to the "on" state by driving current through the gate and cathode terminals. Once the anode current has exceeded a certain "latching" value, the gating current may be terminated without affecting the main anode current flow. The anode current will then cease only when the cathode to anode potential has fallen to such a low value that the anode current becomes less than the minimum or latching value required for continuous operation.

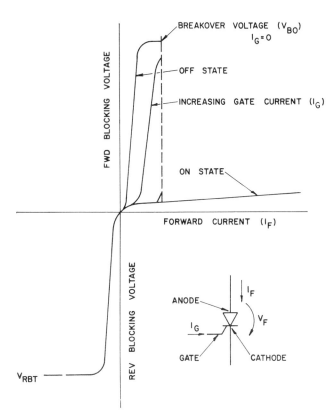

Figure 3-67:
Characteristics of a thyristor. (Operational capabilities are determined by its ability to block applied voltage in forward and reverse directions, and when gated to conduct current in forward direction with low forward voltage drop.)

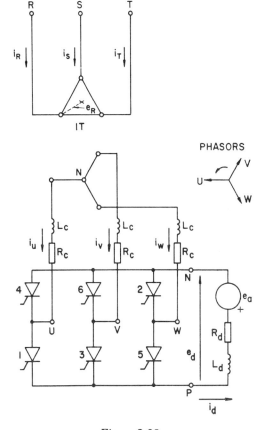

Figure 3-68:
Six-phase double-way circuit for thyristor power supply.

Many a-c to d-c converters have been developed using these solid-state devices. The 6-phase double-way circuit (3-phase bridge) has evolved as the preferred arrangement for thyristor power supplies and is shown schematically in Figure 3-68. The transformer is delta-wye connected since in this arrangement the flux ripple of the third harmonic in the core is eliminated. The main purpose of the transformer is to adjust the line voltage to the proper level, to provide isolation and to introduce inductance into the converter current path. (The inductance is required to control the rates of currents during commutation and faults.) The d-c output side of the converter, classified as a unidirectional or single converter, is connected to the load circuit consisting of a counter electro-motive force (emf), a resistance and an inductance. Gating pulses are applied to the pairs of the gate electrodes of the thyristor at the proper phasing with respect to the 3-phase a-c supply.

The d-c output voltage is controlled by the phasing α of the gating pulses. If a gating pulse is applied as soon as the corresponding anode becomes positive ($\alpha = 0$), then the output is a maximum and the thyristors act like single diodes. Increasing the gate's angle decreases the d-c output since the output voltage is proportional to $\cos \alpha$. When $\alpha = 90°$, therefore, the d-c output is zero. Thus, for values of α between $0°$ and $90°$, the flow of power is into the load. If the load is generative, then for values of α in excess of $90°$, power can be transferred from the load back into the line and the circuit, instead of being a rectifier, becomes an inverter.

If current reversal is required, a second converter can be added and connected to the first one in an antiparallel circuit as shown in Figure 3-69. This circuit is classified as a bidirectional or dual converter. Basically there are three ways to control such a circuit. One requires that the forward and reversing gating angles (α_F and α_R) total 180 degrees, and a second involves the netire suppresion of gating pulses to the converter intended to be inoperative. However, the most convenient method is known as offset control with the phasing of the gating pulses being such that

$$\alpha_F + \alpha_R = 240°$$

Figure 3-69: Thyristor Circuit for Current Reversal.

This, however, results in a dead band between the control characteristics of the forward and reverse converter and this is undesirable for current reversal. Fortunately, these dead zones, or nonlinearities, in the system may be essentially removed by means of a negative feedback loop as illustrated schematically in Figure 3-70. It also has the advantage of linearizing the load characteristics of the converter at light loads.

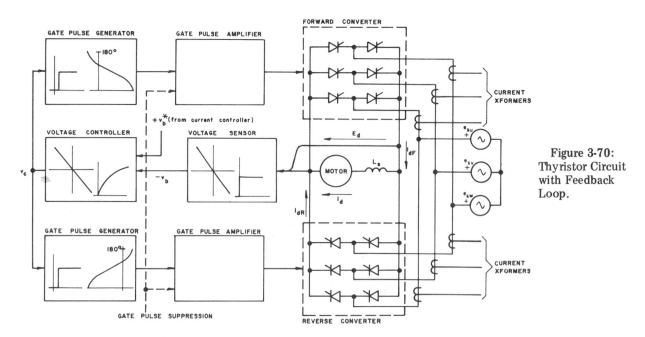

Figure 3-70: Thyristor Circuit with Feedback Loop.

COMPONENTS OF COLD ROLLING MILLS

3-19 Static Power Systems for A.C. Motors

One important limitation of the d-c motor is the commutator which not only restricts the motor ratings but, because of the brushes, is subject to wear and must be serviced. To avoid this limitation, a 3-phase induction motor with a squirrel cage rotor may be used in conjunction with a 3-phase power supply of variable frequency and amplitude.◈ ◊ ♦

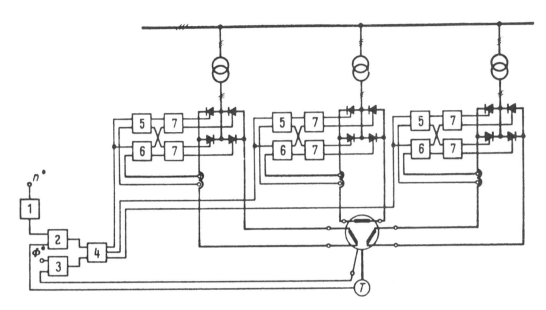

Figure 3-71:
Diagram of Static Power System for 3-Phase A-C Motor.

The frequency of the 3-phase voltage supply is converted and the voltage varied simultaneously by electronic components in such a way that the speed of the motor and the torque it provides can be regulated just as simply and accurately as that of the d-c commutator motor. This may be achieved either by the use of silicon controlled rectifiers or grid controlled mercury arc rectifiers as illustrated in Figure 3-71. The major parts of the circuit are: (η) speed reference signal, (φ) flux reference signal, (1) acceleration controller, (2) speed controller, (3) field controller, (4) frequency and amplitude control, (5) current controller, (6) control command units for commutation without circulating current and (7) control units. In the arrangement, each phase winding of the motor is fed by a 6-pulse, 3-phase/single converter which consists of two 3-phase bridge circuits in nonparallel connection arranged for zero circulating current. The actual speed of the motor is measured by a tachometer generator and the torque (related to the magnetic air gap flux) is measured by measuring the magnetic flux density. Signals relating to these two parameters are fed back to the speed and field controllers for regulation purposes.

3-20 Idler and Bridle Rolls

Idler rolls are often used at the entry and exit sides of the mill stand to ensure that the strip being rolled is maintained on the proper passline of the mill without appreciably changing the tensions in the strip, as illustrated in Figure 3-72. Being undriven, such rolls should not skid with respect to the roll surface and should, therefore, possess low inertia and "frictionless" bearings.

◈ D. F. Grubb, "Theory and Application of A-C Static Adjustable Speed Drives", Iron and Steel Engineer Year Book, 1969, pp. 463-471.

◊ J. Ullmann, "Variable-Speed Reversing Drive for Rolling Mills Features Static Control", Iron and Steel Engineer Year Book, 1965, pp. 519-520.

♦ R. G. Wilson, "Static Drives — State of the Art", Iron and Steel Engineer Year Book, 1966, pp. 441-446.

They should be of sufficiently large diameter (in excess of 100 times the strip thickness) so that no plastic deformation of the strip occurs as it is bent around the roll. Moreover, such rolls should have a very smooth finish and may be chrome plated or otherwise finished to provide a wear-resistant surface.

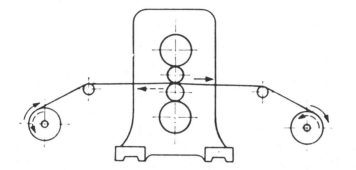

Figure 3-72:
Four-high reversing strip mill (with an idler roll between each reel and the mill stand).

In many cases, however, it is desirable to not only maintain the strip on the passline of the mill but also to change the tensile stresses within it. This is particularly true in the rolling of light gage strip where a relatively high tensile stress may be desired on the exit side of the roll bite but a much lower stress required for coiling (where higher stresses may cause a collapse of the eye of the coil). To provide sufficiently high strip tensions at the roll bite, tension bridles are often used, an example of which is illustrated in Figure 3-73. In such a bridle, one or both rolls may be driven and to facilitate the threading of the bridle with the strip, guides may be provided as shown in the Figure.♦

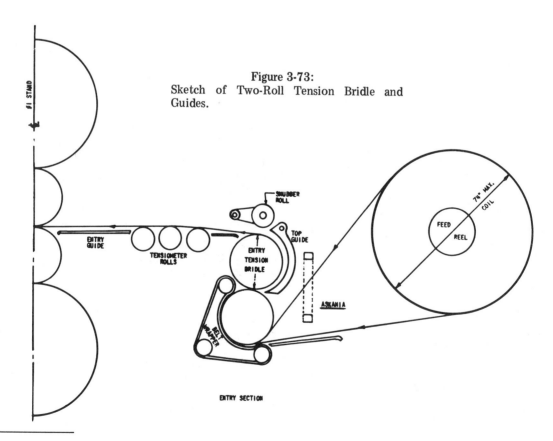

Figure 3-73:
Sketch of Two-Roll Tension Bridle and Guides.

♦ T. A. Bessent and R. F. Long, "Mechanical Features of the 2-Stand Duo-Reducing Mill At Sparrows Point", Iron and Steel Engineer Year Book, 1966, pp. 189-199.

COMPONENTS OF COLD ROLLING MILLS

To understand the effectiveness of these bridles, reference should be made to Figure 3-74, showing a sketch of a single roll with strip passing around it in such a manner that the arc of contact of the strip with the roll subtends an angle a (expressed in radians) at the center of the roll. The relation between the incoming tension T_1 and the outgoing tension T_2 is given by

$$T_1 = T_2 e^{fa}$$

where e = 2.718 (the base of Naperian logarithms) and f is the coefficient of friction between the strip and roll surfaces. The above equation is developed on the basis that the strip is fully flexible and the speed is so low that centrifugal effects are negligible.

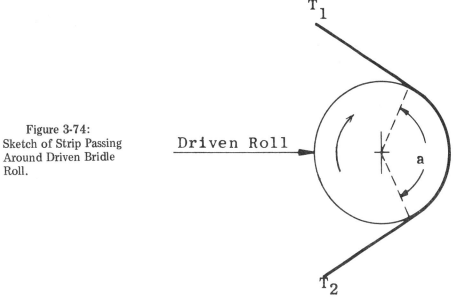

Figure 3-74:
Sketch of Strip Passing Around Driven Bridle Roll.

The value of e^{fa} is known as the amplification factor, and is shown plotted in Figure 3-75 as a function of wrap angle for various coefficients of friction.

Critical in the design of bridle rolls is, therefore, the selection of the roll material and the coefficient of friction it provides with the strip surface. Table 3-I lists the values of the coefficient of friction for a variety of roll surfaces.▼

Table 3-I

Coefficient of Friction Provided by Various Roll Surfaces

Roll Surface	Measured Coefficient of Friction	Design Values of Coefficient
Polished Steel (Dry)	0.31 — 0.45	0.15 to 0.18
Polished Steel (Oiled)	0.10 — 0.23	~ 0.05 to 0.1
Ground Steel (Dry)	0.29 — 0.41	0.15 to 0.18
Ground Steel (Oiled)	0.05 — 0.10	~ 0.05
Specially Compounded Plastic Rubber (Dry)	0.33 — 0.37	0.18 to 0.20
Plastic Rubber (Oiled)	0.18 — 0.26	0.13 to 0.15
Rubber (Dry)	0.52 — 0.64	0.25 to 0.28
Laminated Plastic (Dry)	0.31 — 0.35	0.18 to 0.20

▼ G. C. Turner, "The Design and Application of Bridles for Process Lines", Iron and Steel Engineer Year Book, 1965, pp. 211-218.

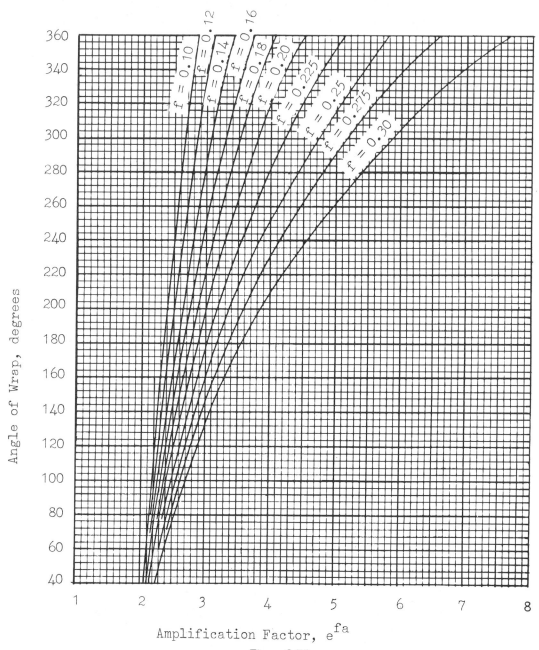

Figure 3-75:
Curves Showing the Variation of Amplification Factor with Angle of Wrap for Various Coefficients of Friction.

A two roll bridle, such as that shown in Figure 3-74, may have a total wrap of at least 310°, and if the coefficient of friction is 0.1, the amplification factor would be at least 1.7.

The theoretical horsepower P required to drive a bridle roll such that its peripheral speed is V feet/minute and the tensions are T_1 and T_2 pounds is simply

$$P = \frac{V(T_1 - T_2)}{33,000} \text{ HP}$$

COMPONENTS OF COLD ROLLING MILLS

3-21 Coil Boxes and Uncoilers

To pay off strip into a cold mill, coil boxes as shown in Figure 3-76 are generally used with respect to the handling of thick strip, such as pickled, hot-band, whereas mandril-type uncoilers are used to feed thinner strip into cold mills at higher speeds.✠

Figure 3-76:
Coil Box at Entry
End of Tandem
Cold Reduction
Mill.

In coil boxes, the feed coil is placed on rollers (one or more of which may be power driven) and the end of the coil fed into the bite of the first stand between guide rolls. In some instances, drums with flanges may be moved inwards to engage the eye of the coil as illustrated in Figure 3-77.

Unbending the end of the strip may be accomplished electromagnetically as shown in Figure 3-78. This uncoiler has a cradle 1 with rolls 2 and 3 onto which the coil is rolled from a conveyor. The coil is turned by the driver roller 3 so that the extended end of the strip is toward the electromagnet 4, and it is then raised by a screw mechanism 5 until it is in line with the axes of the drums 6 of the centering device which grips the edges. The electromagnet then descends, and when raised, it unbends the end of the strip until it touches the spring mounted driver roll 7. Subsequently, the end of the strip is gripped between this roll and roll 8, the latter being raised by a crank mechanism. The engaging of roll 7 after the disconnection of the electromagnet starts the uncoiling which is accompanied by the levelling of strip as it passes between rolls 7 and 8. If necessary, the drums 6 of the centering device can be braked by pneumatic brakes 9 during uncoiling.

On mandril-type uncoilers, the coil is slipped over the mandril and the latter expanded so as to be tight in the eye of the coil. In some cases the mandril may be moved backwards and

---

✠R. C. Conley, "Armco-Middletown's 86-Inch Tandem Cold Mill", Iron and Steel Engineer Year Book, 1972, pp. 349-352.

forwards along its axis so as to center the strip in the mill. The mandril may be powered with a motor that may be used for inching the coil into the mill for threading purposes but which may also act as a drag-generator and provide the necessary back tension during rolling. In reversing mills, the uncoiler will act as the coiler and vice versa in the reversed direction of the mill.

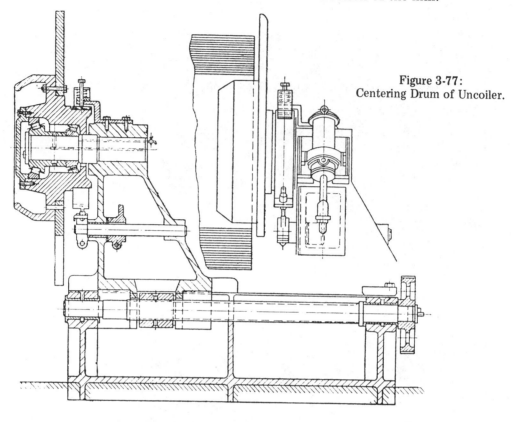

Figure 3-77: Centering Drum of Uncoiler.

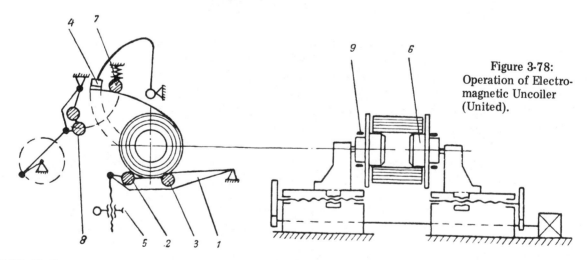

Figure 3-78: Operation of Electromagnetic Uncoiler (United).

3-22 Coilers

For cold mills, drum expanding mandril-type coilers of the type shown in Figure 3-79 are generally used. Such coilers are capable of providing the high forward tensions that may be required in rolling, as the strip is stretch levelled and wound tightly and uniformly on the drum.

Matching the rolling and coiling speeds and maintaining constant tension in the strip are complicated by the changing diameter of the coil as it is being built up. In old mills, some drums

Figure 3-79: Expanding Mandril-Type Coiler.

were driven by the mill motor and speeds were matched by slippage in the drive system (as, for example, in slipping clutches). In modern mills, however, the coilers are driven by variable speed motors which ensures matched speeds at the desired tension level. Moreover, in reversing mills, when the strip is being paid off a coiler, the motor acts as a generator returning energy from the back tension to the power supply.

In the U. S., the diameters of drums (when expanded) are usually about 16 inches. To minimize the possibility of coils collapsing or telescoping, the drum diameter should be as small as possible. However, the diameter should not be so small that the strip is plastically deformed as it is bent around it. Moreover, it must have sufficient strength to bear the weight of the strip, the effect of the strip tension and the radial pressures exerted on it by the strip. Unfortunately, the internal components of the drum required to reduce its diameter for coil removal also diminish the strength of the drum and necessitate its being of the order of 16 inches.

In some instances, additional support for the mandril is provided by an outboard bearing which may be moved aside when the coil is to be removed from the mandril.

Mandril design for cold mills varies considerably ranging from simple drums with no mechanisms for fastening the strip to the mandril to more elaborate types that clamp the head end of the strip. With the former, a common practice is to fasten the end of the strip onto the drum with a piece of adhesive tape and then rotating the drum to accumulate several wraps on it before applying high tension.

One method of wrapping the head end of the coil around the mandril involves the use of a belt wrapper. This is an endless belt of fabric that may be moved around the mandril as a movable guide. Such a device is shown in diagrammatic form in Figure 3-80.

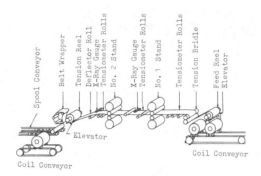

Figure 3-80:
Sketch of 2-Stand Secondary Cold Reduction Mill Showing Location of Belt Wrapper.

An example of the second type of mandril is illustrated in Figure 3-81. Here, the drum consists of three segments, the main segment being rigidly connected to the brake disc 2 and the movable segments 3 connected to the fixed segment by the spindles 4 and to each other by the springs 5. After the end of the strip has been fed into the jaws 6 and 7, the driving shaft 8 is rotated clockwise with the disc 2 slightly braked. At the same time, the driving shaft projections, acting on a lever 9 and rollers 10, ensure that the strip is tightly gripped by the jaws and the segments 3 are moved apart. When a coil is to be removed, the driving shaft 8 is reversed and the projection 11 comes up against the stop in the main segment. The jaws are then automatically opened and the segment 3 are pulled together by the springs, thereby freeing the coil on the mandril.

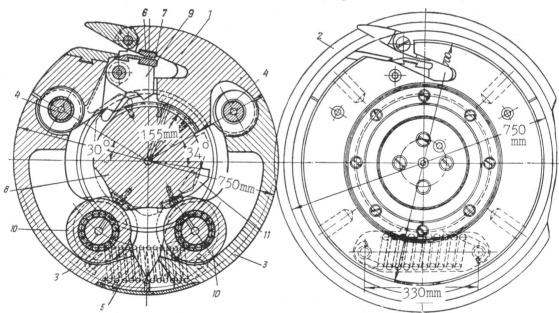

Figure 3-81:
Lateral Cross-section and End View of 750 mm-Diameter Coiling Drum.

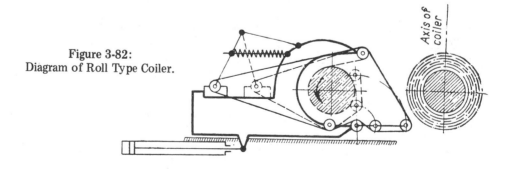

Figure 3-82:
Diagram of Roll Type Coiler.

COMPONENTS OF COLD ROLLING MILLS

The power required to drive a coiler is stated * to be

$$\left(M_{bend} + TR + \Sigma P \mu \frac{d}{2}\right) \frac{\nu}{R\eta} \frac{1}{33{,}000} \text{ HP}$$

Where: M_{bend} is the bending moment of the strip when uncoiling and in practical calculations is often taken as equal to the moment of plastic bending $M_S = \sigma_S S$ (pounds-feet) where σ_S is the yield stress and S is the moment of resistance of the strip to plastic bending.

T is the tension in the strip (lbs).

R is the minimum radius of the coil (practically the radius of the drum) (feet).

$\Sigma P \mu d/2$ is the total moment of friction in the bearings of the drum shaft (allowing for the weight of the drum and the coil) (pounds-feet).

ν and η are the speed of coiling (feet per minute) and the efficiency factor of the transmission.

Roll type coilers of the type shown in Figure 3-82 are sometimes used on cold mills where thick strip is to be coiled. They form coils by bending the strip between rollers instead of by winding it on a drum and have the advantage that strip can be fed into them at full speed.

Before coils are removed from a mandril, the loose end of the strip is usually secured. this may be accomplished by strapping the coil with a steel band, using an adhesive tape, or spot welding to secure the outermost wrap to the wrap immediately adjacent to it.

3-23 Coil Handling Equipment

The movement of coils to and from the mill requires the use of materials handling equipment that can be used effectively and safely without damaging the coil. Several approaches to coil handling are in commercial use; the particular approach chosen must be tailored to the type of mill being used.

To move coils appreciable distances to and from the mills, overhead cranes or mobile coil carriers are generally used. The former may be equipped with the use of "C" hooks which lift the coil by inserting the hook through the eye of the coil, maintaining the eye of the coil in an essentially horizontal position. (See Figure 3-83.) If the eye of the coil is to be maintained in a vertical position, a grab as depicted in Figure 3-84 or an electromagnet as shown in Figure 3-85 may be used. A mobile coil carrier suitable for in-plant use is illustrated in Figure 3-86.+

To move coils onto and off a mandril, an overhead crane may again be used, particularly for small coils such as may be used on a laboratory mill. In this case, a belt-type sling may be wrapped around the coil.

On production mills, a coil buggy may be used in connection with each mandril. Generally, the buggy rides on rails and features a "V"-shaped platform on which the coil rides with its eye horizontal. The buggy is moved towards the mandril and the platform may be raised or lowered so that the coil may be conveniently slid on the mandril. The platform may then be lowered and the mandril expanded. In removing coils, the same equipment may be used with the procedure reversed.

In some commercial mills, a V-shaped coil carrier sunk into the floor is used for coil handling. Here, hydraulic cylinders raise or lower the carrier and move it backwards or forwards in line with the axis of the mandril.

---

* A. I. Tselikov and V. V. Smirnov, "Rolling Mills", Pergamon Press, London, 1965.
+ J. A. Draxler, "Design Factors for Steel Mill Tractors", Iron and Steel Engineer Year Book, 1956, pp. 222-226.

Figure 3-83:
C-Hook.

Figure 3-84:
Grab for Handling Coils.

Figure 3-85:
Electromagnet.

Figure 3-86:
Mobile Coil Carrier.

COMPONENTS OF COLD ROLLING MILLS

3-24 Mill Instrumentation

Although earlier models of cold rolling mills were, as a rule, poorly equipped with respect to instrumentation, the modern mill usually possesses sensors to monitor virtually all of the important process parameters. The more important of the instruments are located, for the convenience of the roller, close to the mill stand on a panel or pendant such as is illustrated in Figure 3-87. Other instruments, often duplicates of those on the control pendant, may be found remote from the mill on the electrical control panels for the mill power supply.

The more important instrumentation of the mill monitors the following:

1. Main mill drive motor voltage and current.
2. Mill speed.
3. Rolling force at each side of the mill housing. (This information is usually presented as a total rolling force and as the unbalance or difference between the two forces.)
4. Screw settings (often as an "average" and an unbalance or difference in settings for the two sides of the mill.
5. Strip tension at the entry and exit sides of the mill stand and between bridles and terminal equipment (such as uncoilers and coilers).
6. Outgoing strip thickness.

Other instrumentation may be used to measure the rolling "solution" pressure and flow rate, the pressures in various hydraulic and lubrication systems, and the temperatures of various fluids used on the mill.

Figure 3-87: View of Cold Mill Showing Instrument Panel at the Last Stand.

Where data accumulation systems are available, the mill data monitored by the various instruments may be converted from analog to digital form and then periodically printed out. In this form, if supplemented with other information relating to the product, the mill crew, etc., the complete mill data may be conveniently analyzed by industrial engineering personnel in establishing the overall performance of the mill.

In more sophisticated mill installations, where computers are utilized for mill control, the information provided by the sensors or transducers is vital to the effective use of the computer. For example, where screw settings are automatically adjusted by the computer, rolling force data are utilized to ensure that mill stands are not overloaded. Similarly if drive motor speeds are automatically controlled, strip tension measurements may be used to ensure that the motor speeds are not excessive.

The various types of transducers utilized on production and laboratory mills is discussed more fully in Chapter 5.

3-25 Operating Controls

To enable the mill operator to handle the rolling process, a control panel is provided near the mill stand (often as a "control pendant") or more remote from the mill as a console or, in more recently built tandem mills, as a control pulpit. (See Figure 3-88.) ♀ Such panels contain all the start-stop buttons associated with the electrical circuits of the mill, the speed controls associated with the individual motors driving the mill rolls, the bridle rolls, the mandrils of the coiler and uncoiler and the screw downs, and the tension-setting controls associated with the coiler and uncoiler drives.

The on-off or start-stop switches are usually associated with the following:

1. The main power to the mill installation.
2. The mill direction selector (in the case of a reversible mill).
3. The line contactors associated with the various motors of the mill.
4. The lubrication and coolant flow systems.
5. The mill screws.
6. The mill and reel jog circuits.

A typical mill speed control circuit is shown in schematic form in Figure 3-89. Here, the operator's control is a reversing switch activating a small reversible d-c motor mechanically coupled to a large rheostat. The rheostat in turn controls the magnitude of the field current in the generator powering the drive motor.◁◾✻

Potentiometers are generally used in connection with controls such as:

1. Tension adjustments related to the coiler and uncoiler.
2. Coil diameter compensation.
3. Draft compensation adjustments.
4. Roll diameter compensation.

♀ Iron and Steel, February, 1973, p. 9.

◁ W. H. Dauberman and S. V. Stickler, Jr., "Outstanding Application Developments in Mill Electrical Drive Systems", Iron and Steel Engineer Year Book, 1962, pp. 33-44.

◾ O. G. Brunner, "Drives and Automation Systems for Wide 5-Stand Cold Mills", Iron and Steel Engineer Year Book, 1965, pp. 939-950.

✻ D. R. DeYoung and T. J. Dolphin, "Recent Trends in Cold Mill Electrical Drive Systems", Iron and Steel Engineer Year Book, 1966, pp. 659-667.

COMPONENTS OF COLD ROLLING MILLS

Figure 3-88: Modern 4-Stand Tandem Mill with Control Pulpit in Foreground.

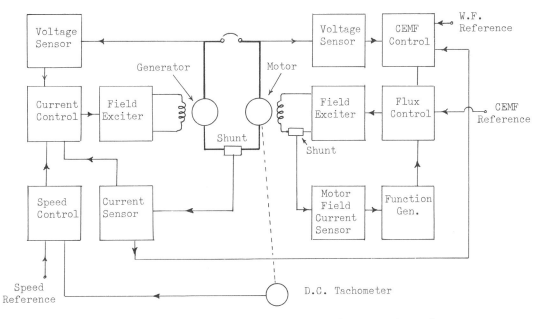

Circuit for speed-regulated mill-stand drive. (Power supply can be either a d-c generator or a thyristor dual converter.)

A typical tension control circuit including items 1 and 2 above is shown in schematic form in Figure 3-90.[a] Draft and roll diameter compensation controls are used to maintain the proper speed synchronization of the terminal equipment (coiler and uncoiler) with the work rolls of the mill.

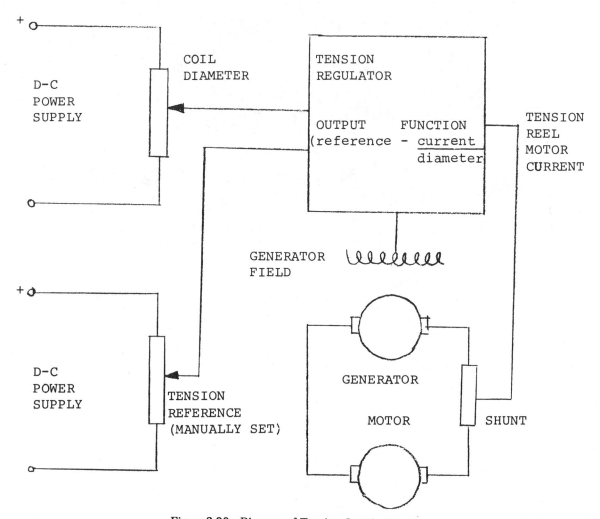

Figure 3-90: Diagram of Tension-Setting Control.

---

[a] W. H. Dauberman and S. V. Stickler, Jr., "Outstanding Application Developments in Mill Electrical Drive Systems", Iron and Steel Engineer Year Book, 1962, pp. 33-44.

Chapter 4 Mill Rolls and Their Bearings

4-1 Introduction

Of all the mill components, none is more critical with respect to the rolling operation than the mill rolls. They must be capable of withstanding the high compressive and shearing stresses needed to deform the strip without themselves undergoing plastic deformation and, though they elastically distort under these stresses, the distortion must be controlled to such an extent that the rolled strip has satisfactory shape, and that the rolling forces are maintained within reasonable ranges. Such stresses and distortions are directly related to the physical properties and the dimensions of the rolls and, accordingly, roll geometries and the metallurgical properties of the rolls are both of considerable importance with respect to mill design and operation.

Mill rolls are manufactured to very diverse specifications. In diameter, they may range from a fraction of an inch (as in the case of cluster mills rolling very thin foils) to several feet (as in the case of large backup rolls used in wide tandem mills). The length of the barrel (or face) may range from a few inches for narrow laboratory-type mills to close to 100 inches for the widest commercial cold strip mills. For small rolls in cluster mills, tungsten carbide is frequently utilized as the roll material, but the rolls of two, three and four-high mills are usually made of cast iron or steel. Most rolls are of single-piece construction, but "built-up" or sleeved rolls are finding increasing use. At the same time, a wide variety of surface finishes are given to the rolls, ranging from a rough, shot-blasted finish to mirror-like finishes with roughnesses as low as one or two micro-inches RMS. Similarly, a wide variety of crowns are used in practice to compensate for roll bending.

Roll necks are usually about one-third to three-quarters of the diameter of the main roll body and are machined to be fitted into bearings accommodated in the roll chocks which position the rolls in the mill housing. Two principal types of bearings are used; the sleeve type or "Morgoil" bearing and the "antifriction" or roller bearing. The former generally features an additional thrust bearing of the roller type and utilizes a constant flow of lubricant, whereas the latter may, in many cases, be packed with grease prior to use or may be mist-lubricated.

This Chapter discusses the design of rolls, their manufacture, the techniques used for finishing and reconditioning and their bearings, primarily as pertaining to four-high mills. Stresses retained in the rolls after fabrication are discussed and stresses impressed into the rolls under mill use are treated insofar as they affect roll life and the longitudinal bending of the rolls. The problems associated with roll life, such as season cracking, spalling and splitting are examined at length and mill practices developed to extend roll life are included in the Chapter. The cooling of mill rolls is treated in Chapter 7, and the relationship between rolling force and the flattening of the rolls is included in Chapter 9.

4-2 Roll Compositions

Experience has shown that both steel and cast-iron rolls are capable of cold rolling sheet and strip products. Forged steel, with a Shore D hardness ranging from 80 to 100, is perhaps the most commonly used material for both work and backup rolls. However, cast steel, with a hardness in excess of 70 Shore, may be used for backup rolls in wide, cold, strip mills and for the various rolls in sheet temper mills. Chilled cast-iron rolls may also be used for work rolls in most cold-rolling operations and flaked-graphite iron rolls, with a hardness in excess of 80 Shore, may be used as work rolls in wide strip mills. Generally speaking, the thinner the strip being processed, the harder should be the mill rolls. However, in four-high mills, the hardness of the backup rolls, which are always steel, should be kept 20 to 40 Shore D lower than the hardness of the work rolls.♦

♦ "Rolling Mill Rolls", Parts I, II and II, Based on the work of Jachem C. Thieme, Balsthal, Switzerland, and Sepp Ammereller, Bochum, Germany, and used by permission of Climax Molybdenum Company, "33"/The Magazine of Metals Producing, January, February and April, 1966.

Most rolls are made of a type of chrome steel similar to that used in roller bearings. In this material, the carbon content ranges from 0.8 to 1.0% and the chromium content from 1.0 to 2.0% depending on the manufacturer and the type of roll. Usually a composition with the maximum carbon and the least chromium is used for small diameter rolls, while the lesser carbon and higher chromium contents (which provide better tempering characteristics) are used for large diameter rolls. However, it is possible to modify the tempering characteristics of the chrome steel by the control of the concentration within it of secondary alloying elements, such as manganese, or by small additions of elements such as molybdenum or nickel.* The effects of varying alloying elements on both alloy irons and steels are presented in Tables 4-1 and 4-2.□

Table 4-1

Effects of Alloying Elements in Iron Rolls

Element	Effects
Carbon	Increases hardness, wear resistance and brittleness. Decreases ductility and depth of chill.*
Silicon	Increases graphite, adds to cleanliness and decreases depth of chill.
Phosphorus	Increases hardness and brittleness.
Sulphur	Increases hardness, brittleness and depth of chill.
Manganese	Reduces chill in lower ranges and increases it in higher ranges; in combination with nickel increases hardness and with chromium increases brittleness.
Nickel	Increases hardness, strength and wear resistance but decreases depth of chill.
Molybdenum	Increases strength, produces grain refinement.
Vanadium	Increases chill depth and strength but lowers ductility.
Chromium	Generally used in combination with nickel and/or molybdenum to increase hardness and depth of chill.
Copper	In small amounts, similar in effect to nickel.
Boron	Carefully used to increase hardness.

*The "chill" is the layer of dense, fine-grained structure at the surface of the roll barrel and results from a more-rapid rate of cooling accomplished by the use of "chills" in the mold.

Rolls for special applications, such as for use in Sendzimir mills, possess a larger-than-normal length-to-diameter ratio which presents certain difficulties with respect to their heat treatment and prevents the use of normal chrome steels.* Such rolls usually contain about 2% carbon with 12 to 14% chromium but occasionally high-speed steels (with 12-18% tungsten, 5-8% molybdenum, about 1% carbon and the usual additions of chromium and vanadium) may be used for their manufacture.

An improved roll composition for conventional mill rolls that has been found to reduce premature failures and increase the overall roll life was reported a few years ago.* This composition is shown in Table 4-3 together with that of a typical roll. It is claimed that it provides, from forged, vacuum-degassed ingots, rolls with an extremely clean steel possessing a fine, uniform sulphur distribution that exhibits low residual stresses and a far better response to hardening practices.

---

*L. Colombier, "Cylindres de Laminoirs à Froid", Technologie des Laminoirs, Cahier I. Irsid, Centre D'études Superieres de la Sidérurgie, 17, Avenue Serpenoise, Metz, 1963.

□"Roll Design and Mill Layout" by Ross E. Beynon, Published by the Association of Iron and Steel Engineers, Pittsburgh, Pa., 1956.

*G. F. Melloy, "Development of Improved Forged, Hardened Steel Roll Composition". Iron and Steel Engineer Year Book, 1965, pp. 451-460.

MILL ROLLS AND THEIR BEARINGS

Table 4-2
Effects of Alloying Elements in Steel Rolls

Element	Effects
Carbon	Increases hardness, brittleness and wear resistance but decreases resistance to shock.
Silicon	In the range 0.20 to 0.35 per cent increases cleanliness of steel. Increases hardness, used as a deoxidizer and to promote sound casting.
Phosphorus	Increases hardness and brittleness but decreases ductility.
Sulphur	Similar to phosphorus and must be carefully used.
Manganese	Increases hardness, brittleness, tensile strength and wear resistance.
Nickel	Increases strength and hardness.
Molybdenum	Increases strength and hardness.
Vanadium	Increases toughness, hardness and susceptibility to heat treatment.
Chromium	Used in combination with nickel and/or molybdenum, increases hardness.
Copper	Effects similar to those of nickel.
Boron	Increases hardness.

Table 4-3
Roll Compositions

Roll Composition	Alloying Elements (Per Cent)					
	C	Mn	Si	Cr	Mo	V
Typical	0.90	0.37	0.27	1.75	0.25	0.09
New Composition	0.70	0.25	0.27	0.77	0.65	0.09

Similar compositions with respect to carbon, chromium, molybdenum and vanadium are used in the manufacture of backup roll sleeves. However, the managanese is sometimes increased, as is evident from data published by Bracht and Bradd* and presented in Table 4-4.

Table 4-4
Roll Sleeve Compositions

Sleeve Composition	Alloying Elements (Per Cent)					Shore "C" Hardness
	C	Mn	Cr	Mo	V	
Type 1	0.60	0.70	1.10	0.44	0.10	60-65
Type 2	0.70	0.75	1.50	0.45	0.05	65-70

4-3 Mill-Roll Dimensions

In mill design, the barrel length (or face) of the mill rolls is established by the maximum width of strip to be rolled, the mill width being usually a few inches greater than the maximum strip width. The diameters of the work and backup rolls are then selected on the basis of a number of considerations, such as the mill width, the type of material to be rolled, the maximum reduction to

---

* N. A. Bracht and A. A. Bradd, "Factors Affecting Backup Roll Life". Iron and Steel Engineer Year Book, 1965, pp. 243-251.

be given to the strip, the mill speed and the provisions for roll cooling. Generally speaking, in the case of four-high mills, the backup roll diameter is established primarily by the mill width, but the diameter of the work rolls has, until recently, been established primarily on the basis of experience. However, as is seen in Chapter 7, valid technical procedures may be adopted with respect to work-roll-diameter selection. Only those parameters which limit the work-roll diameter and the uniformity of specific rolling force are discussed in this Section.

Inasmuch as backup rolls must support the corresponding work rolls, they must exhibit considerable rigidity and are, therefore, generally designed with a roll face-to-roll diameter ratio of approximately unity ("square" rolls). For cold mills constructed since 1945 in the United States, the relationship of backup-roll diameter to face length is illustrated graphically in Figure 2-7.[ø]

To prevent spalling of the rolls in four-high mills, contact stresses at the interface of work and backup roll must be limited to less than 300,000 psi.[+] This means, in effect, that if the work and backup roll diameters are fixed, there is a maximum specific rolling force that must not be exceeded. Such limits may be determined from Figure 2-8. For example, specific rolling forces for a mill with 20-inch diameter work rolls and 40-inch diameter backup rolls should not exceed about 115,000 pounds per inch. Alternately, it means that for a given maximum specific rolling force, the work-roll diameter must not be less than a certain minimum value ascertainable from Figure 2-8. On the other hand, the choice of too large a diameter for the work roll will lead to excessive rolling forces as will be understood more fully from the mathematical models pertaining to rolling force discussed in Chapter 9.

Polukhin and his coworkers have studied the uniformity of the specific rolling force across the width (B) of the strip being rolled in relation to the mill width (L) and the ratio of the diameters of the work and backup rolls (D_W/D_B).[■] They showed that there is an optimum strip width for every ratio of diameters in a given four-high mill such that there is a uniform specific rolling force across the width of the strip in the roll bite and that the deflections of the axes of the two rolls are the same. In general, three forms of inter-roll pressure distribution are possible; a maximum in the middle of the roll barrel, maxima at the edges of the strip and a uniform distribution.

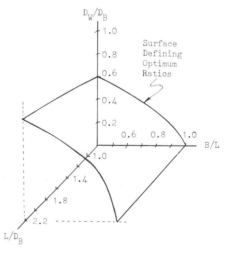

Figure 4-1:
Optimum D_W/D_B, L/D_B and B/L ratios which provide uniform specific rolling force across the width of the strip in the roll bite.

The interrelationship between the basic parameters is shown in a three-dimensional diagram in Figure 4-1. The surface in this diagram is represented by the optimum ratios D_W/D_B; L/D_B and B/L which provide a uniform specific rolling force across the width of the strip in the roll

---

[ø] "Cold Strip Mills in the U.S.", "33"/The Magazine of Metals Producing, Volume 3, No. 9, September, 1965.

[+] J. I. Greenberger, "Rolling of Metals", Iron and Steel Engineer Year Book, 1959, pp. 215-223.

[■] V. P. Polukhin et al. "Optimum Rigidity and Elastic Deformations of the Rolls in Four-High Rolling Mills", Izv. VUZ Chern Met., 1965, (1) pp. 78-84. (British Iron and Steel Institute Translation #5994, October, 1965, published by The Iron and Steel Institute, 4, Grosvenor Gardens, London SW1, England.)

MILL ROLLS AND THEIR BEARINGS

bite. The ratios which provide a maximum value of the specific force at the edge of the strip are located above this surface, whereas, those that provide a maximum at the center of the roll are located below the surface of the diagram.

The relationship between backup and work roll diameters for cold rolling and temper mills installed in the USA between the years 1945 and 1965 is illustrated by Figure 4-2.[ø] Backup rolls with diameters in the range 49-56 inches and work rolls with diameters in the range of 18-24 inches are the most commonly used for sheet and strip production. For seven U. S. mills, Polukhin et. al., have established that the B/L ratio ranges from 0.23 to 0.95.[■]

With respect to the roll necks, these are usually designed to be as large as can be conveniently accommodated in the roll chocks. For the rolls of two-high strip mills, Geleji[♦] suggests that the diameter of the neck should be 0.75 to 0.8 times the roll body diameter and that the length of the neck to its diameter should range from 0.8 to 1.0. For rolls manufactured recently in the USA, however, the average ratio of neck to body diameters appears to be about 0.57.[‡] In the case of the necks of the work rolls of four-high mills, which are not required to carry the heavy rolling loads, the ratio of neck-to-body diameter may be as low as about 0.3.

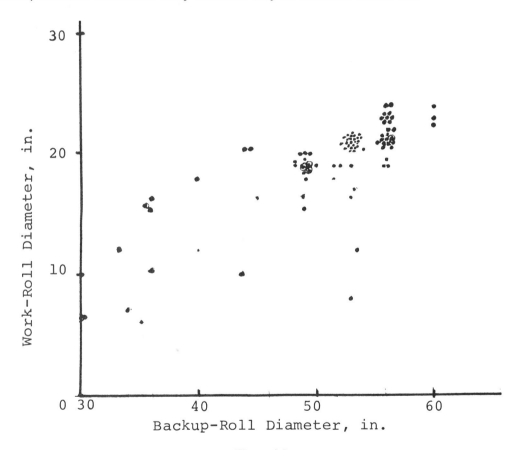

Figure 4-2:
Relationship of Work Roll to Backup Roll Diameters for Four-High Mills
Installed in the U.S.A. Between 1945 and 1965.

[ø] "Cold Strip Mills in the U.S.", "33"/The Magazine of Metals Producing, Volume 3, No. 9, September, 1965.

[■] V. P. Polukhin et al. "Optimum Rigidity and Elastic Deformations of the Rolls in Four-High Rolling Mills", Izv. VUZ Chern Met., 1965, (1) pp. 78-84. (British Iron and Steel Institute Translation #5994, October, 1965, published by The Iron and Steel Institute, 4, Grosvenor Gardens, London SW1, England.)

[♦] "Forge Equipment, Rolling Mills and Accessories" by A. Geleji, Akadémiai Kiado, Budapest — 1967.

[‡] F. A. D'Isa, H. Erzurum and J. Gross, "Stress Concentration Curves for Various Neck Configurations of Steel Mill Rolls", Iron and Steel Engineer Year Book, 1969, pp. 637-641.

4-4 The Design of Sleeved Rolls

Because conventional backup rolls must be frequently removed from service when their diameters have been decreased by about 10% (due to wear and regrinding), sleeves are frequently used on such rolls (see Figure 4-3). Although the cost of a sleeved roll is appreciably more than the cost of a comparable roll of conventional design, the extended life of the main roll body, resulting from the replacement of the sleeves when required, more than compensates for the additional cost of the rolls. Basically, the sleeve should possess physical properties similar to those of the surface of the conventional roll. However, it should be easy to replace when necessary and, at the same time, be shrunk onto the main roll body so tightly that no voids occur at the interface and no slippage occurs during the transfer of power from the sleeve to the roll body or vice versa (as during accelerations or decelerations of the mill or during normal operation of backup roll driven mills).

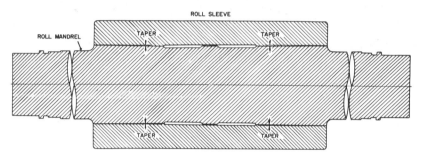

Figure 4-3: Cross Section of a Sleeved Roll.

The optimum thickness for a sleeve remains somewhat in doubt. Sleeves 4-inches thick placed over 25-inch diameter backup rolls have been successfully used on a four-stand sheet mill[•] (and replaced when worn down to 2-3/4-inch thickness), but sleeves ranging in thickness from 5 to 6 inches were found to fail frequently because of splitting resulting from spalls, wreck marks and end rubbing.[□] However, differentially-hardened 5-1/4" thick sleeves were found to be successful on a two-stand tin-temper mill under conditions of light loads and a good protection against roll damage due to cobbles. Later, 8-1/2-inch thick sleeves, with a Shore hardness of 62-65, were satisfactorily used on a four-stand tandem mill.[□]

As a protection against end rubbing, two methods may be conveniently used.[•] One is to shorten the sleeve so that there is more clearance between the end of the sleeve and the housing and the other is to use guard rings at each end of the sleeve, so that if rubbing occurs, it will damage the ring and not the sleeve.

Although sleeves commonly used feature perfectly cylindrical inner surfaces, Skorokhodov and Belevskii [†] have proposed the use of noncylindrical surfaces for the body and sleeve. These include cycloidal (elongated hypocycloid, pericycloid) and equiaxial joined with a noncircular profile in the shape of multiple, sinusoidal curves. Thus, these profiles differ only very slightly from each other but the equiaxial profile has an advantage over the cycloidal profiles in that no special measuring tools are required to check it.

To provide ease of roll changing, the barrel is placed on the shaft with a slight initial fitting clearance, as illustrated in Figure 4-4 for the case of a cycloidal profile. It is then claimed that, under load, the parts satisfactorily center themselves.

As illustrated in Figure 4-5, $D_i/2 + D_e/2 = D_o$, the dimension of the roll body or shaft and $D_e/2 - D_i/2 = 2e$ where e is the eccentricity. At the same time, $D_o + 2e = D_e$ (corresponding to

---

• N. A. Bracht and A. A. Bradd, "Factor Affecting Backup Roll Life", Iron and Steel Engineer Year Book, 1965, pp. 243-251.

□ N. Powell, discussion of above paper.

† N. E. Skorokhodov and L. S. Belevskii, "New Design of Rolls for Sheet Rolling", Izv. VUZ Chern. Met., 1969, (12) pp. 123-125. (British Iron & Steel Institute Translation #8521).

MILL ROLLS AND THEIR BEARINGS

twice the distance between the axis and the furthest point on the profile) and $D_o - 2e = D_i$ (corresponding to twice the distance to the closest point on the profile).

In the case of a cycloidal profile, when torque is applied, the shaft of the roll rotates in the sleeve by an angle corresponding to the clearance between the two, assuming the position of the dotted lines in Figure 4-4.

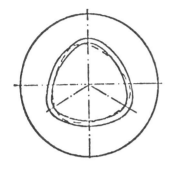

Figure 4-4:
Sketch of Sleeved Roll with
Non-Cylindrical Mating Surfaces.

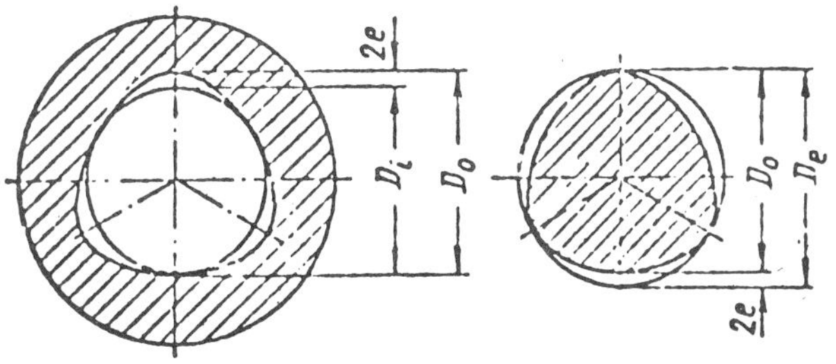

Figure 4-5: Dimensions of Sleeved Roll.

4-5 Molds Used in the Casting of Rolls

Virtually all molds used in roll manufacture are designed for the bottom swirl method patented in 1835.▲ In this method, a circular motion is induced by fast pouring the molten metal down a runner and through a tangential spout, centrifugally throwing the denser, purer, metal to the surface.▲

Two types of molds are in general use. One is the chill type shown in Figure 4-6, used mainly for plain-body iron rolls and, with sand linings, for plain body steel rolls, such as backup rolls. The second, illustrated in Figure 4-7, is known as the split flask and, though generally used for shaped rolls, may also be used for plain-body steel rolls.

In the chill type molds, the sections for the top and bottom necks, as well as for the shrinkhead, are made of foundry sand, whereas the body mold (consisting of one or more cylinders) is made of metal which acts as a chiller for the body of the roll. Due to the more rapid cooling of the roll body, natural segregation within it is minimized and, in the case of iron rolls, this portion of the roll contains a minimum of graphite. However, because of the sand molds for the roll necks, these portions of the rolls exhibit softer and tougher characteristics.

In the split-flask mold, the mold is split lengthwise, each half being filled with a well-tamped, high-silica molding sand.♦ The sand in each half flask is shaped by means of a sweep, which is a metal-edged flat piece of wood with the same outline as the contour of the roll. The molding sand is carefully smoothed down, and runners and gates are sometimes formed in the sand, as illustrated in Figure 4-8. To prevent the sand from being torn loose by the swirling motion of the liquid metal during casting, large headed nails may be driven into the sand. The portions of the mold are then thoroughly dried and hardened in a large oven. They are then clamped together and set on end in the pouring pit. The hot top, which consists of sand-lined cast-iron rings clamped together, is set on top of the end of the mold with its runner hole aligned with that of the mold.

Rolls (both iron and steel) are usually cast slightly oversize to provide stock for subsequent machining and thereby permit defects on the cast roll surface to be conveniently

---

▲ F. H. Allison, Jr. and C. E. Peterson, "Modern Manufacture and Use of Cast-Iron Rolling-Mill Rolls", Iron and Steel Engineer Year Book, 1954, pp. 952-961.

♦ "Rolling Mills, Rolls and Roll Making", Mackintosh-Hemphill Company, Pittsburgh, Pa., 1953.

removed. However, the amount of excess metal on the rough casting must be limited, since the necessity for removing a considerable thickness of the metal would seriously diminish the thickness of the hardened layer produced either by casting or heat treatment.□

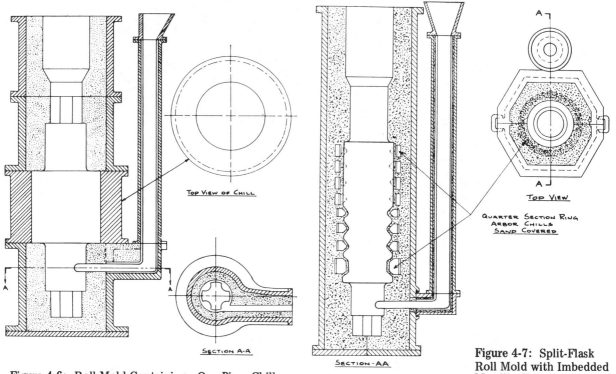

Figure 4-6: Roll Mold Containing a One-Piece Chiller.

Figure 4-7: Split-Flask Roll Mold with Imbedded Metallic Chillers.

4-6 Melting and Pouring Techniques in the Casting of Rolls

For melting the metal for iron rolls, reverberatory air furnaces, fired with powdered coal, are generally used, whereas for chill roll metals and frequently the center metals of double-poured rolls, a cupola is employed. Metal for cast steel rolls is generally melted in acid open-hearth furnaces capable of reaching temperatures of 3150°F, although basic open-hearth and electric furnaces are occasionally used.

To produce hot metal for cast iron rolls, a substantial percentage (at least 40%) of scrap rolls should be added to the furnace charge to ensure the proper relationship between combined carbon and graphite and to attain the required denseness of the surface metal.▲ However, cast steels used for rolls present no problem in this respect.

In the case of iron rolls, the molten metal is tapped from the furnace at a temperature as close as possible to that prescribed and run into one or more lipped ladles. The slag is then carefully skimmed off its surface and, at the correct pouring temperature (ranging from 2400° to 2550°F), the iron is decanted into the runner. A casting rate of about 50 tons per hour is maintained to create the desired swirling motion in the mold.

Cast steel and alloy cast steel rolls are poured in a similar manner except that the pouring temperature is higher (2600° to 2700°F depending on the carbon content) and the rate is

---

□ "Roll Design and Mill Layout" by Ross E. Beynon, Association of Iron and Steel Engineers, Pittsburgh, Pa., 1956.

▲ F. H. Allison, Jr. and C. E. Peterson, "Modern Manufacture and Use of Cast-Iron Rolling-Mill Rolls", Iron and Steel Engineer Year Book, 1954, pp. 952-961.

MILL ROLLS AND THEIR BEARINGS

appreciably higher (150 to 180 tons per hour).♦ Cast steel and alloy cast steel require a considerably longer solidification time in the mold because of the use of sand instead of cast iron chills.

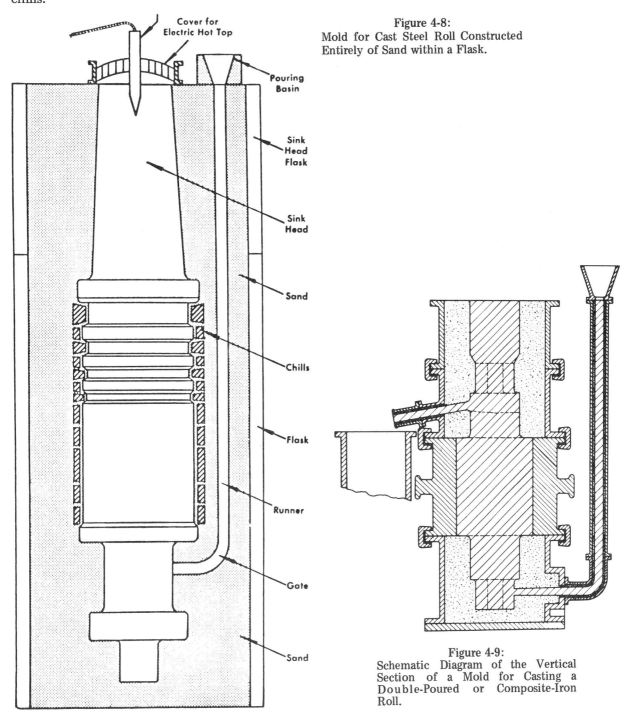

Figure 4-8:
Mold for Cast Steel Roll Constructed Entirely of Sand within a Flask.

Figure 4-9:
Schematic Diagram of the Vertical Section of a Mold for Casting a Double-Poured or Composite-Iron Roll.

In 1936, the double-pour method of casting was introduced to produce rolls with a hard, wear-resistant body surface, yet with body core, necks and wabblers that would be tough and shock resistant. Double pouring is further classified into composite pouring and inoculation

---

♦ "Rolling Mills, Rolls and Roll Making", published by the Mackintosh-Hemphill Company, Pittsburgh, U.S.A., 1953.

pouring. In the former, the filled mold is permitted to stand for a predetermined time which is just long enough for the metal in the body to solidify to the proper depth (1 to 4 inches) but short enough so that the metal in the sand molds (necks and wabblers) remains molten. A second molten alloy that is softer but tougher than the first is now poured into the top of the mold replacing the still molten metal and forcing it back out through the runner into a ladle for reuse. As soon as the color difference shows that the newer, hotter alloy is coming back out through the runner, the second pour is stopped and the whole casting allowed to cool naturally. Alternatively, composite pouring may be accomplished by permitting the proper time to elapse after the first pour, then pouring the second alloy into the same runner forcing the first metal into an overflow at the base of the upper neck of the roll (see Figure 4-9). The overflow is stopped up when the second, hotter metal begins to flow from it and the pouring is then continued until the upper neck and the hot-top are completely filled.

In the inoculation-pouring procedure, the original molten metal is poured through the runner only until it reaches the top of the roll body. After permitting solidification in the roll body to proceed for a certain time, the second metal is poured into the runner. This metal is poured until the total metal of both pours reaches the top of the hot top on the mold and then the casting is allowed to cool naturally. This method mixes the much less rich alloy with the original rich alloy which is still molten in the sand molds so that the combination of the two alloys gives the softer but tougher body core, necks and wabblers as desired.

As reported by Allison and Peterson,[▲] much skill is needed in the production of the double-poured roll, not only with respect to pouring; but also in producing the required shell depth and in balancing the composition between shell and fill metal for different roll applications. Moreover, because of the two separate melts and the extra equipment required, it is more expensive to make double-poured than single-poured rolls.

Since the top of the roll casting (the top or cope neck) is the last to solidify, it contains the most segregation. It is, therefore, apt to be weaker than the bottom or drag neck and should not be used for connection to the spindles of the mill. In rolls with identically shaped ends, many roll makers mark the ends with a "C" or a "D" (so that the user can place the "D" or drag neck in the drive position).

4-7 The Production of Forged Rolls and Sleeves

In the preparation of molten steel intended for forged rolls and sleeves, particular care must be exercised in the choice of materials and the melting, deoxidation and degassing practices. Basic electric arc furnaces or induction furnaces are generally used for melting and care is taken to keep the sulphur and phosphorus content of the steel to a minimum. Similarly, care must be exercised in the deoxidation practice so that inclusions formed of complex oxides of manganese, silicon and aluminum are kept at as low a level as possible.[■]

Hydrogen inevitably enters the molten steel from water vapor in the atmosphere but unless the hydrogen content of the subsequently forged steel is maintained below certain levels (on the order of 2 parts per million), "flaking" occurs. This is characterized by the development of numerous small internal ruptures within the steel, often at some distance away from its surfaces (see Figure 4-10). As will be seen in Section 4-26, flaking is the cause of "season cracking" or the explosive splitting of rolls even before they are put into service.

Hodge, Orehoski and Steiner [⊕] have shown that, for steels similar to those used in roll manufacture, there are threshold values of hydrogen content below which the susceptibility to

---

[▲] F. H. Allison, Jr. and C. E. Peterson, "Modern Manufacture and Use of Cast Rolling Mill Rolls", Iron and Steel Engineer Year Book, 1954, pp. 952-961.

[■] A. A. Bradd, "Material and Design Defects in Forged Steel Rolls", Iron and Steel Engineer Year Book, 1961, pp. 31-44.

[⊕] J. M. Hodge, M. A. Orehoski and J. E. Steiner, "Effect of Hydrogen Content on Susceptibility to Flaking", Transactions of the Metallurgical Society of AIME, Vol. 230, August, 1964, pp. 1182-1193.

MILL ROLLS AND THEIR BEARINGS

flaking is very low and above which the susceptibility increases abruptly and markedly as the hydrogen content increases. This is illustrated in Figure 4-11 where the ordinate relates to the number of flakes found on an etched surface one square foot in area.

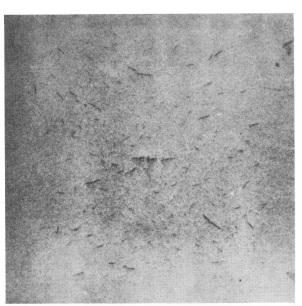

Figure 4-10: Flakes in Hood-Cooled Ni-Mo-V Steel.

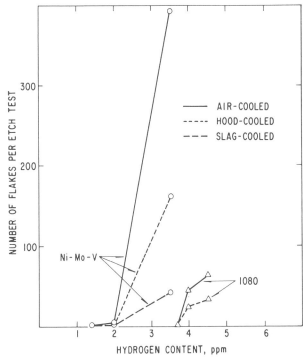

Figure 4-11: Effects of Hydrogen Content on Flaking of Ni-Mo-V and 1080 Steels.

One manufacturing technique that may be employed to minimize the occurrence of flaking involves vacuum degassing the steel with a facility such as that shown in Figure 4-12. However, to obtain hydrogren concentrations in the steel of the order of 1 ppm, pressures are on the order of 1 mm Hg are required in the vacuum tank.

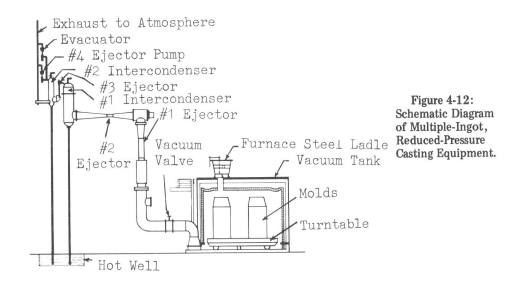

Figure 4-12: Schematic Diagram of Multiple-Ingot, Reduced-Pressure Casting Equipment.

The ingot, usually cast in a mold with a roughly circular, fluted cross section, must be carefully reheated after solidification, particularly with respect to the maximum temperature and the temperature distribution attained. Reheating must be conducted slowly, preferably in direct-fired, car-bottom type furnaces. Since the thermal conductivity of the steel is relatively low, rapid reheating might result in too great an expansion of the outer layers of the steel relative to the cooler, internal regions with the consequent occurrence of internal cracks.✠

Step heating may be employed; that is, the work piece may be held at one or more temperature levels below forging temperature and then allowed to equalize before proceeding to a higher temperature level. It has been found that, after its temperature has been equalized at a point slightly above the upper critical temperature (about 1475°F), steel can be heated at a rate of 40° to 60°F per hour until forging temperature is attained. This cycle results in heating times corresponding to approximately 3/4 to 1 hour per inch of diameter of thickness of the ingot or forging. In general, carbon steels containing over 0.50 per cent of carbon and alloy steels require slower rates of heating than carbon steels with less than 0.50 per cent carbon.

The forging temperature is selected to provide the best condition for hot working a given steel. Although the final properties of a finished forging are established largely by the heat treatments given to it subsequent to hot working, the temperature at which it is hot worked influences to varying degrees, depending on the grade, what heat treatments are necessary as well as the final mechanical properties of the steel. In general, lower finishing temperatures result in a finer-grained microstructure when forging is completed. Moreover, the finer-grained structures respond better to heat treatment than do coarser-grained structures. However, the finishing temperature must be kept high enough to prevent the occurrence of forging bursts (internal ruptures) that may result from severe stresses induced by working large masses of steel at too low a temperature.

If the temperature of the forging is too high, "burning" may occur in which actual fusion and oxidation take place at the austenite grain bounderies. Under these circumstances, a hot shortness ensues resulting in badly torn surfaces and internal ruptures during hot working. Such damaged steel cannot be salvaged. On the other hand, the effects of overheating the forging (but not to the extent to cause burning) can be largely removed by hot working but more severe overheating can cause low ductility in forgings tested after final heat treatment.

During forging, the cross section of the work piece is reduced by a factor of 3 or 4, this reduction being necessary to achieve the desired grain size, the homogeneity of structure and the absence of defects. However, the deformation should not be too large, otherwise excessive elongation of the inclusions might intensify their effectiveness in causing roll breakage. In addition, careful attention must be paid so that any surface laps or cracks are removed before proceeding with the forging.

Figure 4-13 shows the steps involved in the production of a sleeve for a cold-reduction mill backup roll. Such sleeves are generally made from a high carbon (0.50 to 0.65 per cent), nickel-chromium or chromium-molybdenum steel. With respect to Figure 4-13, the sleeve has a 42-5/8-inch O.D., a 28-inch I.D. and a 42-7/8-inch face. Three sleeves of this size are produced from a 48-inch diameter ingot weighing 72,800 lbs. The round, corrugated ingot is first forged to a 44-inch octagon, then sheared into three pieces each 42 inches in length. The pieces are reheated and each is upset forged to 36 inches in length, after which a 16-1/2-inch diameter hole is punched through the longitudinal axis of each. After reheating, each piece is forged on a 16-inch diameter expanding bar to increase the hole to 20-inch I.D. Each expanded piece is again reheated and forged on a 19-1/2-inch diameter bar to a 39-1/4-inch octagon and the hole is enlarged to 21 inches. Another reheating is followed by forging the ends of each piece to a level contour and, after a final reheating, the sleeves are each forged on a 19-inch diameter bar to 26-inch I.D., 44-5/8-inch O.D. and approximately 53 inches in length.

✠ "The Making, Shaping and Treating of Steel", edited by H. E. McGannon, U. S. Steel Corporation, 8th Edition, 1964, Chapter 37.

MILL ROLLS AND THEIR BEARINGS

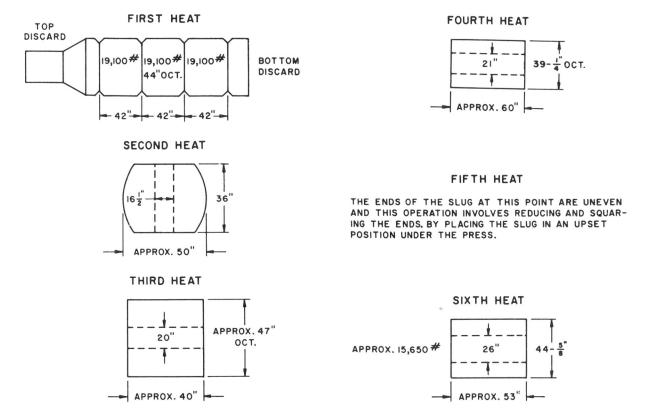

Figure 4-13:
Steps Involved in Forging a Sleeve for a Backup Roll for a Cold Reduction Mill.

After forging, a slow cooling is imperative. In many cases, it is desirable to replace the roll or sleeve in the furnace and to follow well-known treatments intended to prevent flaking. These include sustained heating for long periods and very slow cooling permitting the diffusion and elimination of hydrogen.

4-8 Heat Treating Cast-Steel Rolls

To obtain the desired properties of a cast-steel roll, such as strength to resist breakage and hardness to resist wear, cast-steel rolls are usually heat treated. The treatment given to a particular roll depends on its composition and size and, as a consequence, heat treatments differ considerably with respect to rate of heating, the maximum temperature, the rate and method of cooling and the quenching medium.

In single annealing, the roll is heated at a rate of 15 to 50°F per hour to a temperature high enough to change the "as-cast" structure, to remove the residual stresses and to soften the roll. The roll is then maintained at its maximum temperature for a minimum of one hour for each inch of diameter and then cooled down to 200°F in a furnace. The time required for this processing is approximately one week, after which the roll is relatively soft but strain free.

In the double-annealing practice, the roll is heated at the same rate to a considerably higher temperature than that used for a single anneal and is held at the maximum temperature for the same length of time. After a slow cool in the furnace, the roll then undergoes the same heating cycle as for a single anneal except that it is cooled in the furnace at the same rate at which it was heated. The first heating breaks up the as-cast structure and the second heating refines the grain structure. This processing, which may take as long as two weeks, produces the softest roll of a given composition but its fine grain structure provides it with a high degree of resistance to breakage.

In normalizing, the roll is subject to a heat treatment similar to single annealing except that it is cooled in air. As a consequence, normalizing increases the strength and hardness of a roll but also increases its brittleness.

When a roll is to be hardened, it is first single annealed, reheated to the proper temperature for the composition, and held there for a period dependent upon its size. After withdrawal from the furnace, the roll is then quenched in the proper medium but the roll necks and wabblers are protected from the quench. The severity of the quench establishes the degree of hardness and, after the roll temperature has reached a certain value, the roll is replaced in the heat treating furnace and tempered to the specified hardness. This hardening operation consumes one to three weeks but it provides a high degree of wear resistance to the roll.

In the production of a differentially heat-treated roll, between the annealing, quenching and tempering cycles described above, the roll is placed in a specially constructed reheat furnace capable of heating the roll at a very rapid and uniform rate so that proper hardening temperatures can be attained in the surface area of even large rolls without a long soak time.§ Following this heating operation, the roll is quenched and the hardened shell is then refined through a tempering operation as is standard with conventionally heat-treated rolls.

4-9 Heat Treating Forged Steel Rolls

To obtain the requisite hardness and wear resistance of the roll bodies, forged steel rolls must be heat treated. This constitutes a difficult manufacturing procedure, particularly with respect to cold mill work rolls.× Many different techniques and pieces of equipment are used for this

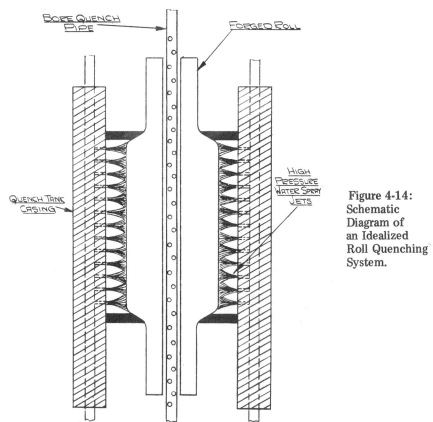

Figure 4-14: Schematic Diagram of an Idealized Roll Quenching System.

---

§ J. M. Dugan, "Differential Heat Treatment Tailors Roll Structure to Satisfy Performance Demands", "33" Magazine, November, 1973, pp. 44-47.

× J. Dugan, "Application of Progressive Induction Heating to Forged Roll Hardening Operations", Iron & Steel Engineer Year Book, 1958, pp. 789-800.

purpose, but basically, the practice can be described in general heat-treating terms. The roll is first heated above the critical temperature and then quenched rapidly through the transition temperature range so as to produce a martensitic surface.

Figure 4-14 presents a sketch of an idealized, conventional roll-hardening quench treatment. The roll has been heated to a temperature above the upper critical and is being quenched on all points of the roll surface with high pressure, cold water sprays. A spray arrangement is also used in a center bore to remove heat from the center of the roll forging so that the surface quenching may be more effective. After being subjected to such drastic quenching, the roll is then immediately placed in a stress-relieving furnace in order to avoid failure due to bursting in the quenched, stressed condition.

Typical microstructures in a 21-inch diameter roll obtained by this general method of heat treating are illustrated in Figure 4-15 and the corresponding Shore scleroscope hardness values (Standard "C" scale) are shown in Figure 4-16. It is to be noted that the martensitic and martensitic-bainitic bands are located on the outer and inner surfaces of the roll and Shore hardnesses as high as 100 may be obtained on the outer surface.

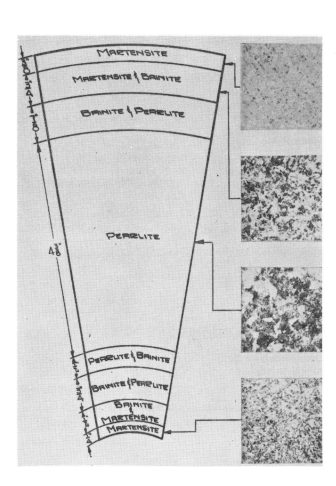

Figure 4-15:
Typical Microstructures
in a 21-Inch Diameter
Roll Given a Conventional
Roll-Hardening Quench
Treatment.

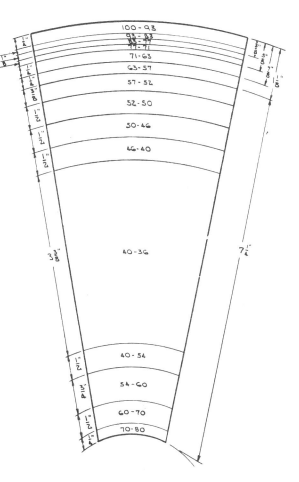

Figure 4-16:
Typical Hardness
Values in a 21-Inch
Diameter Roll
Given a Conventional
Roll-Hardening
Quench Treatment.

Although the induction heating of rolls began to be practiced in the 1930's, the progressive induction heating and quenching of rolls was not developed until the 1950's. Using a coil around only a small portion of the roll barrel, as illustrated in Figure 4-17, the heating of the roll can be localized in a manner dependent on the power output and frequency (usually about 1200 Hz) of the electrical generator, the gap between the inductor coil and the roll and the restivity of the roll metal. By locating a ring of sprays near the inductor, as shown in Figure 4-17, and by lowering the roll slowly down through the coil and the spray ring, it is possible to progressively heat and quench the roll body surface. Temperatures which are developed on the roll surface during the hardening operation are indicated in Figure 4-18. However, the downward movement of the roll must be carefully controlled so that adequate time above the critical temperature is allowed for each point on the roll surface. The direction of the water jets, the volume of water used, and its pressure level must all be coordinated. While changes in the water temperature in the range 50-80°F do not greatly affect the resulting hardness, it has been found that brine can give slightly higher surface-hardness values.

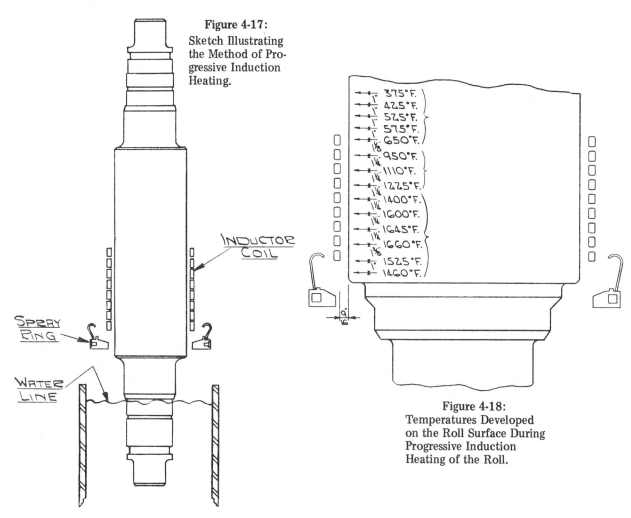

Figure 4-17: Sketch Illustrating the Method of Progressive Induction Heating.

Figure 4-18: Temperatures Developed on the Roll Surface During Progressive Induction Heating of the Roll.

Typical microstructures obtained by the progressive-hardening method are shown in Figure 4-19 with the corresponding hardnesses in Figure 4-20. It is to be noted that there is no appreciable hardening at the surface of the center bore in the roll and thus quenching in this region of the roll is unnecessary.

It should be noted, however, that rolls which are to be hardened by induction heating must receive prior conditioning heat treatments in order to have a microstructure which will

MILL ROLLS AND THEIR BEARINGS

properly respond to induction heating and become fully austenitized. Stresses in the roll may be removed by a high-temperature anneal after the conditioning treatment but before the final induction-hardening operation.

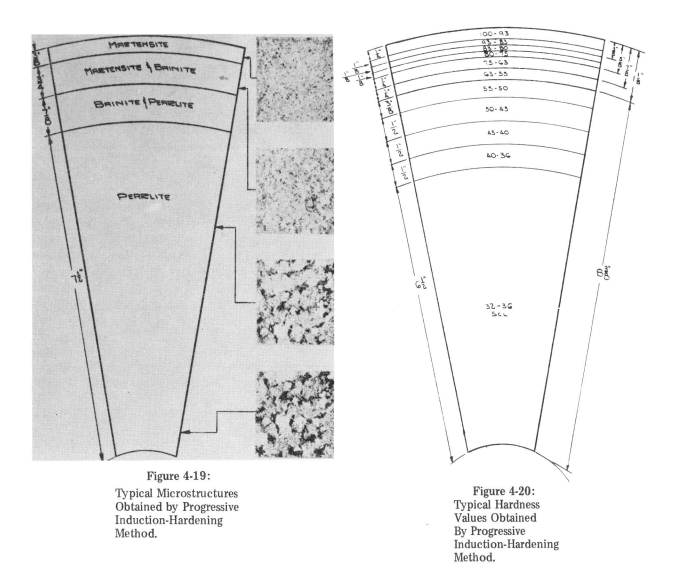

Figure 4-19:
Typical Microstructures Obtained by Progressive Induction-Hardening Method.

Figure 4-20:
Typical Hardness Values Obtained By Progressive Induction-Hardening Method.

    The hardening mechanisms with respect to conventional rolls may be understood with reference to the TTT (time-temperature-transformation) diagram shown in Figure 4-21 Here it is seen that the bainite start temperature is about 850° to 900°F, and that there is a strong ferrite and pearlite reaction (indicated by the separation of the steep ferrite curve labeled Fo_s). On cooling, the material near the surface of the roll (corresponding to the cooling curves labeled 1/8 and 1/4 inch from the quench) transforms directly from austenite to the hard martensite, whereas, that at greater depths from the surface contains appreciable quantities of ferrite and pearlite. The improved roll composition discussed in Section 4-2 exhibits a TTT diagram shown in Figure 4-22.* Here the ferrite reaction is almost entirely suppressed, the pearlite reaction eliminated, and the bainite start level raised to temperatures close to 1100°F. Because of these different characteristics, the composition permits roll hardening to a greater depth.

---

* G. F. Melloy, "Development of an Improved Forged, Hardened Steel Roll Composition", Iron and Steel Engineer Year Book, 1965, pp 451-460.

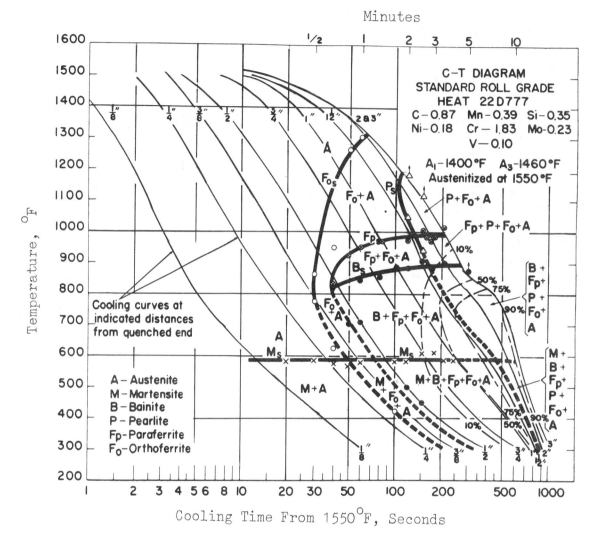

Figure 4-21:
Time-Temperature-Transformation Diagram for Standard Roll Grade of Steel.

4-10 Reconditioning Forged Steel Rolls

When forged steel mill rolls have been worn below optimum hardness levels, it is frequently possible to recondition them into top-grade rolls and use them either in the same mill or one that will accommodate smaller diameter rolls.° Such rolls must first receive a tempering treatment to remove all of the stresses created by the original hardening or imposed on the roll through mill service. This may be accomplished by an installation as sketched in Figure 4-23 in which only the body of the roll is subjected to the high tempering-temperature treatment. The temperature of the roll journals is kept low so that no warpage occurs in them. Following this tempering treatment, the roll is machined to provide a defect-free surface on the body and then rehardened, as for example, by the progressive induction method discussed in the preceding section ⸸. A forged-roll refinishing complex of modern design is illustrated in Figure 4-24.

° "Worn Forged Rolls Can be Reconditioned for Use in a Smaller Mill", Iron and Steel Engineer, November, 1965, p. 162.

⸸ "Roll Reclamation Route Reaps Savings for Steelmakers", "33" Magazine, November, 1973, pp. 48-51.

MILL ROLLS AND THEIR BEARINGS

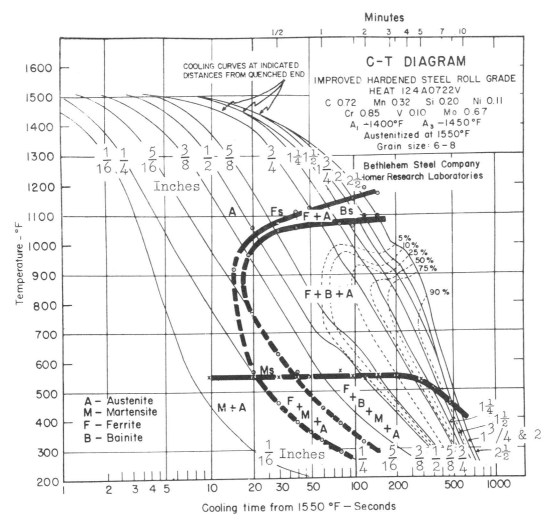

Figure 4-22:
Time-Temperature-Transition Diagram for Improved Roll Composition.

It is frequently believed that better service life is received from rehardened rolls than from the original ones.× Under some circumstances, rehardened rolls do show a statistical advantage in this respect resulting from the following causes:

1. Inherent defects in rolls, such as unrelieved quenching stresses usually become evident during the period of the original use of such rolls;

2. The original hardening treatment may have created a dispersion of carbides which may be a superior structure for the second or subsequent hardening treatments, and better service life in such cases may be expected from rehardened rolls.

Although rolls can be rehardened and returned to service, the reduction in the roll diameter due to wear and regrinding finally reaches a point where the roll cannot be used. As a consequence, it is not unusual for 90% of the original roll weight to be scrapped.◈ To remedy this situation, some rolls may be hard surfaced, reground, and returned to service at a fraction of the cost of new rolls. In this reclamation process, the rolls are carefully cleaned and the roll body

---

× J. Dugan, "Application of Progressive Induction Heating to Forged Roll Hardening Operations", Iron and Steel Engineer Year Book, 1958, pp. 789-800.

◈ I. W. Evans, "Hardsurfacing of Mill Rolls and Mill Equipment", Iron and Steel Engineer Year Book, 1962, pp. 643-648.

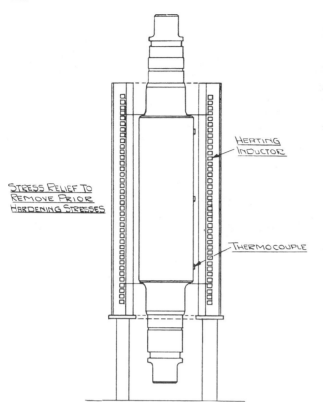

Figure 4-24: Forged Roll Refinishing Complex.

Figure 4-23: Induction Tempering Furnace.

Figure 4-25: Preheating a Roll at the Welding Fixture.

chemistry checked. The rolls are preheated at the welding fixture to a temperature of about 600°F, as shown in Figure 4-25, and then an overlay of a suitable composition is welded onto the roll surface. In the case of a roll, of say 0.77% carbon and 1.10% manganese, a relatively low carbon "buttering" layer should be first used to dilute the carbon in the roll and make it more compatible with respect to the wear- and impact-resistant overlay. The desirable overlay weldment might contain 0.55% carbon, 0.90% manganese, 1.10% silicon, 5.30% molybdenum, 5.30% nickel with the balance being iron.

MILL ROLLS AND THEIR BEARINGS

The welding procedure, particularly with respect to the width of the bead and the overlap of each bead is quite critical since most hard surfacing materials are inclined to be hot-short. Though considerable cost savings are claimed by welding roll surfaces, many welding problems still remain to be solved.○

4-11 Roll Grinding

To provide the surface finish and the precise dimensions required with respect to mill rolls, roll grinders, such as that shown in Figure 4-26, are used. In such a machine, an abrasive wheel is rotated at high speed against the surface of the roll turning in the grinder, much like a cutting tool is impressed upon a work piece held in a lathe.

Figure 4-26: Roll Grinding Machine.

However, grinding is much less predictable than machining operations such as turning and milling. Whereas it is possible to preprogram the operation of a lathe to provide the desired dimensions and surface finishes of the work piece, the wear of the grinding wheel renders the attainment of a specified dimension and finish much more difficult.

Grinding wheels consist of particles of abrasive materials, such as aluminum oxide, bonded together with resins of other binders. The finish of a ground roll depends primarily on the coarseness of these particles but also on such parameters as grinding speed, the roll composition, the nature of the coolant used, and the accuracy with which the wheel is balanced. Wheel grit sizes and typical surface roughnesses they produce are presented in Table 4-5.

○G. D. Reis, "Problems in Steel Mill Roll Salvage", Iron and Steel Engineer Year Book, 1969, pp. 608-617.

Table 4-5
Approximate Relationship of Finish Roughness to Wheel Grit Size

Grit Size (Wires per inch in screening mesh)	Surface Roughness (Microinches R.M.S.)
25	200-350
40	100-250
80	50-150
120	< 50

In recent years, fully-automatic, numerically-controlled roll grinders have been developed and the schematic diagram of such a machine installation is shown in Figure 4-27.◆ẟ♦ The operator's control console is at one end of the machine and an electronic calipering device is positioned over the roll mounted in the machine. The machine features a 36-inch diameter by 4-inch wide grinding wheel with a precision feed control.

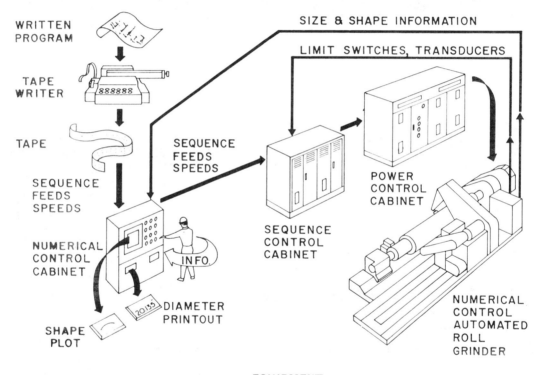

Figure 4-27: Numerically-Controlled Roll Grinder.

In operation, the stock removal rate is automatically controlled by accurately maintaining the wheel motor current close to a preset value. Under these circumstances, it is possible to produce low-microinch finishes without feed lines or chatter. Automatic cycles required to carry out taped grinding programs are initiated at the operator's console, the dials for presetting

---

◆ W. G. Nichol, "Automatic Roll Grinders for Flat Products Mills", Iron and Steel Engineer Year Book, 1967, pp. 530-539.

ẟ "Programmed Roll Grinding", Iron and Steel, November, 1968, p. 502.

♦ T. E. Shipley, Jr. and Q. B. Danzig, "Precise Rolls, Original and Redressed, Through Final Stand", Iron and Steel Engineer Year Book, 1968, pp. 272-277.

MILL ROLLS AND THEIR BEARINGS

the crown and the roll-diameter reduction being located at the console. The peripheral speed of the grinding wheel is maintained at a relatively constant value in spite of the wear of the wheel. This is accomplished by coupling the wheel speed regulation with the wheel dressing diamond which, in turn, is positioned by the wheel head probe. In Figure 4-28, the grinding wheel has been removed to show the wheel dressing diamond just to the left of the wheel spindle while to the right of the spindle is the wheel head probe.

Figure 4-28:
The Grinding Unit With the Wheel Removed (The wheel-dressing diamond is located to the left of the wheel spindle while to the right is the wheel-head probe which contacts the roll and actuates the wheel infeed control.)

With the exception of the manual loading and unloading of the rolls on the grinder and such actions as setting the dials for crown and roll diameter decrease, all the operator functions are broken down and each is assigned a tape function. Since each function is individually coded, the sequence of operations can be changed and any steps can be omitted at will. The tape, therefore, controls the entire series of grinding operations including periodic functions such as wheel dressing and the number of passes at specified head stock, wheel and carriage speeds at fixed wheel loads for each pass. In addition, the functions of roll alignment check and diameter calipering and printout are also directed by the tape program.

Assuming that the rolls are mounted correctly on the grinder, the supporting gibs are in good condition and the roll necks have correct tolerances, the body of a roll can be ground to a roundness of 0.0001 inch. On a straight roll, the diameter across the face will not vary over 0.0002 inches from a straight line and the taper from end to end of a roll will not exceed 0.0005 inches. Moreover, for crowned rolls, the deviation from a target curve will not exceed 0.0005 inches. Such tolerances are normally well beyond what could be accomplished by manual grinding methods, even if executed with considerable care.

To achieve economies in roll grinding operations, cold mill rolls may be ground while still in their chocks.[a] In this practice, the roll is placed in the grinder, lined up so that its axis is parallel to the grinder axis and driven through universal couplings (see Figure 4-29). When antifriction roller bearings are used in the chocks, these bearings are lubricated during the roll grinding operation. Under these conditions, with the roll rotating at a fixed speed and a grinding wheel traversing the roll at a constant rate, turning at a constant rate and exerting a constant pressure, a concentric round roll is produced whether it is turning in its bearings or on journal rests.

---

[a] R. V. Wardle and T. H. Kimes, "Grinding Rolls in Their Chocks", Iron and Steel Engineer Year Book, 1966, pp. 225-229.

Figure 4-29:
Entire Top Backup Roll Assembly in Position on Roll Grinder.

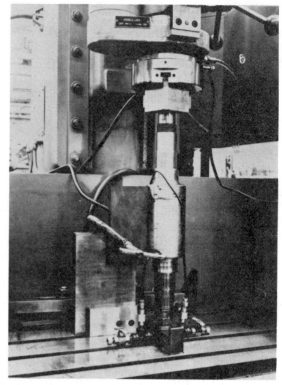

Figure 4-30:
Experimental Equipment for EDM Finishing a Work Roll.

Although this roll grinding practice involves additional capital investment in chocks and bearings, this investment may be compensated by the longer life of split rings, collars and seals, lesser grease wastage through bad seals, longer bearing life and reduced manpower, space and crane service.

4-12 Producing Rough Roll Surfaces

Rolls with relatively rough surfaces are used in certain mills, particularly temper mills, where a thin layer on each surface of the product is to be work hardened or where the final product is specifically intended to exhibit a rough surface. In the former case, a two stand temper mill may be equipped with rough work rolls in the first stand and highly finished rolls in the second, the rough rolls in the first stand "kneading" the surface layer and subjecting it to more severe deformation than the body of the strip. In the latter case, the surface condition is required for better paint adherence or for some other purpose and the finish roughness is frequently specified in terms of microinches RMS.

Although roll grinding techniques can be used to produce rolls with relatively rough finishes (up to about 350 microinches), rough grinding leads to a roll surface distinctly different in the circumferential direction than in the axial direction. A more desirable surface may be achieved by other methods such as grit blasting and electrical discharge machining. The former is the method generally used at present whereas the latter is still in the experimental stages.

In grit blasting operations, steel grit of a specified size is projected at high speed against the surface of the body of the roll, the roll necks being protected by rubber jackets or by other means. In some machines, the grit is propelled by air pressure whereas in others it is thrown from a rapidly rotating wheel. In the air blasters, the roll finish will depend on (a) the number of passes, (b) the grit size, (c) the air pressure, (d) the nozzle speed and (e) the grit flow rate and, in the rotating wheel type, on such parameters as (a) the number of passes, (b) the grit size, (c) the wheel speed, (d) the car speed and (e) the rotational speed of the roll. Grit sizes usually range from G25 (used to give finishes in the range 300-350 microinches RMS) to G80 (used to give finishes in the range of 60-90 microinches).

MILL ROLLS AND THEIR BEARINGS

In operation, the grit deteriorates so that, if a blaster is to be consistent in its performance, the grit should be continually screened and the finer particles removed. Moreover, the effectiveness of the grit depends on the hardness of the roll being processed,[◙] and the quality of the finished surface is sensibly dependent on the operator's skill. For these reasons, electrical discharge machining is currently being studied as an alternate method of roll finishing.

In electrical discharge machining (EDM), an electrode is placed in close proximity to the roll surface and the intervening space filled with a fluid dielectric (as illustrated in an experimental setup shown in Figure 4-30). Electrical current impulses at a frequency in the range of 18 to 250 kilohertz are passed between the electrode and the roll surface. These discharges erode minute particles from the roll surface and, the lower the frequency, the rougher the finish achieved. Moreover, the hardness of the roll surface has no significant effect on the efficiency of the process and the type of finish is readily controllable.

4-13 Measuring and Matching Roll Diameters

Although operators frequently use a steel tape to measure roll diameter, saddle micrometers are far more accurate. Moreover, they can detect any ovularity in the roll section. However, on modern roll grinders, as discussed in Section 4-11, roll diameters may be measured electronically with a very high degree of accuracy.

Many operators insist that work roll diameters match to within a few mils. The necessity of such close matching is only justified where a very small reduction is given to the strip, where the forward slip of the strip is very close to zero or where the rolls are to be run together on face. Such conditions occur in temper mills[✕] and sometimes in secondary cold mills where a mill stand is used for flattening the strip or providing a desired finish to the strip after reduction. Under such circumstances, appreciable spindle torque unbalance may occur with mismatched rolls and pinion gears may be damaged.

In primary cold mills, however, where reductions are usually in the range of 20 to 50 per cent, slips may range from 5 to 15 per cent. Under such circumstances, so long as the rolls are not run on face in the absence of strip, the roll diameters need not be matched with any great degree of accuracy, since a mismatch results in a lesser forward slip with respect to one roll and a greater forward slip with respect to the other.

The mismatch of roll diameters under these circumstances, causes only a slight unbalance in the torques of the two drive spindles. For mills where the spindles are not mechanically synchronized but where the separate drive motors are speed matched a similar spindle torque unbalance would occur. For mills where the separate motors are load balanced and where identical frictional conditions exist in the roll bite for each side of the strip, then the spindles would assume a rotational speed ratio inversely proportional to the ratio of the diameters of the rolls which they drive.

Where the surfaces of work rolls press on each other beyond the edges of the strip being rolled, and where the spindles are mechanically coupled by pinion gears, a torque unbalance again ensues. This unbalance, due to one spindle trying to drive the other, is dependent only on the total force exerted by the one roll on the other over the areas of contact, the diameter of friction of the surfaces in contact and the mean roll diameter. It should be noted that this component of torque unbalance, although significant, is sensibly independent of the degree of mismatching.

The frictional energy dissipated at the roll surfaces in contact does depend on the degree of mismatch in roll diameters, but because the relative slipping speed of the two surfaces is quite low, this energy is not usually significant.

◙ D. V. Barney, Jr. and G. C. Robb, "An Analysis of the Characteristics and Preparation of Tin Temper Mill Blast Rolls", Iron and Steel Engineer Year Book, 1969, pp. 520-525.

✕ In some mills, such as older temper mills, a single spindle is used for drive purposes. Under these circumstances, the necessity for matching mill diameters is obviated.

Although it might be thought that mismatching of the rolls would lead to coil set in the rolled product, strip rolled on a laboratory mill with work rolls mismatched by 3% in diameter produced perfectly flat strip.

4-14 Roll Finish and Its Measurement

The types of finish or surface smoothness exhibited by cold mill rolls varies considerably. For tin plate and stainless steel strip production, rolls should usually have a very smooth, mirror-like finish with a surface roughness of a few microinches. However, for other types of steel product, grit blasted rolls are often used to impart a correspondingly rough finish to the sheet. However, where very smooth rolls are used they will tend to get rougher in use; conversely where very rough rolls are used, they tend to become smoother on account of the slippage that occurs in the roll bite.

The relation between roll and strip roughnesses has been investigated by Barney and Robb[⊚] in the production of relatively rough finishes on the rolled strip. They found that the roll roughness wears off rapidly as rolling begins. For the first coil (approximately 15,000 ft. of strip), the relationship between strip finish and the initial roll roughness was found to be as represented by Figure 4-31. Subsequent to the rolling of the first coil, the drop-off in strip roughness with the length of strip rolled was determined to be as shown in Figure 4-32.

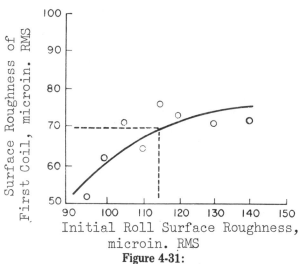

Figure 4-31:
— Relationship between strip roughness of the first coil produced and the initial roll roughness.

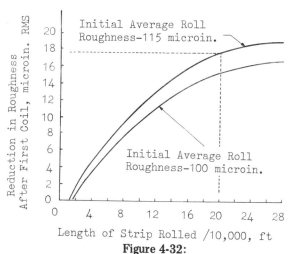

Figure 4-32:
— Approximate drop-off in strip roughness that results as footage in excess of the first coil is produced.

Usually, however, the strip or sheet product should have the same finish on each surface. Accordingly, mill operators generally exercise considerable care in matching the finish of a set of work rolls (usually to within 2 to 3 microinches).

In measuring surface finish or roughness, a fine-pointed stylus of a suitable instrument is drawn slowly across the surface so that it executes a time-displacement curve corresponding to the microscopic profile of the surface, as illustrated diagrammatically in Figure 4-33. In turn, this motion of the stylus is made to generate a corresponding electrical signal which can be analyzed several ways to classify surface roughness. The most common method of analysis involves the measurement of the average roughness height which is described as the arithmetical average deviation from the mean line or level of the relatively finely-spaced surface irregularities. On the

---

[⊚] D. V. Barney, Jr. and G. C. Robb, "An Analysis of the Characteristics and Preparation of Tin Temper Mill Blast Rolls", Iron and Steel Engineer Year Book, 1969, pp. 520-525.

MILL ROLLS AND THEIR BEARINGS

other hand, peak frequency measurements are used to count the number of peaks occurring in a given distance along the surface.

Two commercial instruments find widespread use in surface finish measurement. One is the Surfindicator, illustrated in Figure 4-34, used with its companion the Surfacount Unit shown in Figure 4-35, both being manufactured by Brush Instruments. The other instrument is the Profilometer, manufactured by Micrometrical and shown in Figure 4-36.

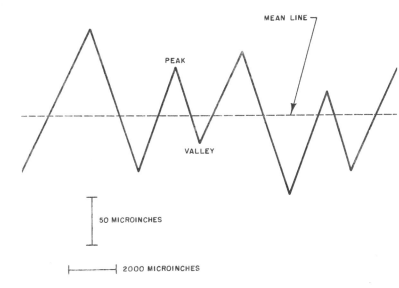

Figure 4-33: Sketch of Surface Profile of 35 Microinches Average Roughness and 300 Peaks Per Inch.

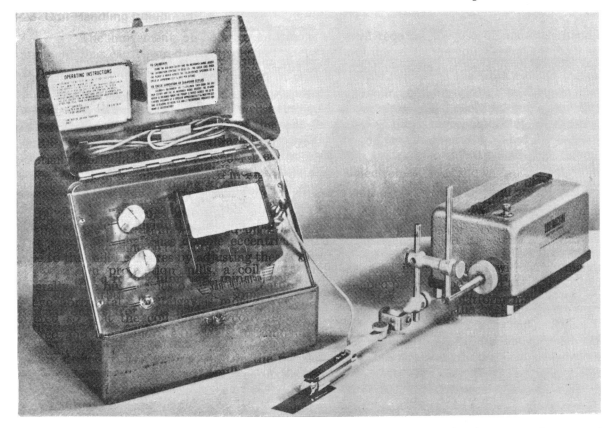

Figure 4-34: Surfindicator.

The Surfindicator incorporates a tracing head and a motorized unit to move it across the surface under study at a constant speed of 1/8 inch per second. The tracing head consists of a diamond stylus attached to the plate of an electron tube through a viscous coupling and a metal bellows. Movement of the stylus on the specimen changes the tube characteristic, which produces a voltage change across the plate-load resistor. This voltage is fed into an amplifier that has controls

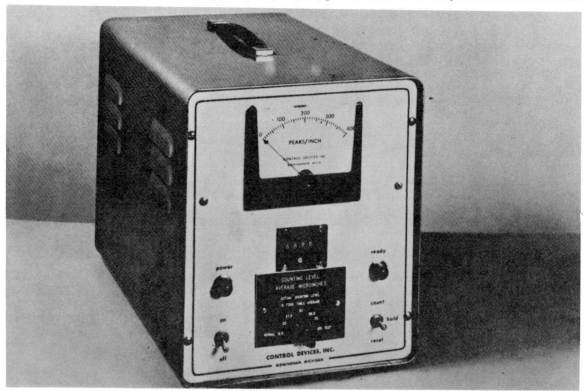

Figure 4-35: Surfacount Unit.

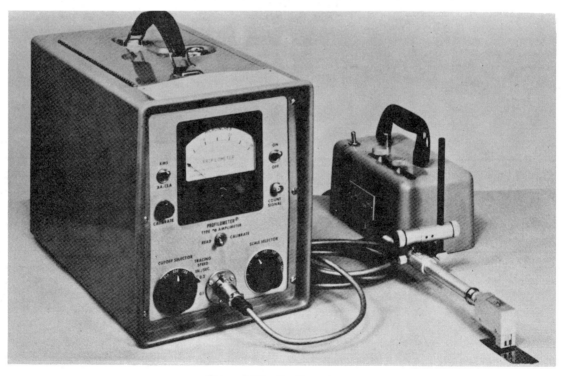

Figure 4-36: Profilometer.

MILL ROLLS AND THEIR BEARINGS

for range (10 to 1000 microinches full scale) and roughness-width cutoff △ (0.030 0.010, and 0.003 inch). The output of the amplifier is rectified and applied to a meter that reads average roughness in microinches. The unit is supplied with precision reference specimens of 125 and 20 microinches average surface roughness so that the unit may be calibrated and the condition of the stylus checked.

The Surfacount Unit shown in Figure 4-35 receives the signal from the Surfindicator tracing head and examines every surface deviation. The Surfacount measures the height of the deviations in the profile to the next reversal of the slope. If the deviation is larger than 50 microinches, peak to valley, the unit switches and looks at the deviation that follows immediately. If any deviation is smaller than 50 microinches top to bottom, the unit does not switch but waits for the next deviation having the same sign on the slope, and, remembering the point at which the previous deviation started, looks for a total height change of 50 microinches or more. The unit registers one count after it has received two deviations the slopes of which are opposite in sign and each greater than 50 microinches in magnitude. This mode of operation is illustrated in Figure 4-37.

The peaks are totaled on an electromechanical counter rated at a maximum of 50 counts per second. An interlock is provided between the motorized tracer and the Surfacount to start counting at the beginning and stop counting at the end of a trace one inch long. The counter therefore reads directly in peaks per inch. A panel meter is provided to give a visual indication of the instantaneous frequency of peaks.

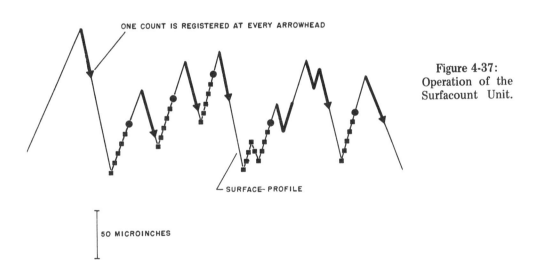

Figure 4-37: Operation of the Surfacount Unit.

The Surfacount is calibrated by switching in a 10-to-1 attenuator, tracing the 125-microinch average (500 microinches from peak to valley) precision reference specimen and adjusting the level control so that the unit counts only some of the peaks. When the attenuator is switched out, the unit then requires deviations of 50 microinches from peak to valley (12.5 microinch average) to trigger it.

The Micrometrical Profilometer consists of two units, a Mototrace Unit and an Amplimeter, as shown in Figure 4-36. A magnetic pickup head is moved at 0.3 inches per second by the Mototrace Unit and it produces a voltage proportional to the vertical velocity of the moving

---

△ The term "roughness-width cut-off" is the maximum width of an irregularity to be counted in the measurement of roughness and is established in the measuring instrument by the manner in which the mean roughness is computed.

stylus. This voltage is fed to the Amplimeter, which has a roughness-width cutoff circuit (0.010, 0.030, and 0.100 inch), a circuit to compensate for the velocity output of the pickup, a range switch (3 to 3000 microinches full scale) and an amplifier, rectifier and meter. A switch on the Amplimeter boosts the meter reading by 11 per cent to provide a root-mean square (rms) average reading. The Amplimeter and tracers are factory calibrated, but a ground glass plate is provided to check the condition of the stylus and the amplifier. The output of the Amplimeter may be fed to a counting unit which will total those signals from the Profilometer which make excursions beyond preset maximum and minimum levels measured with respect to the mean line.

The counter is calibrated by throwing a switch on the Amplimeter that applies an adjustable 60-cycle sine wave to both the roughness meter and the counter. The level of the sine wave is adjusted to give the desired meter reading and then the trigger circuit in the counter is adjusted so that it barely counts. Since the zero-to-peak value of the sine wave is 1.57 times the average value of the rectified wave form read on the meter, the signal from the surface must rise above the mean line and fall below the mean line 1.57 times the calibration level originally set on the meter. A representation of the operation of the counter is shown in Figure 4-38.

The counter is gated by a microswitch located in the Mototrace in such a way that the peaks are counted on the outstroke, the number is displayed on the instroke, and the counter is reset to zero at the start of the next outstroke.

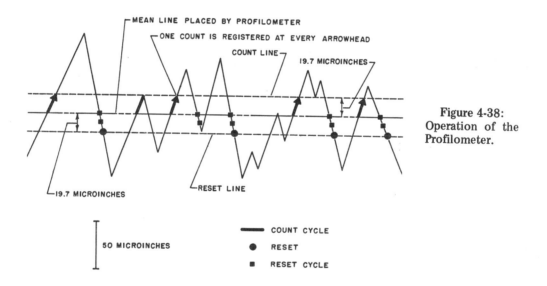

Figure 4-38: Operation of the Profilometer.

4-15 Measuring Roll Surface Hardness

In measuring the hardness of a mill roll, it is generally desirable to avoid the use of any method which visably marks or mars the roll surface. For this reason, the most commonly used instrument for this purpose is the sclerescope manufactured since 1907 by the Shore Instrument and Manufacturing Company, Inc. However, the Rockwell Hardness Tester, which does produce a small indentation of the surface under test, also finds application in roll hardness measurements.

The sclerescope, one model of which is illustrated in Figure 4-39, drops a diamond-tipped hammer from a fixed height and makes only a minute indentation of the metal under test. The hammer rebounds, but not to its original height because some of its energy is dissipated in producing the indentation. The height of the rebound varies in proportion to the hardness of the metal, the higher the rebound the harder the metal. The scale consists of units

MILL ROLLS AND THEIR BEARINGS

which are determined by dividing the average rebound of the hammer from quenched tool steel of ultimate hardness into 100 equal parts. (The scale is carried higher than 100 to cover super-hard metals.) In one version (Model C-2), the height of the hammer rebound is read directly by viewing the hammer as it moves inside a glass tube. In another version (Model D), the rebound height is read directly on a meter, with a scale that is also calibrated in terms of Brinell hardness.

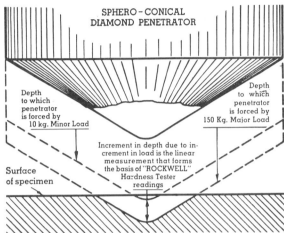

Figure 4-40: Principle of the Rockwell Hardness Test.

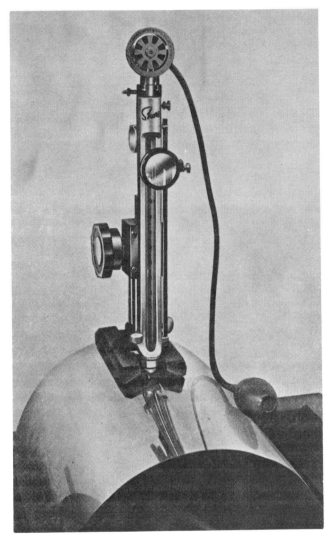

Figure 4-39:
Model C-2 Scleroscope Barrel with Roll Testing Stand.

Table 4-6
Rockwell Hardness Scales

Scale or Prefix	Type of Penetrator	Major Loads in Kgs.
STANDARD SCALES		
B	1/16" Ball	100
C	BRALE Diamond	150
SPECIAL SCALES		
A	BRALE	60
D	BRALE	100
E	1/8" Ball	100
F	1/16" Ball	60
G	1/16" Ball	150
H	1/8" Ball	60
K	1/8" Ball	150

In the Rockwell hardness test, developed in 1921, the hardness number obtained represents the additional depth to which a test ball or sphero-conical diamond test penetrator is driven by a heavy load beyond the depth of a previously applied light load, as illustrated in Figure 4-40. The minor load (10 kgs.) is applied to the penetrator first, the major load (60, 100, or 150 kgs.) applied, the major load is then removed without moving the surface under test, and the hardness number is then automatically indicated on a gage. The entire test requires only five or ten seconds and was designed to measure the hardness of all types of metals, both hard and soft, in any shape or form. The incremental depth of penetration for each point of hardness of the Rockwell scale is 0.00008 inches.

All Rockwell hardness tests, however made, should be designated by the letter A, B, C, etc., or a statement as to the type of penetration used and the magnitude of the load supplied. These designations are given in Table 4-6.

The scale used principally for testing heat-treated and hardened steel is the C scale utilizing a major load of 150 kgs. and a spheroconical diamond penetrator. Hardness is read on the black-figured scale of the dial. Brass, bronze, unhardened steel, etc., in rolled, drawn, extruded, as well as cast metal, are usually tested with the 100-kg. load and a 1/16" diameter steel ball. The readings are then taken from the red-figured scale on the dial, and the letter B is prefixed.

A Rockwell hardness tester adapted for the measurement of roll surface hardness is shown in Figure 4-41.

The "Rockwell" Superficial Hardness Tester was introduced in the early 1930's for testing extremely thin sheet, nitrided steel and lightly carburized steel. It employs lighter loads and a more sensitive depth-measuring system where tests have to be limited to very shallow indentations.

The "Rockwell" Twin Tester, for Rockwell hardness testing and Rockwell superficial hardness testing, introduced in the late 1950's, is a combination instrument designed primarily for use in laboratories, tool departments, maintenance repair shops, inspection departments, small-lot job shops, etc.

Figure 4-41:
Rockwell Tester Adapted for the Measurement of Roll Surface Hardness.

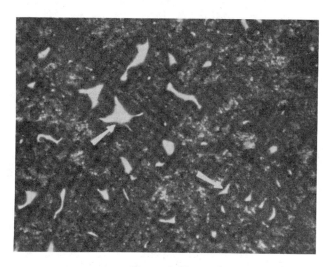

Figure 4-42:
Microscopic Discontinuities on Roll Surface Due to Inclusions.

4-16 Replication Techniques for Studying Roll Surfaces

In recent years, a technique has been developed for studying roll surfaces wherein a thin replica is made of the roll surface and examined under the electron microscope at magnifications up to 200,000X.⊞ ✶ ◄ ♦ Designated electron microfractography, it provides a method of permanently recording roll appearance and has been utilized in investigations pertaining to fracture foci or origins, to material defects (inclusions, stringers, spalls and pits), to surface patterns such as shading, to surface finish variations and to sheet patterns.

⊞ A. Phillips and G. V. Bennett, "The Electron Microscope...A New Tool for Examining Fractures", Metal Progress, May, 1961, p. 97.

✶ R. Pelloux, "An Analysis of Fracture Surfaces by Electron Microscopy", ASM Technical Report. P-19-3-64, Philadelphia, Pa., 1964.

◄ D. Krashers, "Plastic Replicas Help Rate Surface Quality", Metal Progress, September, 1965, p. 98.

♦ L. E. Arnold, "Replicas Enable New Look at Roll Surfaces", Iron and Steel Engineer Year Book, 1966, pp. 625-629.

MILL ROLLS AND THEIR BEARINGS

To make a replica, a strip of cellulose acetate is cut and softened with acetone on one surface. This side of the strip is then pressed against the metallic surface to be studied and, after the strip has rehardened, it is peeled off the surface. Chromium is then vapor deposited on the replica surface at a suitable impingement angle which will emphasize the "highs" and "lows" of the replica. The thin chromium coating may then be studied under a microscope.

This replication technique has been used to study and record details of fracture foci (located by tracing the flow patterns of crack propagation), defects on roll surfaces (attributed to inclusions exemplified by those shown in Figure 4-42), spalls (usually related to high nonuniform contact stresses), pitted surfaces (caused either by material defects, such as inclusions, or by a faulty grinding practice) and surface patterns (due to chatter or other causes occurring during grinding).

The technique has also been used in studying the surfaces of sheet product in attempts to correlate surface defects like "orange peel" with certain variables associated with the rolling practice.

4-17 Stresses in Mill Rolls

Stresses in mill rolls may be categorized on the basis of their origins as (a) residual, (b) thermal and (c) operational. Residual stresses occur as the result of roll manufacturing techniques, particularly those associated with the hardening of roll body surfaces. Thermal stresses are attributable to the existence of temperature gradients within the roll and operational stresses result from the application of external forces to the rolls. Under any given set of circumstances, the existing stresses within the mill rolls result from a combination of these three types of stresses.

Such stresses are important, not only because they affect roll life, but because they affect the operation of the mill and the quality of the rolled product. Damage to rolls occurring prior to or during their use, such as season cracking, may be directly traceable to excessive stresses. Under certain circumstances, stress reversals due to the rotation of the rolls cause fatigue damage and, at the same time, high stresses may result in excessive roll wear and changes in the surface finish of rolls.

The effects of stresses on mill operation and the quality of the rolled strip result from strains or distortions induced in the rolls by the stresses. For example, the radial stresses induced in the work rolls at the region of contact with the sheet or strip flatten the rolls so that, in effect, they exhibit a larger diameter (the "deformed diameter"). For a given reduction incurred by the work piece, therefore, the area of each work roll surface in contact with the work piece is larger than if the rolls were perfectly rigid. This, in turn, implies that the rolling force is increased by the roll flattening.

As well as the radial distortion or flattening of the work rolls, the stresses existing in both the work and backup rolls induce them to bend in a plane containing their axes. Unless this bending is compensated for by the use of roll crowns, roll bending (either work or backup) or by some other method, the rolled strip will exhibit over-rolled or full edges.

In the succeeding sections, residual and thermal stresses are discussed in greater detail. The stresses resulting from roll-to-roll and roll-to-workpiece contacts are examined mainly from the roll bending viewpoint (and its correction by crowning) since the effects of roll flattening on mill operation are treated more fully in Chapter 9.

4-18 Residual Stresses in Rolls

In the rapid quenching of steel such as frequently occurs in roll manufacture, residual stresses may be created both by slight plastic deformation induced in the metal as well as by localized changes occurring in its microstructure. Plastic deformation results from the existence of large temperature gradients and the thermal expansion of the steel so that, at some point in the quenching (or heating) operation, the induced stresses exceed the yield point of the steel. However, if the yield point is not exceeded and only elastic straining occurs, no residual stresses attributable to this cause remain after cooling to a uniform temperature. Stresses arising from microstructural changes are related to both the rate of cooling of the workpiece and the dilation of the steel in its transformation from the face-centered cubic structure of austenite to the body centered tetragonal structure of martensite.[*] (See Figure 4-43).

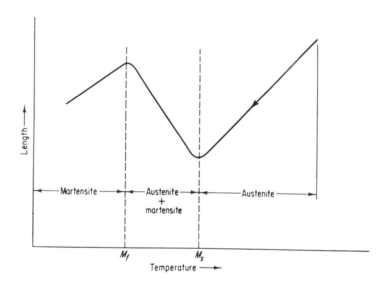

Figure 4-43: Schematic Dilation Curve for Martensite Transformation.

To understand how residual stresses are developed in a roll, it is convenient to consider a cylindrical steel bar quenched from an austenitizing temperature of about 1600°F disregarding, at first, any microstructural changes that occur. Time-temperature curves pertaining to the surface and center of the bar and a point midway between the two are shown in Figure 4-44. At the end of 10 seconds, the surface has cooled to about 700°F while the center is at 1500°F. Because of the thermal expansion characteristics of the steel, after 10 seconds of cooling, the surface should have contracted much more than the inside. However, because the surface and inside of the bar are attached to each other, the interior of the bar will prevent the surface from contracting as much as it should. As a consequence, during cooling, the surface will be in tension in the longitudinal and tangential directions and the interior of the bar in compression as indicated in Figure 4-45. If the stress exceeds the ultimate strength of the steel, cracking will occur. However, in the case of steel, thermal stresses alone very rarely lead to cracking but, if the stress exceeds the yield strength, the surface layer will be plastically deformed or permanently elongated. If yielding occurs, then when the bar has reached room temperature, the surface will have a residual compressive stress and the interior, a residual tensile stress. Thus, if the specimen was originally cylindrical, it will become slightly barrelshaped.

Considering now the effects of the martensitic transformation (with a volume change of about 4.6 per cent), it is desirable first to examine the case of a cylindrical workpiece transformed across its entire cross section and then the case where the hardening extends only to a shallow depth.

For the through-hardened case, the TTT curves and the surface and center cooling curves are illustrated in Figure 4-46. During the first stage, up to time t_1, the stresses present are due to the radial temperature gradient. The surface, prevented from contracting as much as it should by the center, will be in tension while the center will be in compression. During the second stage, between times t_1 and t_2, the surface, having reached the martensite-start (M_s) temperature, transforms to martensite and expands. The center, however, is undergoing normal contraction due

---

[*]"Introduction to Physical Metallurgy", by Sidney H. Avner, McGraw Hill Book Company, New York, New York, 1964.

MILL ROLLS AND THEIR BEARINGS

to cooling and it prevents the surface from expanding as much as it should. The surface, therefore, tends to be in compression while the center will tend to be in tension. After t_2, the surface has reached room temperature and will be a hard, brittle, martensitic structure. During the third stage, the center finally reaches the M_S temperature and begins to form martensite and expand. The center, as it expands, will try to pull the surface along with it, putting the surface in tension. The stress condition of the three stages is summarized below:

Stage	Stress Condition (in longitudinal and tangential directions)	
	Surface	Center
1st (Cooling of austenite)	Tension	Compression
2nd (Transformation to martensite at surface)	Compression	Tension
3rd (Transformation to martensite at center)	Tension	Compression

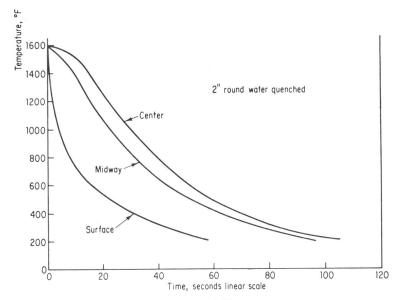

Figure 4-44:
Cooling Curves at the Surface, Midway on the Radius, and the Center of a 2-inch Diameter Bar when Water-Quenched.

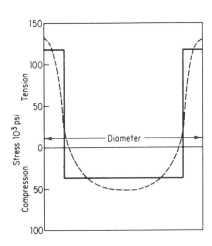

Figure 4-45:
Schematic Representation of the Stress Distribution Across the Diameter of a Bar During Quenching.

During the first stage, if the tensile stress in the surface is high enough, the steel will deform plastically thereby relieving the stress. In the second stage, the compressive stress will also tend to produce deformation rather than cracking, but in the last stage, the tensile stresses exerted on the hard, unyielding martensite pose the greatest danger of cracking. For this reason, tempering should immediately follow hardening, as this practice will impart some ductility to the surface before the center transforms.

The surface and center cooling curves for a shallow-hardened steel cylinder are shown in Figure 4-47. During the first stage, up to time t_1, the stresses present are due only to the radial temperature gradient and, as in the through-hardened condition, the surface will be in tension while the center will be in compression. During the second stage, between times t_1 and t_2, both the surface and center will transform. The surface will transform to martensite while the center will transform to a softer product, like pearlite. The entire piece is expanding but, since the expansion resulting from the formation of martensite is greater than that resulting from the formation of pearlite, the surface tends to expand more than the center. As a consequence, the center becomes subject to tensile stresses whereas the surface becomes compressed. After t_2, the center will contract on cooling from the transformation temperature to room temperature. The surface, being martensitic and having reached ambient temperature much earlier, will prevent the center from contracting as much as it should. This will result in higher tensile stresses in the center. The stress condition during the three time periods may, therefore, be summarized as follows.

Stage	Stress Condition (in longitudinal and tangential directions)	
	Surface	Center
1st (Cooling of austenite)	Tension	Compression
2nd (Transformation to martensite at surface and to pearlite at the center)	Compression	Tension
3rd (Cooling of center to ambient temperature)	Greater Compression	Greater Tension

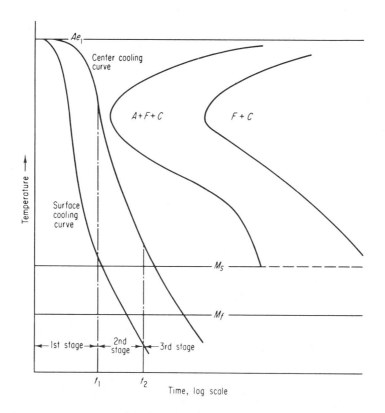

Figure 4-46: Center and Surface Cooling Curves Superimposed on the TTT Curves for the Steel (Through-Hardened Condition).

As in the through-hardened case, in the first stage, the surface will yield rather than crack. In the second stage, compressive stresses in the hardened casing will prevent cracking. However, during the third stage, the interior is under tension. Accordingly, it may exhibit cracks but these would be more difficult to detect because they are subsurface.

It must be remembered that the residual stresses discussed to this point have pertained only to the longitudinal (parallel to the cylinder axis) and tangential directions. Stresses in the radial direction are important, however, as high tensile stresses in this direction may cause hardened layers to separate from the roll body. Generally speaking, if the longitudinal and tangential stresses are compressive, the radial stresses will be tensile and vice versa. Thus, for example, a roll sleeve shrunk fit on the roll body will, after cooling, be in tension in the longitudinal and tangential directions, but exert a compressive stress on the roll body in the radial direction.

MILL ROLLS AND THEIR BEARINGS

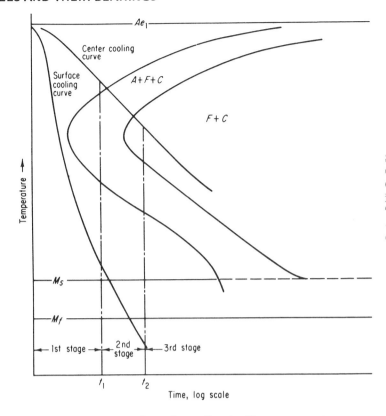

Figure 4-47: Surface- and Center-Cooling Curves Superimposed on the TTT Curves (Shallow-Hardened Condition).

4-19 Residual Stress Computations from Strain Measurements

Measurements of residual stresses in double-poured, cast iron work rolls have been made by DiGioia and Peck.◆ They utilized a relaxation method whereby strain gages were mounted on rolls while they were still intact and then the portions on the rolls on which they were mounted were machined away from the body of the roll. The relaxation strain in each such portion of the roll surface was then measured using an electrical bridge circuit.

The gages were mounted tangentially and longitudinally on each of four rolls (as illustrated in Figure 4-48), having diameters of 12 inches, 18 inches (two rolls, one broken) and 22 inches. Stresses were then computed using the expression

$$S_1 = \frac{E}{(1-\gamma^2)} (\epsilon_1 + \gamma \epsilon_2)$$

where S_1 is the stress in the direction of 1, ϵ_1 is the strain in the direction 1, ϵ_2 is the strain at right angles to the direction 1, E is the modulus of elasticity and γ is Poisson's ratio. The moduli of elasticity ranged from 16×10^6 to 28×10^6 psi and the value of Poisson's ratio was found to be 0.28.

On the basis of the experimental data, both the longitudinal and tangential stresses on the roll surfaces were found to be compressive (as would be expected on a cast iron roll). The longitudinal stresses ranged from a low value of 1,120 psi (for the 12-inch diameter roll) to 29,000 psi (for the unbroken 18-inch diameter roll) while the tangential stresses ranged from 4,590 psi (for the 12-inch diameter roll) to 49,157 psi (for the broken 18-inch diameter roll). The relative longitudinal stress distribution in the body of the roll has been computed by Peck▼ and is shown in Figure 4-49.

◆ A. M. DiGioia, Jr. and C. F. Peck, Jr., "Residual Stresses in Double-Pour, Cast-Iron Work Rolls", Iron and Steel Engineer Year Book, 1956, pp. 182-185.

▼ C. F. Peck, Jr., "A Proposed Method of Calculating Residual Stresses in Iron Rolls", Iron and Steel Engineer Year Book, 1956, pp. 1047-1050.

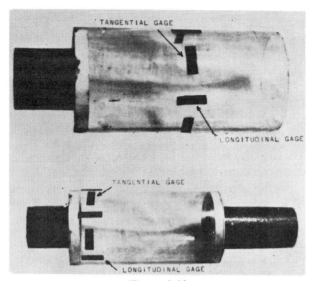

Figure 4-48:
Placement of Strain Gages on Rolls for the Measurement of Residual Stresses.

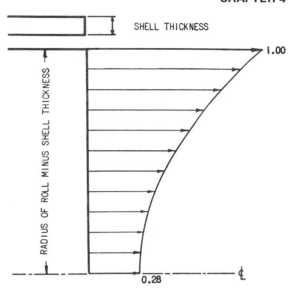

Figure 4-49:
Relative Longitudinal Stress Distribution in the Body of a Work Roll. (Peck)

In computing the radial stresses in the roll, Peck assumed that the chilled region, or shell of the roll, was analogous to a tank and that the internal stress (acting like a fluid pressure in the tank) could be determined from the calculated tangential stress in the shell (S_{TAN}), the internal diameter of the shell (D') and the thickness of the shell (t). Referring to Figure 4-50, the radial stress S_{RAD} is given by the equation

$$S_{RAD} = \frac{2t\, S_{TAN}}{D'}$$

Using this formula, a chart may be drawn up as shown in Figure 4-51, showing the variation of the ratio of the maximum radial tensile stress to the surface compressive stress with both the roll diameter and the depth or thickness of the shell. On this basis, Peck established radial tensile stresses as low as 1,530 psi (for the 12-inch diameter roll) and as high as 9,830 psi (for the broken 18-inch diameter roll).

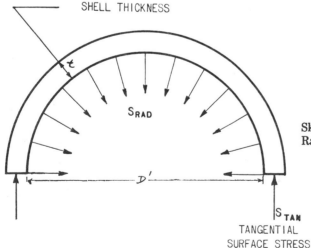

Figure 4-50:
Sketch Showing Method of Computing Radial Stress.

In addition to the foregoing stresses, surface shearing stresses (as illustrated in Figure 4-52) exist, due to the compressed stresses in the shell.

MILL ROLLS AND THEIR BEARINGS

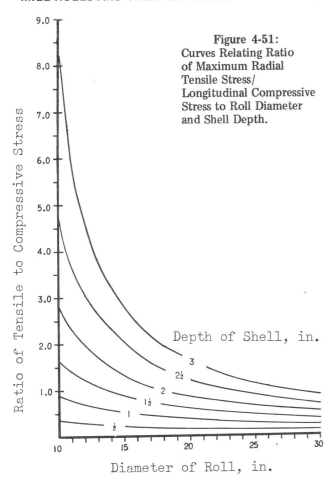

Figure 4-51:
Curves Relating Ratio of Maximum Radial Tensile Stress/Longitudinal Compressive Stress to Roll Diameter and Shell Depth.

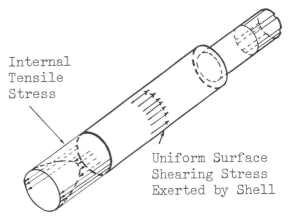

Figure 4-52:
Sketch Illustrating Shearing Stresses in a Roll Body.

4-20 Thermal Stresses in Mill Rolls

Just as stresses due to temperature gradients are induced in mill rolls during their manufacture and heat treatment, stresses are similarly created by temperature gradients that occur during roll usage in the mill. In particular, two conditions are sometimes encountered that give rise to undesirable thermal stresses. One is the rapid heating of rolls during the initial "warm-up" period of the mill and the other is the rapid surface chilling of rolls once they have become heated. In both cases, the effects of the heating or chilling can be readily understood if the surface layers of the roll are thought of as a shrunk-fitted sleeve.

Where the surface of a roll is rapidly heated, longitudinal and tangential compressive stresses are generated in the roll surface accompanied by radial tensile stresses. Thus, the surface layers of the roll, though exhibiting no tendency to crack, will tend to pull away from the roll body, just as a heated sleeve would tend to become loose on an arbor. Thus, spalling could occur during a too-rapid warming up of a roll.

To minimize thermal stresses arising during warm-up, Stepanek[♀] suggests that rolls should be given one of the following treatments:

(1) Placed in the vicinity of heating furnaces for at least 48 hours prior to use so that they can acquire a more or less uniform temperature approximately equal to the ambient temperatures in the vicinity of the mill.

(2) Placed in a tempering pit or bath so that they acquire a temperature of 30°C.

[♀] R. Stepanek, "Principles of Limit Cooling in Rolling Mill Rolls", Hutnik, 1969, Vol. 19, (4), pp. 132-137, (British Iron and Steel Industry Translation BISI #7673).

(3) Heated by an induction heating equipment to a temperature of approximately 60ºC.

(4) Subjected to a less severe usage immediately after placement in the mill.

An 8-year project of the AISE and RMI concerned with roll spalling led to the belief that, if the surface temperature of a roll put into a mill cold does not exceed 170ºF in the first six minutes of warm up and is not allowed to increase in temperature faster than 8ºF per minute for the next eight minutes, the roll is not likely to break due to temperature stresses.◊ Under any circumstances, it is desirable to flood mill rolls with hot water or the roll coolant for a brief period prior to the initiation of rolling operations.

The sudden cooling of a hot roll creates circumferential tensile stresses in the surface of the roll and radial compressive stresses within it (see Figure 4-53).▼ Under these conditions, the tensile stresses may lead to the development of surface cracks which may propagate well into the roll body. Peck and Mavis▼ cite an instance where, inadvertently, cooling water had not been applied to a roll and, when it was suddenly applied, the roll split. Had the roll been allowed to cool slowly instead of being quenched, the splitting may have been averted. Under any circumstances, however, when rolls (particularly backup rolls) are pulled from a mill due to a cobble, it is advisable to place the rolls within a protective blanket or a shielding enclosure to eliminate the danger of mill personnel being injured by flying roll fragments.

4-21 Stresses at the Work Roll — Backup Roll Contact

It has been pointed out by Stone◁ that, if two mill rolls of different diameters are pressed together with their axes parallel under a uniform specific contact force, then a complex system of stresses is set up both at the region of contact as well as throughout the cross section of the rolls. These are in addition to the usual bending and shear stresses associated with the beam action of the rolls which transmit the total rolling load through their neck bearings to the mill housings. Hertz,✪ who was the first to study these stresses, recognized that, at the region of contact, a local elastic flattening takes place over which a semi-elliptical distribution of contact compressive stresses exists as indicated in Figure 4-54. The width (b) of this contact area and the magnitude of the maximum compressive stress (P_{Max}) existing along a line through the roll centers were given by the equations

$$b = 1.52 \sqrt{P' \times \frac{d_1 d_2}{(d_1 + d_2)} \times \frac{(E_1 + E_2)}{E_1 E_2}}$$

$$\text{and } P_{Max} = \frac{4P'}{\pi b} = 1.27 \frac{P'}{b}$$

$$= 0.83 \sqrt{P' \times \frac{(d_1 + d_2)}{d_1 d_2} \times \frac{E_1 E_2}{(E_1 + E_2)}}$$

where P' is the force per unit length exerted by one cylinder on the other, d_1 and d_2 are the diameters of the two cylinders and E_1 and E_2 are their respective moduli of elasticity. As noted in Section 4-3, the value of P_{Max} should not exceed 300,000 psi, otherwise roll spalling may occur.

◊ S. Stasko, "Evaluation of Mill Factors", Iron and Steel Engineer Year Book, 1966, pp. 152-155.

▼ C. F. Peck, Jr. and F. T. Mavis, "A Study of Failures in Iron Work Rolls", Iron and Steel Engineer Year Book, 1955, pp. 567-578.

◁ M. D. Stone, "Benefits, Improvements and Ideas from Roll Research Program", Iron and Steel Engineer Year Book, 1966, pp. 143-163.

✪ H. Hertz, Crelle's Mathematical Journal (Berlin), Vol. 9, 1881.

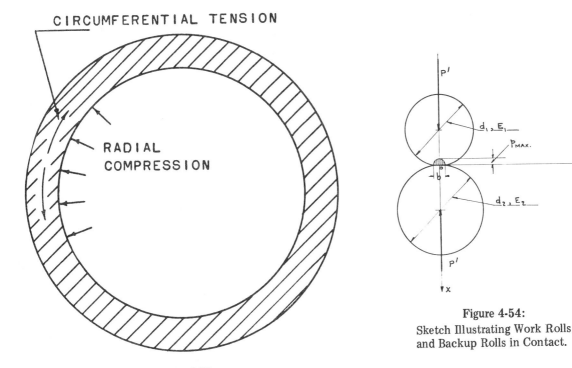

Figure 4-53:
Sketch Illustrating the Development of Stresses in a Roll Due to Sudden Surface Cooling.

Figure 4-54:
Sketch Illustrating Work Rolls and Backup Rolls in Contact.

In addition to the normal stress at the region of contact, other body stresses may occur in the rolls. Figure 4-55 shows the principal types of shear stresses. τ_{45} are the major shear stresses that occur along the line connecting the roll centers and are oriented at 45 degrees to this line. They vary from zero at the point of contact (0) increasing to a maximum value at A and attenuating again with increasing depth below the roll surface. Point A occurs at a distance 0.39b below the flattened roll surface and the value of this maximum shear stress is τ_{45} (max.) = 0.304 P_{max}.[⊙] As the two rolls rotate, the stresses at this point increase from zero to a maximum and decrease to zero again as the point passes under the contact line.

Radzimovsky[■] investigated the τ_{xy} stresses, which exist parallel to the x and y axes at any point within the roll body. He found that the maximum value of these stresses occurs in a plane (actually a cylindrical surface) at a distance equal to 0.25b below the contact point 0 and that the maximum value of this shear stress is 0.256 P_{Max}. Of particular significance is the fact that this stress undergoes its maximum value at some distance away from the x axis (connecting the roll centers), going through zero at the x axis and reaching its same maximum value again (but of opposite sign) on the other side of the x axis. Hence, the term "maximum reversed shear stress" was given to this stress which occurs at a depth below the surface somewhat less than that for the so-called "maximum shear stress".

The transverse distribution of stresses along the region of contact has been investigated by MacNaughton and Bradd.[□] They used flat plastic replicas of rolls made of 1/4-inch thick Catalin cast resin 61-893, with the backup rolls (and, in the case of 2-high configurations, the work rolls) loaded on the journals. Figure 4-56 shows the interference pattern obtained with a uniformly

---

⊙ N. Belyayev, "Bulletin of Engineering Inst. of Ways and Communications", (St. Petersburg), 1917.

■ E. I. Radzimovsky, "Stress Distribution and Strength Conditions of Two Rolling Cylinders Pressed Together", University of Illinois Engineering Experimental Station, Bulletin 408, 1953.

□ L. MacNaughton and A. A. Bradd, "Photoelastic Study of Stress Distribution in Rolls", Iron and Steel Engineer Year Book, 1962, pp. 265-272.

loaded 4-high mill with a narrow strip between the rolls. The stress concentrations at the ends of the contact region between work and backup rolls are evident.

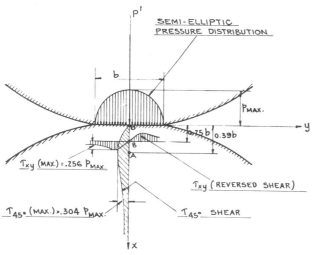

Figure 4-55:
Sketch showing the Magnitude and Distribution of Major Stresses at the Zone of Contact Between Work and Backup Rolls.

Figure 4-56:
Stress Pattern Resulting from the Insertion of a Narrow Strip into a Uniformly-Loaded Four-High Mill.

4-22 Deflection of a Roll Due to Bending and Shear

In the operation of 4-high mills, the backup rolls are designed to provide the rigidity required. Accordingly, in computing roll deflections, they are treated similarly to work rolls in a 2-high mill.

Roll deflections occur as a result of both bending and shear strains, with the former usually significantly larger than the latter. However, in the expressions derived from roll deflections presented in this section, both types of strain have been taken into consideration, and the derivations of these expressions are as given by Larke[a] and Tselikov and Smirnov.[b]

Referring to Figure 4-57, the following symbols will be used to represent various parameters associated with the rolls.

F = total rolling force
w = width of roll face
b = width of strip being rolled
L = separation of the centers of the roll neck bearings
D = diameter of the barrel
d = diameter of the roll neck (assumed to be cylindrical)
n = distance from the center of a roll neck bearing to the closer end of the roll barrel

---

[a] "The Rolling of Strip, Sheet & Plate", by E.C. Larke, The MacMillan Company, New York, 1957.
[b] "Rolling Mills", A. I. Tselikov and V. V. Smirnov, Pergamon Press, London, 1965.

MILL ROLLS AND THEIR BEARINGS

e = distance from the center of the roll neck bearing to the closer edge of the strip
y = deflection of the roll axis normal to the axis and in a plane containing both roll axes
x = distance along the roll axis from the center of a roll neck bearing

Considering only the deflection due to the bending forces, in the region $0 \leq x \leq n$, the deflection may be denoted by y_1 and may be derived from the equation

$$\frac{A d^2 y_1}{dx^2} = -\frac{Fx}{2}$$

where $A = EI_d$, E representing the modulus of elasticity and I_d the moment of inertia of the cross section of the roll neck. Similarly, in the region $n \leq x \leq e$, the deflection y_2 may be derived from the equation

$$\frac{B d^2 y_2}{dx^2} = -\frac{Fx}{2}$$

where $B = EI_D$, I_D representing the moment of inertia of the cross section of the roll body. Finally, for the continuously supported region $e \leq x \leq L/2$, the deflection y_3 is associated with the equation

$$\frac{B d^2 y_3}{dx^2} = -\frac{F}{2}\left(x - \frac{(x-e)^2}{b}\right)$$

The double integration of each equation yields

$$-A y_1 = \frac{Fx^3}{12} + Z_1 x$$

$$-B y_2 = \frac{Fx^3}{12} + Z_2 x + Z_3$$

$$-B y_3 = \frac{Fx^3}{12} - \frac{P(x-e)^4}{24b} + Z_4 x + Z_5$$

where Z_1 to Z_5 represent constants of integration. After these have been established from the boundary conditions

$$y_1 = CFx[\beta + 3n^2(\alpha - 4) - \alpha x^2] \qquad (1)$$

$$y_2 = CF[\beta x - 4x^3 + 2n^3(\alpha - 4)] \qquad (2)$$

and $$y_3 = CF\left[x(\beta - 4x^2) + \frac{2(x-e)^4}{b} + 2n^3(\alpha - 4)\right] \qquad (3)$$

where
$$C = \frac{4}{3\pi ED^4}$$

$$\beta = 3L^2 - b^2$$

$$\alpha = 4(D/d)^4$$

$$e = (L-b)/2$$

If it is assumed that the width of the strip b is equal to the length of the roll face w and that only the deflection of the roll axis relative to the deflection at the end of the roll face is of concern, then n should be substituted for x in equation 1, (L−2n) should be substituted for b in equation 2, and the first equation subtracted from the second. Then

$$\Delta y_B = \frac{Fc(x-n)}{24EI}\left[\frac{(x-n)^3}{w} + L^2 + 2wn - 2x(x+n)\right]$$

where Δy_B = the deflection (−) or the camber (+) at any point along the roll barrel due to the bending stresses and F_c = rolling force uniformly distributed across the face of the roll

At the center of the roll, the maximum deflection $\Delta y_B(\text{max.})$ is derived by substituting $L/2$ for x and is

$$\Delta y_B(\text{max.}) = \frac{F_c w^2}{6 \pi E D^4} (5w + 24n)$$

Considering now the deflection due to shearing stresses and using similar notation, it can be shown that

$$N \frac{dy_1}{dx} = \frac{2F}{\pi d^2} \tag{4}$$

$$N \frac{dy_2}{dx} = \frac{2F}{\pi D^2} \tag{5}$$

and $$N \frac{dy_3}{dx} = \frac{2F(L-2x)}{b \pi D^2} \tag{6}$$

where N is the modulus of rigidity, which is related to the modulus of elasticity as follows

$$N = \frac{E}{2(1+\gamma)}$$

where γ is Poisson's ratio. If γ is assumed to be 0.25

$$N = 2E/5$$

Integration of equations 4, 5 and 6 yields

$$Ny_1 = \frac{2Fx}{\pi d^2} \tag{7}$$

$$Ny_2 = \frac{2Fx}{\pi D^2} + Z_6 \tag{8}$$

$$Ny_3 = \frac{2Fx(L-x)}{b \pi D^2} + Z_7 \tag{9}$$

where Z_6 and Z_7 are constants of integration. Substituting n for x in equations 7 and 8 gives

$$Z_6 = \frac{2Fn}{\pi D^2} \left[\frac{D^2}{d^2} - 1 \right]$$

Similarly, substituting e for x in equations 8 and 9 yields

$$Z_7 = \frac{2F}{b \pi D^2} \left[\left(\frac{D^2}{d^2} - 1 \right) bn - e^2 \right]$$

Using these values for Z_6 and Z_7, in conjunction with equations 7, 8 and 9, it can be shown that

$$y_1 = \frac{2Fx}{N \pi d^2} \tag{10}$$

$$y_2 = \frac{2F}{N \pi D^2} \left[x + n \left(\frac{D^2}{d^2} - 1 \right) \right] \tag{11}$$

and $$y_3 = \frac{2F}{Nb \pi D^2} \left[x(L-x) + \left(\frac{D^2}{d^2} - 1 \right) bn - e^2 \right] \tag{12}$$

MILL ROLLS AND THEIR BEARINGS

Putting n for x in equation 10 and subtracting the resultant expression from equation 12, it can be shown that the shearing deflection

$$\Delta y_s = \frac{F_c}{2\pi ND^2}\left[\frac{4x(L-x)-(L-b)^2-4bn}{b}\right]$$

Assuming that the strip width is equal to the length of the roll face,

$$\Delta b = L - 2n$$

and

$$\Delta y_s = \frac{2F_c}{\pi ND^2}\left[\frac{x(L-x)-n(L-n)}{w}\right]$$

At the center of the roll where $x = L/2$, the maximum deflection $\Delta y_s(\text{max})$ relative to the deflection at the ends of the roll face is therefore given by

$$\Delta y_s(\text{max.}) = \frac{F_c w}{2\pi ND^2}$$

The total roll deflection (Δy TOT) at the center of the roll relative to the ends of the barrel is equal to the sum of the deflections due to bending and shear.

Thus
$$\Delta y_{TOT} = \Delta y_B(\text{max.}) + \Delta y_s(\text{max.})$$

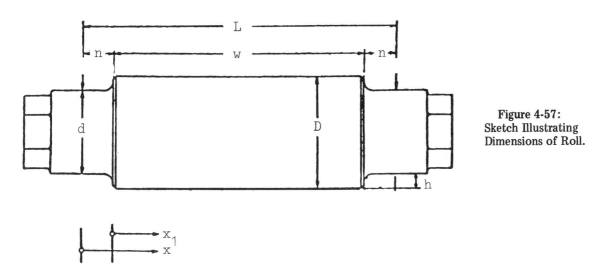

Figure 4-57:
Sketch Illustrating
Dimensions of Roll.

4-23 Compensating for Roll Deflection

To compensate for roll deflections due to bending and shearing stresses, one or more of the following techniques are used:

(a) Rolls are "cambered" or "crowned", that is, given a slightly barrel-shaped profile.

(b) Rolls are distorted by temperature gradients within them which either occur naturally or which have been induced by the selective heating or cooling of the rolls.

(c) Rolls, either work or backup rolls, are bent through the action of external forces applied to their roll necks.

The method of selectively heating and/or cooling of the mill rolls and work and backup roll bending are discussed in Chapter 11. Accordingly, only the grinding of crowns on mill rolls will be treated in this section.

In 4-high mills with backup rolls that are approximately "square", the temperature gradients in the rolls are such that very little crowning of the rolls is required. Thus, for example, a typical secondary cold reduction mill may use crowns[⊢] of 0.001 inch on the work rolls and 0.002 inch on the backup rolls. Because of the difficulty in computing the temperature gradients and the crowning effects they produce, the crowns established by grinding practices are usually selected on the basis of experience. Wusatowski[Δ] states that, for cold rolling, rolls are usually dressed to a convex camber up to 0.10 mm depending on the length of the roll barrel and the type of metal rolled, the convexity increasing with roll length and the hardness of the workpiece.

Barrels of cold rolls are usually given a positive crown of constant curvature, but to avoid the work rolls touching each other beyond the edges of the strip, the curvature of the surface is lesser at the center of the roll than at its ends.

Larke[■] has shown that crowns calculated on the basis of a completely filled roll bite are satisfactory for narrow widths provided lower rolling forces are employed. For example, a roll of 20 inches diameter and 48 inches face under a uniformly distributed load of 574 tons[ᴿ] would suffer a total deflection at its midpoint due to bending and shear of 0.0151 inches. However, a camber designed to offset this deflection will be satisfactory for rolling 24-inch wide strip under a rolling force of 412 tons.[ᴿ]

To facilitate the grinding of rolls, Larke proposed a parabolic curve represented by

$$cy = \frac{F_c}{3\pi E D^4}\left[10w + 48n + \frac{15 D^2}{w}\right](w - x_1)x_1$$

where x_1 is zero at the end of the roll barrel and equal to $w/2$ at the midpoint of the face, the other symbols being as shown in Figure 4-57. In this expression, used for calculating camber on the basis that the strip occupies the whole roll bite, the term $15D^2/w$ is equivalent to the deflection due to shear. The more general expression, when $b \neq w$, is

$$my = \frac{F_m}{3E\pi D^4 w^2}\left[16w^2(w + 3n) - 2b^2(4w - b) + 15D^2(2w - b)\right](w - x_1)x_1$$

where F_m is the actual rolling load.

Prior to the advent of numerically-controlled machines, roll grinders were sometimes equipped with a cam (corresponding to the roll crown desired) and a linkage system that would automatically move the wheel away from and back toward the roll axis as it traversed the length of the roll face. With the modern grinders discussed in Section 4-11, the designed profile may be programmed on a tape.

4-24 Stress Concentration in The Roll Necks

Because of the abrupt change in diameter from that of the barrel to that of the roll neck at each end of the roll body, stresses tend to be intensified at the ends of the necks close to the body. As a consequence, roll necks are frequently broken at these locations when severely overloaded.

To minimize the stress-concentration factor (K) occurring at the necks,[⊖] a smooth fillet is usually provided as illustrated in Figure 4-58. This fillet is machined to one or more radii. In the

---

[⊢] A roll crown is defined as the excess of its mid-face diameter over its face-end diameter. Negative crowns indicate the roll has the least diameter at its midsection.

[Δ] "Fundamentals of Rolling", by Z. Wusatowski, Pergamon Press, London, 1969.

[■] "The Rolling of Strip, Sheet and Plate", by E. C. Larke, The MacMillan Company, New York, 1957.

[ᴿ] English or long tons (2,240 lbs.) are implied here.

[⊖] The stress-concentration factor K may be defined as the ratio of the maximum safe stress that may exist generally in the body of the roll to the maximum safe stress that may exist at the juncture of the body and neck.

MILL ROLLS AND THEIR BEARINGS

simplest case with one radius, the stress concentration is dependent on the ratio h/d (the height of the shoulder/neck diameter) and the ratio r/d (radius of the fillet/neck diameter). The relationship is illustrated in Figure 4-59.

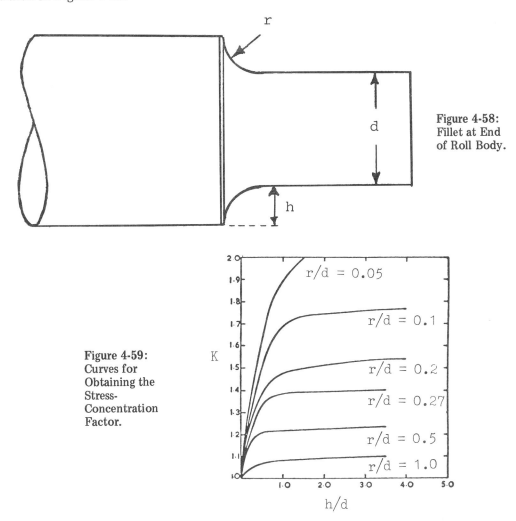

Figure 4-58: Fillet at End of Roll Body.

Figure 4-59: Curves for Obtaining the Stress-Concentration Factor.

Stress-concentration factors for necks of a more complicated design have been measured by D'Isa, Erzurum and Gross.‡ The basic geometry of these roll necks is illustrated in Figure 4-60 and four types (designated 2C, CSC, CSCT and CSCM with relative dimensions as indicated in Figure 4-60) were studied in the form of small steel models. For the CSC, CSCT and CSCM types, strain gages were placed at the tangent points T_1 and T_2 as indicated in the figure and, for the 2C series, one gage was placed on the curve of radius R_2 at point G and another at the tangent point of the two circles having radii R_1 and R_2.

The stress-concentration factors, as measured by the gages closer to the supports are shown plotted against the ratio R_2/d in Figure 4-61. Similarly, the stress-concentration factors as measured by the other gages are shown in Figure 4-62. The investigators believed that the data for the 2C and CSC configurations shown in Figure 4-62 are reasonably accurate provided $1.5 \leq D/d \leq 2$, $1 \leq a/b \leq 2$ and $0.1 \leq a/d \leq 0.4$. The curves for the other configurations were believed to be indicative but less reliable due to the availability of only limited data.

---

‡ F.A. D'Isa, H. Erzurum and J. Gross, "Stress Concentration Curves for Various Neck Configurations of Steel Mill Rolls", Iron and Steel Engineer Year Book, 1969, pp. 637-641.

2C series: $R_1 + R_2 = 2a$
$R_1 = ((\sqrt{3} + 1)b - (\sqrt{3} - 1)a)/2$
$R_2 = ((\sqrt{3} + 3)a - (\sqrt{3} + 1)b)/2$
$L = 0$

CSC series: $R_1 + R_2 = \frac{2}{3}(a + b)$
$R_1 = (9b - a)/12$
$R_2 = (9a - b)/12$
$L = (150a^2 - 268ab + 150b^2)^{\frac{1}{2}}/12$

CSCT series: $R_1 + R_2 = 2a$
$R_1 = (4b - a)/3$
$R_2 = (7a - 4b)/3$
$L = (a - b)/3$

CSCM series: $R_1 + R_2 = a$
$R_1 = (5b + 3a)/16$
$R_2 = (13a - 5b)/16$
$L = (238a^2 - 476ab + 366b^2)^{\frac{1}{2}}/16$
$W = 1.05a$

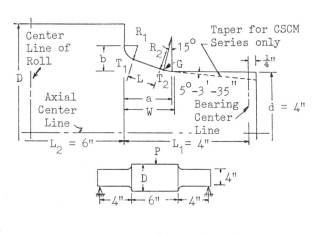

Figure 4-60: Roll Neck Configuration With Two Radii.

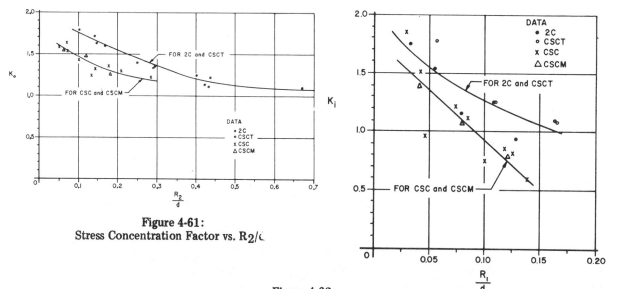

Figure 4-61:
Stress Concentration Factor vs. R_2/d.

Figure 4-62:
Stress-Concentration Factor vs. R_1/d.

4-25 The Failure of Mill Rolls Before and During Use

Although mill rolls are generally replaced due to wear, they frequently suffer damage of a more drastic nature, even during the period prior to their first use in a mill. This premature failure, known as "season cracking", discussed in the following section, results in the explosive fracture of a roll whereas the more common failures occurring during mill use may be categorized as:

(1) Breaks in the main body of the roll, both perpendicular and angular.

(2) Breaks in the roll necks.

MILL ROLLS AND THEIR BEARINGS

(3) Spalls.

(4) Splits in the main body of the roll.

With respect to hot strip mills, the first three constitute the more serious modes of failure.▼ A typical vertical, perpendicular break in a roll body is shown in Figure 4-63, a diagonal break in Figure 4-64, a break in the neck fillet in Figure 4-65, a horizontal split (probably caused by sudden surface cooling) in Figure 4-66, and a spalled roll in Figure 4-67.

Figure 4-63:
Typical Perpendicular Break in Roll Body.

Figure 4-64:
A Diagonal Break in the Main Body of a Roll.

Figure 4-65:
A Break in The Roll Neck Fillet.

Figure 4-66:
Horizontal Split Failure in a Roll Probably Caused by Sudden Surface Cooling.

▼C. F. Peck, Jr. and F. T. Mavis, "Reducing Roll Breakage", Iron and Steel Engineer Year Book, 1956, pp. 177-181.

In many instances, failures are due to improper utilization or handling of the rolls. However, other failures, including season cracking, are due primarily to the defective processing of the roll during manufacture. Yet, throughout the years, the causes of many roll failures were the subject of speculation and the source of argument between roll manufacturer and roll user. In 1956, when the need for intensive studies of roll fractures had become apparent, the Association of Iron and Steel Engineers (AISE) and the Roll Manufacturers' Institute (RMI) undertook a cooperative investigation into roll spalling, a project which extended over eight years. During the course of this work, several articles were published,[⊛] and a review and summary of the work were presented in 1966.[▲]

Early in this work, it was reported that, in typical strip mill operations, spalling occurred in 19% of the work rolls, resulting in a loss of 6% of the overall life and, in the case of backup rolls, 78% spalled, resulting in a 17% loss of overall life. (In the case of backup rolls, the ratio of spalling to loss of overall life was lower because spalls occurred later in their life.) Roll spalling was found to be attributable to two causes; "accidental", due to bruising or mishandling of the rolls and "chronic", related to mill characteristics and operating practices. The AISE-RMI research was directed to a better understanding of the latter, and is discussed briefly in Sections 4-27 and 4-28.

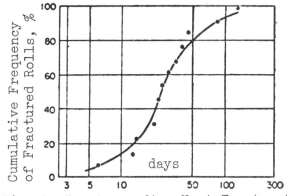

Figure 4-67:
Spalled Portion of a Backup Roll.

Figure 4-68:
Cumulative Frequency of Roll Cracking as a Function of the Time Subsequent to Heat Treatment (Sakabe).

[⊛] "Roll Wear in Finishing Trains of Hot Strip Mills", by Dr. G. Sachs, J. V. Latorre and M. K. Chakko, p. 1015, Iron and Steel Engineer Year Book, 1961, pp. 1015-1036.

[⊛] "Investigation of Spalling of Work and Backup Rolls in Hot and Cold Strip Mills", by J. V. Latorre and M. K. Chakko, p. 899, Iron and Steel Engineer Year Book, 1962, pp. 899-922.

[⊛] "Mechanical Properties of Work and Backup Roll Materials", by J. V. Latorre, p. 297, Iron and Steel Engineer Year Book, 1963, pp. 297-305.

[⊛] "Influence of Wear on Pressure and Stress Distribution in Rolls of 4-High Mills", by Dr. K. N. Tong, M. Sadre and M. K. Chakko, p. 339, Iron and Steel Engineer Year Book, 1963, pp. 339-345.

[⊛] "Contact Pressure Distribution and Comparative Tests for Evaluation of Resistance to Spalling of Roll Materials", by Dr. K. N. Tong, Dr. M. K. Chakko and J. V. Latorre, p. 555, Iron and Steel Engineer Year Book, 1963, pp. 555-566.

[⊛] "Predictions of Roll Spalling in 4-High Mills Based on Fatigue Strength of Roll Materials and Wear Pattern of Rolls", by Dr. K. N. Tong and M. K. Chakko, p. 539, Iron and Steel Engineer Year Book, 1964, pp. 539-569.

[⊛] "Evaluation of Resistance to Spalling of Roll Materials", by Dr. M. K. Chakko and Dr. K. N. Tong, p. 911, Iron and Steel Engineer Year Book, 1965, pp. 911-924.

[▲] F. H. Allison, Jr.; L. J. Spivak; D. Yorke; S. Stasko; M. D. Stone; H. E. Muller, Jr. and J. M. Dugan, "Benefits, Improvements and Ideas from Roll Research Program", Iron and Steel Engineer Year Book, 1966, pp. 143-163.

MILL ROLLS AND THEIR BEARINGS

4-26 Season Cracking

Season cracking may occur within a certain period subsequent to the heat treatment given to a roll during manufacture. Thus it may occur during maching or grinding, while it is being stored for future use or just after its insertion into a mill.ε However, from all tests and inspections made during fabrication of these rolls, nothing about their production appears abnormal.

The cumulative frequency of cracking, as a function of time elapsed from the last heat treatment until failure, is shown in Figure 4-68. Here it will be seen that roughly 80% of those liable to fail will do so between 10 and 50 days after heat treatment. Moreover, if season cracking is to be avoided during mill operation, the rolls must be stored for at least 200 days prior to use.

Figure 4-69 shows the incidence of season cracking during a year, showing none in July and August, rising to a peak in November and then decreasing again. Thus the phenomenon occurs most frequently during the cold season of the year (in the northern hemisphere). Because of this, Sykes' and Jones, et al.^ attributed season cracking to excessive residual stresses including additional heat stress set up by variations of the ambient temperature. They, therefore, advocated avoiding fluctuations of air temperature during roll storage, perhaps by using heat insulating materials around the rolls in winter and not letting the roll temperature fall below 10°C. Kawaguchi et al. ⋔ attributed the cracking to the increase of residual stress that might ensue from the gradual decomposition of residual austenite contained in the quench-hardened layer after hardening.

Sakabe ε argued that ambient temperature variations are more or less constant (about 8°C) the year round, and that these are relatively small compared to the temperature rise of the rolls (about 50°C) that occurs during warm up of the rolls in the mill. With respect to the effects of the austenite-martensite reaction, Imai et al. found that, for forged steel hardened rolls, the magnitude of the residual stress and the frequency of occurrence of quenching cracks were more dependent on the rate of progress of the martensitic transformation than on the quantity of martensite produced during hardening, and that further decomposition would not affect the stresses already present.

Sakabeε then correlated the frequency of cracking with the hydrogen content of the steel just before tapping the molten metal from the acid open hearth furnace. He found that season cracking is liable to occur when there is a high hydrogen concentration in excess of 3.5 ppm.

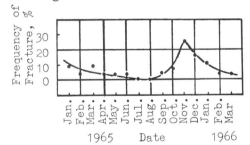

Figure 4-69: Seasonal Frequency of Cracking.

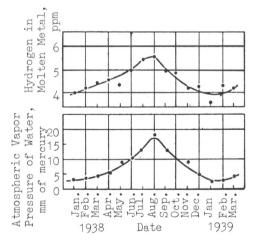

Figure 4-70:
Variation of the Hydrogen Content of Hot Metal and Moisture in the Atmosphere Throughout the Year. (Sakabe).

ε K. Sakabe, "Considerations on the Premature Failure of Fully Hardened Rolls", Tetsu to Hagane, 1967, 53 (6), pp. 611-628, (British Iron and Steel Industry Translation #6393).

ı C. Sykes, Steel Processing, 1954, 3, p. 168.

^ F. W. Jones, Steel and Coal, 1962, August 24, p. 354.

⋔ Kawaguchi, Shisaki: Nippon Seiko Tech. Rep., 1959, 1, p. 23.

That the hydrogen content of the metal at tapping is related to the moisture content of the atmosphere has been amply demonstrated. Figure 4-70 shows the data published by Kobayashi,[11] and it can be seen that these curves closely resemble that related to the frequency of cracking until the month during which the steel was produced (Figure 4-71).

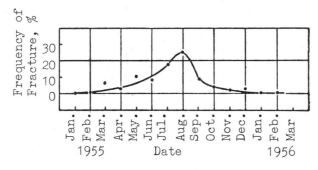

Figure 4-71:
Monthly Variation of
Roll Cracking Frequency.

On the basis of this and other work, Sakabe states that "season cracking of fully hardened rolls must be attributed to a delayed failure effect caused by hydrogen occluded in the roll and the stress conditions, and not to the hitherto accepted idea of simple quench cracks due to excessive residual stress during hardening, or else a failure caused by additional stresses set up by progressive decomposition of residual austenite". Since residual stresses cannot be avoided during hardening, the removal of the hydrogen from the melt by vacuum degassing (Section 4-7) is the best assurance of obviating the season cracking of rolls.

4-27 Roll Spalling

In the AISE-RMI investigation into roll spalling referred to in Section 4-25, a general review of roll pressure distribution theories was undertaken. Some of the data established for typical mills is shown in Figure 4-72, the shear stresses (the two lowest curves) being important since they are the most frequent cause of roll failure. The third highest curve is important inasmuch as it shows that the maximum value of the shear stress occurs below the roll surface and that failures from fatigue could originate at this location.

The investigation was also concerned with the mechanical properties of the shell and core materials of double-poured, cast-iron work rolls and with the surface layers of cast-steel backup rolls. The measured values of some of these properties are shown in Table 4-6 (the designations A, C, D, E, F, T and 73 referring to different rolls under test).

Since there was no agreement between the data obtained from standard fatigue tests, roll-on-roll fatigue tests were made on a modified mill. This was a 6-inch, 4-high mill converted to a 2-high configuration (as shown in Figure 4-73) so that it could accommodate test discs up to 4.75 inches wide and 12 inches in diameter mounted on arbors. Hydraulic cylinders on top of the mill housings exerted loads up to 114,000 lbs. so that, when operated with a 1-inch face width, Hertz pressures up to 440,000 psi and maximum reversed shear stresses up to 110,000 psi could be developed.

The mill was equipped with a 50 hp motor and a variable speed gear reducer driving the lower roll enabling speeds ranging from 70 to 250 rpm to be obtained. Various combinations of rolls were used simulating most of the combinations found in commercial mills.

The data was plotted as the number of cycles to failure against the maximum reversed shear stress in the material as illustrated in Figure 4-74. It is to be noted that these curves resemble standard fatigue curves, in that the number of cycles to failure increases as the mill load decreases. In certain tests, it was also found that a hard, cast-iron roll (with a Shore hardness in excess of 70) caused a flow of metal to occur in a cast-steel roll of about 45 Shore hardness with no sign of

---

[11] Kobayashi: Nippon Gakujutsu Shinkokai, 19th Steelmaking Committee Report, 1940, p. 209.

MILL ROLLS AND THEIR BEARINGS

spalling. (This same ductile behavior of cast steel is actually observed in mill operations, the surface of cast steel backup rolls having been found to flow both circumferentially and laterally.)

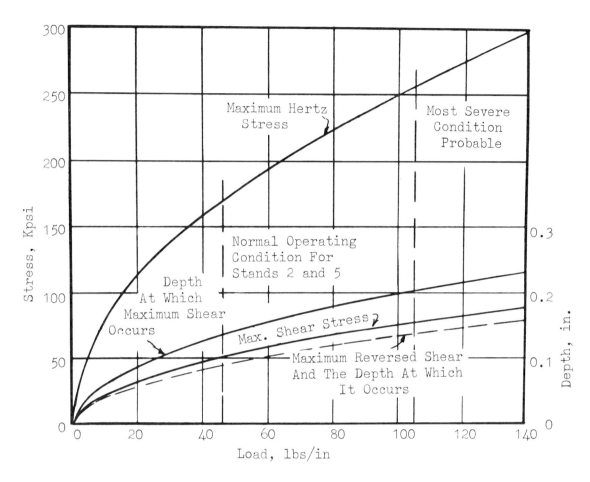

Figure 4-72: Magnitude and Location of Maximum Stresses in Roll.

Table 4-6

Tensile-Compression Properties

	Double pour, shell A	Double pour, shell C		Double pour, core D		Cast steel, E	Cast steel, F	Cast steel, T	Cast steel, 73
Yield strength, psi	—	—		—		105,000	—	90,000	86,000
Ultimate tensile strength, psi	53,000 (5) [1] L [2]	63,600 (5) L [2]		28,600 (6) L		158,300 (6) L	146,000 L	128,000 L	128,200
		69,400 (3) R [3]		36,550 (2) R		157,700 (3) R			
Compression strength, psi	—	314,000		134,000 (3)		—	—	—	280,000
Hardness, Rockwell C	51-59	53-56 L	45-51 R	22-28 L	23-25 R	37-39	30	32-33.5	—
Shore	68-80	71-74 L	60-68 R	35-41 L	36-37 R	50-52	43		
Modulus (tensile) $\times 10^6$, psi	21.9 (3)	26.6 L	30.0 R	13.6 L	15.4 R	31.0 L	—	31.2	29.5
						31.0 R	—		
Modulus (compression) $\times 10^6$, psi	—	25.0 L	25.0 R	20.0 L	20.0 R	29.4 L	28.0	—	30.0
						29.4 R		29.4 R	

[1] Number of tests.
[2] Longitudinal orientation.
[3] Radial orientation.

Figure 4-73:
Experimental Equipment Used in Roll Spalling Tests.

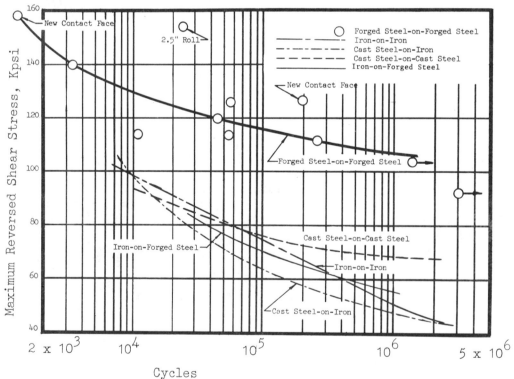

Figure 4-74:
Cycles-to-Failure Data for Various Roll Materials.

MILL ROLLS AND THEIR BEARINGS

Some of the data obtained with the mill is shown in Figure 4-75. The single point occurring at 71 Ksi and 10^6 cycles represents a test using an air blast for cooling instead of water normally employed. Although only one point, it tends to support the argument that rapid damage to rolling surfaces is caused by hydraulic pumping of the coolant into microscopic surface fissures.

Other experimental data obtained from the mill is shown in Figure 4-76. They relate to the rolling of a steel roll on an iron roll and the numbers associated with each point represent the lengths of the roll faces in inches (the face length of the steel roll being given first). From the consistency of the data, it must be concluded that the edge effects in the experimental mill were not large.

Depth hardness tests made on samples rolled on the experimental mill showed a considerable amount of cold work in the surface layers with a maximum occurring at a depth below the surface corresponding to the location of the calculated maximum shear. (See Figure 4-77.) As a consequence of these observations, it appears desirable to remove the cold-worked surface layer before it becomes a source of fatigue failure and spalling. Increased backup roll life resulting from the implementation of this recommendation has been claimed by a number of mill operators.

Figure 4-78 shows the variation of material hardness in the roll at the location of maximum shear stress with the number of cycles of operation.

The maximum depths of the spalls developed in the experimental mill rolls is shown in Figure 4-79 plotted as a function of cycles for three types of rolls. It should be noted that forged rolls had deeper spalls at a small number of cycles and appear to be load dependent. Conversely, the spalls in the iron and cast steel rolls became deeper with increasing roll usage under relatively constant loads.

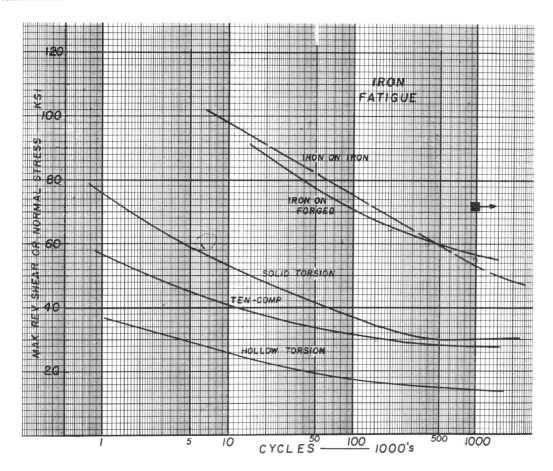

Figure 4-75: Cycles-to-Failure Curves for Iron Rolls Compared with Fatigue Data (Three lowest curves).

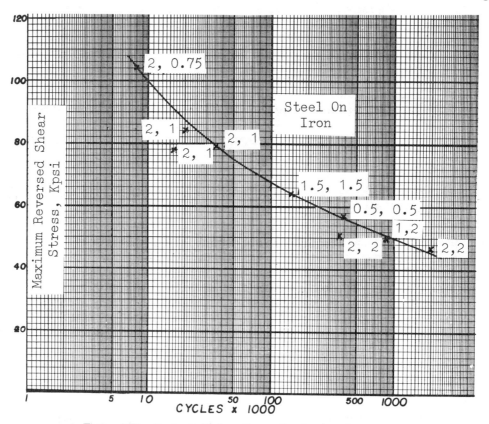

Figure 4-76: Cycles-to-Failure Curves for Steel on Iron Rolls.

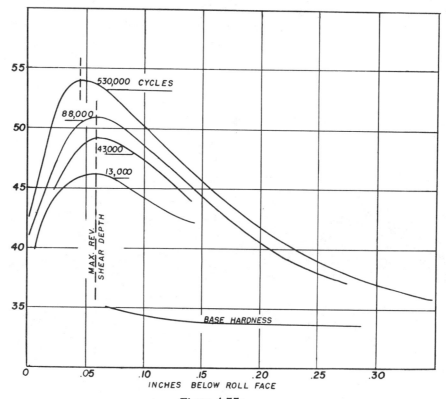

Figure 4-77:
Curves Showing Variation of Hardness with Depth Below Roll Face for Various Maximum Reverse Shear Stress Levels.

MILL ROLLS AND THEIR BEARINGS

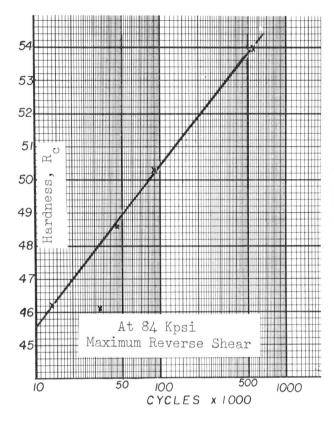

Figure 4-78: Curve Showing Relationship Between Hardness and the Number of Cycles at the Location of 84,000 psi Maximum Reverse Shear.

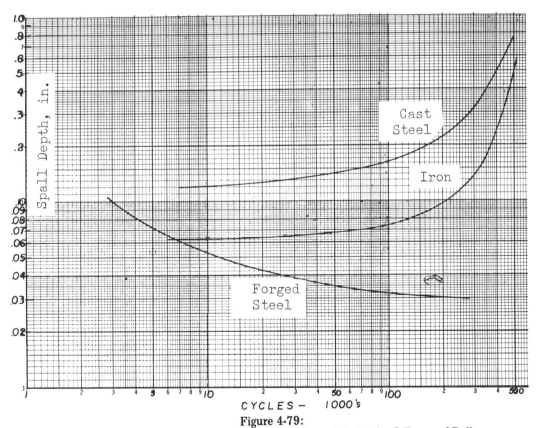

Figure 4-79:
Relationship of Maximum Spall Depth to Number of Cycles for 3 Types of Rolls.

To minimize the occurrence of roll spalling, Lankau [=] recommends the following measures:

(a) The use of roll materials with a low susceptibility to the formation of grinding fissures and a high fatigue strength.

(b) Proper hardening to ensure the lowest levels of residual stresses and residual austenite.

(c) Very careful regrinding of the rolls.

(d) Use of the most effective rolling lubrication.

(e) Limiting the usage of the rolls between regrinds.

(f) Maintaining, by preheating the rolls, the lowest possible temperature gradients within them when they are placed into service.

More recently, however, Melford et al.[<] have reached the conclusion that roll spalling on a mill used for rolling strip for ERW tube making was attributable to a hydrogen injection in the rolls resulting from a corrosion reaction with a soluble oil coolant. They alleviated the problem considerably by changing the concentration of the rolling oil from 5 to 10 per cent and replacing the solutions more frequently.

4-28 The Fatigue Failure of Rolls Initiated at the Bore

Fatigue cracks may occur in the roll bore soon after it has been put into use in the mill, resulting in an elliptical deformation of the roll and sometimes its complete disintegration.[ε] Figure 4-80 shows a sketch of such a crack originating at the bore and extending in a spiral toward the roll surface.

The radial stress in the roll during operation is illustrated in Figure 4-81. In the a-a section, a compressive stress acts at the roll face and a tensile stress in the surface of the bore. In the b-b section, rotated 90° from the a-a section, there is a tensile stress in the roll face and a compressive stress in the surface of the bore. Thus, in the rolling operation, the bore wall is subject to cyclical stresses. In addition, however, there are residual stresses as discussed in Section 4-18 and these decrease the amplitude of the allowable stress φ_a in accordance with the Goodman relationship.

$$\varphi_a = \varphi_w (1 - \varphi_m / \varphi_B)$$

where φ_w is the alternating stress fatigue limit, φ_m is the mean residual stress and φ_B the tensile strength of the roll material.

The reduction in the allowable alternating stress φ_a accounts for fatigue failure in the bore and the tendency to failure is increased by poor surface finishes in the bore, by decarburization of the bore surface, by excessive rolling loads, and by thermal stresses induced during the warm up of the roll. Fortunately, repeated grindings of the outer roll surface tend to decrease the residual stresses and so the frequency of fatigue failure initiated at the bore decreases with increasing roll life.

[=] G. Lankau, "Shell Fractures (Spalling) of Cold Rolls, Their Formation and Interpretation of Their Origin", Neue Hutte, 1966, 11, July, pp. 404-412 (in German). (Available in the English Translation BISI 9163, published April, 1971, by the British Iron and Steel Translation Service.)

[<] D. A. Melford, V. B. Nileshwar, R. E. Royce and M. E. Giles, "Influence of Hydrogen Pick-Up on the Spalling Behavior of Work Rolls in a Cold Rolling Mill", Journal of the Iron and Steel Institute, March, 1972, pp. 163-167.

[ε] K. Sakabe, "Considerations on the Premature Failure of Fully Hardened Rolls", Tetsu to Hagane, 1967, 53 (6), pp. 611-628, (British Iron and Steel Institute Translation #6393.)

MILL ROLLS AND THEIR BEARINGS

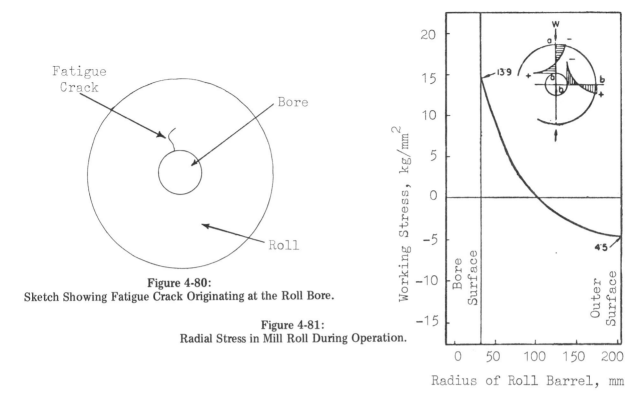

Figure 4-80:
Sketch Showing Fatigue Crack Originating at the Roll Bore.

Figure 4-81:
Radial Stress in Mill Roll During Operation.

4-29 The Wear of Mill Rolls

In the normal operation of rolling mills, work and backup rolls, even if undamaged by cracking, spalling or in any other manner, must be frequently changed as the result of wear. In general, wear has two detrimental consequences, these being (a) the changes in the surface finish of the rolls (very smooth rolls tend to become rougher and grit-blasted rolls tend to become smoother) and (b) changes in the surface profile (or crown). In rolling operations where a very uniform finish and/or high luster is demanded in the rolled product, work rolls must be frequently changed to ensure product quality.

The wear rate of work rolls is subject to a number of variables. On the basis of tons rolled, it would be expected that small rolls would require more frequent regrinds than larger diameter rolls. However, assuming the use of similar rolling lubricants and drafting practices, smaller rolls are subject to lesser peak pressures (see Chapter 9). Better lubricants, higher mill speeds and lower rolling forces all tend to diminish roll wear.

Accurate statistics relative to work roll wear rates are difficult to establish in practice, primarily because rolls are often changed because they are:

(a) Slightly damaged due to defective surfaces on the incoming strip, and/or

(b) believed to be responsible for defective shape conditions, such as crossbow or twist (see Chapter 11).

However, Figure 4-82 shows the life dispersion curve for the work rolls of a 4-high tandem cold mill that were scrapped during a certain year.[*] From this curve, it is seen that roll life should be stated in statistical terms inasmuch as one roll was scrapped after rolling about 1,400 tons, whereas, another rolled in excess of 16,000 tons before being discarded. With the increasing use of data logging systems in conjunction with cold mills, it is anticipated that more reliable roll life data will soon be available for analysis.

---

[*] N. A. Bracht and A. A. Bradd, "Factors Affecting Backup Roll Life", Iron and Steel Engineer Year Book, 1965, pp. 243-251.

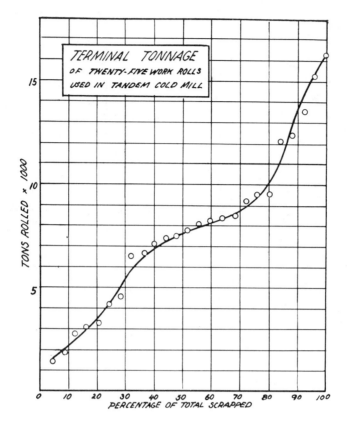

Figure 4-82: Life Dispersion Curve for Work Rolls Scrapped During a Period of One Year.

Backup rolls are removed from the mills less frequently than the work rolls and, as a consequence, their wear rates are perhaps better established. With respect to sleeved, backup rolls, Bracht and Bradd* have published data relative to their use in a 4-stand sheet mill. The sleeves initially possessed diameters of 33 inches, with a 32-inch face and a Shore sclerescope hardness in the range 55-70. The best sleeve usage occurred during a certain year when 195,052 tons of sheet product were rolled requiring 63 roll dressings. These data correspond to an average diameter decrease of 0.131 inch per dressing or 23,610 tons per inch change in roll diameter. However, the following year, this rate of roll usage had, for a variety of reasons, approximately tripled (7,450 tons per inch change in diameter). It was also found that harder sleeves (65-70 Shore "C"), with a soft bore, were superior to through-hardened but softer sleeves (60-65 Shore "C"), in that the former required a diameter change of 0.083 inches per dressing whereas the latter required 0.14 inch per dressing.

4-30 Practices Developed for Conserving Backup Roll Life

As a consequence of backup roll life studies, mill operators have increased roll life by grinding the surface to a specified depth after a specified number of turns in the mill. Bracht and Bradd* reported that, with respect to a 4-stand sheet mill, each backup roll is pulled after 40 turns of operation and, on a 10 turn per week basis, this means one backup roll change per week. At the same time, backup roll changes are made when requested by the operators but, since the adoption of this special practice, unscheduled roll changes are rare. Such changes in no way interfere with the scheduled roll removal practice. With respect to the redressing operation, a minimum diameter reduction of 0.050 inches is taken at each dressing except every third, when a minimum 0.100 inch is removed. However, more material is, of course, removed when necessary but all dressing is by grinding (except where the reduction in diameter is more than 0.250 inch).

---

* N. A. Bracht and A. A. Bradd, "Factors Affecting Backup Roll Life", Iron and Steel Engineer Year Book, 1965, pp. 243-251.

MILL ROLLS AND THEIR BEARINGS

For a three-year period following the introduction of this practice, the average diameter reduction per dressing was 0.110 inch, and the tons per inch of diameter change was 14,439. These data are presented in greater detail in Table 4-7.

Table 4-7 — Backup Roll Life Statistics

Year	Roll Type	Total Stock Removal (In.)	Dressings	Average Removal Per Dressing (In.)	Tons Rolled	Tons Rolled per Inch Diameter Change	Remarks
1956	Cast	15.984	52	0.307	121,631	7,610	Indefinite Time Between Grinds
1957	Rolls &	15.798	62	0.255	131,826	8,340	
1958	Sleeves	19.601	66	0.247	129,933	6,620	
1959		8.261	63	0.131	195,052	23,610	
1960		21.962	84	0.261	163,537	7,450	
Totals		81.606	327	0.249	741,979	9,090	
1961	Sleeves	10.031	91	0.110	160,289	15,980	Rolls Changed on Schedule With 0.050-0.100 in. Minimum Reduction
1962	Only	9.802	101	0.907	142,555	14,540	
1963		11.371	91	0.125	147,695	12,981	

A similar practice to prevent spalling and thereby extend hot-mill backup roll life has been reported by Muller.± In this regular "grind-and-cut" program, the reduction specified was 0.090 inches on each grind for three successive dressings with turning each fourth time to remove 0.25 in.

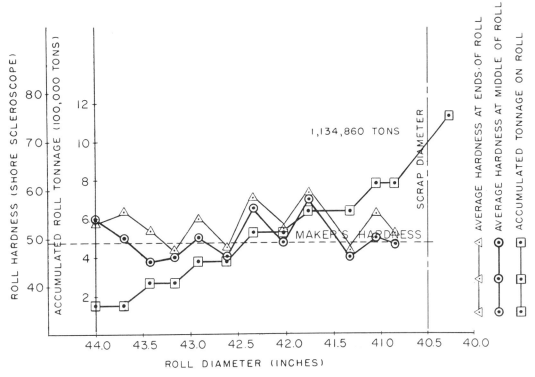

Figure 4-83:
Plot of Successive Roll Reduction vs. Hardness and Accumulated Tonnage Rolled.

± H. E. Muller, Jr., "Practice Developed for Conserving Backup Roll Life", Iron and Steel Engineer Year Book, 1966, pp. 160-163.

Figure 4-83 shows plots of roll hardness at the center and ends of rolls after regrinds, together with plots of the accumulated tonnage rolled. These curves demonstrate that full roll life can be obtained by continually removing the work hardened material so that the original hardness is maintained. Figure 4-84 shows that failure to reduce the hardness to the proper level resulted in a spall. Similar results have also been reported by Butylkina.◈

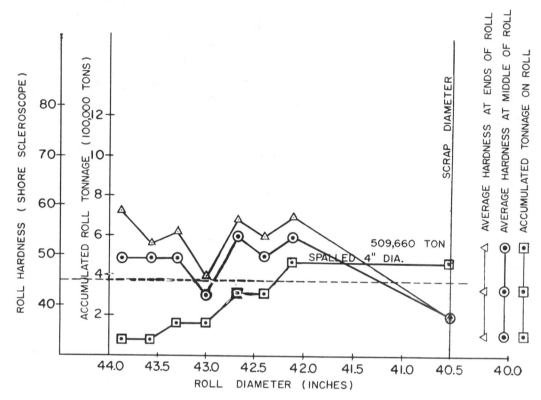

Figure 4-84:
Plot of Successive Roll Reductions vs. Hardness and Accumulated Tonnage for Spalled Roll.

4-31 Extending Roll Life by Burnishing and Chrome Plating

Belkin and Benzhega• have investigated the effects of burnishing on roll life, particularly in relation to the contact fatigue strength of the roll material. They processed specimens of hardened 9X steel on a 3-roll burnishing machine mounted on the carriage of a lathe, the specimens being copiously lubricated during the surface treatment. It was found that burnishing could appreciably increase the hardness of the specimens (see Figure 4-85) as well as their resistance to spalling. The investigators believed the increased contact fatigue strength was associated with the increased uniformity in the microstructure of the surface layers. Roller burnishing apparently evens out the stress gradients so that work hardening and subsequent changes in specific volume, structure and plasticity vary gradually from point to point. At the same time, some of the residual austenite in the roll surface is also transformed in the 0.3 mm thick work hardened layer.

As a consequence of their laboratory work, 30 work rolls of a 12-roll mill used to roll brass were burnished to determine the effectiveness of this roll processing technique under

---

◈ L. I. Butylkina, "Hardness and Life of Backup Rolls for Cold Rolling Mills", Stal in English, October, 1969, p. 903.

• M. Ya Belkin and A. S. Benzhega, "Increased Wear Resistance in Rolls for Cold Rolling Obtained by Roller Burnishing", Izvest. VUZ. Chem. Met. 1967, 10, 9, pp. 120-123, (British Iron and Steel Industry Translation #6232).

MILL ROLLS AND THEIR BEARINGS

production conditions. It was established that burnishing extended roll life by 40 per cent and, as a consequence, this technique has been adopted as the standard practice at the Staro-Kramalorsk Engineering Works in the USSR.

Gurin[G] and his coworkers have reported that electroplated chromium coatings on work rolls of cold mills has provided increased hardness and wear resistance of the rolls and was particularly beneficial in the rolling of precision steel strip 0.15 mm thick. The type of coating found to be most satisfactory was a hard chrome coating obtained at a temperature of 53-56°C and a current density of 52-58 A/dm^2. This type of coating provided a four- to five-fold increase in wear resistance. The optimum coating thickness was found to be 5 μm for a temper mill and use of the coated rolls significantly decreased the number of surface defects on the rolled strip.

4-32 Roll Chocks

Roll chocks, which accommodate the roll neck bearings for the mill rolls, are usually steel castings designed to fit into the windows of the mill housing. As such, they are intended to maintain accurate positioning of the rolls and, in the case of backup roll chocks, to transmit the rolling force from the housing to the rolls. Because of the larger size bearings they must accommodate, backup roll chocks are considerably larger than the work roll chocks, so much so that the latter are usually designed to fit inside the "wings" or projections of the former as illustrated in Figures 3-5 and 3-25.

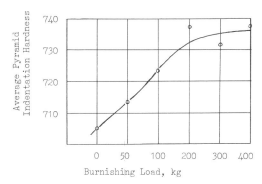

Figure 4-85:
Changes in Surface Hardness of Hardened 9X Steel, Roller Burnished Under Various Loads. (After Belkin and Benzhega)

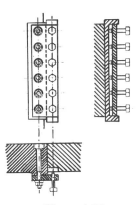

Figure 4-86:
Roll Chock Guide with Abrasion Rail.

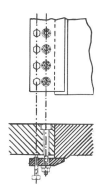

Figure 4-87:
Sketch Showing Guide Abutting Chipping Strip of Chock.

To prevent appreciable movement of the chocks in the housing in a direction parallel to the roll axes, guides or adjusting plates, which are securely attached to the housing on the operator's side only, are used as shown in Figure 4-86.♦ In this instance, the guides are fitted with adjusting bolts which should be secured by nuts. The bolts are not in direct contact with the chock, but with an abrasion rail placed in a groove of the chock. The rail abuts against the guide with its ends lapped above and below so that it is not moved by any displacement of the chock. Hence, the chock does not slide on the adjusting bolts but on the abrasion rail. In Figure 4-87, the guide abuts directly on to a chipping strip on the chock and is fastened to the housing by means of bolts. Fine adjustment can be achieved by adjusting screws. Figure 4-88 illustrates how the guides may be fitted to the housing.

Another method of securing the chocks is by means of claws, shown in Figure 4-89, which slide into flat recesses of the chock and the housing in order to prevent being turned.♦ On

---

[G] S. M. Gurin, A. F. Pimenov and Z. M. Shvartsman, "Chromizing Steel Work Rolls for Cold Rolling Mills", Stal in English, October, 1969, pp. 901-903.

♦ "Forge Equipment, Rolling Mills and Accessories", by A. Geleji, Akadémiai Kiadó, Budapest, 1967.

the drive side of the mill, the chocks are not usually locked in position. This enables the chocks to slide slightly in the axial direction of the rolls to compensate for thermal expansion occurring in the rolls.ʎ

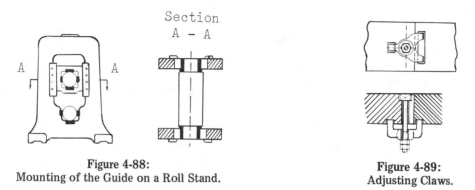

Figure 4-88:
Mounting of the Guide on a Roll Stand.

Figure 4-89:
Adjusting Claws.

To prevent unnecessarily large bending stresses in the backup roll chocks, it is essential that the chocks be self-aligning in order to prevent misalignment of the bearings. Self-alignment of the top chock is effected by the spherical surfaces of the screws. However, for the self-alignment of the bottom chocks, they may be designed to rest on rounded or short supports as shown in Figure 4-90.

The surfaces of the backup roll chocks contacting the mill posts are frequently faced with liners. Maintenance of these liners is essential for good bearing life since they govern the vertical alignment of all chocks.ᴊ Very loose or "sloppy" chocks result in crossed rolls and damage to bearings. Figure 4-91 shows one side of a backup roll chock machined to carry two independent liners.

To enable the backup rolls to be kept apart for work roll changing purposes, hydraulic cylinders are often incorporated in the lower work and backup roll chocks as illustrated in Figure 3-25. In such cylinders, the hydraulic fluid pressure is usually 100-150 kg/cm^2 (1420-2130 psi).▶

Where backup roll chocks incorporate lubricated, sleeve-type bearings, such as Morgoil bearings, they also feature an inlet and an outlet for the lubricant as indicated in Figure 4-92.

Work roll chocks, in general, exhibit many of the features of backup roll chocks. Small, counterbalancing hydraulic cylinders are used in the lower set of chocks, as illustrated in Figure 3-25. However, where they use roller bearings prepacked with grease, they require no lubricant connections.

4-33 Roller Bearings

The tapered roller bearing is the commonest bearing of this type employed in cold reduction mills, finding almost universal use as work roll bearings in 4-high mills and increasing use as backup roll bearings. In use since 1926,♦ these bearings are significantly more expensive than the sleeve or plain bearings discussed in Section 3-3, but exhibit significantly less friction and may, in many cases, be prepacked with lubricant, thus simplifying the construction of both the bearing and the roll chock.

ʎ D. G. Gibson and W. E. McCoy, "Tapered Roller Bearings as Applied to High Speed Rolling Mills", Iron and Steel Engineer Year Book, 1970, pp. 597-608.

ᴊ D. Lomax, "Rolling Mill Chock, Maintenance at DOFASCO", Iron and Steel Engineer Year Book, 1966, pp. 631-638.

▶ "Rolling Mills", A. I. Tselikov and V. V. Smirnov, Edited by W. J. McG. Tegart, Pergamon Press, London, 1965.

♦ "Forge Equipment, Rolling Mills and Accessories" by A. Geleji, Akademiai Kiado, Budapest, 1967.

MILL ROLLS AND THEIR BEARINGS

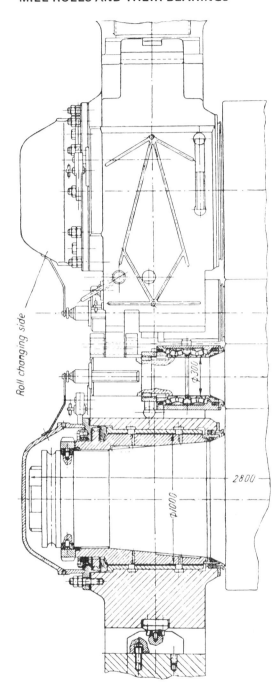

Figure 4-90:
General View of Chocks in Wide, 4-High Mill.

Figure 4-91:
Backup Roll
Chock Machined
to Carry Two
Independent
Liners.

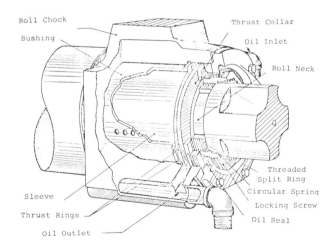

Figure 4-92: Oil Film Bearing (Morgoil Type).

Two main types of roller bearings have been developed for large rolling mills; these being the barrel-shaped roller bearings shown in Figure 4-93 and the tapered roller bearings shown in Figure 4-94. Besides these two types, there are also cylindrical roller bearings used for smaller rolling mills.

A four-row tapered roller bearing mounting, designated type TQO, used for straight bore, loose roll neck fits is illustrated in Figure 4-95.[λ] Loose fits are used between the cone bore and the roll neck to facilitate assembly and disassembly of the bearing and chock with respect to the roll neck. With a loose fit between the inner race and the roll neck, a relative rotation occurs of one with respect to the other, resulting in wear (particularly of the roll neck because it is softer than the race). This can be minimized by improved lubrication between the neck and the race and some necks are drilled so that grease can be periodically injected through a swivel connection on the end of the roll while the mill is in operation.

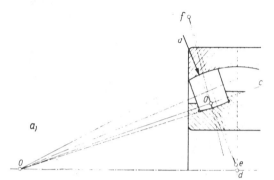

Figure 4-93:
Barrel-Shaped Roller Bearing.

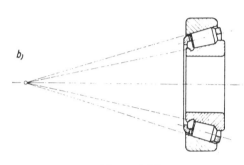

Figure 4-94:
Tapered Roller Bearing.

Oil mist provides a superior roll neck cone bore lubrication than grease, provided the sealing is adequate. With mist, the lubricant is continually being replenished while the mill is operating.

Work rolls with loose fitted type TQO roll neck bearings, similar to those shown in Figure 4-95, have been operated at mill speeds up to 7,000 fpm or a bearing speed of 1240 rpm. With work rolls, roll neck wear, in general, is not a problem with either mist or grease lubrication during the normal life of the rolls. Where grease is used, adequate lubrication can usually be maintained on the roll neck even on high speed mills because frequent roll changes afford an opportunity for periodic relubrication of the roll neck at the time of bearing assembly. A factor favorable to work rolls used in 4-high mills is that, generally speaking, the bearings are relatively lightly loaded.

Backup rolls with loose fitted straight bore roll neck bearings have been in operation for a number of years with mill speeds in the range 1500-2000 fpm. In recent years, aluminum foil mills with this design have been successfully operated in the 2000-3000 fpm range and occasionally at speeds up to 5000 fpm.

The wear or scuffing of the straight bore loose roll neck fits is obviated by the tapered bore, interference roll neck fits illustrated in Figures 4-96 and 4-97.[∞] This bearing, designated Type TQIT, and consisting of a 3-piece inner race (cones) and a 2-piece outer race (cups) with a cup spacer, was first used on a temper mill in 1951. During assembly, the bearing is first installed in the chock and the bearing and chock are slipped onto the roll neck. A hydraulic jack backed up by a split clamp ring pushes the bearing up the roll neck (as indicated in Figure 4-98), thereby creating an interference fit. The axial travel of the bearing and, consequently, the degree of interference, is determined by the length of the fillet ring. Each fillet ring is individually sized for each roll neck such that the roll neck diameter at the face of the fillet ring is practically identical for all rolls at a given installation. This means that all bearings are interchangeable on all roll necks at each mill installation.

[λ] D. G. Gibson and W. E. McCoy, "Tapered Roller Bearings as Applied to High Speed Rolling Mills", Iron and Steel Engineer Year Book, 1970, pp. 597-608.

[∞] It should be noted that the orientations of the rollers in the TQIT bearings differ from those of the TQO bearings.

MILL ROLLS AND THEIR BEARINGS

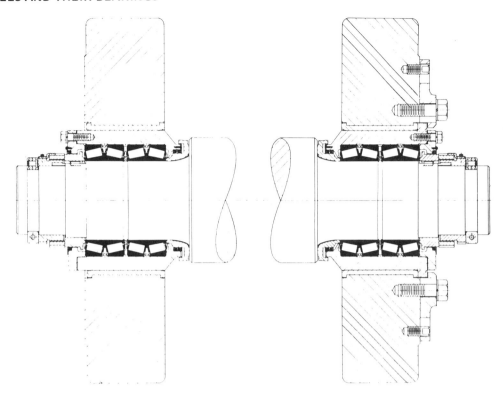

Figure 4-95: Type TQO Four-Row Tapered Backup Roll Bearing Mounting.

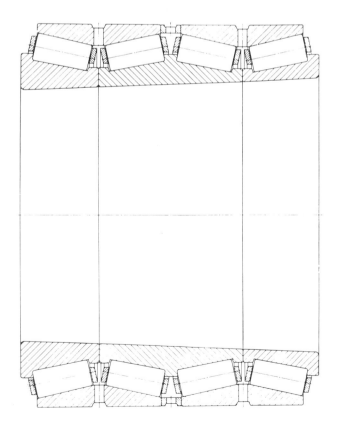

Figure 4-96: Type TQIT Tapered Bore Bearing.

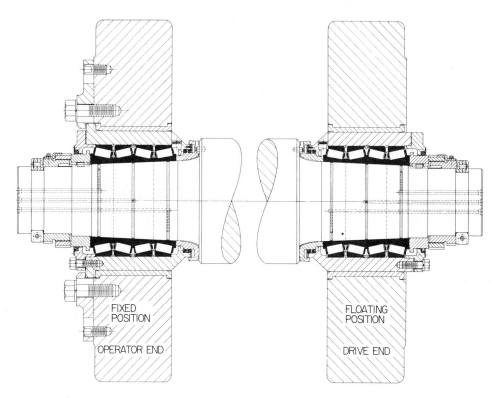

Figure 4-97: Type TQIT Tapered Bore Bearing in Assembled Mounting.

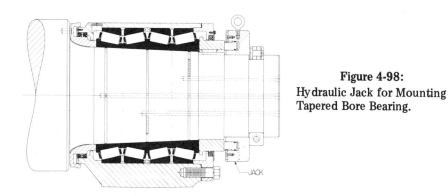

Figure 4-98: Hydraulic Jack for Mounting Tapered Bore Bearing.

The roll neck taper has a ratio of 12:1 on diameter which provides a self-locking taper and allows for a high degree of accuracy regarding the interference fit.

Extensive experience has indicated that it is necessary to mount the fillet ring with an interference fit on the roll neck and pinned to the roll barrel to prevent rotation. To prevent shearing the pin or having the pin back out of the roll barrel, even when welded to the fillet ring, a loose-fitted slot in the fillet ring is recommended.

A straight bore, four-row tapered roller bearing of the TQO type, modified to a tapered bore bearing, is designated type TQOT. In the TQOT mounting (Figure 4-99) the cups, by necessity, must be securely clamped into the chock at both the fixed and floating positions. The cones are also rigidly clamped to the roll neck. While there appears to be a close similarity in the axial location of the cups and cones regarding the straight bore, loose fitted type TQO bearing and the type TQOT design, there is a significant difference. The type TQO bearings, at both the fixed and floating positions, are not clamped to the roll neck With loose fitted bearings, it is recommended that the clamping nuts be backed off up to 0.040-inch, depending on the locking device, to allow the cones to freely creep on the roll neck to prevent excessive frictional heat and fire cracking on the cone faces. With the TQO mounting, while the chock on the drive side of the mill is allowed axial float, most of the axial thermal expansion can be taken through the loose cone fit on the roll neck. With the TQOT mounting, the axial roll expansion is accommodated by the chock on the drive side of the mill sliding on what is generally a nonlubricated surface between the chock and the breaker block or the rocker plate. How these axial forces react on both the TQOT and TQIT designs is discussed later in this Section.

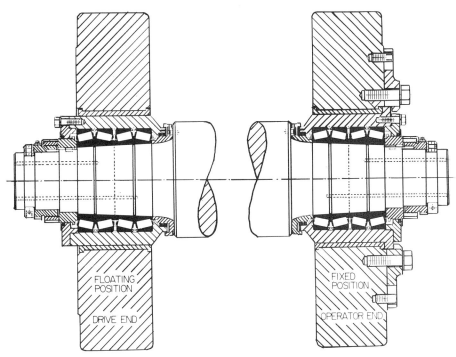

Figure 4-99:
Straight Bore, Four-Row Tapered Bearing Design Converted to Tapered Bore.

Figure 4-100 shows a comparison of the two designs. The effective center of a tapered roller bearing is considered as being that point located by the intersection of the horizontal axis of the bearing and a line drawn perpendicularly through the midpoint of the cup's ID. It will be noted that the direction of the rollers in the TQIT bearing below the centerline is reversed in comparison to the direction of the rollers in the TQOT bearing above the centerline. It can be seen that there are, in effect, two centers for the latter and three for the former and, since the effective spread or stability of the bearing is indicated by the distance between these centers, the stability of the TQIT design is therefore significantly greater than the TQOT design.

The hydraulic removal of the TQOT bearing and chock from the roll neck is somewhat more difficult than for the TQIT bearing illustrated in Figure 4-101. In the latter case, each of the three inner races or cones can be hydraulically removed in sequence independently of each other. With the TQOT design, hydraulic pressure is applied to both cone bores simultaneously or by alternately applying pressure separately to each cone. (See Figure 4-102). An alternate disassembly procedure is to loosen the bolts of the chock cover plate that clamps the cups in place. This will allow removal of the outer cone and then the inner cone, in that order.

In general, it would appear that the type TQIT bearing has distinct operating advantages as compared with the TQOT design and this difference should be reflected in a more favorable life experience for the TQIT bearings.

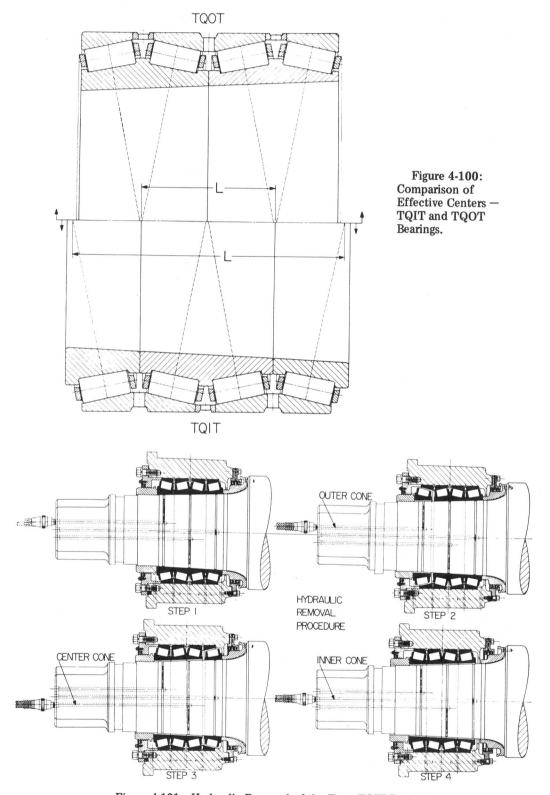

Figure 4-100: Comparison of Effective Centers — TQIT and TQOT Bearings.

Figure 4-101: Hydraulic Removal of the Type TQIT Bearing.

MILL ROLLS AND THEIR BEARINGS

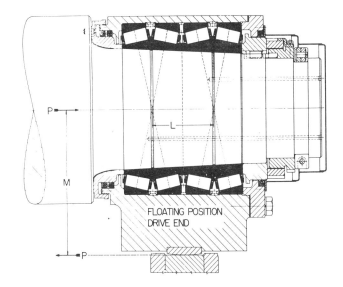

Figure 4-102:
Type TQOT
Bearing Design.

4-34 Roller Bearing Lubrication

Although work roll bearings of the roller type may frequently be lubricated with an extreme pressure grease (either prepacked or dispersed through a centralized lubrication system), oil-mist lubrication of such bearings is being more extensively applied to modern mills.[1] It has seven basic advantages; viz., no product staining, no tramp oil from the bearing contaminating the rolling lubricant, no churning of the lubricant in the bearing, low oil consumption, a continuously controlled oil flow to the bearing, low cost with respect to installation and use and reliability of operation.[2]

The quantity of mist required per bearing for adequate lubrication, (expressed in cubic feet per minute), is obtained by multiplying the product of the bearing bore (in inches) and the number of rows of rollers by the factor 0.05. For a four-row bearing, two nozzles are generally used in addition to a single nozzle for a roll seal.

The mist is generated by dispersing the oil into very small droplets so that it can be carried by the air stream. One or two large misting units supplying the complete mill stand are generally used in conjunction with monitoring systems which give assurance that the equipment is operating satisfactorily. Three different types of mist generating units are available, these being (a) those without any heaters, (b) those with oil heaters and (c) those with oil and air heaters. Those utilizing heated air have the advantages of providing higher flow rates for oils with higher viscosities regardless of the ambient temperature.

Delivery pipe size is determined by the quantity of mist required and should be such that the flow rate of the mist is less than 24 feet per second and preferably about 15 feet per second. The delivery pipe should be sloped so that the oil which is condensed within it flows to the bearing or returns to the mist unit.

Figure 4-103 shows a schematic diagram of the mist distribution system with one large mist unit supplying the mill bearings. The same figure shows the flow rates and the pipe sizes at various points in the system. A hose of carefully selected size is generally used to connect each chock to the delivery pipe and the couplings must also be carefully chosen so that they do not restrict the mist flow rate.

---

[1] C. H. West and H. S. Struttmann, "Guide Lines for Mist Lubrication of Roll Neck Bearings", Iron and Steel Engineer Year Book, 1967, pp. 821-827.

[2] W. E. McCoy, C. W. West and P. E. Wilks, "New Mist Lubrication Concepts for Tapered Roller Bearings Used on High Speed Rolling Mill Backup Rolls", "Tribology in Iron and Steel Works", The Iron and Steel Institute, ISI Publication 125, 1970, pp. 106-111.

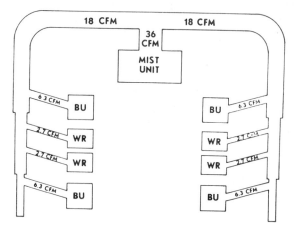

Figure 4-103:
Misting System for 4-High Mill.

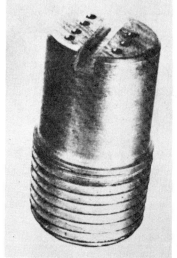
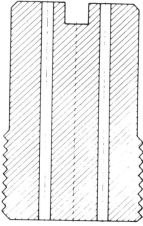

Figure 4-104:
Straight-Through Nozzle for Mist Application.

Since the best distribution of oil mist is obtained with a minimum of back pressure, free venting is recommended. In larger mills, operators prefer carrying the vented mist away from the mill area by sufficiently large hoses attached to the vent connections on the chocks.

The mist nozzles or orifices not only assure the desired distribution of the lubricant, but they also change the oil droplet size so that it will readily condense. When operating at a pressure of 20 inches of water, the conventional, straight-through nozzle (Figure 4-104) will condense about 75 per cent of the oil.

In original installations, the nozzles were located on the face of the chock but, in modern designs, they are located in radially-drilled holes near the bearing (Figure 4-105). This design has the advantage of a single connection at the face of the chock and the condensation of the oil closer to the bearing.

Oil selection is one of the critical aspects of mist lubrication since the stray mist should be at the lowest level and must be nontoxic. The desired characteristics are presented in Table 4-8.

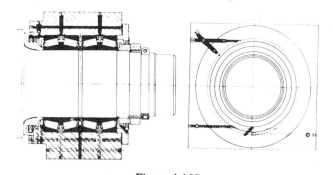

Figure 4-105:
Mist Nozzles Located in Radially-Drilled Holes Near Straight-Bore Backup Roll Bearing.

Table 4-8
EP Mist Oil Characteristics

Oil	High quality, high viscosity index, solvent refined petroleum oil
Viscosity	750 SSU (minimum)
EP	35 lb. Timken (minimum)
Mistability	Good total output
	Minimum stray mist
	No nozzle clogging
Other	Good oxidative and thermal stability
	Nonfoaming
	Good rust prevention in presence of water

In the development of oils specifically formulated for misting, emphasis has been given to the reduction of stray mist. However, a lower oil/air ratio is frequently encountered with such oils. Certain additives and insufficiently refined oils can produce nozzle clogging and to prevent this, oils with precipitation numbers (ASTM D91-61) exceeding 0.1 should be avoided.

MILL ROLLS AND THEIR BEARINGS

Extreme pressure (EP) additives are necessary ingredients in lubricants for heavily loaded, high-speed bearings. They prevent scoring type damage of asperities on both mating roller ends and the cone rib. These surfaces involve sliding motion and require careful consideration in selecting lubricants. There are many EP additives of different chemical types and most are satisfactory for mist applications.

4-35 Sleeve Type ("Morgoil") Bearings

Successfully employed for over 30 years, the Morgoil bearing is a hydrodynamic design which distributes bearing load over a large area with no concentration points. The film of oil on which the bearing operates has immense load carrying capacity and is, in fact, the stiffest load carrying element in the mill. Because of its continuous, unbroken nature, the film eliminates wear caused by metal-to-metal contact. This oil film is constantly maintained by the hydrodynamic action of the rotating sleeve to which a surplus of oil is presented at controlled temperatures.

The Morgoil bearing is uniquely compact. Its essential parts are a rotating sleeve which fits over the tapered roll neck, and a nonrotating bushing mounted in the protective chock. Because these components are so compact and the oil film so thin, chock space is conserved so that the largest possible roll neck can be accommodated. These Morgoil characteristics contribute in a major way to mill stiffness. Furthermore, the compact Morgoil design permits use of a rigid chock while leaving ample room in the backup chock for roll bending devices.

Referring to Figure 4-106, the radial load is carried by the film of oil between the sleeve and the bushing. The thrust assembly receives no radial load. For bearing sizes 21" and larger, a roller thrust assembly is employed. For bearing sizes below 21", a ball thrust unit is used.

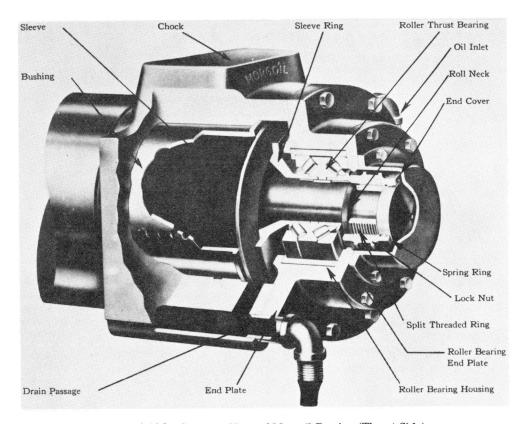

Figure 4-106: Cutaway View of Morgoil Bearing (Thrust Side).

A recent refinement has now speeded up and simplified removal of the entire bearing assembly from the roll, reducing bearing change time by 50 per cent. Key to this feature is the bearing locknut which, when backed off, pulls the bearing components (all linked with a series of engaging flanges) away from the nonlocking roll neck taper. Threaded ring and locknut remain an integral part of the bearing assembly, thus eliminating handling and affording protection for precision parts.

The complete range of sizes and ratings of "Morgoil" bearings is shown in Table 4-9, the dimensions A, E, K, G and Z being as illustrated in Figure 4-107.

Table 4-9

	SERIES						DIMENSIONS COMMON TO ALL SERIES			
	60		72		90					
BRG. SIZE	BEARING RATING	A	BEARING RATING	A	BEARING RATING	A	MIN. E	K	MIN. G	Z
8 1/4″	75,000	3 13/16″	91,700	4 1/4″	....	......	8″	6.212″	9 3/4″	5 9/16″
9 1/2″	91,000	4 1/4″	106,800	4 5/8″	....	......	9 1/2″	6.921″	11 1/4″	6.27″
10″	110,000	4 3/8″	131,000	4 13/16″	....	......	9 3/4″	7.480″	12″	6 3/4″
11″	131,000	4 11/16″	156,200	5 3/16″	194,400	5 15/16″	10 7/8″	8.283″	13″	7.43″
12″	158,000	4 15/16″	190,000	5 1/2″	236,000	6 15/16″	11 5/8″	9.100″	14″	8 1/8″
14″	212,000	5 3/8″	257,000	6 1/16″	318,000	7″	13 3/8″	10.512″	16″	9 3/8″
16″	274,000	5 7/8″	334,000	6 11/16″	414,000	7 3/4″	15 1/8″	12.050″	18″	10 3/4″
18″	352,000	6 1/2″	426,000	7 3/8″	525,000	8 9/16″	17″	13.587″	20″	12 1/8″
21″	475,000	7 7/16″	574,000	8 7/16″	714,000	9 7/8″	20 1/2″	15.912″	23″	14 1/8″
24″	622,000	8 3/4″	754,000	9 15/16″	936,000	11 9/16″	23 1/2″	18.855″	26″	16 1/8″
26″	736,800	9 9/16″	888,300	10 13/16″	1,109,100	12 5/8″	25 1/2″	20.355″	28″	17 5/8″
28″	848,700	10″	1,027,800	11 3/8″	1,281,000	13 5/16″	27 1/2″	21.548″	30″	19″
30″	976,000	10 13/16″	1,169,000	12 3/16″	1,468,000	14 5/16″	29 3/8″	23.457″	32″	20 1/2″
32″	1,104,300	11 3/8″	1,327,500	12 7/8″	1,658,000	15 1/16″	31 1/4″	24.923″	34″	21 7/8″
34″	1,256,700	11 7/8″	1,512,600	13 1/2″	1,879,500	15 13/16″	33 1/4″	26.298″	36″	23 1/4″
36″	1,400,700	12 1/2″	1,692,900	14 1/4″	2,101,500	16 11/16″	35 1/4″	27.844″	38″	24 3/4″
38″	1,570,500	13″	1,891,500	14 13/16″	2,346,000	17 3/8″	37 1/4″	29.219″	40″	26 1/8″
40″	1,732,500	13 9/16″	2,092,500	15 1/2″	2,592,000	18 3/16″	39 1/4″	30.719″	42″	27 5/8″
42″	1,900,500	14 1/4″	2,304,000	16 5/16″	2,850,000	19 1/8″	41 1/4″	32.583″	44″	29 1/8″
44″	2,091,000	14 3/4″	2,526,000	16 7/8″	3,111,000	19 3/4″	43″	33.867″	46″	30 1/2″
46″	2,283,000	15 5/16″	2,763,000	17 13/16″	3,435,000	20 15/16″	45″	35.515″	48″	31 7/8″
48″	2,499,000	15 13/16″	3,015,000	18 1/8″	3,741,000	21 3/8″	46 1/2″	36.924″	50″	33 3/8″
50″	2,700,000	16 7/16″	3,270,000	18 7/8″	4,050,000	22 1/4″	48 1/2″	38.344″	52″	34 3/4″
52″	2,913,000	16 15/16″	3,528,000	19 1/2″	4,377,000	23″	49 1/2″	39.708″	54″	36 1/4″
54″	3,161,000	17 9/16″	3,825,000	20 3/16″	4,750,000	23 7/8″	51″	41.265″	56″	37 5/8″
56″	3,498,000	18 3/8″	4,191,000	21″	5,247,000	25″	53″	42.640″	58″	39″
60″	3,983,000	19″	4,760,000	21 3/4″	5,950,000	26″	56 1/2″	46.549″	60″	43″
64″	4,500,000	20″	5,400,000	23″	6,750,000	27 1/2″	60″	49.549″	66″	46″

4-36 "Morgoil" Bearing Lubrication

The modern "Morgoil" lubrication system delivers a controlled supply of oil to the bearings. When the roll is turning, the sleeve, fitted over the roll neck and secured by a key to prevent slippage, turns within the stationary bushing which is secured by a lockpin to the chock. Bearing load is distributed over the entire load zone and the continuity of the oil film is ensured by delivery of the oil through internal passages within the chock to the journal. Under conditions of hydrodynamic lubrication, satisfactory starting under normal conditions is achieved.

When extremely low starting torques are necessary, hydrostatic lubrication can be added, as illustrated in Figure 4-108. Hydrostatic pressure takes over when the mill speed drops

MILL ROLLS AND THEIR BEARINGS

below a predetermined value, thereby maintaining an oil film in the load zone. Thus, there is no time limit for the operation of the mill at low speed or for when it is stopped. Moreover, the mill, after being stopped, can be restarted without raising the screws.

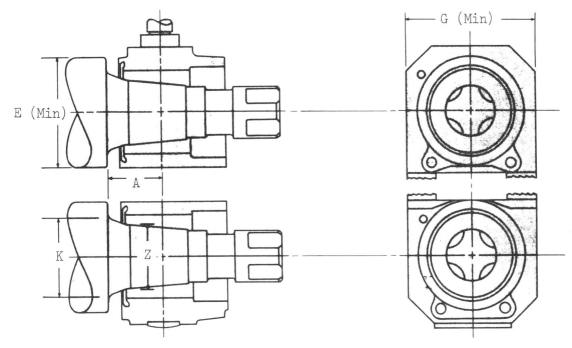

Figure 4-107: Dimensions of "Morgoil" Bearings (For Use With Table 4-9).

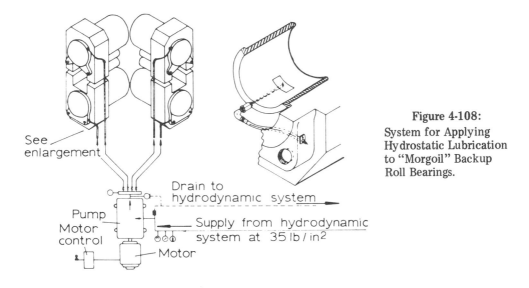

Figure 4-108: System for Applying Hydrostatic Lubrication to "Morgoil" Backup Roll Bearings.

The first major mill designed and built for hydrostatic bearings was a three-stand secondary cold reduction mill commissioned in 1962.[1,2] In this instance, the hydrostatic system operates from 0-150 fpm mill speed and then cuts out, with hydrodynamic lubrication taking over at speeds in excess of 150 fpm.

---

[1] R. W. Barnitz and W. L. Cooper, "J&L's 3-Stand Cold Reducing Mill for Thin Tin Plate", Iron and Steel Engineer Year Book, 1963, pp. 265-273.

[2] A. E. Cichelli, "Backup Roll Bearings in Cold Reducing Mills", "Tribology in Iron and Steel Works", ISI Publication 125, The Iron and Steel Institute, 1970.

The hydrostatic system is designed to be "fail safe". If for any reason it should malfunction, the bearings will perform in their normal hydrodynamic manner and no mill downtime should be incurred.

One advantage of the normal oil film bearing is that the hydrodynamic lubrication within it provides roll neck cooling. For this reason, the roll neck equilibrium temperature is quickly reached and a roll-body-end temperature is maintained which is approximately equal to, or lower than, the temperature of the roll body center. It is therefore claimed that the hydrodynamic lubrication provided in the "Morgoil" bearing tends to stabilize the roll temperature profile and therefore facilitate the attainment of good shape in the rolled product. (See Section 7-4.)

4-37 Roll-Neck Bearing Seals

For satisfactory operation, roll-neck bearings must be protected from the ingress of water, the rolling solution and other contaminants. Seals, usually made from rubber or plastic, are used for this purpose.

With respect to roller bearings, a completely sealed TQIT type is illustrated in Figure 4-109. It features an extension of the cone adjacent to the roll barrel which functions as a sealing surface. As reported by Gibson and McCoy,[λ] the design has the following advantages:

(1) It provides a hardened and ground seat for the seal.

(2) The chock and the bearing are a sealed unit.

(3) The lips of the seal point outwards for maximum effectiveness.

(4) An extension of the cone protects the seals during handling.

(5) The need for an expensive fillet ring is eliminated.

(6) The roll neck rigidity is increased since the bearing centerline is brought closer to the roll body.

(7) Shorter, less expensive rolls may therefore be used.

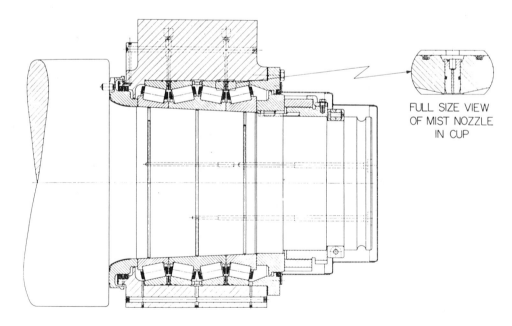

Figure 4-109: Completely Sealed Type TQIT Bearing and Chock Assembly.

[λ] D. G. Gibson and W. E. McCoy, "Tapered Roller Bearings as Applied to High Speed Rolling Mills", Iron and Steel Engineer Year Book, 1970, pp. 597-608.

MILL ROLLS AND THEIR BEARINGS

This extended cone bearing design, using three lip-type seals, has been used successfully on both "wet" and "dry" mills operating at speeds less than 3,000 fpm. At higher mill speeds, wear, abrasion, heat and seal eccentricity constitute more critical problems. However, these problems may be ameliorated by the use of a flinger ring pressed onto the roll shoulder as illustrated in Figure 4-110 for the case of a 56-inch diameter backup roll. This ring effectively eases the burden of the rubbing seals and protects them from the inevitable seal damage that may result from strip breakage and mill wrecks. Furthermore, with the use of the flinger ring, only two lip-type seals are necessary on the main bearing seal, as shown in Figure 4-110.

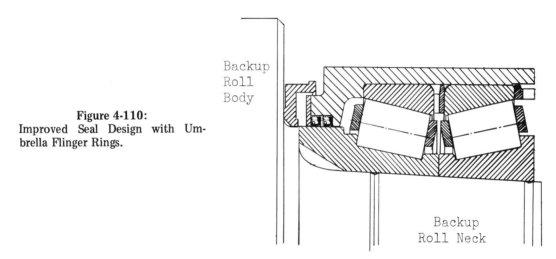

Figure 4-110: Improved Seal Design with Umbrella Flinger Rings.

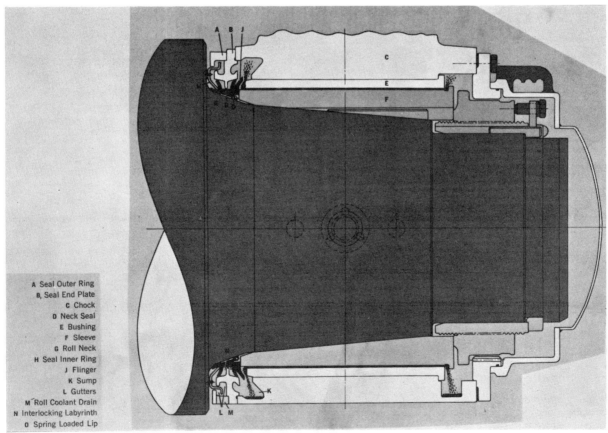

A Seal Outer Ring
B Seal End Plate
C Chock
D Neck Seal
E Bushing
F Sleeve
G Roll Neck
H Seal Inner Ring
J Flinger
K Sump
L Gutters
M Roll Coolant Drain
N Interlocking Labyrinth
O Spring Loaded Lip

Figure 4-111: Cross Section of "Morgoil" Bearing.

In the case of "Morgoil" bearings, the seals are located at each end of the bearing as shown in Figure 4-111 and Figure 4-112. Referring to the latter figure, the flinger portion (1) of the molded synthetic rubber neck seal diverts oil into a deep sump (2) within the chock, from which it passes to the drain through internal passages. A one-piece neck seal (3) with no joint and a static, spring-loaded lip (4) afford positive sealing against leakage from the sleeve and the roll neck interface. At the same time, a nonrubbing series of rotating and stationary gutters (5) provide ample capacity for draining away the rolling solution or coolant. Interlocking, deep, labyrinthine seal legs and the end plate dam (6) also prevent seepage of contaminants into the bearing. The inner seal ring (7) is made of high-strength aluminum alloy for corrosion resistance and low inertia, and the seal end plate (8) features a chrome-plated surface.

For "dry" mills, such as temper mills, rubbing seals are unnecessary. Figure 4-113 illustrates a suitable seal wherein the spring-loaded lip and flinger retain the lubricant in the bearing.

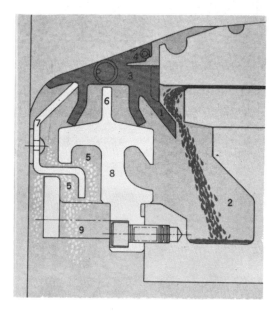

Figure 4-112:
Details of "Morgoil" Neck Seal.

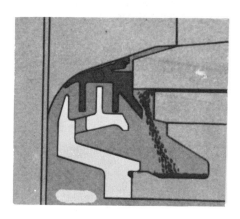

Figure 4-113:
Seal for "Morgoil" Bearing
Used on a "Dry" Mill.

Chapter 5 The Instrumentation and Automatic Control of Cold Rolling Mills

5-1 The Utility of Adequate Instrumentation

Adequate mill instrumentation is desirable for a number of reasons. First, for the protection of the mill equipment itself, as, for example, in the prevention of excessive motor loads or rolling forces on the roll necks and the mill housings; second, to ensure that the strip is undamaged in processing by virtually eliminating strip breaks and cobbles due to improper tensions and rolling forces; third, to provide the desired quality in the rolled product with respect to such properties as gage and shape; fourth, to facilitate the general operation of the mill, particularly during its acceleration and deceleration so that the amount of off-gage strip at the head and tail end of a coil is held to a minimum; fifth, to assist in the proper maintenance of the rolling facility by indicating when changes must be made with respect to various components of the mill such as rolls and lubricants; sixth, to provide, with the assistance of automatic and computer control systems utilized on the mill, reliable gage control and mill set-up; last, to acquire experimental data for use in studies relating to rolling theories, strip behavior, mill design and the like.

Mill instrumentation has improved dramatically in recent years. Early rolling mills were almost devoid of instrumentation and, prior to World War II, even such instrumentation as was supplied related mainly to the electrical drives of the mills. Operation of such mills to produce quality strip was indeed an art. However, during the last two decades, high-speed tandem cold mills have been equipped with many different types of instrumentation and sometimes with sophisticated automatic and computer control systems to affect rapid and accurate changes to the operating conditions of the mill.

This chapter discusses the general characteristics of the more commonly used instruments, data accumulators and control systems. Some of the more common instruments, such as voltmeters and ammeters, are undoubtedly familiar to the reader but have been included for the sake of completeness. Instruments relating to rolled strip shape, will, however, be omitted from this chapter and are described in Chapter 11.

5-2 The Location of Instruments and Controls

For the convenience of mill operators, controls and instruments have always been provided on panels or pendants located close to the mill stands, (see Figure 3-87). In recent years, tandem cold mills have been constructed with pulpits (Figure 3-88) that provide the head roller with an overview of the mill and virtually a complete duplication of the instrumentation and controls on each of the operator's pendants.

Some instrumentation is located remotely from the mill, however. Rooms housing motor generator sets and electrical control equipment are often some short distances from the mill and sometimes below ground level. For the convenience of maintenance personnel, such equipment usually features enough instrumentation to ensure its satisfactory control and operation; this instrumentation being frequently located on cabinet doors, as seen in Figure 5-1. Similarly, most of the components of hydraulic and rolling oil recirculating systems are located in so-called oil basements or cellars. Again, for the convenience of maintenance personnel and mill operators (who have to make up rolling solutions and keep them adjusted with respect to concentration, temperature, cleanliness and emulsion stability), a minimum number of instruments relating to these fluid systems is essential.

Computers, data accumulation systems, and sophisticated automatic control systems used with cold mills are usually housed in clean, airconditioned rooms, such environmental control being necessary to ensure their reliability. However, where coil data is to be fed into and read out of such equipment, certain peripheral equipment, such as data input panels and printout machines, must be located close to mill-side for the convenience of operators, as illustrated by Figure 5-2.

Figure 5-1:
View of Operational Amplifiers
and Thyristor Excitors Used
for Speed Regulation of 52-in.
Tandem Mill.
(Note the Instrumentation
on the Panel Doors)

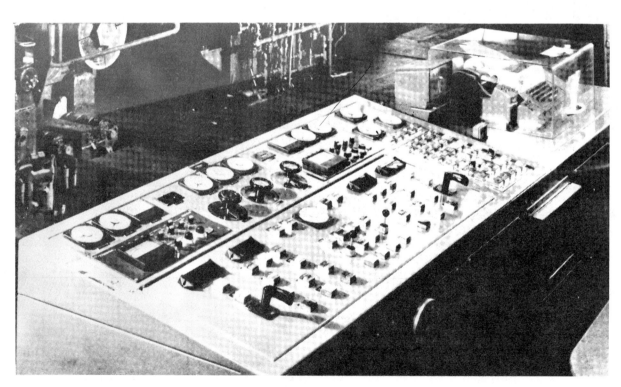

Figure 5-2:
Data Entry and Print-Out Equipment Located on Control Console of a Cold Reduction Mill.

AUTOMATIC CONTROL OF COLD ROLLING MILLS

5-3 Voltmeters and Ammeters

D.C. voltmeters currently used on pendants and controls are usually similar to that shown in Figure 5-3. They generally utilize a rugged, d'Arsonval type movement and require a direct current of the order of 10 mA to provide full scale deflection. For different full-scale voltages, different series resistors are utilized, either internally or externally, with the same basic meter movement.

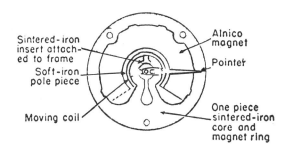

Figure 5-3:
Appearance and Component Parts of a D. C. Voltmeter.

In the D.C. voltmeter shown in Figure 5-3, a one-piece core and ring of sintered iron is die cast around an alnico magnet to form a solid mass with a uniform air gap. The radially-magnetized magnet, with its soft iron pole piece, provides uniform flux distribution and is effectively shielded from stray magnetic fields. The moving shaft rotates in jewel bearings mounted in a one-piece, die cast aluminum frame and provision is made for easy replacement of the moving element assembly.

In recent years, "edgewise" instruments have found increasing use on control panels. Utilizing the same basic movements as the conventional meters, these units are designed for both vertical and horizontal mounting, as shown in Figure 5-4.

Figure 5-4:
Horizontal and Vertical Edgewise Instruments.
(a) A-C ammeter, iron-vane type.
(b) D-C microammeter, d'Arsonval type.

As voltmeters, the foregoing instruments are utilized in reading the outputs of D.C. generators or rectifier units supplying electrical power to the various D.C. motors associated with the mill installation. However, they are frequently used in circuits where their scales are calibrated in units other than volts. For example, a tachometer generator providing a D.C. output will use a D.C. voltmeter with its scale calibrated in terms of revolutions per minute or feet per minute. In other instances, meters may have a center-zero scale, as for example, in connection with a thickness gaging system where they show the deviation, either positive or negative, from a set point value.

A.C. voltmeters are generally similar in appearance to their D.C. counterparts but commonly use a basic movement as shown in Figure 5-5. This movement comprises a fixed vane and two adjustable vanes of carefully determined configuration mounted inside of a field coil and an armature vane on a pivoted shaft. With an A.C. current flowing through the coil, the shaft rotates to a position where magnetically attractive and repulsive forces acting on the armature vane are in balance. Cup shaped shields fitted over the coil ends increase the sensitivity of the instrument and can be rotated to adjust its scale reading. Scales on these voltmeters are, however, usually characterized by nonlinear calibrations. A.C. instruments of this type are generally used to monitor A.C. power line voltages, but they may also find application where the output of an electrical circuit provides an A.C. output signal as, for example, in the case of a Pressductor load cell.

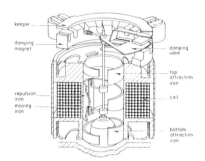

Figure 5-5:
Non-Linear Scale of Iron-Vane A-C Voltmeter and Drawing Illustrating its Construction.

The D.C. and A.C. voltmeters described above are essentially current monitoring indicators and may be simply used as such for currents within the basic range of the instrument. However, currents to be monitored in mill circuits are usually several orders of magnitude greater than the maximum current permissible for the basic movement.

For this reason shunts, as illustrated in Figure 5-6, are generally used in the measurement of large D.C. currents and current transformers, as shown in Figure 5-7, for monitoring A.C. currents. A D.C. voltmeter is connected across a shunt and an A.C. voltmeter connected to the secondary of the current transformer. These arrangements permit the voltmeters, the scales of which are calibrated in terms of current, to be installed at appreciable distances from the conductors carrying the large currents.

Figure 5-6:
Shunts of Various Ranges Used in Monitoring D-C Currents.

Figure 5-7:
Current Transformer (The high current passes through the copper bar and the a-c meter is connected to the terminals on the top of the unit).

D.C. current transformers may also be used for the remote indication of D.C. currents up to about 10,000 amperes. A transformer of this type, as illustrated in Figure 5-8, employs two saturable reactors to control the current in an A.C. circuit in accordance with the magnitude of the D.C. current in a bus magnetically coupled to the two reactors. The strip-wound cores of the two saturable reactors possess essentially rectangular magnetization or B-H curves, and the effective inductance of each reactor is proportional to the permeability of the core material. Both reactors are designed so that the cores are saturated with no A.C. excitation applied and with only a small percentage of rated D.C. current flowing in a bus centered within the cores.

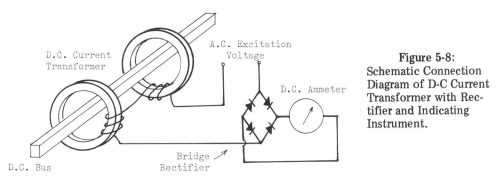

Figure 5-8: Schematic Connection Diagram of D-C Current Transformer with Rectifier and Indicating Instrument.

Because the A.C. coils are connected in series opposition, the instantaneous A.C. polarity of one saturable reactor is always opposite to the polarity of the other when A.C. excitation is applied. The A.C. flux in one core opposes the D.C. flux from the primary bus and tends to desaturate the core, while, in the other, the A.C. flux will aid the D.C. flux and further saturate that core.

The resultant current drawn from the A.C. line has a substantially square wave form and is proportional to the magnitude of the D.C. bus current. Since A.C. instruments are calibrated for sine wave rather than square wave currents, the output current should be rectified and measured with an average reading D.C. ammeter.

It is important to note that the D.C. transformer is not sensitive to the direction of current flow in the primary bus. For most mill applications, however, this does not present a serious drawback.

5-4 Tachometers

Tachometers are used for measuring rotational speeds or, if associated with rolls of known diameter, such as the work rolls, they may monitor linear speeds, such as strip speeds. The

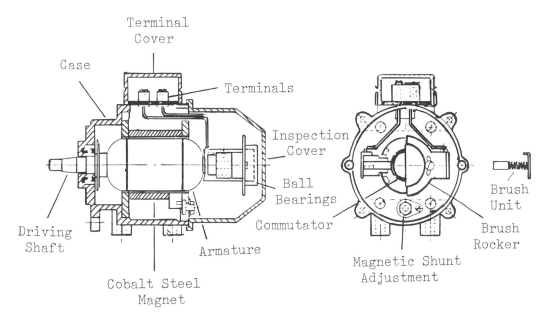

Figure 5-9: Generator for Use with a Voltmeter for Speed Measurements.

most commonly used type of tachometer is a D.C. generator which provides an output potential essentially proportional to rotational speed and a polarity related to the direction of rotation. A typical generator of this type is shown in Figure 5-9 and is intended for use with a D.C. voltmeter, as discussed in Section 5-3.

Where a high order of accuracy is required, digital tachometers are used. These take a variety of forms but all provide a sequence of voltage impulses (the recurrence frequency being directly and accurately related to rotational speed), and have the advantage of not using slip rings. For lower speed applications, magnetic or electromagnetic units may be used. In the case of the former, a wheel with small magnets embedded in its periphery may be used with a reed type switch shown in Figure 5-10. In the case of the latter, however, a detector that senses the proximity of metal may be used in conjunction with a gear wheel or similar rotational unit. For higher frequency operations, optical systems may be used, as illustrated in Figure 5-11.

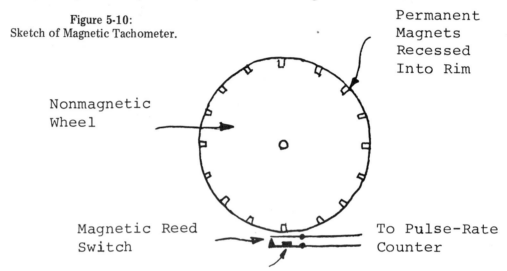

Figure 5-10: Sketch of Magnetic Tachometer.

To measure the recurrence frequency of impulses generated by digital tachometers, electronic counters are used. These are so designed as to admit the pulses to the counter through an electronic "gate" open only for a very accurately controlled period of time, such as a second. The number of pulses entering the counter is then totalled and indicated on a digital type display. On most units, the frequency at which the gage may be opened (the "sampling" rate), is controllable.

5-5 Roll Position or Roll Opening Indicators

Until recently, to provide an indication of the roll positions in a mill stand, it was the practice to mechanically attach one or more large dial indicators to the mill screw drive system. These showed not only the approximate position of the work rolls with respect to each other, but also indicated whether or not the mill screws were moving and in which direction.

Illustrative of the general construction of such indicators is that described for a blooming mill by Tselikov and Smirnov* and shown in Figure 5-12 and Figure 5-13. In this instance, the arms of the roll opening indicator are driven by tapered pinions on one of the screwdowns through gearing. Planetary reducing gear allows the arms to be moved by a 0.15 kw motor independently of the screwdowns. This is essential for adjusting the base reading of the roll opening indicator to make allowance for wear or roll and chock changes. For the provision of an additional remote indicator, a Selsyn transmitter is incorporated into the screwdown system and this unit may be electrically connected to a Selsyn receiver unit driving the remote indicator.

---

* "Rolling Mills" by A. I. Tselikov and V. V. Smirnov (Translated from the Russian by M.H.T. Alford and Edited by W. J. McG. Tegart) Pergamon Press, New York, New York, 1965.

AUTOMATIC CONTROL OF COLD ROLLING MILLS

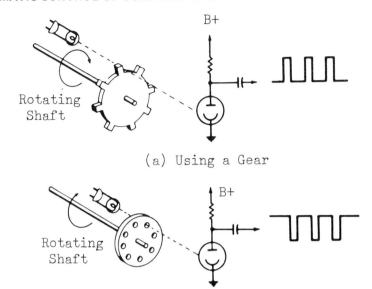

(a) Using a Gear

(b) Using an Opaque Disc with a Number of Openings

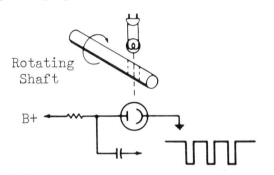

Figure 5-11: Optical Tachometers.

(c) Using a Hole Through the Shaft

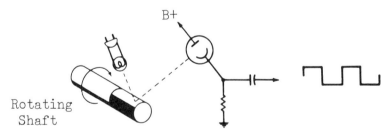

(d) Using a Painted Shaft

In mills of more recent design, however, the large mechanical indicators adjacent to the screws are omitted and sensors, such as Selsyn transmitters, provide the roll position information remotely to consoles, pendants and computers. The electrical signals so provided are used, not only by the operator for the manual control of the mill but also for automatic gage control systems, particularly of the gagemeter type discussed in Section 5-29 and for computers automatically controlling the set-up of the mill, as discussed in Sections 5-36 and 5-37.

In the case of hydraulically loaded mills (which may be devoid of screws), other types of transducers must be used to indicate roll position. An inductive roll-gap transducer has been described by Guetlbauer♦ and is illustrated in Figure 5-14. This unit, in two parts, is located

---

♦ F. G. Guetlbauer, "In-Gap Gaging Improves Strip-Mill Control", Control Engineering, July, 1970, pp. 57-60.

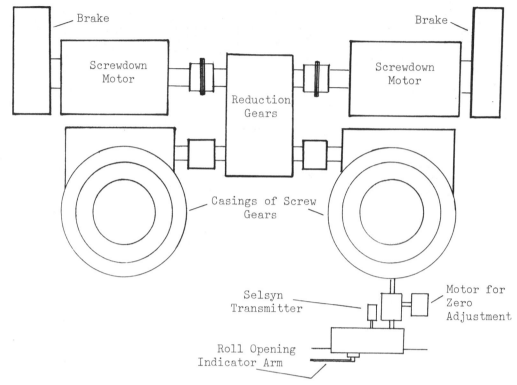

Figure 5-12: Screwdown System Showing Arrangement of Roll Opening Indicator.

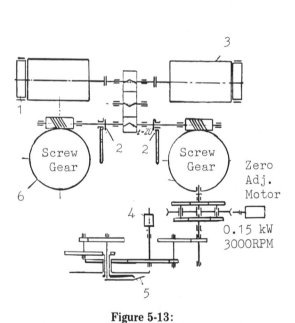

Figure 5-13:
Schematic Diagram of Coupling of Roll Opening Indicator to Screwdown System (1 — Brake; 2 — Connecting Coupling; 3 — Shunt Motor [150 h.p., 490 r.p.m.]; 4 — Selsyn Transmitter; 5 — Roll Opening Indicator Arm; 6 — Screwdown).

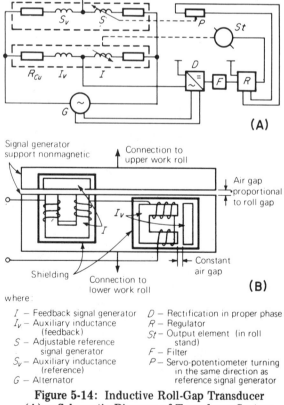

where:
I — Feedback signal generator
I_v — Auxiliary inductance (feedback)
S — Adjustable reference signal generator
S_v — Auxiliary inductance (reference)
G — Alternator
D — Rectification in proper phase
R — Regulator
St — Output element (in roll stand)
F — Filter
P — Servo-potentiometer turning in the same direction as reference signal generator

Figure 5-14: Inductive Roll-Gap Transducer
(A) — Schematic Diagram of Transducer System
(B) — Mechanical Construction of the Transducer.

AUTOMATIC CONTROL OF COLD ROLLING MILLS

between the work rolls using mounting rings as shown in Figure 5-15. Referring to Figure 5-14, the sensing variable inductance (I) with its air gap proportional to the roll gap, is connected in a bridge circuit with a variable reference inductance (S) and two fixed inductances (I_V) and (S_V). All four inductances have the same electrical values and the same geometry. Combining two halves of the bridge system, (I) and (I_V), in one transducer assembly makes the system immune to large variations in coil resistance (represented by R_{cu}) due to temperature changes.

The symmetrical arrangement and the electromagnetic shielding of the transducer assembly reduce variations in inductance and ensure that the relationship between the air gap and the output signal is not disturbed by magnetic interference. Moreover, the operation of the transducer is not affected by the roll coolant, but care must be taken in grinding the rolls that no eccentricity occurs between the roll barrel and the roll shaft face.

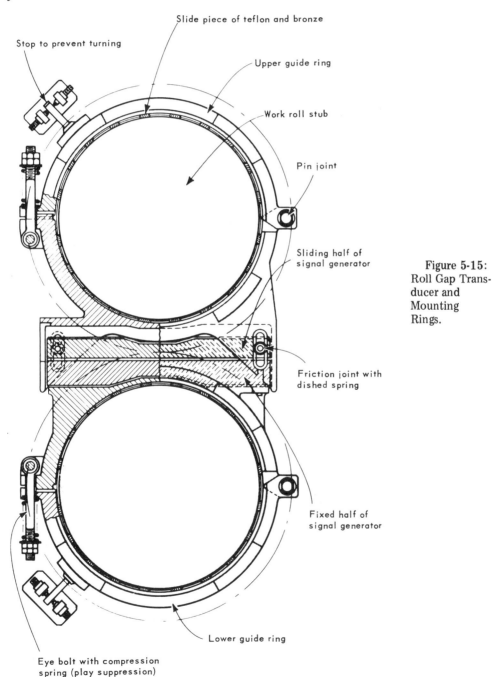

Figure 5-15: Roll Gap Transducer and Mounting Rings.

5-6 Rolling Force Measurement Systems

Until recent years, rugged and reliable transducers for the measurement of rolling force were not available. Such instruments as had been previously installed on mills did not meet with the general approval of operators and were soon either removed from the mills or became unusable through lack of maintenance.

Yet the value of the rolling force is a very important process parameter and operators of mills with reliable rolling force indicators soon learn to depend upon these instruments for proper mill control. For example, in the absence of instrumentation which monitors rolled strip shape, rollers usually find that operating the mill stands with rolling forces within certain ranges constitutes their best assurance in attaining satisfactory shape. Moreover, by avoiding excessive forces, roll wear can be minimized, roll surface spalling and neck breaking can be avoided and bearing life prolonged.

Where hydraulic systems are used instead of screws, as discussed in Section 3-7, the rolling force may be very simply computed from the hydraulic pressure and the areas of the pistons in the cylinders. In fact, a pressure reading indicator, which may be coupled to a pressure sensor in the hydraulic system, may be directly calibrated in terms of rolling force. Related to this method of measuring rolling force is the use of a hydrostatic type load cell placed under the mill screws or located within the roll assembly as shown in Figure 5-16.‡

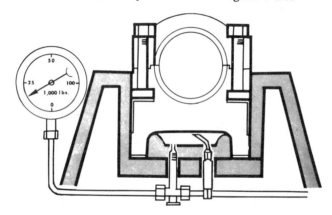

Figure 5-16:
Cross Section of a Hydraulic-Type Load Cell Located within the Roll Assembly.

Force measuring systems for mills with conventional screws generally fall into two categories; those that measure the strain in a stressed component of the mill housing and those that measure the force directly. Of the latter, strain gage load cells and Pressductors find the greatest use. These systems are discussed in greater detail in the next two sections.

Usually, the instruments are electrically connected so as to provide not only the total rolling force but also the difference or the unbalance of the screw forces exerted on the two chocks of each backup roll. This differential screw pressure is useful in aiding operators to level the mill and minimize any tendency to roll product with camber or sweep.

In the use of rolling force measuring systems, however, it must be recognized that some of the rolling force may be shunted around the load cell. For example, load cells placed in proximity to the screws will read a less-than-actual force on the workpiece if binding occurs between the chocks of the upper work or backup roll chocks and the mill housing.

Alternately, the load cells may read higher-than-actual forces on the strip. This may occur when operators use hydraulic jacks, located in various roll chocks, to engage other roll chocks in an effort to control the shape of the rolled strip. In a typical secondary cold mill, for example, the lower backup roll chocks contain two sets of jacks; one set to engage the upper backup roll chocks and the other set to engage the upper work roll chocks. These jacks are installed for the

---

‡ I. G. Orellana, "Application of Load Cells to Strip Mill Operation and Control", Iron and Steel Engineer Year Book, 1968, pp. 25-32.

AUTOMATIC CONTROL OF COLD ROLLING MILLS

purpose of keeping the various rolls separated during the changing of work rolls, and their use during rolling will divert a fraction of the rolling force away from the workpiece.

5-7 Measuring Stresses in the Mill Housing

Although the stresses developed in a mill housing are relatively small, the mill posts are of sufficient height to yield a measurable change in length under load.[◊] One method of doing this utilizes a rod attached at one end of the mill housing, the other end being coupled to a unit that will measure the relative displacement between the end of the rod and the mill housing as illustrated in Figure 5-17. However, such devices are difficult to calibrate and to maintain in good working order. Another method involves the measurement of the strain in the mill screws as shown in Figure 5-18.

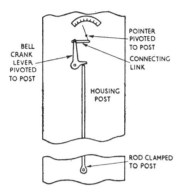

Figure 5-17:
Principle of a Simple Lever Amplifying System for Measuring Mill Housing Strain.

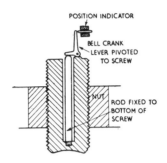

Figure 5-18:
Simple System for Measuring Strain in a Mill Screw.

With the advent of the resistance strain gage about 1940, attempts were made to use this sensor applied directly to mill components. However, to be accurate, measurement of housing strain must be made on the neutral axis of the mill structure. Unfortunately, this axis is difficult to establish correctly and can shift depending on such variables as the direction of rolling and temperature rises in the mill components. Accordingly, this direct utilization of strain gages to mill housings has hitherto been restricted mainly to laboratory or engineering studies. Recently, however, U. S. Steel Corporation developed a reliable rolling load measurement and display (ROLMAX) system which utilizes strain gages mounted on the mill housing in conjunction with electronic circuitry which compensates for temperature effects.

5-8 Load Cells Using Strain Gages

To date, one of the most widely employed types of transducer for the measurement of rolling force has been the load cell utilizing strain gages. Various designs are currently in use but all measure the strains produced in an element, such as a short cylinder compressed under the influence of the rolling load. (See Figure 5-19) Ideally these units should be as compact as possible, particularly where they are to be fitted into existing mills. They may be placed in each mill housing window either between the lowest roll chock and the housing, between the uppermost chock and the screw, or between the mill housing and the screw nut. The strain gages of such load cells are connected in a bridge arrangement similar to that shown in Figure 5-20 with the output indicator calibrated in tons or thousands of pounds.

In recent years, load cells of the "washer" type manufactured by George Kelk, Limited, Ontario, Canada, have found numerous applications in rolling facilities and are conveniently

---

[◊] "The Theory and Practice of Flat Rolling", by C. W. Starling, University of London Press, Ltd., London, England, 1962.

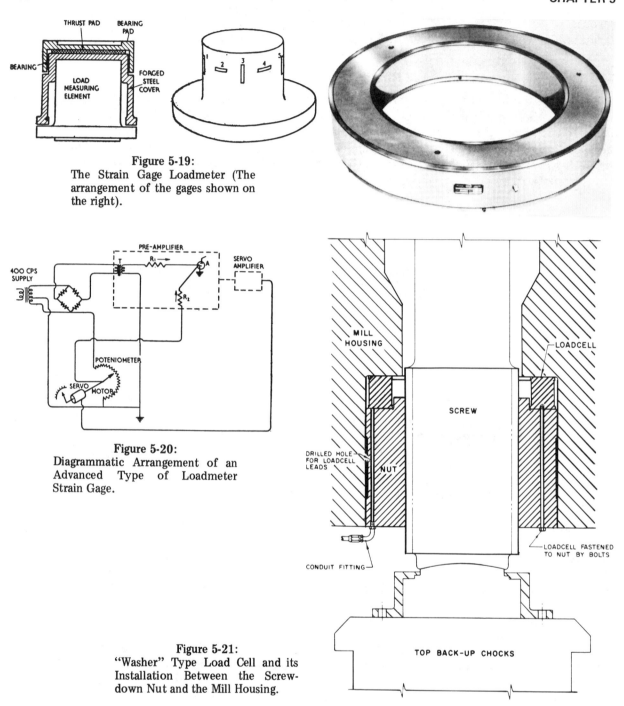

Figure 5-19: The Strain Gage Loadmeter (The arrangement of the gages shown on the right).

Figure 5-20: Diagrammatic Arrangement of an Advanced Type of Loadmeter Strain Gage.

Figure 5-21: "Washer" Type Load Cell and its Installation Between the Screwdown Nut and the Mill Housing.

mounted between the screwdown nut and the mill housing as illustrated in Figure 5-21. This type of transducer utilizes strain gages for the measurement of compressive forces and units are available with load-carrying capacities up to 10 million pounds per cell. The load sensing elements in these units are forged steel rings to which are bonded wire strain bridges.‡

The washer type of load cell has the advantage of complete protection by the mill stand structure and is not affected by the rolling solution. However, the mill nut must be securely held so that no torque is transmitted to the load cell as the mill screws are rotated. In the case of strip mills

---

‡ I. G. Orellana, "Applications of Load Cells to Strip Mill Operation and Control", Iron and Steel Engineer Year Book, 1968, pp. 25-32.

AUTOMATIC CONTROL OF COLD ROLLING MILLS

using electrohydraulic automatic gage control, where the mill nut assemblies are purposely rotated, thrust bearings and suitable nonrotating pressure plates are required between the screwdown nuts and the washer type load cell.

5-9 The Pressductor

The Pressductor, manufactured by ASEA, Vasteras, Sweden, is a transducer which, when mechanically loaded, produces an A.C. output voltage proportional to the applied load. A unit capable of measuring loads up to 1.4 million pounds is shown in Figure 5-22 and its construction in Figure 5-23.

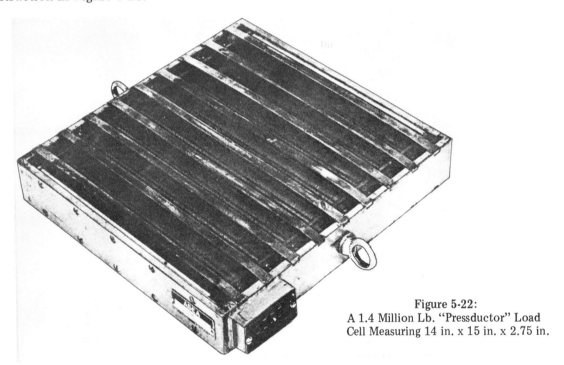

Figure 5-22:
A 1.4 Million Lb. "Pressductor" Load Cell Measuring 14 in. x 15 in. x 2.75 in.

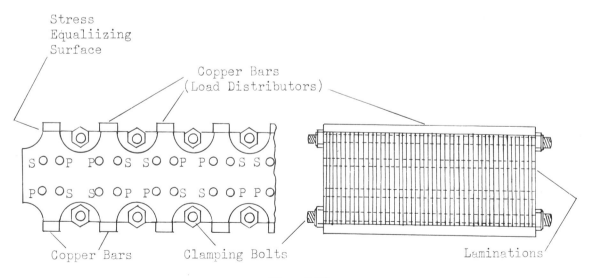

Figure 5-23:
Construction of "Pressductor" Load Cell Shown in Figure 5-22
(P denotes primary winding hole and S denotes secondary winding hole)

The Pressductor consists of a number of stampings of transformer sheet with groups of four symmetrically-punched holes. The laminations are pressed together to form a block and the four channels formed by each set of holes are wound with two crossed windings as indicated in Figure 5-24. The primary winding is fed with an alternating current and the secondary is connected to an indicating instrument. Since the windings are placed at right angles to each other, no voltage is induced in the secondary winding when the Pressductor is unloaded. If the Pressductor is subjected to a compressive force as shown in Figure 5-24, the magnetic permeability of the sheets is reduced in the direction of the applied force. This causes a change in the flux distribution so that the flux now passes partially through the secondary winding, thereby inducing a voltage across the secondary winding.

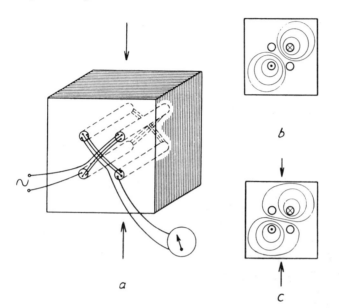

Figure 5-24:
The "Pressductor" Principle for the Measurement of Compressive Force.
a) "Pressductor" Unit.
b) Flux Lines in Unstressed Condition.
c) Flux Lines in Stressed Condition.

In Pressductors suitable for measuring rolling forces, there are a considerable number of laminations, each of which carries a portion of the applied load. Each sheet is provided with extensions in the direction of the force, shaped so that a suitable bearing face is formed and the stresses properly distributed within the transducer.

Because of the impossibility of manufacturing Pressductors with identical characteristics, each unit is "trimmed" to provide the same response characteristics by connecting suitable high stability, wire-wound resistors between the secondary winding and the output terminals. After this trimming procedure, all "Pressductors" for the same rated load and A.C. input voltage have sensitivities and internal impedances within 1 per cent of the aim values.

The block schematic diagram of the circuit utilized with two Pressductors in a rolling mill stand is shown in Figure 5-25. This circuit provides output signals corresponding to both the total rolling force and the unbalance of the rolling force with respect to the two sides of the mill.

These force measuring units have found extensive use in the steel industry, particularly in mills that roll in one direction. For use in reversing mills, however, the circular load cell has been found to be preferable in that it appears to be less affected by the peculiar type of cyclical "rocking" loading forces exhibited by such mills.♦

5-10 Torque Monitoring Devices

For most purposes, the magnitude of the torque in a single shaft can be calculated with sufficient accuracy from the voltage and current readings associated with the drive motor. For this

---

♦ "Instrumentation Confab Covers Rolling Mills", "33" Magazine, November, 1965, p. 59.

AUTOMATIC CONTROL OF COLD ROLLING MILLS

reason, torquemeters have not found extensive use on production mills. However, where pinion stands are used to drive two spindles from a single motor, it has been found that the torques to the two spindles may be quite dissimilar. For this reason, torque measurement in individual spindles is highly desirable both on laboratory and production mills.ᵟ

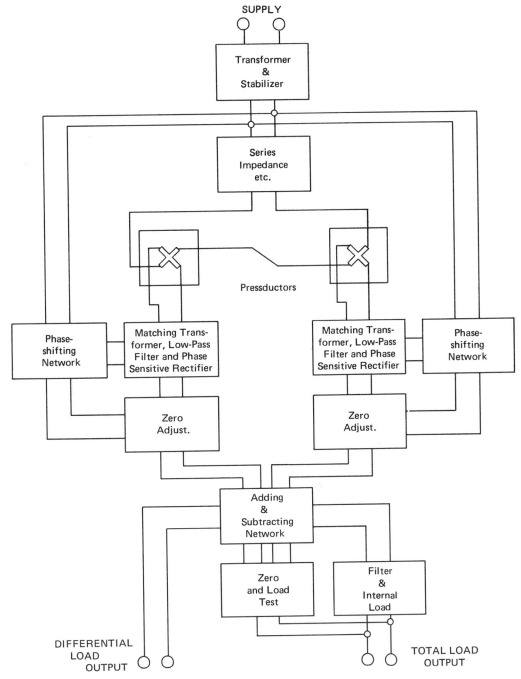

Figure 5-25: Block Schematic Diagram of Circuitry Utilized With a "Pressductor" Load Cell.

Strain gages are usually used for this purpose arranged on the spindle as shown in Figure 5-26 to form part of a bridge circuit as in the case of load cells. Under such circumstances,

---

ᵟ "The Theory and Practice of Flat Rolling", C. W. Starling, University of London Press, Ltd., London, England, 1962.

connections to the strain gages may be made to the rest of the circuitry via slip rings. However, techniques are now available for connecting the whole bridge circuit to a small, frequency-modulated, low-power radio transmitter attached to each spindle, the output signal of each bridge modulating the corresponding transmitter. The signal radiated by each transmitter is then picked up and demodulated by a suitable radio receiver for recording purposes.

Figure 5-26: Mill Drive Spindles with Torque Measuring Cells.

5-11 Monitoring Hydraulic Systems

In modern cold-rolling mills, hydraulic systems are used extensively to adjust or control the positions of many mill components. For example, mill rolls are frequently counterbalanced by hydraulic cylinders (see Section 3-6) and, more recently, hydraulically loaded mills have found increasing use (see Section 3-7). Employing hydraulically actuated mechanisms, coils are moved to and from reels, mandrels are expanded and contracted, belt wrappers are moved into and out of position, guides are correctly located for threading the mill, thickness gages are brought to their operational positions and mill screws may be given very rapid vernier adjustments, to enumerate but a few of the functions performed by such mechanisms.

These mechanisms necessitate the use of one or more hydraulic systems which provide the hydraulic fluid at the pressure and flow rate demanded by the various units. It is the general practice, therefore, to monitor the operation of these systems with respect to pressure, flow rate, fluid level (in the reservoirs) and temperature. The construction and operation of typical instruments used for these purposes are discussed below.

Fluid pressure is generally measured by Bourdon tube type indicating gages, such as that illustrated by Figure 5-27, since they have the advantages of accuracy, ruggedness, reliability, simplicity and low cost. In such a gage, one end of a tube, usually formed into the segment of a circle, is fastened to a socket which connects to the pressure source. The tube is flattened on opposite sides and has an approximately elliptical cross section. When pressure is applied inside the tube, the walls deflect and tend to assume a round cross section. This sets up stresses which increase

AUTOMATIC CONTROL OF COLD ROLLING MILLS

the coiling radius and the free end of the tube moves a small amount which is translated into the rotary motion of an indicating pointer by a linkage and gear arrangement.

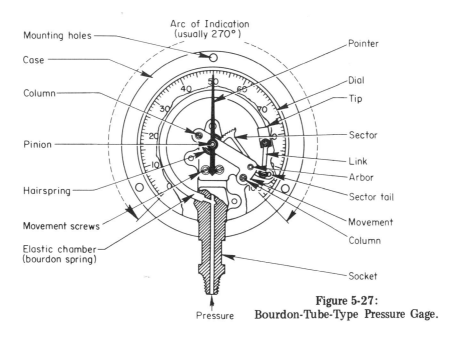

Figure 5-27: Bourdon-Tube-Type Pressure Gage.

For measuring liquid flow rates, a number of instruments, such as Pitot tubes, venturi tubes, and rotameters are available.[⊕] These instruments, illustrated diagrammatically in Figure 5-28, may be read visually or coupled to remote instrumentation. However, where high reliability is required, electromagnetic flow meters offer the advantages of having no moving parts and require no insertions into the flow stream. By producing a strong magnetic field around the stream, these meters develop a voltage proportional to the number of charged particles carried past the instrument by the stream.

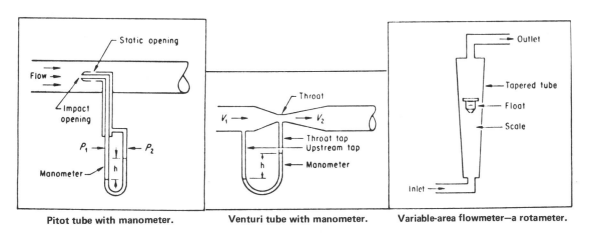

Figure 5-28: Three Types of Flow Meters.

Positive displacement meters offer the ultimate in volumetric accuracy, but require the maximum interference with the flow stream. Consisting of a tight fitting, rotatable gearset or vaned rotor, or a piston/cylinder set, the operating member of this type of meter must be displaced by the

---

[⊕] Machine Design, November 19, 1970, Engineering Department Equipment Reference Issue.

fluid as it passes through the instrument. These meters are, therefore, very similar to the hydraulic motors used in fluid power applications.

The level of a liquid in a reservoir or a tank may be measured by any of a large variety of methods. These include such equipment as sight gages, floats, and pressure tapes. Probes are finding increasing use and these are of several types; namely, capacitative, resistive, thermal and ultrasonic. However, capacitative units, similar to those shown in Figure 5-29, are the most common. In these sensors, the probe functions as one plate in a capacitor, with the fluid serving as the dielectric and the fluid container as the other plate. As the liquid level increases, the capacitance associated with the probe increases and is conveniently measured by a bridge circuit which also provides an output signal for visual display or for recording purposes.

Some capacitative probes need not actually make contact with the fluid. Others are available in lengths over 12 feet, and some are able to withstand pressures up to 600 psi.

Any of a variety of thermometric indicators may be used to measure the temperature of the rolling solution and hydraulic fluids on the mill but the indicator should preferably be in the form of a panel type meter. The circuit for one suitable type is illustrated in Figure 5-30, the sensing element constituting one arm of a Wheatstone bridge, the other arms being fixed resistors with a negligible temperature coefficient. The excitation is provided by the rectification of a constant A.C. voltage and the meters are of the ratio type, permanent magnet, moving coil design. Approximately one volt ampere is required for operation and because of the ratio design, a voltage change of 10 per cent above or below normal will not affect the accuracy materially.

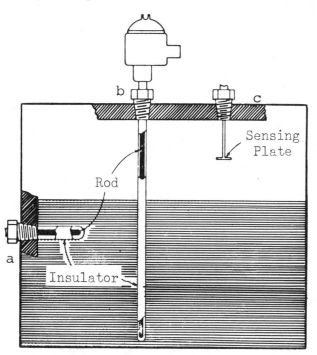

Figure 5-29:
Capacitance-Type Liquid Level Probes.
(a) Single point sensing,
(b) continuous level sensing and
(c) a plate probe.

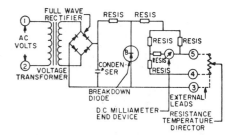

Figure 5-30:
Schematic Diagram of Circuit for Use with Resistance Thermometer.

5-12 Monitoring the Rolling Solution

As will be understood from other sections of this book, the characteristics of the rolling solution vitally affect the whole rolling operation and the quality of the rolled product. Ideally, in the case of tandem mill operations, for example, it would be desirable to know the effective coefficient of friction existing in each roll bite. Except in cases where adequate instrumentation is utilized with digital computers, however, the effectiveness of the lubricant must be determined from indirect measurements, such as the rolling force and the slip. If these measurements indicate poor lubricity, mill operators must then determine the cause of the high friction so as to take proper corrective measures. For example, the temperature of the solution may be outside of the normal operating range thereby preventing the proper plate out of the oil on the strip. Under such circumstances, the continuous measurement of solution temperature would be highly desirable. Alternately, the concentration of the oil in the solution may be too low and additions of oil should be made to the solution. Hence, a continuously reading concentration meter would also be desirable. Instruments for temperature and concentration measurement are available and, in cases of suspected trouble with the solution, assist the operator in deciding whether to "dump" or retain the rolling solution and, in the latter case, what steps he should take to improve it.

Where recirculated rolling solutions are used, frequent additions of oil and water must be made to the solution to compensate for drag out of the lubricant, evaporation of the water and other losses. For this reason, it is desirable to continuously monitor the concentration of the rolling solution. Freeman and his coworkers § at the Homer Research Laboratories of the Bethlehem Steel Corporation have utilized for this purpose a gamma ray liquid density gage installed in the coolant system of a cold reduction mill. The instrument measures the rolling oil content of the coolant with an accuracy of ± 5 percent in the range of zero to 20 per cent oil by volume.

The gage consists of two main components, a measuring element and an electronic unit. The former is attached directly to the feeder pipe carrying coolant to the mill stands and consists of a radioactive caesium 137 source unit and an ionization detector. The gamma rays generated by the source unit pass through the pipe walls and coolant at a rate inversely proportional to the density. The transmitted gamma radiation is received by the ionization detector which produces an electrical signal of corresponding magnitude. The signals are read on a meter or recorded on an auxiliary recorder.

It has been found that coolant flow rate affects the gage reading to some extent, particularly with unstable oil-water emulsions which tend to separate out into their respective oil and water phases under conditions of insufficient agitation. However, this condition only occurs during periods when no rolling is being carried out, the instrument reading returning to its correct value as soon as rolling and the full coolant flow are resumed.

5-13 Thickness Gaging of the Rolled Strip

Since one of the prime requisites concerning the characteristics of the rolled strip is that its thickness must be within proper commercial tolerance, accurate thickness gaging of the strip is of paramount importance. Basically, there are three direct methods of thickness measurement. One method, developed by R. B. Sims, utilizes a knowledge of the mill spring, the rolling force and the mill screw settings to develop a calculated value of the rolled strip thickness.†‡ The oldest method for continuous gaging utilizes flying micrometers with small rollers engaging each side of the strip whereas the most accurate and commonly used method at the present time is based on the use of radiation type gages. The latter gages may be classified into those which incorporate x-ray tubes and those which use radioactive isotopes as radiation sources. These direct approaches to the gaging of strip thickness are discussed in greater detail in the following sections.

In tandem mill operation, it is usually desirable to know the strip thickness at all positions along the pass line of the mill. It is not, however, practical or economical to position radiation type gages at all interstand locations and recourse must therefore be made to an indirect method of gaging. This method is based on (a) an accurate, direct measurement of the final gage at the exit end of the mill, and (b) measurements of the work roll speed at each stand of the mill. At

---

§ J. B. Freeman, R. J. Horst and J. R. Shuman, "Liquid Density Gage for Determining Rolling Oil in Cold Mill Coolant", Iron and Steel Engineer Year Book, 1966, pp. 700-702.

† R. B. Sims, "Measuring Apparatus for Rolling or Drawing Sheet or Strip Material", U. S. Patent 2,726,541.

‡ I. G. Orellana, "Application of Load Cells to Strip Mill Operation and Control", Iron and Steel Engineer Year Book, 1968, pp. 25-32.

each stand, it is assumed that the exit speed of the strip bears a constant relationship to the peripheral speed of the work rolls (constant forward slip) and the interstand strip thickness may therefore be calculated on the fact that, for given rolling conditions, the mass flow of the strip is constant throughout the mill.

Attempts have been made to improve the accuracy of this method by the direct measurement of interstand strip speed by the use of tensiometer or other rolls coupled to tachometers. Unfortunately, such attempts have not been successful, primarily due to the presence of rolling oil on the strip and the rolls, and the inevitable slippage that occurs between them.

5-14 Sims' Method of Gaging

The variation of mill spring (or elongation of the mill housing) with rolling force for a typical 4-high mill is illustrated by the curve in Figure 5-31. It is seen to be nonlinear for the range of small rolling forces (due to the elastic deformation of the rolls and the bearings), but the major position of the curve is linear, its slope corresponding to the mill modulus M. If the mill is to be operated in the linear region of the curve, the mill spring is given by $S_O + F/M$ where S_O is the intercept of the extrapolated linear portion of the curve on the mill spring axis, and F is the total rolling force.

Assuming that the rolls have been separated by a distance S (the initial roll gap), then the roll gap under operating conditions is given by $S + S_O + F/M$. (See Figure 5-32.) Sims equates this roll gap with the rolled strip thickness h so that

$$h = S + S_O + F/M$$

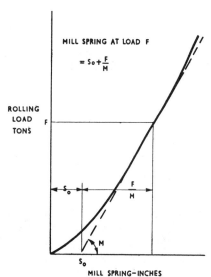

Figure 5-31:
Curve Showing How Mill Spring is Calculated from the Mill Modulus.

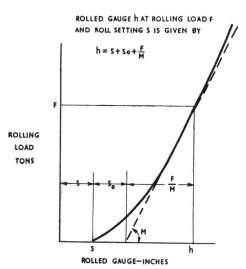

Figure 5-32:
Diagrammatic Curve Showing How Gage is Calculated from the Mill Modulus.

Although this method, also known as the gaugemeter method, appears simple, Starling[◊] points out three factors which influence its accuracy. One is the curvature of the mill spring force curve, although this will contribute relatively small inaccuracies if the mill is usually operated within a restricted range of rolling force. A second is the increase in roll diameter and other mill dimensions with increasing temperature and the fluctuations of roll and housing temperatures during actual rolling operations. Thirdly, eccentricity between the backup roll neck and the barrel and any ovality of the rolls causes cyclical variations of the roll gap even with a constant rolling

---

◊ "The Theory and Practice of Flat Rolling", by C. W. Starling, University of London Press, London, England, 1962.

AUTOMATIC CONTROL OF COLD ROLLING MILLS

force. In spite of these difficulties, automatic gage control systems based on this method have been built and are used to a limited extent on cold mills.

A simplified schematic diagram of the Gaugemeter circuit is shown in Figure 5-33. Once the elastic constant M has been determined, a signal proportional to F/M may be obtained from the load cells (4) placed in the mill housing. A voltage proportional to the value of S_o (the passive roll gap) is obtained from the slide wire (2) attached to the mill screws.↓±° Both the loadmeter (4) and the slidewire (2) may be supplied with D.C. potentials as shown and a potential proportional to the desired gage (h) may be established by the slider of the potentiometer (3). Thus the output of the D.C. amplifier (5), as read by the meter (6), is proportional to the deviation from aim gage.

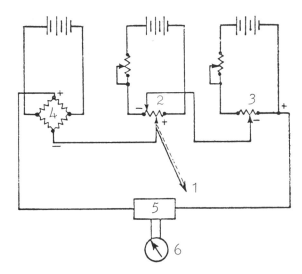

Figure 5-33:
Simplified Diagram of Gaugemeter Circuit.
(1 = connection to mill screw;
2 = slidewire providing e.m.f. proportional to S_o;
3 = slidewire providing e.m.f. proportional to h;
4 = loadmeter;
5 = D.C. amplifier;
6 = centre-zero meter indicating $\triangle h$)

5-15 Continuous Contact Gages

The earliest type of gage used for the on-line measurement of strip thickness was the flying micrometer, examples of which are the "Magnetic Continuous Gage" shown in Figure 5-34 and the "Electrolimit Continuous Gage" depicted in Figure 5-35, both instruments being manufactured by Pratt and Whitney. The latter instrument, which uses a combination of mechanical metering and electrical amplification, is set by means of a counter to the desired thickness of the strip to be checked, and the deviation of gage from the aim value is read out on the indicating meter. The contact gage stands about 17 inches high and projects about 10 inches from its mounting on the mill stand. Though normally installed on a constant pass line, the gage has the ability to float on the strip and, with special arm attachments, it can tolerate pass line variations up to 5-1/2 inches. The gage normally contacts the strip within 2 inches from the edge but measurements at a greater distance from the edge may be made with a special throat.

The higher speed and wider cold mills have relegated this type of gage to use as a backup gage on newer mills. The economy of this type of instrument has always made it attractive, however, especially for use as a standby unit.

The major weakness of a continuous contact gage is its exposure to cobbles and wrecks.■ Because of the necessity of contact, any malfunction in the process line, such as strip

---

↓ R. B. Sims and P. R. A. Briggs, "Control of Strip Thickness in Hot and Cold Rolling by Automatic Screwdown", Sheet Metal Industries, 1954, Vol. 31, p. 181.

± R. B. Sims, "Gaugemeter for Strip Mills", Engineering, 1953, Vol. 175, p. 33.

° W. C. F. Hessenberg and R. B. Sims, "Principles of Continuous Gage Control in Sheet and Strip Rolling", Proc. Inst. Mech. Eng., 1952, 166, p. 75.

■ "Contact Thickness Gaging: Back-up for Non-Contact Units", "33" Magazine, May, 1967, p. 77.

breakage or cobbling, can damage the gage. However, by fabricating the arms supporting the gage of cast aluminum which fracture under impact, the gage may be prevented from being twisted out of alignment.

Figure 5-34: Components of the Magnetic Continuous Gage (Pratt and Whitney).

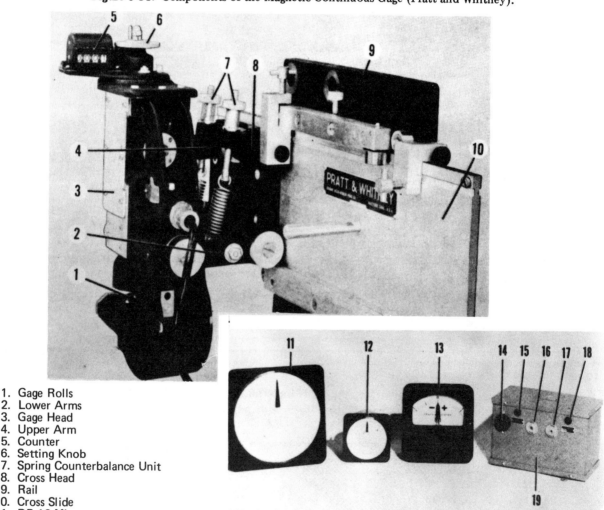

1. Gage Rolls
2. Lower Arms
3. Gage Head
4. Upper Arm
5. Counter
6. Setting Knob
7. Spring Counterbalance Unit
8. Cross Head
9. Rail
10. Cross Slide
11. DB-16 Microammeter
12. DB-18 Microammeter
13. DD-15U Microammeter
14. Gage Head Receptacle
15. Meter Cable Hole
16. Fine Magnification Adjustment
17. Coarse Magnification Adjustment
18. Power Cable Hole
19. Power Unit

Figure 5-35: Electrolimit — Continuous Gage (Pratt and Whitney).

AUTOMATIC CONTROL OF COLD ROLLING MILLS

5-16 X-Ray Thickness Gages

Because of the desirability of using noncontacting gages, the flying micrometer has given way to the use of radiation type gages, particularly on high speed mills. Both x-ray and radio-isotope gages are employed but since, in general, radioactive isotopes emit very penetrating radiation, x-ray tubes are frequently used to generate the radiation. Although originally somewhat erratic in performance, in recent years they have been developed and engineered into quite reliable instruments.+

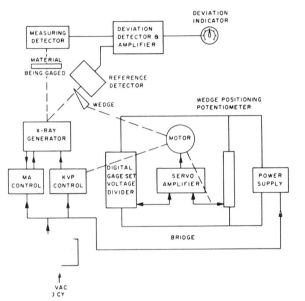

Figure 5-36: G. E. Raymike 600 Thickness Gaging System.

pe of x-ray gage (the Raymike 600 unit), manufactured by General Electric, is, in in one, utilizing one source and two ion-chamber detector heads (Figure 5-36). ed through a finely-calibrated wedge and the rays enter a detection chamber. Rays d through a strip in the jaw of the gage into another detection chamber. Any ignals developed by the two ion chambers is read out as a deviation from the set me the strip has a different aim thickness, the wedge is moved into a new position by ol.

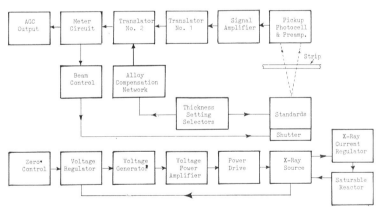

Figure 5-37: Daystrom Gaging System.

Another x-ray gage (the Xactray, manufactured by Daystrom) uses a standards magazine to guarantee its accuracy, (see Figure 5-37). On setting the gage, the operator automatically inserts a calibrated sample of the desired thickness into the beam and zeroes his gage deviation indicator. The

---

+W. R. Baarck, "The Generation and Behavior of X-Rays in Thickness Measurements", Iron and Steel Engineer Year Book, 1963, pp. 868-870.

calibrated sample consists of a number of strips of thicknesses based on a binary system, the exact thickness being chosen by the selection of the appropriate strips. The detector of this unit is made up of a crystal (which converts the x-rays to light waves), and a photomultiplier tube.

Yet another unit (the Sheffield Measuray), shown in Figure 5-38, also converts x-radiation into light energy and uses a higher frequency (360 cps) x-ray power source, thereby achieving a faster rate of response and a lower noise level.

In the use of x-ray gages, high-carbon sheets introduce gaging errors in the range of 0.2 per cent with the deviation indicator reading light. Alloys and stainless steel introduce errors of far greater magnitude depending upon their composition.

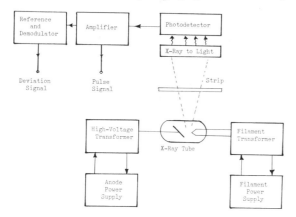

Figure 5-38:
Sheffield Measuray Gaging System.

Figure 5-39:
Accuray Gage in Use on Tandem Cold Reduction Mill.

5-17 Radio Isotope Thickness Gages

Both beta-ray and gamma-ray sources are used for thickness gaging and have the significant advantage over x-ray type gages in that the half life of the radioactive source is generally considerably longer than the useful life of x-ray tubes (six months to five years). In addition, the accuracy of beta-ray gages is far less dependent on the composition of the steel being monitored than is the case with x-ray gages. However, these advantages are to some degree offset by the fact that nuclear gages are generally more expensive than the corresponding x-ray units.[✠]

Beta-ray gages utilize as sources radioactive elements such as strontium-90 (half life, 25 years) or ruthenium-106 (half life, approximately one year). The maximum thickness for this type of gage, however, is limited to 0.8 mm. (0.0315 inches) using strontium-90 and 1.2 mm. (0.0472 inches) using ruthenium-106 when monitoring steel strip.

The "Accuray" (produced by Industrial Nucleonics), shown in Figure 5-39, utilizes different sources for different strip thicknesses and, once the gage is calibrated, compositional changes in the strip have a negligible effect on the accuracy of the gage.

Because of the variety of gaging applications, there are no standard radioisotope source-detector geometries.[†] They are, in fact, custom designed for each application, but experience with many installations has led to the development of several basic geometries. These are composed of standard components, which are specifically arranged for any given application. One geometry, designated type TLK, is suitable for foil thicknesses in the range of 0.0001 to 0.005 inches and has been used for monitoring strip thicknesses on Sendzimir mills where dynamic gage

---

[✠] T. Mizukoshi, S. Umeda and M. Shimizu, "Gamma Ray Thickness Gage for Strip Processing Lines", Iron and Steel Engineer Year Book, 1968, pp. 33-37.

[†] J. L. Griffith, "Progress Report — Radio-Isotope Thickness Measurement Systems and Techniques", Iron and Steel Engineer Year Book, 1967, pp. 153-165.

AUTOMATIC CONTROL OF COLD ROLLING MILLS

control accuracies of better than ± 5 microinches are being achieved. This geometry uses a rather low energy beta emitter, krypton-85, as the source. For greater thickness ranging from 0.001 to 0.030 inches, the higher energy strontium-90 isotope is used in TM-type geometries with source strengths of both 25 and 100 millicuries being available.

A schematic diagram of the basic measuring circuit is illustrated in Figure 5-40. Since the current output from the ionization chamber is in the order of micro-microamperes, a detector load resistor of extremely high value (50 to 500 megohms) is required to develop a usable signal amplitude. A preamplifier feeds this signal to a servosystem whose indicator shows the change in the thickness which caused the original change in the electrometer input signal. In the standard system, the indicator is the pen of a strip chart recorder which provides a permanent analog record of the strip thickness variation.

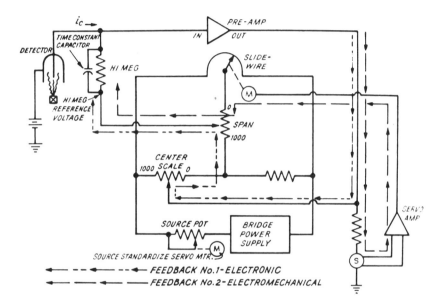

Figure 5-40: Basic Measuring Circuit of Radio-isotope Thickness Measuring System.

As in the case of beta-rays, gamma-rays are absorbed and attenuated as they pass through the material to be measured. By measuring the residual intensity of the attenuated radiation, the thickness of the strip being gaged can be determined. When the energy of the gamma radiation is great (as with cobalt-60), the mass absorption coefficient is independent of the atomic number of the material. However, when the energy is small, more of the gamma radiation is absorbed by the heavier elements than the lighter elements. This can be seen in Figure 5-41 where the mass absorption coefficient is shown plotted against atomic number for gamma-rays of different energies.

In recent years, a gage has been developed which uses the soft gamma-ray emission of the radioactive isotope-241 of the element americium as its source. This element is a metal (specific gravity, 10 to 11) of the ultra-uranium group and is produced by the bombardment of uranium with high energy helium nuclei. The 241 isotope's half life is 458 years, and it radiates gamma rays of 60 kev. energy.

The absorption curves of this radiation with respect to copper, aluminum and iron are shown in Figure 5-42. Note that the upper limit of measurement is approximately 4.5 mm. (0.177 inches) for iron and 2.5 mm. (0.098 inches) for copper. There is a small difference in the absorption curve caused by a difference in composition (as, for example, in the case of stainless steels) and, if a heavy metal, such as molybdenum, is contained in the strip, the absorption becomes greater.

The schematic diagram of the americium-241 thickness gage is shown in Figure 5-43. It is to be noted that the gamma-rays which penetrate the material to be measured are converted to an electric current in the ionization chamber and that this current is a function of the thickness of the material. During this process, by feeding the backing voltage to the feedback circuit of the amplifier, the gage becomes a deviation measuring device capable of very fine measurements. Correction of source attenuation can easily be accomplished by adjustment of the amplifier gain and need be made only once every year.

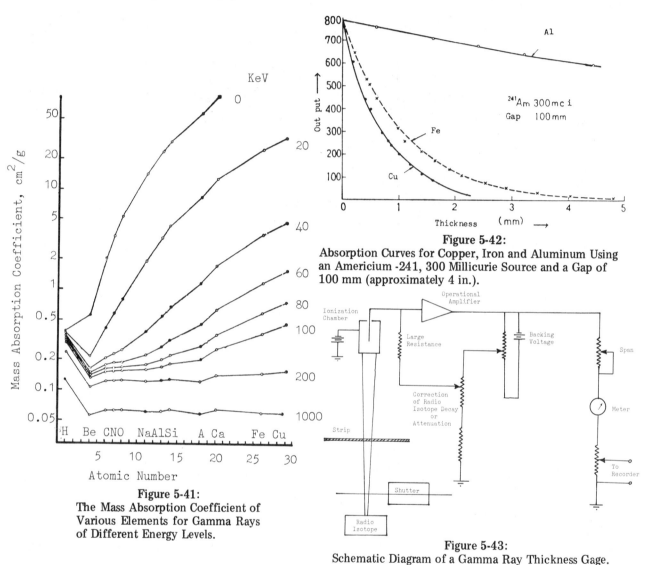

Figure 5-41:
The Mass Absorption Coefficient of Various Elements for Gamma Rays of Different Energy Levels.

Figure 5-42:
Absorption Curves for Copper, Iron and Aluminum Using an Americium-241, 300 Millicurie Source and a Gap of 100 mm (approximately 4 in.).

Figure 5-43:
Schematic Diagram of a Gamma Ray Thickness Gage.

5-18 Extensometers

In some rolling operations, as in temper rolling, it is desirable to continuously monitor the small extension or reduction (usually about 1 per cent) throughout the rolling operation, this extension being more important than the final gage. For this purpose a number of instruments, known as extensometers, have been developed. The most accurate of such units utilize pulse type tachometers attached to tension bridles or other rolls on the entry and exit sides of the mill so as to provide pulse trains with recurrence frequencies proportional to the speed of the strip before and after reduction. The pulse trains are fed to digital counters and electronic circuits which periodically measure the ratio of the two pulse recurrence frequencies to a high degree of accuracy and visually display the ratio in terms of the percentage extension of the strip. A simplified diagram illustrating the installation of an extensometer on a temper mill is presented in Figure 5-44, and the use of this type of monitoring equipment for the automatic control of extension is discussed in Section 5-31.

AUTOMATIC CONTROL OF COLD ROLLING MILLS

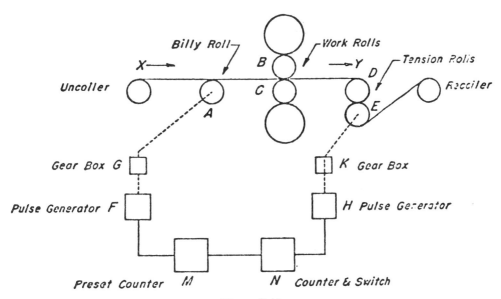

Figure 5-44:
Diagram Illustrating the Application of an Extensometer to a Single-Stand Temper Mill.

Instead of using billy rolls, idler rolls or bridle rolls for the measurement of strip speed, Baltz ⌗ proposes the use of tachometers attached to the entry and exit reels of a reversing mill as illustrated in Figure 5-45 and 5-46. In Figure 5-45, tachometers and diameter sensing rollers provide a product signal proportional to the strip speed for each reel, and the reduction is computed by ratioing these signals. In Figure 5-46, the contacting rollers of Figure 5-45 have been replaced by light sources and photocells to provide a measurement of coil diameter.

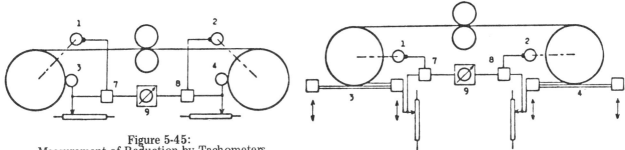

Figure 5-45:
Measurement of Reduction by Tachometers and Diameter Sensing Rollers.
(1, 2 — Tachometers;
3, 4 — Diameter Sensing Rollers;
7, 8 — Signal Multipliers;
9 — Ratiometer).

Figure 5-46:
Measurement of Reduction by Tachometer and Electronic Diameter Scanning.
(1, 2 — Tachometers;   7, 8 — Signal Multipliers;
3, 4 — Photosensing Units;   9 — Ratio Meter).

A more modern type of extensometer gage measures the increase in distance between magnetic patterns induced in the strip. Output signals from an alternating current generator driven by the mill rolls are fed to two "printing" coils, one on each side of the mill. The coils are displaced relative to each other so that the magnetic patterns appear side by side on the strip. Two pickup coils detect the patterns and record how much they are out of phase. The position of the ingoing

---

⌗ W. E. Baltz, "Measurement of Reduction in Cold Rolling Mills", Blech, 1968, 15, February, pp. 80-82 (In German).
(Also available in English as BISI 6837, a translation published by the British Iron and Steel Industry Translation Service, May, 1967).

5-19 Tensiometers

Knowledge of the tension in the strip at different positions in the mill is very desirable if strip breaks and cobbles are to be avoided.[◆] Loss of tension may cause a looping of the strip with the attendant danger of cobbles. On the other hand, excessive tensions result in strip tears and breaks which are equally undesirable from a production viewpoint. Alternatively, high tensions may result in gross slippage of the strip in the mill and undesirable surface conditions on the rolled strip.

Although in some instances, as in the case of tension bridles, strip tensions may be computed with a reasonable degree of accuracy from drive motor currents, instrumentation specifically designed to measure strip tension is now generally used. Instruments for this purpose are designated tensiometers and they basically measure the force exerted on a roll by a strip partially wrapped around it, the angle of wrap being accurately controlled. The force may be measured by any type of load cell such as a strain gage type or a Pressductor.[+][♦]

In installations where the degree of wrap is controlled by other components, such as the work rolls and tension bridle rolls, only a single roll tensiometer need be used. Referring to Figure 5-47, it will be seen that the force F_V exerted on the load cells beneath the bearings of the rolls is given by

$$F_V = S \sin \alpha + S \sin \beta + F_t$$

where S is the strip tension, α and β are the angles the strip form with the horizontal and F_t is the weight of the roll and its bearings. A single roll tensiometer for use with a fixed pass-line is shown in Figure 5-48. Where only one angle would be fixed in the pass-line, however, a second roll must be utilized to fix the angle of the strip on the other side of the tensiometer roll as shown in Figure

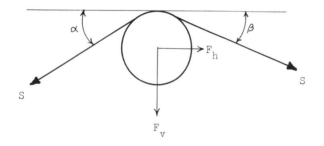

Figure 5-47:
Sketch Illustrating the Strip Pass-Line Over a Tensiometer Roll.

$$F_{vmax} = S_{max} \sin \alpha + S_{max} \sin \beta + F_t$$

where
 F_{vmax} = Force component parallel to cell axis (Tensiometer rating)
 S_{max} = Maximum working strip tension
 α and β = Deflection angles
 F_t = Weight component of roll and shaft

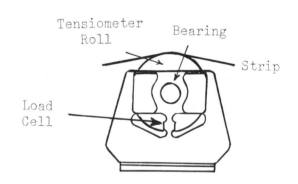

Figure 5-48:
Single-Roll Tensiometer.

[□] J. D. Heys, "Control of the Cold Rolling Process", Sheet Metal Industries, April, 1959, pp. 281-299.
[◆] V. O. Sorokin, "Design and Operation of Tensiometers", Iron and Steel Engineer Year Book, 1961, pp. 458-462.
[+] E. Angeid, "Strip Stiffness as the Deciding Factor in the Choice Between Different Tensiometer Principles", Iron and Steel Engineer Year Book, 1964, pp. 281-291.
[♦] K. A. Petraske and R. M. Sills, "Developments in Drive Systems and Gage Control for Reversing Cold Mills", Iron and Steel Engineer Year Book, 1961, pp. 991-997.

AUTOMATIC CONTROL OF COLD ROLLING MILLS

5-49. Where neither angle would be accurately fixed by the mill components, a three-roll unit must be used as illustrated in Figure 5-50. Where two ranges of tensions are to be measured, a dual-range unit, such as that manufactured by ASEA (see Figure 5-51), may be employed.

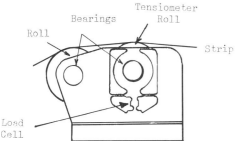

Figure 5-49: Two-Roll Tensiometer.

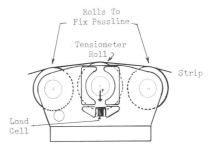

Figure 5-50: Three-Roll Tensiometer.

Since tensiometers utilize two load sensors, one at each bearing of the roll, it is possible to provide a differential output signal as well as one corresponding to the total tension. The differential signal is used by many mill operators as a rather crude method of monitoring strip shape (see Section 11-17).

5-20 Measuring Slip in the Roll Bite

"Slip" or "relative slip" is defined as the excess speed of the strip on exit from the mill over that of the work rolls, expressed either as a ratio or a percentage. Thus, if the exit speed of the strip is V_S and the peripheral speed of the rolls, V_R, the slip, n, as a percentage, is given by

$$n = \left(\frac{V_S - V_R}{V_R}\right) \times 100\%$$

The measurement of this parameter utilizes a circuit arrangement very similar to that used for measuring extension. In this case, however, one pulse tachometer is attached to a drive shaft associated with the work rolls of the mill and another to the shaft of a roll on the exit side of the mill, the peripheral speed of which is equal to the speed of the strip. The ratio of the pulse recurrence frequencies is measured by digital counters and gating circuits which are adjusted so that a ratio of unity is indicated when the peripheral speed of the work roll matches that of the other roll. Because of work roll changes, however, and the use of rolls of different diameters, readjustment of such a slip indicator must be made after every roll change.

Because of the difficulties associated with the accurate measurement of strip speed in cold mills (except temper mills), operational mills seldom carry instrumentation to measure slip.

5-21 Rolled Coil Measurement Systems

In cold rolled sheet manufacture, temper rolling is often the final processing step. Accordingly, coils may be shipped directly from the temper mill to the customer and, therefore, must be of specified size. A coil diameter measuring system (Figure 5-52) accurately monitors the size of the coil being wound on the takeup mandrel.[△] This enables the operator to stop the mill when a coil of the desired size has been rolled.

The measurement system operates from the outputs of two digital tachometers, one associated with a delivery bridle roll and the other with the coiling mandrel. These outputs enable the circumference of each wrap and, hence, the diameter of the coil to be computed at any instant.

[△] D. E. Rea and F. G. Johnson, "Cold Rolled Sheet Temper Mills", Iron and Steel Engineer Year Book, 1966, pp. 347-353.

The system determines the diameter for each revolution of the reel and its value is independent of all previous measurement. Automatic stopping of the mill when the prescribed coil diameter has been reached may also be provided.

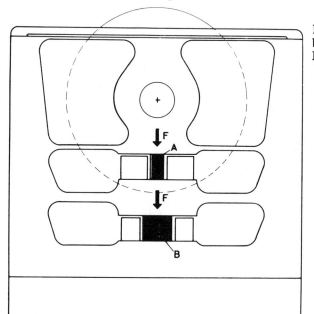

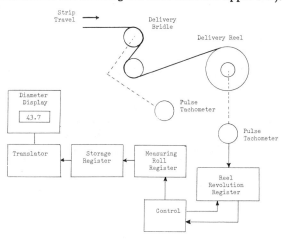

Figure 5-51:
Dual-Range Tensiometer. (At heavier loads, steel blocks [A] transfer load to the lower, heavier, Pressductor cell relieving the more sensitive upper cell).

Figure 5-52: Coil Diameter Measuring System.

In recent years, buyers of steel sheet products have shown a greater interest in purchasing coiled, rolled steel on the basis of area rather than weight.☐ Since the strip width is accurately specified and known, the length of the coil must be measured with precision. A number of systems for length measurement have been developed, principally for use on sheet temper mills, but the one shown in Figure 5-53 is perhaps the simplest. In this system, a 1 ft. circumference, rubber-coated roll makes direct contact with the strip or the surface of an idler roll. This roll and its associated pulse tachometer generates a pulse for each foot of steel processed. The accumulator is reset to zero at the start of each coil and it continually displays the length of coil rolled.

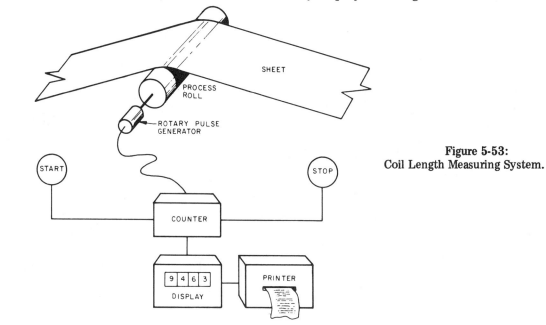

Figure 5-53:
Coil Length Measuring System.

☐ "Taking Guesswork out of Steel Measuring", Iron Age, June 4, 1970, p. 77.

AUTOMATIC CONTROL OF COLD ROLLING MILLS 217

The accuracy of the system is determined by the linear footage wheel. This is ground to an accuracy of $\pm$ 0.005 per cent but, allowing for wear or slippage, an accuracy of 0.1 per cent is generally attainable.

Boos and Achey[○] have recently described a unique sheet length measuring method based on computing the length of every wrap of sheet in the coil. By subjecting the individual wrap length measurements to statistical analysis, this method corrects for slippage between the idler or process roll and the sheet product passing over it.

5-22 Recording Mill Data

In both commercial and experimental rolling operations, data must usually be recorded. For the former, such records are necessary for industrial engineering, production scheduling, quality control, accounting and other purposes. For the latter, the information recorded is usually of a technical nature and is collected for such purposes as the verification of rolling theories, evaluating rolling lubricants, assessing the "rollability" of various steel compositions and to obtain criteria for mill design.

The earliest method of recording such data was by handwritten entries into a log book or by notations made on production tickets and the information so noted generally related to the products rolled, such as the thickness and width of the product and when it was rolled. As mills became better instrumented, it was not unusual for operators to make observations regarding the mill operation, such as the rolling force required and the speed of mill operation.

The advent of continuously recording meters (or chart recorders, as they are generally termed), which provide an ink trace on a roll of paper, simplified the data recording process to a certain extent. Such recorders could be attached to virtually any sensor providing an electrical output, but in commercial rolling operations, they found perhaps the greatest use in continuously recording the gage of the rolled strip, particularly where the strip thickness was monitored by radiation type gages. In the laboratory, however, such recorders, especially the multi-channel type, were used to continuously monitor a number of rolling parameters such as roll force, strip tensions, spindle torques and strip speeds.

Continuously recording meters are relatively slow-speed devices and, where transient signals are to be monitored, oscilloscopes or other instruments with faster response characteristics must be used. Yet because of their continuing use on production mills, recorders are discussed in greater detail in Section 5-23.

The outputs of most mill sensors, being in continuous or analog form, are not directly applicable to digital computer systems, to teleprinters or to certain types of information storage systems. Accordingly, analog signals must be converted to digital form by analog-to-digital converters. Such units periodically sample the analog signal and translate its amplitude and polarity into conventional units, as described in Section 5-24.

Data accumulation systems are designed to utilize the outputs of analog-to-digital converters and to permanently record such data on punched or magnetic tapes, cards or printed format. Where digital computers are used for mill control, the peripheral equipment associated with such computers readily performs the data accumulation function and, in addition, the computer may be programmed to perform routine calculations with the acquired data, such as totalizing the mill production and calculating the mill energy requirements in terms of horsepower hours per ton of steel rolled. A relatively simple data accumulation system used in conjunction with two laboratory type cold mills is discussed in Section 5-25.

[○] R. T. Boos and F. A. Achey, "New Method of Measuring Coil Sheet Lengths", Iron and Steel Engineer, January, 1974, pp. 86-90.

5-23 Strip Chart Recorders

Most industrial analog recorders of good accuracy with response characteristics of less than a few Hertz, are of the self-balancing, potentiometer type with either a circular or a strip chart output as illustrated in Figure 5-54.[⊕] Both ink and inkless pens (the latter being used with a special paper) are available in these instruments and, in most cases, the recording pen is linked to an indicating pointer to provide a simultaneous, visual output. Chart speeds range from 0.5 inches per hour to 3 inches per second, and the charts may be driven by spring motor units as well as by electric motors.

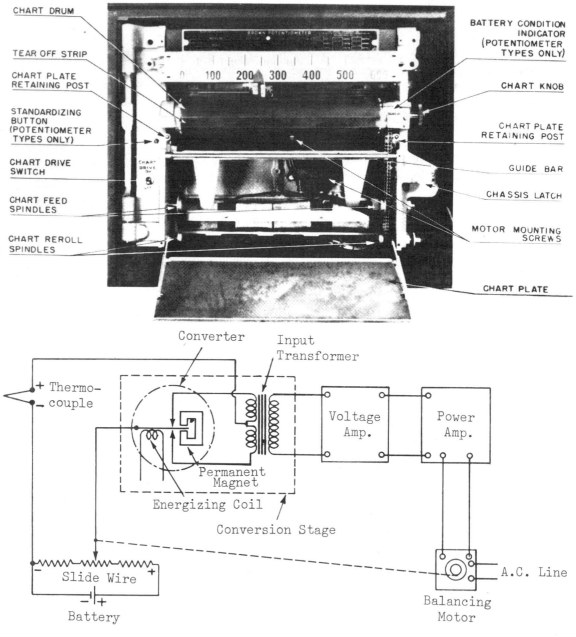

SCHEMATIC DIAGRAM OF BROWN CONTINUOUS BALANCE SYSTEM
(USED IN POTENTIOMETER CIRCUIT)

Figure 5-54: Photograph of Strip Chart Recorder and Schematic Diagram of its Circuit.

[⊕]Machine Design, November 19, 1970, Engineering Department Equipment Reference Issue, p. 86.

AUTOMATIC CONTROL OF COLD ROLLING MILLS

Multiple pen units provide four or more simultaneous records in various colors on the same chart. Multipoint units use a single pen in a scanning arrangement to sequentially record up to 24 variables, while a printing wheel stamps the chart to identify each point. For each point so printed, the time interval may range from 1 to 30 seconds.

5-24 Analog-to-Digital Conversion

To be used with digital displays or digital computers, the analog signals from the various mill sensors must be converted to digital form. Additional benefits of such conversion ensue from the fact that the signal transmission is less susceptible to noise, interference and errors due to the attenuation occurring in long cable runs. Before conversion, the analog signal must be reduced to correspond to a shaft angle, a frequency or a voltage. Shaft angle can be digitized with electromechanical converters, usually coded wheels directly read by photocells or mechanical brushes. Frequency signals are converted using counters operating on a fixed time basis, whereas voltages may be digitized by several methods. One involves pulse-width modulation wherein the analog voltage determines the width of the pulse that operates a gating circuit which, in turn, controls the number of evenly-spaced pulses of a "clock-pulse" generator.

To avoid the necessity of using many signal channels, multiplexing is frequently used. This involves the rapid switching or scanning of the various sensors or transducers in rotation so that the digital signals they produce after analog-to-digital conversion are transmitted sequentially. Relay switching devices of this type have a sampling rate of a few hundred per second which is usually adequate for slowly changing transducer signals. Where rapid transients are involved, however, solid state switching devices, such as that illustrated in Figure 5-55, are used. Early solid state units of this type required high level input signals (5 to 10 volts) and, therefore, an expensive array of input amplifiers but newer, solid state multiplexers switch at low signal levels and use a single amplifier stage for all channels.

Figure 5-55: Analog-to-Digital Converter.

5-25 Data Accumulation Systems

Data accumulation systems associated with cold rolling facilities are usually used in conjunction with digital computers. In the laboratory, the accumulated data may be analyzed, routine calculations performed, graphs automatically drawn and the information presented in printed form. If desired, a single, data-accumulation system may be used with two or more mills.

An example of a laboratory type of data accumulation system is that used by the Research Laboratory of the U. S. Steel Corporation in connection with two experimental cold mills. A photograph of this unit is shown in Figure 5-56, and it is represented in block diagrammatic form in Figure 5-57. The system has fourteen strain-gage channels, eight analog channels, four pulse counters, and twenty-four thumb wheel inputs. The strain gage, analog, and thumb wheel inputs are permanently assigned to each mill but the pulse counting channels are shared by each mill. The output of this system is a punched paper tape.

Figure 5-56: Data Accumulation Equipment for Experimental Rolling Mills.

For the larger mill (a 4-high mill with 6-1/2-inch diameter work rolls of 14-inch face), the system provides seven strain-gage channels, six of which are connected to the rolling-force load cells, the rolling torque transducers, and the entry and exit tensiometer load cells. Each strain-gage channel is provided with its own bridge power supply balancing potentiometer, calibration circuit, and variable gain (1 to 1000) amplifier. An analog channel assigned to the mill is used to measure thickness deviation from a nominal thickness set on the contacting micrometers that measure the entry strip thickness. This channel is also equipped with a variable gain (1 to 1000) amplifier. Three pulse counters are used to measure entry strip speed, mill speed, and exit strip speed from the outputs of the pulse tachometers.

The preset thumb wheels are used to input test data, strip width, work roll diameter, nominal setting on the entry side contacting micrometers, and numerical code information required by the computer for data processing. There is also an automatically sequenced test number counter.

The outputs of the strain gage amplifiers and analog amplifiers are changed from an analog signal to a digital signal with a digital-to-analog converter. The digital information is then used to punch the test data into paper tape using a high-speed punch.

The system is activated by the test engineer at a remote station located at the instrumentation cabinet of the experimental mill. Upon activation, all strain-gage channels, analog

AUTOMATIC CONTROL OF COLD ROLLING MILLS

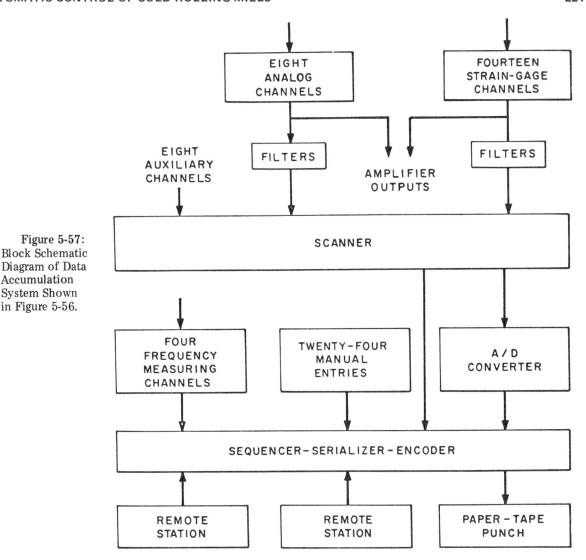

Figure 5-57: Block Schematic Diagram of Data Accumulation System Shown in Figure 5-56.

channel, pulse counters, and thumb wheel preset information are punched in digital form on the tape. The tape is then processed on a tape reader and data-processing computer in accordance with a computer program previously written by the researcher. A flow chart illustrating the steps involved in the acquisition and processing of the data is shown in Figure 5-58.

A data logging system on a production mill reduces the amount of manual effort required to record information about the coil being processed on a mill at a certain time. In addition, it provides the following benefits:

1. The information is recorded in the same place and in the same way despite changes in personnel operating the equipment.
2. Coil identification information can be prepunched in a card and the card read into the system. This eliminates manual recording or entry of a group of numbers on a form or into a data system.
3. The information collected on the coil rolled is recorded in numerical form with the proper decimal point location.
4. The data logging system has its own real time clock so that the time that a coil was processed is available and is printed out with other data.

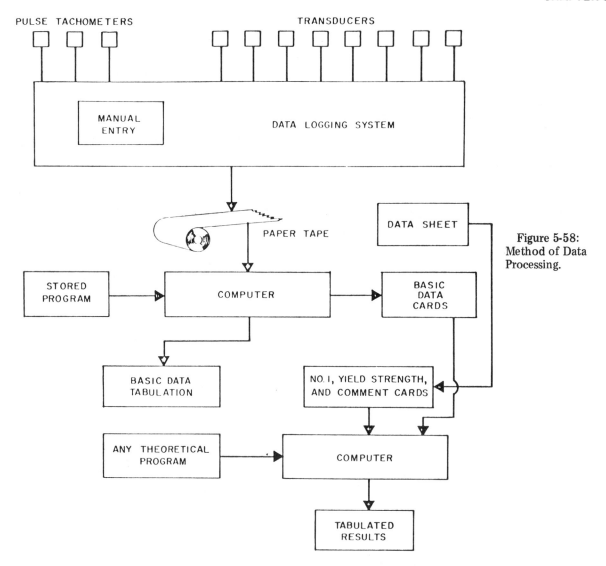

Figure 5-58: Method of Data Processing.

A typical data system applied to a temper mill is shown pictorially in Figure 5-59 and in block diagrammatic form in Figure 5-60. This system records information identifying the coil being rolled and accumulates data on its physical characteristics.○△

Input information for coil identification is obtained from either the card reader or a manual entry station. The other sources of input information provide a measure of product quality which is recorded as aim gage, total footage, percentage of strip on gage and average extension. The system generates the time at which the coil was processed, a number indicating coil sequence and a number identifying the facility processing the coil.

The data assembled for a particular coil is transmitted to a typewriter which prints a ticket that accompanies the coil. The information is also punched into a paper tape which is read into a plant-wide data collection system. In this installation, the mill operator's console features a typewriter, a card reader and a manual entry station. Such a system can be expanded with the addition of peripheral equipment to accumulate and log engineering information.

○T. E. Bryan and J. E. Horn, "Application of Computers to Temper Mill Control", Iron and Steel Engineer Year Book, 1968, pp. 405-416.

△D. E. Rea and F. G. Johnson, "Cold Rolled Sheet Temper Mills", Iron and Steel Engineer Year Book, 1966, pp. 347-353.

AUTOMATIC CONTROL OF COLD ROLLING MILLS

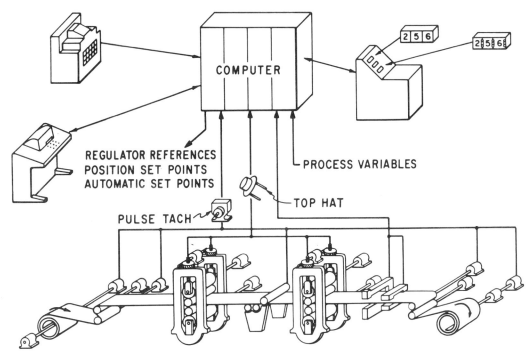

Figure 5-59: A Computer Control System Applied to a 2-Stand Temper Mill.

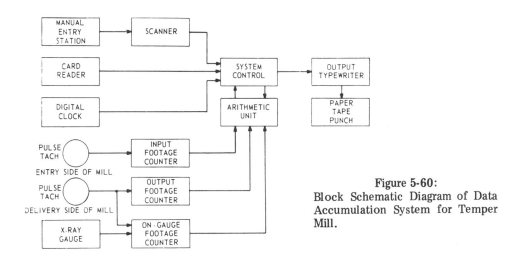

Figure 5-60:
Block Schematic Diagram of Data Accumulation System for Temper Mill.

5-26 Automatic Control Systems

During the last two decades, the availability of reliable transducers, electronic amplifiers and electro-mechanical and electro-hydraulic actuators has made possible the application of a number of important automatic control systems to cold rolling mill operation. Perhaps the best known are the automatic gage control (AGC) systems discussed in Sections 5-28 to 5-30, but systems for automatically decelerating and stopping mills, providing a prescribed extension to the strip (as in temper mills), ensuring the proper feed angle for the strip entering a roll bite and guiding the strip along the desired passline are also used, and are discussed in later sections of this chapter.

In its simplest form, an automatic control system, (1) measures the parameter (such as strip thickness) to be controlled, (2) subtracts the signal corresponding to this measurement from a signal corresponding to the desired value (such as the aim gage), (3) uses the difference or error signal to effect a change in the operating conditions (such as strip tensions or screw settings) so as to decrease the value of the error signal. Such systems are often called "feedback" control systems, the error signal being used to correct a situation already in existence. However, feed-forward systems can be used where an error signal can be used to correct an anticipated problem situation.

Care must be exercised in the design of automatic control systems to ensure they exhibit the necessary stability. Failure to do so could result in cyclical overcorrections or "hunting". Systems are more prone to exhibit this tendency when some delay time (often called a transport delay) exists between the corrective action and the corresponding measurement.

If two or more ostensibly independent automatic control systems are applied to the same mill, care must be exercised so that interactive effects are held to a minimum. For example, in the operation of a tandem mill, an automatic control system intended to load the mill motors uniformly could well interfere with the operation of an AGC system. It is also conceivable that an automatic system designed to produce strip of consistently good shape (as, for example, by maintaining the rolling force constant at the last stand) could also interfere with an AGC system.

Computers, both analog and digital, can be used to provide many of the functions demanded of control systems, in addition to other functions such as mill set-up and data accumulation. Ideally, such computers are programmed with one or more mathematical models representing the rolling operation of the mill they control. Such models take into account the interactions of the various mill parameters and are discussed more fully in Chapters 9 and 10.

5-27 Automatic Decelerating and Stopping Equipment for Reversing Mills

In reversing cold mills, the scrap ends remaining on the reels between passes may be reduced to a minimum by rolling the strip as closely as possible to its ends. However, on the one hand, the pass marks at each end of the rolled length of strip must not enter the roll bite and, on the other hand, the time required for advancing a pass mark towards the roll bite must be kept to a minimum.▲*

The time required to make one pass, which is determined by the length of the workpiece, rolling speed, final strip thickness and by the periods of acceleration and deceleration, can be reduced considerably, particularly on small and medium sized coils, by using automatic decelerating and stopping equipment. Thus the braking period exhibited by the mill drive is minimized. If braking is initiated too early, time is wasted as the pass mark moves too slowly towards the roll bite and, if braking is left too late, the pass mark may enter the bite damaging the rolls and bearings.

To ensure that the strip is rolled to the ends without incurring wastage of time or damage to the rolls, the automatic control equipment must correctly establish the instant braking is to begin from a knowledge of (1) rolling speed, (2) length of strip, (3) reduction per pass and (4) strip thickness.

The same equipment must control the rolling speed until the mill has completely stopped. In addition, the resolution of this equipment must be high enough to allow the determination of the positions of the pass marks, so as to ensure that the mill does not stop before the distance between the pass mark and the roll bite is minimized.

A further advantage of the automatic decelerating and stopping equipment is that it allows more stock to be rolled at full speed, thus improving strip shape and reducing thickness variations due to conditions of improved thermal stability in the roll gap.

▲ O. Steinbrecher, "Programmed Cold Rolling Reversing Mills", Iron and Steel, October, 1969, pp. 303-317.

* "Fully Automatic Cold Rolling — Description of Control System Installed on a Four-High Reversing Mill at J. J. Habershen and Sons, Ltd.", Sheet Metal Industries, October, 1959, pp. 657-659, 662.

AUTOMATIC CONTROL OF COLD ROLLING MILLS

A block schematic diagram of an automatic decelerating and stopping equipment is shown in Figure 5-61. This equipment stores the number of revolutions made by the coiler during the first rolling pass and stops the mill automatically from the second pass on. Tachometer generators are attached to the reversing reels and to the mill drive to monitor the reduction per pass, the strip thickness and the resultant coil buildup. The outputs of the tachometers are fed to the analog computer to effect the breaking operation.

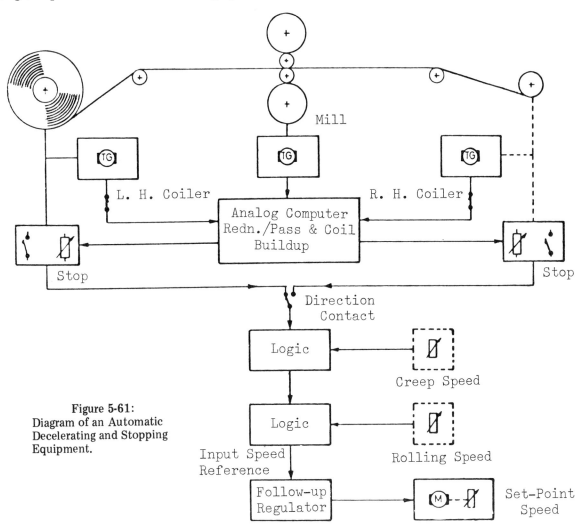

Figure 5-61: Diagram of an Automatic Decelerating and Stopping Equipment.

From the second pass on, the automatic decelarating and stopping equipment takes over to slow down the mill near the coil end. When reaching a constant creep speed at the end of each pass, the control equipment generates the order to stop the mill. If the mill is stopped when a pass is being rolled and the automatic decelerating and stopping equipment is still switched on, the number of revolutions made by the counter attached to the coiler remains intact so long as the strip remains under tension at all points along the pass line. Therefore, after the mill has been restarted under such circumstances, stopping is effected automatically when the coil end approaches. However, if the strip tension is relaxed during a stoppage, the operation of the control equipment must be reset.

5-28 Automatic Gage Control (AGC) Systems

Because of the commercial demand for more accurate and consistent dimensions of flat rolled products, it was inevitable that the gage of the rolled strip should be automatically controlled. Accordingly, in the midfifties, automatic gage control (AGC) systems were introduced on cold mills and since that time a great variety of such systems have been built.[1] Basically, they are closed-loop feedback systems but, unlike process controls, AGC circuits are basically vernier control loops continually adjusting the original mill settings made either by an operator or a computer.

All AGC systems involve the use of an accurate gaging unit and, while most mills today utilize radiation type gages, some of the older, slower mills still use flying micrometers and some use the rolling force as an indirect measure of gage (Sims' method). To make the necessary adjustments to gage, changes are made either to strip tensions and/or to roll positions.

On lighter strip, adjustment of the back tension is widely used as a means of control. However, limits are usually set to the permissible tension that can be utilized so that if the tension tends to swing outside the prescribed limits, adjustments to the screws are made. Conventional screwdown control is of two types. In one type, the screws are jogged with a jog time proportional to the magnitude of the error signal. Alternately, if a feedback signal corresponding to screw position is available, it may be balanced against the gage error signal. To prevent the system from "hunting" or "overcorrecting", a time delay may be introduced if there is a substantial transport lag.

The automatic control of gage has been improved through the use of hydraulic-roll positioning systems, since such systems have an appreciably faster speed of response to control signals than conventional mill screws.[2,3] However, as pointed out by King and Sills,[4] in the case of hydraulic mills, there is a need for continuous feedback and control of the main roll positioning cylinders because of the absence of the braking and locking action of conventional screws.

In foil rolling mills, it is possible to use strip speed as one of the controlling variables.[5] In effect, variation of the mill speed creates a change in the effective coefficient of friction in the roll bite, the greater the rolling speed, generally speaking, the less the coefficient of friction. Consequently, if the rolled strip is too thick, the rolling speed should be increased and vice versa.

Automatic gage control on modern tandem mills has considerably lessened the tolerance limits on strip thickness.[6] Most tandem mills claim 85 per cent of rolled sheet to be within half sheet tolerances ($\pm$ 10 per cent). Coils show little thickness deviation throughout their entire length and control circuit data indicate that once the sheet comes "on gage", the AGC has very little to do. Deviation will therefore stay within the error discrimination, or dead band, so the gage will not initiate any signals to the tension control.

In a British installation on a four-stand cold rolling mill, with one AGC arrangement, 96.5 per cent of the rolled product was within one mil of nominal and 92 per cent was within half a mil. In a digital control system for revamped two-stand tandem mills to be used as double cold reduction mills, 59 per cent of the product was within 1/2 per cent of the nominal aim thickness (0.0066 inches) compared with 27 per cent obtained under the manual control of the mill. Automatic control produced 96 per cent of the output within 1-1/2 per cent of nominal compared to only 68 per cent under operator control.[7] A similar improvement in gage uniformity has also been reported by Druzhinin, et. al., for a Russian four-stand sheet mill.[8]

---

[1] J. W. Wallace, "Fundaments of Strip Mill Automatic Gage Control Systems", Iron & Steel Engineer Year Book, 1964, pp. 753-762.

[2] M. D. Stone, "Hydraulic vs. Electric A.G.C. Systems — A Comparison of Speed of Response", Iron and Steel Engineer Year Book, 1972, pp. 346-348.

[3] M. D. Stone, "Hydraulic Automatic Gage Control Mills", Iron and Steel Engineer Year Book, 1973, p. 313.

[4] W. D. King and R. M. Sills, "New Approaches to Cold Mill Gage Control", Iron and Steel Engineer Year Book, 1973, pp. 187-198.

[5] O. Steinbrecher, "Automatic Gauge Control for Cold Rolling Mills", Iron and Steel, June, 1969, pp. 174-179.

[6] "Thickness Gaging: Where the Customer's Mike is Always Right", "33" Magazine, May, 1967, pp. 67-78.

[7] "AGC: Putting Thickness Measurement to Work", "33" Magazine, May, 1967, pp. 61-67.

[8] N. N. Druzhinin, A. G. Mirer, V. M. Kolyadich, A. N. Druzhinin and B. A. Shikhanovich, "Automatic Strip Thickness and Tension Control Systems for Continuous Cold Rolling Mills", Stal in English, April, 1969, pp. 396-399.

AUTOMATIC CONTROL OF COLD ROLLING MILLS

Generally speaking, the purpose of AGC systems is to maintain constancy in the gage of the rolled strip. However, with the advent of the truly continuous cold mills discussed in Section 2-5, the necessity has arisen of changing "on-the-fly" in tandem cold mills. This problem has been solved through an analysis of the dynamic behavior of tandem mills and the computer simulation of their operation. On the basis of a mathematical model developed during the analytical work, gage changes "on-the-fly" are now made on a fully continuous commercial mill by the simultaneous adjustment of the roll positions and roll speeds at all stands while maintaining constant tension between stands.▼

5-29 AGC Systems for Single-Stand Mills

The simplest but most effective form of gage control for a single-stand mill involving automatic screw positioning and tension adjustment is shown in Figure 5-62.⊗ It uses plug-in modular electronic equipment and operates from either a contactless measuring device, such as an x-ray gage, or a flying micrometer.

For small gage variations, the thickness error is used to alter the back tension within prescribed limits by varying the current or tension regulator reference for the uncoiler. Outside the tension limits, the error signal pulses the screws and, at the same time, the tension is reset to its original value. For large gage errors, such as occur at the beginning of a pass or at a weld, the screws are pulsed until the gage error is within tolerance. The pulses are adjusted so that the screwdown "on" time is a function of both the gage error and the setting of the preset hardness control (which takes account of the metallurgical hardness) and the cross section of the material being rolled. The spacing between pulses is an inverse function of the mill speed, thereby allowing a change in strip gage due to movement of the screws to reach the measuring device in time to determine whether further corrective action is necessary.

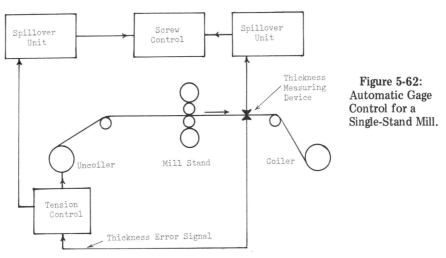

Figure 5-62: Automatic Gage Control for a Single-Stand Mill.

An AGC system for a single-stand, hydraulically loaded mill has been described by Heys ◻ and is illustrated schematically in Figure 5-63. The system is capable of operation from an error signal derived from a thickness gage monitoring the strip on the exit side of the mill or a similar gage monitoring the entry thickness of the strip. In the first case, the system can compensate for virtually any factor causing a change in gage whereas, in the second case, only a general improvement is effected since the system cannot correct for changes in the hardness of the strip or for changes in mill speed.

▼ T. Arimura, M. Okado and M. Kamata, "NKK Applies Computer Controls to Hot and Cold Strip Rolling Facilities", "33" Magazine, September, 1973, pp. 38-42.
⊗ D. W. Draper and J. C. Maxey, "Cold Mills", AEI Engineering, Metal Industries Supplement, 1967, pp. 79-87.
◻ J. D. Heys, "Control of the Cold Rolling Process", Sheet Metal Industries, April, 1959, pp. 281-299.

Referring to Figure 5-63, it will be seen that the signal from one of the flying mikes is fed to an amplifier which, in turn, is connected to a solenoid-operated pressure control valve. Within the range employed, the delivery pressure of the variable pressure pump is proportional to the current flowing through the solenoid and hence to the magnitude of the error signal. The variable pressure pump controls the pressure in the rams, or cylinders, and hence the rolling force, increasing the latter if the outgoing or incoming strip is thicker than specified.

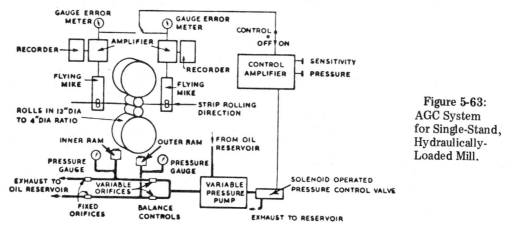

Figure 5-63: AGC System for Single-Stand, Hydraulically-Loaded Mill.

A variable orifice is fitted to the pressure line to each ram to allow for balancing of load for shape control. The pressure to each ram can be dropped to about 15 per cent below the pump delivery pressure and this adjustment can be made while the control is in operation.

5-30 AGC Systems for Tandem Mills

An AGC system for Kaiser Steel's 3-stand cold reducing mill for lightweight tin mill products has been described by Thomson and Coleman [1] and is briefly reviewed here as an example of an AGC system for secondary cold mills. This mill is used for the production of 55- and 60-pound basis weight strip and is operated as follows: tension payoff reel and entry bridle provide 4,000- to 5,000-pound tension into stand No. 1; the tension between stands No. 1 and No. 2, lies between 5,000 and 9,000 pounds, between stands No. 2 and No. 3 — 4,000 to 6,000 pounds and 1,000- to 1,500-pound tension on the delivery side of stand No. 3. Stand No. 1 is operated as a dry mill with from 200 to 400 tons total rolling force with very little reduction at that stand. With rolling solution applied at stand No. 2 and with a force ranging from 500 to 1,000 tons, reductions ranging from 25 to 30 per cent are obtained. Stand No. 3 is also operated with a lubricant and, with a force in the range of 200 to 400 tons, provides a reduction of 5 to 10 per cent.

The AGC system underwent modification after installation until it assumed the form shown schematically in Figure 5-64. As will be seen, the inputs to the AGC system are the output of the thickness gage after the last stand, the output of the tensiometer located between stands 1 and 2 and the rate of change of gage signal from the thickness gage also located between stands 1 and 2. The outputs of the system were fed to the stand 1 speed regulator and to the screws of stand 2, the latter being necessary to prevent the tension between stands 1 and 2 deviating by more than plus or minus 15 per cent from the set points. A tension limit circuit was installed to allow the gage free control of stand 1 speed as long as the tension remained within plus or minus 15 per cent of the set tension. The gain of the AGC system is progressively reduced when the tension error exceeds the set limits, limiting the magnitude of the gage error that can be compensated for by tension alone without waiting for the screws to assist.

---

[1] O. B. Thomson and V. S. Coleman, "Kaiser Steel's 3-Stand Cold Reducing Mill for Light Weight Tin Mill Products", Iron and Steel Engineer Year Book, 1967, pp. 543-551.

AUTOMATIC CONTROL OF COLD ROLLING MILLS

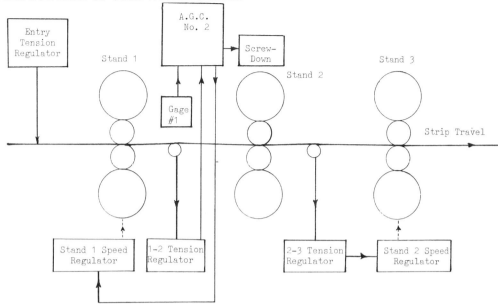

Figure 5-64: AGC System for 3-Stand Secondary Cold Reduction Mill.

An automatic gage control system used on an English 4-stand cold strip mill has been described by Gifford and Bray,[⊠] and is illustrated in Figure 5-65. This mill utilizes four-high stands with 10.5 inch diameter driven work rolls and 22.5 inch diameter backup rolls. The mill is capable of rolling strip 18 inches wide at a speed of 2,000 fpm and provides total reductions up to 85 per cent. Gage control is provided on the first and last stand with the gagemeter method used at both stands for measuring the thickness of the strip. As seen in Figure 5-65, signals proportional to roll force and screw position are summed at stands 1 and 4 and from the totals are subtracted signals proportional to the aim gage. The resultant signals are then compared against the output of the flying mike to provide error signals which adjust the screw position at stand 1 and the field of the main drive motor at stand 4 so as to change the gage by tension control. However, if the gage error exceeds 0.0005 inches, the screws on stand 4 are energized to correct the error. It should be added that, since the AGC system cannot operate during acceleration, it is switched on automatically at approximately full rolling speed.

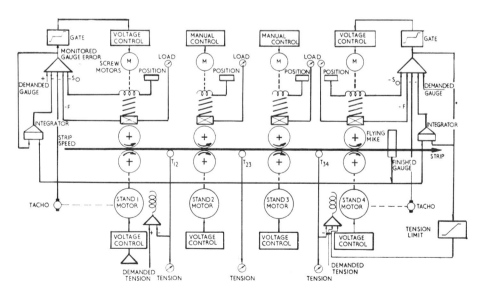

Figure 5-65: AGC System for 4-Stand Mill.

[⊠] J. D. Gifford and D. B. Bray, "Control of a Multi-Stand Cold Strip Mill", Steel Times, March, 1969, pp. 171-179.

A computerized AGC system for a 5-stand, 80-inch sheet mill has recently been described by Lapham.⊖^ This is illustrated in diagrammatic form in Figure 5-66. It will be noted that it utilizes six x-ray gages and it is unique in that it provides tension rather than speed regulation. The tension regulators (indicated by the blocks labelled TR) along with their associated speed regulaters (labelled SR) are hard-wired functions.

Interstand tensions are moved within preset limits to compensate for changes in gage. The AGC system sends control signals to the tension regulators based on the incoming x-ray gage deviation. The stand screw positions change in response to incoming gage deviations (CAL 0) or to interstand tension limitations between stands 3 and 4 and stands 4 and 5. Screwdowns for stands 1-4 are all regulated by the AGC system.

The error signal derived from the stand 2 x-ray gage feeds the AGC tension control for stands 2 and 3 to achieve a constant gage out of stand 3 (a feed forward system). (The more conventional feed-back system is shown by the dotted lines of Figure 5-66.)

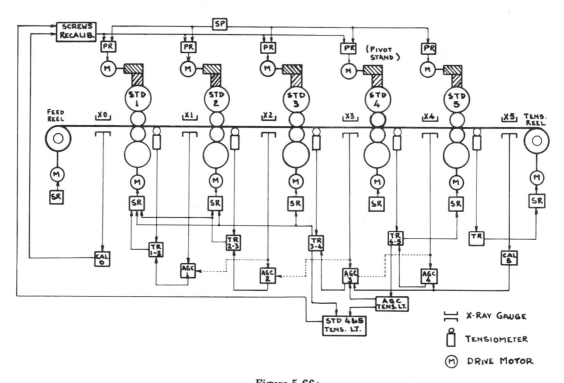

Figure 5-66:
Block Schematic Diagram of AGC System for 5-Stand Sheet Mill (SR = Speed Regulator;
TR - Tension Regulator; PR = Position Regulator; SP = Screw Position).

The computer control of a Japanese 5-stand, 68 inch wide hydraulic mill has recently been described by Togashi, et. al.* The control scheme for the mill is shown in Figure 5-67 and, at high mill speeds, the hydraulic cylinders are activated by the x-ray thickness gage on the exit side of stand 1 and the tension control is activated by the deviation signal from the x-ray thickness gage on the exit side of stand 5. In addition, the AGC system compensates for the variation in the thicknesses of the oil films in the backup roll bearings.

⊖ G. C. Lapham, "Computerized Cold Reduction: Inland's 80-Inch Tandem Mill", Iron and Steel Engineer Year Book, 1972, pp. 577-582.

^ See also O. G. Brunner, "Drives and Automation Systems for Wide 5-Stand Cold Mills", Iron and Steel Engineer Year Book, 1965, pp. 939-950.

* N. Togashi, K. Takemura, A. Kishida and S. Kitao, "Computer Control of a 5-Stand Cold Mill at Mizushima Works", Iron and Steel Engineer, February, 1974, pp. 47-51.

AUTOMATIC CONTROL OF COLD ROLLING MILLS

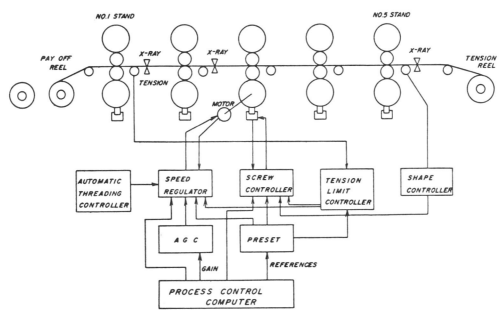

Figure 5-67: AGC System Used with 5-Stand Hydraulic Mill.

5-31 Automatic Extension Control for Temper Mills

The digital type extensometer described in Section 5-18 may be used with the automatic control arrangement shown in Figure 5-68.△ Depending on the strip thickness, primary control can be accomplished either by entry tension or stand 1 screw adjustment or both. When operating with entry tension as the primary controller and the strip tension exceeds upper or lower tension limits, stand 1 screws take over and continue to operate until the tensions are again within limits. On the heavier gages, the tension control is turned off and only the screws used for automatic control of extension. An off timer allows for the transport time from stand 1 to the exit pulse tachometer and, after a correction has been completed, it will delay the start of another extension sample, the off-time interval being varied inversely as the strip speed.

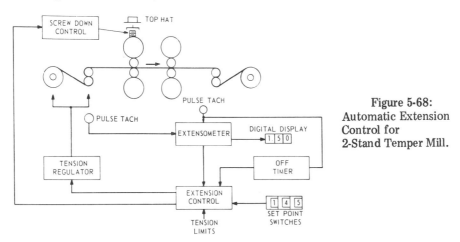

Figure 5-68: Automatic Extension Control for 2-Stand Temper Mill.

A digital screw reset system complements the extension control system by returning the screws to the stored setup position for the next coil after they have been moved for extension correction and/or mill threading.

△D. E. Rea and F. G. Johnson, "Cold Rolled Sheet Temper Mills", Iron and Steel Engineer Year Book, 1966, pp. 347-353.

5-32 Constant Strip Entry Angle Control

Some temper mills are provided with electrical and mechanical equipment designed to provide a constant strip entry angle to the mill as strip is uncoiled from the payoff reel.△ This equipment is intended to reduce the surface defects caused by excessive bending of an annealed strip over the entry deflector roll.

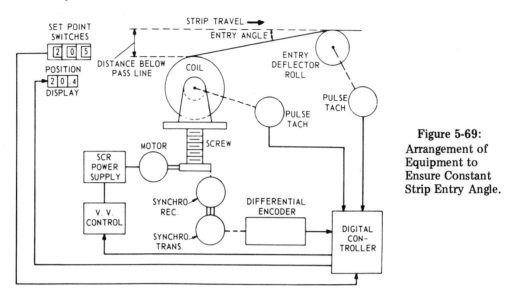

Figure 5-69: Arrangement of Equipment to Ensure Constant Strip Entry Angle.

One method of maintaining a constant entry angle is illustrated in Figure 5-69 wherein the payoff reel is capable of being moved vertically. The lifting mechanism is equipped with a digital positioning system and the controller ensures that the top of the coil is at a constant distance below the mill passline. To obtain high accuracy and fast response over a wide range of mill speeds, coil sizes and material thicknesses, the platform is moved by a screw mechanism driven by an SCR-controlled D.C. motor.

The coil diameter is measured using a differential encoder gated by a photoelectric limit switch as the coil moves into position for the coil hoist to put it on the uncoiler mandrels. This measurement is first used to properly position the mandrel for automatic loading. Then it is used to set the counter emf rheostat to a position based on incoming coil diameter. This measurement is also used in the system to initially position the coil to the required distance below the passline for proper entry angle.

To maintain the entry angle after the start of rolling, use is made of the outputs of two tachometers, one coupled to the mandrel and the other to the entry deflector roll. The control system generates a positioning signal based on the set point and the computed coil diameter at the particular time.

5-33 Strip Guidance Systems

In the rolling of thin strip where edge guides are not practical, the positioning of the strip in the mill during rolling may be accomplished automatically by edge guidance equipment. Such equipment senses the position of one edge of the strip at a given position along the passline prior to entry into the first roll bite and automatically adjusts the axial position of the coil being unwound so as to maintain an essentially constant strip edge position. A similar arrangement may be used in association with the takeup reel so that the wound-up coil maintains a desirable geometry.

△ D. E. Rea and F. G. Johnson, "Cold Rolled Sheet Temper Mills", Iron and Steel Engineer Year Book, 1966, pp. 347-353.

AUTOMATIC CONTROL OF COLD ROLLING MILLS

For sensing the position of the strip edge, optical or pneumatic methods may be employed. A diagrammatic illustration of the former is shown in Figure 5-70, and of the latter in Figure 5-71.[T~]

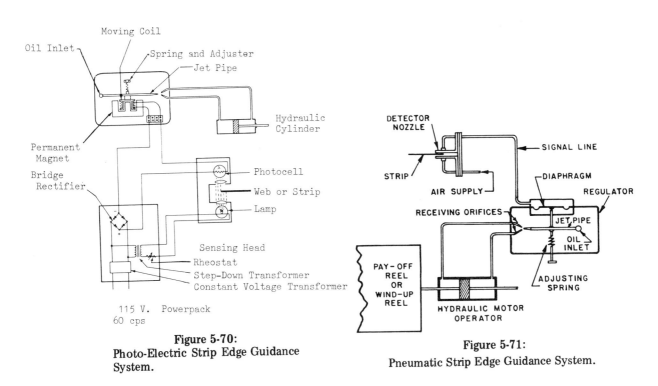

Figure 5-70:
Photo-Electric Strip Edge Guidance System.

Figure 5-71:
Pneumatic Strip Edge Guidance System.

For adjusting the coil position, hydraulic cylinders may be utilized to move the whole reel assembly horizontally on its bed, as illustrated in Figure 5-72 for the case of a payoff reel. The sensing nozzle is in a fixed position corresponding to the required edge position and the strip is moved to the nozzle. The first idler roll must be rigidly affixed and move with the payoff reel. When using the pneumatic sensing nozzle, this allows for a uniform passline through the sensing nozzle and insures sufficient wrap on the moving idler roll, eliminating lateral slippage of the strip from maximum to minimum coil diameter. The sensing head is located as close as mechanically possible to the moving idler roll. The instant the reel is displaced, the strip must move and the sensor must be aware of this movement. Moreover, satisfactory edge position control cannot be maintained unless there is a constant tension on the strip beyond the pivot point.

The use of a photo-electric sensor provides greater accuracy and eliminates the necessity for an idler roll. The gap between the light source and the photocell is established by the maximum coil diameter so the sensor can detect the edge position of the strip during the change in passline as the coils pay out.

On the wind-up reel application, the reel follows the lateral strip travel and places the reel in the correct position to receive the strip. As shown in Figure 5-73, the sensor is rigidly fastened to the reel and moves with it. As in the case of the payoff reel, the photoelectric system

---

T J. M. Deering, "New Concepts for Automatic Strip Guiding for the Metals Industry", Iron and Steel Engineer Year Book, 1965, pp. 975-982.

~ See also, J. J. O'Brien and G. H. Mattke, "Strip Guiding — A Major Factor in Productivity", Iron and Steel Engineer, February, 1974, pp. 31-40.

provides greater accuracy in wind-up reel applications than the pneumatic system and can be readily adjusted for all strip speeds and tensions.

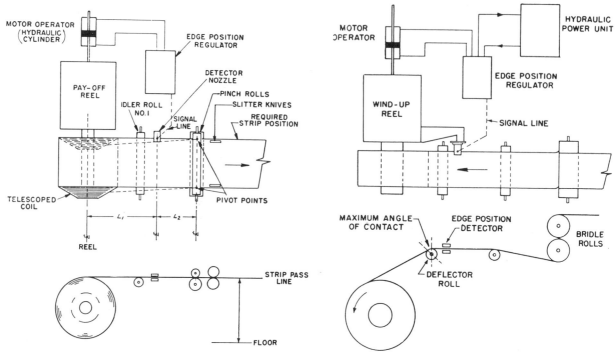

Figure 5-72:
Edge Guidance System Applied to a Pay-Off Reel.

Figure 5-73:
Edge Guidance System Applied to a Coiler.

In primary cold mills of modern design, centering of the coil with respect to the mill pass line is accomplished through the automatic operation of the stub mandrils in the coil box. To perform this operation, the coil width is first determined through the use of phototubes.[9]

5-34 Preprogrammed Control Systems

Where the setup of a mill is accomplished through the use of electrically-operated controls, it is possible to utilize preprogrammed control systems for this purpose. Slabbing mills were the first to be controlled in this manner in the 1950's using punched cards to establish work roll openings and other rolling conditions. In the 1960's, preprogramming was applied to cold mills, primarily those large tandem mills subject to digital computer control. In this case, the computer could, with relative ease, ascertain from its stored information the optimum mill setup to use for rolling a given coil and, at the appropriate time, ensure that the mill was correctly set up on this basis.

For single-stand reversing mills, simpler preprogramming controls may be used. Steinbrecher[▲] has described such controls (shown diagrammatically in Figure 5-74) that automate a reversing mill with respect to setting or actuating (1) the rolls for each pass, (2) the thickness gages, (3) the nominal front and back tensions for each pass, (4) the speed for each pass, (5) the switches controlling the auxiliary equipment, such as the rolling solution sprays, (6) the rolling direction, (7) the rolling sequence after setting all nominal values and switching on all necessary circuits, (8) the operation of the automatic gage control equipment and (9) the stopping of the mill

---

[9] R. C. Conley, "Armco-Middletown's 86-In. Tandem Cold Mill", Iron and Steel Engineer Year Book, 1972, pp. 349-352.

[▲] O. Steinbrecher, "Programmed Cold Rolling Reversing Mills", Iron and Steel, October, 1969, pp. 303-317.

AUTOMATIC CONTROL OF COLD ROLLING MILLS

by automatic braking equipment at the ends of the coil. The system was designed so that the nominal thickness of the strip could be set by means of potentiometers and other manual controls or, preferably, by punched cards. It is possible, for example, for the first coil of a series to be rolled under manual control when the roller actuates the potentiometers and hand-operated units on his console. However, the settings for the individual passes preset in this way are stored for the next coil. When the direction of rolling is reversed, the settings of the potentiometers and the manual units for the next pass are put into operation by stepping switches. In this way, all the passes are preprogrammed and, after the first coil of a series has been rolled, the total program has been stored.

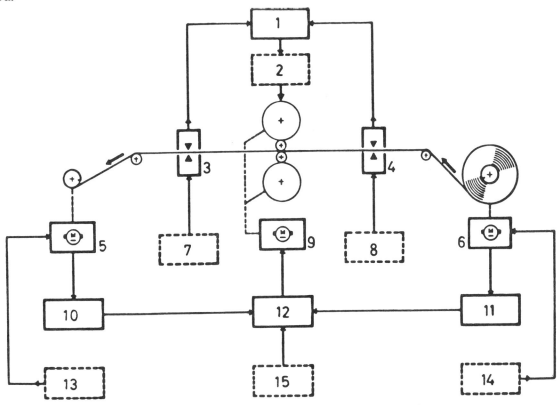

Figure 5-74: Diagrammatic representation of a process control system.

1	= Gauge control system	9	= Mill drive
2	= Screwdown programming	10 and 11	= Automatic decelerating and stopping equipment
3 and 4	= Thickness gauges	12	= Speed control
5 and 6	= Reel drives	13 and 14	= Strip tension programming
7 and 8	= Thickness gauge programming	15	= Speed programming

5-35 Computer Control Systems

Inasmuch as a certain degree of confusion exists as to what exactly is meant by computer control, a few definitions are in order. Generally speaking, where the term "computer" is used, a digital machine is implied capable of being programmed in a variety of ways. However, as is seen in Section 5-37, analog units may also be used for mill control. Yet regardless of the type of computer used, it must be capable of performing various mathematical operations. The term "on-line" has also been misused in the past and perhaps the best definition is that employed by BISRA.◄ϕ BISRA states that "an on-line computer system is one in which a computer is connected

◄ "A Review of On-Line Digital Computer Applications in the British Steel Industry", BISRA, 1965.
ϕ J. F. Roth, "On-Line Computer Systems in the Metals Industry — A Survey", Iron and Steel, 25, May, 1966, pp. 251-255.

to a process and the computer system responds to external stimuli as they occur, the method of communication between the process and the computer in no way affecting the process itself. Thus information describing the state of the process is supplied to the computer, which is so organized that none of this information can be inadvertently overlooked. After the information has been assimilated by the computer, data regarding the control of the process or its state are generated. These data can be presented to the operator so that, if he wishes, he may make manual adjustments to the process. Alternately, the signals so generated by the computer may be used for the direct control of the process. Under these circumstances, the on-line computer must work at a speed dictated by the process itself".

On-line computers used in industrial applications may perform process control and/or sequence control. In the majority of applications, particularly in the area of cold reduction, only the first need be considered since computers attempt to control the process so that the best possible, or optimum, performance is achieved. Measurements are made describing the physical state of the process and this information is used by the computer in conjunction with some form of mathematical model of the process so as to determine what adjustments have to be made to the operating conditions to attain and maintain the required performance.

Usually the mathematical models of a process, such as that associated with a tandem cold mill, are very complicated and the calculations to be performed by the computer are necessarily complex.[∞] In any specific application, the complexity of the model and the calculations define the type of computer to be used. However, it must be emphasized that satisfactory operation of the process can only be achieved when the process and all the boundary conditions (or constraints) are clearly specified in mathematical terms.

5-36 Digital Computer Control Systems

The use of digital computers for the control of primary, tandem cold reduction mills began in 1967,[φ] but the number of mill installations equipped with this type of control is still rather limited. Although the basic objective of this type of control is to produce the maximum tonnage of rolled strip of acceptable quality at minimum cost, computers have not really attained this objective. This is not to imply that they cannot do so, nor that the investment involved in their purchase is not justified. For they have already enabled relatively untrained mill operators to satisfactorily operate high-speed tandem mills by providing proper mill setup. In other words, they have replaced the "little black books" that skilled rollers carried around with them and which contained the results of years of mill experience.

Digital computers used for control purposes may be thought of as possessing six principal operating characteristics or capabilities.[⋔] These are to

    (1) Scan data from sensors, transducers and contact devices.

    (2) Store such information as is necessary for control purposes.

    (3) Perform arithmetic computations.

    (4) Make decisions or perform logic operations.

    (5) Communicate with peripheral devices, such as typewriters, printers and consoles.

    (6) Produce output signals in the form of analog signals and contact closures necessary for control purposes.

[∞] "Automation of Tandem Mills", Edited by G. F. Bryant, The Iron and Steel Institute, 1 Carlton House Terrace, London SW1Y5DB, 1973.

[φ] B. C. Bradley, M. O. Smith, Jr., and H. N. Cox, "U. S. Steel — Fairfield Works 6-Stand Computer Controlled Cold Mill", Iron and Steel Engineer Year Book, 1970, pp. 257-264.

[⋔] "Application of On-Line Digital Computers to Iron and Steelmaking Processes", "33" Magazine, February, 1965, pp. 48-62.

AUTOMATIC CONTROL OF COLD ROLLING MILLS

Figure 5-75 shows in schematic form the utilization of a digital computer with a 5-stand cold mill.[8] Before the coil to be rolled is threaded into the mill, the coil information provided to the computer by cards or other means enables it to calculate references for screwdowns and side-guide position regulators, for speed and tension regulators and for x-ray gage set points.

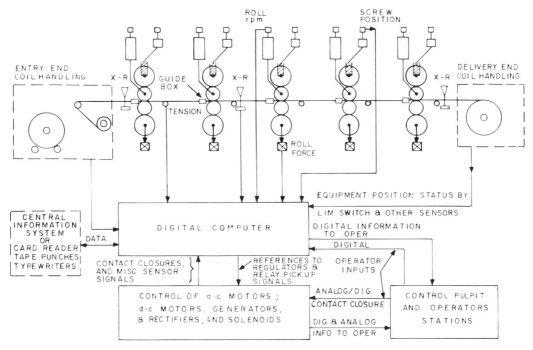

Figure 5-75: Digital Computer Control of a 5-Stand Tandem Mill.

When a coil is to be rolled, the computer must control:

(a) The transfer of the coil from the handling conveyor to the handling equipment at the entry end of the mill.

(b) The measurement of the coil dimensions, including diameter and width.

(c) The correct positioning of the coil with respect to the mill centerline.

(d) The removal of the coil band and the flattening of the head end of the coil.

(e) The positioning of the coil so that the pay-off reel mandrils may be properly inserted into the eye of the coil.

On threading the strip into the first stand, the sprays (often only water sprays) are turned on and the guides are closed. From a measurement of the rolling force at the first stand, the "rollability" of the strip is assessed and any necessary adjustments are made to the computed mill setup.

While the head end of the strip is moving towards the second stand, the operator will level the screws at the first stand to straighten out the delivery of the strip from that stand. If the positions of the strip edges could be satisfactorily sensed, this levelling operation could easily be performed by computer.

On entry into the guides of stand 2, the guides are closed so as to be only slightly wider than the strip. This ensures that the centerline of the strip coincides as closely as possible with the centerline of the mill.

After entry into the second stand, not only are the top guides closed and the rolling solution sprays turned on, but the mill thread speed is adjusted to give the desired strip speed

---

[8] H. N. Cox and A. S. Norton, "Computer Set-up and Control of Tandem Cold Mills", Iron and Steel Engineer Year Book, 1968, pp. 479-486.

between stands 2 and 3. So as to obtain the "run-speed" gage from stand 2, the speed of stand 1 is reduced. At the same time, the screws of stand 2 are adjusted to maintain the desired tension between stands 1 and 2. This sequence of control steps is repeated for each stand on the entry of the head end of the coil.

After complete threading of the mill, the motor loads, rolling forces, tensions, strip thicknesses and speeds are monitored by the computer to determine if conditions are normal. If not, corrective or informative actions are provided by the computer and the mill may not be accelerated.

At the delivery end of the mill, the establishment of proper tension at the coiler initiates

(a)   the classification of strip thickness for data logging,

(b)   movement of the guides to the "run-speed" positions,

(c)   the operation of the automatic gage control system,

(d)   the retraction of the belt wrapper,

(e)   the acceleration of the mill, and

(f)   the automatic control of interstand tensions.

During the rolling operation, data from all the many analog and digital sensors are utilized by the computer for the production of engineering logs as well as to modify the mathematical model utilized by the computer to compensate for changes in the rolling solution, roll diameters and other parameters. The computer also aids in establishing slight changes to the reduction schedule necessary to maintain acceptable strip shape. Moreover, as the end of the coil approaches, the computer calculates the proper time to begin deceleration utilizing such information as mill speed, drafting, deceleration and "tail out" speed. During "tailing out", the computer initiates closing of the entry guide boxes at all stands and sprays are turned off so as to maintain the thermal roll crowns. At the same time, references are reestablished for the stand. The coil diameter on the winding reel is measured so that the coil may be properly positioned for reel stripping and the many subsequent operations involved in placing the coil on the delivery conveyor.

In the many control functions outlined above, direct digital control consists of time sharing the computer to close the loop of a number of digital regulators. Each function being so controlled, has its own error storage and feedback hardware, as illustrated in Figure 5-76. Reference data, derived from a compilation of punched cards or direct manual input are stored in the computer memory, and the stored program logic dictates that the drives are moved to their reference positions. In an interval of a few milliseconds, the input multiplexer selects the appropriate feedback sensor and the computer reads the sensor position. A comparison is then made between the actual and reference positions and the error, or difference, is computed. The appropriate error buffer storage hardware is sleected, the error signal is stored in this buffer and the drive system moves in a direction to reduce the error. After having serviced all the other regulators, the computer turns its attention again to this particular loop and repeats the operations.

For adjustable voltage drives, the position signal is a proportional analog signal with correct polarity for direction of movement. For constant potential drives, the error signal calls for fast or slow speed corrective response in the proper direction. For A.C. drives, the error signal calls for direction of movement and, in the case of hydraulic or pneumatic drives, high or low speed and the appropriate direction.

Another 5-stand tandem mill that is computer controlled is the No. 2 cold rolling mill at Fuji steel's Nagoya Works which began operation in December 1967. Its computer provides data logging and automatic presetting of the mill controls. The mill is also equipped with work roll bending devices on all stands, and it features a tension-type AGC system with the gauge-meter type AGC installed on stands 1 and 2.

"Fully Automated Cold Tandem Mill in Operation at Nagoya Works", Fuji Steel News, Vol. 5, No. 1, January-February, 1969, p. 7.

AUTOMATIC CONTROL OF COLD ROLLING MILLS

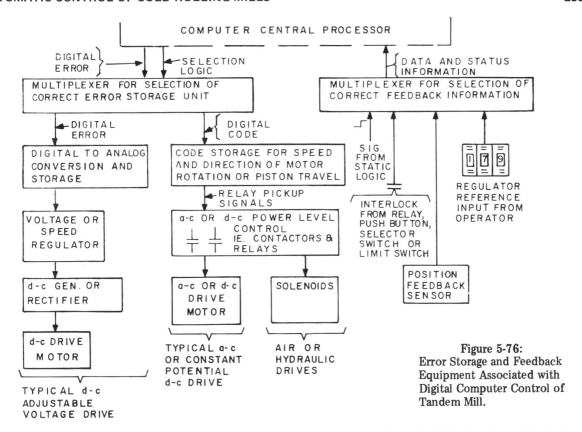

Figure 5-76:
Error Storage and Feedback Equipment Associated with Digital Computer Control of Tandem Mill.

A 6-stand mill with computer control installed at the Fairfield Works of the U. S. Steel Corporation is described in Section 2-14.ɸ A similar mill under computer control is installed at the Gary Works of the U. S. Steel Corporation and, in this case, the computer has been programmed utilizing a simplified mathematical model of the rolling process.ᶧ⁺

Five-stand sheet mills under computer control have been placed in service recently. One, a 68-inch mill at the Mizushima Works of Kawasaki Steel Corporation was put into operation in March 1971, and its computer control system is shown in simplified, diagrammatic form in Figure 5-77.*

Another 5-stand, 84-inch computer-controlled sheet mill was installed at the Irvin Works of the United States Steel Corporation. It features automatic gage control and utilizes the computer for setting interstand tension and stand speed.ᵛ

5-37 Analog Computer Control Systems

In some instances, analog computers have been satisfactorily used for tandem mill control. One example has been described by Schmidt relative to a 5-stand mill at Hoesch A.G.** This

---

ɸ B. C. Bradley, M. O. Smith, Jr., and H. N. Cox, "U. S. Steel — Fairfield Works' 6-Stand Computer-Controlled Cold Mill", Iron and Steel Engineer Year Book, 1970, pp. 257-264.

ᶧ⁺ W. L. Roberts, "A Simplified Cold Rolling Model", Iron and Steel Engineer Year Book, 1965, pp. 925-937.

* N. Togashi, K. Takemura, A. Kishida and S. Kitao, "Computer Control of a 5-Stand Cold Mill at Mizushima Works", Iron and Steel Engineer, February, 1974, pp. 47-51.

ᵛ C. J. Labee, "A Century of Steelmaking — The Edgar Thomson-Irvin Works of U. S. Steel Corp.", Iron and Steel Engineer Year Book, 1972, pp. 495-506.

** H. Schmidt, "Pilot Model of an Analogue Programme Computer For a Five-Stand Tandem Rolling Mill", Stahl und Eisen, 1967, 87, Dec. 28, pp. 1565-1571 (In German).
(English Translation BISI 6485, published August 1968, available from the British Iron and Steel Industry Translation Service.)

mill, with a width of 1.42 m, rolls sheet to a finish gage of 0.5 and 1.5 mm and thin strip down to a thickness of 0.12 mm. The maximum rolling speed is 1800 m/minute and the total drive power is 17.9 megawatts. The mill is equipped with the gaugemeter or Sims' method for gage control at the first stand and an x-ray gage for controlling the speed (and hence the entry tension), at the last stand.

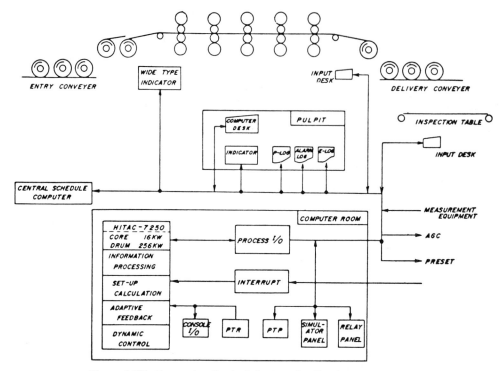

Figure 5-77: Computer Control System for Tandem Sheet Mill.

At the end of 1964, it was decided to equip the mill with a computer which would facilitate changeover of the mill during program changes. The computer was to monitor the various control loops and calculate the desired values for (1) rotational speeds of the drive motors, (2) the screw positions, (3) the interstand tensions and (4) the interstand strip thicknesses. This information was to be derived from (1) the desired finish thickness, (2) the incoming strip thickness, (3) the strip width and (4) the deformation resistance of the strip.

The advantages anticipated by this computer control were (1) a reduction in the quantity of off-gage product, (2) better utilization of the rolling capacity, and (3) a reduction in the time required for mill setup.

Basically there are two possible methods of programming a computer of this type. One involves the storage of information relating to all possible rolling conditions, recalling this data and making necessary interpolations when required. The other involves the analog storage of the functional relationships between the strip data and the mill data, such as screw settings, so that mill settings may be calculated on receipt of the desired strip parameters. The second option was chosen because of its greater economy.

Since the relationship between the input energy to the mill per ton of steel rolled and the total deformation (Figure 5-78) is relatively insensitive to changes in rolling lubrication, etc., it was decided to base the mathematical model on energy requirements. This decision was validated by experiments showing that, in accordance with theory, the energy utilized by the mill per ton of steel rolled is dependent upon the elongation of the strip but practically independent of the entry thickness of the strip. Moreover, in the case of the rolling of strip of medium thickness, the energy consumption when rolling with mineral oil emulsion was similar to that when palm oil was used.

AUTOMATIC CONTROL OF COLD ROLLING MILLS

Initially, therefore, no adjustment was made for the type of rolling lubricant used although it was found later that improved operation could be obtained by making slight adjustments to the stored data in the computer to correspond with different lubricants.

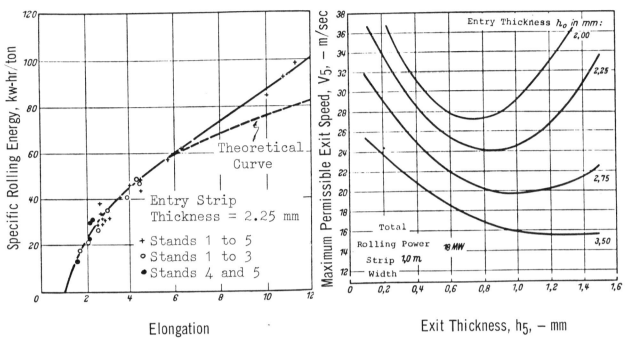

Figure 5-78:
Specific Energy Requirement in Cold Rolling in Relation to Elongation (Including Motor Losses).

Figure 5-79:
Exit Speed in Relation to Entry and Exit Strip Thicknesses.

Figure 5-79 shows the relationship, for a low carbon steel, between the exit thickness and the maximum permissible exit speed at stand 5 for certain selected entry thicknesses and a strip width of one meter. This diagram was based on a total power consumption of 18 megawatts. If the data corresponding to these curves are stored in the computer, it is apparent that the maximum coiling speed can be calculated for any given values of entry and exit strip thicknesses. Since the energy requirements are essentially proportional to the product of strip width and the deformation resistance, the maximum coiling speed must be modified accordingly. Moreover, if the maximum speed of any mill motor is exceeded, the maximum coiling speed must be reduced accordingly.

The desired strip speeds at the various stands can also be deduced from the energy curve shown in Figure 5-80. If the ordinate is divided up in proportion to the power available at each stand, the resulting abscissa will give the corresponding elongation of the strip (relative to its initial length) as it emerges from the stand. From this elongation, it is simple to compute the entry and exit thickness of the strip at each stand, and hence, with reference to the coiling speed, the speed of each stand may be computed as illustrated in Figure 5-80. (Forward slip of the strip at each stand is disregarded.)

The functional relationship between each motor speed and the total elongation of the strip is stored in the analog computer using circuits of resistors and biased diodes. Data corresponding to the initial and final values of strip thickness are fed into the computer digitally and converted to analog form.

The desired value of the strip thickness at the exit of stand 1, as a direct function of the initial and final strip thicknesses, is stored in the computer. For the rest of the stands, the rolling force is taken into account in computing roll gaps.

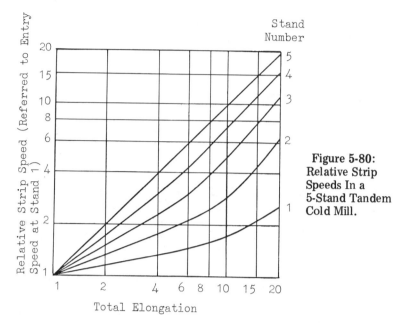

Figure 5-80: Relative Strip Speeds In a 5-Stand Tandem Cold Mill.

During a change in program, not only the desired roll gap setting for the new rolling program but the desired setting from the previous program is calculated, the difference for the two values being used to control the extent of the screw adjustment at each stand. This incremental approach was considered more desirable than moving the screws into fixed positions (as is normal in automatic screwdown systems) so as to compensate for temperature changes in the rolls.

Since there were no sensors available to monitor the amount of screw travel, this factor was determined from the on-time of the screws taking accelerating and decelerating effects into account. This method was found to give adequate accuracy. After a roll change, the rolls are set simply by feeding certain strip dimensions into the computer and then adjusting the rolls manually to an appropriate preliminary roll force. The automatic adjustment to a desired rolling program can then be carried out.

Resetting of motor speeds and screwdowns takes 2 to 3 seconds on the average. The first strip of a new program can be fed into stand 1 as the tail end of the preceding coil is leaving stand 2 provided side guides do not have to be adjusted.

After two years of operation with this computer, it has been found that the length of off-gage strip could be halved on sheet product and reduced by two-thirds on very thin strip. At normal production rates, this increased the yield by about 50 tons per month. Shortening of changeover times was achieved and could have been reduced even further by automatic side guide adjustment. Rolling speeds (and therefore throughput of the mill) were also increased. Schmidt concluded from this operational experience that the analog computer was adequate for the purpose and a more costly digital computer unnecessary.

Chapter 6 Cold Rolling Lubrication

6-1 Introduction

Except for temper rolling, virtually all other cold reduction operations associated with the flat processing of steel products involve the use of rolling lubricants. These materials, usually in the form of fatty or mineral oils, applied either neat or in emulsion form, greatly facilitate the reduction of the strip in that they considerably reduce the rolling forces required for deformation. By so doing, they also (1) make the attainment of acceptable "shape" or flatness in the rolled product much easier, (2) lessen roll wear, (3) reduce roll and strip temperatures, and (4) prevent rusting of the reduced strip.

In this chapter, an attempt has been made to discuss, in general terms, friction in the roll bite under dry and wet rolling conditions, the types of rolling lubricants that are commonly used, their methods of application, how they may be tested and how their frictional characteristics may be established from rolling mill data. Without a knowledge of the effectiveness of the rolling lubricants used, the satisfactory operation of commercial rolling facilities is more difficult to achieve. Moreover, it becomes difficult to use mathematical models of the cold rolling process (such as those presented in Chapter 9) for engineering purposes (such as mill design) and for the computer control of existing mills.

From an economic viewpoint, it is desirable to use the cheapest material that will provide adequate lubricity in the roll bite. Generally speaking, lubricant consumption ranges from 3 to 8 lbs. per ton of steel rolled. From a practical viewpoint, reduction in the quantity of rolling lubricant used is not always feasible because of the "drag out" of the lubricant on the surface of the rolled strip, because of the skimming of the lubricant tanks and because of the other operations necessary to maintain the aqueous dispersion of the lubricant in a satisfactorily clean condition. Accordingly, it is often easier to achieve economies by changing the nature of the rolling lubricant rather than by reducing its consumption.

Although cold rolling lubricants have been extensively used since the early 1930's, research into their behavior was not undertaken to any significant extent until the late 1940's. Even then, for a variety of reasons, the research data was often difficult to interpret or relate to other rolling situations. Moreover, because of incorrect assumptions relating to rolling models, computed values of the coefficient of friction in the roll bite were often of questionable accuracy.

Except in dry rolling (that is, rolling without the use of a liquid lubricant), frictional effects in the roll bite appear to be associated with boundary or thin film lubrication at low rolling speeds and with hydrodynamic lubrication at high mill speeds. Initially, it was believed that the former type of lubrication was prevalent under virtually all rolling conditions and that a constant coefficient of friction could be used to characterize it. However, the realization that the viscosity of the lubricant and the rolling speed profoundly influenced the frictional conditions in the roll bite led to the belief that hydrodynamic effects were also present. However, even at very high mill speeds, it is unlikely that the lubrication is fully hydrodynamic or the oil film continuous in nature.

Various attempts have been made to analyze the frictional conditions in the roll bite on a mathematical basis. Assuming rigid work rolls, a perfectly rigid/plastic material and a lubricant of constant viscosity, it can be shown from energy considerations that the neutral point (where roll and strip surface speeds are the same) lies close to the exit plane and that, contrary to experience, it would be moved towards the entry plane of the roll bite as the entry or back tension is increased. More complex mathematical models of the rolling process have been developed by Cheng[‡] and

---

[‡] H. S. Cheng, "Friction and Lubrication in Metal Processing", ASME, New York, 1966, pp. 69-89.

Bedi and Hillier.□ In both models, a "friction hill" (discussed in Chapter 9) is envisaged, but different approaches are taken with respect to the oil film. Cheng allows for a variation of lubricant viscosity as a function of pressure and temperature, whereas Bedi and Hillier assume a lubricant film constant both in thickness and viscosity. Unfortunately, both models predict a neutral point close to the center of the arc of contact, whereas it should be (and actually is) close to the exit plane.

Thus much remains to be learned about rolling lubrication. However, because of the progress made in understanding the dynamic behavior of metals as they undergo deformation and because of the availability of more extensive and accurate rolling data, a clearer insight is gradually being gained into the frictional phenomena occurring in the roll bite, as will be seen later in this chapter.

6-2 The Necessity of Friction in the Roll Bite

In the cold rolling process, friction along the arcs of contact at the roll-strip interfaces is necessary for the transmission of deformation energy from the work rolls to the strip. If the frictional forces are too small, the peripheral speed of the roll will exceed the exit speed of the strip or, in other words, the rolls will skid.* With the minimum coefficient of friction in the roll bite, the two speeds will be closely matched, whereas larger coefficients will result in a forward or positive slip of the strip (as evidenced by the fact that the exit speed of the strip is in excess of the peripheral speed of the rolls).

If tensile stresses in the strip on the entry and exit sides of a mill stand are disregarded and if the work rolls are considered to be perfectly rigid (i.e., not sustaining any "flattening" or elastic deformation), for conditions of minimum permissible friction, the specific rolling force may be approximated by

$$f \simeq \sigma_C \sqrt{\frac{Dtr}{2}} \qquad (6\text{-}1)$$

where σ_C is the compressive stress required to deform the strip, D is the work roll diameter, t is the entry thickness of the strip and r is the reduction (expressed as a decimal fraction) as illustrated in Figure 6-1. The total frictional force per unit width of strip dragging the strip into the roll bite is

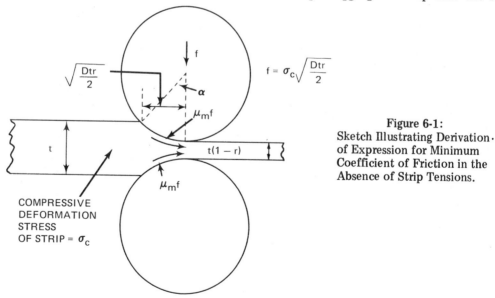

Figure 6-1:
Sketch Illustrating Derivation of Expression for Minimum Coefficient of Friction in the Absence of Strip Tensions.

□ D. S. Bedi and M. J. Hillier, "Hydrodynamic Model for Cold Strip Rolling", Proc. Inst. Mech. Eng., (London), 182, 1967-1968, pp. 153-162.

*Rolling is still often possible even when the roll surfaces are moving faster than the rolled strip. This condition of negative slip is sometimes encountered in the rolling of coated strip (see Section 6-32).

COLD ROLLING LUBRICATION

twice the product of the minimum coefficient of friction, μ_m, and the specific rolling force, namely $2\mu_m f$. Resisting the entry of the strip of unit width into the bite is a force approximately equal to the product of the draft, tr, and the compressive stress, σ_c, namely, tr σ_c. Considering only the horizontal forces acting on the strip of unit width, and disregarding acceleration effects, for conditions of equilibrium, as an approximation

$$2\mu_m f \simeq tr\, \sigma_c \qquad (6\text{-}2)$$

or

$$\mu_m = \sqrt{\frac{tr}{2D}} \qquad (6\text{-}3)$$

where strip tensions exist, however, this relationship is modified. Referring to Figure 6-2, horizontal equilibrium is now attained when

$$tr\, \sigma_c' + t\, \sigma_1 \simeq 2\mu_m f + t(1-r)\, \sigma_2 \qquad (6\text{-}4)$$

where σ_c' is the new value of the compressive deformation stress and σ_1 and σ_2 represent the entry and exit tensile stresses in the strip. Thus,

$$\mu_m \simeq \left(1 + \frac{\sigma_1 - (1-r)\sigma_2}{r\sigma_c'}\right)\sqrt{\frac{tr}{2D}} \qquad (6\text{-}5)$$

From this expression, it will be noted that (1) increasing the entry stress σ_1 increases the value of μ_m and (2) increasing the exit stress σ_2 decreases the value of μ_m.

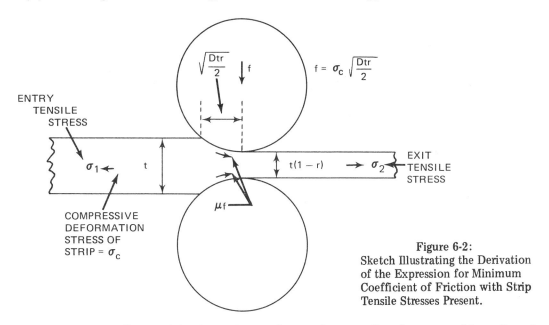

Figure 6-2:
Sketch Illustrating the Derivation of the Expression for Minimum Coefficient of Friction with Strip Tensile Stresses Present.

Referring to Figure 6-1, the entry angle α, when small and expressed in radians is given by

$$\alpha = \sqrt{\frac{Dtr}{2}} \bigg/ (D/2) = \sqrt{\frac{2\,tr}{D}} \qquad (6\text{-}6)$$

Under these circumstances

$$\mu_m \simeq \alpha/2 \qquad (6\text{-}7)$$

and similarly, equation (6-5) may be expressed

$$\mu_m \simeq \left(1 + \frac{\sigma_1 - (1-r)\sigma_2}{r\sigma_c'}\right)\frac{\alpha}{2} \qquad (6\text{-}8)$$

In practice, however, the work rolls elastically deform as illustrated in Figure 6-3 so that their effective or deformed diameter, D', may be two or three times the actual diameter D. The value of D' may be related to D by the Hitchcock [■] relationship (as discussed in Chapter 9)

$$D' = D\left(1 + \frac{4.63f}{Etr}\right) \qquad (6\text{-}9)$$

where E is the elastic, or Young's modulus of the rolls, or by a modification of the relationship

$$D' = D\left(1 + 2\sqrt{\frac{f}{Etr}} + \frac{2f}{Etr}\right) \qquad (6\text{-}10)$$

(also discussed in Chapter 9).[●]

For practical purposes and for calculating the frictional characteristics of a lubricant (see Section 6-31), the value of D' may be considered as double the actual work roll diameter for typical cold rolling situations. Accordingly, in the absence of strip tensions

$$\mu_m \simeq \frac{1}{2}\sqrt{\frac{tr}{D}} \qquad (6\text{-}11)$$

and, in the presence of strip tensions,

$$\mu_m \simeq \frac{1}{2}\left(1 + \frac{\sigma_1 - (1-r)\,\sigma_2}{r\sigma_c}\right)\sqrt{\frac{tr}{D}} \qquad (6\text{-}12)$$

In commercial cold rolling operations using 4-high mills, perhaps the largest value of μ_m would be experienced in a mill rolling hot band, say, 0.25 inches thick with rolls about 25 inches in diameter. Assuming a 25 per cent reduction to be required, μ_m (as derived by equation 6-11) would have to be at least 0.025. In the rolling of light gage strip and foil products, the value of μ_m is considerably smaller and frequently less than 0.010. Fortunately, the use of effective lubricants on high speed mills enables coefficients to be attained which are not significantly larger than the required minimum value, as will be seen in later sections of this chapter.

Friction is also important in the roll bite in that it affects the sideways spread of the workpiece as it undergoes deformation. In the rolling of thin wide strip, the increase in width is negligible so that the assumption that the deformation is characterized by plane strain is justified. However, where relatively thick, relatively narrow strips are cold rolled, some lateral flow of the metal does occur as demonstrated by Siebel and Lueg[◇] and Capus and Cockroft.[□] For the same reduction in the pass, the larger the diameter of the work rolls, the greater the spread whereas, for the same diameter, spread increases with the length of the arc of contact to strip width ratio.[†] Numerous observations have been made to show that reduced friction also reduces spread, the metal in effect finding it easier to elongate in the rolling direction than in the transverse direction.

Increasing the friction in rolling increases the degree of nonuniformity in the deformation of the workpiece as indicated by the curvature of initially straight, vertical lines on the sides, or a longitudinal section, of the workpiece as illustrated in Figure 6-4. A reversal of the curvature of grid lines near the surfaces of a workpiece has been observed by Crane and Alexander[▭] indicating a reversal of frictional forces during the passage of the workpiece through the roll bite.

[■] J. Hitchcock, "Roll Neck Bearings", ASME Research Publication, Appendix 1, 1935.

[●] W. L. Roberts, "Choice of Work Roll Diameter in Cold Rolling Mill Design", Paper Presented at the AISI Regional Technical Meeting, Chicago, Illinois, October 16, 1969.

[◇] E. Siebel and W. Lueg, "Investigations into the Distribution of Pressure at the Surface of the Material in Contact with the Rolls", Arch. Eisenhuttw., 15, 1933, pp. 1-14.

[□] J. M. Capus and M. G. Cockroft, "Relative Slip and Deformation During Cold Rolling", J. Inst. of Metals, 90, 1961-62, pp. 289-296.

[†] "Metal Deformation Processes/Friction and Lubrication", Edited by J. A. Schey, Marcel Dekker, Inc., New York, 1970.

[▭] F. A. A. Crane and J. M. Alexander, "Friction in Hot Rolling", J. Inst. of Metals, 91, 1962-1963, pp. 188-189.

COLD ROLLING LUBRICATION

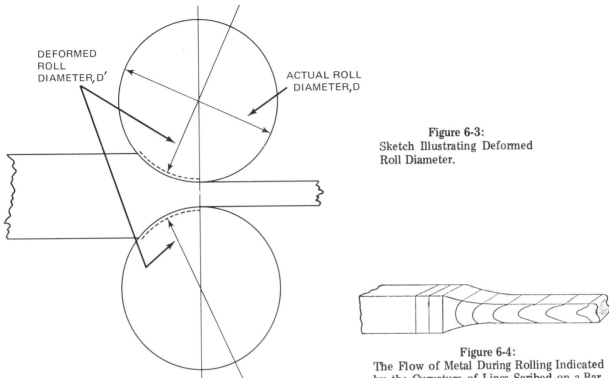

Figure 6-3:
Sketch Illustrating Deformed Roll Diameter.

Figure 6-4:
The Flow of Metal During Rolling Indicated by the Curvature of Lines Scribed on a Bar.

6-3 The Effects of Excessive Friction in the Roll Bite

As seen in Section 6-2, the minimum theoretical coefficient of friction in the roll bite should generally be less than 0.025 for conventional 4-high mills. However, in the rubbing together of dry, steel surfaces, coefficients of friction are developed which are, at least, an order of magnitude larger (generally in the range 0.3 to 0.9). Such coefficients are tolerable only in special circumstances as, for example, in temper rolling, where only a very light reduction is given to the workpiece. With the use of lubricants, however, the effective coefficients are reduced to much lower values thereby making cold rolling a commercially feasible process.

Even with the use of lubricants, however, there is generally an excess of friction in the roll bite. This results in (1) increased rolling forces, (2) slightly increased rolling power requirements, and (3) a forward slip of the strip as it leaves the roll bite. If the frictional effects are considerably in excess of those corresponding to minimal frictional requirements, then difficulties may be encountered in obtaining satisfactory rolling in that the rolling forces may be so large that the accompanying roll bending gives the rolled strip a poor "shape" or an inadequate degree of flatness. Moreover, the dissipation of the excessive frictional energy may result in abnormally high roll and strip temperatures. In the case of the former, nonuniform heating of the rolls may affect the ease of attaining satisfactory rolled strip shape, whereas the latter may be detrimental with respect to the quality of the coiled strip if used for such critical applications as tin plate.

Increased friction also produces a more lustrous finish on the rolled strip, on account of the increased buffing action of the rolls on the strip surfaces (see Section 6-16). Under some circumstances, the higher luster may be desirable as in the rolling of stainless steels. However, if the lubricant is not equally effective at all points in the roll bite, the rolled strip may exhibit streaking or, on a smaller scale, a so-called "staining".

The effect of friction on the rolling force may be understood by reference to Figure 6-5. Disregarding work hardening, the pressure exerted by each roll on the strip along the arc of contact may be regarded as essentially constant (except for very short regions at the end of each arc where elastic deformation of the strip occurs) for the case where the coefficient of friction assumes the minimum value. However, as the coefficient is increased, the neutral point (where roll and strip surface speeds are equal) moves towards the entry plane of the roll bite. The pressure along the arc of contact is no longer constant but assumes the form of two exponential curves creating a "friction hill" first postulated by Siebel♦ (and discussed in greater detail in Chapter 9). The specific rolling force (or the rolling force per unit width of strip) is obtained by integrating the area under the "friction hill" and, as the neutral point is shifted towards the entry plane due to increasing friction, the height of the "friction hill" and, consequently, the specific rolling force both increase.

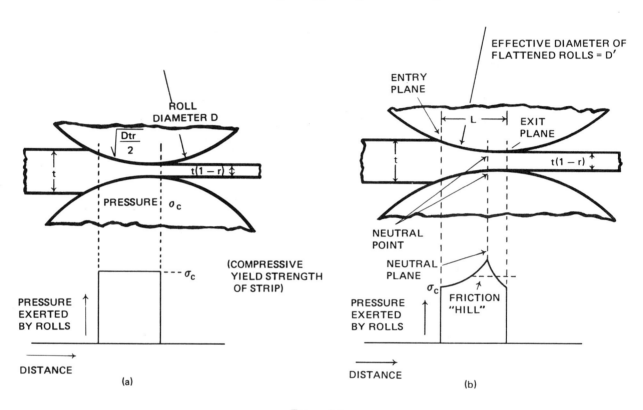

Figure 6-5:
Sketches Illustrating the Hypothetical Pressure Distribution Along the Arc of Contact.
a) Under Minimal Frictional Conditions; b) With the Existence of a "Friction Hill".

The effect of friction in the roll bite is perhaps most dramatically demonstrated when strip of minimum gage is being rolled (see Chapter 9). According to Stones'‖ § well-known formula, the lightest gage strip that can be rolled on a mill will have a thickness t_m given by

$$t_m = \frac{3.58 \, D \mu \, \sigma_c}{E} \quad (6\text{-}13)$$

where D is the diameter of the work rolls
μ is the coefficient of friction
σ_c is the stress required to deform the strip under compression and
E is the elastic modulus of the rolls.

♦ E. Siebel, "Berichte des Walzwerksausschusses", Verein deutscher Eisenhuttenleute, No. 37, 1924.
‖ M. D. Stone, "The Rolling of Thin Strip — Part I", Iron and Steel Engineer Year Book, 1953, pp. 115-128.
§ M. D. Stone, "The Rolling of Thin Strip — Part II", Iron and Steel Engineer Year Book, 1956, pp. 981-1002.

Although other minimum gage formulae have been developed,♦ they all essentially show a direct relationship between final gage and the effective coefficient of friction in the roll bite. For this reason, in rolling foils with thicknesses close to the minimum obtainable, the use of the most effective lubricant becomes a matter of critical economic importance.

The increased rolling force attributable to excessive friction in the roll bite increases the frictional energy dissipation along each arc of contact as well as the bearing losses associated with the mill rolls. To some extent, however, this increased energy supplied by spindles is offset by a slightly increased throughput of the mill stand due to the increased forward slip of the strip as it emerges from the roll bite.

6-4 Frictional Conditions in Dry Rolling

In temper rolling, in which high coefficients of friction are encountered in the roll bite, the principal objectives are the attainment of the desired metallurgical properties in the rolled strip and the desired finish of its surfaces. It is a process not intended to significantly change the thickness of the processed strip and it occurs subsequent to its annealing and frequently prior to the application of a coating, such as tin, to its surface. Consequently, it is generally undesirable to utilize a lubricant on a temper mill since the bright lustrous finishes often sought for tin plate products would not be so readily obtainable, the lubricants would have to be removed and the extent of the elongation (or reduction) would be more difficult to control.

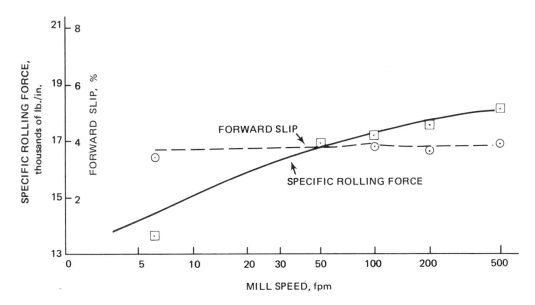

Figure 6-6:
Variation of Forward Slip and Specific Rolling Force with Mill Speed
(Annealed Low Carbon Steel Strip Rolled without a Lubricant).

The indications are that the frictional effects in the roll bite of a "dry" temper mill are attributable to simple metal-to-metal sliding contact and comply with Amanton's law. For example, it has been found that the frictional effects do not change significantly with such processing conditions as mill speed and reduction. This is evidenced by the constancy of the forward slip (which is indicative of the coefficient of friction in the roll bite) as shown in Figure 6-6 where the specific rolling force and the slip are shown plotted against mill speed in the rolling of black plate

---

♦ W. L. Roberts, R. J. Bentz and D. C. Litz, "Cold Rolling Carbon Steel Strip to Minimum Gage", Iron and Steel Engineer Year Book, 1970, pp. 413-420.

on a small laboratory mill.[♦] Nor is the coefficient significantly changed by a change in reduction (or the corresponding change in specific rolling force) as illustrated in Figure 6-7[⊕] where the coefficient of friction appears to be close to 0.3 over the range of reduction given to the black plate. The same value for the coefficient was obtained by Shutt[▲] in the deformation of a cylinder and the same type of frictional effects were observed in the dry rolling of galvanized strip but, in this case, the coefficient was reduced to a value of about 0.25 (Figure 6-8).[⊕]

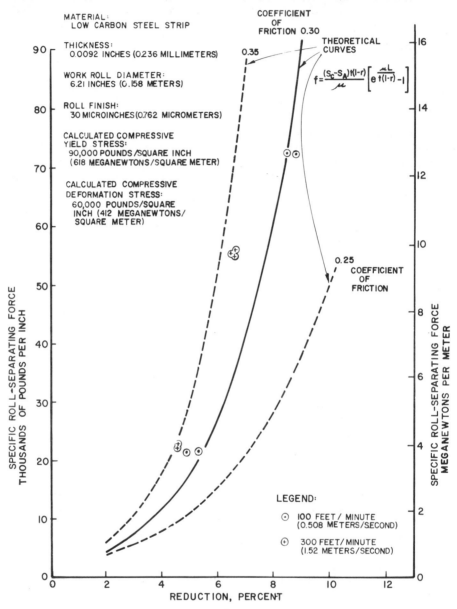

Figure 6-7: Variation of Specific Roll-Separating Force with Reduction for Low-Carbon Steel Strip Processed on a Laboratory Mill.

[♦] R. J. Bentz and W. L. Roberts, "Speed Effects in the Second Cold Reduction of Steel Strip", Mechanical Working and Steel Processing V, Proceedings of the Ninth Mechanical Working and Steel Processing Conference, AIME, Gordon and Breach, New York, New York, 1968, pp. 193-222.

[⊕] W. L. Roberts, "An Approximate Theory of Temper Rolling", Iron and Steel Engineer Year Book, 1972, pp. 530-542.

[▲] A. Shutt, "Note: On the Measurement of Friction Under Yield Conditions", Inst. J. Mech. Sci., Pergamon Press, Ltd., 1966, Vol. 8, pp. 509-511.

COLD ROLLING LUBRICATION

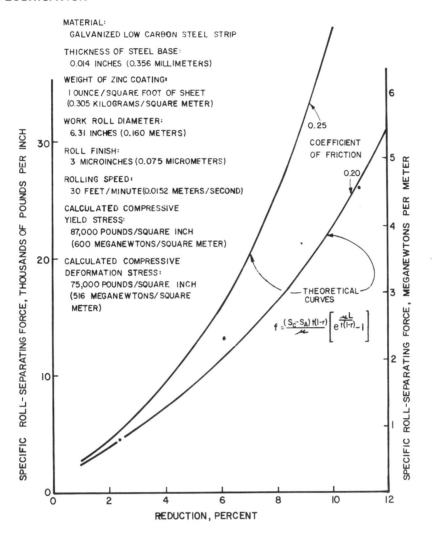

Figure 6-8:
Variation of Specific Rolling Force with Reduction for Galvanized Strip Processed on a Laboratory Mill.

Starchenko[⊞] and his colleagues found that in the reduction (42 per cent) of low carbon steel, 2.7 mm thick, with dry clean rolls of 290 mm (11.42 inches) diameter, the coefficient of friction increased slightly with roll surface speed, the effect being more marked in the range 4.5 — 15.7 m/sec. (885 — 3090 f.p.m.). This is indicated in Table 6-1.

Although dry rolling has certain desirable features, it has the disadvantage of generating metallic dust abraided from the rolls and strip. This can accumulate on the mill housings and produce a "dirty" mill with the danger that parts of the accretion may fall onto the strip and be rolled into it. For this reason, there is a growing tendency to use "wet" temper rolling where a lubricating "solution", albeit a rather poor one from a lubrication standpoint, keeps the mill clean.

In some instances, dry temper rolling of light-gage black plate is carried out prior to secondary cold rolling. This practice appears to break up and remove very slight residues of electrolytes left on the strip from cleaning operations (as well as any residual oxides). Such residues would otherwise impair the effectiveness of the roll bite lubricity on secondary cold reduction mills and necessitate slower rolling speeds.

⊞ D. I. Starchenko, V. I. Kuz'min and S. A. Obernikhin, "Friction in the High Speed Cold Rolling of Metals", Izvest. VUZ — Churn. Met. Vol. 10, 1967, #8, pp. 86-91, (Henry Brutcher Translation HB #7272).

Table 6-1
Variation of Calculated Coefficient of Friction with Mill Speed in the Rolling of Low Carbon Steel without a Lubricant

Speed M/S	Calculated Coefficient of Friction
4.5	0.095
9.5	0.103
15.7	0.105
30.1	0.118

6-5 The Characteristics of An Ideal Rolling Lubricant

The desirable characteristics of a rolling lubricant may be listed as follows: ◈◀•

(1) It must provide a lubricity and a corresponding rolling force so that satisfactory strip shape is obtained.

(2) It should be easy to apply, either neat or in the form of an emulsion (generally, but incorrectly, termed a rolling solution).

(3) It should not seriously impede the heat transfer from the rolls and strip to the coolant.

(4) It should not make mill "housecleaning" more difficult by excessive "plating out" on the mill.

(5) It should be consistent in performance.

(6) It should provide the desired degree of protection from rust in the rolled product.

(7) It should be easily removed from the strip after rolling.

(8) It should be nontoxic and not cause dermatitis or other health problems on contact.

(9) It should be economical with respect to usage.

The lubricity provided by a rolling lubricant is generally assessed in terms of the effective coefficient of friction it provides in the roll bite. Initially, it was believed that the closer the coefficient could be brought to the minimum value required for rolling the better. However, a reduction in the coefficient to this value may not be desirable from the viewpoint of strip shape. For example, if a mill operator has learned to crown his rolls so that, with a lubricant of average effectiveness, he produces satisfactory shape, then when he uses a superior lubricant (that is, one which exhibits a lower coefficient of friction) the rolled strip will tend to be full centered (see Section 11-20). Hence, unless roll bending or roll crowning provisions are made on the mill, the optimum lubricity is that which develops a rolling force such that flat strip is produced.

To be effective, a lubricant must coat the rolls and/or the strip prior to the entry of the latter into the roll bite. Though solid lubricants have been used experimentally, they are usually difficult to apply on high speed mills and, for this reason, virtually all lubricants (even though some are solid at room temperature) are applied as liquids. The melting or pour points must not be too high, otherwise the cooling of the rolls will not be as efficient as desirable and the "housekeeping" of the mill will be made more difficult. When used neat, the lubricants are usually applied as a mist

---

◈ J. A. Schey, "Purposes and Attributes of Metalworking Lubricants", Lubrication Engineering, May, 1967, pp. 193-198.

◀ J. G. Wistreich, "Note on Research into Lubrication in Cold Rolling", The British Iron and Steel Research Association Report MW/A/72/56.

• K. Saeki, "Oils for Steel Strip Rolling", Yukagaku, 1961, 10 (2), pp. 83-88 (British Iron and Steel Industry Translation Service BISI 2657).

COLD ROLLING LUBRICATION

to the rolls and/or the strip. Most commonly, however, lubricants are applied as aqueous emulsions with concentrations in the range 3-15 per cent. Ideally, it would be desirable for lubricants to be so dispersed in water without significant quantities of emulsifiers since the latter act like detergents with respect to roll and strip surfaces and tend to minimize the quantity of oil "plated out" on them.

The "plated out" films of oil on the roll and strip surfaces tend to act as thermal insulators by lowering the coefficient of heat transfer at the surfaces. Ideally, this effect should be kept as small as possible so as to ensure satisfactory roll and strip cooling.

At the present time, most rolling lubricants are sold as proprietary brands, the formulations of which are closely guarded commercial secrets. Since the market prices of the ingredients frequently fluctuate, there is a tendency on the part of the lubricant vendors to make slight changes in the formulations of these materials. As a consequence, the rolling performance of such lubricants tends to be inconsistent, a condition which is aggravated if the local water supply also exhibits seasonal changes in hardness and/or contamination. Accordingly, an ideal lubricant is one which exhibits a consistent rolling performance.

In spite of the high pressures between the roll and strip surfaces encountered in cold rolling, a residual film of lubricant, usually a few millionths of an inch thick, remains on the surfaces of the strip after it emerges from the roll bite. This film is useful in retarding the rusting of the strip and should be as effective as possible in this respect. However, prior to the subsequent processing of the strip the oil film must be completely removed (see Section 12-2). This is usually accomplished, in the case of sheet product, simply by vaporization of the lubricant from the surfaces of the strip during annealing but, in tin plate manufacture, lubricant removal is generally effected by electrolytic action in cleaning tanks. In the former, no carbonaceous or other residues must be left on the strip and, in the latter, the lubricant must be completely removed so that, if the strip is immersed in water, the entire surface is wetted.

A rolling lubricant should, of course, be acceptable from safety and hygienic standpoints. It should not be toxic and its vapors should offer no health hazard. Nor should it act as a culture for bacteria, otherwise it may promote dermititis or other skin problems. At the same time, it should not exhibit rancidity or other objectionable characteristics. For these reasons, rolling oils frequently contain bactericides and anti-oxidants.

The usage rate of a rolling lubricant depends upon its mode of application and the rolled thickness of the strip but usually it lies in the range of 3 to 8 lb. per ton of steel rolled. Accordingly, it is desirable, from a production standpoint, to minimize this cost by using the least expensive materials and by minimizing their usage rates.

6-6 Water as a Rolling Lubricant

Although it is not often used as a rolling lubricant, water can provide coefficients of friction in the roll bite less than those prevailing under conditions of dry rolling. Hence, it finds occasional application in wet temper rolling or where rolling conditions are not severe and where strip cleanliness must be preserved. Without contamination from oil, it provides superior roll cooling but obviously no rust protection as far as the coiled product is concerned.

As in the case of oil emulsions, the coefficient of friction provided by water increases with increasing draft to work roll diameter ratio but, unlike aqueous emulsions of fatty materials, the coefficient increases with increasing rolling speed. The former effect, relating to the so-called first frictional characteristic, is discussed in Section 6-31, whereas the latter, relating to the second frictional characteristic, is evidenced by the increased rolling forces at the higher mill speeds encountered in the rolling of black plate (see Figure 6-9). This increasing friction with increasing mill speed is also encountered in the case of true solutions (not emulsions) of concentrations up to 20 per cent or higher (even where the solute may provide good lubricity in neat form) and frequently in the case of aqueous emulsions of mineral oils.

Figure 6-9:
Curves Showing the Variation of Specific Rolling Force with Reduction for Water and a Substitute Palm Oil at Two Rolling Speeds in the Rolling of Box Annealed Black Plate. (Initial Strip Thickness 0.0095 in.; Roll Finish 30 Microin. R.M.S.; Strip Tensile Stresses, Approx. 15,000 psi)

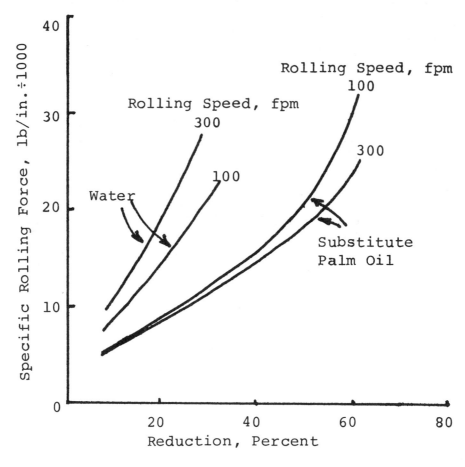

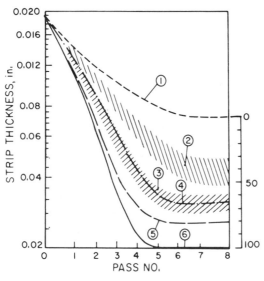

Figure 6-10:
The Efficiency of Various Lubricants in Rolling Low Carbon Steel Strip (1:water; 2:Commercial Emulsions; 3:Improved Emulsions; 4:Undiluted Concentrate; 5:Best Emulsion; 6:Palm Oil)

In the evaluation of rolling lubricants, for the sake of convenience, water has sometimes been used as a reference lubricant at the lower end of a scale with palm oil as a similar reference material at the upper end of the scale. When such a comparison is made, specific rolling forces are measured for a given reduction and are used for the basis of comparison. In the absence of force measuring equipment, however, Billigmann[⊗] evaluated lubricants with the same reference standards but on the basis of the limiting strip thickness achieved and the number of passes required to attain it (Figure 6-10). With water ranking as 0 and neat palm oil as 100, most oil emulsions ranked 50, with tallow base and palm oil emulsions ranking 80-90. The ranking was found to be unaffected by a change in the quality of the mild steel strip used. It did correlate, however, with the free fatty acid content, the saponification number and the potassium hydroxide neutralization number of the oil but not with its viscosity.

Where the lubricity in the roll bite is provided either by a soft metallic or organic coating, or the pickle line oil, water may be used as the roll coolant. However, where metallic coatings are used and relatively large reductions are given to the strip, the metallic coating may be "picked up" by the rolls. This can be usually prevented by the use of a substitute palm oil emulsion as the roll coolant instead of water.

6-7 Commonly Used Rolling Lubricants

Although rolling lubricants are sold commercially under brand names, basically their constituents are selected from oils, fats and waxes. Prior to World War II, palm oil was the most widely used rolling lubricant but the shortage of the imported lubricant spurred the search for substitute materials. This search led to the use of tallows and a large number of other organic materials presently in use today and commly referred to as "substitute palm oils", many of which are usually blended with other materials such as emulsifiers, defoamers, antioxidants, and rust inhibitors. These additives, particularly emulsifiers, if used in excessive amounts, diminish the effectiveness of the rolling lubricant in that it is not so readily "plated out" on the rolls and strip from the rolling solution.

Generally speaking, substitute palm oils (derived from vegetable and animal oils), for use as cold reduction lubricants, are esters of the trihydric alcohol, glycerol, with such saturated fatty acids as lauric, myristic, palmitic, and stearic. These esters are commonly called glycerides. In addition, esters of unsaturated acids of the acrylic series, such as oleic acid, are usually present in the substitute palm oils.

When a fat contains a relatively large proportion of palmitic or stearic acid, it is solid and comparatively hard, like tallow, at ordinary temperatures. But, when it contains a relatively large proportion of combined oleic acid, it is soft and pasty, like lard, or liquid, like olive oil.

The chemical analyses of palm oil and two commercially available substitute palm oil lubricants are shown in Table 6-2.[∞] The substitute palm oils contain less of the palmitic acid but appreciably more of the stearic acid components.

Table 6-2
Analysis of Palm Oil and Two Substitute Palm Oils

Fatty-Acid Component	Palm Oil	Substitute Palm Oils	
		A	B
Palmitic (%)	32.3 − 40.0	27.6	26.1
Linoleic (%)	5.0 − 11.3	3.9	2.9
Oleic (%)	39.8 − 52.4	45.9	39.4
Stearic (%)	2.2 − 6.4	17.1	20.4
Myristic (%)	1.0 − 5.9	2.9	3.2

[⊗] J. Billigmann, "Untersuchungen uber die Schmierwirkung von Walzolen besonders von Walzolemulsionen", Stahl und Eisen, 1955, 75, Dec., pp. 1691-1705.

[∞] W. L. Roberts and R. R. Somers, "The Cold Rolling Lubrication of Steel Strip", Lubrication Engineer, 18, 1962, pp. 362-368.

The optimum level of free fatty acid in a rolling lubricant is still unresolved.[▶] Experiments conducted on a laboratory mill show, on the basis of rolling force, that commercial acidless tallow blended with the methyl esters of tallow fatty acids (with a total free fatty acid content less than 1 per cent) can perform quite satisfactorily as a rolling lubricant, as evidenced by Figure 6-11. However, Neckervis and Evans[⊗] have reported that 16 per cent free fatty acid in Nigerian palm oil provided better lubricity than the same oil refined until its free fatty acid content was reduced to 0.55 per cent. Moreover, the extracted acid was found to be a more effective lubricant than the original oil. However, other investigators have found that materials with only 1 per cent free fatty acid can perform well as rolling oils.[°,+] Lueg et al,[♦] found that additions of free fatty acids to neutral fats (rapeseed oil, tallow and lanolin) provided only slight benefits and drawing experiments conducted by Johnson and his co-workers[ᗺ] showed the addition of free fatty acid to tallow to be either ineffective or harmful.

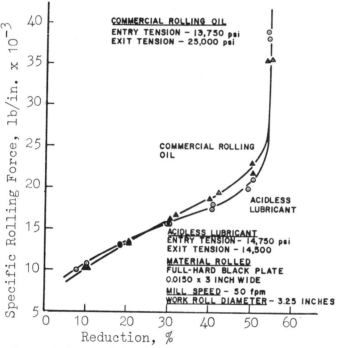

Figure 6-11: Rolling Performance Curves for Acidless Tallow and Commercial Rolling Oil.

Saturated animal fats such as tallow have long been used successfully as substitutes for palm oil whereas fish oils, containing larger proportions of unsaturated fatty acid esters, are generally not as effective as rolling lubricants. Of the fatty oils derived from plants, castor oil and cottonseed oil have been found to equal palm oil in performance.[✦] Rapeseed oil is inferior and linseed and similar oils containing unsaturates are normally undesirable because of their tendency to oxidize.

[▶] W. L. Roberts, "Recent Developments in Rolling Lubrication", Blast Furnace and Steel Plant, May, 1968, pp. 382-394.

[⊗] R. J. Neckervis and R. M. Evans, "Lubricants for Cold Rolling Steel", Iron and Steel Engineer Year Book, 1948, pp. 752-761.

[°] J. C. Whetzel and S. Rodman, "Improved Lubrication in Cold Strip Rolling", Iron and Steel Engineer Year Book, 1959, pp. 238-247.

[+] R. C. Williams and R. K. Brandt, "Experimental Evaluations of Cold Rolling Lubricants for Steel Strip", Lubrication Engineer, 20 (2), 1964, pp. 52-56.

[♦] W. Lueg, P. Funke and W. Dahl, "Untersuchung von Walzolen und Walzolemulsionen im Kaltwalzversuch", Stahl und Eisen, 77, 1957, pp. 1817-1830.

[ᗺ] W. R. Johnson, H. Schwartzbart and J. P. Sheehan, "Substitutes for Palm Oil in the Cold Rolling of Steel", Blast Furnace and Steel Plant, 43, April, 1955, pp. 415-423.

[✦] Y. Iwao, H. Hirano and I. Kokubo, J. Jap. Soc. Technol. Plast. 8, 1967, pp. 248-255.

COLD ROLLING LUBRICATION

Experimental data relating to the use of soybean oil (containing 88 per cent unsaturated acids) are illustrated in Figure 6-12 in comparison with a commonly used commercial rolling lubricant. However, use of such materials may be undesirable for other flat processing operations. Their tendency to polymerize or form resinous materials if uninhibited may make it difficult to store the lubricant prior to use and also to remove the residual oil from the rolled strip.

Under certain conditions, waxes have been found to be efficacious as lubricants in laboratory tests as indicated in Figure 6-13, but when used in the recirculating systems of production mills, they have been found to be ineffective.

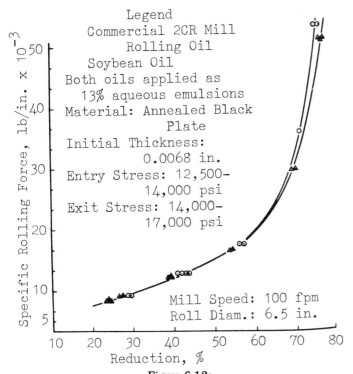

Figure 6-12:
Rolling Performance Curves for Soybean Oil and a Commercial 2-CR Mill Lubricant.

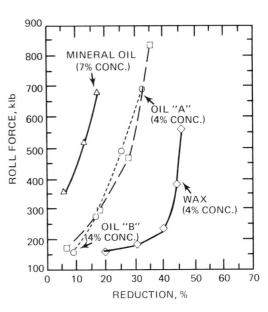

Figure 6-13:
Curves Showing Variation of Rolling Force with Reduction for Full-Hard Low-Carbon Steel Strip Using Various Lubricants.

Petroleum or mineral oils usually do not have the rolling lubricity possessed by animal and vegetable type materials.◊ This is evident from Table 6-3. Seemingly, the chemical constituents of the oil must account for this behavior, the less chemically active, petroleum oil molecules adhering more loosely to the steel surfaces and therefore being unable to withstand the shearing stresses exerted in the rolling process. However, in spite of the higher coefficients they provide, they are frequently used on mills with small diameter work rolls, as diluents for better rolling lubricants, and for the rolling of thicker sheet products when the frictional requirements are not so severe.

It appears that the effectiveness of a mineral oil improves as its viscosity increases,♦✦ but studies by Iwao,✧ et al, have shown that predominantly naphthenic oils are less effective than their paraffinic counterparts of the same viscosity.

◊ W. L. Roberts, "Recent Developments in Rolling Lubrication", Blast Furnace and Steel Plant, May, 1968, pp. 382-394.

♦ J. C. Whetzel and S. Rodman, "Improved Lubrication in Cold Strip Rolling", Iron and Steel Engineer Year Book 1959, pp. 238-247.

✦ S. F. Chisholm, "Some Factors Governing the Choice of Lubricants in the Rolling of Steel", Proc. Inst. Mech. Eng. (London) 179, Pt. 3D (Iron and Steel Works Lubrication), 1964-65, pp. 56-64.

✧ Y. Iwao, H. Hirano and I. Kokubo, J. Jap. Soc. Technol. Plast., 8, 1967, pp. 248-255.

The desirability of using additives in mineral oils is debatable. Certain additives such as lead oleate, oleic acid and stearic acid have been shown to improve their lubricity.※ However, Whetzel and Rodman° found that a 5 per cent addition of stearic acid to a mineral oil with a viscosity of 600 SUS did not improve its performance (see Table 6-3).

Table 6-3
Effect of Lubricant on Reduction Obtained in the Rolling of Steel Strip

Lubricant Type	Characteristics	Reduction, %
Mineral oils	70 SUS at 100°F (38°C)	16.0
	600 SUS at 100°F	24.0
	600 SUS +5% stearic acid	24.0
Fatty oils	Typical animal fat	32.5
	More viscous fat	39.0
	Stearic acid	32.5
	Oleic acid	30.0
	Palm oil	32.5
	Tallow oil (0.25% free fatty acid)	32.5
"Water soluble"	Experimental	28.0—59.0

(0.0063 x 0.250 in., cold rolled SAE 1010 strip; 4 in. diameter rolls at 300 fpm, 38°C; rolls and strip flooded, roll force 3800 lb.)

Extreme pressure additives such as zinc thiophosphate and chlorinated paraffin have been found to be totally ineffective when rolling at low speeds ⊥ although a sulphur containing mineral oil was claimed to be more effective at high speeds.⊞

In recent years attempts have been made to use both water soluble and highly emulsified materials as rolling lubricants in the hope that cleaning the rolled strip might be simplified and that a shinier strip might be produced. Generally, these types of rolling solutions are not satisfactory because of the higher frictional conditions existing in the roll bite and they can be used only in so-called "wet temper" rolling or in rolling situations in which only a light reduction is to be taken.

6-8 Methods of Applying Rolling Lubricants

Generally speaking, rolling lubricants are applied either neat or in the form of aqueous emulsions. In the former method of use, the oil may be applied in minimal quantities to satisfy only the lubrication requirements (roll and strip cooling being carried out separately, if necessary) or it may be applied copiously to act additionally as a coolant and a bearing lubricant at the same time (as in certain Sendzimir mills). When used in the form of aqueous emulsions (usually with concentrations in the 3 to 15 per cent range), the oil "plates out" on the strip and roll surfaces to provide the desired lubricity in the roll bite, but the emulsion also acts as a coolant for both the rolls and the strip.

When applied neat to the mill rolls or strip, the lubricant may be atomized, or misted, using appropriate equipment. One type of applicator, known as a Grammer Oiler, is shown in Figures 6-14 and 6-15 and this type of equipment has been successfully used in applying cottonseed oil to steel strip rolled into foils at speeds in excess of 1,000 f.p.m. on a production mill. It has the

---

※ H. Yamanouchi and Y. Matsuura, Rep. Castings Res. Lab., Waseda Univ. No. 8, 1957.

° J. C. Whetzel, Jr. and S. Rodman, "Improved Lubrication in Cold Strip Rolling", Iron and Steel Engineer Year Book, 1959, pp. 238-247.

⊥ Y. Iwao, H. Hirano and I. Kobubo, J. Jap. Soc. Tech. Plast., 8, 1967, pp. 248-255.

⊞ D. I. Starchenko, V. I. Kuz'min, S. A. Obernikhin and V. I. Kaplanov, "Investigating High Speed Cold Rolling of Low-Carbon Steel with Different Lubricants", Stal in English, 1966 (2), pp. 128-131.

COLD ROLLING LUBRICATION

advantage of applying a very thin uniform film of oil thereby affording economy of lubricant usage and eliminating the problems of streaking or mottle attributable to the nonuniformity of "plate out" sometimes experienced with aqueous emulsions.

Figure 6-14:
Grammer Oiler for
Experimental Mill.

Alternately, a mist type applicator may be used for coating the mill rolls with oil so that the oil is transferred to the roll bite by the rotation of the rolls. This mode of application, often used in hot mill lubrication, is illustrated in Figure 6-16. However, care must be exercised so that when the mill rolls are cooled, the lubricant supply to the roll bite is not interrupted.

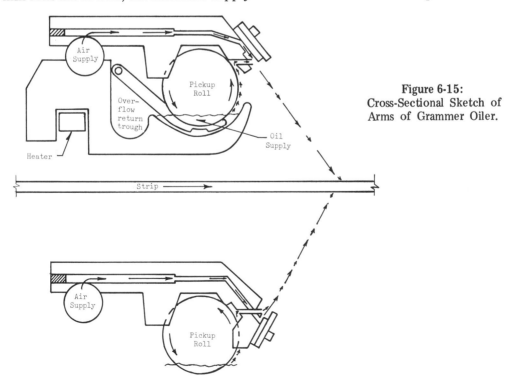

Figure 6-15:
Cross-Sectional Sketch of
Arms of Grammer Oiler.

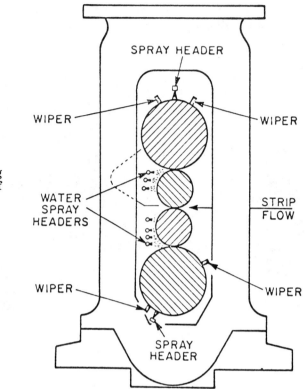

Figure 6-16:
Sketch Illustrating the Application of Sprays to Backup Rolls.

Bentz and Somers + used an electrostatic oiler for applying oil to steel strip in their studies relating to the minimum oil film thickness necessary for adequate lubricity for a 4-high mill with 3-1/4 inch diameter work rolls. They rolled annealed and full hard black plate, as well as tinplate, with cottonseed oil films of various thicknesses deposited on the strip surfaces with the oiler. In addition, they rolled the same materials using a 10 per cent cottonseed oil emulsion sprayed into the roll bite in the conventional manner. Some of their data for tinplate and annealed black plate are reproduced in Figures 6-17 and 6-18, and the dependence of measured rolling force

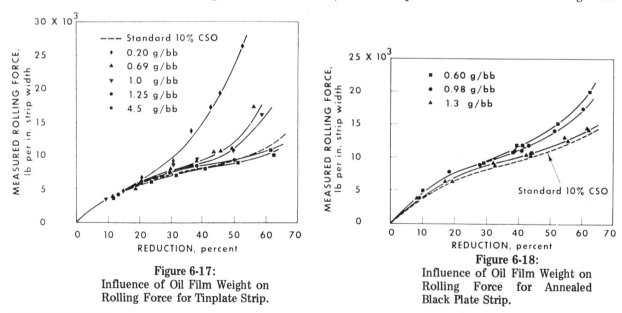

Figure 6-17:
Influence of Oil Film Weight on Rolling Force for Tinplate Strip.

Figure 6-18:
Influence of Oil Film Weight on Rolling Force for Annealed Black Plate Strip.

+ R. J. Bentz and R. R. Somers, "Determination of Minimum Oil Film Weight for Rolling Steel Strip", Lubrication Engineering, February, 1965, pp. 59-64.

on oil film weight for a 50 per cent reduction given to the three types of workpieces, is illustrated in Figure 6-19.

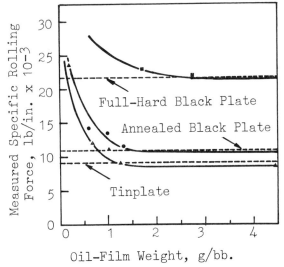

Figure 6-19: Influence of Strip Properties on Minimum Oil-Film Weight Required for 50 Per Cent Cold Reduction.

In most instances, and especially where conventional 4-high mills are concerned, the rolling lubricant is applied as an aqueous emulsion. In the emulsion, the dispersion of the oil in the water is usually achieved by a chemical emulsifier, although the mechanical agitation of the oil and water phases also contributes to the state of emulsification. If the emulsion is stable, the oil droplets will remain dispersed in the water for an indefinite period whereas, if it is unstable, the two phases of the emulsion tend to separate rapidly. Generally speaking, in recirculating systems, it is desirable to use a quasi-stable emulsion in which the oil will remain dispersed for periods longer than the recycle time of the emulsion, otherwise the oil may separate out in the reservoir so that only the water phase would be circulated by the lubrication system of the mill. In direct application systems, less stable emulsions may be used, since the oil need remain dispersed only for the period associated with the transfer of the emulsion from the reservoir to the strip.

The quantity of emulsifier used in effecting the dispersion of the oil in the solution is therefore critical with respect to the lubricity provided by the emulsion. Whetzel and Rodman° found that, whereas 0.5 per cent emulsifier added to a natural fat improved its lubricity in the rolling of steel, a level of 4.0 per cent worsened its performance. Such behavior is related to the effect of the emulsifier on the "plating out" characteristics of the emulsion, the greater the concentration of the emulsifier, the less the tendency for the oil to "plate out" or the thinner the "plated out" film of oil on the roll and strip surfaces.

The stability of recirculated emulsions is affected by a number of factors which include (a) temperature, (b) hardness or contamination of the water used, and (c) chemical changes occurring in the oil. If the temperature of the solution is too low, the lubricant phase may tend to solidify and hence be difficult to maintain in dispersion. Conversely, if the temperature is too high, the effectiveness of the emulsifier may be diminished. Thus a reasonably narrow temperature range exists in which emulsions are generally used, this being about 100-150°F. Chemicals which dissolve in the water phase which affect its hardness may exercise a detrimental effect on the stability of the emulsion and, for this reason, some production mills utilize deionized or purified water for making up emulsions.± Where particulate contamination occurs, for example, due to the presence of iron fines, the oil tends to be pulled out of dispersion through its tendency to adhere to the particles. Chemical changes occurring in the oil phase usually decrease the pH value of the water phase

---

° J. C. Whetzel, Jr. and S. Rodman, "Improved Lubrication in Cold Strip Rolling", Iron and Steel Engineer Year Book, 1959, pp. 238-247.

± A. T. Slyusarev, et al, "Increasing Emulsion Stability in the Cold Rolling of Sheet Metal", Metallurg. 1966 (11), pp. 29-30 (British Iron and Steel Translation Service BISI 5693)

(making it more acidic) and, under these circumstances, the effectiveness of an emulsifier (which is generally intended to be used only in a narrow range of pH values) may be adversely affected.

To maintain the lubricity of an aqueous emulsion, additives are usually blended into the emulsion to prevent (1) chemical changes (oxidation) from occurring, (2) the growth of bacteria, (3) foaming and (4) corrosive attack on the mill components. However, as Schey† has pointed out, some emulsions develop their best performance after decomposition has produced more active constituents within them or, possibly, changed their stability to provide faster "plate out" of the oil.

Although the aqueous emulsion is colloquially called the rolling solution, true solutions are seldom, if ever, used as rolling lubricants primarily because their low viscosity results in poor lubricity. This is evidenced by Table 6-4 showing the effectiveness (in terms of the thickness of the rolled strip) of mineral oil, oleic acid, a detergent and glycerine, both neat and as 5 per cent aqueous dispersions. Glycerine, quite effective as a neat lubricant, virtually loses its lubricity when dissolved in water.♦

Table 6-4
Lubricating Performance of Substances and Their Aqueous Dispersions

Lubricant	Molecular Weight	Neat Lubricant		Water + 5% Lubricant	
		Exit Thickness in.	Viscosity at 38°C, cs	Exit Thickness, in.	Viscosity at 38°C, cs
Water	18	0.018	0.66	—	—
Mineral Oil	250	0.013	5	0.013	1
Oleic acid	282	0.009	21	0.009	1
Detergent	750	0.007	40	0.010	1
Glycerine	92	0.007	224	0.016	1

6-9 The Oiling of Pickled, Hot Rolled Strip

Subsequent to pickling, rinsing and drying, hot rolled strip is oiled prior to coiling. The oil used, designated a "pickler oil" or a "pickle line oil", serves three purposes. It prevents the rusting of the freshly pickled, steel surfaces, it protects the wraps of the coil from scuffing each other as the coil is unwound for further processing, and it acts as a rolling lubricant for one or more passes in a cold reduction mill. Commonly, in the application of the oil, the strip is passed between a set of oiling or wringer rolls (see Figures 6-20 and 6-21) which cover both surfaces with a thin film of oil, the quantity of oil applied being controlled by adjustment of its temperature (and therefore its viscosity).◊

The physical and chemical characteristics of the pickler oil should be contingent on the manner in which the strip is subsequently processed. If the pickled coil is to be fed from an uncoiler, where the danger of scuffing is minimized, a moderate (200 SUS at 100°F) viscosity oil may be satisfactorily used. However, if coil boxes with rollers are used for the subsequent

---

† "Metal Deformation Processes/Friction and Lubrication", Edited by J. A. Schey, Marcel Dekker, Inc., New York, New York, 1970.

♦ K. A. Lloyd, (Discussion of a paper by J. M. Thorpe, "Mechanism of Lubrication in Cold Rolling") Proc. Inst. Mech. Eng. (London) 175, 1961, pp. 593-603, Proc. Inst. Mech. Eng. (London) 175, 1961, pp. 614-615.

◊ J. Douglas, "80-Inch HCl Pickling Line and Regeneration Plant at the Steel Company of Canada", Iron and Steel Engineer Year Book, 1967, pp. 21-29.

COLD ROLLING LUBRICATION

unwinding of the coil, a more viscous (up to 450 SUS) oil is desirable.[δ] Gregory[o] reports that one of the most recent coating media to be introduced is the dry, or wax coating. This is applied as a 2-3 per cent aqueous emulsion, usually during or immediately after the hot water rinse section of the pickle line. The water is evaporated at the air driers, leaving a fine wax-like coating on the strip surface.

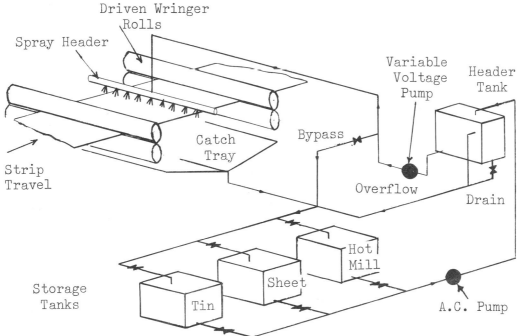

Figure 6-20: Diagrammatic Sketch of Lubricating System for a Pickle Line.

Figure 6-21: Oiling System at End of a Pickle Line.

[δ] H. J. Drake, "Rolling Oils — Their Influence on Productivity and Quality", Iron and Steel Engineer Year Book, 1965, pp. 1064-1070.

[o] W. Gregory, "Effects of Friction on the Cold Rolling Process", "Tribology in Iron and Steel Works", The Iron and Steel Institute, 1970, (ISI Publication 125).

When the pickled strip is to be cold rolled, the pickler oil must provide adequate lubricity, at least for the first pass. For example, when the strip is to be rolled to tinplate gages on a 6-stand tandem mill, the pickler oil and the cooling water provide the only lubricity at the first stand.♥ Moreover, because of the draft-to-roll diameter ratio at this stand, the lubricity that is required of the pickler oil is only minimal. However, the pickler oil tends to be washed off the strip at subsequent stands so that it mixes with the recirculating emulsion used at these stands. For this reason, the pickler oil should be compatible with the lubricant used on the mill. For this reason also, palm oil is sometimes used both on the pickle line and the tandem mill.

In the rolling of sheet product, the pickler oil is frequently the only lubricant used in the cold reduction operation. Even if this is not the case, there is evidence that the pickler oil is not completely removed from the strip surfaces during rolling. Accordingly, the pickler oil must be such that it burns off cleanly during annealing leaving no stains or smut on the product. For this reason, Schey† reports that the fatty oil and fatty acid contents of the pickler oil are often reduced (to a saponification value of 32-40 and a free fatty acid content of 1-2 per cent) by the addition of a mineral oil, preferably of a low staining variety. Drake ⸰ reports that contaminated or reclaimed oils are seldom satisfactory as pickler oils and Gregory⊙ reports that, to prevent the formation of carbon deposits, the carbon residues have to be very low, 0.2% by weight (as determined by the Ramsbottom or Conradson carbon coke method) being typical values. Gregory⊙ also states that the fat content of the oil has to be critically controlled, with the saponification value preferably being in the range 95-115, and the free fatty acid content about 12 per cent.

6-10 Rolling Lubricants for Sheet Mills

Of rolling lubrication requirements, the least severe generally pertain to the operation of sheet mills. However, it is convenient to classify these requirements on the basis of increasing lubricity as (1) slow mills producing thick sheet products, (2) slow mills rolling light gage sheet steel and (3) high-speed mills rolling the lighter gages.

As stated in Section 6-9, the pickler oil (usually a mineral oil) may be utilized as the sole lubricant in the least severe requirement listed above with water applied separately as the coolant for the rolls and strip. For the second, more severe requirement, the mineral oil applied at the pickle line may be blended with fatty oils or a so-called "soluble oil" mixed with a fatty-type lubricant may be applied at the mill. For the most severe requirements, palm oil or palm oil substitutes are generally applied in the form of a moderately stable emulsion. Under such conditions, adequate lubricity is assured but recirculation and filtration are possible so that contamination of the emulsion is kept to a minimum.† These looser emulsions based on mineral oils and compounded with some fatty additives not only provide very economical usage of the lubricant (0.75 to 1.5 lb. per ton of steel rolled) but are capable of being kept clean almost indefinitely if compatible with the pickler oil.⊞ If contamination does occur, the pickler oil will separate out carrying the iron fines and the iron oxides with it.

To facilitate the production of clean sheet steel, a detergent instead of a lubricant may be applied at the last stand of a sheet mill. This procedure reduces the thickness of the residual oil film on the sheet product and hence minimizes the tendency for staining or smut formation during annealing. However, the detergent should have sufficient lubricity to minimize the further

---

♥ R. J. Bentz and W. L. Roberts, "Predicting Rolling Forces and Mill Power Requirements for Tandem Mills", Blast Furnace and Steel Plant, August, 1970, pp. 559-568.

† "Metal Deformation Processes/Friction and Lubrication", Edited by J. A. Schey, Marcel Dekker, Inc., 1970, New York, New York.

⸰ H. J. Drake, "Rolling Oils — Their Influence on Productivity and Quality", Iron and Steel Engineer Year Book, 1965, pp. 1064-1070.

⊙ W. Gregory, "Effects of Friction on the Cold Rolling Process", "Tribology in Iron and Steel Works", The Iron and Steel Institute, 1970, (ISI Publication 125).

⊞ H. Pannek, "Die Brauchbarkeit Verschiedener Emulsionen beim Kaltwalzen von Bandstahl", Stahl und Eisen 75, 1955, pp. 767-769.

COLD ROLLING LUBRICATION

generation of iron smut from the shot blasted rolls commonly used in the last stand of a sheet mill.[*]

6-11 Rolling Lubricants for Tandem Tin Mills

In primary cold rolling operations, the most severe requirements for rolling lubricants occur in the modern high-speed tandem mills rolling strip or tinplate gage. In such mills, the reductions taken per stand are reasonably large (averaging 30 to 35 per cent) and hence excellent lubricity is required. Moreover, the coiling temperature of the strip must not exceed a certain limit (usually in the range 300 to 350°F) because higher temperatures would adversely affect the quality of the tinplate subsequently produced from the rolled strip. Hence the rolling oil emulsion must be satisfactory as a coolant for the strip as well as for the mill rolls.

Quasi-stable emulsions with oil concentrations usually 3 to 5 per cent, and at temperatures 100 to 150° are recirculated on these tandem mills at very high flow rates (3000 to 5000 g.p.m.). Originally palm oil was commonly used for this purpose but the substitute palm oils (generally of a tallow base) employed at the present time provide lubricity superior to that given by palm oil. The analysis of palm oil and two substitute palm oils are given in Table 6-2, and the ranges of the physical and chemical properties of lubricants generally used on tin mills are presented in Table 6-5.[∞]

Table 6-5
Ranges of the Physical and Chemical Properties of Steel Rolling Lubricants

Property	Range
Melting (or pour) point, °C	5-45
Viscosity	50-850 SUS at 100°F (38°C)
	45-200 SUS at 210°F (99°C)
Viscosity index	130-160
Saponification value	125-200
Iodine value	40-75
Free fatty acid, %	3-20

The melting or pour point is important in that an oil that is fluid at ambient temperatures is easier to handle in a recirculating system and provides easier "housecleaning" than one that is solid at such temperatures. In this respect, it should be noted that the titer or melting points of saturated acids are higher than those of unsaturated acids but they may be lowered by the addition of mineral oils. However, the lubricity afforded by mineral oils is distinctly poorer than that provided by fatty materials (see Section 6-7). Accordingly, the use of mineral oils to lower the pour point of a lubricant is accompanied by a degradation in rolling performance. To ensure effective hydrodynamic lubrication, the viscosity should be as high as possible and the viscosity index (which indicates the sensitivity of the viscosity to a change in temperature) should be as large as possible (indicating the least degree of sensitivity). The saponification value of fatty oils, such as glycerides, decreases with increasing molecular weight but, since mineral oils have a zero saponification value, if mineral oils are blended with the fatty oils, the saponification value is lowered and hence this parameter is only a quality indicator in a general sense. Similarly, the iodine value is a measure of the degree of unsaturation of the fatty materials but it, too, is dependent on the molecular weights of the constituents of the oil. A high iodine value, however, indicative of the presence of unsaturated fatty compounds is, in general, undesirable because polyunsaturated acids readily polymerise into resinous substances and tend to become rancid as the result of oxidation. As

---

[*] A. Ohm, "Tribology of the Cold Rolling of Steel Strip, Including Consideration of Hydrodynamic Lubrication, Part I", "Tribology in Iron and Steel Works", The Iron and Steel Institute, London, 1970 (ISI Publication 125).

[∞] W. L. Roberts and R. R. Somers, "The Cold Rolling Lubrication of Steel Strip", Lubrication Engineer 18, 1962, pp. 362-368.

discussed in Section 6-7, the importance of the free fatty acid content of a rolling oil is debatable but the ranges indicated for the substitute palm oils in Table 6-2 are typical of those found in many lubricants.

Generally speaking, the lubricant consumption on tandem tin mills ranges from 1 to 3 lb. per ton of steel rolled and the lubricant emulsion is generally used for 1 or 2 weeks before it is discarded and the system cleaned of sludge. In high-speed mills, the oil in the system must be periodically replenished and water added (the latter at rates of about 1000 g.p.h.) to replace that lost by evaporation. In addition, it should be noted that proper filtration must be used with a recirculation system to remove solid contaminants.

6-12 Lubricants for the Cold Reduction of Stainless and Silicon Steels

Stainless steel, frequently silicon steel, and other relatively hard-to-roll products are generally reduced on Sendzimir mills described in Section 2-11. In earlier versions of these mills, the rolling lubricant also functioned as the roll-bearing lubricant. However, in mills of more recent design, the bearing lubricant has been applied separately, in some cases, as an airborne type of lubrication.◘ Such a method of lubrication (referred to as "Mistlube") makes possible the use of heavier bearing oils with a high viscosity and load-carrying capabilities and, at the same time, allows a choice of a rolling oil appropriate to the material being rolled and the reduction schedule being followed.

Because of the relatively small diameter rolls used in such mills (and hence the relatively high values for the minimum permissible coefficient of friction) and because the lubricants must not stain or mar the surfaces of products like stainless steels, mineral oil containing additives is usually used for flooding the mill housing.* This lubricant is frequently heated, filtered and then passed through oil ways to the top and bottom saddles and the needle bearings of the back-up rolls. However, care and attention are necessary to prevent deterioration of the oil through thermal decomposition, condensation and other factors.

The poorer types of rolling lubricants generally used in Sendzimir or cluster type mills (particularly those of older design) enhances the luster of the rolled strip.× This is illustrated in Figure 6-22, showing how the reflectivity increases as the effective coefficient of friction increases. Frequently, however, the visual appearance of strips rolled with fluids of different lubricity are appreciably more distinctive than would be indicated by glossmeter or reflectometer readings. Murphy, et al,ϒ have obtained a U. S. Patent relating to the production of dull surfaces on stainless steel by using rolling lubricants with a viscosity in the range of about 400-4000 SUS at a temperature of 100°F. Figure 6-23 reproduced from the patent shows the effect of viscosity and the degree of reduction on the surface finish of type 434 stainless steel and Figure 6-24 shows the effect of rolling speed on the reduction required to provide certain gloss levels on the same material using a lubricant with a viscosity of 1750 SUS. Naphthenic base mineral oils were used to provide viscosities of 1750 SUS and lower, and a paraffinic oil provided a lubricant with a viscosity of 4128 SUS. At the same time, the surface finish of the rolls was such as would produce a high degree of luster with the poorer, conventional lubricants.

The effectiveness of various lubricants in the rolling of stainless steel has been published by Whetzel and Rodman° and their data, relative to the rolling of 0.0052 inches by 0.25-inch

---

◘ H. W. Ward, "The Continuing Development of Sendzimir Multi-Roll Cold Mills in World-Wide Strip Production", Sheet Metal Industries, May, 1969, pp. 407-411.

* H. H. Scholefield, J. E. Riley, E. C. Larkman and D. W. Collins, "Some Experiences in Cold Rolling Thin Alloy Strip", Journal of the Institute of Metals, Vol. 88, 1959-60, pp. 289-295.

× W. L. Roberts, "Influence of Rolling Lubricant on Sheet and Strip Quality", "Tribology in Iron and Steel Works", The Iron and Steel Institute, London, 1970 (ISI Publication 125)

ϒ H. L. Murphy, J. P. Bressanelli, and K. E. Pinnow, "Method for Rolling Stainless Steel", U. S. Patent 3,496,746, issued February 24, 1970.

° J. C. Whetzel, Jr., and S. Rodman, "Improved Lubrication in Cold Strip Rolling", Iron and Steel Engineer Year Book, 1959, pp. 238-247.

COLD ROLLING LUBRICATION

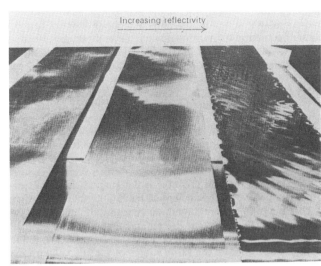

	Computed Coefficient of Friction	
(Low)	(Medium)	(High)
0.037	0.050	0.081
	Reflectivity	
33	37	38

Figure 6-22:
Changes in the Reflectivity of Annealed Low Carbon Sheet Strip Reduced from 0.0090-Inch to 0.0058-Inch at 100 Feet/Minute on a Mill with 6.5-Inch Diameter Work Rolls.

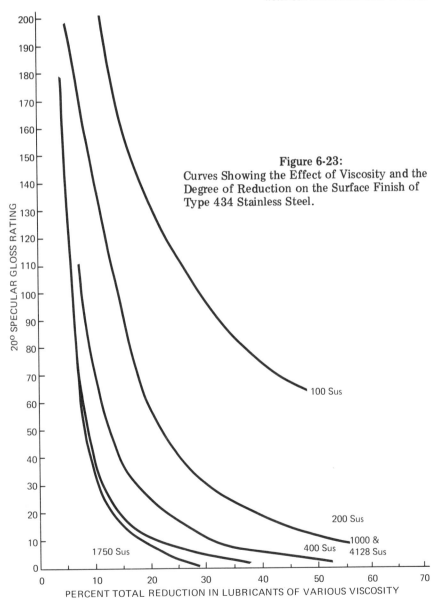

Figure 6-23:
Curves Showing the Effect of Viscosity and the Degree of Reduction on the Surface Finish of Type 434 Stainless Steel.

annealed Type 302 strip at 300 f.p.m. with 4-inch diameter work rolls and a rolling force of 6,000 lbs. are shown in Table 6-6. As indicated in this table, reductions increase with increasing viscosity of the mineral oil and of the animal fats. The experimental "water soluble" oils were viscous pastes containing high molecular weight fatty oil derivatives that could be rinsed off with water. Whetzel and Wyle▼ found that reductions close to 60 per cent could be obtained in a single pass at speeds exceeding 250 f.p.m. when rolling with the most viscous of these experimental lubricants.

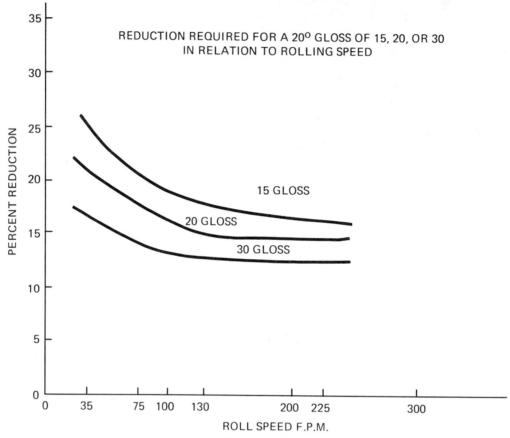

Figure 6-24:
Curves Showing the Effect of Rolling Speed on the Reduction Required to Provide Various Gloss Levels in the Rolling of Stainless Steel Type 434 Using a Lubricant with a Viscosity of 1750 SUS.

Table 6-6
Effect of Lubricant on Reduction Obtained in Rolling Stainless Steel

Lubricant Type	Characteristics	Reduction, %
Mineral oils	70 SUS at 100°F (38°C)	12.5
	600 SUS at 100°F	15.0
	600 SUS ÷ 5% stearic acid	15.0
Fatty oils	Typical animal fat	20.0
	More viscous fat	25.0
	Stearic acid	18.0
"Water soluble"	Experimental, highly viscous fatty oil derivatives	16.0–38.5

▼ J. C. Whetzel, Jr., and C. Wyle, 'Some Effects of Lubricants in Cold Rolling Thin Strip", Metal Progress 70 (6), 1956, pp. 73-76.

COLD ROLLING LUBRICATION

Additional data have been published by Tokar and Chamin[※] and are reproduced in part in Figure 6-25. These curves pertaining to the rolling of 18% nickel — 9% chromium stainless steel strip, 16 inches wide and 0.04 inches thick were obtained by the use of a mineral oil emulsion (1), stearin (2), glyceryl tristearate (tallow) (3), table fat (4) and castor oil (5).

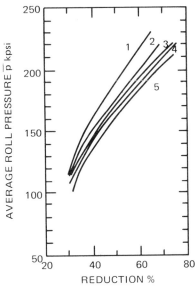

Figure 6-25:
The Effectiveness of Various Lubricants in Cold Rolling Stainless Steel (1:Mineral Oil Emulsion; 2:Stearin; 3:Tallow; 4:Table Fat and 5:Castor Oil).

Where hard to roll strips, such as the silicon steels, are processed on reversing 4-high mills with relatively large work rolls, better lubricants may be employed. These usually contain fatty materials and generally are similar to those used on tandem cold mills.

6-13 Rolling Lubricity in the Reduction of Coated Products

In the rolling of steel foils, it was soon found that a tin coating of adequate thickness on the surfaces of the strip considerably facilitated its reduction.[△][♀] In recent years, plastic coatings have also been found to aid the effectiveness of the rolling lubricant.[♦]

The relative effect of unmelted, electrolytically deposited tin coatings of thicknesses up to 15 g/m^2 on the calculated coefficient of friction in the rolling of steel strip at two speeds and two reductions is illustrated in Figure 6-26. It is seen that, provided the tin coating is thick enough, the calculated coefficient of friction is reduced. In the case of electrolytically deposited coatings that have been melted or reflowed (as on FERROSTAN lines), the effectiveness of the coating is seriously reduced. This is probably attributable to (1) the fact that the reflowing process has diminished the surface roughness of the strip (and therefore reduced the quantity of lubricant it can drag into the roll bite, and (2) the existence of hard iron-tin allow layers at the steel-tin interfaces. As a consequence, strip with light (about 0.1 lb. per base box) melted tin coatings are more difficult to roll than uncoated strips with the same roughness on the steel surface. This effect is further illustrated in Figure 6-27 showing the minimum thicknesses that can be obtained from 0.007-inch black plate and tinplate with various types of coatings in a single pass on a 4-high mill with 6.5-inch

---

[※] I. K. Tokar and I. A. Chamin, "New Greases for the Cold Rolling of Sheet", Metallurg. 1960 (4), pp. 28-29 and Metallurg. 1960 (4) pp. 161-163.

[△] W. L. Roberts, "Friction in the Cold Rolling of Steel Strip", "Friction and Lubrication in Metal Processing", Edited by F. F. Ling, R. L. Whitely, P. M. Ku and M. B. Peterson, Published by the American Society of Mechanical Engineers, New York, New York, 1966, pp. 103-121.

[♀] D. C. Litz, "Rolling of Light Gage Strip on a Laboratory Mill", Blast Furnace and Steel Plant, November, 1967 pp. 1027-1035.

[♦] W. L. Roberts, R. J. Bentz and D. C. Litz, "Cold Rolling Low-Carbon Steel Strip to Minimum Gage", Iron and Steel Engineer Year Book, 1970, pp. 413-420.

diameter work rolls. It is to be noted that melted tin coatings of 0.1 and 0.25 lb./base box are more difficult to roll than the uncoated black plate. Zinc, terne metal and aluminum coatings (as applied by the hot-dip process) exhibit a similar effectiveness in reducing the friction in the roll bite. As discussed in Section 6-32, the major reason for the enhanced lubricity of such coatings appears to be the increased values of their second frictional characteristics which are related to their ability to draw the lubricant into the roll bite at high mill speeds.

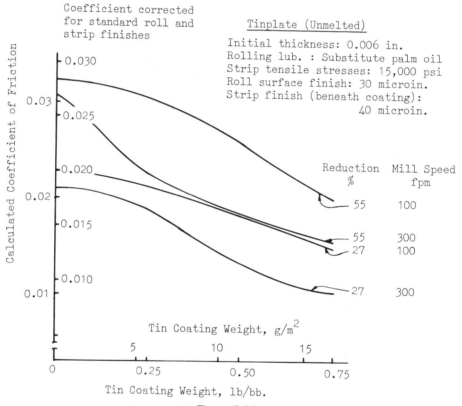

Figure 6-26:
Curves Relating Calculated Coefficient of Friction with Coating Weight for Unmelted Tinplate.

Organic coatings on steel strip surfaces have also been found to provide additional lubricity. Figure 6-28 illustrates the effectiveness of an alkyd amine coating in the rolling of full hard 0.015-inch thick black plate with 6.5-inch diameter work rolls using water as a rolling lubricant.

Similarly, Figure 6-29 shows the effectiveness of epoxy and alkyd amines in the rolling of Type 304 stainless steel 0.008 inches thick on the same mill, again using only water as a lubricant in conjunction with the coatings.

6-14 Roll Wear and the Use of Roll Coatings

As in hot rolling, mill rolls become worn through the processing of strip so that, if they are initially rough, they tend to become smoother approaching a finish in the general range of 20-25 microinches. Conversely, smooth rolls gradually become rougher and also approach a similar final surface condition. Good lubricants, of course, minimize the rate of the wear and also assure its uniformity. With poor lubricants, detritus may become embedded in the rolls (roll pick-up) which mars the finish of the rolled strip.

COLD ROLLING LUBRICATION

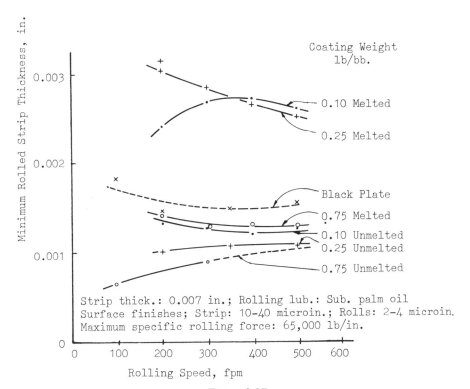

Figure 6-27:
Variation of Minimum Rolled Strip Thickness with Rolling Speed for Black Plate and Tinplate.

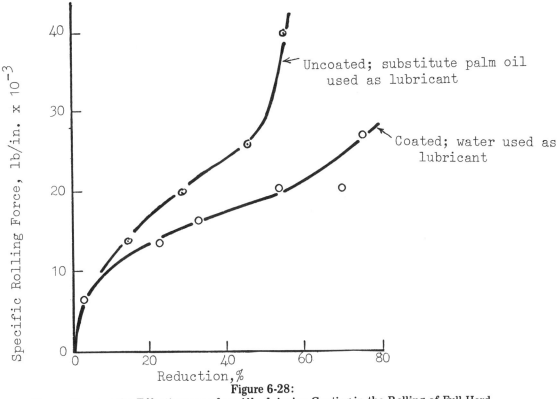

Figure 6-28:
Curves Showing the Effectiveness of an Alkyd Amine Coating in the Rolling of Full-Hard 0.015-Inch Thick Black Plate.

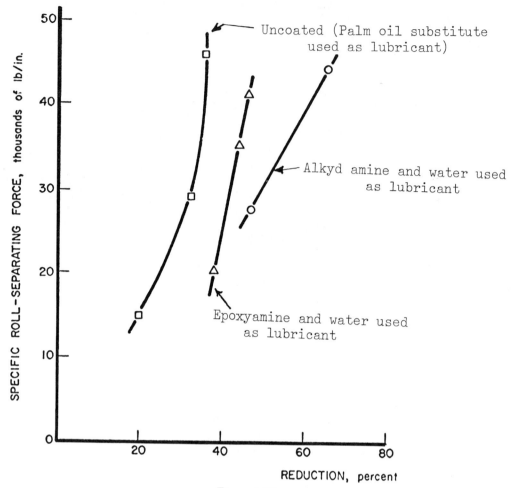

Figure 6-29:
Curves Showing the Effectiveness of Epoxy and Alkyd Amines in Rolling Type 304 Stainless Steel.

Schey[†] reports that few attempts have been made to increase roll life by the choice of a roll coating of high wear resistance. However, Spenceley[#] found lower wear rates and a more consistent strip surface finish when the rolls of 4-high and Sendzimir mills were electrolytically coated with a hard chrome layer of less than 0.001-inch thickness.

On tandem cold mills, work rolls generally wear at a greater rate in the middle stands. However, the rolls in the last stand are the most critical with respect to the finish of the rolled strip. Thus, in the operation of a 5-stand tandem tin mill, using a typical aqueous emulsion of a fatty oil as a lubricant, the work rolls in the last stand may be changed after rolling 150 tons, those in the penultimate stand after every 250 tons and those in the first 3 stands after every 1,000 tons.[▽] In the rolling of sheet products, greater tonnages may be processed between roll changes. Thus, for example, throughputs ranging from 450 to 1,500 tons have been cited between roll changes in a single-stand reversing mill.[⊞]

[†] "Metal Deformation Processes/Friction and Lubrication", Edited by J. A. Schey, Marcel Dekker, Inc., 1970, New York.

[#] G. D. Spenceley, "The Use of Hard Chrome Plated Rolls for Cold Rolling", Sheet Metal Industries, 42, 1965, pp. 408-416.

[▽] "The Making, Shaping and Treating of Steel", Edited by H. E. McGannon, U. S. Steel Corporation, Pittsburgh, Pa., 7th Edition, 1957, p. 607.

[⊞] H. Pannek, "Die Brauchbarkeit verscheidener Emulsionen beim Kaltwalzen von Bandstahl", Stahl und Eisen, 75, 1955, pp. 767-769.

COLD ROLLING LUBRICATION

Nikolaev and his co-workers[◁] have established that a work roll surface defect known as "orange peel" or rippling develops after the rolling of 30 to 50 tons on a reversing 4-high cold mill. They established that the dendritic structure and carbon heterogeneity in the roll metal led to different wear rates on the surface on a microscopic scale, forming a micro-uneveness of the roll surface. This effect was worse on the lower work roll because of poorer lubricity and cooling.

6-15 Changes Occurring in Recirculated Rolling Emulsions

In the continued use of recirculated rolling emulsions, it is not uncommon to experience gradual changes in the lubricity they provide, usually in the direction of decreased effectiveness. To restore the emulsion to its normal performance, mill operators will frequently add oil so as to increase the oil-water ratio of the emulsion. While such a procedure may be satisfactory in some instances, in others the addition of the fresh lubricant aggravates the situation in that the emulsion becomes quite unstable with the oil separating out from it.

Such changes are attributable to physical and chemical reactions taking place in the emulsion. Among the physical changes, particulate and solute contamination may occur, the former due largely to iron fines dislodged from the rolls and the strip and the latter to the continuous evaporation of the water from the roll and strip surfaces. Chemical changes may result from the oxidation of the oils (with the iron fines acting as a catalyst) or from bactericidal action occurring in the emulsion.

The effect of sludge concentration on the lubricity of a rolling solution has been measured in the laboratory and it is seen that even small percentages of sludge noticeably change the coefficient of friction.[∞] (See Figure 6-30.)

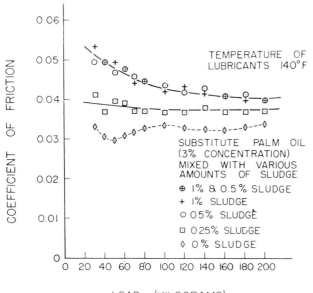

Figure 6-30:
The Effect of Sludge Concentration on the Lubricity of a Rolling Solution as Measured on a Bench-Testing Machine.

Ludwig,[*] in a thorough investigation of changes occurring in the rolling emulsion of a 5-stand tin mill during a week's usage, found that the insoluble inorganic material (fines) in the emulsion did not appear to be a function of the elapsed time of use (or age) of the emulsion. He

---

[◁] V. A. Nikolaev, et al, "Wear on the Surfaces of the Work Rolls of a Reversing Cold Rolling Mill", Stal in English, 1969, pp. 196-198.

[∞] W. L. Roberts and R. R. Somers, "The Cold Rolling Lubrication of Steel Strip", Lubrication Engineer, 18, 1962, pp. 362-368.

[*] J. R. Ludwig, "Changes in Cold Reduction Lubricant Solution During Rolling", Blast Furnace and Steel Plant, August, 1969, pp. 641-651.

made semiquantitative emission spectrographic analyses of the fines for metallic elements and elemental carbon. The results of these analyses are listed in Table 6-7 and they show that the major element in all cases was iron, with copper, silicon and carbon being the most abundant minor elements. The presence of copper could possibly have been due to the abrasion of brass guides and other copper and brass fittings on the mill. The silicon was probably accumulated from dissolved and suspended siliceous materials in the make-up water as well as from substances carried in on the strip. The carbon content of the fines showed a continuous increase during the week. The presence of the carbon could possibly be caused by thermal degradation of the lubricant at the roll bite. The 12.4 percent carbon content in the final skim sample on 7/29-3 in particular could have resulted in this manner. On the turn preceding the one on which this sample was taken, thin-gage strip (0.007-inch) was being rolled. The thin-gage strip requires greater roll forces, resulting in greater frictional effects and thus higher roll bite temperatures.

Table 6-7
Semiquantitative Emission Spectrographic
Analyses of Fines Recovered from a
Cold-Rolling Lubricant Solution

	Composition, percent			
Element	7/25* 3rd Skim	7/27 1st Skim	7/27 3rd Skim	7/29 3rd Skim
Aluminum	0.3	0.3	0.3	0.5
Calcium	0.2	0.2	0.2	0.2
Chromium	0.03	0.03	0.04	0.04
Copper	1.1	0.9	0.9	0.7
Iron	Major	Major	Major	Major
Magnesium	0.7	0.7	0.3	0.7
Manganese	0.3	0.2	0.3	0.2
Molybdenum	0.04	0.04	0.05	0.05
Sodium	N.D.***	N.D.***	N.D.***	N.D.***
Nickel	0.1	0.1	0.2	0.1
Lead	0.08	0.06	0.1	0.1
Silicon	1.2	1.0	1.3	2.1
Tin	0.07	0.07	0.07	0.05
Titanium	0.01	0.01	0.01	0.01
**Carbon	2.1	2.2	5.6	12.4

*1st line, date; 2nd line, turn; 3rd line, type of sample.

**Elemental carbon analyses determined by a combustion — gas chromatographic technique.

***None Detected, (<0.01).

Ludwig, in the same study, also investigated changes occurring in the aqueous phase of the emulsion. During the rolling of strip pickled with sulphuric acid, the sulphate concentrations were in excess of 100 ppm with the chloride values in the range 30 to 40 ppm. However, later in the life of the emulsion, when hydrochloric acid-pickled strip was being rolled, the chloride values ranged from 50 to 60 ppm and the sulphate concentrations decreased to 40-50 ppm. During the complete life of the solution, the detection limit of 0.3 ppm of soluble iron was never exceeded. This absence of iron in the aqueous phase may possibly have been due to the reaction of any dissolved iron with the free fatty acids in the emulsion. The iron soaps resulting from this reaction would be insoluble in water but, being soluble in organic materials, would accumulate in the oil phase. During the life of the emulsion, the total hardness (as ppm of calcium carbonate) ranged from 60 to 260 ppm and the total soluble material from 110 to 720 ppm.

In studies of the oil phase, Ludwig concluded that the analyses suggested that appreciable deterioration of the lubricant due to thermal decomposition and oxidation occurs in rolling emulsions. Some of these changes, such as the formation of soluble metallic soaps and the build-up of soluble inorganic materials may undoubtedly affect the stability of the emulsion.

COLD ROLLING LUBRICATION

The presence of metal soaps could cause other problems, also, because metal soaps, metal oxides, and metals themselves have been shown to be effective catalysts for the oxidation of both hydrocarbon and fatty lubricants.◊ By using radioactive tracers, Berezin, Berezkina, and Nosova ⅃ have shown that stearic acid and its esters undergo oxidative decarboxylation with atmospheric oxygen with the formation of lower molecular weight acids and esters. This could provide an explanation for the appearance in the rolling solution of C_{17}, C_{15}, and the lower-molecular-weight species which were detected. These materials could appreciably affect both the stability of the solution and the lubricity of the oils. It is not known, however, whether all the changes which occur in the system are detrimental, since some of the oxidation and thermal degradation products may be providing beneficial properties to the solution.

6-16 The Effect of the Lubricant on the Surface Finish of the Rolled Strip

As discussed in Section 6-12, a correlation exists between the general appearance (reflectivity, gloss, luster or shininess) of the rolled strip and various rolling parameters including the effective coefficient of friction in the roll bite. This results from the fact that the surface finish is, in part, dependent on the amount of buffing the strip receives from the roll surfaces prior to its emergence from the roll bite. In the case of a good lubricant, the neutral point (or that point along each arc of contact at which the roll and strip surface speeds are equal) is close to the exit plane, as illustrated in Figure 6-31. There is, therefore, little slippage of the strip with respect to the roll surface after the neutral point has been passed and, accordingly, very little buffing of the surface, so that the strip surface has a dull or matte finish. On the other hand, when a poor lubricant is used, the neutral point is moved towards the entry plane and the slip between the outgoing strip and each roll surface is increased with the result that more buffing occurs to produce a brighter strip.

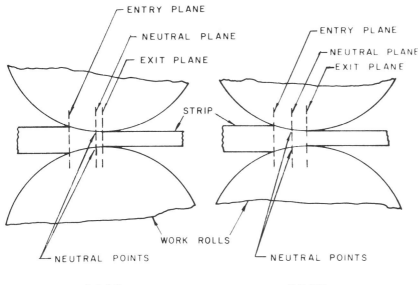

Figure 6-31: Sketches Illustrating How the Position of the Neutral Point is Influenced by the Effectiveness of the Rolling Lubricant.

To some extent, the strip surface appearance can be changed by changing the strip tensions. This is due to the fact that the neutral point is not only shifted by changes in the coefficient of friction but also by changes in the ratio of the input to the output strip tensile stresses. Increasing the ratio of the entry to the exit tensile stress moves the neutral point towards the exit plane (thereby decreasing the slip) and tends to produce duller strip. Conversely, decreasing

---

◊ E. Fedeli, "Riv. Ital. Sostanze Grosse", 43, 99-103 (1966).
⅃ N. M. Emanuel, "The Oxidation of Hydrocarbons in the Liquid Phase", The MacMillan Company, New York, New York, 1965.

the ratio and increasing the slip will tend to produce brighter strip as the neutral point is moved in the opposite direction. In a laboratory experiment, the reflectivity of strip rolled with a slip of 2.3 per cent was found to have a reflectivity of 16 per cent, whereas strip rolled with a slip of 14.2 per cent exhibited a reflectivity of 19.5 per cent, the lubricant used in both cases being a mixture of a substitute palm oil and a mineral oil.

The nonuniform application of the rolling lubricant may result in a surface defect commonly referred to as "staining" or "mottle" as illustrated in Figure 6-32. It occurs primarily with the use of oil-water emulsions, where the degree of "plate out" of the oil on the steel surface may vary from point to point. Where the "plate out" has been adequate, the strip assumes a duller appearance than the other portions that were subjected to more of a burnishing action by the rolls. Careful analyses of the surface by microprobes have shown that, chemically, the mottled strip surface is uniform, but interference microscopy shows that the duller areas are indeed rougher or more pitted.

Figure 6-32: Samples of Black Plate with Mottled and Uniform Surfaces.

Figure 6-33: Photomicrograph Showing Many Small Pits in the Streaked Area on the Surface of Mottled Black Plate. Unetched. X50.

COLD ROLLING LUBRICATION

Figure 6-33 is a photomicrograph of a dull, streaked area showing a high concentration of small dark pits. The horizontal lines on the magnified surface in the rolling direction are typical of the normal strip finish with a surface roughness of approximately five microinches obtainable with work rolls of a No. 7 finish. When the magnification of the "stained" or mottled area is increased, the dark pits appear as small depressions in the surface as illustrated in Figure 6-34. A cross section of these depressions at a magnification of 750 is shown in Figure 6-35. From interference photomicrographs, such as that illustrated in Figure 6-36, the depths of the pits are found to range from 10 to 80 microinches.

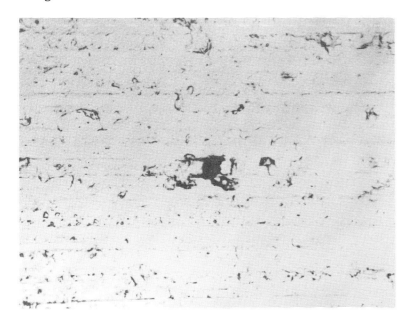

Figure 6-34: Photomicrograph of Pits on the Unpolished Surface of Black Plate. Unetched. X500

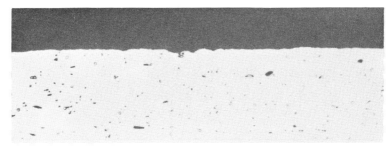

Figure 6-35: Cross Section Through Pits on Mottled Black Plate. Picral Etch. X750.

To ensure the complete absence of staining or mottle it is advisable to use neat lubricants. However, where emulsions are used, the effect may be minimized by reducing the rolling speed. Alternatively, the stability of the emulsion may be decreased thereby facilitating plate out of the oil on the strip surfaces.

Under certain conditions, the presence of light mottle may be tolerated. For example, in the production of tinplate, the streaked appearance of the strip is largely concealed by the coating of tin, especially if the coating of the tin is sufficiently thick, say 0.5 lb./base box or heavier. On the other hand, very light coatings of chromium, as in the case of tin-free steel, tend to enhance the visual appearance of the mottle.

Another surface defect, associated in part with the lubricant, but also with the shape of the strip entering the roll bite, is often designated as "tiger stripes". It gives the appearance of broad bands of alternating levels of reflectivity running approximately parallel to the direction of rolling. If on one surface the band is shiny, on the other surface of the strip there is a band with a dull finish, and vice versa. The cause of the defect would appear to be a tendency for the strip to enter the bite in a very slightly corrugated manner, as shown in Figure 6-37, where there are regions on each surface with a higher-than-average coating of lubricant entering the roll bite.

Figure 6-36: Interference Photomicrograph of Surface Under Thallium Light, 0.27 Microns Between Bands. Unetched X312.

Figure 6-37: Sketch Illustrating Possible Cause of "Tiger Stripes".

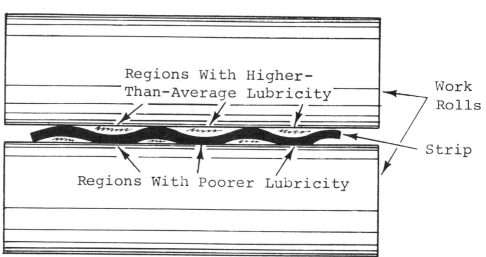

6-17 The Effect of the Lubricant on the Shape of the Rolled Strip

In cold rolling operations associated with conventional 4-high mills, the mill rolls are crowned by grinding and/or by a nonuniform temperature distribution within them, so as to compensate for their bending during the operation of the mill and thereby provide an essentially-uniform roll gap. For a given set of roll crowns, or cambers, there is ideally one value (or a small range of values) of rolling force under which the exact compensation occurs. Since, as is shown in Chapter 9, the rolling force for a given set of rolling conditions is dependent on the value of the effective coefficient of friction in the roll bite, the relationship between roll bite lubricity and the shape of the rolled strip becomes apparent.

Assuming that flat strip is being produced under a given set of rolling conditions, if the roll crowns remain the same, then any change in conditions that lessens the total rolling force will tend to produce "full-centered" strip. Such changes include (a) a decrease in the yield strength of the strip, (b) a decrease in the width of the strip, (c) a decrease in the reduction given to the strip by

COLD ROLLING LUBRICATION

the rolling pass, (d) increases in the strip tension, (e) an increase in the surface roughness of the incoming strip, and (f) a decrease in the work roll surface roughness. To compensate for any of these changes, the effective coefficient of friction should be increased. Conversely, if any of the changes are in the reverse direction so as to increase the rolling force, the tendency to produce strip with over-rolled edges may be best compensated for by the use of a better lubricant.

The influence of the lubricant on strip shape has been demonstrated on a 4-high experimental rolling mill with 6.5-inch diameter rolls and 14-inch face [x] as illustrated in Figure 3-1. Annealed black plate strip could be produced with a full center, good shape or with full edges by varying the lubricity of a blend of oils applied neat to the strip prior to reduction. Examples of such strips are shown in Figure 6-38 together with the measured rolling force and the calculated coefficient of friction in the roll bite for each sample. It should be noted that, as the coefficient of friction provided by the lubricant is increased, the shininess, or brightness of the strip is also increased as evidenced by the markedly darker appearing, full edge strip at the right.

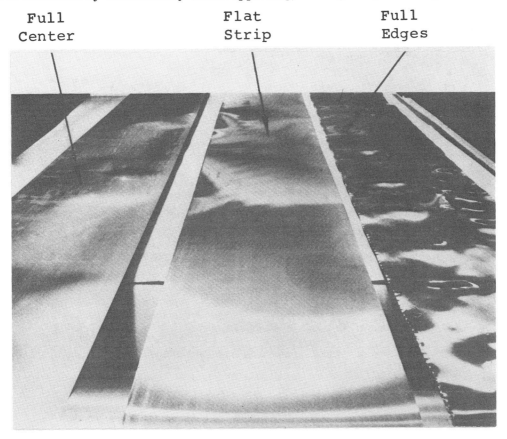

| Full Center | Flat Strip | Full Edges |

Specific Rolling Force

| 16,000 lb./in. | 18,600 lb./in. | 27,100 lb./in. |

Computed Coefficient of Friction

| 0.037 | 0.050 | 0.081 |

Figure 6-38:
Annealed Low-Carbon Steel Strip Rolled from 0.009-Inch to 0.006-Inch in a Single Pass on a Mill with 6.5-Inch Diameter Work Rolls Using Three Different Lubricants. (Backup Roll Crowns 0.002-Inch; Work Roll Crowns 0.001-Inch.)

---

[x] W. L. Roberts, "Influence of Rolling Lubricant on Sheet and Strip Quality", "Tribology in Iron and Steel Works", Iron and Steel Institute, London, 1970, (ISI Publication 125)

If the degree of reduction given the strip is changed, then, with the same crowns and lubricant, there is a good possibility that the shape of the rolled strip will be adversely affected. This is illustrated in Figure 6-39, which shows how strip that is rolled flat with a 49 per cent reduction has an over-rolled center when given a 14 per cent reduction and over-rolled edges when given a 56 per cent reduction. To obtain flat strip at the reductions other than 49 per cent, the coefficient of friction should be changed so that the rolling force is maintained approximately constant, regardless of the reduction given. This is illustrated in Figure 6-40, which shows flat strip obtained after 22, 28 and 40 per cent reductions with essentially the same values of specific rolling force but with different lubricants providing different coefficients of friction.

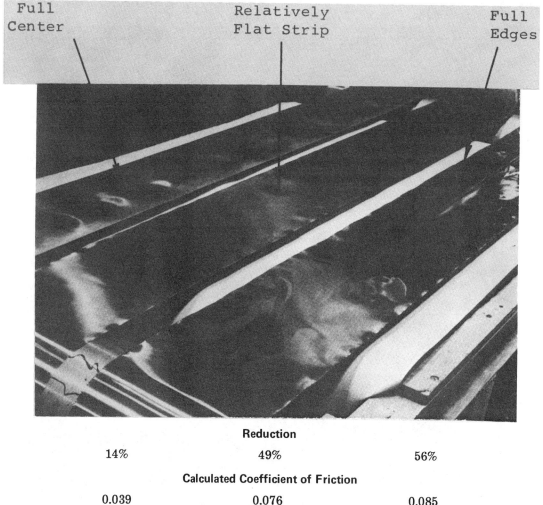

	Reduction	
14%	49%	56%
	Calculated Coefficient of Friction	
0.039	0.076	0.085
	Measured Specific Rolling Force	
10,900 lb./in.	21,000 lb./in.	37,600 lb./in.

Figure 6-39:
Annealed Low-Carbon Steel Strip 0.009-Inch Thick Given Three Different Reductions on a Mill with 6.5-Inch Diameter Work Rolls Using the Same Lubricant. (Backup Roll Crowns 0.002-Inch; Work Roll Crowns 0.001-Inch.)

6-18 Residual Oil Films on the Rolled Strip

When rolled with either a neat lubricant or an aqueous emulsion of an oil, the strip emerges from the roll bite with its surfaces covered with a very thin film of the lubricant of the order of a few microinches thick. The actual thickness of this film depends on the rolling conditions, such as the reduction given to the strip and the work roll diameter. However, it is interesting to note that, even if an emulsion is used, the water phase is generally completely prevented from passage through the roll bite.

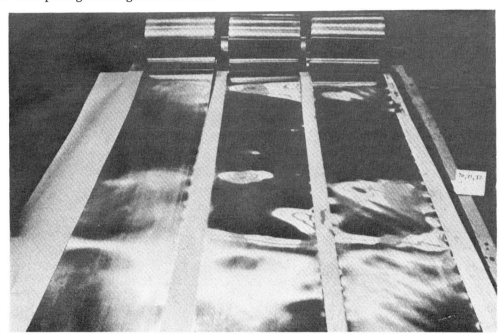

	Reduction	
22%	28%	40%
Calculated Coefficient of Friction		
0.076	0.060	0.041
Measured Specific Rolling Force		
18,200 lb./in.	18,600 lb./in.	18,000 lb./in.

Figure 6-40:
Annealed Low-Carbon Steel Strip 0.009-Inch Thick Given Three Different Reductions at 100 Feet/Minute on a Mill with 6.5-Inch Diameter Rolls Using Three Different Lubricants. (Backup Roll Crowns 0.002-Inch; Work Roll Crowns 0.001-Inch.)

The residual oil film affords a certain degree of rust protection to the rolled product during the period the coil awaits further processing (such as annealing or tinning) or during shipment from the plant to a customer. In some instances, the residual oil film is not detrimental in subsequent operations (such as annealing or drawing operations) but, in others, its removal is necessary. Where removal occurs either by burn off in box annealing furnaces or by electrolytic action in cleaning tanks, no residues should be left on the strip or, in other words, the oil should possess good "cleanability" (see Section 12-2).

If the residual oil is of the type that tends to polymerize and the film is thin enough, it may be possible in can manufacture to lacquer and lithograph the oil coated strip. Under such circumstances, it may be necessary to ensure that the lubricant meets the requirements of the Food and Drug Administration * with respect to content and coating weight. In other cases, such as the production of steel foil which is to be laminated to paper or other materials, the residual oil film must be such that it does not preclude the use of conventional bonding adhesives.

---

*FDA Amendment Published in the Federal Register, April 27, 1962, 27F.R. 4014 Subpart F — Food Additives, page 31.

In tests conducted on a foil mill, using a proprietary brand rolling oil in neat form, it has been found that residual oil film weight

(a) decreases as the specific rolling force is increased,

(b) increases as the mill speed increases to a certain value and then decreases with further increases in rolling speed,

(c) decreases as the reduction increases and,

(d) where the strip is precoated with a lubricant thickness greater than 10 g/base box, the residual oil film thickness is relatively insensitive to the thickness of the precoat.

When using the same lubricant as an aqueous emulsion, however, the residual oil film weight

(a) is lower than that obtained when rolling with neat oil,

(b) decreases with increasing rolling force and increased mill speed, and

(c) decreases as the reduction is increased.

Film thicknesses obtained in these tests range from 0.91 to 2.1 g/base box in the case of neat oil and from 0.7 to 1.3 g/base box in the case of emulsions. Figure 6-41 illustrates the range of residual oil film weights encountered in the rolling of 0.0061-inch thick black plate at a mill speed of 200 f.p.m. using a heavy precoat of a commercial lubricant. Such data is in general agreement with that published by Whetzel and Rodman° who found that the oil film left on strip processed at 4,000 f.p.m. on a 5-stand tandem mill had a thickness of 7-9 microinches (or 80-160 molecules).

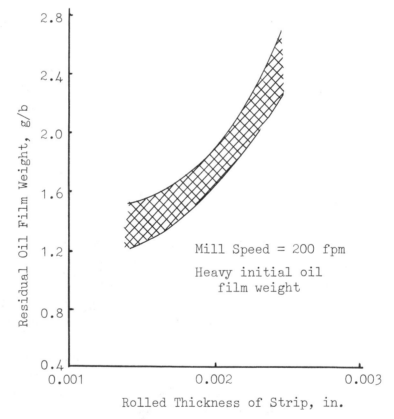

Figure 6-41: Variation of Residual Oil Film Weight with Rolled Thickness of Strip. (Black Plate Initially 0.0061-Inch Thick.)

° J. C. Whetzel, Jr. and S. Rodman, "Improved Lubrication in Cold Strip Rolling", Iron and Steel Engineer Year Book, 1959, pp. 238-247.

COLD ROLLING LUBRICATION

That edible type materials (which meet with FDA approval) can satisfactorily perform as rolling lubricants is shown in Figure 6-42. The data was obtained in reducing low carbon, full hard (80 $R_{30\text{-}T}$ hardness) black plate 0.0149 inches thick at 100 f.p.m. on an experimental mill with 3.25-inch diameter work rolls. The lubricants were used as 4 per cent aqueous emulsions and were sprayed onto the strip at the entry to the roll bite at a temperature of 140°F.

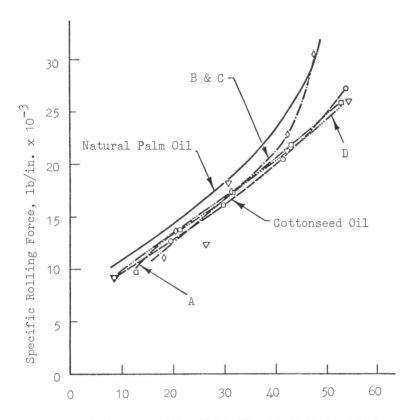

Figure 6-42: Rolling Performance of Four Edible Materials (A, B, C and D) Compared with That of Palm Oil and Cottonseed Oil.

6-19 Criteria for the Evaluation of Lubricants

From the foregoing sections, it becomes apparent that a rolling lubricant must exhibit a number of qualities that facilitate the rolling and subsequent processing of the strip, and that it is desirable to evaluate a lubricant on the basis of all pertinent properties. Otherwise, situations may arise, for example, where a lubricant, used on the basis of the excellent lubricity it provides in the roll bite, cannot retain its lubricity for long periods as an emulsion or cannot be easily removed from the strip after rolling. Consequently, in assessing the merits of a lubricant, the following criteria should be established:

(1) the lubricity it provides in the roll bite (in terms of an effective coefficient of friction),

(2) how its lubricity is affected by the physical conditions associated with the rolling operations.

(3) The stability of an aqueous emulsion of the oil and the factors that effect it,

(4) its cleanability and burn off characteristics, and

(5) the extent to which it affords rust or corrosion protection.

All of the above criteria may be established in quantitative terms even though, at times, only a ranking of the lubricants with respect to some particular property may be required. Although lubricants may be evaluated on full size production mills, it is not always desirable, for reasons of economy, to utilize a mill trial of a lubricant that may have little chance of being successfully used on a production basis. Accordingly, small sized experimental mills may be conveniently used to investigate the properties of rolling lubricants and bench-type testing machines may also be used to advantage.

Basically, the logical purpose of lubricity tests is to be able to predict exactly how the lubricant will perform under specified circumstances on a production mill and the behavior of the residual oil film in subsequent processing operations. Ideally, from measurements made on the properties of the lubricant, it should be possible to compute such parameters as the rolling force to be anticipated at any rolling stand and the charge density required to remove the oil from the strip surfaces in an electrolytic cleaning line. In addition, the data should permit predictions relative to the behavior of an aqueous emulsion of the lubricant over extended time periods and whether or not the rolled strip will exhibit mottle.

In later sections of this chapter, the various tests that may be performed on lubricants are discussed in detail and typical data pertaining to various lubricants are presented.

6-20 The General Assessment of Lubricants on Experimental Mills

The simplest method of using a mill to evaluate lubricants involves presetting the positions of the rolls by means of the screws, feeding into the roll bite samples of the strip (all of the same dimensions) coated with the lubricants under test and measuring the lengths of the samples after emergence from the roll bite. Basically, the poorest lubricant will be that used to roll the strip with the shortest final length, the reason being that the lubricant creates the largest roll separating force and hence the greatest elastic deformation of the rolls and housing and, therefore, the greatest separation of the work rolls. Conversely, the best lubricant provides the greatest length of rolled strip. Care must be taken to clean the rolls between the rollings of the various samples but, so long as this precaution is observed, the method has the obvious advantage of not requiring any instrumentation on the mill.

Although the method does not provide a quantitative assessment of lubricant performance, it is simple to undertake and enables lubricants to be ranked in order of merit with respect to the conditions of the test. Schey[†] points out that the sensitivity of the technique is increased if reductions are taken far enough to reach the condition of minimum gage. Under these conditions, the minimum attainable thickness will be directly proportional to the effective coefficient of friction in the roll bite (see Chapter 9).

This method has been used by Dedek, et al,[‡] for evaluating four new types of cooling and lubricating emulsions listed in Table 6-8 using a small 2-high mill. However, all the emulsions produced strip of the same final thickness and therefore provided comparable lubricity under the conditions of the test. Lueg and Funke[*] also used this technique for investigating the various lubricants listed in Table 6-9 and their data is shown graphically in Figure 6-43, and as a histogram in Figure 6-44. It is to be noted that dry rolling was used as the reference in this instance, and in order of increasing lubricity were ranked mineral oils, fats and palm oil.

[†] "Metal Deformation Processes/Friction and Lubrication", Edited by J. A. Schey, Marcel Dekker, Inc., New York, New York, 1970.

[‡] V. Dedek, et al, "Operation Tests of New Type Oil Emulsions for Cold Strip Rolling", Hutn. Listy. 1964, 19 (2), pp. 102-108. (British Iron and Steel Translation Service BISI 7880)

[*] W. Lueg, P. Funke and W. Dahl, "Untersuchung von Walzolen und Walzolemulsionen im Kaltwalzversuch", Stahl und Eisen, 77, 1957, No. 25, p. 1817.

COLD ROLLING LUBRICATION

Table 6-8
Constitution of Oil Emulsions Tested
By Dedek Et Al

Type of Lubricant	Component	Percentage Composition	Function of Component
EOT	stratum oil B1	84.3	basic lubricant
	ethylene	10.0	free fatty acid and emulsifier
	ditriethanolamine	5.7	emulsifier and corrosion prevention additive
Akvol	sulphoricinated oil	48.2	lubricant
	ethylene	15.0	fatty acid
	triethanolamine	24.6	corrosion-prevention additive
	cutting oil P	8.46	lubricant
	lanolin	0.8	emulsifier
	denatured spirit	2.5	stabilizer
	thymol	0.3	bactericide
	fuchsin, tannin		dye and bactericide
	$AgNO_3$, $CaCO_3$	*	
PP 607	mineral oil K420	72.3	basic lubricant
	oxyethyl naphthenic acid	13.0	emulsifier
	oxyethyl octyl phenol	5.8	fatty acid and emulsifier
	ethylene	3.2	fatty acid
	50% solution of iron sulpho-succinate	1.3	emulsifier
	H_2O	3.2	
	triethanolamine	1.2	corrosion-prevention additive
K1	mineral oil K420	78.0	basic lubricant
	ethylene	12.0	fatty acid
	triethanolamine	7.0	emulsifier and corrosion prevention additive
	oxyamide 2	3.0	emulsifier

*Quantity unknown.

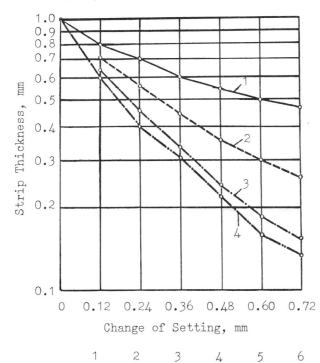

Figure 6-43:
Curves Showing the Influence of the Lubricant on the Final Rolled Thickness. (1:Dry Rolling; 2:Mineral Oils; 3:Fats; 4:Palm Oil)

Table 6-9
Lubricants and Emulsions (investigated by Lueg and Funke)

Group	Lubricant Number	Type of Lubricant	Composition	Content of Free Fatty Acids, %
A	1 2 3	Natural fats	Palm oil Rape oil Castor oil	1.5-13
B	4	Natural fats with addition of fatty acids	Rape oil Animal fat	3.0-6.0
C	7 12	Pure mineral oil, mineral oil with an increasing content of unsaturated fatty acids	Mineral oils, mineral oils with addition of fatty acids, fatty acids	
D	13 14	Mineral oils with addition of saturated fatty acids	Mineral oils and stearic acid	3.0-6.0
E	15 16	Mineral oils with 50% of saturated and 50% of unsaturated fatty acids	Mineral oil and fatty acid	2.8-5.0
F	17 18	Mineral oil with a considerable addition of saturated fatty acid	Mineral oil with stearic and palmitic acids	25-30
G	19 20	Lubricants with a very low content of fatty acids	Animal fat Cottonseed oil	2.2-4.2
H	21 26	Lubricants of G group with addition of fatty acids	Animal fat, cottonseed oil and stearic acid, palmitic acid, fatty acids with C_8–C_{10} and C_{10}–C_{18} chains	3.0
I	27 28	Lubricants with a constant consistency	Animal fat, rape oil, fatty acid	3.0
K	29 32	Lubricants of the G group with addition of fatty acids	Animal oils, cottonseed oil, fatty acids, emulsifiers	20.0
L	33 34	Commercial emulsions	Not given	12.0

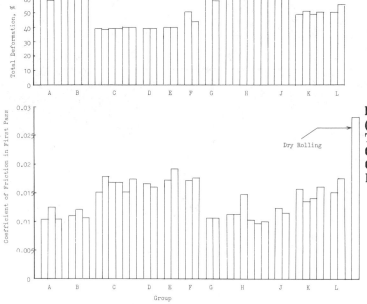

Figure 6-44: Histograms Showing (a) Change of Strip Thickness and (b) Coefficient of Friction Calculated for the First Pass.

COLD ROLLING LUBRICATION

Gorecki and Madejski[▷] used the same method for the evaluation of a number of rolling oils taking precautions to maintain the rolls and the mill housing at virtually constant temperature. They also used a 2-high mill with 260 x 350 mm rolls to reduce strip initially 1.0 mm thick at a speed of 17.5 m/min. Rolling was accomplished in 3 passes with the same settings of the roll gap adjusted to give a 45 per cent reduction to the strip when rolling dry. Their data for five lubricants is illustrated in Figure 6-45. After the first pass, a dry-rolled strip (5) was 83 per cent and, after the third pass, 26.8 per cent thicker than strip rolled with a mixture of rapeseed oil and wax (1) (this mixture showing properties superior to palm oil in the laboratory).

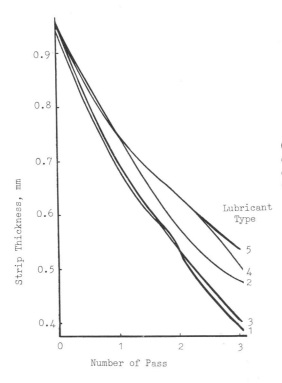

Figure 6-45:
Curves Showing Thickness of Rolled Strip as Function of Lubricant Type (1:Rape Oil + Wax; 2:Spindle Oil; 3:Palm Oil; 4:Kerosene; 5:Dry Rolling)

It should be remembered that these tests, though useful for screening candidate lubricants, provide a ranking of lubricants only under the test conditions. As will be seen later in this chapter, the behavior of lubricants may differ appreciably when the rolling speed is changed so the ranking of lubricants developed on the basis of low speed rolling may not correspond with their relative performances on a high-speed production mill.

6-21 The Direct Measurement of the Coefficient of Friction During Rolling

If the pressure and the shearing stress could be measured at all points along an arc of contact in the roll bite, it would then be possible to calculate the coefficient of friction for any position along the arc and to study its variation from the entry to the exit plane of the bite. A number of attempts to accomplish this have been made by various investigators who inserted transducers in one of the work rolls of the mill capable, in some instances, of measuring the rolling pressure normal to the roll surface and, in others, the shear stresses tangential to the roll surface as

---

▷W. Gorecki and J. Madejski, "Selection of Emulsion Lubricants for Cold Reduction of Steel Strip", Journal Prace Institutow Hutniczych 17, (1965), pp. 303-313.

well. Even if only the pressure distribution along the arc of contact is measured, the coefficient of friction may be calculated from its rate of variation.×

Seibel and Lueg◇ conducted some classic experiments in the 1930's in which they measured the pressure distribution along the arc of contact using the type of equipment shown diagrammatically in Figure 6-46.○ In this transducer, two quartz crystals (X) were inserted in the hole drilled into an accurately made circular segment which acted as the bottom roll. One end of a metal pin, of small cross-sectional area, contacted the face of the top crystal, the length of the pin being such that the other end was made to fit flush with the circumferential surface of the segment. During rolling, pressure was transmitted through the pin to the quartz crystal which developed potentials proportional to the pressure.

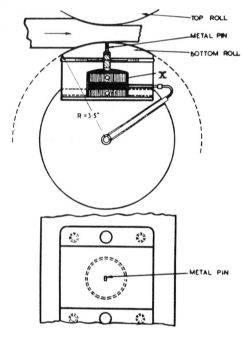

Figure 6-46:
Seibel and Lueg's Apparatus for the Direct Measurement of Rolling Pressure.

More recent experiments of this nature have been conducted at the Cavendish Laboratory, Cambridge, by Smith, Scott and Sylwestrowicz◻ with respect to the rolling of copper. In their method, a photoelastic dynamometer inside the lower roll measured the normal force exerted on a radial pin inserted into the roll, the surface of the pin being flush with the roll surface.

Van Rooyen and Backofen= refined the method and used a special roll containing two pins installed at right angles to the roll axis, one in a radial direction and the other inclined to the first in an oblique direction. Measurements were made of the stress along the axis of each pin as it travelled over the arc of contact and, from the ratio of stress in the oblique and radial directions, a coefficient of friction could be computed as a function of position in the roll gap. Figure 6-47 illustrates schematically the directions of stresses measured by the pins and the following symbols are used in the mathematical treatment. Van Rooyen and Backofen developed corrections for the effect of friction between the pins and the roll so as to enhance the accuracy of the method.

× E. Orowan, "The Calculation of Roll Pressure in Hot and Cold Flat Rolling", Proc. Inst. Mech. Eng., (London), 1943, Vol. 150, pp. 140-167.

◇ E. Seibel and W. Lueg, "Investigations into the Distribution of Pressure at the Surface of the Material in Contact With the Rolls", Mitt. K. W. Inst. Eisenf. 1933, 15, p. 1.

○ "The Rolling of Strip, Sheet and Plate", by E. C. Larke, The MacMillan Company, New York, 1957.

◻ C. L. Smith, F. H. Scott and W. Sylwestrowicz, "Pressure Distribution Between Stock and Rolls in Hot and Cold Flat Rolling", Journal of the Iron and Steel Institute, April, 1952, pp. 347-358.

= G. T. Van Rooyen and W. A. Backofen, "Friction in Cold Rolling", Journal of the Iron and Steel Institute, June, 1957, pp. 235-244.

COLD ROLLING LUBRICATION

σ = radial or normal stress between rolls and strip at any point M

τ = friction-induced shear stress on the surface at any point M

p_r = stress in radial pin

$p+\phi$ = stress in oblique pin inclined at an angle $+\phi°$ measured from the radial direction towards the entrance point O

$p-\phi$ = stress in oblique pin inclined at an angle $-\phi°$ measured from the radial direction towards the exit point N such that $|+\phi°| = |-\phi°|$

μ = coefficient of friction between strip and roll; $\mu = \tau/\sigma$

σ_Y^* = yield stress of the strip in plane-strain compression

τ_Y^* = yield stress in shear, $\sigma_Y^*/2$

A = cross-sectional area of pins

Determination of p_r and $p+\phi$ with radial and oblique pins respectively could be translated into a value of μ. Since

$$p_r A = \sigma A \qquad (6\text{-}14)$$

$$p+\phi A = \sigma \frac{A}{\cos\phi} \cos\phi + \tau \frac{A}{\cos\phi} \sin\phi \qquad (6\text{-}15)$$

$$\frac{p+\phi}{p_r} = 1 + \mu \tan\phi \qquad (6\text{-}16)$$

therefore

$$\mu = (p+\phi/p_r - 1) \cot\phi . \qquad (6\text{-}17)$$

The friction coefficient is thus a function of the ratio of the stress in the oblique pin to that in the radial pin. The above equation is based on the assumption that there is no interference with pin response by friction between the pins and the roll in which they are installed ($p_r = \sigma$).

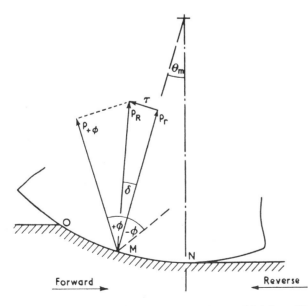

Figure 6-47: Sketch Illustrating Directions of Stresses Measured by Pins (Van Rooyen and Backofen).

Their experiments were carried out on a 2-high mill with hardened steel rolls of 6-inch diameter and 10-inch face at a rolling speed of 25 f.p.m. The material rolled was 2S-18H aluminum in the form of strips 2 inches wide by 9 inches long by 0.25 inches thick. Reduction was fixed at 50 per cent and the roll surface preparation was varied to change frictional conditions. To explore the coefficient of friction each side of the neutral point, tests were made with rolling carried out in both directions (see Figure 6-47).

A cross-sectional view of the experimental roll, showing only the oblique pin, is presented in Figure 6-48. Both pins were incorporated in the lower roll so that they terminated at the roll surface, with their centers spaced 7/8-inch apart on the same line parallel to the roll axis. Electrical strain gages were mounted on the square body of the pin to form a Wheatstone bridge circuit which provided compensation for both temperature and any bending stresses imposed on the pin. A schematic diagram of the electrical circuit is illustrated in Figure 6-49, the d-c supply voltage to the two weigh bars being chopped by a rotary commutator fitted with silver contacts. A complete cycle of chopping consisted of applying a d-c voltage to the radial pin gages in 1/3 of the cycle, applying a d-c voltage to the oblique pin gages in the next 1/3 cycle and then isolating the circuits from the d-c voltage. Thus it was possible to obtain both signal traces as well as a zero line on the oscilloscope screen.

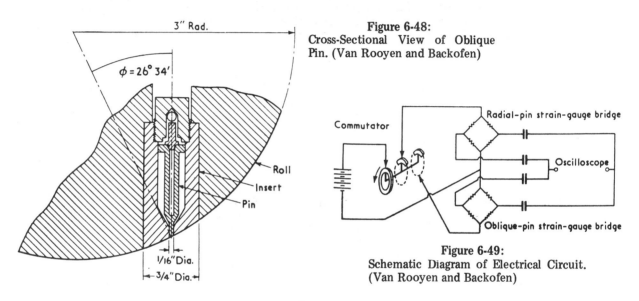

Figure 6-48: Cross-Sectional View of Oblique Pin. (Van Rooyen and Backofen)

Figure 6-49: Schematic Diagram of Electrical Circuit. (Van Rooyen and Backofen)

Typical test results are shown in Figures 6-50 and 6-51. In the former, the reduction is given to aluminum strip with sandblasted (60 microinch RMS) rolls without a lubricant whereas, in the latter, smooth (11 microinch RMS) rolls are used to reduce the strip with the aid of a rolling lubricant.

6-22 Measuring the Coefficient of Friction Under Conditions of Zero Slip

Whitton and Ford◇ developed a method for measuring the friction in the roll gap under typical rolling conditions, yet without making use of any theory of the distribution of roll pressure along each arc of contact. Referring to Figure 6-52, consider an element on the roll surface of length dl and of unit width making an angle ϕ with the vertical (the roll axes being assumed horizontal) subjected to a normal stress s. The resultant vertical force on the element is sdl and the specific rolling force f is given by

$$f = \int_0^L s \, dl \qquad (6\text{-}18)$$

since l is equal to zero at the exit plane and L at the entry plane.

◇ P. W. Whitton and H. Ford, "Surface Friction and Lubrication in Cold Strip Rolling", "Research on the Rolling of Strip", A Symposium of Selected Papers, 1948-1958, The British Iron and Steel Research Association.

COLD ROLLING LUBRICATION

The torque per unit width T is given by

$$T = \mu R \int_0^L s\, dl \qquad (6\text{-}19)$$

and therefore, to a close approximation,

$$\mu = T/fR \qquad (6\text{-}20)$$

This last equation is strictly valid only when the forward slip is zero or negative.

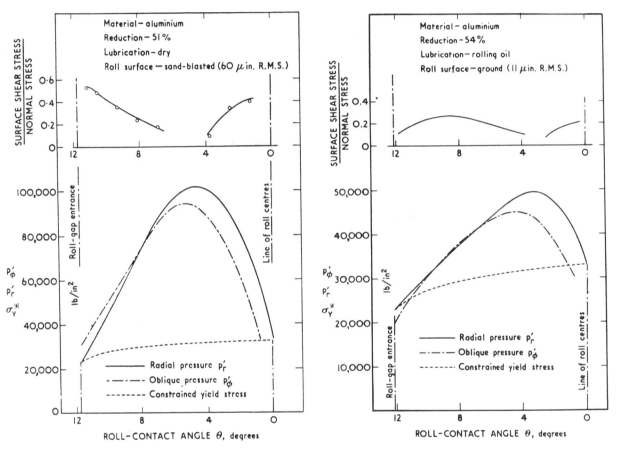

Figure 6-50: Ratio of Surface Shear Stress to Normal Stress with Sandblasted Roll.

Figure 6-51: Ratio of Surface Shear Stress to Normal Stress with Ground Roll and Oil.

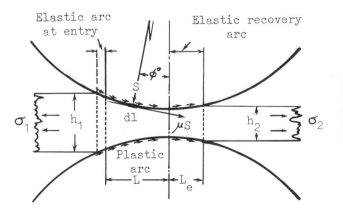

Figure 6-52: Sketch Illustrating Method of Determining the Coefficient of Friction (Whitton and Ford).

The experimental technique required of this method necessitates the measurement of the spindle torques, the rolling force and the slip of the strip, with respect to the roll surface speed, on exit from the roll bite during the rolling operation. The strip is entered into the roll bite, rolled at a steady speed at any suitable pass reduction and the back tension is gradually increased until the neutral point is brought to the exit plane. The rolling force and torque are then accurately measured and substituted in equation 6-20 above.

Table 6-10
Coefficients of Friction for Various Lubricants as Measured Under Conditions of Zero Slip

Test No.	Lubricant	Pass number	Nominal pass reduction, per cent	Coefficient of friction, μ
	Mild steel — initially in annealed condition			
5	None, Rolls and strip clean and dry	First	15.0	0.085
4	Paraffin oil	First	16.5	0.080
15	Paraffin oil	Second	17.0	0.068
19	Paraffin oil	Third	22.0	0.060
12	Paraffin oil +1 per cent stearic acid	First	16.7	0.075
13	Paraffin oil +1 per cent stearic acid +0.6 per cent sulphur	First	17.0	0.071
11	Paraffin +5 per cent copper stearate	Second	16.8	0.063
24	Paraffin +5 per cent sodium stearate	Third	24.0	0.060
9	Paraffin +5 per cent lead stearate	Second	17.3	0.058
10	Paraffin +5 per cent lead oleate	Second	17.4	0.058
26	Paraffin +1 per cent lauric acid	Third	24.3	0.053
30	Paraffin +1 per cent lauric acid	Second	18.8	0.052
25	Paraffin +5 per cent sodium oleate	Fourth	23.0	0.049
28	Paraffin +1 per cent palmitic acid	Third	22.0	0.043
6	68/615 graphite in oil	First	15.5	0.072
20	615 graphite in oil	Fourth	24.5	0.047
7	Vac. R.O. 546	First	15	0.070
8	Vac. R.O. 950	First	15.6	0.069
29	Vac. R.O. 40A	First	17.0	0.061
16	Shell P.E. 6	Third	23	0.050
21	Shell P.E. 6	Fourth	27.9	0.053
18	Esso Baywest	Fourth	27.5	0.050
31	Esso Pale 885	Third	24.0	0.052
17	Esso Paranox 108	Second	21.4	0.054
22	Esso Paranox 108	Third	25.9	0.056
14	Olive oil	Second	18.1	0.057
23	Castor oil	Fourth	23.0	0.045
32	Lanoline	Fourth	26.5	0.041
27	Camphor flowers	Fourth	27.2	0.038

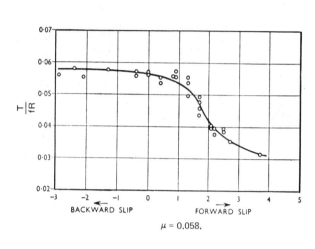

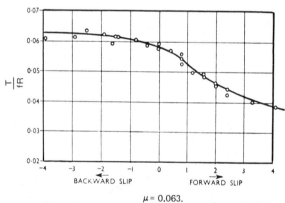

Figure 6-53:
Curves Showing Variation of T/fR with Slip for Two Typical Tests (Whitton and Ford).

Whitton and Ford utilized a 2-high mill with 4-inch diameter rolls of 6-inch face operating at a constant rolling speed of 32 f.p.m. The rolling force was measured by force meters accurately located between each screw and the top roll chock, rolling torques were measured by resistance strain gages mounted on the spindles (the data being corrected for bearing losses) and the slip was measured by cinephotography. A commercial quality rimming steel (0.08 per cent C, 0.01 per cent Si, 0.03 per cent P, 0.03 per cent S and 0.35 per cent Mn) was used as fully annealed strip 0.07-inch thick and 1.5-inch wide. Figure 6-53 illustrates the results for test numbers 9 and 11 listed in Table 6-10. The table presents the calculated coefficients of friction for a variety of lubricants used with smooth (< 9 microinch RMS) mill rolls.

COLD ROLLING LUBRICATION

Yamanouchi and Matsuura[※] adopted the same technique and taking into account the elastic deformation of the rolls developed the expression where α is the angle subtended by the arc of contact at the center of the work roll, P_O and T_O are the force and torque with no tension applied, P and T are the force and torque with sufficient back tension to cause slipping and R' is the deformed roll radius.

$$\mu = \left[1 + \left(1 - R/R'\right)\left(\frac{RP_O\alpha}{2T_O}\right)\right]\left[\frac{T}{PR}\right] \qquad (6\text{-}21)$$

Unfortunately, this method of applying a high back tension to the strip entering the roll bite has the disadvantage of creating an artificial situation in the roll gap, in that the neutral point cannot assume its normal position (determined by the balance of forces in the roll bite). Nevertheless, the method is simple and it does provide values of the coefficient related directly to the rolling process.

6-23 Calculating the Coefficient of Friction Under Conditions of Finite Slip

Using measurements of rolling force, torque and forward slip, it is possible to derive an approximate value of the coefficient of friction for virtually any rolling situation. The method assumes a circular arc of contact corresponding to the deformed roll diameter, a friction hill with geometrical simularity each side of the neutral point and a constant value μ for the coefficient of friction along the arc of contact.

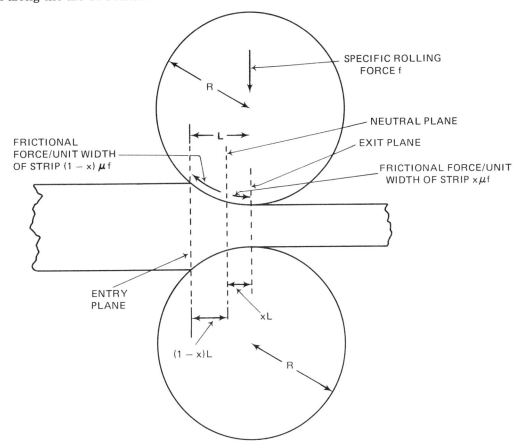

Figure 6-54: Sketch of Section of Roll Bite.

[※] H. Yamanouchi and Y. Matsuura, Rep. Castings Res. Lab., Waseda University, No. 8, 1957.

Referring to Figure 6-54, if the distance between the exit plane and the neutral point represents a fraction X of the length of the arc of contact, then the frictional force exerted by each roll surface ahead of the neutral point in forcing strip of unit width into the mill is (1-X) μ f where f is the specific rolling force. Similarly, the frictional force retarding its exit from the bite is $X\mu f$. Thus the net tangential force on the surface of each roll is (1-2X) μ f and this can be equated to the specific torque T for the corresponding spindle by

$$T = (1-2X)\mu fR \qquad (6-22)$$

or

$$\mu = \frac{T}{fR(1-2X)} \qquad (6-23)$$

If the slip is ν and the peripheral speed of the work roll is V, then the exit speed of the strip is $V(1+\nu)$. Since its thickness at entry is t and at exit is t(1-r), its thickness at the neutral point is $(1+\nu)t(1-r)$. Thus the strip is $\nu t(1-r)$ thicker at the neutral point than at exit and tr thicker at entry than at exit from the bite. From geometrical considerations, these excess thicknesses are proportional to the square of the corresponding distance from the exit plane. Thus

$$\frac{X^2}{1^2} = \frac{\nu t(1-r)}{tr} \qquad (6-24)$$

and therefore

$$X = \sqrt{\frac{\nu(1-r)}{r}} \qquad (6-25)$$

Substituting this expression for X in equation (6-23) the relationship for the coefficient of friction becomes

$$\mu = \frac{T}{fR\left(1-2\sqrt{\frac{\nu(1-r)}{r}}\right)} \qquad (6-26)$$

This method has been used for computing the coefficient of friction in rolling experiments in the laboratory and in the operation of secondary cold reduction mills. It has the distinct advantage of not requiring a knowledge of either the strip tensions or the dynamic yielding behavior of the workpiece. Figures 6-55 and 6-56 show data obtained by this method in the rolling of tinplate and black plate on a laboratory mill with 3.25-inch diameter work rolls.

In the derivation of equation 6-26, the neutral point position was calculated from the slip and an assumed roll bite geometry. Capus and Cockroft[□] developed a relatively simple method for establishing the position experimentally. They utilized the marks left on the strip by the roll-surface asperities when the mill was quickly stopped and opened up. Where there is no lateral movement of the strip in the bite, a point type asperity would, in its contact with the strip between the entry and neutral planes, produce a straight scratch parallel to the rolling direction. At the neutral point, the asperity would coincide with the end of the scratch further into the bite. After passing the neutral point, the asperity would begin to retrace its path along the scratch and it would be impossible to establish the neutral point position. However, where a lateral movement of the strip occurs, either naturally due to the sideways spread of the workpiece or due to an impulsive movement given to a work roll in an axial direction, the scratches assume the forms shown in Figure 6-57 so that, on exit from the bite, they appear like hooks. The neutral point coincides with the position on the strip where the tangent to the hooked-shaped scratch lies parallel to the axis of the rolls.

□ J. M. Capus and M. G. Cockroft, "Relative Slip and Deformation During Cold Rolling", Journal of the Institute of Metals, 1961-62, Vol. 90, pp. 289-297.

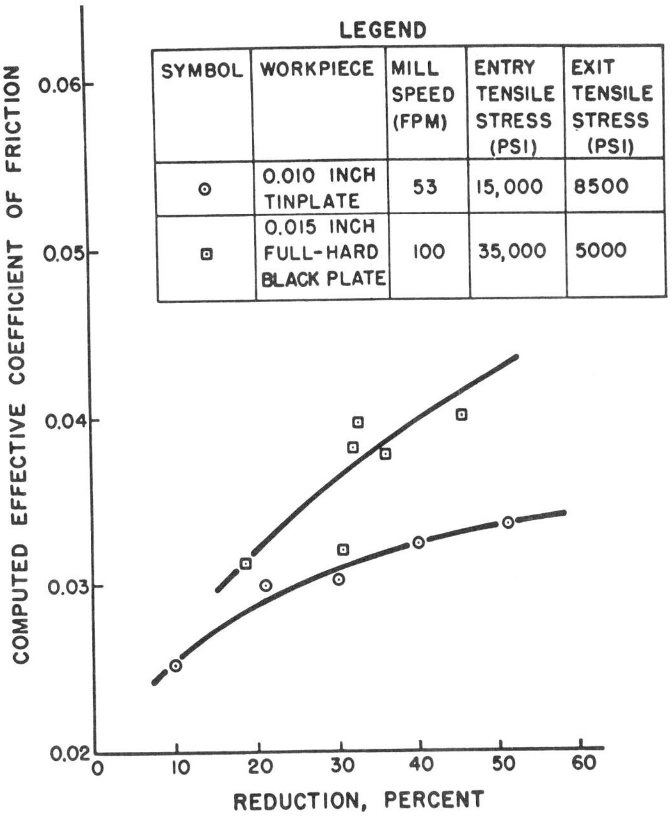

COLD ROLLING LUBRICATION

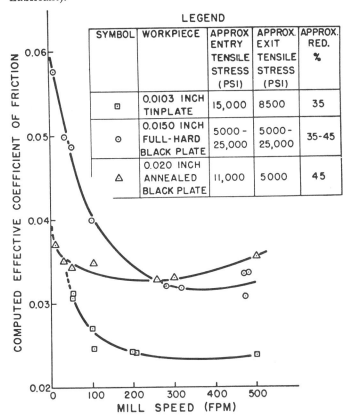

Figure 6-55:
Variation of the Computed Coefficient with Reduction for Two Types of Strip Rolled on a 3.25-Inch Diameter Work Roll Mill (Substitute Cottonseed Oil Used as the Lubricant).

Figure 6-56:
Variation of the Computed Coefficient with Mill Speed for Two Types of Strip Rolled on the 3.25-Inch Diameter Work Roll Mill (Substitute Palm Oil Used as the Lubricant).

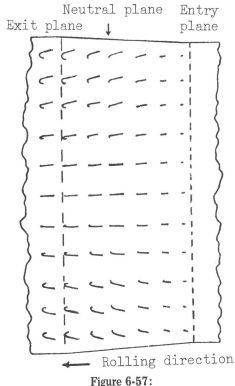

Figure 6-57:
Distribution of Scratches on the Contact Area of a Partially Rolled Strip. (Schematic; Not to Scale.) (Capus and Cockroft).

6-24 Calculating the Coefficient of Friction from Rolling Force and Strip Data

Providing the strip tensions and the dynamic yielding characteristics of the strip are known, it is possible to calculate the coefficient of friction solely from measurements of the rolling force (without a knowledge of the slip). The ease with which this may be accomplished depends upon the nature of the mathematical model which must be employed for this purpose. A simple algebraic model * of the rolling process is the most convenient type to use and such a model is presented below (and its derivation in Chapter 9).

Referring to Figure 6-58, the mathematical model consists basically of the following relationships:

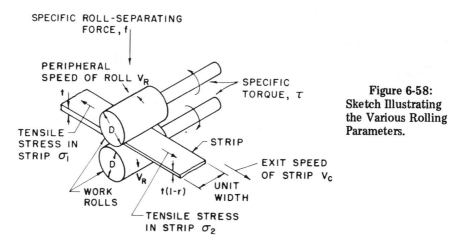

Figure 6-58: Sketch Illustrating the Various Rolling Parameters.

(1) The expression for strain rate $\dot{e}$ (seconds^{-1})

$$\dot{e} = 0.178\ V_R \sqrt{\frac{r}{Dt}} \qquad (6\text{-}27)$$

where V_R is the rolling speed (fpm) and t is the entry thickness of the strip (inches);

(2) the expression for the constrained yield stress σ_c

$$\sigma_c = 1.155\ (\sigma_{yt} + a\ \log_{10} 1000\dot{e}) \qquad (6\text{-}28)$$

where σ_{yt} is the yield strength in tension as conventionally measured and

a is a constant with a value about 6,250 psi per decade change of strain rate;

(3) the expression for the deformed roll diameter D'

$$D' = \left(1 + 2\sqrt{\frac{f}{Etr} + \frac{2f}{Etr}}\right) D \qquad (6\text{-}29)$$

where f is the specific rolling force and E is the elastic modulus of the roll material;

(4) the expression for the coefficient of friction μ

$$\mu = 2\sqrt{\frac{2t}{D'\ r}} \left(\frac{f\ (1-r)}{(\sigma_c - \sigma_1)} \sqrt{\frac{2}{D'\ tr} - 1} + \frac{5r}{4}\right) \qquad (6\text{-}30)$$

where σ_1 is the average of the tensile stresses in the strip on entry into and exit from the roll bite.

From any set of rolling data, the value of the strain rate determined by equation 6-27 can be substituted in equation 6-28 to yield the constrained yield stress σ_c which, in turn, may be

---

* W. L. Roberts, "Computing the Coefficient of Friction in the Roll Bite from Mill Data", <u>Blast Furnace and Steel Plant</u>, June, 1967, pp. 499-508.

COLD ROLLING LUBRICATION

used in equation 6-30. The solution of the simultaneous equations 6-29 and 6-30 therefore yields the coefficient of friction μ.

For the graphical solution to equations 6-29 and 6-30, it is perhaps most convenient to use equation 6-30 in the form

$$f = \frac{(\sigma_c - \sigma_1)}{(1-r)} \sqrt{\frac{D'tr}{2}} \left\{ 1 - \frac{5r}{4} + \frac{\mu}{2}\sqrt{\frac{D'r}{2t}} \right\} \quad (6\text{-}31)$$

Then ascribing fixed values to $(\sigma_c - \sigma_1)$ and t, the following procedure may be adopted;

(1) For various reductions (r = 0.1, 0.2, 0.3, etc.) a family of curves relating f to D on the basis of Equation 6-29 may be drawn up. Such a family is illustrated in Figure 6-59.

(2) For the same values of r, and for various values of μ (μ = 0.02, 0.03, 0.04, etc.), a similar family of curves for Equation 6-30 may be drawn up on semitransparent graph paper using the same scales for the axes as in Step 1. Figure 6-60 represents some of the curves, the remainder being omitted for the sake of clarity.

(3) By superpositioning the two graphs, the solutions for f may be read off from the intersections of the appropriate curves for various values of μ and r.

(4) From the data acquired under Step 3, a family of curves relating f to r for various values of μ may then be prepared as illustrated in Figure 6-61.

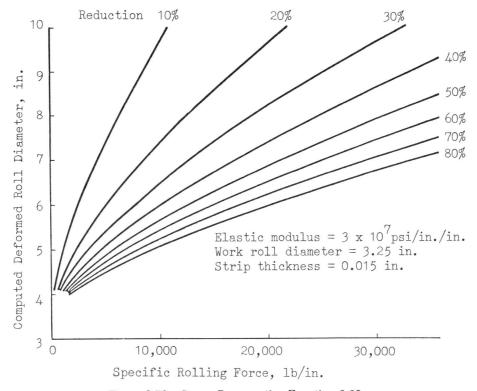

Figure 6-59: Curves Representing Equation 6-29.

Experimentally determined values for the specific rolling force may now be plotted directly on curves such as those shown in Figure 6-61, and the values of the coefficient can be determined without any further computation.

It will be noted that, for a small diameter work roll mill, such as that considered in the computation, the curves in Figure 6-61, corresponding to different values of the coefficient, are

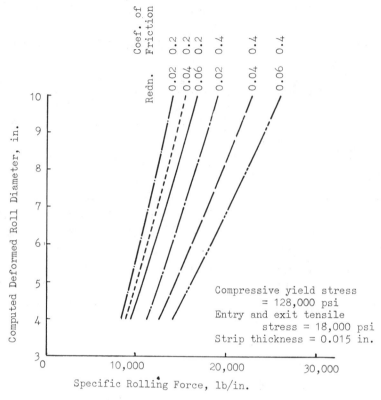

Figure 6-60: Curves Representing Equation 6-30.

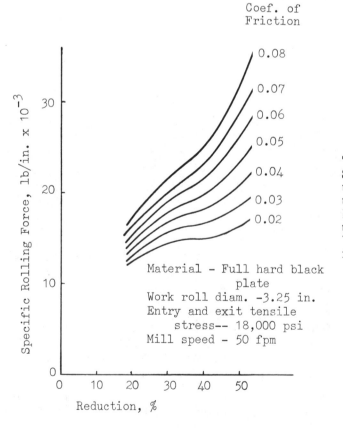

Figure 6-61: Theoretical Curves Showing the Relationship Between the Specific Rolling Force and Reduction for Various Values of the Coefficient of Friction.

COLD ROLLING LUBRICATION

closely grouped, especially at low reductions. This implies that the dynamic yield strength at rolling speed must be accurately known and the specific roll separating force must be carefully measured if a reliable value of the coefficient is to be obtained.

Where the use of a computer is available, it is possible to print out sets of data relating to any roll diameter, D, entry thickness t and the minimum or "frictionless" deformation pressure ($\sigma_c - \sigma_1$). A typical page from such a set of data is illustrated in Figure 6-62, and it is a simple procedure to determine the value of the coefficient μ from such tables.

Roll Diameter — 21.000 (in.) Entry Thickness — 0.0121 (in.) Pressure — 95,000 (psi)

Reduction, Percent	Specific Rolling Force (lb./in.)											
	m 0.010	m 0.015	m 0.020	m 0.025	m 0.030	m 0.035	m 0.040	m 0.045	m 0.050	m 0.055	m 0.060	m 0.065
6	20693	22337	24137	26133	28450	31023	33956	37265	41107	45508	50701	56683
8	21717	23527	25496	27678	30230	33051	36239	39898	44128	49015	54727	61354
10	22726	24651	26781	29154	31951	35005	38513	42525	47166	52541	58874	66432
12	23618	25699	28005	30660	33616	36956	40783	45166	50248	56185	63138	71621
14	24467	26697	29183	32069	35273	38900	43062	47844	53433	59987	67597	77099
16	25261	27648	30394	33457	36916	40847	45369	50608	56722	63912	72530	82949
18	25990	28557	31522	34822	38556	42810	47745	53455	60136	67992	77619	89257
20	26713	29493	32632	36176	40202	44828	50156	56394	63731	72504	83096	96112
22	27381	30358	33720	37527	41863	46861	52665	59458	67478	77214	89002	103607
24	28014	31198	34796	38881	43548	48971	55268	62694	71416	82251	95429	111897
26	28615	32016	35863	40244	45290	51124	57976	66099	75838	87725	102482	120990
28	29203	32816	36926	41622	47051	53378	60844	69651	80490	93716	110311	131621
30	29745	33599	37989	43022	48887	55729	63864	73616	85511	100295	118982	143514
32	30276	34369	39055	44473	50788	58217	67040	77801	91021	107616	128946	157227
34	30763	35125	40149	45961	52765	60834	70517	82367	97060	115770	140234	173312
36	31240	35871	41235	47469	54826	63602	74232	87333	103785	125049	153301	192412
38	31653	36625	42339	49051	57018	66500	78271	92810	111364	135649	168665	215591
40	32080	37357	43483	50697	59331	69765	82657	98919	119906	147976	186993	244310
42	32480	38100	44655	52439	61765	73228	87519	105757	129742	162463	209330	280880
44	32852	38823	45859	54244	64439	77031	92925	113440	141177	179864	237165	329159
46	33194	39562	47103	56174	67309	81197	99009	122470	154660	201129	272925	395706
48	33506	40283	48394	58201	70423	85857	105885	132880	170924	227562	320613	493031
50	33771	41024	49764	60468	73865	91052	113834	145244	190899	262408	387242	647878
52	34017	41770	51201	62875	77684	96988	123108	159948	216031	308964	486863	928297
54	34225	42524	52677	65495	81935	103814	134096	178525	249043	375102	650734	
56	34392	43306	54356	68404	86800	111769	147422	201878	293924	476333	965277	
58	34514	44084	56125	71652	92355	121228	163909	232644	358724	649408		
60	34584	44880	58046	75316	98867	132701	185012	275076	460540			

Figure 6-62:
Portion of Table for Determining Specific Rolling Force from Other Process Parameters (m Denotes the Coefficient of Friction).

In modern cold rolling research facilities, the transducers used on an experimental mill for the measurement of rolling force, spindle torques, strip tensions and other parameters may be coupled directly to analog-to-digital converters, the outputs of which may be fed to a data accumulation system such as that shown in diagrammatic form in Figure 5-1, which provides the experimental information in the form of a punched tape. The data on the tape may then be transferred automatically to punched cards, which in turn may then be used with a computer to provide a printout of the data in any desired form such as that shown in Figure 6-63. Calculations, such as those pertaining to coefficient of friction and deformed roll diameter, may be made at the same time and curves relating the various parameters may be automatically drawn (Figure 6-64).

The mathematical model of the rolling process discussed in this section has been utilized in studies relating to the operation of both primary and secondary cold reduction mills in addition to experiments conducted on laboratory mills.♦►♯♯ Needless to say, other algebraic models could be used with a corresponding degree of convenience, the accuracy of the computed values of the coefficient depending on the validity of the model used.

		WORK ROLL DIA.	6.303	STRIP WIDTH 10.00		TEST DATE		120
MATERIAL ALUM COATED BP		LUBRICANT		CODE NUMBER		0		
TEST NUMBER		144	145	146	147	148	149	150	151
ENTRY GAGE	IN	.0313	.0307	.0303	.0305	.0305	.0306	.0307	.0299
EXIT GAGE	IN	.03002	.02926	.02890	.02788	.02785	.02597	.02635	.02407
REDUCTION	PCT	4.1	4.8	4.8	8.6	8.9	15.3	14.3	19.5
FORC DRIVE	LB	47350	46600	46300	84950	85150	142150	142400	187050
FORC OPER	LB	31000	29750	29350	54600	54600	103400	103450	143550
TOTAL FORCE	LB	78350	76350	75650	139550	139750	245550	245850	330600
SPEC FORCE	LB/IN	7835	7635	7565	13955	13975	24555	24585	33060
TOP TORQUE	FT-LB	347.0	363.0	387.0	615.0	629.0	1081.0	1063.0	1420.0
BOT TORQUE	FT-LB	284.0	283.0	263.0	519.0	518.0	1011.0	1038.0	1467.0
TOT TORQUE	FT-LB	631.0	646.0	650.0	1134.0	1147.0	2092.0	2101.0	2887.0
SP TORQ	FT-LB/IN	63.1	64.6	65.0	113.4	114.7	209.2	210.1	288.7
ENTRY TENSION	LB	3217	3267	3275	3327	3328	3369	3364	3376
ENT STRESS	PSI	10278	10642	10809	10908	10911	11010	10958	11291
EXIT TENSION	LB	2976	2877	2858	2755	2742	2693	2682	2662
EXIT STRESS	PSI	9913	9833	9889	9882	9846	10370	10178	11059
ENT SPEED	FPM	81.7	81.9	81.9	79.4	79.3	75.1	75.2	72.4
MILL SPEED	FPM	85.4	85.2	85.1	85.3	85.3	85.3	85.4	85.4
EXIT SPEED	FPM	85.3	86.1	86.1	87.0	87.1	88.7	87.9	90.0
FORWARD SLIP	PCT	-0.00	1.00	1.10	1.90	2.00	4.00	2.80	5.40
MAX SLIP ERROR		.22	.22	.22	.22	.22	.22	.22	.21
HPOWER SPINDLE		6.22	6.35	6.38	11.16	11.29	20.59	20.70	28.45
HPOWER ENT	REEL	7.96	8.11	8.13	8.00	8.00	7.67	7.67	7.41
HPOWER EXIT	REEL	7.69	7.51	7.46	7.26	7.24	7.24	7.14	7.26
HPOWER TOTAL		5.95	5.75	5.71	10.42	10.53	20.16	20.18	28.30
ENERGY	HP-HR/TON	2.28	2.24	2.25	4.21	4.26	8.58	8.54	12.81

Figure 6-63: Portion of Computer Printout of Mill Data.

A number of investigators have used rolling mills instrumented to measure the rolling force but have used their equipment basically to rank or compare lubricants under test with the behavior of materials such as palm oil or water. Jackson and Guy * compared the performance of a group of rolling oils in this manner using a mill with 6-inch diameter work rolls to reduce mild rimming steel strip initially 0.029-inch thick at speeds up to 1,000 f.p.m. Emulsions with concentrations close to 4 per cent at a temperature of 35 ± 2°C were fed to the roll bite at a rate of 1,150 g.p.h. by means of single jets which gave a flat spread about 8 inches wide. Four main passes were given to the strip alternated with skim passes as indicated in Table 6-11.

Some of the experimental data is presented in Table 6-12 which shows the loads required for a particular drafting in each pass at a speed of 1,000 f.p.m. It will be noted that the differences in performance of the 5 oils became very noticeable during the third pass, and the order of merit was established as palm oil, Z, W, X and Y.

♦ R. J. Bentz and W. L. Roberts, "Predicting Rolling Forces and Mill Power Requirements for Tandem Mills", Blast Furnace and Steel Plant, August, 1970, pp. 559-568.

► W. L. Roberts, "Recent Developments in Rolling Lubrication", Blast Furnace and Steel Plant, May, 1968, pp. 382-394.

♯♯ W. L. Roberts, "Frictional Characteristics of Rolling Lubricants", Proceedings of the 11th Mechanical Working and Steel Processing Conference, AIME, 1969, New York, New York, pp. 299-316.

* G. Jackson and V. H. Guy, "Progress Report on the Assessment of New Rolling Oils Intended for Use in High Speed Tandem Mills", British Iron and Steel Research Association MW/A/14/60.

COLD ROLLING LUBRICATION

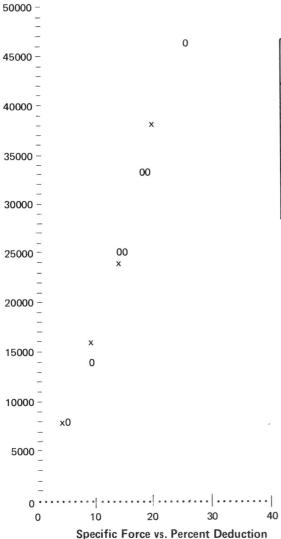

Figure 6-64:
Rolling Data Plotted by Computer.

Table 6-11
Rolling Schedule (Jackson and Guy)

Pass No.	Strip Thickness (in.) Ingoing	Strip Thickness (in.) Outgoing	Speed (ft./min.)	Tensions (tons) Front	Tensions (tons) Back
1	0.029	0.018 − 0.021	1000, 800	1.25	1.0
Skim	0.018 − 0.021	0.017	1000	1.25	1.0
2	0.017	0.013 − 0.009	1000, 800 600, 400	1.0	0.75
Skim	0.013 − 0.009	0.008	1000	0.75	0.5
3	0.008	0.007 − 0.005	1000, 800 600, 400	0.5	0.4
Skim	0.007 − 0.005	0.004	1000	0.4	0.3
4	0.004	0.004 − 0.002	1000	0.3	0.25

Table 6-12
Experimental Rolling Data (Jackson and Guy)

LUBRICANT	Total Rolling Load (tons) for Ingoing and Exit Gauges Shown (in.)			
	0.029 0.018	0.017 0.009	0.008 0.005	0.004 0.0035
W	47.5	53.0	54.0	150
X	53.0	51.0	56.0	− *
Y	52.5	51.0	68.0	− *
Z	51.5	51.0	47.0	150
Palm Oil	40.0	48.0	46.5	40

*Pass 4 was not completed on these two oils as the limiting reduction was found to be above 0.004 in.

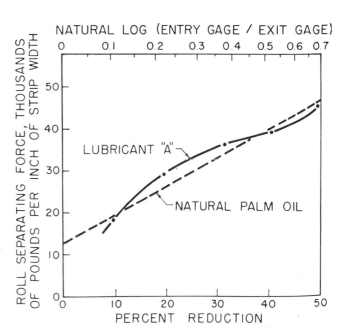

Figure 6-65:
Variation of Rolling Force with Reduction for a Lubricant Under Test and for Palm Oil.

Vojnovic and his co-workers⊖ conducted similar tests on a number of commercially available lubricants with full hard black plate 0.010, 0.0209 and 0.030 inches thick on a mill with 9-inch diameter work rolls operating at a speed close to 100 f.p.m. Figure 6-65 shows the data for lubricant A and that for natural palm oil. At reductions of 10, 20, 30, 40 and 50 per cent, the ratios for the specific rolling force for the lubricants under test to the specific rolling force for palm oil (designated the "rolling indices") were computed and found to be as listed in Table 6-13. The average rolling index relates to the average of the rolling indices for the various reductions (10 to 50 per cent). The significance of the data presented in the other columns of Table 6-13 is discussed in later sections of this chapter.

Table 6-13
Bench Machine and Rolling Mill Data Relating to
Cottonseed Oil, Palm Oil and Various Commercial Lubricants

Lubricant	Concentration of aqueous mixture (per cent)	Thickness of strip rolled (10^{-3} in.)	Free-fatty-acid content (per cent)	Saponification value	Viscosity Index	Coefficient of friction (μ)	Load-bearing capacity, P_f, psi	Ratio, μ/P_f (sq. in. per lb.)	RI 10	RI 20	RI 30	RI 40	RI 50	Average Rolling Index, RI_A	Rating of lubricant
A	4	20.9	–	–	–	0.061	289 x10^3	2.11 x 10^7	0.95	1.12	1.08	0.98	1.00	1.03	Medium
B	4	30.0	8 ± 0.5	193	164	0.048	318	1.51	1.10	1.04	1.00	0.98	0.96	1.00	Good
C	4	20.9	8 ± 0.5	195	141	0.047	281	1.67	1.04	1.00	1.03	0.98	0.94	1.00	Good
D	4	10.0	10	190	91	0.046	289	1.59	1.10	1.04	1.00	0.98	0.96	1.00	Good
E	4	30.0	10	198	157	0.047	274	1.71	1.16	1.08	1.05	1.01	1.15	1.09	Medium
F	4	30.0	3	60	72	0.061	289	2.11	1.08	1.08	1.14	1.15	1.32	1.16	Poor ▽
G	7	10.0	–	42	88	0.077	274	2.81	2.40	2.40	*	*	*	2.40	Poor ▽
H	4	10.0	–	42	–	0.050	289	1.73	0.72	0.47	0.32	–	–	–	Good †
I	4	30.0	2 ± 0.5	65	–	0.057	289	1.97	1.08	1.15	1.14	1.16	*	1.13	Poor
J	4	30.0	8 ± 0.5	194	164	0.045	250	1.80	1.21	1.16	1.08	1.05	*	1.13	Poor
K	4	20.9	–	100	122	0.061	237	2.57	1.16	1.23	1.14	1.08	1.04	1.15	Poor
L	4	20.9	–	–	–	0.086	227	3.70	–	–	1.17	*	*	–	Poor
M	4	20.9	11	150	140	0.046	296	1.55	1.18	1.15	1.02	1.00	1.17	1.10	Medium
N	4	20.9	–	183.5	–	0.047	289	1.63	1.13	1.09	1.00	0.95	1.11	1.06	Medium
O	4	20.9	–	177 ± 8	193	0.041	260	1.58	1.16	1.15	1.07	1.00	1.35	1.15	Poor
P	15	20.9	–	196	–	0.044	311	1.42	–	–	1.10	1.07	–	1.09	Medium
Q	4	20.9	7.5 ± 0.5	196	149	0.046	281	1.64	1.21	1.08	1.03	1.00	1.04	1.07	Medium
R	4	20.9	0.4	196	155	0.047	296	1.59	1.26	1.23	1.15	1.05	*	1.17	Poor
S	4	20.9	10 ± 0.5	197	–	0.043	341	1.26	1.00	1.00	0.94	0.89	0.85	0.94	Good
T	4	20.9	11	194	–	0.045	301	1.50	1.00	1.04	0.94	0.93	0.91	0.96	Good
U	4	20.9	10 ± 1	192	–	0.045	311	1.45	1.21	1.15	1.00	1.00	1.11	1.09	Medium
V	4	20.9	13	192	125	0.045	207	2.17	1.29	1.35	1.21	1.08	1.35	1.26	Poor
W	4	20.9	–	190	142	0.047	341	1.38	1.00	1.00	0.97	0.94	0.89	0.96	Good
X	4	20.9	15	190	157	0.048	267	1.80	1.18	1.15	1.06	1.03	1.30	1.14	Poor
Cottonseed Oil	10	20.9	2 ± 1	194	–	0.052	333	1.56	–	–	1.00	1.00	–	1.00	Good
Palm Oil	4	20.9	9.5	192	–	0.047	348	1.35	1.00	1.00	1.00	1.00	1.00	1.00	Good

▽ The low saponification values appeared responsible for high rolling index.
* Rolling force too high to achieve the specified reduction.
– Indicates data not available.
† This material acted as a solid lubricant during rolling.

6-25 Measuring the Coefficient of Friction and the Load Bearing Capacity on Bench Testing Machines

For reasons of economy and convenience, many efforts have been made to assess the lubricating properties of rolling oils on bench type friction testing machines such as the Shell 4-Ball, the Falex and the Timken machines. None of these attempts have been wholly successful but such machines are useful in screening materials since those which perform poorly on these machines also perform poorly as rolling lubricants. However, the converse is not always true since many materials appear to be good lubricants when tested on these machines but fail to perform satisfactorily under the different conditions existing in a roll bite.

To determine the dynamic coefficient of friction, a laboratory test should simulate the rolling operation as closely as possible and should, therefore, be carried out under conditions of sliding and rolling friction. The Amsler Wear Testing Machine, Type A-135, shown in Figure 6-66, has been found to be suitable for this purpose. Under conditions of relatively light loads and low speeds, it gives the value of the coefficient of friction that exists between the peripheries of two test discs as they roll and slide on each other. Through a sequence of gears and shafts, a 1.4-hp motor drives two lubricated discs, 1-inch and 0.786-inch in radius, in peripheral contact, in such a manner

⊖S. N. Vojnovic, R. R. Somers and W. L. Roberts, "The Laboratory Evaluation of Lubricants Used in the Cold Rolling of Steel Strip", Iron and Steel Engineer Year Book, 1964, pp. 333-339.

COLD ROLLING LUBRICATION

that the circumferential speeds are 210 and 150 fpm. The two discs have rounded edges with radii of curvature equal to the maximum radii of the discs. They are therefore central segments of spheres, as may be seen in Figure 6-67. They are made of 4140 steel hardened to a value of 59 Rockwell C, with a finish of 4 to 10 microinches, and are held in contact by means of a spring and screw arrangement. By this means, one disc exerts a force of 440 lb. on the other to provide an average pressure at the zone of contact of 250,000 psi.

Figure 6-66: The Amsler Wear Testing Machine.

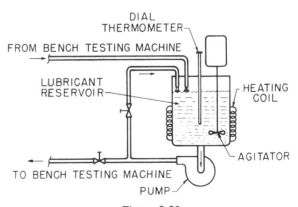

Figure 6-67: Close-up View of Discs on Amsler Machine and Lubricant Directed on Them.

Figure 6-68: Recirculating System for Use With Amsler Machine.

A stream of lubricant is directed into the bite of the discs (Figure 6-67). To ensure a constant and uniform consistency, the lubricant is maintained at a constant temperature in the recirculating system illustrated schematically in Figure 6-68. This system, which provides a flow rate of 0.125 gal. per min., is also used with the Shell four-ball machine, described later.

By means of a pendulum arrangement, the frictional torque transmitted by one disc to the other may be measured and is recorded on a chart recorder attached to the machine. It is desirable to operate the machine for a period of at least 10 minutes so that the recorded torque value becomes constant. From this value, the coefficient of friction may be computed.[8]

Typical values for the coefficients of various lubricants as derived from the Amsler tests are given in Table 6-14.

Table 6-14
Coefficients of Friction Exhibited by Various Rolling Lubricants
(As measured on the Amsler Machine)

Lubricant	Coefficient	Lubricant	Coefficient
Di-isoactyl diphenate	0.078	High Boiling Xylyl Palmitates	0.052
Bis (o-cresyl) isosebacate	0.065	Neat Tar Acid Oleate	0.052
Motor oil SAE 10	0.065	Rapeseed Oil	0.050
X-methylnaphthyl-y-methyl- −2 ethylhexoate	0.063	High-Boiling Xylyl Stearate	0.050
2, 2-Biphenyl dipelargonate	0.062	O-Cresyl Palmitate	0.050
Motor oil SAE 50	0.062	O-Cresyl Stearate	0.049
Neutral Mineral Oil (Viscosity 90° at 100°F)	0.062	Ethylene Glycol	0.048 − 0.051
		Tallow (acidless, refined)	0.046
Kerosene	0.058	Soybean Oil	0.045
Cresyl pelargonate	0.057	Di-Octyl Adipate	0.044
Metalign	0.053	Natural Palm Oil	0.042
		Butyl Stearate	0.041
Cottonseed Oil	0.052	Methyl Stearate	0.037 − 0.039

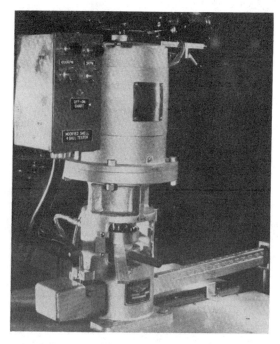

Figure 6-69:
The Shell 4-Ball Machine.

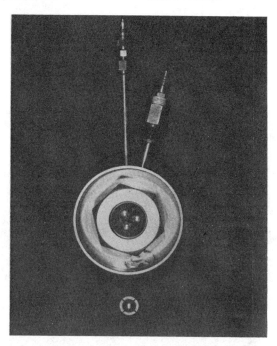

Figure 6-70: Tetrahedronal Arrangement of Balls in the Shell 4-Ball Machine.

---

8 "Wear and Mechanical Properties of Railroad Bearing Bronzes at Different Temperatures", H. J. French, S. J. Rosenberg, W. Le C. Harbaugh, and H. C. Cross, Bureau of Standards Journal of Research, Vol. 1, Research Papers No. 1-36, July to December, 1928, pp. 357-362.

COLD ROLLING LUBRICATION

Because of the high pressures encountered in the cold rolling of steel, the lubricant should have a high load bearing capacity. This property is evaluated on a machine, normally designated as the Shell Four-Ball E.P. Tester, illustrated in Figure 6-69, the tests being conducted in accordance with a standard test procedure. As may be seen from Figure 6-70, in this machine the upper ball of a tetrahedronal arrangement of four balls is rotated while the other three balls, which are held rigidly in a pot, are made to exert a force against it. A 2-hp motor drives the top ball at a speed of 1800 rpm and the loading is provided by a lever arm associated with the ball pot.

To test the load bearing capacity of a lubricant, the machine is operated generally in accordance with Federal Test Method Standard No. 791,[ħ] with the rolling lubricant pumped through the ball pot. The loading is set up on the machine and the top ball is rotated at 1800 rpm for 10 sec. The ball pot is then removed from the machine and the average scar diameter on the three fixed balls is computed from observations of the scars with a microscope. The data relating scar diameter to load are then plotted as illustrated in Figure 6-71. For a particular type of ball bearing, the corresponding value of the mean pressure may be read directly from the chart for each point plotted.

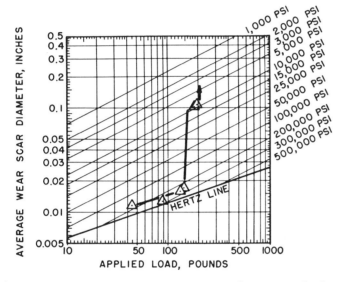

Figure 6-71: Data from Shell 4-Ball Machine Plotted on Special Chart.

For a typical lubricant, it will be noted that, as the load is incrementally increased, the scar diameter increases uniformly until, at a certain loading, the scar diameter increases abruptly. The pressure at which this abrupt change takes place is regarded as the load bearing capacity of the lubricant.

Loads that caused failure on the Shell four-ball machine ranged from 126 to 205 lb. and produced scar diameters of 0.0146 to 0.0197-inch. Rolling lubricants which perform well in mills have been found to exhibit load bearing capacities (or pressures of failure) P_f, as measured by the machine, in excess of 280,000 psi.

In the belief that a satisfactory rolling lubricant should provide both a coefficient of friction, μ, and a high load bearing capacity, P_f, the ratio of these two physical properties has been plotted against the average rolling index, RI_A, discussed in Section 6-24. The resultant graph is shown in Figure 6-72 and as may be seen, the ratio RI_A increases with increasing values for μ/P_f.

As a consequence of their work conducted in 1963, Vojnovic, et al,[⊖] expressed the belief that, with a reasonable degree of confidence, the rolling characteristics of a lubricant may be

---

[ħ] Federal Test Method Standard 791 A; Lubricants, Liquid Fuels, and Related Products; December 31, 1961 Edition; U. S. Government Printing Office. Obtainable from General Services Administration, Business Service Center, Washington, D. C.

[⊖] S. N. Vojnovic, R. R. Somers and W. L. Roberts, "The Laboratory Evaluation of Lubricants Used in the Cold Rolling of Steel Strip", Iron and Steel Engineer Year Book, 1964, pp. 333-339.

predicted from measurements of its coefficient of friction and its load bearing capacity as obtained from the Amsler and Shell testing machines. In addition, knowledge of a lubricant's saponification, free fatty acid and viscosity index values is important. In their opinion, a satisfactory lubricant should possess the following characteristics:

1. A coefficient of friction less than 0.055.
2. A load bearing capacity in excess of 280,000 psi.
3. A μ/P_f ratio less than 1.45×10^{-7} sq. in. per lb.
4. A saponification value between 175 and 200.
5. A free fatty acid content not greater than 10 per cent.
6. A viscosity index greater than 140.

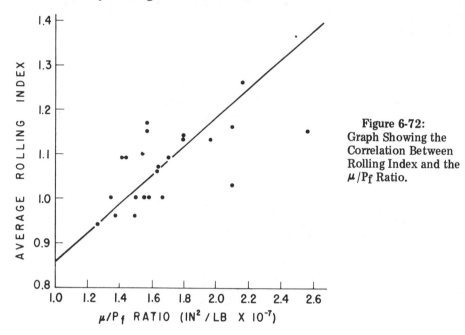

Figure 6-72: Graph Showing the Correlation Between Rolling Index and the μ/P_f Ratio.

6-26 Pour Point, Viscosity and Viscosity Index

It is the common practice to measure a number of other properties, besides the coefficient of friction and the load bearing capacity, relating to the lubricant, or its emulsion, some of which (such as its viscosity) may relate directly to its lubricating properties whereas others (such as its pour point) may relate more to the effectiveness of the emulsion as a coolant. Clear cut correlations between many of these properties and the general effectiveness of the lubricant have not been established but they are frequently of value in identifying the main constituents of lubricants and their contaminants. In this section, pour point, viscosity and viscosity index are discussed and their methods of measurement are described.

(a) <u>Pour Point</u>

Ideally, the rolling oil should be liquid at ambient temperature (as indeed many oils are), since no special techniques need be employed in using the rolling solution. Under these circumstances, the solution can be employed at a lower temperature, thus improving its effectiveness as a coolant.

Lowering the melting or pour point of the rolling oil is usually accomplished by the addition of paraffinic compounds of shorter chain lengths. However, care must be exercised in this, because such additions will frequently decrease the lubricity of the solution and its load bearing capacity.

To determine the pour point, a sample of the oil is poured into a test jar up to the level of an etched mark (2 to 2.25 inches from the bottom) and the jar closed tightly by a cork carrying a test thermometer (see Figure 6-73) positioned so that the beginning of the capillary of the thermometer is 1/8-inch below the surface of the oil. A ring gasket is placed around the jar and it is placed in a jacket in a freezing bath containing crushed ice and salt or solid carbon dioxide and acetone. Beginning at a temperature 20°F before the expected pour point, at each thermometer reading that is a multiple of 5°F, the jar is removed from the jacket and tilted to ascertain whether or not there is a movement of the oil. As soon as the temperature is reached when the oil does not flow, the jar is held in the horizontal position for 5 seconds. If movement occurs, the test is repeated at a 5°F lower temperature until no flow is observed in 5 seconds. The pour point is then regarded as being 5°F above the temperature where solidification is noted.

(b) Viscosity

In a cold reduction mill where an oil-water emulsion is used as the rolling lubricant, the oil is the phase that "plates out" on the rolls and the strip and provides the lubrication. Therefore, in rolling operations, it is primarily the viscosity of the oil and not that of the emulsion that is important.

The viscosity of a liquid is affected by the temperature, the higher the temperature, the lower the viscosity of the fluid. Accordingly, it is general practice to measure viscosity at a definite temperature (100 or 210°F), and also to compute the viscosity index (see under (c) below) which defines the variation of viscosity with temperature.

The viscosity of a fluid is also affected by pressure; some liquids become less viscous, others solidify under pressure. It is not the usual practice, however, to measure this effect of pressure, especially at pressures corresponding to the yield strength of steel, since such measurements would not be easy to make.

The relationship between viscosity and the effectiveness of the lubricant is discussed in Section 6-29. Accordingly, only the methods of measurement will be described in this section.

The Saybolt viscosimeter (or viscometer), which is often used to measure the viscosity of an oil, is illustrated in Figure 6-74 and basically the property is expressed in terms of the time in seconds (Saybolt Universal Seconds or SUS Units) required for 60 ml. of the oil to flow through a standard orifice under a standard falling head of pressure at a given temperature, 100 and 210°F being the common temperatures for reporting viscosity.

In using the Saybolt viscometer, the following procedure is adopted. A draft-free room is used with its temperature maintained between 68 and 86°F and the controls for the viscometer are adjusted so that the desired bath temperature is attained. Previous to the introduction of oil into the oil tube, the tube is cleaned with an effective solvent such as xylol, and all excess solvent is removed from the tube and gallery. The oil is shaken thoroughly before it is heated in an aluminum pot and is not heated to more than 3°F above the temperature of the test. The oil is then transferred to the oil tube, straining it through a 100-mesh wire strainer, with sufficient oil to overflow into the gallery. The bath temperature is adjusted until the oil temperature remains constant. After removal of the oil tube thermometer, the oil is removed from the gallery until the oil level in the gallery is below that of the oil tube proper. The receiving flask is placed below the orifice and a timer is started as soon as the cork is removed and stopped as soon as the bottom of the meniscus in the flask reaches the 60 ml. mark. The results are recorded to the nearest 0.1 second below 200 seconds and the nearest whole second for values of 200 seconds or above.

Capillary-type viscometers are now generally preferred for the measurement of kinematic viscosity. The American Society for Testing Materials (ASTM) and the British Institute of Petroleum have jointly issued a method (ASTM D 445-64, IP 71/65) which prescribes the standard procedure for measuring the kinematic viscosity of oils. This method describes 17 capillary viscometers of which a few are illustrated in Figure 6-75. The first six are used for transparent liquids, the Cannon Master for calibration work, and the last three are designed chiefly for opaque liquids. In this method, a fixed amount of oil is allowed to flow through a capillary tube under the force of gravity. The unit of kinematic viscosity is the stoke or centistoke, and the viscosity in these units may be obtained by multiplying the flow time by a constant depending on the type of tube used. The testing procedure is illustrated in Figure 6-76, water or standard oils being used as references.

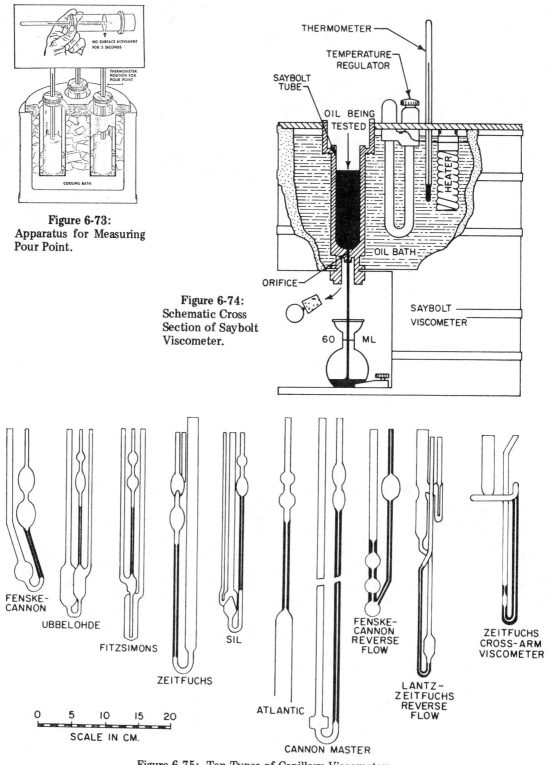

Figure 6-73: Apparatus for Measuring Pour Point.

Figure 6-74: Schematic Cross Section of Saybolt Viscometer.

Figure 6-75: Ten Types of Capillary Viscometers.

COLD ROLLING LUBRICATION

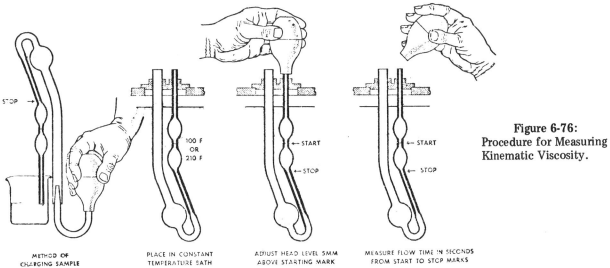

Figure 6-76: Procedure for Measuring Kinematic Viscosity.

Table 6-15
CONVERSION OF KINEMATIC VISCOSITY TO SAYBOLT UNIVERSAL VISCOSITY

Kinematic Viscosity, cs	Equivalent Saybolt Universal Viscosity, sec		Kinematic Viscosity, cs	Equivalent Saybolt Universal Viscosity, sec	
	At 100°F Basic Values	At 210°F		At 100°F Basic Values	At 210°F
2.0	32.6	32.9	31.0	145.7	146.8
2.5	34.4	34.7	32.0	150.2	151.2
3.0	36.0	36.3	33.0	154.7	155.8
3.5	37.6	37.9	34.0	159.2	160.3
4.0	39.1	39.4	35.0	163.7	164.9
4.5	40.8	41.0			
5.0	42.4	42.7			
6.0	45.6	45.9	36.0	168.2	169.4
7.0	48.8	49.1	37.0	172.7	173.9
8.0	52.1	52.5	38.0	177.3	178.5
9.0	55.5	55.9	39.0	181.8	183.0
10.0	58.9	59.3	40.0	186.3	187.6
11.0	62.4	62.9	41.0	190.8	192.1
12.0	66.0	66.5	42.0	195.3	196.7
13.0	69.8	70.3	43.0	199.8	201.2
14.0	73.6	74.1	44.0	204.4	205.9
15.0	77.4	77.9	45.0	209.1	210.5
16.0	81.3	81.9	46.0	213.7	215.2
17.0	85.3	85.9	47.0	218.3	219.8
18.0	89.4	90.1	48.0	222.9	224.5
19.0	93.6	94.2	49.0	227.5	229.1
20.0	97.8	98.5	50.0	232.1	233.8
21.0	102.0	102.8	55.0	255.2	257.0
22.0	106.4	107.1	60.0	278.3	280.2
23.0	110.7	111.4	65.0	301.4	303.5
24.0	115.0	115.8	70.0	324.4	326.7
25.0	119.3	120.1			
26.0	123.7	124.5	Over 70.0	Saybolt seconds = centistokes x 4.635	Saybolt seconds = centistokes x 4.667
27.0	128.1	129.0			
28.0	132.5	133.4			
29.0	136.9	137.9			
30.0	141.3	142.3			

NOTE: To obtain the Saybolt Universal viscosity equivalent to a kinematic viscosity determined at $t\,°F$ multiply the equivalent Saybolt Universal viscosity at 100°F by $1 + (t - 100)\,0.000064$; for example, 10 cs at 210°F are equivalent to 58.9 x 1.0070 or 59.3 sec Saybolt Universal at 210°F.

For conversion from Saybolt Universal viscosity to kinematic viscosity and vice versa, Table 6-15 may be utilized.

Kinoshita⊖ developed the following equation for converting kinematic viscosity ν in centistokes, determined at temperature T(°F) into Saybolt Universal seconds (SUS)

$$\text{SUS} = \frac{(4.605 + 0.000297T)\nu}{\left(1 - 10^{-0.07445\nu^{0.9538}}\right)} \quad (6\text{-}32)$$

This equation is satisfactory for kinematic viscosities greater than 2cs and temperatures between 70 and 300°F.

(c) <u>Viscosity Index</u>

The viscosity index is an emperical number indicating the rate of change in the viscosity of an oil with temperature within a given temperature range.∞ A low viscosity index signifies a relatively large change of viscosity with temperature while a high viscosity index signifies the opposite. The viscosity index, VI, may be obtained from the relationship

$$\text{VI} = \frac{L - U}{L - H} \times 100 \quad (6\text{-}33)$$

where U is the viscosity in SUS at 100°F of the oil whose viscosity index is to be calculated; L is the viscosity in SUS at 100°F of an oil of zero viscosity index having the same viscosity at 210°F as the sample under test, and H is the viscosity in SUS at 100°F of an oil with a viscosity index of 100 having the same viscosity at 210°F as the sample. For convenience, however, a chart such as that shown in Table 6-16 may be used for conversion purposes as follows. The column corresponding to the nearest SUS value at 210°F should be selected, the column should then be scanned to find the nearest number to the SUS value at 100°F, and the viscosity index should be read off from the bold face column at either side. For example, if the SUS values at 210 and 100°F were 110 and 1856 seconds, respectively, the viscosity index would be 75.

6-27 The Surface Tension of the Oil and the Quenching Properties of the Emulsion

Two other physical characteristics relating to the oil are sometimes considered in connection with its use as an aqueous emulsion, these being (a) the surface tensions of the oil, the water and the oil-water interface and (b) the quenching properties of the emulsion. Their influence on the cold reduction process are discussed below.

(a) <u>Surface Tension Values</u>

The relative values of the surface tensions of an oil and water largely determine the ease with which the oil may be emulsified and the stability of the resulting emulsion. The surface tensions of oils usually lie in the range 20 to 30 dynes per centimeter, whereas water lies in the range 60 to 75, the exact values for both depending on temperature. The addition of an emulsifier (or detergent) to water decreases its surface tension to a value closer to that of the oil, and thus improves the stability of the emulsion. The ability of the oil to plate out on the rolls and strip is therefore influenced to some extent by the surface tension values of the two phases of the rolling solution. However, the rapidity with which the oil plates out and the thickness of the resultant oil film have not been correlated with the surface tension values; so that the effect of surface tension values on the lubricity of the rolling solution remains a matter of speculation.

⊖M. Kinoshita, J. Inst. of Pet. <u>43</u>, May, 1957.
∞For an excellent discussion of this subject see "Lubrication", Vol. 52, 1966, No. 3. Published by Texaco, Inc.

COLD ROLLING LUBRICATION

Table 6-16
CHART FOR DETERMINING VISCOSITY INDICES

Procedure: (1) Scan the bold faced type across the top row for the nearest value to the SUS at 210°F value.

(2) Scan the corresponding vertical column for a number closest to the SUS at 100°F value.

(3) Read off the viscosity index at either end of the horizontal row.

	40	**45**	**50**	**55**	**60**	**65**	**70**	**75**	**80**	**85**	**90**	**95**	**100**	**105**	**110**	**115**	**120**	**125**	**130**	**135**	**140**	**145**	**150**	**155**	
0	138	265	422	596	781	976	1182	1399	1627	1865	2115	2375	2646	2928	3220	3524	3838	4163	4498	4845	5202	5570	5959	6339	**0**
5	136	261	414	584	763	953	1153	1364	1585	1816	2059	2311	2573	2846	3129	3423	3727	4042	4366	4701	5046	5402	5768	6145	**5**
10	135	256	405	570	745	930	1124	1329	1543	1767	2002	2246	2500	2765	3038	3323	3616	3920	4233	4557	4890	5234	5587	5950	**10**
15	133	252	397	557	727	907	1095	1294	1502	1718	1946	2182	2427	2683	2947	3222	3505	3799	4101	4413	4734	5065	5406	5756	**15**
20	132	247	389	545	710	884	1066	1259	1460	1670	1889	2117	2355	2601	2856	3121	3394	3677	3968	4269	4578	4897	5225	5562	**20**
25	130	243	380	532	692	861	1038	1224	1418	1621	1833	2053	2282	2520	2765	3021	3284	3556	3836	4125	4423	4729	5044	5368	**25**
30	129	238	372	519	674	837	1009	1188	1376	1572	1776	1989	2209	2438	2674	2920	3173	3434	3703	3981	4267	4561	4863	5173	**30**
35	127	234	364	506	656	814	980	1153	1334	1523	1720	1924	2136	2356	2583	2819	3062	3313	3571	3837	4111	4392	4682	4979	**35**
40	126	230	355	493	638	791	951	1118	1293	1474	1663	1860	2063	2274	2492	2718	2951	3191	3438	3693	3955	4224	4501	4785	**40**
45	124	225	347	480	621	768	922	1083	1251	1425	1607	1795	1990	2193	2401	2618	2840	3070	3306	3549	3799	4056	4320	4590	**45**
50	123	221	339	468	603	745	893	1048	1209	1377	1551	1731	1918	2111	2311	2517	2729	2948	3173	3405	3643	3888	4139	4396	**50**
55	121	216	330	455	585	722	864	1013	1167	1328	1494	1667	1845	2029	2220	2416	2618	2827	3041	3261	3487	3719	3957	4202	**55**
60	119	212	322	442	568	699	835	978	1125	1279	1438	1602	1772	1948	2129	2316	2507	2705	2908	3117	3331	3551	3776	4007	**60**
65	118	207	314	429	550	676	806	943	1084	1230	1381	1538	1699	1866	2038	2215	2396	2584	2776	2973	3175	3383	3595	3813	**65**
70	116	203	305	416	532	653	777	908	1042	1181	1325	1473	1626	1788	1947	2114	2285	2462	2643	2829	3019	3215	3414	3619	**70**
75	115	199	297	403	514	630	749	873	1000	1132	1268	1409	1553	1703	1856	2014	2175	2341	2511	2685	2864	3046	3233	3425	**75**
80	113	194	288	391	497	606	720	837	958	1083	1212	1345	1480	1621	1765	1913	2064	2219	2378	2541	2708	2878	3052	3230	**80**
85	112	190	280	378	479	583	691	802	916	1035	1155	1280	1408	1539	1674	1812	1953	2098	2246	2397	2551	2710	2871	3036	**85**
90	110	185	272	365	461	560	662	767	875	986	1099	1216	1335	1457	1583	1711	1842	1976	2113	2253	2396	2542	2690	2842	**90**
95	109	181	263	352	443	537	633	732	833	937	1042	1151	1262	1376	1492	1611	1731	1855	1981	2109	2240	2373	2509	2647	**95**
100	107	176	255	339	426	514	604	697	791	888	986	1087	1189	1294	1401	1510	1620	1733	1848	1965	2084	2205	2328	2453	**100**
105	106	172	247	326	408	491	575	662	749	839	930	1023	1117	1212	1310	1409	1509	1612	1716	1821	1928	2037	2147	2259	**105**
110	104	167	238	314	390	468	546	627	707	790	873	958	1043	1131	1219	1309	1398	1490	1583	1677	1772	1869	1966	2064	**110**
115	103	163	230	301	372	445	517	592	666	741	817	894	970	1049	1128	1208	1287	1369	1451	1533	1616	1700	1785	1870	**115**
120	101	159	222	288	355	422	488	557	624	693	760	829	898	967	1037	1107	1176	1247	1318	1389	1460	1532	1604	1676	**120**
125	99	154	213	275	337	399	460	522	582	644	704	765	825	886	946	1007	1066	1126	1186	1245	1305	1364	1423	1482	**125**
130	98	150	205	262	319	375	431	486	540	595	647	701	752	804	855	906	955	1004	1053	1101	1149	1196	1242	1287	**130**
135	96	145	197	249	301	352	402	451	498	546	591	636	679	722	764	805	844	883	921	957	993	1027	1061	1093	**135**
140	95	141	188	236	284	329	373	416	457	497	534	572	606	640	673	704	733	761	788	813	806	859	880	899	**140**

An instrument used for obtaining surface tension values is illustrated in Figure 6-77. It measures, in dynes per centimeter, the force required to pull a ring of platinum wire from the surface of the liquid. Where interfacial tension is to be measured, it is determined from the force required to pull the ring from the surface of the water into a layer of floating oil.

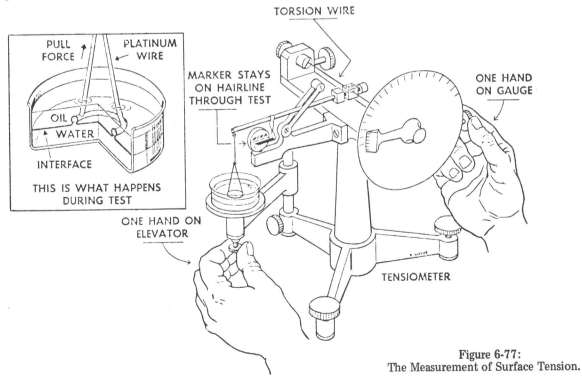

Figure 6-77:
The Measurement of Surface Tension.

(b) The Quenching Properties of the Solution

To determine the effectiveness of a rolling solution as a coolant, the Standard of Indiana quench test apparatus may be used as illustrated in Figure 6-78. This equipment, by the use of automatic electric timers, measures the time required to cool a thermocouple from preset maximum to preset minimum temperatures. Results are expressed in terms of seconds, and are the times required to quench from the higher to the lower temperature. Three bath temperatures, 40, 130, and 150°F are generally used; the temperature ranges from which the thermocouple is quenched are 1500 to 500°F and 800 to 150°F. Table 6-17 shows typical results for water, pure palm oil, and a used rolling solution.

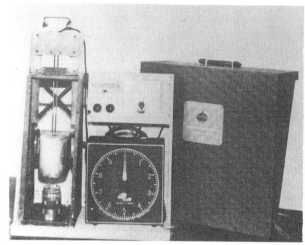
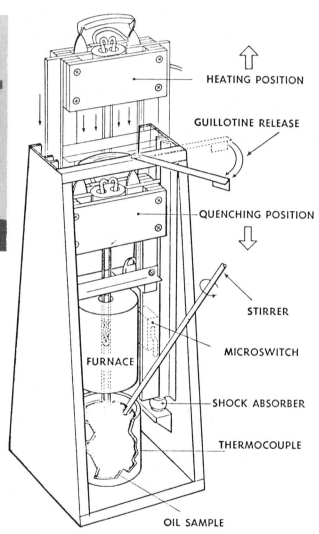

Figure 6-78: Standard of Indiana Quench Test Apparatus. (The Quenching Unit is Shown in Greater Detail to the Right.)

Table 6-17

Typical Quenching Data
(As Determined by the Standard of Indiana Quench Test)

	Quench Time (seconds)			
	Bath 130°F		Bath 150°F	
Coolant	1500 to 500°F	800 to 150°F	1500 to 500°F	800 to 150°F
Water	1.8	2.9	3.3	3.5
Palm Oil (neat)	5.5	4.4	7.4	5.6
Used Rolling Solution	3.4	4.2	5.0	6.0

COLD ROLLING LUBRICATION

Table 6-17 indicates that the quenching properties of the rolling solution lie between those of the water and the neat oil. For ideal cooling, the quenching values of the solution should be as close as possible to those for water.

6-28 The Saponification, Iodine, Peroxide and pH Values and the Free Fatty Acid Content

The five principal chemical tests to which rolling oils and their emulsions are subjected are those relating to the saponification, iodine, peroxide and pH values and the free fatty acid content. These are important with respect to establishing the constituents of the lubricants and the deterioration of emulsions during prolonged use. The significance and method of measurement of each parameter are discussed below.

(a) <u>Saponification Value</u>

When a natural fat or oil is suitably treated with an aqueous solution of caustic potash or soda, soap and glycerin are formed. This process is called saponification and the number of milligrams of caustic potash combining with one gram of fat or oil is equivalent to the saponification value of the fat or oil. Mineral oils are not attacked by an alkali and therefore the saponification value can be used to indicate the ratio of natural fats and oils to mineral oils in a lubricant.

Where mixtures of mineral and natural oils are used, the saponification value cannot, however, give an absolute analysis of the mixture. This is because for pure glycerides the saponification value will depend on the molecular weight of the glyceride; the higher the molecular weight, the lower the saponification value. Accordingly, the saponification value of a rolling lubricant is of interest only in a general sense, with commonly used rolling oils having values in the range of 125 to 200.

(b) <u>Iodine Value</u>

Organic compounds characterized by double or unsaturated valency bonds, viz. $-\overset{|}{C}=\overset{|}{C}-$, may be made to react with iodine under specific conditions. Accordingly, the degree of unsaturation of such compounds may be indicated by measuring the "iodine value" or the weight in grams of iodine chloride, ICl, expressed in terms of iodine, which combines with 100 grams of the oil or fat. Again, the measured values are only of general significance since the iodine values for pure unsaturated fatty compounds depend on their molecular weights.

From rolling practice and from laboratory measurements of lubricity, it has been established that saturated fatty acids (such as stearic, palmitic, lauric and myristic) provide better lubricity than unsaturated acids (such as oleic and linoleic). Moreover, unsaturated acids, under higher than ambient pressures and temperatures, tend to polymerize into resinous substances that are more difficult to remove from the rolled strip. Furthermore, they are also prone to become rancid as a result of atmospheric oxidation.

Accordingly, because of the disadvantages of unsaturated fatty acids, the iodine value of a rolling oil should be as low as possible.

(c) <u>The Peroxide Value</u>

Hydrocarbons which contain tertiary carbon atoms react with molecular oxygen to form peroxides of the general formula ROOH and are therefore alkyl derivatives of hydrogen peroxide H_2O_2.^ A knowledge of the peroxide content of the oil is therefore of value in detecting chemical changes occurring within the oil and the extent to which it is oxidized.

The peroxide content of oils is measured by the peroxide value or number which is the milliequivalents of constituents per kilogram of sample that will oxidize potassium iodide. This

---

^ "Textbook of Organic Chemistry" by A. Geo, John Wiley and Sons, Inc., New York, New York, 1963, p. 110.

value is determined as follows.~ The oil sample is dissolved in carbon tetrachloride and brought into contact with aqueous potassium iodide solution. The peroxides present are reduced by the potassium iodide and an equivalent amount of iodine is liberated, this being quantitatively determined by titration with sodium thiosulphate solution.

(d) The pH Value of the Emulsion

During usage of a rolling solution, changes in its alkalinity or acidity often occur associated with oxidative reactions occurring within it. Conceivably, these are related to changes in the free fatty acid content and tend to enhance the hydrolysis of the rolling oil, produce saponification and possibly change the effectiveness of the emulsifier. Accordingly, it is good mill practice to monitor the pH value of the rolling emulsion at frequent intervals with an instrument such as that shown in Figure 6-79. With this instrument, pH determination is accomplished by measuring the potential developed by an electrical cell consisting of two electrodes — a glass electrode and a reference electrode — immersed in the emulsion. Ideally, it would be desirable to continuously monitor the pH of the rolling emulsion but the contamination of the emulsion by sludge has so far prevented the development of a reliable instrument of this type.

Figure 6-79:
A Laboratory-Type pH Meter.

(e) Free Fatty Acid Content

Although triglycerides are the main constituents of natural fats and oils, some uncombined or free fatty acids are also present. The percentage by weight of such acids is determined by titrating the oil or fat with aqueous sodium hydroxide.

The relationship between free fatty acid content and the effectiveness of a lubricant is discussed in Sections 6-7 and 6-11.

6-29 Lubricant Properties Affecting the Coefficient of Friction

As would be expected, the chemical and physical characteristics of the lubricant profoundly influence the coefficient of friction it provides in the roll bite. Of particular significance with respect to the chemical properties of the lubricant are (1) the carbon chain lengths of its organic constituents and (2) its molecular structure, in particular, whether or not it contains unsaturated hydrocarbons. These chemical characteristics affect the physical properties of the lubricant in that higher viscosities and higher melting points are generally associated with longer carbon chain lengths, whereas increasing the degree of unsaturation has the opposite effect.

As Schey † points out, if hydrodynamic lubrication plays a significant part in the rolling process, increasing lubricant viscosity should lower friction. This has been confirmed by a number

---

~ Tentative Method of Test for Peroxide Number of Petroleum Wax — ASTM Designation D1832-61T.

† "Metal Deformation Processes/Friction and Lubrication", Edited by J. A. Schey, Marcel Dekker, Inc., 1970, New York, New York.

COLD ROLLING LUBRICATION

of investigators. For example, Whetzel and Rodman[°] state that, for a variety of fatty lubricants used to roll steel, friction decreased proportionally with the logarithm of the viscosity increase. Using three paraffins of the same series (hexane: C_6H_{14}; dodecane: $C_{12}H_{26}$ and hexadecane: $C_{16}H_{34}$) Thorpe[Φ] found that, at high rolling speeds, the coefficient of friction decreased linearly with the kinematic viscosity of the lubricant on a log-log plot (see Figure 6-80).

A correlation between the effectiveness of a rolling lubricant and its viscosity was established by Williams and Brandt[+] who used a 4-high mill with 1-5/8-inch diameter work rolls fixed in a preset position. Soft, low carbon steel strips measuring 0.010 inch by 2.0 inches by 24 inches dipped into the lubricants at 150°F were fed into the mill operating at a speed of 215 fpm. Effectiveness of the lubricants was gaged by the reductions they provided (as discussed in Section 6-20) and the data for the various lubricants (which were mainly fatty materials) are shown plotted in Figure 6-81.

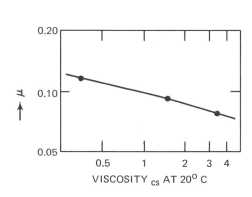

Figure 6-80:
The Effect of Viscosity in Rolling Low Carbon Steel with High Purity Paraffins.

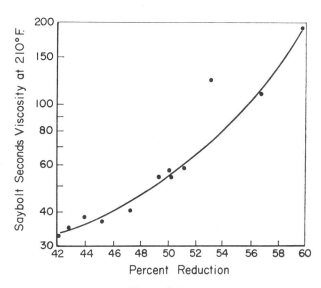

Figure 6-81:
Rolling Reductions Versus Viscosities of a Series of Oils Tested on a Laboratory Mill. (Williams and Brandt)

However, it must be remembered that the viscosity of the lubricant under the considerable pressure and the higher temperatures encountered in the roll bite may be quite different from the viscosity measured under typical laboratory conditions.

Using the Amsler machine discussed in Section 6-25, it has been found that aqueous emulsions of the saturated fatty acids (myristic, palmitic and stearic) exhibited coefficients of friction which decreased with increasing carbon chain length (see Table 6-18). Two unsaturated fatty acids (oleic and linoleic) comparable in carbon chain length to palmitic acid, provided markedly higher coefficients of friction. Calcium and zinc stearate provided low coefficients but sodium stearate, sodium oleate and potassium stearate (all of which are soluble in water) provided high coefficients.

Inasmuch as most rolling lubricants are applied in the form of an aqueous emulsion, it would be expected that the concentration of oil in the emulsion and the stability of the dispersion

---

[°] J. C. Whetzel, Jr. and S. Rodman, "Improved Lubrication in Cold Strip Rolling", Iron and Steel Engineer Year Book, 1959, pp. 238-247.

[Φ] J. M. Thorpe, "Mechanism of Lubrication in Cold Rolling", Proc. Inst. Mech. Eng. (London), 175, 1961, pp. 593-603.

[+] R. C. Williams and R. K. Brandt, "Experimental Evaluations of Cold Rolling Lubricants for Strip Steel", Lubrication Engineering, 20 (2), 1964, pp. 52-56.

would affect the lubricity provided in the roll bite. Generally speaking, so long as an emulsion is applied to the strip and the rolls in copious quantities, emulsion concentrations of 3 to 4 per cent appear to be adequate and greater concentrations afford little increase in lubricity. Emulsion stability is discussed in greater detail in Section 6-33 but, as a general principle, it may be stated that only the minimum concentration of the emulsifier required to maintain satisfactory performance of the solution should be used.

Table 6-18
Rating of Lubricants According to Type of Chemical

Lubricant	Chemical Formula	Coefficient of Friction
Fatty Acids — Saturated		
Stearic acid	$C_{18}H_{35}O_2$	0.0224
Palmitic acid	$C_{16}H_{32}O_2$	0.0268
Myristic acid	$C_{14}H_{28}O_2$	0.0403
Fatty Acids — Unsaturated		
Oleic acid	$C_{18}H_{34}O_2$	0.0372
Linoleic acid	$C_{18}H_{32}O_2$	0.0444
Soaps and Salts		
Calcium stearate	$Ca(C_{18}H_{35}O_2)_2$	0.0229
Zinc stearate	$Zn(C_{18}H_{35}O_2)_2$	0.0255
Sodium stearate	$Na(C_{18}H_{35}O_2)$	0.0442
Sodium oleate	$Na(C_{18}H_{33}O_2)$	0.0476
Potassium stearate	$K(C_{18}H_{35}O_2)$	0.0558
Miscellaneous		
Cetyl alcohol	$C_{16}H_{34}O$	0.0335
Tristearin	$(C_{17}H_{35}COO)_3C_3H_5$	0.0403
Tripalmitin	$(C_{15}H_{31}COO)_3C_3H_5$	0.0409

6-30 The Dependence of the Coefficient of Friction on Various Rolling Parameters

The careful analysis of mill data has shown that the computed coefficient of friction in the roll bite is not determined solely by the physical and chemical properties of the lubricant but by various other parameters associated with the rolling process as well.[♦] In particular, the following rolling conditions are known to affect roll bite lubricity:

(a) the draft or reduction given to the workpiece,

(b) work roll diameter,

(c) the mill speed,

(d) the surface roughness of the strip,

(e) the surface roughness of the rolls,

(f) the presence of metallic coatings on the strip surfaces, and

(g) the yield strength of the strip prior to reduction.

In general, when the calculated coefficient of friction has been plotted against the reduction given by the rolling pass, the coefficient has been found to increase with reduction, particularly in the case of reductions in excess of 10 or 20 per cent (see Figure 6-82 relating to the rolling of annealed strip). In some instances, calculated values of the coefficient have been found to increase with decreasing reduction for small reductions (less than 10 per cent) but this trend appears to reflect the inadequacy of the mathematical model used for computing the coefficient. The principal effect of increasing coefficient with increasing reduction may be partially explained by the

---

[♦] W. L. Roberts, R. J. Bentz and D. C. Litz, "Cold Rolling Low Carbon Steel Strip to Minimum Gage", Iron and Steel Engineer Year Book, 1970, pp. 413-420.

following; (a) with larger reductions, the oil film thickness entering the bite is decreased due to the lower entry speed of the strip and the larger entry angle at the bite and (b) the oil film entering the bite is decreased in thickness to a greater extent as it passes through the bite.

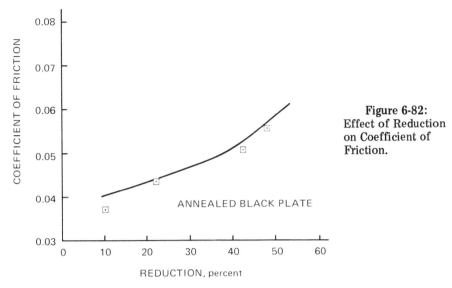

Figure 6-82: Effect of Reduction on Coefficient of Friction.

It has also been shown that, for the same lubricant, type of workpiece, rolling speed and roll finishes, the computed value of the coefficient decreases with increasing work roll diameter. Table 6-19 shows values of the coefficient computed for rolling annealed black plate approximately 0.010-inch thick at speeds of 500 fpm on mills with work roll diameters of 3-1/4, 6-1/2 and 21 inches. The reduction taken was 30 per cent and a good rolling lubricant was used. The effect is presumably due to the easier entrapment of oil in the bite by rolls of larger diameter.

Table 6-19
Effect of Work-Roll Diameter on Coefficient of Friction

Diameter (in.)	Coefficient of Friction
3¼	0.056
6½	0.023
21	0.017

The effect of mill speed on roll bite lubricity was noticed soon after the introduction of high-speed cold mills but the first systematic study of the effect was carried out by Ford[m] and by Sims and Arthur.[:] The decreased rolling forces encountered at the higher mill speeds could only be satisfactorily explained on the basis of improved lubricity in the roll bite, presumably due to hydrodynamic effects. Ford conducted his experiments on a 2-high mill equipped with 10-inch diameter rolls at speeds ranging from 5 to 300 fpm using 0.2 per cent carbon steel and high purity copper as strip materials. He found the effect to be more pronounced the thinner and harder the strip entering the mill. Sims and Arthur used the same mill and they found that using oil as a rolling lubricant, the rolled strip thickness decreased from 0.035 inch to 0.027-inch as the rolling speed was increased from 10 to 250 fpm. Their computed values for the coefficient of friction decreased with mill speed as illustrated in Figure 6-83. On the other hand, when rolling strip with dry graphite lubricant, the outgoing thickness of the strip remained constant at 0.038-inch.

---

[m] H. Ford, "The Effect of Speed of Rolling in the Cold Rolling Process", J. Iron and Steel Institute (London), 156, 1947, pp. 380-398.

[:] R. B. Sims and D. F. Arthur, "Speed Dependent Variables in Cold Strip Rolling", J. Iron and Steel Institute (London), 172, 1952, pp. 285-295.

Billigmann and Pomp[Ω] found that, while the rolling force increased with speed under dry rolling conditions, with a lubricant the gage of the strip invariably decreased with increasing speed but, like Ford, they found the speed effect occurred only if the ratio of the rolled gage to the roll diameter was smaller than a certain limiting value, the magnitude of which was dependent on the nature of the workpiece, the reduction per pass and the surface finish of the rolls.

When a straight soluble oil is used as a lubricant, the speed effect is virtually absent. This was shown by Whetzel and Wyle[▼] whose data is reproduced in Figure 6-84. However, they showed a dramatic decrease for the coefficient provided by liquid fat, verifying calculations made earlier by Stone[ᵛ] and later by Pawelski.[⊥]

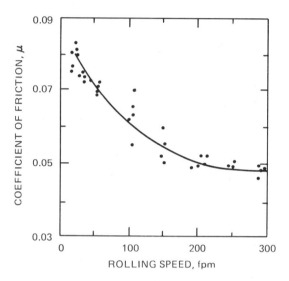

Figure 6-83:
Variation of Coefficient of Friction With Rolling Speed. (Sims and Arthur)

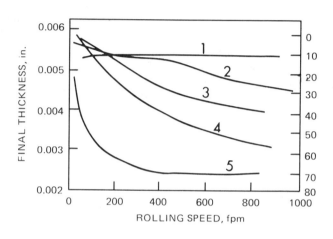

Figure 6-84:
Speed Effect in Experimental Rolling of Low Carbon Steel Strip (1: "Soluble Oil"; 2: Mineral Oil; 3 and 5: Experimental Lubricants; 4: Liquid Fat). (Whetzel and Wyle)

When strip is prelubricated with a thin oil film, the speed effect is not significant. This was demonstrated by Bentz and Somers[+] who obtained the data shown in Figure 6-85 during the rolling of 0.010-inch thick, annealed black plate coated with cottonseed oil to a weight of 1.3g/base box.

More recently, the variation of rolling force with mill speed has been studied in connection with the secondary cold reduction of box annealed black plate (see Figure 6-86).[♦] The initial drop in force with increasing mill speed was found to result from a decreasing coefficient of friction and the subsequent increase in force was attributed to the strain rate effect on the

---

[Ω] J. Billigmann and A. Pomp, "Untersuchungen uber den Einfluss der Walzgeschwindigkeit auf den Walzdruck, die Festigkeit-Seigenschaften und die Banddicke beim Kaltwalzen von Bandstahl", Stahl und Eisen 74, 1954, pp. 441-461.

[▼] J. C. Whetzel and C. Wyle, "Some Effects of Lubricants in Cold Rolling Thin Strip", Metal Progress, 70 (6), 1956, pp. 73-76.

[ᵛ] M. D. Stone, "Rolling of Tin Strip", Iron and Steel Engineer Year Book, 1953, pp. 115-128.

[⊥] O. Pawelski, "Investigation of the Coefficient of Friction in Metal Forming", Paper 24, Third Annual Meeting of the Lubrication and Wear Group, Institution of Mechanical Engineers (London), Cardiff, 1964.

[+] R. J. Benz and R. R. Somers, "Determination of Minimum Oil Film Weight for Rolling Strip", Lubrication Engineering, February, 1965, pp. 59-64.

[♦] R. J. Bentz and W. L. Roberts, "Speed Effects in the Second Cold Reduction of Steel Strip", "Mechanical Working and Steel Processing V", Proceedings of the Ninth Mechanical Working and Steel Processing Conference, AIME, Gordon and Breach, New York, New York, 1968.

COLD ROLLING LUBRICATION

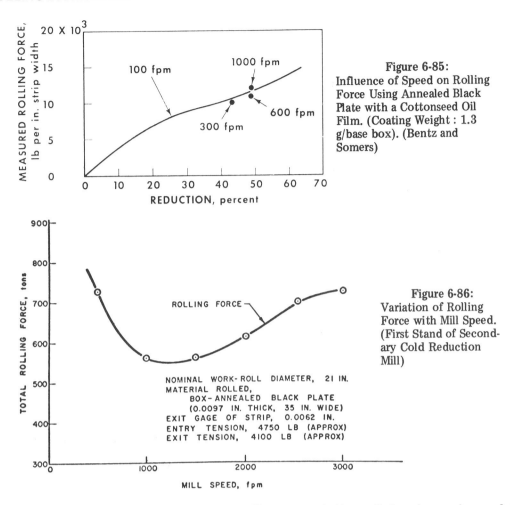

Figure 6-85: Influence of Speed on Rolling Force Using Annealed Black Plate with a Cottonseed Oil Film. (Coating Weight: 1.3 g/base box). (Bentz and Somers)

Figure 6-86: Variation of Rolling Force with Mill Speed. (First Stand of Secondary Cold Reduction Mill)

constrained yield strength of the workpiece. However, at times it has been observed that the computed coefficient of friction increases with increasing speed across virtually the whole speed range of the mill as shown in Figure 6-87. This is probably attributable to an insufficient quantity of lubricant applied to the roll bite or an insufficient time for the lubricant to plate out properly on the roll and strip surfaces.

The fact that, under dry lubrication, the speed effect is not observable has led to the belief that the effect is due to hydrodynamic phenomena, analogous to those encountered in a

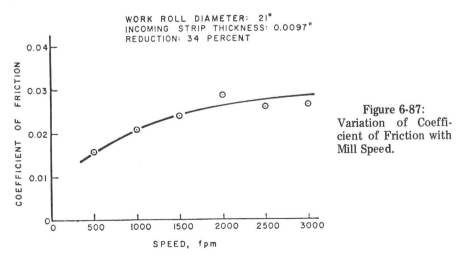

Figure 6-87: Variation of Coefficient of Friction with Mill Speed.

journal bearing. Sims** believed the explanation of the effect should be based on the viscous flow properties and the attachment of the lubricant to the strip surface and he postulated that, at low rolling speeds, more of the attached lubricant is squeezed out of the bite.

Rolling lubricity is generally improved as the surfaces of the strip are increased in roughness. Presumably, the additional lubricant entrapped in the pores of the surface and dragged into the roll bite is released as the metal is deformed and is able to provide a thicker-than-normal lubricant film. As an example of this effect, it has been found that, in the rolling of full hard black plate 0.0061-inch thick on a commercial mill with 21-inch diameter rolls, a maximum reduction of only nine per cent was obtained with a strip having a smooth (5 to 10 microinches) surface finish as compared with a reduction of 32 per cent obtainable with a rough surface finish (30 to 40 microinches). These data, relating to the use of a substitute cottonseed oil, are shown in Figure 6-88.

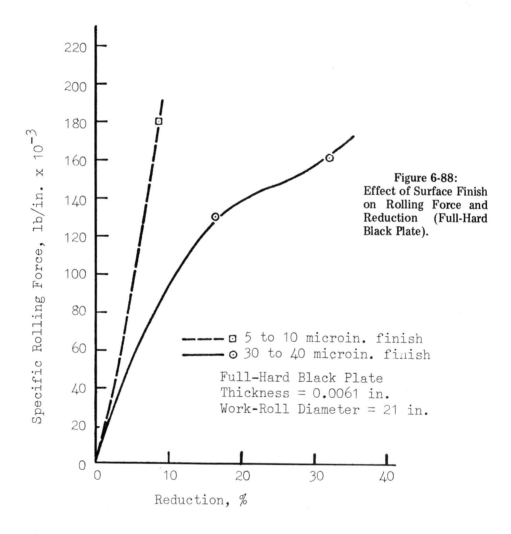

Figure 6-88: Effect of Surface Finish on Rolling Force and Reduction (Full-Hard Black Plate).

On the other hand, rougher roll surfaces provide poorer lubricity, primarily because the entrapped lubricant cannot be released to be effective. This effect has been discussed by Witt ⊖ who found that a change from a 1 to an 11 microinch finish on work rolls produced a measurable

---

**R. B. Sims, J. Inst. Petrol. 40, 1953, pp. 314-318.

⊖F. A. Witt, Jr., "Comparison Between Theoretical and Experimental Cold Rolling Forces", (Discussion of Paper), Iron and Steel Engineer Year Book, 1965, pp. 531-538.

COLD ROLLING LUBRICATION

influence on torque and separating force. Similar data have been published more recently[♦] and is presented in Figure 6-89. Thorpe and Guy[⊢] found, however, that at low mill speeds, rolls transversely ground (that is, with grinding marks parallel to the axis of the rolls) provided lower frictional effects than conventionally ground rolls when used with a typical lubricant.

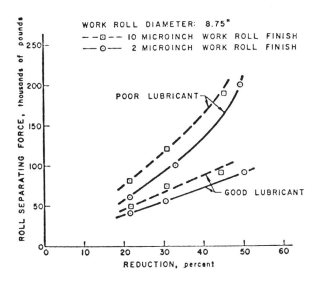 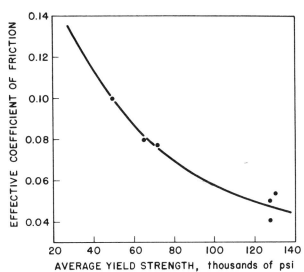

Figure 6-89:
Effect of Roll Finish on Rolling Force.

Figure 6-90:
Variation of Effective Coefficient of Friction with Average Yield Strength. (Witt)

Softer metallic coatings generally enhance the lubricity provided by a conventional lubricant unless the coating has been made to alloy with the steel to form a harder, more brittle, alloy layer. Under such circumstances, the coefficient of friction may be increased. (See Section 6-32.)

The effect of workpiece hardness (and therefore rolling pressure) on the coefficient of friction has been studied by a number of investigators who have drawn varied, and sometimes contradictory, conclusions. Schey[◈] found the coefficient to increase slightly with pressure for low carbon steel rolled with a mineral oil. LeMay and Vigneron[<] obtained similar results and Zhuchin and Pavlov,[■] in the rolling of a soft magnetic alloy on a commercial 4-high mill, found the coefficient to increase with increasing entry strip thickness and with work hardening. Witt[⊖] found that, to obtain a fit between experimental data, the effective coefficient would have to vary with the average yield strength as indicated in Figure 6-90. His data is generally in agreement with other experimental data shown in Figure 6-91.[✻]

♦ W. L. Roberts, R. J. Bentz and D. C. Litz, "Cold Rolling Low Carbon Steel Strip to Minimum Gage", Iron and Steel Engineer Year Book, 1970, pp. 413-420.

⊢ J. M. Thorpe and V. H. Guy, "Second Progress Report on Improved Lubrication in Cold Rolling Using Transversely Ground Rolls", B.I.S.R.A. Report #112, September, 1960.

◈ J. A. Schey, "The Nature of Lubrication in the Cold Rolling of Aluminum and Its Alloys", J. Inst. Metals 89, 1960-61, pp. 1-6.

< I. LeMay and F. R. Vigneron, "Initial Studies on the Use of Rapeseed Oil as a Cold Rolling Lubricant", Lubrication Engineer 21, 1965, pp. 276-281.

■ V. N. Zhuchin and I. M. Pavlov, "The Coefficient of Friction During Cold Rolling", Stal in English, 1963 (3), pp. 201-203.

⊖ F. A. Witt, Jr., "Comparison Between Theoretical and Experimental Cold Rolling Forces", Iron and Steel Engineer Year Book, 1965, pp. 531-538.

✻ W. L. Roberts, "Computing the Coefficient of Friction in the Roll Bite From Mill Data", Blast Furnace and Steel Plant, June, 1967, pp. 499-508.

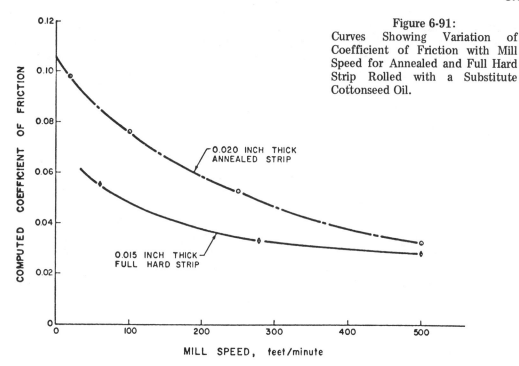

Figure 6-91:
Curves Showing Variation of Coefficient of Friction with Mill Speed for Annealed and Full Hard Strip Rolled with a Substitute Cottonseed Oil.

Increasing strip tension has virtually the same effect as decreasing the yield strength of the strip. Accordingly, it would be expected that the effective coefficient for high strip tension would be higher than for lower tensions (and higher rolling forces). Data relating to the rolling of 0.0135-inch thick silicon steel strip at 30 fpm on a small mill verifying this expectation are illustrated in Figure 6-92. However, since all such values for the coefficient are computed using a mathematical model of the rolling process, it must be remembered that these values depend on the nature of the model, the various assumptions made (as, for example, with respect to work hardening) and other parameters, such as roll and strip surface roughnesses and mill speed.

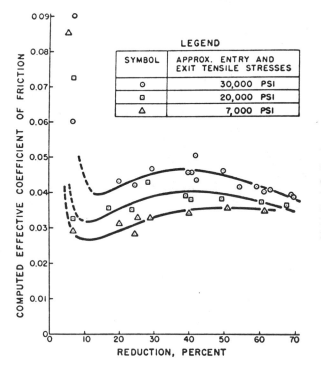

Figure 6-92:
Variation of the Computed Coefficient of Friction with Reduction for Silicon Steel Strip for Three Strip Tension Conditions.

COLD ROLLING LUBRICATION

6-31 The Frictional Characteristics of a Lubricant

Since it has been shown in the preceding section that the lubricity in the roll bite is affected by changes in (1) mill speed, (2) work roll diameter, (3) the draft given to the workpiece, (4) the type of material being rolled and (5) the surface finishes of the rolls and strip, attempts to define the general effectiveness of a lubricant in terms of a single coefficient of friction is of questionable value. However, it has been established empirically that, by evaluating a lubricant in terms of two parameters, designated the first and second frictional characteristics (K_1 and K_2), the coefficient of friction may be readily computed for any given set of rolling conditions.✠ How these characteristics are derived, their physical meanings and how they may be used for computing the coefficient are discussed in this section.

In examining the geometry of the roll bite (see Figure 6-93), it would appear reasonable to expect that, for comparable roll surface and strip entry speeds, the coefficient of friction would be dependent on the angle A formed by the plane containing the tangent to the roll surface at entry into the roll bite and the plane containing the surface of the strip. As the angle A is decreased, on the basis of hydrodynamic lubrication, the amount of lubricant dragged into the roll bite should increase with a corresponding diminution in the coefficient of friction. For a rigid roll, it can be shown that the angle A expressed in radians is equal to the square root of the ratio of twice the draft (d) divided by the roll diameter D, namely $\sqrt{\frac{2d}{D}}$. Thus it would be expected that the calculated value of the coefficient of friction (μ) could be directly correlated with the dimensionless parameter $\frac{d}{D}$ for comparable rolling speeds.

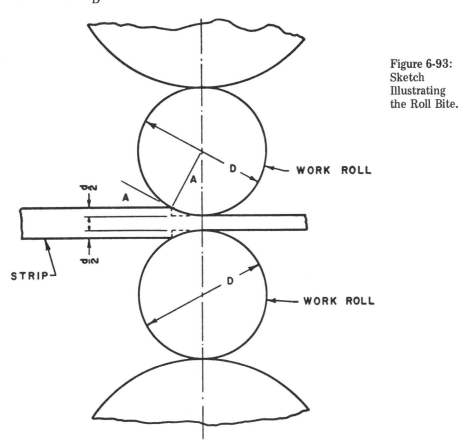

Figure 6-93:
Sketch
Illustrating
the Roll Bite.

✠W. L. Roberts, "The Frictional Characteristics of Rolling Lubricants", "Mechanical Working and Steel Processing VII", Edited by McGrann, Murphy and Richardson, AIME, 1969.

Empirically, it has been established that, for a constant rolling speed V, a direct proportionality exists between the calculated coefficient and the parameter $\sqrt{\frac{d}{D}}$, designated the "entry angle factor", as illustrated in Figure 6-94. Thus

$$\mu = K \left(\sqrt{\frac{d}{D}} \right)_V \qquad (6\text{-}34)$$

As the rolling speed V decreases, the constant of proportionality K increases. When extrapolated to zero mill speed and referred to rolls and strip with surface finishes of 10 microinches, the constant is designated K_1, the first frictional characteristic.

Because, in the rolling operation, friction is necessary to transmit the deformation energy from the work rolls to the strip, it can be shown that, when the entry and exit strip tensile stresses are equal or small enough to be negligible, the minimum coefficient of friction (μ_m) required in the deformation of strip by elastic rolls is $\sqrt{\frac{d}{2D'}}$, where D' is the effective deformed diameter of the work rolls in the roll bite. As a reasonable approximation for most rolling operations, the deformed roll diameter D' is about twice the actual roll diameter D, so that

$$\mu_m \simeq 0.5 \sqrt{\frac{d}{D}} \qquad (6\text{-}35)$$

Thus, there is a minimum frictional curve corresponding to equation 6-35, as shown in Figure 6-94, that, in reality, represents very high (or infinite) rolling speed with a constant of proportionality of approximately 0.5. Now the manner in which the constant of proportionality K in equation 6-34 varies with rolling speed between the extreme limits of K_1 (the first frictional characteristic corresponding to zero mill speed) and 0.5 (corresponding to very high mill speeds) is expressed by the second frictional characteristic K_2.

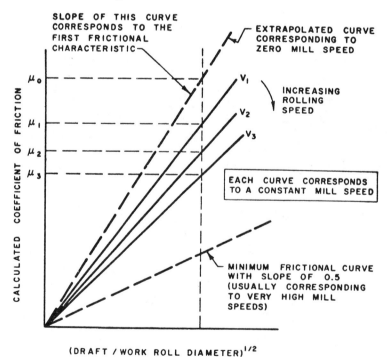

Figure 6-94: Variation of Coefficient of Friction with Entry Angle Factor for Various Mill Speeds.

COLD ROLLING LUBRICATION

Empirically, it has been established that when the logarithm of the difference between the computed and the minimum values of the coefficient $[\log_e(\mu - \mu_m)]$ is plotted against the rolling speed, a straight line curve is obtained. The magnitude of the slope of this curve is designated the second frictional characteristic (K_2) and has the dimensions of reciprocal speed. This is illustrated in Figure 6-95. Thus

$$\log_e(\mu - \mu_m) = \log_e(\mu_o - \mu_m) - K_2 V \qquad (6\text{-}36)$$

or

$$\mu = \mu_m + (\mu_o - \mu_m)e^{-K_2 V} \qquad (6\text{-}37)$$

where μ_o is the coefficient of friction extrapolated to correspond with zero mill speed.

Thus

$$\mu = \sqrt{\frac{d}{D}}\left(0.5 + (K_1 - 0.5)e^{-K_2 V}\right) \qquad (6\text{-}38)$$

Given, for any rolling situation, the values of the draft-to-roll diameter ratio (d/D), the frictional characteristics K_1 and K_2, and the mill speed V, the effective coefficient μ may therefore be readily calculated from equation 6-38. Experimentally, it has been found that values of the frictional characteristics determined for an experimental lubricant on a mill with 3.25-inch diameter work rolls could be used to satisfactorily predict rolling forces on mills with 6.5 and 21-inch diameter work rolls.

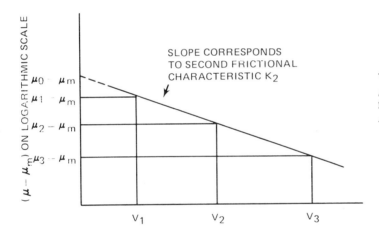

Figure 6-95: Variation of $(\mu - \mu_m)$ with Mill Speed for Constant Entry Angle Factor.

Assuming that the work rolls of a mill could be suitably crowned to provide flat strip with any value of rolling force, it would be desirable from the viewpoint of rolling force to maintain a coefficient of friction close to the minimum coefficient required for rolling. Under these circumstances, K_1 should have a value close to 0.5 and K_2 should be approximately zero. In practice, however, typical lubricants have K_1 values in excess of unity (even as high as 3), and it is therefore desirable for the K_2 values to be such that, at the operational speed of the mill, the coefficient of friction is as close as possible to the minimum value. Typical K_2 values range from 0.5×10^{-3} to 2.0×10^{-3} (fpm)$^{-1}$.

The well-known speed effect is, however, directly related to the second rolling characteristic, the larger the K_2 value, the more pronounced is the speed effect. Generally speaking, large speed effects make gage and shape control more difficult in the rolling operations and should be avoided as much as possible.

In practice, however, rolls have specific crowns established both by initial grinding and by nonuniform temperature distributions within them. Under the circumstances, as will be discussed in Chapter 11, the preferred value of the coefficient of friction is that which provides a uniform roll gap or flat strip.

As previously mentioned, the K_1 and K_2 values are based on roll and strip surface finishes of 10 microinches. When other surface finishes are encountered, scaling factors must be used to correct the coefficient of friction as derived from equation 6-38. If μ_N is the corrected coefficient

$$\mu_N = B_S B_R \mu \qquad (6\text{-}39)$$

where B_S and B_R are the scaling factors for the strip and roll finishes, respectively. Figure 6-96 presents a chart for the determination of the two factors in terms of surface roughness. On an approximate basis the following expressions may be used for computing the B_S and B_R factors

$$B_S = \frac{1}{(0.8 + 0.02\phi_S)} \qquad (6\text{-}40)$$

and

$$B_R = (0.8 + 0.02\phi_R) \qquad (6\text{-}41)$$

where ϕ_S and ϕ_R are the surface roughnesses (in terms of microinches) of the strip and rolls, respectively.

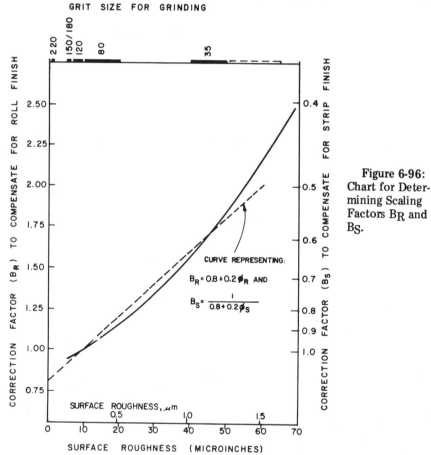

Figure 6-96: Chart for Determining Scaling Factors B_R and B_S.

6-32 The Effect of Metallic Coatings on the Frictional Characteristics

For a given rolling oil of the substitute palm oil variety, a soft metallic coating on the strip surface changes the values of the frictional characteristics K_1 and K_2 normally provided by the lubricant. The corresponding changed values are designated K_{1c} and K_{2c}. The first frictional characteristic may be increased or decreased but the second characteristic is generally increased. The proportional changes in these parameters attributable to the coating are represented by scaling or

COLD ROLLING LUBRICATION

proportionality factors b_1 and b_2 (affecting K_1 and K_2, respectively). Thus, for a given lubricant with constant rolling conditions

$$K_{1c} = b_1 K_1 \qquad (6\text{-}42)$$

and

$$K_{2c} = b_2 K_2 \qquad (6\text{-}43)$$

Tin coatings deposited electrolytically, but unmelted, (so as to provide a matte finish with no alloy coating), considerably enhance the roll bite lubricity provided by a conventional rolling oil. As would be expected, the thicker the coating, the greater its effect, as evidenced by Figures 6-97 and 6-98 showing the variation of K_1 and K_2 and b_1 and b_2 with tin coating weight in the case of a substitute palm oil. When the tin is melted or reflowed (as in the case of conventional tinplate manufactured on FERROSTAN lines), the effectiveness of the coating is dramatically changed. For tinplate with a 0.75-lb./base box coating weight (melted), K_1 is increased and K_2 decreased with respect to the unmelted coating. (See Table 6-20.) This is further evidenced by the minimum gages obtainable with black plate and tin plate (both melted and unmelted) shown as functions of mill speed in Figure 6-27. Remembering that the less the coefficient of friction, the thinner the rolled strip, it can be seen from the figure that 0.1 and 0.25-lb./base box coatings which have been melted provide poorer lubricity than the uncoated black plate. A melted 0.75-lb./base box coating provides a lubricity comparable to a 0.25-lb./base box unmelted coating. An 0.75-lb./base box unmelted coating, on the other hand, provides excellent lubricity.

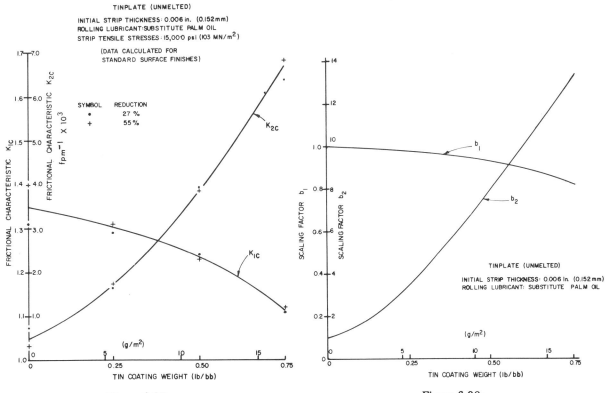

Figure 6-97:
Curves Showing Variation of K_{1c} and K_{2c} With Coating Thickness for Unmelted Tinplate.

Figure 6-98:
Curves Showing Variation of b_1 and b_2 With Coating Weight for Unmelted Tinplate.

The behavior of terne metal and zinc coatings are also indicated by the corresponding b_1 and b_2 values listed in Table 6-20. A terne metal coating of 0.09-oz./square foot (0.045-oz./square foot on each side) enhances the lubricity in a manner comparable to 0.75-lb./base box melted tin coating whereas the galvanized coating is not as effective.

Table 6-20
Scaling Factors
for
Various Coating Materials

Coating Material	Coating Weight	b_1	b_2
Aluminum	1.2 oz./ft.² of sheet (366 g/m²)	0.54	48
Tin (unmelted)	0.75 lb./bb[8] (16.8 g/m²)	0.82	13.4
Terne metal[7]	0.09 oz./ft.² of sheet (27.5 g/m²)	1.08	5.95
Tin (melted)	0.75 lb./bb (16.8 g/m²)	1.37	9.2
Zinc	0.94 oz./ft.² of sheet (287 g/m²)	1.46	4.94

The effectiveness of the aluminum coating (1.2-oz./square foot) is also given in Table 6-20. Of all the metallic coatings investigated, it appeared the most beneficial from a lubricity viewpoint.

Reference to the table shows that the coatings (with the exception of the first two coatings) generally increase the value of K_1 thereby worsening the lubricity in the roll bite at very low speeds. The benefit of such coatings therefore arises from larger values of K_2 (or increased speed effects) so that the rolling advantages afforded by the coatings are to be obtained at the higher mill speeds.

To calculate the coefficient of friction μ_s under "standard" conditions (i.e., with roll and strip surfaces with a 10 microinch finish) the following formula should be used

$$\mu_s = \sqrt{\frac{d}{D}} \left\{ 0.5 + (b_1 K_1 - 0.5) \, e^{-b_2 K_2 V} \right\} \quad (6\text{-}44)$$

where nonstandard roll and strip surface conditions exist, the coefficient of friction μ_N is given by

$$\mu_N = B_S B_R \mu_S \quad (6\text{-}45)$$

where the factors B_S and B_R are determined as discussed in the preceding section.

6-33 The Emulsion Stability Index and Its Measurement

The effectiveness of a rolling solution is closely associated with its stability as an emulsion. If the aqueous mixture is highly emulsified (a "tight" emulsion), the oil phase does not readily plate out on to the strip with the result that high values of the coefficient of friction may be encountered. On the other hand, an emulsion that is too "loose" may permit separation out of the oil in the storage tanks with the result that the application system may be pumping a solution of very low concentration to the mill. Accordingly, a quasi-stable emulsion is desirable such that ensures no separation in the recirculation system but a ready plate-out on the strip.

To express emulsion stability in quantitative terms, an "emulsion stability index" (ESI) is used. This is determined by allowing the emulsion to stand for 8 minutes in a separatory funnel and then separating the lowest and uppermost quarters of the emulsion. After centrifuging these

---

▶ W. L. Roberts, "Recent Developments in Rolling Lubrication", Blast Furnace and Steel Plant, May, 1968, pp. 382-394.

COLD ROLLING LUBRICATION

portions, the ratio of the oil content of the lowest to the uppermost fractions constitutes the emulsion stability index. An ESI of unity denotes a perfectly stable emulsion, whereas an ESI close to zero denotes a highly unstable emulsion. In rolling operations, an ESI of about 0.75 usually seems to be satisfactory on 2CR mills and about 0.65 on multistand tandem mills rolling light gage strip.

6-34 Chemical Stability Tests for Emulsions

The rate of utilization of a lubricant in a recirculating system is such that a quantity of lubricant equal to the initial quantity of oil in the emulsion is usually added to the system every 10 to 20 hours of mill use. Consequently, it can be said that the oil is kept in the system for a number of days and is exposed to a number of mill and atmospheric contaminants. Thus oxidation and hydrolysis of the lubricant frequently occur and corresponding changes take place in the pH, iodine, free fatty acid, and peroxide values. Such chemical changes may also affect the stability of the emulsion and hence its lubricity. Experience has established that the desirable, preferred values and the maximum permissible changes in one week for the four parameters are shown in Table 6-21.

Table 6-21
Preferred Values and Maximum Permissible Weekly Changes in Various Characteristics of Rolling Solutions

Characteristic	Preferred Value	Maximum Permissible Change In One Week
Free fatty acid	0-10%	Should not increase by more than an additional 3%
Iodine value	40-70	Should not decrease more than 10%
pH value	6-8	Should not decrease more than 2%
Peroxide value	0-50	Should not increase more than 150%

To evaluate the resistance of the lubricant to oxidation, the solution at its normal operating temperature should be recycled over iron filings in an apparatus such as that illustrated in Figure 6-99 and shown diagrammatically in Figure 6-100. Periodically, the solution should be sampled and measurements made in the conventional manner of the free fatty acid content, the iodine value, the solution pH, and the peroxide value.[Ψ] Figures 6-101 and 6-102 illustrate the experimentally determined variations of these characteristics with time for a satisfactory commercial lubricant.

6-35 The Cleanability and Burn-Off Characteristics of the Lubricant

It would be desirable to leave as thin a film of residual lubricant on the strip after rolling as possible because, where the strip is to be processed further, such films must be either removed or reduced to a low level. However, residual oil film thicknesses depend on a number of factors such as rolling speed, type of lubricant, method of application (whether applied neat or as an emulsion), and the reduction taken. Usually the film thickness lies in the range of 1 to 10 grams of oil per base box (approximately 1 to 10 microinches in thickness).

Removal of the oil film may be accomplished by several methods. In the case of sheet products, the film may be partly removed by detergent sprays located at the exit end of the mill and then the remainder removed by evaporation during box anneal. For strip used in tinplate and black plate manufacture, removal of the oil is usually accomplished by electrolysis using an alkaline

---

[Ψ] F. D. Snell and F. M. Biffin, "Commercial Methods of Analysis", Chemical Publishing Company, Inc., New York, New York, 1964.

cleaning solution. With respect to the latter method, the lower the charge density on the strip surface (coulombs/square foot) required to remove the lubricant, the more desirable the lubricant becomes from a cleaning standpoint.

Figure 6-99: Lubricant Recirculation Equipment.

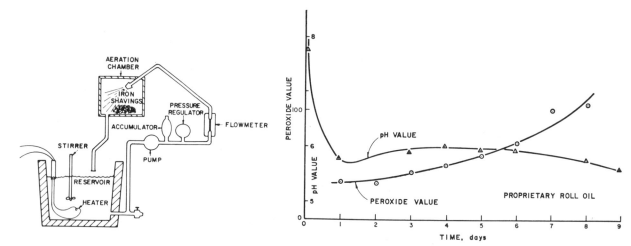

Figure 6-100:
Schematic Diagram of Recirculation Equipment.

Figure 6-101:
Changes in Peroxide and pH Values During Recirculation Tests.

The ease with which a lubricant may be electrolytically removed from a surface is usually evaluated in terms of the electrical charge per unit area of the strip required to produce a clean surface. The test is conducted as follows. Strip rolled with an emulsion of the lubricant under test is stored in a normal atmosphere for three days and is then cleaned electrolytically, using a

COLD ROLLING LUBRICATION

3-ounce-per-gallon solution of a mixture of 80 per cent sodium hydroxide and 20 per cent sodium tripolyphosphate. The cell used is shown in Figure 6-103 with the strip samples being approximately 3 inches wide and 9 inches long immersed to a depth of about 7 inches. The solution in the cell is kept at a temperature of 180 to 190°F.

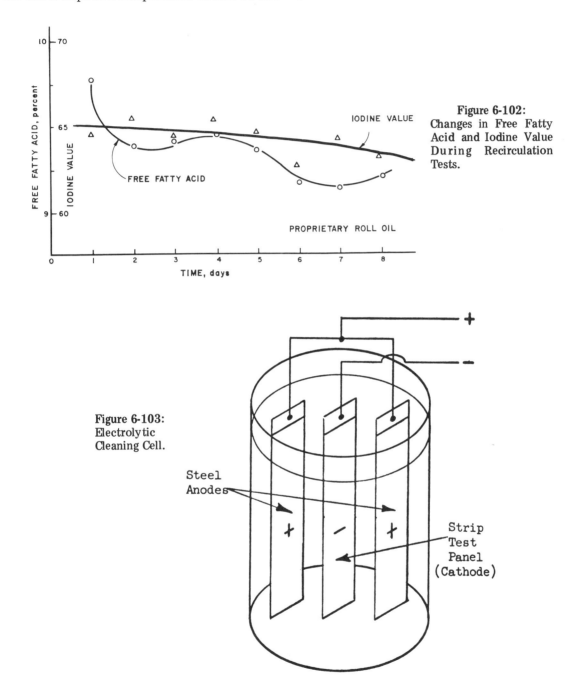

Figure 6-102: Changes in Free Fatty Acid and Iodine Value During Recirculation Tests.

Figure 6-103: Electrolytic Cleaning Cell.

The test procedure involves subjecting the panels to brief impulses of current and the number of coulombs per square foot of surface area required to remove the oil layer is measured using the "water break" test as an indication of satisfactory oil removal.

Lubricants necessitating charge densities of less than 480 coulombs/square foot are rated as "good", those needing 480 to 620 coulombs/square foot as "fair" and those in excess of 620 coulombs/square foot as "poor".

Chapter 7   Thermal Aspects of the Cold Rolling Process

7-1 Introduction

Modern high-speed cold rolling facilities are currently being designed with drives exceeding 40,000 hp or, in terms of specific total power, about 700 hp per inch width of the mill.♦⊕§ If utilized to capacity, therefore, the rate of energy input to such mills corresponds to more than a million Btu's per minute. Since the electrical power input to the rolling process is converted mainly to thermal energy, it is apparent that the adequate cooling of the mill components (primarily the rolls) and the rolled products is of considerable concern to mill operators.

A number of operational problems may ensue from inadequate cooling of the mill rolls. As far as the rolls themselves are concerned, improper cooling may develop large enough stresses within them to cause spalling and, in the case of sleeved backup rolls, the sleeve may slip in a localized area thereby creating an eccentricity in the roll. Moreover, higher roll temperatures may also result in increased roll wear.

Other operational problems ensue from abnormally high roll temperatures or from undesirable thermal gradients within them. With respect to the former, perhaps the most significant consequence is the decreased effectiveness of the rolling lubricant. This, in turn, results in increased frictional energy dissipation in the roll bite, correspondingly increased roll and strip surface temperatures and a progressively aggravated situation. Under such circumstances, decreasing the mill speed is the quickest remedy available to the mill operators. Undesirable thermal gradients in the rolls, on the other hand, create roll profiles or crowns such that flat strip, or satisfactory shape in the rolled product, is difficult to attain. Ideally, the thermal expansion of the rolls, combined with their initial, ground-in crowns, should, under the rolling conditions encountered, provide a roll gap geometry matching the cross-sectional profile of the strip entering the roll bite. Under such circumstances, strip flatness may be preserved. Otherwise, the strip may be overrolled at either the edges or the center.

Improper cooling of the rolled product may be equally undesirable. For example, if the strip attains too high a temperature, an oxide film may form on its surface of such a thickness as to impair the corrosion resistance of tinplate subsequently produced from the strip. As a consequence, mill operators try to maintain the coiling temperatures at less than 350°F, if necessary, by slowing down the speed of the mill. Even so, coiling temperatures as high as 400°F have been encountered on some mills.✢

Failure to provide the mill and the rolled products with satisfactory cooling may therefore lead to decreased productivity as a result of (1) decreased mill speeds, (2) increased number of roll changes, and (3) decreased yields due to unsatisfactory shape. It is therefore of paramount importance that, when a mill is designed, the thermal aspects associated with the rolling process be fully considered so that optimum sizes of mill rolls and drive motors may be selected and adequate roll and strip cooling systems be engineered.

It is the intent of this chapter to identify the sources of heat generation in the roll bite and the modes of heat transfer affecting roll and strip temperatures. Mathematical models are presented for the computation of these temperatures for single and multistand mills and, lastly, typical mill coolant systems are described.

♦ "Developments in the Iron and Steel Industry During 1971", Iron and Steel Engineer Year Book, 1972, pp. 31-82.
⊕ "Developments in the Iron and Steel Industry During 1972", Iron and Steel Engineer Year Book, 1973, pp. 1-48.
§ "Development in the Iron and Steel Industry During 1973", Iron and Steel Engineer, January, 1974, pp. D1-D32.
✢ "Making, Shaping and Treating of Steel", Edited by H. E. McGannon, Ninth Edition, U. S. Steel Corporation, 1971, p. 968.

7-2 Energy Dissipated in The Plastic Deformation of Steel Strip

Basic to the thermal considerations of cold mill design and operation are data relating to the theoretical energy involved in the plastic deformation of steel. Until relatively recently, the only data available were in the form of graphs, as exemplified by Figure 7-1 showing the total energy per ton dissipated in the reduction of hot rolled strip on a tandem mill.★

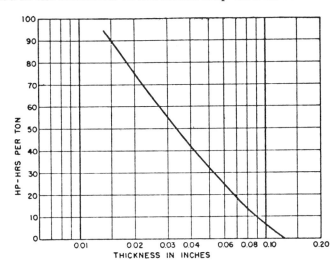

Figure 7-1: Energy Required to Reduce Low Carbon Steel 0.125-Inch Thick.

Although such data serve a useful purpose in the computation of mill drive requirements, to achieve a better understanding of the cold reduction process, it is necessary to know in detail how this total energy is dissipated by the mill and the strip.

If steel strip could be regarded as an isotropic material exhibiting constant mechanical properties during its deformation, the computation of its deformation energy would be considerably simplified. However, the following factors must generally be taken into account in calculating deformation energies:

(a) the initial yield strength of the strip as conventionally measured in tension at low strain rates, (σ),

(b) the strain rate during the deformation process, ($\dot{e}$),■

(c) the strain rate effect, (a), and

(d) the constraint on the sideways flow of the strip in the roll bite.

The dynamic constrained yield stress of the workpiece under compression (σ_c) is then given by

$$\sigma_c = 1.155 \, (\sigma + a \log_{10} 1000\dot{e}) \qquad (7\text{-}1)$$

Now the initial yield strength σ is dependent on both the yield strength (σ_o) of the strip in the fully annealed condition and the total prior reduction (r_p) given to it. The relationship between (σ) and the other parameters may be expressed by an empirical expression of the form

$$\sigma = \sigma_o + a_1 r_p + a_2 r_p^2 + a_3 r_p^3 \qquad (7\text{-}2)$$

where a_1, a_2 and a_3 are coefficients expressed in terms of stress (psi).

The deformation given to a workpiece may be denoted in terms of the strain (e). This parameter is related to the reduction (r) (expressed as a decimal fraction) by the relationship

$$e = \ln \left(\frac{1}{1-r} \right) \qquad (7\text{-}3)$$

★ J. E. Peebles, "Power Requirements and Selection of Electrical Equipment for Reversing Cold Strip Mills", Iron and Steel Engineer Year Book, 1956, pp. 1028-1046.

■ The strain rate associated with the rolling process is discussed more fully in Chapter 9.

It has been shown that, once the steel strip is undergoing deformation, work hardening does not appear to occur as plastic flow proceeds.* Thus, assuming no redundant work to occur during the deformation, the energy E expended per unit volume is given by

$$E = \sigma_c e = \sigma_c \ln\left(\frac{1}{1-r}\right) \qquad (7\text{-}4)$$

The magnitudes of the parameters σ_0, a, a_1, a_2, and a_3, as defined above, are dependent basically on the physical properties of the workpiece. Moreover, the parameter a is dependent on the prior reduction to which the strip has been subjected, decreasing with increasing prior reduction as indicated by the slopes of the curves of Figure 7-2. A typical low carbon steel may have values for the parameters referred to above as presented in Table 7-1.

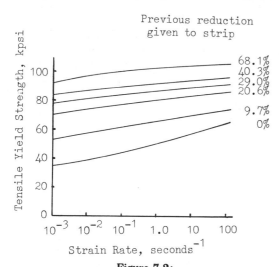

Figure 7-2:
Variation of Tensile Yield Strength with Strain Rate (Annealed Low-Carbon Steel Strip).

Table 7-1

Physical Parameters for a Typical Low Carbon Steel

Parameter	Symbol	Value
Yield strength at low strain rates when fully annealed	σ_0	40,000 psi
Strain rate effect	a	4,460 psi
Work hardening coefficient	a_1	177,300 psi
Work hardening coefficient	a_2	−292,000 psi
Work hardening coefficient	a_3	195,000 psi

The single-pass deformation energy (in terms of hp-hours/ton) for this steel (when fully annealed) as a function of reduction and for various strain rates, is as shown in Figure 7-3. Assuming the specific gravity of steel is 0.283 lb./inch^3 and the specific heat is 0.11 Btu/lb.-F, then a deformation energy of 1 hp-hour/ton (assumed to be all converted to heat) would produce a temperature rise close to 11.6°F in the steel. It should also be noted that, where a reasonably large reduction is to be given to annealed strip, less deformation energy is incurred when the reduction is made in a single pass rather than a plurality of passes. This is evidenced by Table 7-2 showing the deformation energy required by a 50 percent overall reduction given to annealed low carbon steel at a strain rate of 100 seconds^{-1} in 1, 2, 3 and 4 passes. Similarly where a given deformation is to be produced in a fixed number of passes, a drafting schedule with decreasing reductions will require slightly less energy than a schedule with increasing reductions, as indicated in Table 7-3.•

In the foregoing discussion of deformation energy, it should be noted that the only constraint pertaining to the flow of the strip in the roll bite has been assumed to be that in the lateral direction (preventing sideways spread of the strip) and that, in the direction of rolling, the strip is assumed to be free to elongate in the absence of frictional effects. This is an idealistic situation and the energy of deformation based on such conditions constitutes the minimum theoretical energy required to reduce the strip by the rolling process. In actual rolling operations,

---

* R. J. Bentz and W. L. Roberts, "Speed Effects in the Second Cold Reduction of Steel Strip", Mechanical Working and Steel Processing V, Proceedings of the Ninth Mechanical Working and Steel Processing Conference, AIME, Gordon and Breach, New York, New York, pp. 193-222.

• W. L. Roberts, "Thermal Considerations in Tandem Cold Rolling Operations", Iron and Steel Engineer Year Book, 1968, pp. 362-370.

THERMAL ASPECTS OF COLD ROLLING

friction existing between the roll and strip creates restraint to the flow of metal in the direction of rolling as well as sideways and therefore pressures larger than σ_c must be exerted by the rolls on the strip if the latter is to be reduced in thickness (see Chapter 9).

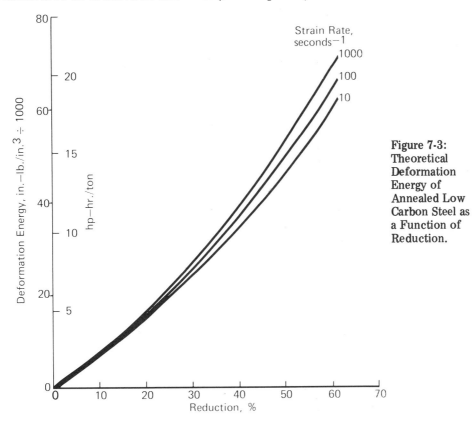

Figure 7-3: Theoretical Deformation Energy of Annealed Low Carbon Steel as a Function of Reduction.

Table 7-2

Theoretical Deformation Energy Required for a 50 Per Cent Reduction of Annealed Low-Carbon Steel in 1, 2, 3, and 4 Passes

No. of Passes	Reduction per Pass (%)	Theoretical Deformation Energy (In.-lb./in.^{3})
1	50.0	49,900
2	29.3	62,700
3	20.6	70,500
4	15.8	75,300

Table 7-3

Theoretical Deformation Energy of Low Carbon Steel as Related to Drafting Practice on a 5-Stand Tandem Mill

Reduction Schedule	Stand 1		Stand 2		Stand 3		Stand 4		Stand 5		Total Energy (In.-lb./in.^{3})
	Red. (%)	Energy (In.-lb./in.^{3})	Red. (%)	Energy (In.-lb./in.^{3})	Red. (%)	Energy (In.-lb./in.^{3})	Red. (%)	Energy (In.-lb./in.^{3})	Red. (%)	Energy In.-lb./in.^{3})	
1	48	47,050	42	63,980	36	71,750	30	71,800	26	71,690	326,270
2	37	33,240	37	52,180	37	71,130	37	90,080	37	109,020	355,650
3	26	21,650	30	37,960	36	64,100	42	100,520	48	148,900	373,130

7-3 Frictional Energy Dissipation in the Roll Bite

Friction between the surfaces of the work rolls and the strip is necessary to transmit deformation energy from the former to the latter. As a consequence of the shearing forces established at the roll surfaces by this friction and because of the relative slippage that must occur between the rolls and the strip, there is a frictional energy dissipation at each roll-strip interface (or arc of contact). The fraction of the total energy delivered by the rolls that is dissipated as frictional energy (and therefore as heat) is dependent on both the reduction given to the strip in the rolling operation and the effective coefficient of friction along each arc of contact. For a given reduction, the least frictional energy is dissipated when the minimum value of the coefficient of friction occurs (see Section 6-2), or when the neutral plane of the roll bite coincides with the exit plane and the delivery speed of the strip is equal to the peripheral speed of the work rolls.

This minimum frictional energy dissipation, expressed as a fraction of the total energy delivered by the rolls, may be derived on an approximate basis as follows. Referring to Figure 7-4, it

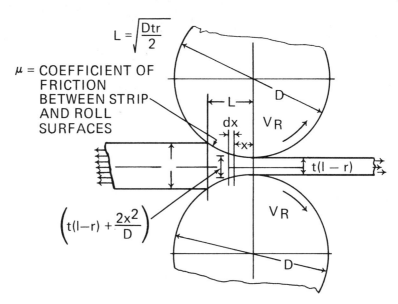

Figure 7-4:
Diagrammatic Sketch of the Roll Bite.

is assumed that (1) the strip is of unit width, (2) the roll-strip interfaces are circular arcs of contact, and (3) the deformation pressure (p) is of a constant value along each arc of contact. An element of dimension (dx) at a distance (x) from the exit plane has a thickness $t(1-r) + 2x^2/D$ where t is the entry thickness of the strip, r is the reduction expressed as a decimal fraction and D is the deformed roll diameter. If the peripheral speed of the roll is V_R, then the speed of the element is $V_R/(1 + 2x^2/D't[1-r])$, and the relative slipping speed between the roll and strip surfaces is

$$V_R \left(\frac{1}{\frac{D't(1-r)}{2x^2} + 1} \right)$$

The frictional power dissipated on both the top and bottom ends of the element is therefore

$$2\mu p V_R \left(\frac{1}{\frac{D't(1-r)}{2x^2} + 1} \right)$$

The total frictional energy dissipation is therefore

THERMAL ASPECTS OF COLD ROLLING

$$2\mu pV_R \int_0^L \frac{x^2 \, dx}{\left(\frac{D't(1-r)+x^2}{2}\right)}$$

$$= 2\mu pV_R \sqrt{\frac{D'tr}{2}} \left[1 - \sqrt{\frac{(1-r)}{r}} \tan^{-1}\sqrt{\frac{r}{(1-r)}}\right]$$

since $L = \sqrt{\frac{D'tr}{2}}$ (7-5)

Now the total energy delivered by the two work rolls is $2\mu pV_R \sqrt{\frac{D'tr}{2}}$ so that the fraction of this energy converted to frictional heat is

$$1 - \sqrt{\frac{(1-r)}{r}} \tan^{-1}\sqrt{\frac{r}{(1-r)}}$$

This may be shown to be approximately equal to 0.358r. However, since in practice, the neutral plane is not at the exit of the roll bite, it is generally of sufficient accuracy to assume that the fraction of the energy dissipated is 0.5r, and that this energy is converted to heat along each arc of contact where it is equally partitioned between the rolls and the strip. (See Section 7-6.)

7-4 Roll Bearing and Other Mill Stand Losses

In the normal operation of a mill, power losses are incurred at the roll bearings (particularly the backup roll bearings) and at the regions of contact between the work and backup rolls. It is not, however, convenient to measure these losses separately so the usual mill practice is to measure the drive power required to operate the mill stand at various speeds under various screw forces with no workpiece in the roll bite (or with the rolls "below face"). Mill stand power loss for a typical commercial 4-high mill stand equipped with sleeve bearings is illustrated in Figure 7-5.

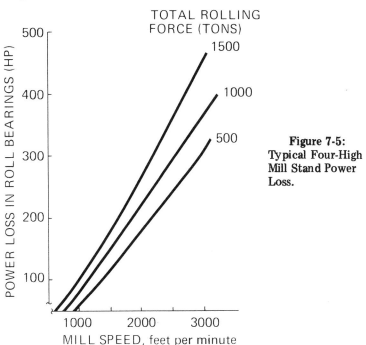

Figure 7-5: Typical Four-High Mill Stand Power Loss.

It will be noted that, for a given screw force, the power dissipation increases in an approximately linear manner with mill speed. Such losses are not as sensitive to screw force, however, since at mill speeds in the range 2000-3000 fpm, tripling the screw force increases the power loss by about 50 per cent.

When the mill stand losses are measured as described above, a fraction of the total energy relates to frictional energy developed when one work roll is in contact with the other. Under normal rolling conditions, if the work rolls touch each other (as they may in the rolling of thin strips or foils) they do so only at the ends of the rolls and this component of the mill stand energy loss is considerably reduced. In cases where the rolls are not below face this component disappears completely. Accordingly, it must be remembered that the mill stand losses measured without a strip in the mill may be slightly larger than would occur under typical rolling conditions.

Probably the largest mill stand frictional losses are encountered at the work roll — backup roll contacts. Because of the complexity of the phenomena associated with rolling friction, however, it is difficult to write quantitative laws pertaining to it (as has been done in the case of sliding friction). Coefficients of rolling friction are generally in the range 10^{-5} to 5×10^{-4},[▲] and if the mill stand losses shown in Figure 7-5 were to be attributed solely to rolling friction between work and backup rolls, the coefficient would have to be close to 10^{-3}.

Some of the energy dissipated by rolling friction is undoubtedly due to elastic hysteresis losses exhibited by the roll materials. In the majority of cases, the effect is believed to make a predominant contribution to the total rolling coefficient.[×‡]

With certain types of roll bearings, the heat dissipated in the bearing may be removed, in part, by the bearing lubricant. For example, as discussed in Section 4-34, an oil mist used to lubricate the bearing may be vented away from it. However, in sleeve-type bearings, the oil is pumped through the bearing, as described in Sections 4-35 and 4-36, thereby enabling the equilibrium temperature of the roll necks to be quickly reached. This is verified by roll temperature profile measurements made on dry mills with different types of bearings.[◊] Profiles for two of these mills are illustrated in Figure 7-6.

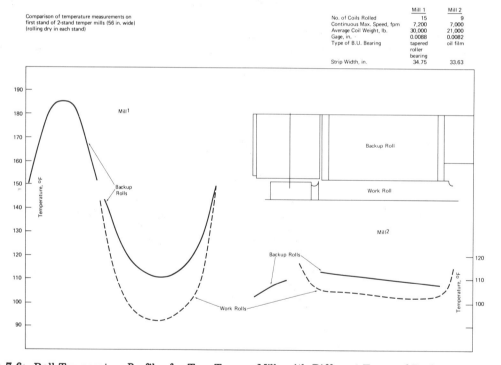

Figure 7-6: Roll Temperature Profiles for Two Temper Mills with Different Types of Backup Roll Bearings.

▲ "Friction and Wear of Materials", by E. Rabinowicz, John Wiley and Sons, Inc., New York, N.Y., 1965, pp. 82-85.
× D. Tabor, "The Mechanism of 'Free' Rolling Friction", Lubrication Engineering, (12) 1956, pp. 379-386.
‡ R. C. Drutowski, "Energy Losses of Balls Rolling on the Plates", pp. 16-35 of "Friction and Wear", Edited by R. Davies, Elsevier, Amsterdam, 1959.
◊ Data kindly supplied by the Morgan Construction Company, Worcester, Mass.

THERMAL ASPECTS OF COLD ROLLING

7-5 Peak Temperatures Encountered in The Roll Bite

The dissipation of frictional energy along each arc of contact in the roll bite must, under conditions of high-speed cold rolling, create relatively high transient temperatures (particularly on a microscopic basis) on the surfaces of the rolls and strip. In many instances, such as in the rolling of strip with soft metallic coatings and in roll-wear studies, it is desirable to know the peak temperatures reached along the arcs of contact. Such temperatures occurring under normal rolling conditions may be computed from analytical considerations and may be measured using a recently developed temperature sensing transducer.

Blok[⊙] used the following approximate treatment in developing an expression for the maximum flash temperature occurring at the interface of two bodies sliding one over the other. He considered a band-shaped heat source with some distribution of the heat flux q (having an average denoted by q_{av} and moving at a uniform speed v along a solid body bounded by a plane surface (see Figure 7-7).

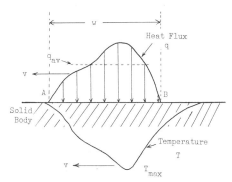

Figure 7-7: Moving Band-Shaped Heat Source. (Blok)

The flash temperatures, which have to be superimposed on the initial bulk temperature of the body, will naturally be highest at the surface swept by the heat source. The maximum flash temperature, $T_{max.}$, will as a rule occur somewhere between the middle and the trailing edge, B, of the heat source. However, in the case of a uniformly distributed heat flux, the maximum flash temperature will occur at the trailing edge.

Where the sliding speed, v, is high, the maximum flash temperature, $T_{max.}$, can be related to various other parameters by means of the expression

$$T_{max.} = \frac{A\, q_{av}}{b} \frac{w^{1/2}}{v^{1/2}} \qquad (7\text{-}6)$$

where w is the width of the source and v is its velocity (relative to the other surface). The parameter b denotes the square root of the product of the thermal conductivity, k, the density, ρ, and the specific heat, c. Therefore,

$$b = (k.\rho.c)^{1/2} \qquad (7\text{-}7)$$

Hence

$$T_{max.} = A \frac{q_{av}}{(k\rho c)^{1/2}} \frac{w^{1/2}}{v^{1/2}} \qquad (7\text{-}8)$$

The factor A is a form factor in that its value depends only on the form of the distribution q over the width w. For instance, for a uniform distribution with $q = q_{av}$ over the entire width,

$$A = 2/\sqrt{\pi} = 1.13 \qquad (7\text{-}9)$$

and for a semi-elliptical distribution

$$A = 1.11 \qquad (7\text{-}10)$$

[⊙] H. Blok, "The Flash Temperature Concept", Wear, 6 (1963), pp. 483-494.

Regarding the contact of the two sliding bodies considered by Blok to be analogous to the interface between a work roll and the strip in a roll bite, then the formula for T_{max} may be expressed in terms of various rolling parameters. Now the heat flux q_{av} may be approximately equated to $\mu f V r / 4 J L$ ◊ where μ is the coefficient, f is the specific rolling force, V is the rolling speed, r is the reduction (expressed as a decimal fraction), J is the mechanical equivalent of heat and L is the length of the arc of contact. Similarly, the width of the heat source may be regarded as being equivalent to the length of the arc of contact and v as being equal to the mean sliding speed $Vr/2$. Thus

$$T_{max.} = \frac{A \mu f r^{1/2} V^{1/2}}{4 J L^{1/2} (k \rho c)^{1/2}}$$

$$= \frac{A \mu f}{4 J} \left(\frac{rV}{L k \rho c} \right)^{1/2} \qquad (7\text{-}11)$$

If the specific rolling force f is approximated by σL where σ is the compressive flow stress of the strip, then

$$T_{max.} = \frac{A \mu \sigma}{4 J} \left(\frac{LrV}{k \rho c} \right)^{1/2} \qquad (7\text{-}12)$$

Archard ⊞ developed a similar formula for the temperature θ of a point on a plane surface of a semi-infinite solid, viz.,

$$\theta = \frac{2 q t^{1/2}}{(\pi k \rho c)^{1/2}} \qquad (7\text{-}13)$$

where q is the rate of supply of heat per unit area and t is the time for which it has been supplied. In terms of rolling parameters, q may again be equated on an approximate basis to $\frac{\mu \sigma V r}{4J}$ and the time t roughly approximated by L/v. Thus

$$\theta \simeq \frac{\mu \sigma r}{2J} \left(\frac{LV}{\pi k \rho c} \right)^{1/2} \qquad (7\text{-}14)$$

From the above formulas, which must be regarded only as approximate and applying basically to nonlubricated dry surfaces, it would appear that the peak temperature (considered as a temperature increment above ambient) encountered on the strip surface in the roll bite is directly proportional to

(1) the coefficient of friction (μ)
(2) the constrained compressive flow stress of the workpiece (σ),
(3) the square root of the length of the arc of contact (L),
(4) the square root of the rolling speed (V), and inversely proportional to the square roots of
(5) the thermal conductivity of the workpiece (k),
(6) its density (ρ) and
(7) its specific heat (c).

For steel, typical values of the parameters k, ρ, and c are as follows: k = 25 Btu/hr-ft-F; ρ = 487 lb_m/ft^3; c = 0.113 Btu/(lb_m-F), and J = 778 ft-lb/Btu. Thus, if conventional units are used, formula 7-12 for the flash temperature in degrees F becomes

$$T_{max.} = 0.0028 \, A \, \mu \sigma \, (LrV)^{1/2} \quad (°F) \qquad (7\text{-}15)$$

◊ The frictional heat is considered to be equally partitioned between the rolls and the strip.
⊞ J. F. Archard, "The Temperature of Rubbing Surfaces", Wear 2, (1958/59), pp. 438-455.

THERMAL ASPECTS OF COLD ROLLING

where σ is expressed in psi, L in inches and V in fpm. In a typical rolling situation, where $\mu \sim 0.03$, $\sigma \sim 75,000$ psi, $L \sim 0.75$-inch, $r \sim 0.3$ and $V \sim 2000$ fpm, then T_{max} would be about $148°$ F.

Thus, in conventional cold rolling, assuming formula 7-15 to be valid where rolling lubricants are used, the flash temperatures are relatively low and would be less than the melting points for coating metals such as tin and aluminum. In dry temper rolling where μ may be close to 0.3, the value of the reduction r is about 0.01. Accordingly, it is doubtful if significantly high temperatures are reached even in this operation.

Measurements of the temperature distribution along the arc of contact in the rolling of annealed, low carbon steel strip 0.010-inch thick on an experimental mill with 3-inch diameter work rolls have recently been made by Kannel and Dow[1] of Battelle's Columbus Laboratories. For this purpose, a narrow, (~ 0.002-inch), thin (~ 3-microinch) strip of titanium is vapor deposited on top of a silica substrate on the roll surface parallel to the roll axis, protected by a thin overlay of silica or alumina. The titanium strip is connected to a bridge circuit via slip rings on the roll neck and the change in resistance of the strip is measured by means of a high-speed oscilloscope. A typical oscillogram is illustrated in Figure 7-8 where a peak temperature about $46°$F above ambient is recorded. Such a temperature rise is about 3 to 4 times larger than would be calculated from equation 7-15 and it is to be suspected that the protective layer of silica or alumina provides a relatively high coefficient of friction (~ 0.11) in spite of the fact that fatty materials were used as rolling lubricants.

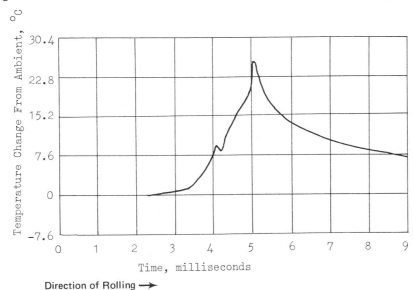

Figure 7-8: Temperature Distribution Along Arc of Contact (Kannel and Dow) 12.1 Per Cent Reduction of Annealed, Low Carbon Steel 0.010-inch thick at 350 fpm. Scope Sweep Speed = 1 millisec/large division. Vertical Calibration = $7.6°$ C/large division.

7-6 Partitioning of the Frictional Energy Between the Rolls and the Strip

In developing a thermal model of the cold rolling process, it is desirable to know how the frictional energy dissipated along each arc of contact is shared between the rolls and the strip. Generally speaking, because the rolls and strip are both steel (albeit of different grades), it is commonly assumed that there is an equipartitioning of this frictional heat. However, there are many instances where the rolls and workpiece are of quite dissimilar materials as, for example, in the rolling of austenitic stainless steel grades with steel rolls and the rolling of electrical steels on Sendzimir mills with tungsten carbide rolls.

---

[1] J. W. Kannel and T. A. Dow, "The Evolution of Surface Pressure and Temperature Measurement Techniques for Use in the Study of Lubrication in Metal Rolling", paper presented at the ASME Lubrication Symposium in Key Biscayne, Florida, June 17-19, 1974.

Blok[*][△] made the assumption (subsequently verified by Symm[○]) that the heat flow is partitioned in such a manner that the peak or flash temperatures of the two surfaces are the same. If a dimensionless parameter α represents the fraction of the heat entering surface 1, $(1-\alpha)$ is the fraction entering surface 2, and if

$$\beta_1 = \frac{A}{(k_1 \rho_1 c_1)^{1/2} v_1^{1/2}} \tag{7-16}$$

and

$$\beta_2 = \frac{A}{(k_2 \rho_2 c_2)^{1/2} v_2^{1/2}} \tag{7-17}$$

where A is a form factor, then

$$T_{max.1} = \alpha q w^{1/2} \beta_1 \tag{7-18}$$

and

$$T_{max.2} = (1-\alpha) q w^{1/2} \beta_2 \tag{7-19}$$

where q is the total frictional heat flux ($= 2 q_{av}$)

For equality of flash temperatures

$$\alpha \beta_1 = (1-\alpha) \beta_2 \tag{7-20}$$

or

$$\alpha = \frac{1}{\left(1 + \beta_1/\beta_2\right)} \tag{7-21}$$

If the materials are the same but the speeds v_1 and v_2 are different, then

$$\alpha = \frac{1}{1 + \left(v_2/v_1\right)^{1/2}} \tag{7-22}$$

and the flash temperature for the interface of the two dissimilar metals is given by

$$T_{max.} = \frac{2 q_{av} w^{1/2} \beta_1}{\left(1 + \beta_1/\beta_2\right)} \tag{7-23}$$

With respect to equation 7-15 of Section 7-5,

$$T_{max.} = 0.0056 \, A \mu \sigma \, (LrV)^{1/2} \, \frac{\beta_1}{\left(1 + \beta_1/\beta_2\right)} \tag{7-24}$$

Under normal rolling conditions, the average strip speed in the roll bite is slightly less than the peripheral speed of the rolls so there should be slightly more frictional heat energy entering the rolls than the strip if their thermal properties were identical.

Stainless steel (18Cr, 8Ni) exhibits a thermal conductivity about 2-1/2 times less than carbon steels so that, if it were to be reduced with steel rolls, more heat would enter the rolls than the strip. Conversely, if aluminum were to be rolled, its thermal properties are such that appreciably more frictional energy would enter the strip than the rolls. Yet again, rolls with relatively poor thermal conductivities, such as carbide rolls, would ensure that the larger fraction of the frictional energy would enter the strip being rolled.

7-7 Preheating Work Rolls

The creation of heat as discussed in Sections 7-2 and 7-4 occurs during the rolling process. If the listing of heat sources is to be complete, however, the preheating of work rolls prior

---

[*] H. Blok, General Discussion of Lubrication Institution of Mechanical Engineers, (1937) 2, pp. 222-235.
[△] "Basic Lubrication Theory", by A. Cameron, Longman Group, Ltd., London, 1970, pp. 167-169.
[○] G. T. Symm, Quarterly Journal Applied Mathematics and Mechanics, (1967) 20, pp. 381-391.

THERMAL ASPECTS OF COLD ROLLING

to their use in a mill should not be overlooked. This practice, under development in recent years, is intended to offer two advantages to mill operators; (1) a reduction in the "warm up" time of the mill, and (2) the prevention of thermal shock to the rolls and the possibilities of spalls or cracks occurring within them during the warm-up period.

Liputkin□ and his co-workers have published information concerning the preheating of work rolls for cold mills as carried out in the USSR. Induction heating, used to preheat the work rolls for hot mills, was found to be unsatisfactory for cold mill rolls in that it did not provide a sufficient uniformity of temperature. However, it was found that preheating could be accomplished using the mill's rolling solution and, at the Cherepovets Metallurgical Works, the steam-heated solution was flooded on to 4 pairs of rolls which were then air cooled so as to achieve the desired temperature distributions in the rolls. The equipment used for this purpose is shown in Figure 7-9.

Figure 7-9: Work Roll Heating Equipment.

To establish heating practices suitable for a tandem mill, the temperature distributions in rolls under operating conditions were investigated. The surface temperatures were measured by a contact chromel-copel thermocouple in the intervals between the rolling of coils and after the rolls have been removed from the mill. The crowns of the rolls were measured before insertion into the mill and after removal from the stands.

It was established that the surface temperature of the rolls in the four stands varied between 40 and 60°C depending on the grade of steel rolled, the rolling rate and the flow rate of the rolling solution. In rolling metals 0.5-1.0-mm thick, the roll temperature was found to be 5-10°C higher than when rolling thicker sheet. The temperatures of the upper work rolls were 6-10°C higher than that of the lower rolls because some of the solution applied to the upper rolls also fell on the lower rolls and cooled them also. The temperature differential over the length of the roll barrel depended on the width and thickness of the rolled stock. When rolling sheet more than 1200-mm wide, it amounted to 1-6°C; with narrow sheet product with a thickness of 0.5-1.0 mm, the differential reached 7-15°C and for narrow strip of thicknesses more than 1.0 mm, it was 5-10°C.

Heating schedules for the work rolls were developed using a specially prepared roll 490-mm in diameter. Holes were bored in the annealed roll in four cross sections at different depths (Figure 7-10) and the temperatures in the interior of the roll were measured by calibrated chromel-alumel thermocouples inserted into the holes and packed with asbestos. The variation in the temperature of the roll when heated by feeding the emulsion in the center of the barrel (Schedule A) and then air cooled is shown in Figure 7-11, and the surface and internal temperature distributions of the roll are presented in Figure 7-12 for rolls heated in accordance with Schedule A and for uniform heating along the roll face (Schedule B).

□ Y. V. Liputkin, et al, "Preheating the Work Rolls of a 1700 Cold Rolling Mill", Stal, 1971 (4), pp. 346-347. (British Iron and Steel Translation Service, BISI, 9458.)

344

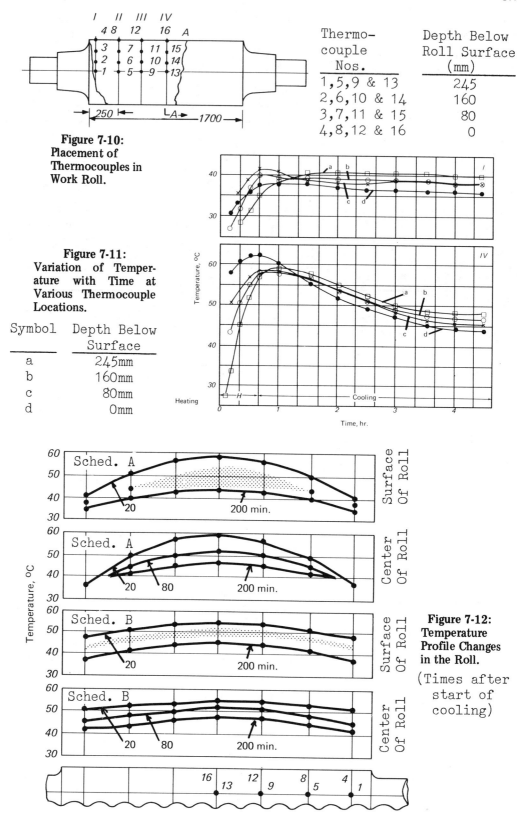

Figure 7-10: Placement of Thermocouples in Work Roll.

Thermocouple Nos.	Depth Below Roll Surface (mm)
1, 5, 9 & 13	245
2, 6, 10 & 14	160
3, 7, 11 & 15	80
4, 8, 12 & 16	0

Figure 7-11: Variation of Temperature with Time at Various Thermocouple Locations.

Symbol	Depth Below Surface
a	245mm
b	160mm
c	80mm
d	0mm

Figure 7-12: Temperature Profile Changes in the Roll. (Times after start of cooling)

Referring to Figure 7-11, the difference in temperature between the surface and the center of the roll at its midsection after a heating cycle of 40 minutes is about 9°C with a temperature at the center being 54°C. Thirty minutes later in the cooling cycle, the surface and center temperatures are approximately equal (57°C) and with further cooling the surface temperature becomes 3 − 4°C less than the center temperature. After 4 hours of cooling, the cooling rate is seen to have decreased from 4 to 2 − 2.5°C per hour.

When the rolls were heated in accordance with Schedule A for 40 minutes, the maximum temperature along the roll axis at point 13 (Figure 7-12) was found to be 57°C with the minimum temperature at point 1 being 34°C. For Schedule B these temperatures were 59°C and 49°C, respectively. On cooling, the temperature inside the roll increases at first. At point 13, for example, it rises to 59°C, falling to 47°C 3 hours after the flow of solution has been stopped. The temperature differential from the center to the end of the face amounts to 8°C in the case of Schedule A and 5°C in the case of Schedule B. Consequently, when heated and air cooled, a high degree of temperature uniformity is ensured across the roll section and a satisfactory degree of thermal crowning is attained.

In practice, the preheated rolls were essentially at their steady state condition while rolling the first coil. By obviating the warming up period of a 1700-mm wide sheet mill, it was claimed that productivity was increased 2% and the downgrading of sheet product 0.5 − 1.0-mm thick attributable to shape defects was decreased by a factor of 1.5 to 2.

7-8 Fundamentals of Heat Transfer

Two basic and distinct types of heat transfer processes exist, these being conduction and radiation.[†] The former may be defined as the transfer of heat by molecular motion between one part of a body to another part of the same body, or between one body and another in physical contact with it. On a macroscopic basis, the thermal conductivity, k, of a material mathematically expresses a proportionality between a heat flux and a temperature gradient.[B][♦]

$$\left(\frac{q}{A}\right)_y = k \frac{\delta T}{\delta y} \qquad (7\text{-}25)$$

where q is the rate of heat transfer across a planar elemental area A normal to which exists a temperature gradient $\delta T/\delta y$. The macroscopic theory of radiation, on the other hand, is based on the Stefan-Boltzmann law which expresses a proportionality between the energy flux (e_b) emitted by an ideal radiator (black body) and the fourth power of its absolute temperature T,

$$e_b = \sigma T^4 \qquad (7\text{-}26)$$

the constant of proportionality σ being called the Stefan-Boltzmann constant. In the case of surfaces other than ideal radiators, the law is modified by several empirical factors.

Generally speaking, conduction and radiation occur simultaneously but, in most heat transfer problems associated with the rolling process, the fraction of the heat transferred by radiation is usually negligible.

(a) Thermal Conductivity

The magnitude of the thermal conductivity k ranges from about zero for highly evacuated spaces to about 7000 Btu/hr-ft-F for copper crystals at low temperatures.

Kinetic theory leads to the following approximate relation for gases

$$k = a \nu c_v \qquad (7\text{-}27)$$

where ν is the viscosity, c_v is the specific heat at constant volume and a is a constant with an approximate value of 2.45 for monatomic, 1.90 for diatomic and 1.70 for triatomic gases. For most

---

[†] "Heat, Mass and Momentum Transfer", by W. M. Rohsonow and H. Choi, Prentice-Hall, Inc., Englewood Cliffs, N. J., 1961, p. 87 ff.

[B] J. B. Biot, Bibliotheque Britannique, 27, 310 (1804) and Traite de Physique, 4, 669 (Paris, 1816).

[♦] J. B. J. Fourier, "Theorie Analytique de la Chaleur", Paris, 1822, English Translation by Freeman, Cambridge, 1878.

gases, the values of k range from 0.005 to 0.05 Btu/hr-ft-F and for most liquids (except liquid metals) from 0.05 to 0.50 Btu/hr-ft-F. For fluids, the variation of the thermal conductivity with temperature is far more significant than its variation with pressure.

The thermal conductivity of crystalline solids varies approximately as the reciprocal of the absolute temperature whereas, the conductivity of amorphous (glasslike) substances increases with temperature. A wide range of thermal conductivities for solids therefore exists as shown in Figure 7-13. Tabulated values of the thermal properties of some metals and alloys are shown in Table 7-4.

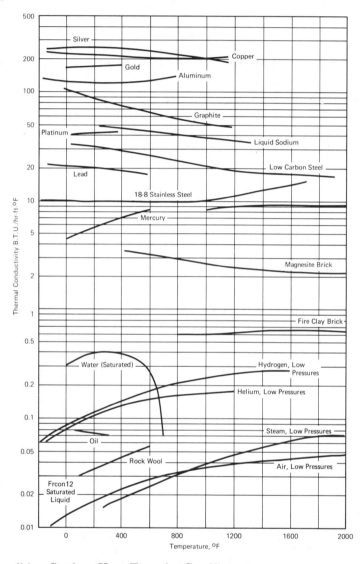

Figure 7-13: Variation of Thermal Conductivity with Temperature for Various Materials.

(b) Surface Heat Transfer Coefficient

For convenience in engineering computations, an empirical surface coefficient of heat transfer, h, is utilized. This is defined by Newton's equation°

$$h = \frac{q/A}{T_s - T_f} \qquad (7\text{-}28)$$

where q is the rate of heat transfer through a surface on the solid of area A, and where T_s and T_f are the solid and fluid temperatures, respectively. The units of the coefficient, h, are Btu/hr-ft^2-F and

---

° I. Newton, Trans. Roy. Soc., (London) 22, 824, (1701).

THERMAL ASPECTS OF COLD ROLLING

its value depends upon the properties of the fluid (including its velocity across the surface) and the roughness of the surface.

Table 7-4
Property Values of Certain Metals and Alloys

Material	Density p (lb.$_m$/ft.^{3}) (68 F)	Specific Heat (constant pressure) c_p (Btu/lb.$_m$ F) (68 F)	Thermal Conductivity k (Btu/hr. ft. F) (68 F)	(212 F)	(1112 F)	Thermal Diffusivity a (ft.^{2}/hr.) (68 F)
Aluminum, pure	169	0.214	118	119		3.665
Brass (70% Cu, 30% Zn)	532	0.092	64	74		1.322
Constantin (60% Cu, 40% Ni)	557	0.098	13	13		0.237
Copper, pure	559	0.0915	223	219	204	4.353
Iron, pure	493	0.108	42	39	23	0.785
cast (C ~4%)	454	0.10	30			0.666
wrought (C <0.5%)	490	0.11	34	33	21	0.634
Lead, pure	710	0.031	20	19.3		0.924
Magnesium, pure	109	0.242	99	97		3.762
Molybdenum	638	0.60	71	68	61	2.074
Nickel, pure (99.9%)	556	0.1065	52	48		0.882
impure (99.2%)	556	0.106	40	37	32	0.677
Silver, pure	657	0.056	242	240		6.601
Steel, mild, 1% C	487	0.113	25	25	19	0.452
Stainless steel (18 Cr, 8 Ni)	488	0.11	9.4	10	13	0.172
Tin, pure	456	0.054	37	34		1.505
Tungsten	1208	0.032	94	87	65	2.430
Zinc, pure	446	0.092	64.8	63		1.591

(c) Dimensionless Parameters in Heat Transfer

In the analysis of heat transfer problems, it has proven convenient to utilize certain dimensionless groupings of the various factors involved. These are usually assigned names to honor earlier investigators in the field of fluid mechanics and heat transfer and are listed in Table 7-5. The symbols used in the third column of the table are presented in Table 7-6.

Table 7-5
Common Dimensionless Groups

Symbol	Name	Group
Bi	Biot number	hr/k
Fo	Fourier number	$k\theta/\rho c r^2$
Gz	Graetz number	wc/kL
Gr	Grashof number	$D^3 \rho^2 g \beta \Delta t/\mu^2$
Nu	Nusselt number	hD/k
Pe	Peclet number	DGc/k
Pr	Prandtl number	$c\mu/k$
Re	Reynolds number	DG/μ, $D\mu p/\mu$
Sc	Schmidt number	$\mu/\rho k$
St	Stanton number	h/cG

In dealing with the thermal problems associated with cold rolling operations, the three dimensionless groupings of interest are the Nusselt, Grashof and Reynolds numbers. The Nusselt number N_u is of interest in that it relates the heat transfer coefficient h to the thermal conductivity k of the cooling fluid and the diameter D of a roll, viz.

$$N_u = \frac{hD}{k} \qquad (7\text{-}29)$$

The Grashof number Gr is more complex and is related to the various parameters by

$$Gr = D^3 \rho \, g \beta \Delta T / \mu^2 \qquad (7\text{-}30)$$

where D is the roll diameter, ρ is the density of the fluid (such as air), g is the acceleration due to gravity, β is the coefficient of volumetric expansion of the fluid, ΔT is the temperature difference between the surface temperature of the solid (e.g., mill roll) and the temperature of the fluid (such as air) and μ is the viscosity of the fluid.

The Reynolds number R_e is represented by

$$R_e = \frac{D u \rho}{\mu} \qquad (7\text{-}31)$$

where D is the diameter, u is the velocity of the fluid, ρ is its density and μ is its viscosity.

Table 7-6
Dimensions and Units

Dimensions: Force = F, heat = H, length = L, mass = M, temperature = T, time = θ. The force-pound is the poundal, the force-gram is the dyne

Symbol	Quantity: consistent engineering and metric units	Dimension
g	Acceleration of gravity, ft/hr^2, cm/sec^2	L/θ^2
A	Area or surface, ft^2, cm^2	L^2
K_H	Conversion from kinetic energy to heat	$ML^2/H\theta^2$
K_M	Conversion from force to mass	$ML/F\theta^2$
ρ	Density, lb/ft^3, g/cm^3	M/L^3
D	Diameter, ft, cm	L
k_d	Diffusivity (volumetric), ft^2/hr, cm^2/sec	L^2/θ
F	Force, force-pound (poundal), force-gram (dyne)	F
H	Heat, Btu, cal	H
L	Length, ft, cm	L
M	Mass, lb, g	M
W	Mass flow, lb/hr, g/sec	M/θ
G	Mass velocity, lb/(hr)(ft^2), g/(sec)(cm^2)	$M/\theta L^2$
J	Mechanical equivalent of heat, (force-lb) (ft)/Btu, (force-g) (cm)/cal	FL/H
P	Pressure, force-lb/ft^2, force-g/cm^2	F/L^2
P_o	Power, (force-lb) (ft)/hr, (force-g) (cm)/sec	FL/θ
r	Radius, ft, cm	L
c	Specific heat, Btu/(lb) (°F), cal/(g)(°C)	H/MT
υ	Specific volume, ft^3/lb, cm^3/g	L^3/M
τ	Stress, force-lb/ft^2, force-g/cm^2	F/L^2
σ	Surface tension, force-lb/ft, force-g/cm	F/L
T	Temperature, °F, °C	T
k	Thermal conductivity, Btu/(hr) (ft^2) (°F/ft), cal/(sec) (cm^2)(°C/cm)	$H/LT\theta$
α	Thermal diffusivity, ft^2/hr, cm^2/sec	L^2/θ
β	Thermal coefficient of expansion, 1/°F, 1/°C	$1/T$
R_t	Thermal resistivity, (°F)(ft)(hr)/Btu, (°C)(cm)(sec)/cal	$LT\theta/H$
θ	Time, hr, sec	θ
u	Velocity, ft/hr, cm/sec	L/θ
μ_σ	Viscosity (force-lb)(hr)/ft^2, (force-g)(sec)/cm^2	$F\theta/L^2$
μ	Viscosity (abs), lb/(ft)(hr), g/(cm)(sec)	$M/L\theta$
M	Mass, lb, g	M
w_o	Work, (force-lb)(ft), (force-g)(cm)	FL

7-9 Heat Transfer From the Strip to the Rolls

In the rolling of steel foil, it is possible to obtain large reductions ($\sim 80 - 90$ per cent) in a single pass. The heat created by such deformations would, if retained in the foil, raise its temperature several hundred degrees F. The surface of the foil exiting the roll bite does not appear to have been subjected to such temperature rises and the belief arose that a considerable fraction of

THERMAL ASPECTS OF COLD ROLLING

the deformation energy of the strip was being conducted away by the work rolls of the mill. To verify this belief, G. T. Pallone[✠] developed a simplified steady state, heat transfer model of the cold rolling process.

A schematic diagram of the rolls and the strip configuration in the area of the roll bite is shown, together with the nomenclature used in Figure 7-14. The area considered lay between the entry and exit planes and the problem involved deriving the steady state temperature distributions in both the rolls and the strip. The solution had to be obtained by thermally coupling the strip and one roll at the common boundary; the strip-roll interface.

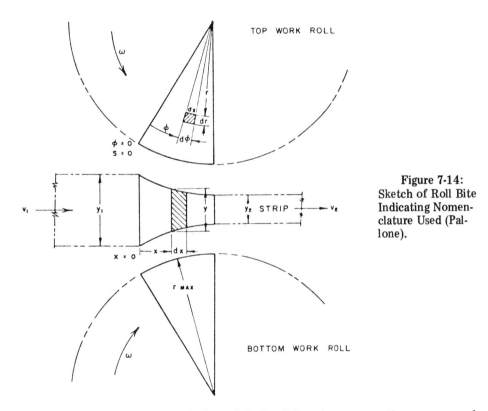

Figure 7-14: Sketch of Roll Bite Indicating Nomenclature Used (Pallone).

In developing the mathematical model, the following assumptions were made:

(1) The rolls and strip were homogeneous solids with constant thermophysical properties.
(2) The energy released by both the deformation and frictional energy dissipation was generated within the strip and was uniform along its length.
(3) A constant coefficient of thermal contact conductance existed between the roll and the strip.
(4) There was no variation in temperature in the width direction in either the roll or the strip.
(5) There was very little variation of temperature through the strip thickness.
(6) Conduction in the circumferential direction in the roll and in the direction of rolling in the strip was small compared to the conductive flux between strip and roll.
(7) Only a thin outer layer of the roll displayed a periodic temperature fluctuation due to heating in the roll bite and cooling during the remainder of its rotation.

[✠] G. T. Pallone — Unpublished work.

(8) Heat conduction within the plane of the strip was negligible compared to the heat flux through the interface.

Disregarding the curvature of the rolls in the vicinity of the roll bite, the thermal transient equation for the roll could be written

$$\frac{\delta T_R(x,r)}{\delta x} = \frac{\alpha(1+FS)}{v_2} \frac{\delta^2 T_R(x,r)}{\delta r^2} \qquad (7\text{-}32)$$

where T_R is the roll temperature

  x is the distance of an element of strip from the entry plane

  α is the thermal diffusivity of the roll

  FS is the forward slip of the strip on leaving the roll bite

  v_2 is the velocity of the strip on leaving the roll bite, and

  r is the distance radially outward in the roll.

For the strip, referring to the element shown in Figure 7-14, the heat flux q_{in} into the element due to motion is

$$q_{in} = \rho\, cvy T_s \qquad (7\text{-}33)$$

where, with respect to the strip:

  ρ is its density,

  c is its specific heat,

  y is its instantaneous thickness, and

  T_s is its temperature

The heat flux q_{out} out of the element is

$$q_{out} = \rho c\left[vyT_s + \frac{d}{dx}(vyT_s)\,dx\right] \qquad (7\text{-}34)$$

The heat conducted into the rolls $q_{contact}$ by contact with the strip is

$$q_{contact} = 2h\,[T_s(x) - T_R(x,r_{max.})]\,dx \qquad (7\text{-}35)$$

and the heat generated $q_{gen.}$ in the element may be expressed

$$q_{gen.} = q_G y\,dx \qquad (7\text{-}36)$$

where q_G is the internal heat generation rate in unit volume of the strip.

Applying an energy balance to the element of the strip and noting that the continuity equation requires that the product vy remain constant, then

$$\frac{dT_s(x)}{dx} = \frac{1}{\rho cv}\left\{q_G - \frac{2h}{y(x)}\left[T_s(x) - T_R(x,r_{max.})\right]\right\} \qquad (7\text{-}37)$$

For the roll, the boundary conditions are

(1) $\quad T_R(0,r) = f(r) \qquad (7\text{-}38)$

where the distribution $f(r)$ is assumed to be known,

(2) $\quad \dfrac{\delta T_R(x,0)}{\delta r} = 0 \qquad (7\text{-}39)$

(3) $\quad \dfrac{\delta T_R(x,r_{max.})}{\delta r} = \dfrac{h}{k}\left[T_s(x) - T_R(x,r_{max.})\right] \qquad (7\text{-}40)$

For the strip, the boundary condition is

$$T_s(0) = T_1 \qquad (7\text{-}41)$$

where T_1 is the known strip temperature at the entry into the roll bite.

THERMAL ASPECTS OF COLD ROLLING

Figure 7-15 shows a plot of predicted strip and roll surface temperatures in the rolling of tinplate foil from a thickness of 0.0065 (with a reduction of 79 per cent) to 0.0013-inch at 500 fpm on a mill with work rolls 6.5 inches in diameter. The contact conductance coefficient h was assumed to be 40,000 Btu/hr-ft^2-F. The innermost roll position that experienced a change in temperature was close to 0.02-inch.

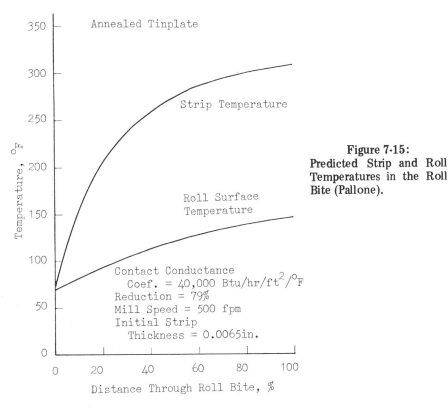

Figure 7-15: Predicted Strip and Roll Temperatures in the Roll Bite (Pallone).

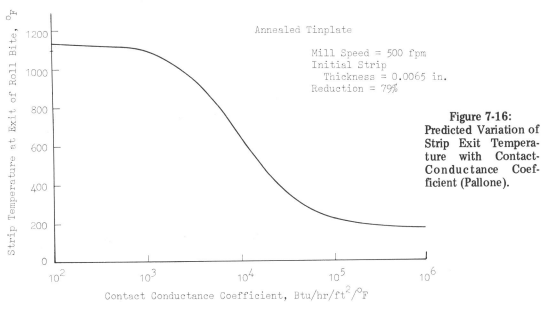

Figure 7-16: Predicted Variation of Strip Exit Temperature with Contact-Conductance Coefficient (Pallone).

Figure 7-16 illustrates the effect of the contact conductance coefficient between the roll and strip on the predicted strip temperature on exit from the roll bite. Since measured values of the exiting foil temperature generally range from 200 to 250°F, h values in the range 40,000 to 100,000 are believed to exist along the arcs of contact.

It should be remembered, however, in connection with the above model that the frictional energy was assumed to be dissipated entirely within the strip instead of being partitioned as discussed in Section 7-6. Accordingly, the estimated values of the coefficient h may be somewhat too large.

7-10 Air Cooling of Stationary Mill Rolls

In dry rolling operations, such as temper rolling, the mill rolls are air cooled. It is therefore desirable to examine first the heat transfer from the rolls under such conditions, considering first the air cooling of stationary rolls and then the air cooling of rotating rolls. Necessary to these studies is a knowledge of the thermal properties of the air and these are presented in Table 7-7.

Table 7-7
Physical Properties of Air at Atmospheric Pressure †

Temperature (°F)	Density (lb.$_m$/ft.^{3})	Specific Heat (Const. Press.) (Btu/lb.$_m$F)	Thermal Conductivity (Btu/hr. ft. F)
0	0.0863	0.239	0.013
60	0.0763	0.240	0.015
100	0.0709	0.241	0.016
200	0.0601	0.242	0.018

Temperature (°F)	Absolute Viscosity (lb.$_m$/hr. ft.)	Kinematic Viscosity (ft.^{2}/hr.)	Thermal Diffusivity (ft.^{2}/hr.)
0	0.040	0.457	0.633
60	0.043	0.560	0.799
100	0.046	0.648	0.919
200	0.052	0.864	1.250

Temperature (°F)	Prandtl Number	$g\beta/\nu^2$* (1/ft.^{3} F)
0	0.722	4.2 × 10^6
60	0.712	2.5 × 10^6
100	0.706	1.8 × 10^6
200	0.694	0.86 × 10^6

$$*g\beta/\nu^2 = \frac{\text{(Gravitational Acceleration)} \times \text{(Thermal Coef. of Volume Expansion)}}{\text{(Kinematic Viscosity)}^2}$$

The total heat flux q from the surface of a roll (regarded here simply as a cylinder) is equal to the sum of the convective and radiative components, viz.,

$$q = h_c (T_s - T_\infty) + \sigma\epsilon (T_s^4 - T_\infty^4) \tag{7-42}$$

where h_c is the heat transfer coefficient

T_s is the surface temperature of the cylinder

T_∞ is the ambient temperature of the surroundings

σ is the Stefan-Boltzmann constant and

ϵ is the emissivity of the surface.

McAdams◊ has shown that the heat transfer coefficient of horizontal cylinders in air may be computed from the following equation

$$h_c = 0.27 \left(\frac{T_s - T_\infty}{D}\right)^{0.25} \tag{7-43}$$

---

† "Heat, Mass and Momentum Transfer", by T. S. Rohsenow and S. W. Choi, Prentice-Hall, Inc., Englewood Cliffs, N.J., 1961.

◊ "Heat Transmission", by W. H. McAdams, McGraw Hill Book Company, Inc., New York, 1954.

THERMAL ASPECTS OF COLD ROLLING

where D is the external diameter of the cylinder.

It is convenient to express

$$(T_s^4 - T_\infty^4) = 4 T_f^3 (T_s - T_\infty) \qquad (7\text{-}44)$$

where T_f is the film temperature approximating the arithmetic average of T_s and T_∞, or

$$T_f = 1/2 (T_s + T_\infty) \qquad (7\text{-}45)$$

Thus $4 \sigma \epsilon T_f^3$ may be regarded as a "radiation" heat transfer coefficient (h_r) with the emissivity of the surface dependent on a number of factors but generally lying in the range 0.2 to 0.3. Thus if h_m is regarded as the total or measured heat transfer coefficient equal to $q/(T_s - T_\infty)$ then

$$h_m = h_c + h_r \qquad (7\text{-}46)$$

Hogshead,▼ using a cylinder about 6.9 inches in diameter with an electrical heating ribbon wrapped around it measured the total heat flux lost from its surface and calculated its emissivity. His data are presented in Table 7-8 and it is interesting to note that, for stationary conditions the radiation heat transfer coefficient h_r is about a third to a half of the value of the convection heat transfer coefficient h_c.

Table 7-8
Data Pertaining to Air Cooling of Stationary Cylinder (Hogshead)

	Average Surface Temp. T_S	$\Delta T = T_S - T_\infty$	h_m Measured Heat Transfer Coefficient $\left(\dfrac{Btu}{hr\text{-}ft^2\text{-}°F}\right)$	h_c Convection Heat Transfer Coefficient $\left(\dfrac{Btu}{hr\text{-}ft^2\text{-}°F}\right)$	h_r Radiation Heat Transfer Coefficient $\left(\dfrac{Btu}{hr\text{-}ft^2\text{-}°F}\right)$	Emissivity ϵ
1	107	24	1.040	0.694	0.346	0.295
2	133	50	1.155	0.827	0.328	0.261
3	131	49.5	1.184	0.827	0.357	0.287
4	170	82	1.266	0.936	0.332	0.236
5	164	83	1.277	0.939	0.338	0.249
6	222	143	1.503	1.076	0.427	0.274
7	270	188	1.646	1.151	0.495	0.281

In the data presented in Table 7-8, the average surface temperature was calculated as the mean of the temperature measured at the top ($\alpha = 0°$) side ($\alpha = 90°$) and bottom ($\alpha = 180°$) of the cylinder. The data pertaining to the individual temperature measurements are shown in Table 7-9 and it is seen that the Nusselt number corresponding to the bottom of the cylinder is approximately 50 per cent larger than the corresponding value at the top and that the Nusselt number for the side is about 30 per cent larger than the value of the top.

Eckert and Soehngen◆ measured the temperature field around a stationary cylinder using interference photographs and estimated a 400 per cent increase in the Nusselt number from top to bottom. Their experimental results, together with the theoretical values of Herman✝ and the experimental data of Hogshead▼ are shown in Figure 7-17.

▼ T. H. Hogshead, "Heat Transfer and Temperature Distribution in the Rolling of Metal Strip", PhD Thesis, Carnegie-Mellon University, 1967.

◆ E. R. G. Eckert and Soehnghen, "Studies on Heat Transfer in Laminar Free Convection with the Zehnder-Mach Interferometer", USAF Tech. Report 5747, December, 1948.

✝ R. Herman, NACA TM 1336, Translation of German Paper, November, 1954.

Table 7-9
Variation of Heat Transfer Parameters Around Periphery of Stationary Cylinder (Hogshead)

Test	T_S	$\alpha(^\circ)$	T_∞	h_m	h_r	h_c	$Gr \times 10^{-6}$	Nu	$\dfrac{Nu}{Gr^{1/4}}$
1	112	0	83	0.860	0.319	0.541	9.65	19.83	0.356
	106	90		1.085	0.313	0.771	7.87	28.26	0.534
	104	180		1.188	0.312	0.876	7.22	32.11	0.621
2	141	0	83	0.995	0.344	0.650	17.46	23.09	0.357
	131	90		1.203	0.335	0.867	14.98	31.18	0.501
	127	180		1.312	0.332	0.980	14.06	35.47	0.580
3	139	0	81.5	1.019	0.341	0.678	17.63	24.38	0.376
	128	90		1.260	0.331	0.928	14.78	33.59	0.542
	125	180		1.347	0.329	1.018	14.07	36.84	0.601
4	177	0	88	1.166	0.383	0.783	22.80	27.13	0.393
	NR	90		NR	NR	NR	NR	NR	NR
	163	180		1.384	0.369	1.014	20.19	35.58	0.531
5	174	0	81	1.139	0.373	0.766	24.67	26.87	0.382
	161	90		1.324	0.361	0.963	22.44	34.00	0.495
	157	180		1.394	0.357	1.037	21.74	36.61	0.536
6	236	0	79	1.369	0.432	0.936	33.46	31.48	0.414
	217	90		1.557	0.413	1.143	31.48	38.91	0.520
	214	180		1.592	0.410	1.181	31.05	40.20	0.538
7	294	0	82	1.459	0.501	0.958	36.47	30.78	0.397
	264	90		1.700	0.467	1.233	34.37	40.52	0.530
	252	180		1.820	0.454	1.346	33.69	44.50	0.584

NR = Not Recorded

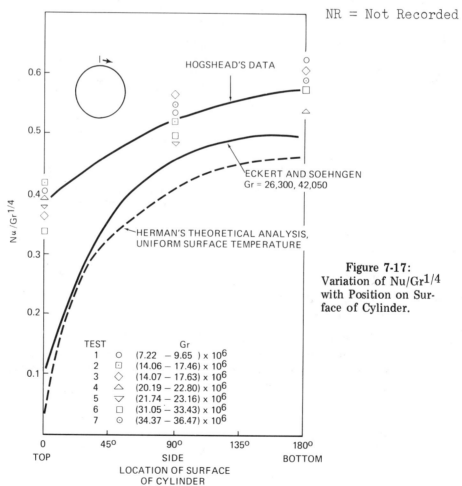

Figure 7-17: Variation of $Nu/Gr^{1/4}$ with Position on Surface of Cylinder.

THERMAL ASPECTS OF COLD ROLLING

7-11 Air Cooling of Rotating Rolls

The heat transfer by convection from a heated horizontal cylinder rotating about its axis in air was measured by Anderson and Saunders.[□] Their apparatus consisted simply of an electrically-heated horizontal cylinder which could be rotated, the surface temperature being recorded by thermocouples and the energy input measured electrically. Connections were made to the heating elements and the thermocouples by means of slip rings and separate end sections were used on the cylinders to minimize end losses. Radiation losses were computed and allowed for and a check on the emissivity of the surface was made by comparing the heat transfer under stationary conditions with established free convection values. In these experiments, three cylinders with diameters 1.0, 1.82 and 3.9 inches in diameter, each being 2 feet long, were used at temperatures up to about 140°F higher than ambient.

The results obtained by Anderson and Saunders are shown in Figures 7-18 and 7-19 in which the Nusselt number is shown plotted against the Reynolds number. The values of k (the thermal conductivity of the air), μ (its viscosity) and ρ (its density) were taken at the mean of the surface and surrounding air temperatures. In Figure 7-18, the relationship between Nu and Re is shown for the 1.82-inch diameter cylinder operated at 3 temperatures, whereas Figure 7-19 presents data for different diameters and gas pressures. (It should be noted that Figure 7-19 includes data published by Micheev.[♦]) In every case, the same trend was observed, the Nusselt number being roughly independent of the Reynolds number up to a critical value, beyond which Nu increases with Re. In some cases, the heat transfer just below the critical value of Re was found to be slightly less than for free convection. This critical value for the Reynolds number occurred when the tangential velocity of the surface of the cylinder becomes approximately equal to the velocity of the vertical free convection flow at the side of a stationary horizontal cylinder. Above the critical Reynolds number, the relationship between Nu and Re was found to be

$$Nu = 0.10 \, Re^{2/3} \qquad (7-47)$$

Similar results were obtained by Dropkin and Carmi[⊗] using rotating nickel-plated copper cylinders 3.25 and 4.5 inches in diameter. They also found that, up to a certain value of Reynolds number, rotation had no effect on the heat transfer coefficient. Above this critical value, the heat transfer coefficient increased as the speed of rotation increased.

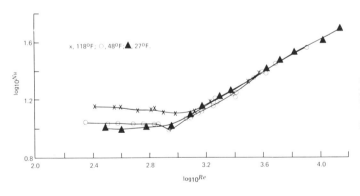

Figure 7-18:
Heat transfer from rotating horizontal cylinder, diameter 1.82-inch Temperature difference.

The results obtained by Dropkin and Carmi are shown in Figure 7-20. The curves indicate that, up to a Reynolds number of approximately 15,000, both free convection and convection due to rotation influence the value of the Nusselt number. Above this number, free

---

[□] J. T. Anderson and O. A. Saunders, "Convection from an Isolated Heated Horizontal Cylinder Rotating about Its Axis", Proceedings of the Royal Society of London, England, Series A, Vol. 217, 1953, pp. 555-562.
[♦] M. A. Micheev, 1951, Investiga Akad, Nauk. U.S.S.R. Otdel Tekh. No. 8, 1259.
[⊗] D. Dropkin and A. Carmi, "Natural-Convection Heat Transfer From a Horizontal Cylinder Rotating in Air", Transactions of the ASME, May, 1957, pp. 741-749.

convection becomes relatively unimportant and only the speed of rotation affects the Nusselt number. Under these circumstances

$$Nu = 0.073 \, Re^{0.7} \tag{7-48}$$

and this relationship applies equally well to both film and bulk properties of the air.

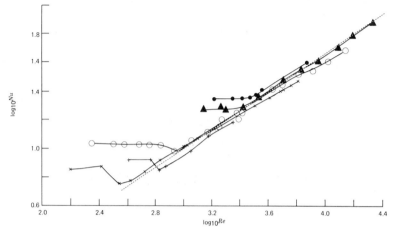

Figure 7-19: Heat transfer from rotating horizontal cylinders of different diameters including results at 4 atm. pressure.

Legend: x, diam. 1·00 inch, atmospheric pressure; ○, diam. 1·82 inch, atmospheric pressure; ▲, diam. 3·90 inches, atmospheric pressure; ●, diam. 1·82 inch, 4 atm pressure; +, (Micheev) diam. 1·1 inch, atmospheric pressure; − − − −, theoretical, $N = 0.10 \, Re^{2/3}$, calculated from analogy with free convection from a heated horizontal surface facing upwards.

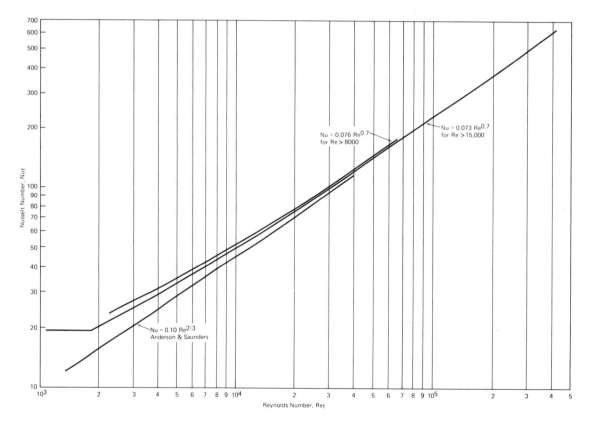

Figure 7-20: Nusselt Number Versus Reynolds Number for Cylinders Rotating in Air (Dropkin and Carmi) (Based on Properties of the Air Film)

THERMAL ASPECTS OF COLD ROLLING

A straight line plot for the region in which the Nusselt number is influenced by rotation is given in Figure 7-21. In this plot Nusselt number is drawn against the empirical parameter $(0.5\ Re^2 + Gr)^{0.35}$.

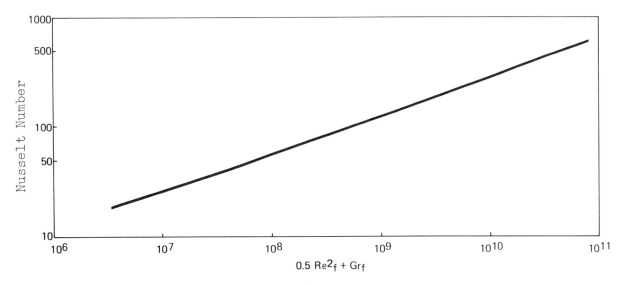

Figure 7-21: Nusselt Number Versus $(0.5\ Re_f^2 + Gr_f)$ for Film Properties. (Dropkin and Carmi)

The region of Reynolds numbers between 0 and 2500 was investigated by the use of titanium tetrachloride which produces a visible smoke. Figure 7-22 shows the smoke pattern for four different Reynolds numbers. In Figure 7-22 (a), the cylinder is stationary but in Figures 7-22 (b), (c) and (d), the cylinder is rotating in a clockwise direction. It is seen that, as the speed of rotation increases, the breakaway point shifts in the direction of rotation until a tangential point is reached. These photographs indicate that the rotating cylinder opposes the free convection currents on the downward side and aids on the upward moving side. The net effect of this, in the regions of low Reynolds numbers, is to keep the heat transfer coefficient the same as for a stationary cylinder. Above the critical value of the Reynolds number, turbulence is indicated around the cylinder and an increase in rotational speed increases the coefficient of heat transfer.

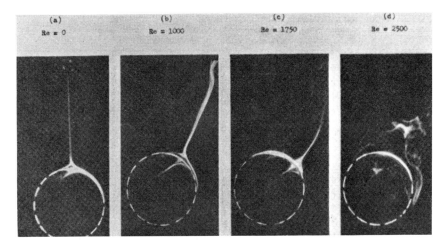

Figure 7-22:
Air-Flow Patterns Around Test Cylinder at Various Reynolds Numbers (Dropkin and Carmi).

Hogshead,▼ using the same equipment as mentioned in Section 7-10, also investigated the heat transfer associated with a rotating cylinder and his data are presented in Table 7-10. It is to be noted that the measured heat transfer coefficient h_T ranged from about 1 to 11.5 Btu/hr-ft^2-F with Nusselt numbers ranging from about 28 to 400. At high-speed rotation, the radiation heat transfer coefficient becomes almost negligible with respect to the conduction coefficient.

Table 7-10
DATA PERTAINING TO A CYLINDER ROTATING IN AIR
(Hogshead)

RPM	T_S	ΔT	δ''	h_T	h_r	h_c	Re	Gr $\times 10^{-7}$	Nu
450	125	39	159	4.08	0.334	3.75	41,800	1.23	136
900	106	25	162	6.48	0.310	6.17	86,800	0.87	226
195	156	67	163	2.44	0.365	2.07	17,200	1.86	72.6
23	182	104	153	1.47	0.379	1.09	1,975	2.70	37.9
205	100	22	56.6	2.27	0.305	2.26	20,000	0.77	84.3
209	335	250	663	2.65	0.570	2.08	14,450	3.74	65.0
950	196	108	700	6.48	0.402	6.08	79,200	2.59	209
2040	143	58	667	11.5	0.348	11.2	184,000	1.72	398
26	112	28	29.8	1.063	0.302	0.761	2,520	0.97	28.1
2070	147	59	675	11.4	0.360	11.04	185,000	1.83	392

7-12 Removal of Heat From Mill Rolls by Water and Rolling Solution

Because of the relatively poor cooling of the rolls afforded by the atmosphere, water and/or rolling solutions are used for those cold mills where significant reductions are to be given to a workpiece in as much as they provide coefficients of heat transfer of the order of fifty to a hundred times those attainable by air cooling.

Heat transfer coefficients relating to the cooling of a roll by water and an aqueous emulsion of a rolling oil have been measured by Hogshead▼ using essentially the same equipment as he used for the air cooling experiments discussed in Sections 7-10 and 7-11. Hogshead used a film jet directed on to the top of the 6.9-inch diameter horizontal rotating roll and, as a consequence, he encountered nonuniform surface temperatures. Minimum and maximum heat transfer coefficients, $h_{min.}$ and $h_{max.}$, were calculated for the corresponding maximum and minimum temperatures, $T_{max.}$ and $T_{min.}$, respectively. Thus

$$h_{min.} = \frac{Q}{(T_{max.} - T_\infty)} \qquad (7\text{-}49)$$

and

$$h_{max.} = \frac{Q}{(T_{min.} - T_\infty)} \qquad (7\text{-}50)$$

where Q is the average heat flux through the roll surface and T_∞ is the temperature of the coolant.

The rate of heat removal from the cylinder was found to be extremely dependent upon the manner in which the coolant flowed around it. Sketches of the observed flow patterns produced by the film jet 0.010-inch by 9.5 inches long flowing at 4 gpm placed 1 inch above the cylinder is illustrated in Figure 7-23. For this jet, the ratio of peripheral area to impingement area was 410.

Referring first to the flow patterns and to heat transfer coefficient data presented in Figure 7-24, until a flow rate of 2 gpm (not photographed) was reached, the film leaving the jet became bent in the direction of rotation by the circulating air and the spray which is centrifugally

---

▼ T. H. Hogshead, "Heat Transfer and Temperature Distributions in the Rolling of Metal Strip", PhD Thesis, Carnegie-Mellon University, 1967.

THERMAL ASPECTS OF COLD ROLLING

thrown from the surface. Increasing the flow rate to 3 gpm stabilized and strengthened the film and a small puddle developed on the backside of the impingement region. Again, all the water was thrown from the roll in the form of a spray. The basic observable change when a flow rate of 4 gpm was reached was that the puddle on the backside of the impingement region became larger and developed a turbulent counterclockwise circulation. When the flow rate was increased to 5 gpm, the puddle completely disappeared and the water was reflected from the impingement region to the backside in the form of large droplets or globules such as would be called a "rooster tail". Increasing the flow rate to 10 gpm caused an almost complete reflection of the fluid from the impingement region. Very little centrifugal spray was observed and the rooster tail was a concentrated stream with a reflection angle as measured from the horizontal greater than the angle caused by the more diverse stream of droplets leaving the impingement region on the front side. Referring to the curves for the maximum and minimum values for the heat transfer coefficient in Figure 7-24, it is seen that maxima occur at a flow rate of close to 4 gpm (corresponding to the appearance of the puddle near the impingement area). It is important to note that increasing the flow rate above this value may actually cause a decrease in the rate of heat removal from the roll.

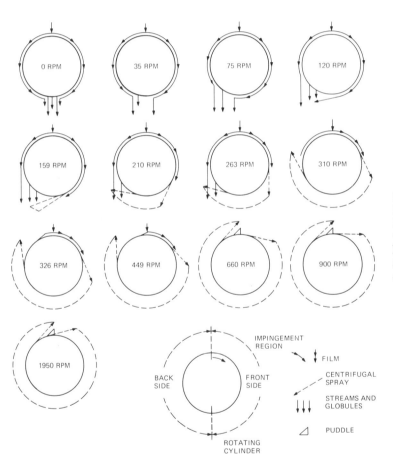

Figure 7-23: Sketches of Observed Flow Patterns, Volume Flow Rate = 4 GPM with Film Jet (Hogshead) (Roll Diameter approximately 6.9 Inches).

Figure 7-25 illustrates how the flow pattern and the heat transfer coefficient varies with roll speed for a constant water flow rate of 4 gpm. For speeds of 35 rpm or less, the flow is very similar to the flow around a stationary cylinder. At 75 rpm, the laminar film adheres to approximately 75 per cent of the roll surface ($-90° < \phi < 180°$) with the streams of fluid leaving the cylinder being largest at $\phi \approx -155°$. At 129 rpm, the streams have completely shifted from the bottom so that they are "hanging" at $\phi = -90°$ while the fluid on the front side has begun to be centrifugally thrown off in a spray near the bottom of the cylinder. At 263 rpm, a slight puddle is formed on the backside of the impingement region and, by 326 rpm, this puddle has become wildly unstable. Further speed increases up to 1950 rpm increase the turbulence of the puddle and decrease the area over which the puddle is in contact with the cylinder. Referring to the curves of Figure 7-25, it is apparent that the heat transfer coefficient exhibits minimum values at about 400 rpm or at a speed somewhat larger than that coinciding with the initial puddle formation.

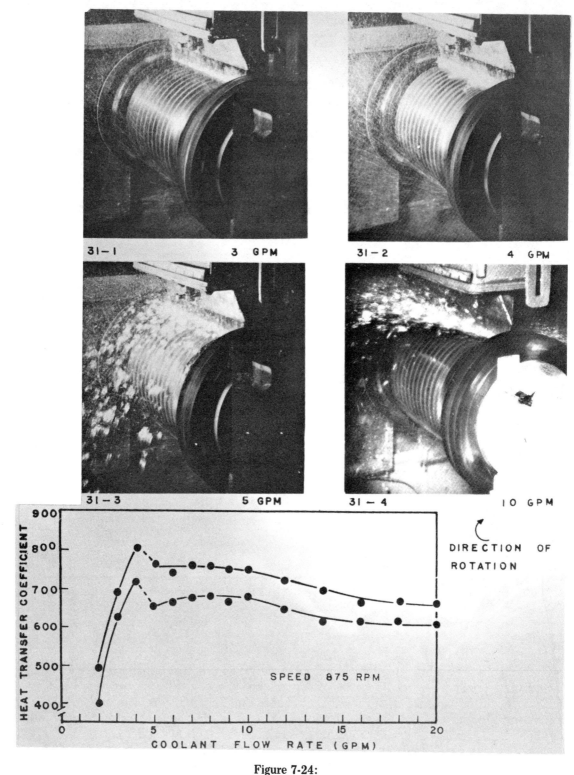

Figure 7-24:
Flow Patterns and Heat Transfer Coefficient Data Relating to a Roll Speed of 875 rpm (Hogshead) (Heat Transfer Coefficient Measured in Btu/hr-ft^2-F).

THERMAL ASPECTS OF COLD ROLLING

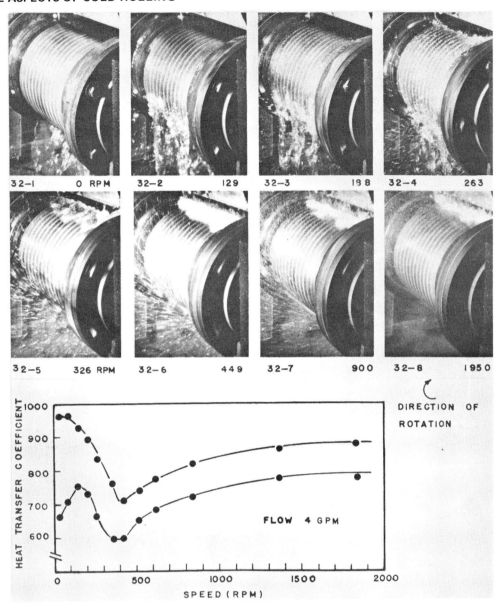

Figure 7-25:
Flow Patterns and Heat Transfer Coefficient Data Relating to a Flow Rate of 4 gpm.
(Heat Transfer Coefficient Measured in Btu/hr-ft^2-F).

It is seen from Hogshead's data that, with adequate flow rates, the coefficient of heat transfer should be at least 600 Btu/hr-ft^2-F and may attain values in the range 800-900 Btu/hr-ft^2-F.

Hogshead also studied the effectiveness of a 90 per cent water − 10 per cent oil emulsion as a coolant, this being applied to the rotating cylinder at a rate of 3 gpm through a 1/4-inch orifice nozzle located 8 inches above the cylinder.

Data obtained with this emulsion at 170°F compared with water at the same temperature is given in Table 7-11. It is seen that the water-oil emulsion is a much poorer coolant than the water; for the same surface heat flux, the rise in surface temperature above the coolant temperature is about 40 to 80 per cent greater than the rise observed with water alone. It is presumed that the "plating out" of the oil on to the roll surface creates a thermal barrier at the roll surface.

Table 7-11

COMPARISON OF WATER AND 10% OIL, 90% WATER EMULSION AS A COOLANT USING 1/4-INCH ORIFICE SPRAY NOZZLE 8 INCHES ABOVE CYLINDER, SPEED = 885 RPM, COOLANT FLOW RATE 3 GPM (Hogshead)

Heat Flux (Btu/hr-ft²)	Temperature (°F)					
	100% Water			90% Water — 10% Rolling Oil		
	$T_{surface}$	$T_{coolant}$	ΔT	$T_{surface}$	$T_{coolant}$	ΔT
16900	192	170	22	204	169	35
	195		25	208		39
27300	209	170	39	227	170	57
	215		45	235		65
38000	222	169	53	255	170	85
	228		59	269		99
51800	231	168	63	—	—	—
	238		70	—	—	—

7-13 Heat Transfer at the Work Roll — Backup Roll Contact

In 4-high and cluster mills, the work rolls are also partially cooled by the backup or intermediate rolls with which they are in contact.[*] The extent of the cooling will depend primarily on (a) the length of the region of contact (as measured in a circumferential direction), (b) the mean temperature difference between the work and backup rolls, and (c) the coefficient of heat transfer across the interface.

The length of the region of contact (b) depends on the diameters of the rolls and the specific rolling force (P' = rolling force per unit width of the mill). In Section 4-21 it was seen that

$$b = 1.52 \sqrt{P' \times \frac{d_1 d_2}{d_1 + d_2} \times \frac{E_1 + E_2}{E_1 E_2}} \qquad (7\text{-}51)$$

where d_1 and d_2 are the diameters of the work and backup rolls and E_1 and E_2 are their respective moduli of elasticity (see Figure 4-54).

The mean temperature difference ΔT between work and backup rolls is more difficult to estimate since it is likely to depend not only on such basic rolling parameters as the draft, mill speed, and roll diameters, but also on the effectiveness of the work roll cooling sprays and their placement, as well as the temperature of the coolant. However, a 10°F temperature differential may be typical in the case of 4-high mills.

The coefficient of heat transfer between two bodies in contact is discussed in Section 7-9 and it is seen that the coefficient may lie in the range 40,000 to 100,000 Btu/hr-ft²-F.

In the case of a 4-high mill with work rolls 20 inches in diameter and backup rolls 50 inches in diameter subject to a specific rolling force of 40,000 lb./in. the length (b) of the region of contact between the rolls would be close to 0.3-inch if the elastic modulus for each roll was considered to be 3×10^7 psi. Assuming a 10°F temperature differential and a heat transfer coefficient of 40,000 Btu/hr-ft²-F, the heat flux through the region of contact would be close to 830 Btu/hr. per inch width of the roll. On the other hand, if the work roll is, say 40°F higher in temperature than the coolant applied uniformly around the roll and the heat transfer is assumed to be 400 Btu/hr-ft²-F, then the heat flux around the roll and the heat transfer is assumed to be 400 Btu/hr-ft²-F, then the heat flux around the roll periphery will be close to 6,960 Btu/hr. per inch width of the roll. Thus about 10 per cent of the roll cooling would be effected through the backup roll contact.

---

[*] D. M. Parke and J. L. L. Baker, "Temperature Effects of Cooling Work Rolls", Iron and Steel Engineer Year Book, 1972, pp. 675-680.

THERMAL ASPECTS OF COLD ROLLING

In the case of cluster mills, such as Sendzimir mills, however, the cooling effect may be appreciably more significant. This is due to the fact that, as the roll sizes are decreased, for the same specific rolling force and elastic moduli, the ratio of the length of each region of contact to the work roll circumference increases. Moreover, there are two zones of contact with backup of intermediate rolls so that it is likely that 20-25 per cent of the work roll cooling may be attributed to conduction to other mill rolls.

7-14 Work Roll Temperature Measurements

Measurements of actual work roll temperature distributions have been made by a number of investigators but in the majority of instances, they have related to hot mills. Tereshko [8] and his co-workers have published some interesting data relating to 500-mm diameter work rolls of a 1700-mm wide continuous mill stand at the Zhdanov Il'vich steel works. In a specially prepared roll to be used in the first stand of the tandem mill, 8-mm diameter holes were radially drilled into the barrel (Figure 7-26) and thermocouple plugs (Figure 7-27) were press-fitted into the holes so that the chromel-copel thermocouples were located at various distances (from 1.5 to 200 mm) from the roll surface. Connections to the thermocouples were made through slip rings and recording galvanometers were used to accumulate the data.

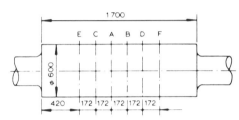

Figure 7-26:
Roll with temperature-measuring transducers, disposed at following distances from surface, mm.
A,*1.5;* B,*5.5;* C,*10;* D,*100;* E,*200;* F,*at surface*
(Other dimensions in mm)

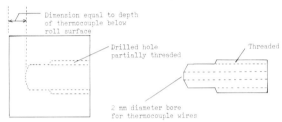

Figure 7-27:
Plug design for mounting thermocouple in roll body at distances h = 1.5; 5.5, 10 mm from roll surface.

The rolling experiments associated with the temperature measurements were performed in several stages. In Stage I, the mill stand was operated below face (without strip) at a force of 100 tons for 7 minutes. In Stage II, strip 1280-mm wide was reduced from 3.0 to 2.4-mm in thickness at a low mill speed (speed 2m/s) for 8 minutes without the use of a rolling solution. Normal rolling at speeds up to 6 m/s was carried out on 10 coils of the same type in Stage III for a period of 67 minutes. Stage IV lasted 23 minutes during which two coils were reduced from 3.8 to 3.0 mm at speeds up to 4 m/s and Stage V involved the removal of the roll from the mill and the recording of its temperatures every 2 hours for a period of 26 hours.

The continuously recorded thermocouple readings taken during the first 4 stages are reproduced in Figure 7-28. The internal layers (100 and 200 mm from the surface) are hardly affected by changes in operational rolling procedures and they do not exhibit the fluctuations experienced by the surface layers of the rolls (curves 1, 2 and 3 of Figure 7-28). During the 7 minute warm-up period, the surface layer temperature increased about 10 degrees C whereas the thermocouples at 100 and 200 mm from the surface recorded hardly any change in temperature.

In the second stage of experiment, when rolling was conducted without a coolant, the surface layer temperatures rose rapidly (8-10°C/min.) and heat penetration into the roll was also recorded by thermocouples 4 and 5.

---

[8] A. K. Tereshko, V. P. Polukhin, V. A. Nikolaev, V. N. Terekhov and I. A. Titarenko. "Experimental Investigation of Work Roll Temperature Distribution During Cold Rolling", Steel in the U.S.S.R., March, 1970, pp. 218-220.

The rate of rise of temperature at the roll surface decreased by a factor of 2 to 2.5 during stage 3 but it is to be noted how sensitive the surface temperature became with respect to rolling practice (intervals between coils, acceleration periods, etc.). Temperature fluctuations due to these causes were as much as 12°C.

The radial temperature distribution curves for different times during the experimental rolling are shown in Figure 7-29. Curve 1 corresponds to the end of Stage 1 before the first coil has been fed into the mill. Curve 2 represents the temperature distribution close to the end of the warm-up period (during which no emulsion was used) and the marked increase in the temperature of the surface layers and the relatively steep gradient across the section of the roll are to be noted. Subsequently, the inner regions of the roll warm-up but the gradients near the surface may still be severe as indicated by curves 3, 4 and 5.

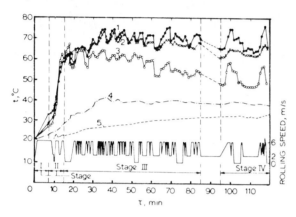

1,*1.5*; 2,*5.5*; 3,*10*; 4,*100*; 5,*200*

Figure 7-28:
Temperature changes in roll layers during mill operation; thermocouples are at distances from roll surface as shown above, mm.

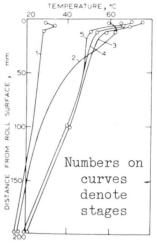

Figure 7-29:
Radial temperature distribution in roll during rolling.

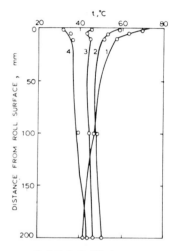

1,*0*; 2,*6*; 3,*15*; 4,*26 h*

Figure 7-30:
Cooling curves for different roll layers on removal from stand and after specified periods, h.

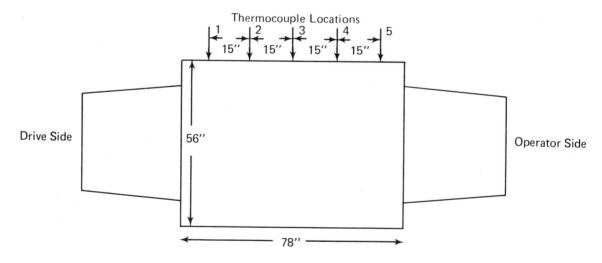

Figure 7-31:
Dimensions of the Backup Roll and the Location of the Thermocouples.

THERMAL ASPECTS OF COLD ROLLING

On the basis of their studies, Tereshko et al, recommended that the emulsion supply to the rolls be cut off between the rolling of coils to minimize the danger of roll cracking and spalling. Their analysis of thermal stresses developed during normal rolling revealed, however, that in the roll contact zones, they would not exceed 4 Kgm/mm² and were therefore not of concern with respect to roll life.

Curves relating to the cooling of the roll after it had been removed from the mill are presented in Figure 7-30. As the figure shows, the internal layers of the roll continue to increase in temperature during the first 5 hours after which all layers in the roll decrease in temperature in an exponential manner. At the later times, when the surface is at a lower temperature than any other portion of the roll, microcracks in the roll surface may open up.

Tereshko and his colleagues also found that, to ensure complete cooling of the roll after service, it must be allowed to stand for 48 hours before regrinding, otherwise its profile may be changed slightly by cooling subsequent to grinding.

7-15 Backup Roll Temperature Measurements

Surface temperatures on the top backup roll of the second stand of a 5-stand sheet mill have been measured by Patula* using sliding contact thermocouples. Temperatures were monitored at 5 locations across the top of the roll face, as indicated in Figure 7-31 with the peripheral speed of the roll being between 1200 and 1500 fpm.

The rolling solution spray system is sketched in Figure 7-32 and it consists of separate headers for each work roll and backup roll. Each header on the backup roll consists of five 15-inch sections with six spray nozzles in each section. The center section, the quarter sections and the end section are separately controllable. Each section, operating at pressures up to 100 psi, could deliver up to almost 100 gpm of coolant at a temperature close to 135°F.

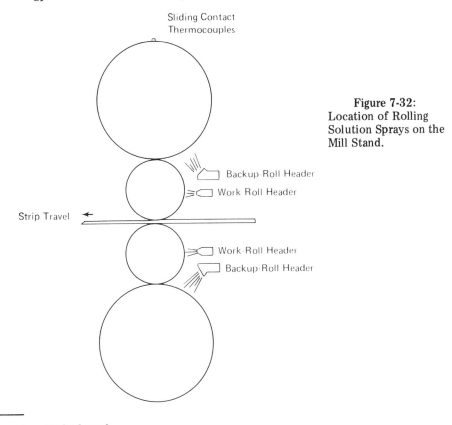

Figure 7-32:
Location of Rolling Solution Sprays on the Mill Stand.

---

* E. J. Patula — Unpublished work.

Temperature measurements were made both for the start-up of a heavy-gage "come down" and during the rolling of light-gage, medium width sheet product of various steel grades. This data is presented in Table 7-12 and it indicates that the maximum surface temperature occurring during the rolling of heavy-gage products was 140°F, dropping to 136°F between coils. For the light-gage products, the corresponding temperatures were 151°F and 140°F, respectively. The data also reveal that the largest temperature difference measured between the center of the roll and its end was 27°F, but after another hour or so of operation, this difference was reduced to approximately 8°F. For heavy-gage sheet products, the maximum roll temperature was 140°F.

Table 7-12

Temperature Profile Across Roll for Heavy and Light Gage Product
(Temperatures Taken on Stand 2 While Rolling)

Time, hrs.	Steel Grade	Band Gage, in.	Final Gage, in.	Width, in.	Backup Roll Temperature (F) Profile				
					(1)	(2)	(3)	(4)	(5)
0	—	—	—	—	72	72	72	72	72
0.25	4202	0.105	0.0335	69.18	92	111	114	90	87
1.0	4862	0.128	0.0543	71.55	104	114	119	103	103
1.75	4202	0.140	0.0599	72.18	110	118	119	110	110
10.0	7021	0.105	0.0357	70.93	132	134	140	131	132
0	—	—	—	—	125	130	132	128	127
0.5	9052	0.090	0.0238	44.00	128	133	133	134	133
1.0	4822	0.075	0.0141	45.62	132	137	139	138	134
2.5	8502	0.075	0.0181	43.62	128	136	132	132	128
4.0	8442	0.075	0.0186	44.75	134	142	151	140	137

By analyzing the roll-surface temperature between coils during start-up, the bulk roll thermal time constant was estimated by Patula to be about 2.0 hours. Thus, approximately 6 hours would be required for the roll to reach 95 per cent of its steady state condition. However, as far as the surface of the roll was concerned, its response time to changes in coolant flow was found to be about 4 seconds.

The depth of heat penetration under steady state conditions was computed to be 0.25-inch and that the roll-surface temperature was essentially uniform around the periphery of the roll (except where cooling is applied).

7-16 The Cooling of the Rolled Strip

In rolling operations where relatively high strip temperatures would be otherwise encountered, it is necessary to cool the strip as well as the mill rolls. This is particularly true in tandem mill operations as, for example, in the rolling of tinplate, otherwise undesirably high coiling temperatures would ensue.[*]

Assuming that the frictional energy is equally distributed between the rolls and the strip and that strip tensions are so small that they may be disregarded in the analysis, the temperature of the strip T_{EX} on exit from the mill, after being given a reduction r, is given by

$$T_{EX} = T_{EN} + \frac{1 - (r/4)}{1 - (r/2)} \frac{\sigma_c \ln\left(\frac{1}{1-r}\right)}{\rho S J} \qquad (7\text{-}52)$$

where:

ρ = the density of the strip
S = the specific heat of the strip
σ_c = compressive yield strength of the strip
J = the mechanical equivalent of heat
T_{EN} = the temperature of the strip at entry into the roll bite

---

[*] W. L. Roberts, "Thermal Considerations in Tandem Cold Rolling Operations", Iron and Steel Engineer Year Book, 1968, pp. 362-370.

THERMAL ASPECTS OF COLD ROLLING

This equation is derived on the supposition that no redundant work occurs in the deformation process. The increase in the temperature of the strip as it passes through the roll bite is illustrated in Figure 7-33.

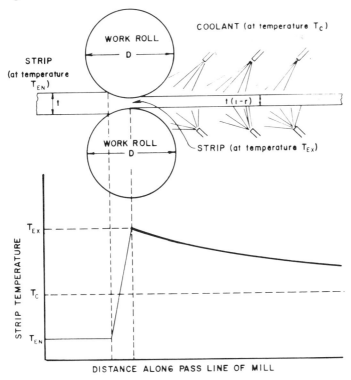

Figure 7-33: Illustration of variation of the temperature of the strip along the passline of the mill.

The heat balance for an element of the rolled strip of unit width in the region of length dx at a distance x from the exit plane of the roll bite may be expressed by the equation

$$vt_r \rho S \left(-\frac{\delta T_x}{\delta x} \right) dx = 2h (T_x - T_c) dx \qquad (7\text{-}53)$$

where:

v	= speed of strip on exit from mill
t_r	= thickness of rolled strip
ρ	= density of strip
S	= specific heat
$\left(-\frac{\delta T_x}{\delta x} \right)$	= temperature gradient
dx	= length of element
2	= constant (number of surfaces)
h	= coefficient of heat transfer
$(T_x - T_c)$	= temperature difference between strip and coolant

thus:

$$\frac{dT_x}{T_x - T_c} = \frac{-2h\,dx}{v\rho S t_r} \qquad (7\text{-}54)$$

and integration gives:

$$\ln(T_x - T_c) = \frac{-2hx}{v\rho S t_r} + C \qquad (7\text{-}55)$$

where:

C = a constant of integration

When x = 0, the value of T_x is T_{EX} and hence equation 7-55 becomes

$$T_x = T_c + (T_{EX} - T_c)e^{-2hx/v\rho S t_r} \qquad (7\text{-}56)$$

In deriving this equation, it is assumed that the strip is a very good thermal conductor, i.e., the Biot modulus is low. (The Biot number or modulus may be expressed as $\bar{h}L/K$ where $\bar{h}$ is the average unit surface thermal conductance, L is the significant length dimension obtained by dividing the volume of the body by the surface area and K is the thermal conductivity of the solid body.) The graphical representation of the expression for T_x is illustrated by the portion of the curve in Figure 7-33 corresponding to the rolled strip on the right-hand side of the roll bite.

When strip is rolled in a tandem mill, its temperature increases in a stepwise manner at each roll bite and decreases (or sometimes increases) exponentially between stands, as illustrated diagrammatically in Figure 7-34. (An exponential temperature increase will occur where the temperature of the coolant is higher than that of the strip.) The entry strip temperatures for the second and successive stands may be computed from the rolling conditions at the preceding stand. It must be remembered, however, that the compressive yield strength increases at successive stands because of the work-hardening effects of the preceding reductions. Under these circumstances, the strip temperature must be calculated in a sequential manner.

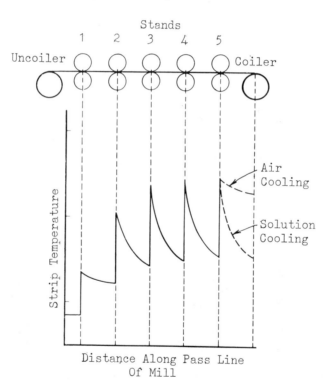

Figure 7-34: Illustration of variation of the temperature of the strip along the passline of the tandem mill.

To establish the general effects of changes on mill parameters on the temperature of strip rolled on a 5-stand mill, computations were made for the following conditions:

1. Incoming material — annealed low-carbon steel strip with work-hardening characteristics as shown in Figure 7-2.
2. Incoming strip thickness — 0.10-inch.
3. Final gage — 0.010-inch.
4. Density of strip — 0.283 lb. per cu. inch.
5. Specific heat of strip — 0.11 Btu per lb.

THERMAL ASPECTS OF COLD ROLLING

6. Initial strip temperature — 50, <u>70</u> and 90°F.
7. Coefficient of heat transfer — 400, <u>600</u>, 800 and 1,000 Btu/hr.–ft.^{2}-F.
8. Coiling speeds — 2000, <u>4000</u> and 6000 fpm.
9. Coolant temperature — 100, 120, <u>140</u> and 160°F.
10. Spacing between stands, and fifth stand and coiler — 8, <u>12</u>, 16 and 20 ft.
11. Drafting practices as follows:

	Reduction, percent		
Stand	Schedule 1	Schedule 2	Schedule 3
1	26	<u>37</u>	48
2	30	<u>37</u>	42
3	36	<u>37</u>	36
4	42	<u>37</u>	30
5	48	<u>37</u>	26

In the computations, only one variable (items 6 to 11) was changed at a time, and the other variables were held constant at the values underlined.

It is to be noted that the coolant was assumed to be applied between the last stand and the coiler. Although this is not usually the case, the computed temperature at the coiler therefore represents a theoretical minimum. On the assumption that no cooling of the strip takes place in this region, then the coiling temperature would be the same as the strip temperature at exit from the fifth stand, this representing the maximum theoretical coiling temperature. Some cooling of the strip does occur between the last stand and the coiler, however, due to the air flow past the strip and the evaporation of rolling solution from it. It is apparent, therefore, that the coiling temperature will lie between the two theoretical extremes, and to minimize coiling temperatures, the most effective strip cooling method should be utilized between the last stand and the coiler.

Effect of initial strip temperature — Reasonable changes in the initial strip temperature, such as plus or minus 20°F, have virtually no effect on coiling temperature (Figure 7-35). This may be readily explained by the fact that the change in the enthalpy of the strip due to a 20°F change in temperature is negligible compared to the heat generated within, and largely removed from, the strip during the rolling process. Accordingly attempts to reduce coiling temperatures by chilling coils prior to rolling would be doomed to failure with respect to the rolling of strip of tinplate gages on 4, 5 and 6-stand mills.

Effect of heat transfer coefficient — As would be expected, the larger the heat transfer coefficient, the lower will be the strip temperatures encountered during the rolling process (Figure 7-36). For most of the computations considered, a value of 600 Btu/hr-ft^2-F has been utilized, being assumed to be a reasonable value. However, it must be realized that the average value of the heat transfer coefficient will depend to some extent on

1. Strip and coolant temperatures.
2. The nature and concentration of the lubricant in the rolling solution.
3. The mode of application of the solution, i.e., the number of sprays, the lubricant flow rate, the velocity and direction of impingement, etc.

The radiation and conduction of energy from the strip has been disregarded, being considered negligible in comparison to the convection cooling under equilibrium conditions.

Effect of heat transfer coefficient for air cooling between the last stand and the coiler — Because of the desirability of coiling strip so that no moisture occurs between the wraps, air cooling is usually used to lower the temperature of the strip after its exit from the last stand. The heat transfer coefficient, under these circumstances, would be expected to lie in the range of 20 to

40 Btu/hr-ft²-F and the drop in strip temperature from the last stand to the coiler would be in the range of 15 to 30°F for strip 0.010-inch thick being coiled at 4000 fpm. This drop in temperature is appreciably less than that encountered by cooling with the rolling solution under which condition the temperature may be reduced in the order of 60 to 70°F.

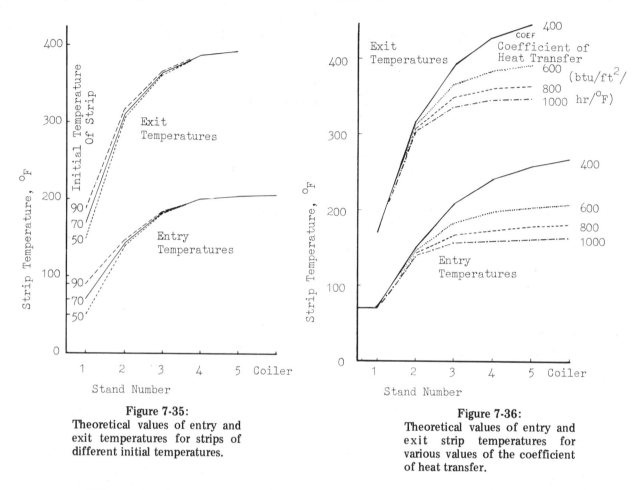

Figure 7-35: Theoretical values of entry and exit temperatures for strips of different initial temperatures.

Figure 7-36: Theoretical values of entry and exit strip temperatures for various values of the coefficient of heat transfer.

Effect of coiling speeds — The faster the mill is operated, the higher will be the coiling temperature (Figure 7-37). These curves, however, were made with the assumption that the frictional conditions remain virtually the same for the three coiling speeds considered. At high strip temperatures, the possibility exists that the effectiveness of the lubricant may diminish and that more energy may be dissipated by friction, with the result that strip temperatures may be higher than those indicated in Figure 7-37.

Effect of coolant temperature — Unlike the initial temperature of the strip, the coolant temperature has a direct effect on strip temperature (Figure 7-38). In the final stages of rolling, a change in coolant temperature will result in an almost identical change in strip temperature at any given point in the strip. Generally speaking, therefore, if lower coiling temperatures are desired, the coolant temperature should be reduced as much as possible. However, it must be remembered that the stability of the rolling solution and the "plate-out" characteristics of the oil may be adversely affected by lowering the solution temperature too far. In any case, however, the solution temperature must be above the melting temperature or the pour point of the oil or lubricant used.

Effect of interstand spacing — In recent years, the trend has been to tightly couple, or more closely space, the stands of tandem cold mills. Unfortunately, as far as coiling temperatures have been concerned, the trend has been in the wrong direction as is evidenced by Figure 7-39. As would be expected from the foregoing mathematical model, the relationship between coiling

THERMAL ASPECTS OF COLD ROLLING

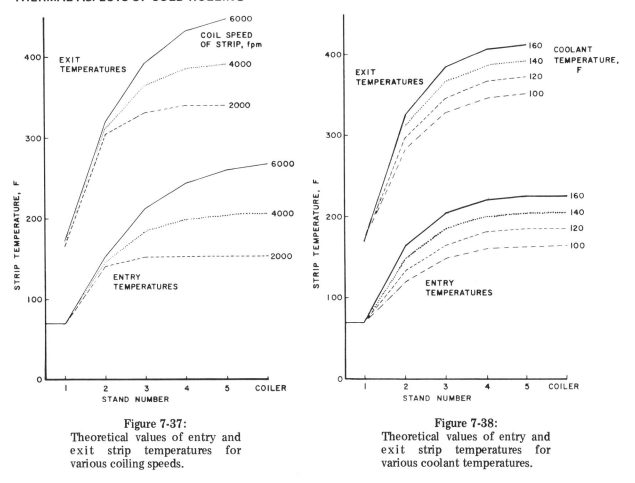

Figure 7-37:
Theoretical values of entry and exit strip temperatures for various coiling speeds.

Figure 7-38:
Theoretical values of entry and exit strip temperatures for various coolant temperatures.

temperature and interstand spacing is not linear, the importance of interstand spacing becoming less at the larger spacings.

Effect of reduction schedule — Because of the work hardening of the strip that occurs as the strip moves from stand to stand and because the efficiency of a mill stand decreases with increasing reduction, coiling temperatures can be changed by changing the reduction schedule for the various stands even though the same overall reduction is maintained. (To a first approximation, mill efficiency may be expressed as $[1 - (r/2)] \times 100$ per cent.) This is illustrated in Figure 7-40, and it will be seen that lower coiling temperatures are attained by a schedule of decreasing reductions. Moreover, the total input energy to the mill may be minimized by the adoption of a schedule of decreasing reductions.

In this examination of the three reduction schedules, no consideration has been given to such problems as the possible slippage of the work rolls or the rolling of strip with unacceptable shape. These problems may be solved by means other than that of rescheduling the reductions at the various stands.

From the foregoing theoretical considerations, if minimum coiling and rolling temperatures and maximum throughput are to be sought as desirable objectives in the operation of a tandem cold rolling mill, the following mill design features and operating practices should be adopted:

1. Work roll diameters should be as large as practicable.
2. The spacing between mill housings should be as large as possible, or means should be provided by rollers or other devices to increase the length of strip between stands.

3. Approximately the same drive power should be available at each mill stand.

4. The coolant temperature should be as low as permissible contingent upon the provision of effective lubrication.

5. The reduction schedule should be such that the largest percentage reduction is taken at the first stand and subsequent reductions are progressively reduced.

6. If the coolant is not to be applied between the last stand and the coiler, the most effective method of air cooling should be used at this location on the mill.

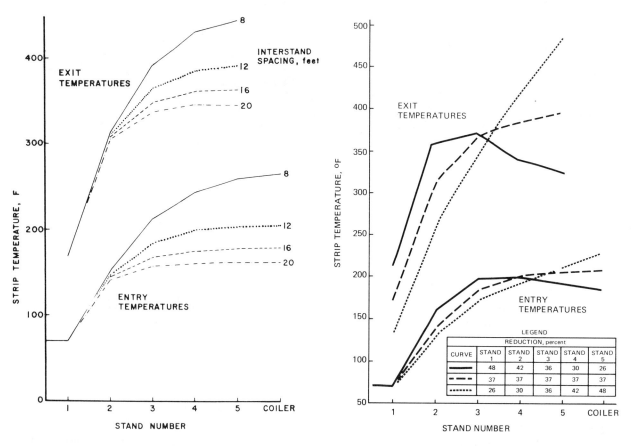

Figure 7-39:
Theoretical values of entry and exit strip temperatures for various interstand spacings.

Figure 7-40:
Theoretical values of entry and exit strip temperatures for various reduction schedules.

7-17 A Thermal Balance of the Cold Rolling Process (Single-Stand Reversing Mills)

An accurate heat balance of a cold rolling operation is difficult to develop on purely theoretical grounds but may become relatively simple to establish if the thermal calculations are based on experimental values. Kiss[=] has described a thermal balance made with respect to the operation of a 135 mm and 500 mm by 500 mm 4-high cold mill carried out by the Institute for Metallurgical Engineering and Metal Forming of the College for Heavy Industry in Hungary. Figure 7-41 shows the diagrammatic sketch of the mill and Figure 7-42 shows the location of the temperature sensors intended to monitor:

[=] E. Kiss, "Heating Conditions During Cold Rolling", Bander Bleche, 1969 10, March, pp. 161-165. (British Iron and Steel Industry Translation Service, BISI, 7309).

THERMAL ASPECTS OF COLD ROLLING

(1) the surface temperature of the work rolls
(2) the temperature of the emulsion in front of the spray nozzles and on return to the tank.
(3) the temperature of the strip entering and leaving the rolls.

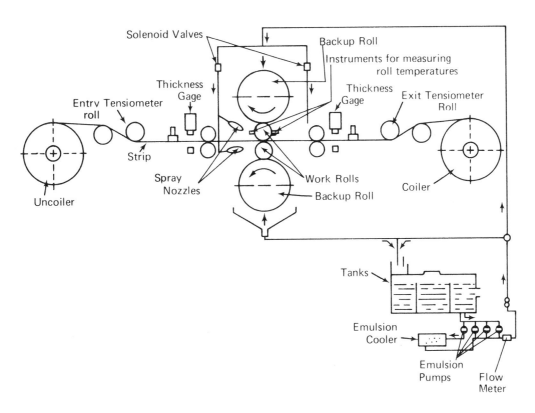

Figure 7-41: Instrumentation to determine rolling factors on a 4-high stand.

The surface temperature of the top work roll was measured with a special temperature-measuring head (Figure 7-43) and the thermocouple (iron-constantan) was located at a constant distance from the surface of the rolls. The temperature measured depended on this distance, and could be accurately determined using a correction factor previously established.

The measuring heads used for simultaneous measurements of strip surface temperature at the exit side of the rolls were of similar design (Figure 7-44).

As an example of the numerous investigations carried out, the temperatures measured during the rolling of dynamo strip under industrial conditions are given in Figure 7-45.

The temperature of the coolant emulsion (5% mineral oil and 95% water) in front of the spray nozzles was 36 to 37°C and in the emulsion tank the temperature fluctuated between 38 and 41.2°C. During the tests, room temperatures between 30.5 and 26.2°C were measured in the immediate environment of the mill.

It was demonstrated that there is a distinct relationship between the measured temperatures and rolling speed, as shown in Figure 7-46.

The equations used by Kiss for the heat balance are as follows:

The heat supplied per second q_{strip}^{in} by the strip entering the roll bite is given by

$$q_{strip}^{in} = cbh_1 \delta V_{in} t_{in}^{strip} \qquad (7\text{-}57)$$

where the following parameters pertain to the strip.

c = specific heat
b = width
h_1 = initial thickness
δ = specific gravity
V_{in} = entry speed into the roll bite
t_{in}^{strip} = temperature of strip entering the roll bite

Similarly, the heat lost per second q_{strip}^{out} by the strip leaving the roll bite may be computed from

$$q_{strip}^{out} = c h_2 b \delta V_{out} t_{out}^{strip} \qquad (7\text{-}58)$$

where h_2 = the final thickness of the strip,
V_{out} = its speed on exit from the mill, and
t_{out}^{strip} = its temperature after rolling

The heat equivalent q_{Def} of the deformation power N_{Def} of the deformation power N_{Def} is given by

$$q_{Def} = 0.24 \, N_{Def} \qquad (7\text{-}59)$$

Now the deformation power is the sum of the drive motor power N_{Mot} (less its losses N_{loss}) and the power associated with the strip tensions, $N_{strip\ tension}$. Thus

$$N_{Def} = N_{Mot} - N_{loss} + N_{strip\ tension} \qquad (7\text{-}60)$$

In turn, the tension power is related to the entry tension, $Z_{braking}$, and the exit tension, Z_{coiler}, and the mean strip speed V_m,

$$N_{strip\ tension} = (Z_{braking} - Z_{coiler}) V_m \times \frac{1}{102} \qquad (7\text{-}61)$$

The heat q_{em} lost through the application of the emulsion is

$$q_{em} = C_{em} \times V_{em} \times \delta_{em} (t_{em2} - t_{em1}) \qquad (7\text{-}62)$$

where the following parameters relate to the emulsion

C_{em} = specific heat
V_{em} = volumetric flow rate
δ_{em} = specific gravity
t_{em2} = final temperature
t_{em1} = initial temperature

The heat balance equation may then be written

$$q_{strip}^{in} + q_{Def} = q_{strip}^{out} + q_{em} + q_{other} \qquad (7\text{-}63)$$

where q_{other} relates to the rate of heat lost by radiation and other means.

Figure 7-47 is a schematic representation (Sankey diagrams) of heat distribution throughout the rolling process (4 passes) and Figure 7-48 illustrates how the heat content of the rolled strip after the fourth pass increases with mill speed.

From his data, Kiss demonstrated:

(1) The heat content of strip leaving the rolls increases proportionally with rolling speed.

(2) The quantity of heat removed with the emulsion (10 to 15%) becomes less as rolling speed increases.

THERMAL ASPECTS OF COLD ROLLING

(3) With smaller strip thickness the heat lost with the emulsion, and other heat losses, are relatively greater.

(4) Cooling of tightly coiled strip is very limited. However, some cooling occurs in the strip entering the rolls, between the pay-off coil and the temperature measuring point.

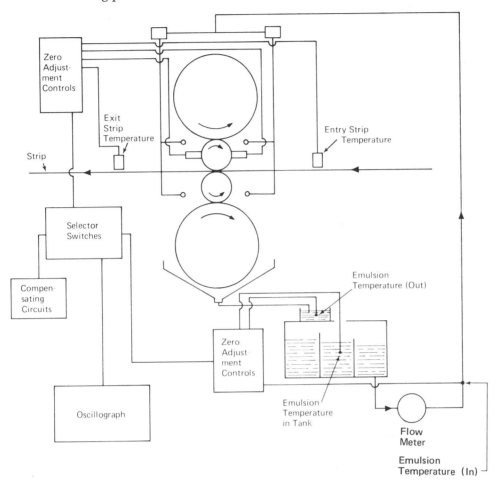

Figure 7-42: Arrangement of Temperature Monitoring Equipment.

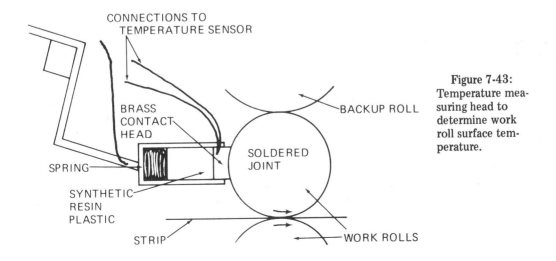

Figure 7-43: Temperature measuring head to determine work roll surface temperature.

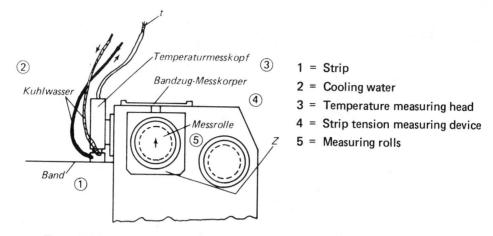

Figure 7-44: Principle of measurement of strip temperature and coiler tension.

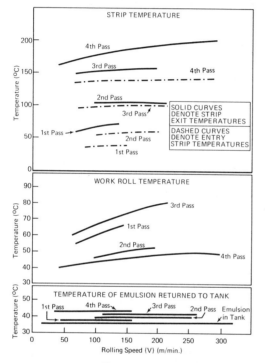

Figure 7-45:
Measured temperatures when finish rolling dynamo strip in four passes.

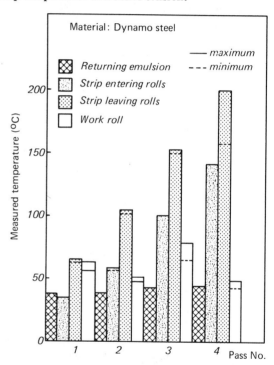

Figure 7-46:
Effect of rolling speed on measured temperatures.

A similar heat balance for a 20-high mill Type ZR 21-61 (see Section 2-11) has been developed by Reihlen.[◘] The general nature of the temperature change occurring in the strip during a rolling pass is sketched in Figure 7-49 and the strip temperature change during a succession of passes is shown in Figure 7-50. The latter illustration refers to the rolling of strip 1050-mm wide from a thickness of 3.6 down to 0.80 mm in 10 passes at various speeds ranging from 100 to 300 m/min.

Data pertaining to the surface temperatures of the rolls in the mill were also published by Reihlen and are reproduced in Figure 7-51. These data were obtained about 100 seconds after

---

[◘] H. Reihlen, "Abschatzung des Warmehaushaltes eines Zwanzigrollenwalzwerkes", Stahl und Eisen 92, 1972, Nr. 5, 2 Marz, pp. 204-209.

THERMAL ASPECTS OF COLD ROLLING

the work rolls had been inserted in the mill in the rolling of 900-mm wide strip at 80 m/min. The mill was operated at full power and the temperature of the strip at the coiler was about 130°C.

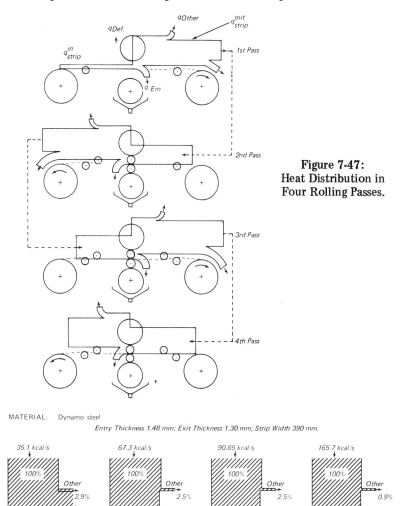

Figure 7-47: Heat Distribution in Four Rolling Passes.

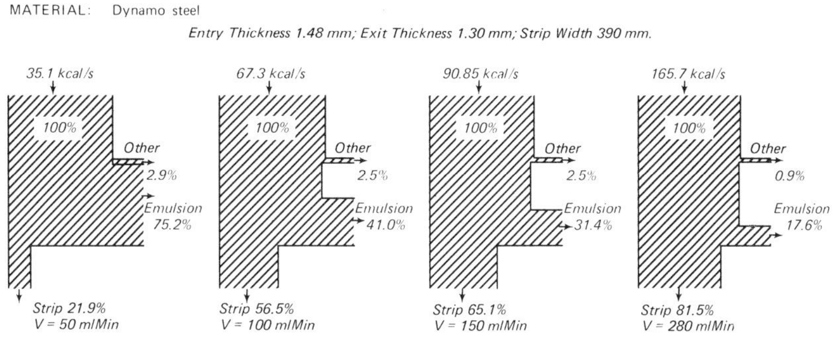

Figure 7-48: Sankey diagrams for the fourth pass when cold rolling dynamo strip (as a function of rolling speed).

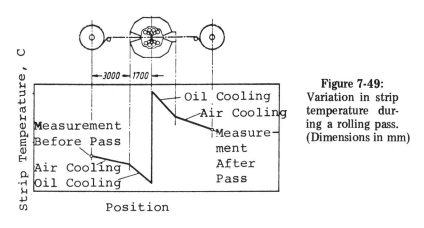

Figure 7-49: Variation in strip temperature during a rolling pass. (Dimensions in mm)

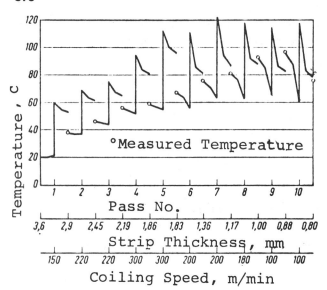

Figure 7-50:
The relationship of strip temperature to pass number for strip 1050 mm in width reduced from 3.6 to 0.80 mm.

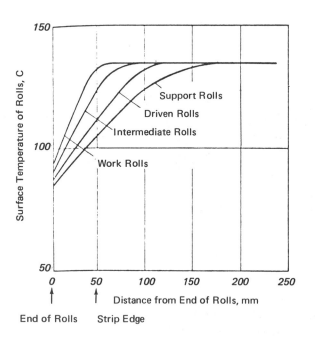

Figure 7-51:
Variation in surface temperature of mill rolls at their ends.

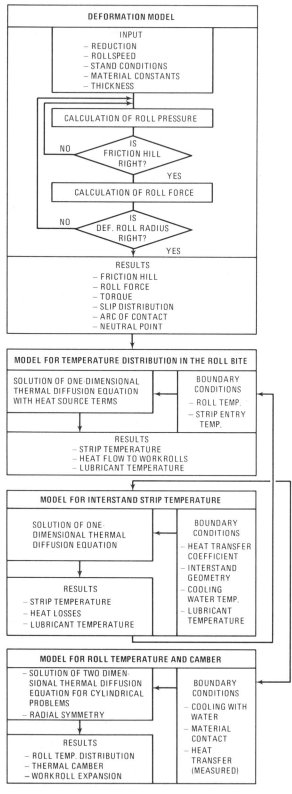

Figure 7-52:
Flow chart pertaining to mathematical models (Koot).

THERMAL ASPECTS OF COLD ROLLING

7-18 Thermal Models for Tandem Mills

Generally speaking, a thermal model for a tandem mill, such as might facilitate the prediction of roll and strip temperatures, can only be developed when more basic models relating to rolling forces and mill drive power requirements have been established. Thus the creation of a satisfactory mathematical model relating the thermal aspects of the tandem cold rolling process to all the other parameters involved becomes a formidable undertaking.

Although a rather elementary model was utilized to prepare the theoretical data concerning strip temperatures discussed in Section 7-16, a much more complex model for a tandem cold mill used for the rolling of tinplate has been developed by Koot.✦ The theoretical data from a strip deformation model constituted the input parameters to the thermal model as illustrated in Figure 7-52. The computer program for the latter was based upon the solution of Fourier's differential equation for heat diffusion in one direction. The integration was accomplished using the numerical implicit finite difference method with the grid system encompassing half the strip thickness, the oxide film on one surface of the strip, a lubricant layer and the shell of the work roll (see Figure 7-53). At points of the grid system inside the strip and in the lubricant layer, heat sources were introduced. The boundary conditions involved symmetry with respect to the center of the strip and a constant temperature for the bulk of the work roll.

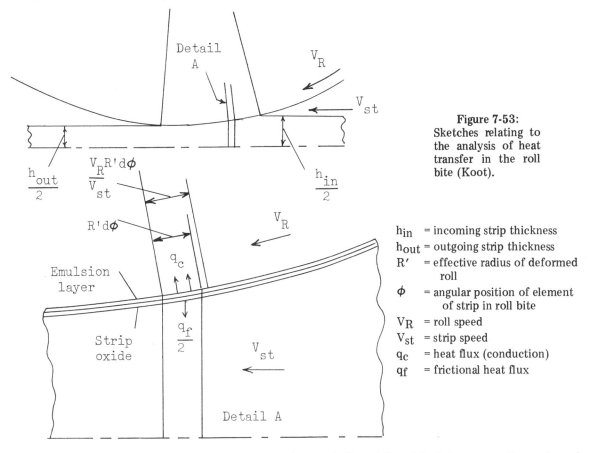

Figure 7-53: Sketches relating to the analysis of heat transfer in the roll bite (Koot).

h_{in} = incoming strip thickness
h_{out} = outgoing strip thickness
R' = effective radius of deformed roll
ϕ = angular position of element of strip in roll bite
V_R = roll speed
V_{st} = strip speed
q_c = heat flux (conduction)
q_f = frictional heat flux

The thermal model (which takes the forward slip of the strip into account) was based on the following assumptions:

(1) The fraction of deformation energy converted to heat was 85 per cent.

(2) Heat was considered to flow only in the thickness direction of the strip.

✦ L. W. Koot, "Process Design Criteria for the Cooling of a Cold Strip Mill", Paper presented at ISI Meeting, "Mathematical Process Models in Iron and Steelmaking", held at Amsterdam, Holland, 19-21, February, 1973.

(3) The deformation energy was homogeneously dissipated over the strip thickness.

(4) The energy associated with friction was dissipated in the lubricant layer.

(5) The temperature of the work roll shell was uniform at the entry plane of the roll bite.

(6) No heat generation resulted from elastic deformation in either the rolls or the strip.

Koot applied his model to the rolling of strip from 2.40 mm to 0.33 mm as indicated in Table 7-13. The most important data so obtained related to the heat flows and temperatures encountered in the various roll bites. As examples, Figure 7-54 shows the variation of heat flow in the bite of stand 4 and Figure 7-55, the corresponding surface temperatures of the strip and the roll, and the temperature of the lubricant film in the same roll bite.

Table 7-13

Data Relating to the Cold Rolling of Strip on a Tandem Mill (Koot)

Stand		1	2	3	4	5	Reel
Entry Thickness	[mm]	2.40	2.00	1.32	0.78	0.49	0.33
Entry tension	[kgf/mm^2]	1.0	10.5	9.3	11.8	13.0	4.0
Roll speed	[ft./min.]	600	1000	1700	2700	4000	
Roll force (Calculated)	[tf]	1090	1315	1680	1610	1590	
Roll power (Measured)	[kw]	300	2250	2670	3150	3680	

Strip width 811 mm.

Work roll diameters 535 — 590 mm

Yield stress 25.2 kgf/mm^2, strain hardening 70 kgf/mm^2 at 85% red.

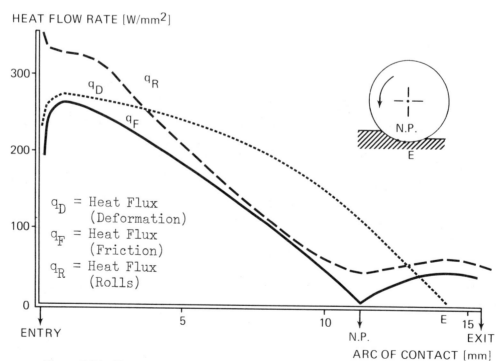

Figure 7-54: Heat flow distribution in the roll bite of the fourth stand (Koot).

(NP = Neutral Point; E = Exit)

THERMAL ASPECTS OF COLD ROLLING

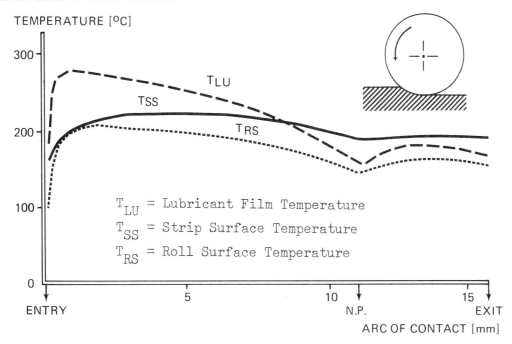

Figure 7-55: Temperature distributions in the roll bite of the fourth stand (Koot).
(NP = Neutral Point)

Strip cooling between stands had to be calculated to determine the strip entry temperatures at successive stands. Again the problem is one dimensional but different conditions exist at different regions between the stands.

On leaving a roll bite, cooling water flows over the top of the strip until the damming rolls are reached. However, on the underside of the strip, the heat loss is mainly by radiation. At the pinch rolls, it is assumed no heat is lost, but beyond these rolls, emulsion covers the strip until the turbulent puddle in front of the next roll bite is encountered. Figure 7-56 shows a typical variation in the temperature of the lubricant on the strip between stands.

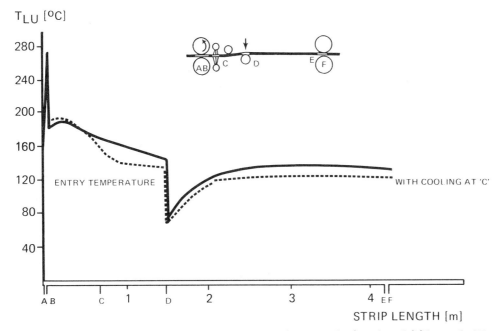

Figure 7-56: Temperature of the lubricant on the strip between the fourth and fifth stands (Koot).

Energy balances for the five mill stands are presented in Figure 7-57 together with a Sankey diagram relating to the fourth stand. Other thermal data including roll and strip temperatures, are listed in Table 7-14.

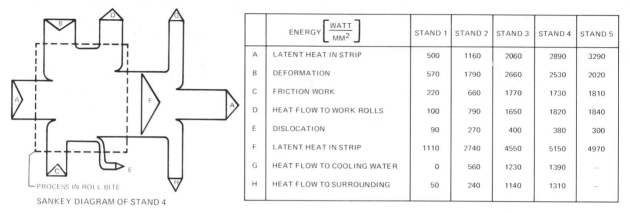

	ENERGY $\left[\dfrac{WATT}{MM^2}\right]$	STAND 1	STAND 2	STAND 3	STAND 4	STAND 5
A	LATENT HEAT IN STRIP	500	1160	2060	2890	3290
B	DEFORMATION	570	1790	2660	2530	2020
C	FRICTION WORK	220	660	1770	1730	1810
D	HEAT FLOW TO WORK ROLLS	100	790	1650	1820	1840
E	DISLOCATION	90	270	400	380	300
F	LATENT HEAT IN STRIP	1110	2740	4550	5150	4970
G	HEAT FLOW TO COOLING WATER	0	560	1230	1390	—
H	HEAT FLOW TO SURROUNDING	50	240	1140	1310	—

SANKEY DIAGRAM OF STAND 4

Figure 7-57: Sankey diagram for the fourth stand and tabulated thermal data for all stands (Koot).

Table 7-14
Energy and Temperature Data Relating to the Operation of a Five-Stand Mill (Koot)

Stand number		1	2	3	4	5
Deformation work	$\left[\dfrac{Watt}{mm}\right]$*	770	1790	2660	2530	2020
Driving power (measured)	$\left[\dfrac{Watt}{mm}\right]$	370	2780	3330	3880	4550
Heat flow to the work rolls	$\left[\dfrac{Watt}{mm}\right]$	160	790	1650	1820	1840
Proportionally to deformation work	[%]	20.8	44.2	62.0	72.0	91.0
Stationary temp. level of the work roll	[°C]	55	63	61	62	69
Flow rate of cooling water	[m³/h]	74	126	200	305	213
Strip temps. at entry and exit of the roll bite	[°C]	20/47	47/11	83/185	117/208	133/202

* All energy data are related values per mm strip width.

7-19 Minimum Work-Roll Diameter to Provide Adequate Cooling

In the design of cold-rolling mills, the choice of work-roll diameter is one of paramount importance.[ǂ] The larger the diameter of the rolls, the greater is the rolling force required to achieve a given rolling operation. Conversely, the smaller the work rolls, the more difficult they are to cool, even at reduced mill speeds. Yet larger rolls exhibit greater rigidity and, in four-high mills, better strip shape may be more easily attained by their use. On the other hand, smaller rolls, depending for their operation on higher frictional levels in the roll bite, are less sensitive to the effects of changes in the rolling lubricant. With these and other considerations in mind, the mill designer faces no simple task in establishing an optimum diameter.

ǂ W. L. Roberts, "Choice of Work Roll Diameter in Cold Rolling Mill Design", AISI Regional Meetings, 1969, pp. 19-41.

THERMAL ASPECTS OF COLD ROLLING

Before the optimum work-roll diameter can be computed for any given rolling situation, consideration must be given to a number of process parameters including (a) the type of rolling lubrication system to be used on the mill, (b) the frictional characteristics of the lubricant to be used, (c) the design of the roll cooling system and (d) the temperature and effectiveness of the coolant.

As discussed in Section 6-8, there are basically three types of lubrication systems:

a. Direct — the neat oil applied as a fine spray or mist

b. Direct — the oil applied as a component of an oil-in-water emulsion

c. Recirculated — an oil-in-water emulsion applied copiously to the roll bite and rolls as a lubricant and coolant; the effluent being reapplied to the strip and rolls.

With respect to the rolling lubricants themselves, dozens of different "rolling oils", as they are generally termed, are commercially available. Because their frictional behavior ranges over wide limits, it is necessary to select a lubricant that will be used on the mill and to establish, by direct experiment if necessary, its frictional characteristics. These frictional characteristics, K_1 and K_2, enable the effective coefficient of friction (μ) on the roll bite to be computed for any rolling condition involving the same type of strip and roll surfaces. (See Section 6-31.)

Since, in many instances, the speeds of mills are limited by thermal effects associated with the work rolls, the cooling system, the temperature and the effectiveness of the coolant must all be specified. In a direct application system, the coolant (water) may be applied separately to each work roll on the exit side of the mill. On the other hand, with recirculation systems, the rolling solution may be applied to the work rolls on the entry or both sides of the mill.

As discussed in Section 7-12, the heat transfer from the roll surface to the coolant depends basically on the difference between the average roll surface and coolant temperatures ($T_R - T_C$). Since, as will be seen later, it is essential to ensure that the average roll temperature (T_R) does not exceed a certain value, in calculating heat removal rates from the rolls, it is also necessary to specify the coolant temperature (T_C).

The effectiveness of the coolant, or the heat transfer coefficient (h) associated with the application of the coolant, must also be known. For the application of water on steel surfaces, the values of h are fairly well established for various types of sprays and coolant flow rates. However, it must be remembered that the rolls will usually be coated with a thin film of oil, thereby lessening the coefficient of heat transfer. Similarly, when an emulsion or recirculated rolling solution is used, the h values are also less than those of pure water on clean steel surfaces. (See Section 7-12.)

As seen in Section 7-3, in the cold rolling operation, if r is the reduction given to the strip a fraction approximately $(1 - \frac{r}{2})$ of the energy feed to the work rolls is utilized to deform the strip and the remaining fraction (r/2) is dissipated by friction at the roll strip interfaces.

Consider a portion of the strip 1-inch wide with a compressive yield strength S_c reduced to a thickness t_f being rolled at a speed of V feet per minute with the frictional conditions in the roll bite such that zero slip occurs. Since the energy of deformation of the strip is $S_c \ln \frac{1}{(1-r)}$ inch-pounds per cubic inch, the power expended in deformation of the strip is $V t_f S_c \ln \frac{1}{(1-r)}$ foot-pounds per minute per inch of width or $\frac{V t_f S_c}{778} \left(\ln \frac{1}{(1-r)} \right)$ Btu/minute/inch of width. the power fed to the rolls must correspond to $\frac{V t_f S_c}{778 (1-r/2)} \left(\ln \frac{1}{(1-r)} \right)$ Btu/minute/inch of width. Assuming an equal distribution of frictional energy between the strip and the roll at each interface, the power dissipated as friction on each roll surface is $\frac{V t_f S_c}{3112 (2-r)} \left(\ln \frac{1}{(1-r)} \right)$ Btu/minute/inch of width. For the work rolls to assume a temperature equilibrium, thermal energy must be removed from the rolls at a rate corresponding to this power dissipation. This thermal balance is illustrated in Figure 7-58, and it assumes that heat removal from the roll is accomplished only by the coolant.

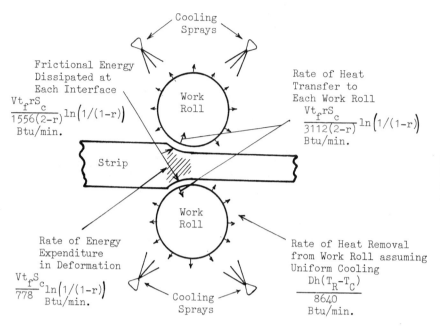

Figure 7-58: Sketch illustrating rate of heat transfer to and from each work roll per unit width of strip.

For a section of each work roll one inch long, the peripheral area of the surface is $\frac{\pi D}{144}$ square feet. If the heat-transfer coefficient is h Btu/hr-ft²-F, the average roll temperature T_R, and the coolant temperature T_C, (both measured in °F) and the coolant is applied essentially to the whole peripheral surface of the sections, the rate of heat removal from the section is $\frac{\pi Dh(T_R - T_C)}{8640}$ Btu/inch/minute. (For cooling systems in which the coolant is applied over only a portion of the roll surface, this expression must be modified accordingly.)

For thermal equilibrium

$$\frac{Vt_f rS_c}{3112(2-r)} \ln\left(\frac{1}{1-r}\right) = \frac{Dh(T_R - T_C)}{8640} \qquad (7\text{-}64)$$

Experience has shown that the average work roll temperature T_R should not exceed a temperature approximately equal to the boiling point of water (212°F). This is presumably not due to any change in the heat transfer coefficient at this temperature but rather to a relatively sharp change in the coefficient of friction in the vicinity of this temperature.

Under these circumstances, if T_R must not exceed 212°F, equation 7-64 may be rewritten

$$D \geq \frac{2.78 \, Vt_f rS_c}{(2-r)\pi h (212-T_C)} \ln\left(\frac{1}{1-r}\right) \qquad (7\text{-}65)$$

This expression showing the linear relationship between the minimum roll diameter and the rolling speed demonstrates the desirability of maximizing the heat transfer coefficient h and minimizing the coolant temperature T_C.

7-20 Spray Nozzles and Header Design

To apply coolants and lubricants to the mill rolls, manifolds or headers equipped with a number of spray nozzles are generally used. In the case of lubricant application, the supply should

THERMAL ASPECTS OF COLD ROLLING

be virtually uniform across the roll face corresponding to the width of the strip. In the case of the coolant, its distribution should be such that a thermal crown is developed that will ensure that the strip rolled is flat. Uniform roll temperatures are not necessarily desirable in achieving product flatness and accordingly headers are often sectionalized (as discussed later) so that by adjusting the coolant flow rates through the various sections, the thermal crowns of the mill rolls may be appropriately controlled.

Numerous types and sizes of spray nozzles are commercially available. On cold mills, nozzles are generally used which provide flat or fan-shaped spray patterns as illustrated in Figure 7-59. The capacity or flow rate of such nozzles is proportional to the square root of the fluid pressure and may range from a fraction of a gallon per minute at pressures of 5 psi for smaller nozzles to rates considerably in excess of 100 gpm for large nozzles operating at pressures around 500 psi.

Spray angle is another characteristic defining a spray nozzle. The angle of the spray increases with fluid pressure and, for certain nozzles operating at maximum pressures, may exceed 150°F.

The size of the droplets in the spray depends upon such factors as nozzle design, capacity, spray angle and pressure. Spray particle sizes become coarser as (1) nozzle capacities increase, (2) spray angles become narrower and (3) pressures decrease. Where fine atomization or misting is required, as for example in the application of neat lubricant, compressed air may be used to break up the liquid in a nozzle such as that shown in Figure 7-60. Certain nozzles of this type feature needle-type valves which not only maintain the nozzles in an internally clean condition, but also permit them to be very rapidly operated (up to 180 "on and off" cycles per minute). The cross section of a typical nozzle of this type is shown in Figure 7-61.

Figure 7-59: Flat spray.

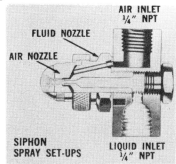

Figure 7-60: Cross section of pneumatic atomizing nozzle.

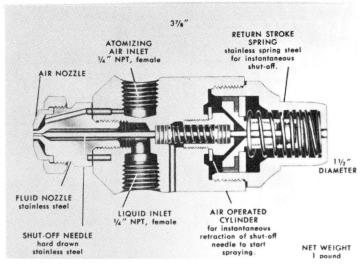

Figure 7-61: Cross section of automatic pneumatic atomizing spray nozzle.

To cool the entire face length of the rolls in a mill stand, the nozzles are fitted into manifolds or headers, an example of which is illustrated in Figure 7-62 and Figure 7-63. In the simplest case, the nozzles are spaced a few inches apart and the manifold is not sectionalized, and the coolant is fed simply into one end of the manifold. Where individual or groups of spray nozzles are to be controlled, a return feed line may be coupled to the manifold, one end being the inlet and the other the outlet.

Figure 7-62:
Spray cooling header (Schaming Industries).

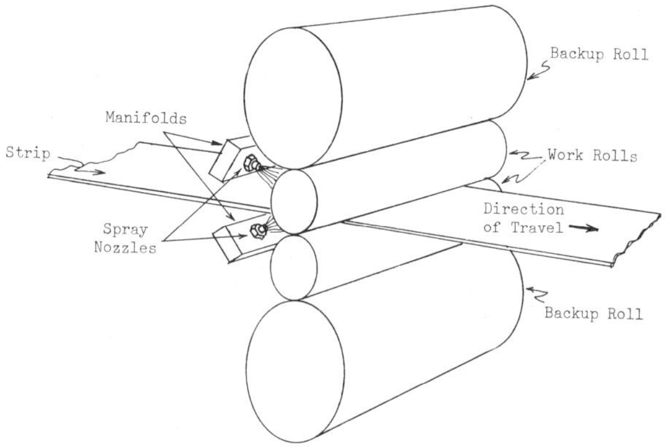

Figure 7-63:
Sketch of spray headers in roll cooling system.

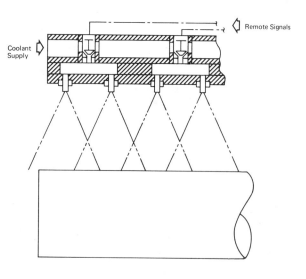

Figure 7-64:
Sketch showing sectionalized header and control of nozzle groups. (Schaming Industries.)

THERMAL ASPECTS OF COLD ROLLING

Such individualized control of spray nozzles, used primarily to achieve the desired thermal crowning of the mill rolls, may be accomplished by electrically or pneumatically operated valves, an example of the latter being shown diagrammatically in Figure 7-64.

A relatively simple cooling system for the application of a recirculated solution to the rolls of a sheet mill is sketched in Figure 7-65. Where a direct application lubrication system is to be used on a tandem sheet mill, the various headers for applying oil, water, air and a cleaning solution may be located on the mill stands as indicated in Figure 7-66.

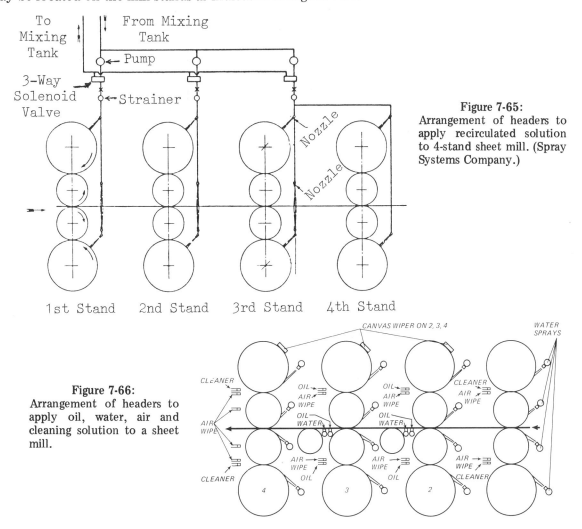

Figure 7-65: Arrangement of headers to apply recirculated solution to 4-stand sheet mill. (Spray Systems Company.)

Figure 7-66: Arrangement of headers to apply oil, water, air and cleaning solution to a sheet mill.

7-21 Direct Application Lubrication Systems

A direct application lubrication system for a 5-stand tandem cold mill used for tinplate production is shown in Figure 7-67.* In this system, the cooling medium (plain water) is applied to the rolls and strip, while the lubricant (palm oil or palm oil substitute) is applied separately. Filtered water is applied at pressures ranging from 150 to 250 psi, usually on the entry and delivery side of each stand.

Since the lubricant is generally a solid at room temperatures, it is necessary to heat the oil to a temperature between 130 and 150°F in a holding reservoir so that it can be pumped and

---

* J. P. Wettach, "The Design of Coolant Systems as Affected by Various Cold Rolled Products", Iron and Steel Engineer Year Book, 1961, pp. 984-990.

properly mixed with a small quantity of water. This initial mixing of the oil with water varies from one part oil and three parts water to one part oil and ten parts water. The ratioing of the oil and water is accomplished by a proportioning pump supplying the oil and hot water to a mixing tank prior to the application of the emulsion to the mill. In some installations, however, the oil and water are metered into the suction of a small centrifugal pump located at each mill stand.

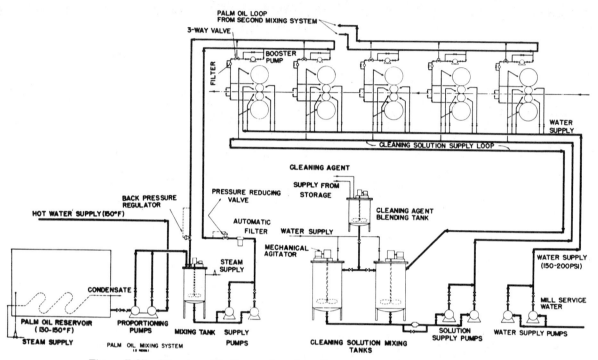

Figure 7-67: Direct application system for a 5-stand tandem cold mill. (Wettach.)

The schematic diagram in Figure 7-67 shows only one application system but tandem tin mills usually require two. This arrangement permits an emulsion richer in oil to be applied to the later stands. Also, it should be noted that the lubricant may not be applied to the first stand as the incoming strip may already be coated with pickle oil.

To maintain the temperature of the oil-water mixture, it is necessary to loop the oil application so that, whether the sprays are being used or not, the solution is continually flowing through the lines back to the mixing tank where it is reheated and continuously agitated. Accordingly, the recirculation capacity is rated above the normal mill requirements and a booster pump is located at the mill to keep the emulsion in motion in the branch lines to the individual mill stands. Remotely operated 3-way, 2 port valves are used to direct the flow of oil to the sprays or to the return header.

As in the case of the water spray headers, the location of the oil spray headers depends on the mill operators. In Figure 7-67, the oil is shown being applied between the backup rolls and the work rolls, the latter carrying it around to the bite. In addition, application is made at the delivery side of each stand to permit the oil to cover the strip uniformly before the next stand is reached.

Automatic edge-type filtration of the emulsion takes place at the discharge of the main supply pumps and another filtering element is located at each individual stand. This arrangement, in order to function properly, must have the necessary pressure reducing valves and back pressure regulators as shown in Figure 7-67.

Corrosion resistant piping is used in the oil loop to minimize the danger of scale developing in the lines and spray nozzles. Provision is also made to steam flush the loop during periods of mill shutdown to eliminate the possibility of clogging the lines with solidified oil.

THERMAL ASPECTS OF COLD ROLLING

Also incorporated into the system is a method of cleaning the strip during the rolling of black plate. In this case, various cleaning agents are mixed in a blending tank before entering other tanks where water and other detergents are mixed. As applied to the rolls and strip, the solution would consist of one part cleaning agent to six to ten parts water.

Other direct application systems have been described by Dowler.[∞] On the five-stand tin mill at the Yorkville plant of Wheeling-Pittsburgh Steel Corporation, the oil is applied with water in a ratio of 4 to 1, and on a three-stand mill, used for the double reduction of tinplate, in a ratio of 20 to 1.

Figure 7-68 shows the system for mixing and supplying oil to this particular five-stand mill. The oil is pumped from field storage where it is maintained at 135°F, to a tank in the tandem mill basement, where the oil temperature is kept at 160° to 165°F. From this tank the oil is pumped by one of four proportioning pumps to a mixing tank; the other three proportioning pumps are for pumping hot water (180°F) to the mixing tank. Oil and water are mechanically mixed and maintained at a temperature of 180°F in this tank. From the mixing tank the oil is circulated through a loop to the mill and back to prevent water and oil separation and at each mill stand there is a diverting valve to supply oil to any given stand as required.

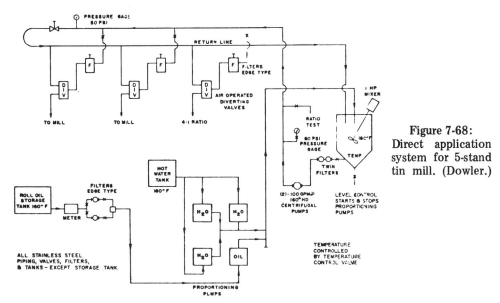

Figure 7-68: Direct application system for 5-stand tin mill. (Dowler.)

The pickler oil acts as the lubricant at stand 1 and oil is used on the other stands. On Stand 3, the oil is applied to the top and bottom work rolls. Between stands 3 and 4 the oil is applied to the top and bottom of the strip, and on stand 4 the oil is applied between the top and bottom backup and work rolls. Oil is also applied to the top and bottom of the strip on the delivery of stand 4 and again to the top of the strip and the bite of the bottom work roll and strip at the entry of stand 5. Oil is applied on the delivery side of stand 5 between the top backup and work roll.

A similar lubrication system for the three-stand, double cold reduction mill is shown in Figure 7-69.

7-22 Recirculating Lubrication Systems

In recirculating systems, the oil-water emulsion functions both as a lubricant and a coolant and is recycled on the mill for continued use. Concentrations of the aqueous emulsions

---

[∞] L. K. Dowler, "Rolling Oil Practice for Tandem Cold Mills and Double Reduced Tin Plate — Characteristics of a Lubricating System", Iron and Steel Engineer Year Book, 1968, pp. 751-754.

generally range from 3 to 20 per cent and they contain various oils and small quantities of emulsifiers. The emulsion is exposed to tramp oil contamination and it is therefore desirable to use an emulsifying agent which does not allow the tramp oil to go into solution. Under these circumstances, the tramp oil rises to the top of a solution storage tank and may be conveniently removed by skimming devices. (The oil removed by skimming may then be centrifuged and used for other purposes.)

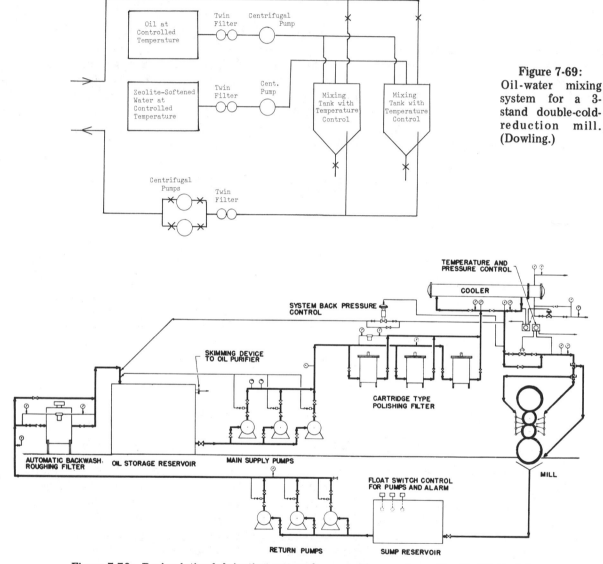

Figure 7-69: Oil-water mixing system for a 3-stand double-cold-reduction mill. (Dowling.)

Figure 7-70: Recirculating lubrication system for use with reversing cold mill. (Wettach.)

Figure 7-70 shows a recirculation system for use with a reversing mill used for cold rolling silicon, carbon and stainless steel strip.[*] In this arrangement, the coolant from the mill pan returns a sump tank before being pumped to the main storage reservoir. Using this method rather than the usual gravity return to the main reservoir, it is possible to place a roughing filter in the main return to the reservoir. This filter, of the automatic continuously back-washing type, will remove contamination down to 0.0025-inch in size. This eliminates a large portion of the heavier

---

[*] J. P. Wettach, "The Design of Coolant Systems as Affected by Various Cold Rolled Products", Iron and Steel Engineer Year Book, 1961, pp. 984-990.

THERMAL ASPECTS OF COLD ROLLING

particles that would normally settle out in the reservoir, and lengthens the time between reservoir cleaning operations. With major contamination removed from the oil before it returns to the reservoir, the final "polishing" filtration is accomplished with the use of cartridge-type filters removing particles down to ten microns in size by the full-flow method before application on the mill.

Since the roughing filter is completely automatic and removes the bulk of the contamination, this lengthens the time between cartridge changes on the polishing filters.

In order to maintain a constant spray pattern on the rolls and strip, it is necessary to have constant pressure and temperature at the mill. All coolant systems should be designed with this in mind. As shown in the installation in Figure 7-70, an automatic pressure control is placed as close to the mill as possible to compensate for pressure drop in the filters, heat exchanger and piping. If the pressure desired at the sprays is 100 psi, the system would be designed with a pump discharge pressure of 175 psi; thus, with a 15 psi drop through the cooler, a 30 psi drop through the filters and a 10 psi drop through the pipe fittings and valves, the pressure drop across the pressure control valve would still be 20 psi, giving the desired 100 psi pressure at the mill. However, since all these pressure drops vary from time to time, especially in the case of the filters, it is necessary to have a pressure controller to maintain a constant spray pattern at the mill.

Since the mill shown in Figure 7-70 is a reversing mill, the operator must have the option of diverting the flow from one side of the mill to the other and also diverting the entire flow to the mill pan for threading or changing rolls.

The heat exchanger used to cool the oil is usually of the shell and tube type with the oil passing through the shell side and the cooling water through the tubes. The cooler is designed to remove 60 to 75 per cent of the heat load generated in the rolling operation based on the horsepower of the mill drive. The remaining heat is considered as being dissipated by such factors as radiation and transmission losses. The coolant system capacity in gpm is also determined on this basis. Thus, a single-stand mill with a drive motor of 1300 hp (figuring 75 per cent of the horsepower for heat load) can develop a heat load of 2,486,250 Btu per hr. The capacity of the system capacity would be: gpm = Btu per hr./(temperature differential X coolant constant). This coolant constant is a factor of the weight of the liquid times the specific heat times the specific gravity. In the above example, using mineral oil as a coolant, the system capacity would be 1200 gpm with a 10°F temperature drop across the cooler or 600 gpm with a 20°F temperature drop across the cooler. The circulating capacity of the same system using soluble oil and water would be 525 gpm with a 10°F drop or 262 gpm with a 20°F drop.

According to Wettach, this method of determining coolant system capacity is generally accepted although there are many other factors involving capacity depending upon the product being rolled, the type of mill and the general experience of the mill operators.

There are numerous installations of dual coolant systems of the same general design as described above being used on a single mill, thus giving greater freedom in use of coolant for the products being rolled. However, on mineral oil systems exposed to the possibilities of oxidation of the oil, it would be necessary to add additional filtering of the Fuller's earth type to absorb the undesirable oxidation contaminants.

Figure 7-71 shows the recirculating lubrication system for a 6-stand tandem mill used for the rolling of tinplate. The rolling solution, at a temperature in the range 130-150°F, is kept constantly agitated while in use. Tramp oils are skimmed from the top of the reservoir by the addition of water up to the overflow level.

Trash screens are located in the reservoir at the return connection from the mill to remove large particles that find their way into the mill pan. Finer screening takes place at the pump suction strainers. Strainers of the double-basket type, which permit cleaning of one basket while the flow is diverted through the opposing basket, are often used. Straining oil with a basket mesh down to 0.02 to 0.015-inch spacing will protect the centrifugal pumps.

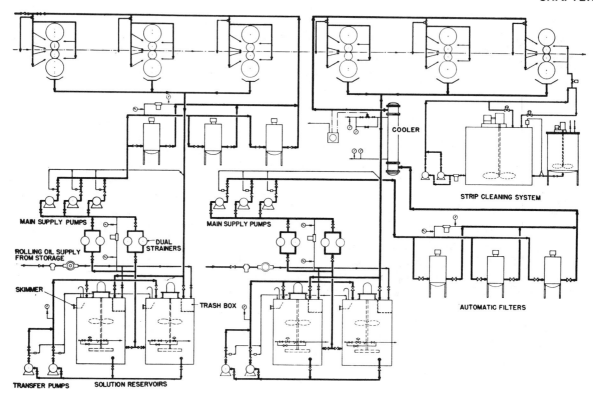

Figure 7-71: Recirculating lubrication system for a 6-stand tin mill.

The main and spare pumps of a typical installation are shown in Figure 7-72. Relief orifices are employed to permit sufficient quantity of coolant solution to pass through the pump even though they might be running against dead shutoff, eliminating the danger of overheating and damaging the pump or impeller.

The main filtering units are of the continuously cleaned edge type with 0.012-inch spacing. It is desirable to have main and spare filtering units to insure continuous filtration of the coolant during maintenance periods on the filter.

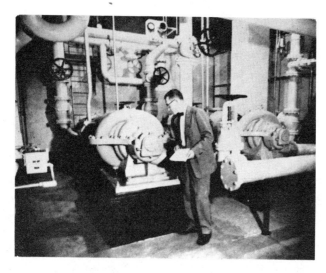

Figure 7-72: Large centrifugal pumps used to recirculate rolling solution on a 6-stand cold mill.

In all coolant systems, it is necessary to have differential pressure switches across strainers and filters to warn of high-differential pressure so that necessary corrections can be made.

THERMAL ASPECTS OF COLD ROLLING

Many other alarms, such as high and low level in the reservoir, high and low temperature and low pressure are desirable.

The method of diverting the flow of solution at the mill varies. In most coolant systems, large capacity centrifugal pumps are used, and if the system is properly designed it is possible to shut off the individual stands and run the pumps against tight shutoff. The method of coolant supply control to individual stands is frequently a remotely-operated 3-way valve, used to dump the entire flow of oil to the mill pan. This has a tendency to flush the mill pan and keep it free of contamination or buildup.

7-23 Filters for Recirculating Lubrication Systems

Although reference to the direct application systems discussed in Section 7-21 reveals their utilization of filters, in recirculating systems filtration becomes considerably more important on account of the contamination of the solution by tramp oils and by the detritus originating from the wear of the surfaces of the rolls and the strip.[Δ] Unless this contamination is effectively removed, not only will the roll and workpiece surfaces deteriorate but spray nozzles will become clogged, pumps will experience severe wear and excessive sludge buildups will occur in the reservoirs.

Oil or water can be filtered relatively easily without the presence of the other, but a mixture of oil and water renders the filtration process quite difficult. A small percentage of water in oil or a small percentage of oil in water will stop the flow. A wide open strainer is usually necessary to strain out (not filter) the contaminants where a mixture of oil and water is present.

There are several ways of applying filters in actual practices. These methods may be classified as: (1) full flow, (2) bypass, (3) shunt and (4) circulation by means of an auxiliary pump and motor. A full flow filter will actually filter 100 per cent of the oil passing to the lubricated part each time the oil passes to that part. A by-pass or bleeder system will bleed off a small percentage of the oil from the pressure line of the lubrication system, pass it through the filter and discharge it back into the reservoir. A shunt-type arrangement allows part of the oil to pass through the filter and then discharge back into the main oil stream so that all of the filtered oil passes directly to the lubricated part. In this case, a means of controlling the pressure, such as a pressure control valve, must be placed in the main lubricating oil line so as to cause a portion of the oil to pass through the filter. The fourth method involves a separate pump and motor to pick up the oil from the oil reservoir, pump it through the filter and discharge it back into the oil reservoir. By this manner of operation the filter may operate entirely independent from the rest of the system. The filter may be run when the rest of the system is not even operating if desired. The filter can be shut off for changing refills, or merely to stand idle if desired.

Filtration may be carried out on a continuous basis or intermittently. Continuous filtration is, however, the most efficient and economical way to operate a filter and it is preferred over intermittent filtration or batch filtration because the contaminants can be removed virtually as they enter the oil.

Particle size removal by the filter is influenced by types of fluid, flow rates, pressure drop, operating viscosity, and type of particle involved. Particle size removal determinations are, however, quite difficult to establish even with the same operator.

Filters can be divided into many types or classifications. These classifications may include depth type, extended area or surface type, edge type, screen type, bulk type, bag type, adsorbent, absorbent, etc. Centrifuge and magnetic separators deserve mention but they are not true filters. Filters are usually designed for one or more specific purposes, but none of them can be applied equally well to all applications.

Δ J. R. McCoy, "Filtration in the Iron and Steel Industry", Iron and Steel Engineer Year Book, 1960, pp. 986-992.

Depth type filters require the oil to flow through a depth of the filter medium. The depth of the filter medium may range all the way from approximately one up to approximately 15 inches. The sediment is distributed throughout the depth of the filter medium. Depth type filters can be made of many different types of filter media, such as fuller's earth,^ cotton, wool, plant fibers or mixtures of these. Depth-type cellulose filters can be made from bulk materials, batted and garneted materials or from rovings. Specially processed rovings can be wound into a cylindrical form into many densities to give the dirt holding capacity and particle size removal desired.

Surface type filters remove their dirt actually on the surface of the filter. Such filters depend on a large surface area for their dirt holding capacity and the larger the surface area, the greater will be their capacity. These filters are usually made of paper or cloth, the filter medium being arranged in such a manner to provide a large surface area in a relatively small space.

Large lubrication systems are usually used with tandem cold mills. Where water is involved, two large settling tanks, ranging up to 10,000 gallons or more each are provided as reservoirs. These tanks should be properly heated and equipped with baffles to facilitate settling of water and scale. Plain unbaffled tanks will allow circulating oil to channel right through the tank and allow relatively little retention time. Accordingly, only a small amount of the oil may actually be in use and circulating, and the rest of the oil may remain stagnant. In such systems, very little settling of water and dirt will occur. In properly designed tanks all of the oil will move slowly and separation and settling of some of the water and solids will result. As the oil is used, water will gradually accumulate and, after a period of time, the water content of the oil may become so great that the oil is no longer fit for service. The standby tank may then be brought into use. Oil in the resting tank may then be heated to 180 to 200°F. This thins the oil, breaks simple emulsions and allows the water to settle and be drained off.

Heat alone is very good in breaking emulsions but, often, stubborn emulsions form and emulsion breakers are necessary. These emulsion breakers are added in the amount of approximately 0.1 per cent by volume. They must be thoroughly mixed with the oil being treated. In combination with heat they are very effective in breaking emulsions and allowing the water to separate and settle. Emulsion breakers are also quite effective in preventing emulsions from forming in the "in-process" oil.

Additional water removing or separating aids are recommended, such as centrifuges, separating screens and filters. As has already been pointed out, these filters operate best when used continuously. It is the practice in some plants to use centrifuges only after the water content of the oil has built up to a rather high level; sometimes, only when the oil is on standby and settling.

Filters used on the lubricating oils of rolling mills are usually of the metal screen, metal edge magnetic bag type or centrifuges. Highly efficient filters, such as are used on hydraulic oils, normally are not used because of the difficulties in filtering high viscosity oils and mixtures of oil and water.

As the oil is used in the lubricating system, it will gradually deteriorate and form products of oxidation, such as acids, asphaltenes, etc. These products of oxidation, as well as finely divided metal and water, act as catalysts to accelerate oil breakdown. They also act as emulsifying agents to emulsify water into the oil and prevent it from separating. In addition, acids may attack bearings and cause pitting.

If the quality of the finished metal is to be improved, the condition of the roll oil must also be improved and be free of contaminants. Some cold rolling processes may render the oil so contaminated that anything better than magnetic separation would be uneconomical. Magnetic separators will take out the large particles and some of the fine particles, but they are not highly efficient. The more viscous the oil, the less efficient the magnetic separator becomes. When high quality processes using mineral oils are involved, filtration is highly desirable and, for maximum

---

^ Fuller's earth is a clay-like substance capable of absorbing (by molecular attraction) oxidation products in the lubricant.

THERMAL ASPECTS OF COLD ROLLING

protection, full flow filters are used. These filters should be capable of removing practically everything down to 20 microns on each pass and they should provide high flow rates at low pressure drops. Diatomaceous earth** precoat filters may be utilized. To maintain the oil in good condition chemically, fuller's earth filters may be used with the oil reservoirs to remove oil deterioration products such as acids, asphaltenes, etc. Fuller's earth filters will also remove undesirable additives which may leak into the oil with bearing lubricants.

Specialized mills, which use the roll oil as a bearing lubricant, require additional filtration facilities to further filter the portion of oil that passes through the bearings.

The flatbed filter, one of the more important and widely used filters, is shown photographically in Figure 7-73 and diagrammatically in Figure 7-74.✪

Figure 7-73:
General view of the flatbed pressure filter, complete with control panel, automatic valves and supervisory instruments.

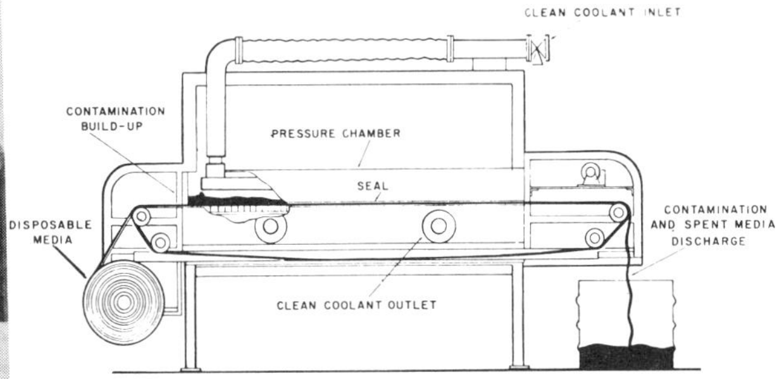

Figure 7-74:
Cross section of a flatbed pressure filter.

In Figure 7-74, a disposable media, intended for filtration of particle sizes from 50 down to 5 microns, is shown. As dirt builds up on the cloth or nonwoven media, it forms a tighter path through which the coolant must pass, thus aiding in the filtration process. When sufficient dirt or contamination has built up on the media to cause a certain pressure drop (e.g., 15 to 20 psi), the actuation of a pressure switch causes the media to index. When this occurs, the filter inlet valve closes, an air valve opens, forcing the remaining coolant through the media, the seal lifts, a new section of the media indexes into the filtration chamber, the seal is seated and filtration resumes. On stable, soluble oil coolants, the filter will generally index once every two to five hours, but on quasi stable emulsions, the cycles will be shorter depending on the type of mill. It should be noted that during indexing, the solution is withdrawn from the clean reservoir so that its supply to the mill is not interrupted.

Where soluble oils are used as roll coolants, such as in the processing of stainless steel, a recirculation system such as that shown in Figure 3-39 may be used. The removal of tramp oils from such a recirculation system is accomplished automatically by the mechanical skimming device

---

**Diatomaceous earth is comprised basically of the siliceous remains of microscopic marine plants known as diatoms. Particles of diatomite are microscopic in size, ranging in some grades down into the colloidal size range. When precoated on a filter screen, the particles have a tendency to interlock, forming a very strong filtering cake with microscopic voids. This, coupled with the porosity of the diatomaceous earth itself, makes it one of the finest filtering mediums available.
Being chemically inert, the filter cake removes only foreign particles in the oil and falls in the classification of mechanical filtration (having no chemical reaction on the oil). Most additives used in mineral oils today will not be removed by this type of filtration.

✪J. P. Wettach, "Recent Advancement in Filtration of Roll Coolants for Cold Rolling Operations", Iron and Steel Engineer Year Book, 1966, pp. 613-619.

shown in Figure 7-75. This unit operates on a continuous basis while the mill is in use, thereby avoiding the possibility of contaminants settling to the bottom of the reservoir during mill shutdowns. Figure 7-76 shows a unit incorporating both filtration (down to 10 microns) and skimming for the removal of fines and tramp oil.

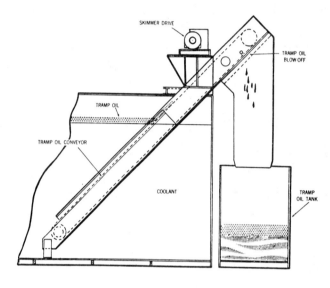

Figure 7-75: Automatic skimmer used for removal of floating tramp oils from a recirculated coolant.

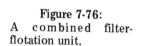

Figure 7-76: A combined filter-flotation unit.

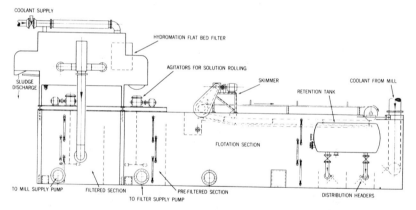

Figure 7-77: Four flatbed pressure filters used in the recirculation system for a tandem tin mill.

In the case of quasi-stable emulsions, such as those used in tandem tin mills, the retention time of the recirculated rolling solution in the system is about 5 minutes and the degree of

THERMAL ASPECTS OF COLD ROLLING

filtration attainable is between 50 and 100 microns. Four flatbed filters designed to handle the rolling solution of a tandem tin mill is shown in Figure 7-77.

The media used in the filters should be sized to suit the specific type of solution. Table 7-15 lists examples of roll coolants used for rolling various products and the approximate degree of filtration possible.

Table 7-15
Examples of Roll Coolants Used for Cold Reduction of Steel Products and Approximate Degree of Filtration Possible (Wettach)

Product	Thickness, in.	Coolant	Degree of filtration possible, microns
Steel sheet	0.187 to 0.142	Soluble oil & water	5-10
Black plate (tin-plate)	0.0149 to 0.0072	Unstable solution	50-100
Double reduced black plate (thin tin)	0.0066 to 0.0033	Meta-stable solution	20-30
Steel foil	0.003 to 0.001	Meta-stable solution	20-30
Stainless steel	Full range	Soluble oil & water	10 (media)
Stainless steel	Full range	Soluble oil & water	1 (diatomaceous earth)
Stainless steel	Full range	Mineral oil	1 (diatomaceous earth)
Silicon steel	Full range	Soluble oil & water	10-20

1. On direct application mills (water and rolling oil applied separately) on steel products, the water phase is filterable down to the 5 to 10-micron range.

2. With the exception of the mineral oil coolant, degrees of filtration possible are based on filtration through the flatbed pressure filter.

Chapter 8   The Alloys of Iron, Their Physical Nature, and Behavior During Deformation

8-1   Introduction

To adequately describe the rolling process in theoretical terms, particularly with respect to rolling force and energy requirements, and to predict the physical condition of the workpiece after it has been rolled, it is desirable to develop an understanding, as complete as possible, concerning the behavior of the workpiece in response to deforming stresses. Accordingly, this chapter is devoted to an examination of the pertinent physical characteristics of various steels and changes that occur in these characteristics as the consequence of cold reduction.

Intrinsically, the physical characteristics of a metal are dependent upon (a) the elements it contains, (b) the crystallographic arrangement of its constituent atoms, (c) the nature of the defects occurring within its crystal lattices and (d) its microstructure. Inasmuch as the principal element involved in steels is iron, this element, its properties, the crystallographic aspects of its allotropes and the general nature of its microstructure are discussed first.

Since the most important alloys of iron are those containing carbon, these are treated in considerable detail. Essentially all carbon steels that are cold rolled have, at some stage in their manufacture, existed in the austenitic form. Thus this form of steel, which is generally unstable at temperatures below those corresponding to the A_3 line (see Figure 8-40 and Section 8-10), is discussed with respect to both its crystallographic form and the products of its decomposition.

The chromium alloys of iron (the stainless steels) are next in economic importance and these are examined in Section 8-15 as well as in Section 12-31. Various magnetic alloys are described in Section 8-16 and the more important silicon steel alloy (grain oriented electrical sheet) is discussed in Section 12-30.

With respect to the elastic and plastic deformation of a strip during rolling, it is desirable to review the concepts of stresses and strains and their relationship in both types of deformation. This is undertaken in Sections 8-17 to 8-23, together with an examination of the physical phenomena associated with plastic deformation, especially crystallographic slip and the development of textures.

Since the yield strength of a material is, from the rolling viewpoint, its principal characteristic, this parameter is studied in greater detail in Sections 8-25 and 8-33, not only with respect to its measurement at various strain rates but also with respect to the influence of various factors, such as temperature and microstructure on its magnitude.

Workhardening is discussed in the concluding sections of this chapter, not only from a theoretical viewpoint but also in connection with its measurement. The apparent failure of low carbon steel to workharden while simultaneously undergoing high-speed deformation is examined, but the importance of this effect in the commercial production of foils is dealt with in Chapter 9. Finally, the energy of deformation and stored energy are discussed, particularly with respect to the release of the latter during recovery and annealing.

8-2   Iron and Its Properties

The element iron occupies the 26th position in the periodic table (Figure 8-1) and its commonest isotope possesses an atomic weight of 56. This metal, together with scandium, titanium, vanadium, chromium, manganese, cobalt and nickel are described as transition elements. The iron atom features 26 protons in its nucleus (each weighing 1.67252×10^{-27} kg), 26 electrons (each weighing 9.1091×10^{-31} kg) in shells around the nucleus and, in the commonest isotope, 30 neutrons (each weighing 1.67482×10^{-27} kg) inside the nucleus.•

• "Physics, Part II", by D. Halliday and R. Resnick, John Wiley and Sons, Inc., New York, Second Printing, 1967.

ALLOYS OF IRON

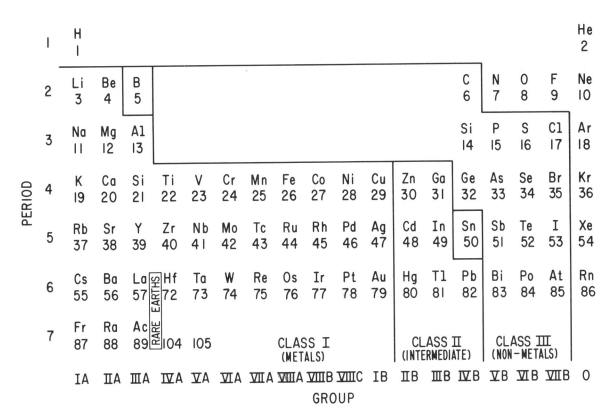

Figure 8-1: Periodic Table of the Elements.

Table 8-1

Configuration of Electrons in Iron Atom

Shell or Group	No. of Electrons	Configuration
K	2	$(1s)^2$
L	8	$(2s)^2 (2p)^6$
M	14	$(3s)^2 (3p)^6 (3d)^6$
N	2	$(4s)^2$

Of the 26 electrons surrounding the nucleus, two are in the K group (closest to the nucleus), 8 in the L group, 14 in the M group and 2 in the N group. On the basis of energy, the quantum numbers of the electrons are as indicated in Table 8-1, and their relative energies are represented schematically in Figure 8-2.

In Table 8-1, the digit inside the parentheses denotes the principal quantum number, the letter s, p or d the secondary quantum number and the superscript to the parenthesis, the number of electrons associated with the two previously-mentioned quantum numbers. Chemically, the element can react in a divalent manner to form ferrous and, in a trivalent manner, to produce ferric compounds.

Iron can exist in both the body-centered cubic (bcc) or the face-centered cubic (fcc) crystallographic form as illustrated in Figure 8-3. The free energy change ΔF^O $(\alpha \rightarrow \gamma)$ accompanying the transformation of pure bcc to fcc iron is shown in Figure 8-4. From this illustration, it is seen that the stable form of iron (corresponding to the lowest free energy) in the temperature range of 910 to 1400°C (1670 to 2552°F) is the γ (or fcc) allotrope while the α (or bcc) form is stable outside of these temperature limits. It is interesting to note that the γ iron would be stable at all temperatures if magnetic phenomena were absent. Moreover, the range of stability of the γ allotrope results from the small difference of the two relatively large free energies associated with the two allotropic forms. Since the presence of alloying elements affects the magnitudes of the free energies of the two forms, the stability of the γ form is readily affected by such elements.

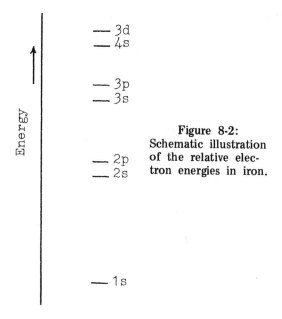

Figure 8-2: Schematic illustration of the relative electron energies in iron.

Table 8-2
The Properties of Ingot Iron

Specific Gravity		7.864
Density		7.868 grams/cm^3
Melting Point		1535°C (2795°F)
Boiling Point		3000°C (5432°F)
Proportional Limit	(annealed, under compression)	19,200 psi
Yield Strength	(annealed, under compression)	20,600 psi
	(hot rolled, under tension)	26,000 – 36,000 psi
	(water quenched, under tension)	30,300 psi
Tensile Strength	(at ambient temperature)	
	(hot rolled, under tension)	42,000 – 48,000 psi
	(cold worked, under tension)	100,000 psi
	(water quenched, under tension)	47,000 psi
Elongation	(hot rolled, under tension)	22 – 28% (in 8 inches)
	(water quenched, under tension)	36% (in 2 inches)
Reduction of Area	(hot rolled, under tension)	65 – 78%
	(water quenched, under tension)	70%
Hardness	(hot rolled) Brinell	82 – 100
	Rockwell B	39 – 55
Fatigue Strength	(annealed, reverse bend)	26,000 psi
Impact	(hot rolled, Izod)	90 ft.-lb. (longitudinal direction)
		56 ft.-lb. (transverse direction)
Resistivity		11 x 10^6 ohm-cm.

In the bcc form at a temperature of 70°F, the lattice parameter is 2.86 Angstroms, the distance of closest approach of the atoms is 2.48 Å, and the density is 7.87 grams/cc (491 lbs./cubic foot). At a temperature of 1800°F, the lattice parameter for the fcc structure is 3.65 Å, the closest approach of the atoms is 2.58 Å, and the density 7.42 grams/cc (463 lbs./cubic foot).

ALLOYS OF IRON

On a normal commercial basis, perfectly pure iron is not available. However, for special purposes, electrolytic iron and carbonyl iron of 99.99 per cent purity are obtainable. In larger quantities, ingot iron, containing about 0.1 per cent of various impurities (about 0.01 per cent carbon),□ though relatively expensive, is sold where its superior ductility, corrosion resistance, electrical conductivity or magnetic permeability are needed. The physical properties of ingot iron are listed in Table 8-2.

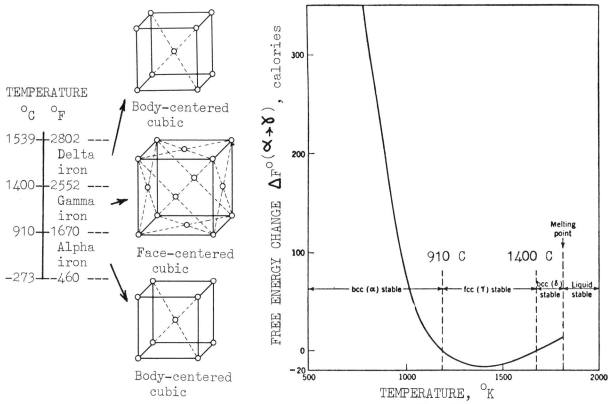

Figure 8-3:
The temperature ranges in which the allotropic forms of iron exist under equilibrium conditions.

Figure 8-4:
The free energy change accompanying the transformation of pure bcc to fcc iron.

Generally speaking, iron and its alloys exist in the polycrystalline state, or as a mass of grains separated from each other by grain boundaries (see Figure 8-5). Each grain represents a portion of the solid phase with a given orientation of the crystal axes and, in passing from one grain to the next, the orientation changes abruptly across the grain boundary. Originally, it was believed that grain boundaries represented regions of amorphous material. However, in the last two or three decades, grain boundaries have been shown to be formed by arrays of dislocations and, in spite of their etched appearance under a microscope, to be extremely thin. They possess stored elastic energy which acts as a driving force in the recrystallization of the metal at high temperatures.

The average size of grains is dependent on the prior history of the iron, particularly that relating to its deformation and heat treatment and, in the case of its alloys, on their composition. The grain size affects the overall metallurgical properties of the metal (such as its yield strength and its rate of recrystallization at annealing temperatures) and its surface appearance after deformation. Grain size is defined by the ASTM grain size number N as expressed by the relationship

$$n = 2^{N-1} \qquad (8\text{-}1)$$

□A typical ingot iron may contain about 0.017 per cent manganese, 0.005 per cent phosphorus, 0.025 per cent sulphur and a trace of silicon.

where n is the number of grains per square inch as seen in a specimen viewed at a magnification of 100 times. The usual range of austenitic grain sizes in steels lies between 1 and 9 with the corresponding number of grains per square inch as viewed at 100 diameters being as shown in Table 8-3.

Figure 8-5:
Appearance of grains in ingot iron. (X100)

Table 8-3
ASTM Grain-Size Numbers

ASTM Grain-Size Numbers	Average Number of Grains per Square Inch as Viewed at 100 Diameters
1	1
2	2
3	4
4	8
5	16
6	32
7	64
8	128
9	256

8-3 The Various Types of Steel Processed in Flat Rolling Operations

Iron alloys readily with a number of other elements to form a variety of irons, many of which are steels of considerable economic importance. It reacts with carbon to form various conventional steels, with silicon to form electrical steels, and with chromium, nickel, and other elements to form various stainless steels. All of these materials are usually cold rolled during production and, for that reason, are discussed in greater detail below.

The iron-carbon equilibrium diagram is shown in Figure 8-6 with the various phases being as shown in the figure. Steels are classified on the basis of carbon content as indicated in

ALLOYS OF IRON

Figure 8-7; low carbon steels containing a maximum 0.2 per cent carbon, medium carbon steels with a carbon content ranging from 0.2 to 0.5 per cent, and high carbon steels containing more than 0.5 per cent carbon. Alternately, steels may be classified as hypoeutectoid (less than 0.8 per cent carbon), eutectoid (0.8 per cent carbon), and hypereutectoid (more than 0.8 per cent carbon). A hypoeutectoid steel contains proeutectoid ferrite and pearlite, while a hypereutectoid steel consists of proeutectoid cementite and pearlite.

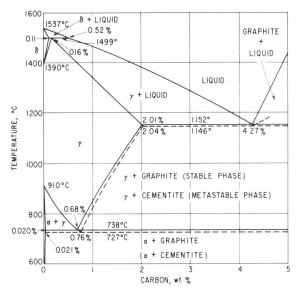

Figure 8-6:
The iron-carbon equilibrium diagram.

Carbon Content (%)	General Classification
Less than 0.2	Low-Carbon *
0.2-0.5	Medium-Carbon *
0.5-0.8	High-Carbon *
0.8	High-Carbon **
More than 0.8	High-Carbon ***

* Hypoeutectoid
** Eutectoid
*** Hypereutectoid

Figure 8-7:
Steel classifications based on carbon content.

Cold-rolled low-carbon steel strip is produced in continuous sheet and strip form generally to a maximum carbon content of about 0.25 per cent (by ladle analysis). The steels so utilized may be described by their method of manufacture as basic open hearth, basic oxygen, acid open hearth, or acid Bessemer steels. At the present time, the basic open hearth steels represent the preponderance of the tonnage and they may be of the rimmed, semikilled, or fully killed variety. The Standard AISI composition ranges of such steels are given in Table 8-4.

Medium and high carbon steel strip (with carbon contents up to about 1.3 per cent) are also cold rolled though usually in relatively narrow widths because of their increased resistance to deformation.

Electrical steels contain low concentrations of carbon with silicon contents ranging from about 0.5 to 4.8 per cent (see Table 8-5). In such steels, the high electrical permeability of the iron is enhanced by the silicon, yet their relatively high resistivity minimizes the core losses. A section of the ternary iron-silicon-carbon equilibrium diagram at the 0.01 to 0.02 per cent carbon level is shown in Figure 8-8 at the left and at the 0.05 to 0.08 per cent carbon level at the right.

Stainless steels (discussed more fully in Section 8-15) are of four general types: (1) ferritic with a bcc structure, (2) austenitic with an fcc structure, (3) martensitic with a bcc tetragonal structure, and (4) the precipitation hardenable or semiaustenitic grades. The chemical composition, limits, and ranges for these sheets are presented in Table 8-6. As far as cold rolling is concerned, only the ferritic (represented by certain of the AISI 400 series) and the austenitic (represented by the 200 and 300 series) grades are of importance.

The approximate densities of iron and various types of steels are presented in Table 8-7. These are of interest in calculating coil sizes and the throughputs of mills.

Table 8-4

Standard A.I.S.I. Composition Ranges of Basic Open Hearth Carbon Steels [1] [2]

A.I.S.I. Number	Chemical Composition Limits, Per Cent				Corresp. SAE No.
	C	Mn	P (Max.)	S (Max.)	
C1005	0.06 Max.	0.35 Max.	0.040	0.050	—
C1006	0.08 Max.	0.25–0.40	0.040	0.050	1006
C1008	0.10 Max.	0.25–0.50	0.040	0.050	1008
C1010	0.08–0.13	0.30–0.60	0.040	0.050	1010
C1011	0.08–0.13	0.60–0.90	0.040	0.050	—
C1012	0.10–0.15	0.30–0.60	0.040	0.050	1012
C1013	0.11–0.16	0.50–0.80	0.040	0.050	—
C1015	0.13–0.18	0.30–0.60	0.040	0.050	1015
C1016	0.13–0.18	0.30–0.60	0.040	0.050	1016
C1017	0.15–0.20	0.60–0.90	0.040	0.050	1017
C1018	0.15–0.20	0.60–0.90	0.040	0.050	1018
C1019	0.15–0.20	0.70–1.00	0.040	0.050	1019
C1020	0.18–0.23	0.30–0.60	0.040	0.050	1020
C1021	0.18–0.23	0.60–0.90	0.040	0.050	1021
C1022	0.18–0.23	0.70–1.00	0.040	0.050	1022
C1023	0.20–0.25	0.30–0.60	0.040	0.050	—
C1024	0.19–0.25	1.35–1.65	0.040	0.050	1024
C1025	0.22–0.28	0.30–0.60	0.040	0.050	1025
C1026	0.22–0.28	0.60–0.90	0.040	0.050	1026
C1027	0.22–0.29	1.20–1.50	0.040	0.050	1027
C1029	0.25–0.31	0.60–0.90	0.040	0.050	—
C1030	0.28–0.34	0.60–0.90	0.040	0.050	1030
C1031	0.28–0.34	0.30–0.60	0.040	0.050	—
C1032	0.30–0.36	0.60–0.90	0.040	0.050	—
C1033	0.30–0.36	0.70–1.00	0.040	0.050	1033
C1034	0.32–0.38	0.50–0.80	0.040	0.050	—
C1035	0.32–0.38	0.60–0.90	0.040	0.050	1035
C1036	0.30–0.37	1.20–1.50	0.040	0.050	1036
C1037	0.32–0.38	0.70–1.00	0.040	0.050	—
C1038	0.35–0.42	0.60–0.90	0.040	0.050	1038
C1039	0.37–0.44	0.70–1.00	0.040	0.050	1039
C1040	0.37–0.44	0.60–0.90	0.040	0.050	1040
C1041	0.36–0.44	1.35–1.65	0.040	0.050	1041
C1042	0.40–0.47	0.60–0.90	0.040	0.050	1042
C1043	0.40–0.47	0.70–1.00	0.040	0.050	1043
C1044	0.43–0.50	0.30–0.60	0.040	0.050	—
C1045	0.43–0.50	0.60–0.90	0.040	0.050	1045
C1046	0.43–0.50	0.70–1.00	0.040	0.050	1046
C1048	0.44–0.52	1.10–1.40	0.040	0.050	1048
C1049	0.46–0.53	0.60–0.90	0.040	0.050	1049
C1050	0.48–0.55	0.60–0.90	0.040	0.050	1050
C1051	0.45–0.56	0.85–1.15	0.040	0.050	—
C1052	0.47–0.55	1.20–1.50	0.040	0.050	1052
C1053	0.48–0.55	0.70–1.00	0.040	0.050	—
C1054	0.50–0.60	0.50–0.80	0.040	0.050	—
C1055	0.50–0.60	0.60–0.90	0.040	0.050	1055
C1060	0.55–0.65	0.60–0.90	0.040	0.050	1060
C1065	0.60–0.70	0.60–0.90	0.040	0.050	1065
C1069	0.65–0.75	0.40–0.70	0.040	0.050	—
C1070	0.65–0.75	0.60–0.90	0.040	0.050	1070
C1075	0.70–0.80	0.40–0.70	0.040	0.050	—
C1078	0.72–0.85	0.30–0.60	0.040	0.050	1078
C1080	0.75–0.88	0.60–0.90	0.040	0.050	1080
C1084	0.80–0.93	0.60–0.90	0.040	0.050	—
C1085	0.80–0.93	0.70–1.00	0.040	0.050	1085
C1086	0.82–0.95	0.30–0.50	0.040	0.050	1086
C1090	0.85–0.98	0.60–0.90	0.040	0.050	1090
C1095	0.90–1.03	0.30–0.50	0.040	0.050	1095

SILICON – When silicon is required, the following ranges and limits are common for standard basic open-hearth steel grades: up to and excluding C1015, silicon limit is 0.10 max.: C1015 up to and including C1025, silicon content may be 0.10 max., 0.10, 0.20: or 0.15, 0.30: over C1025, silicon ranges may be 0.10, 0.20 or 0.15, 0.30.

COPPER AND LEAD – When required, copper and lead are specified as added elements to a standard steel.

---

[1] Hot-rolled carbon-steel bars and semifinished products not exceeding 200 square inches cross-sectional area.

[2] From Steel Products Manuals, American Iron and Steel Institute; "Carbon-Steel – Semifinished for Forging", "Hot-Rolled and Cold-Finished Carbon-Steel Bars", and "Wire and Rods, Carbon Steel".

ALLOYS OF IRON 405

Table 8-5 ★

Typical Mechanical Properties of Fully Processed Electrical Sheets

Grade*		Thickness (In.)	Approximate Si Content (Per Cent)	Approximate Resistivity (Microhms per cm³)	Yield Point Longitudinal (Lb. per Sq. In.)	Tensile Strength Longitudinal (Lb. per Sq. In.)	Per Cent Elongation in 2 In. Longitudinal	Rockwell Hardness B	Erichsen Cup (mm.)	Amsler Bends
Electrical	(CR)	0.025	1.60	31	37,000	56,000	28	72	—	21
Motor	(CR)	0.025	2.80	45	51,000	68,000	24	80	7.0	10
Dynamo	(CR)	0.0185	3.25	50	55,000	69,000	16	82	5.4	10
Transformer 72	(CR)	0.0185	3.25	50	59,000	68,000	16	84	5.2	9
Transformer 65	(HR)	0.014	4.00	58	59,000	66,000	12	70	4.0	7
Transformer 58	(HR)	0.014	4.25	61	60,000	61,600	4	75	3.5	5
Transformer 52	(HR)	0.014	4.50	64	60,000	69,500	2	74	2.7	3
Grain-Oriented Transformer 60	(CR)	0.014	3.25	50	48,200	55,500	10	77	—	—

*(CR) indicates cold-reduced product; (HR) indicates hot-rolled product.

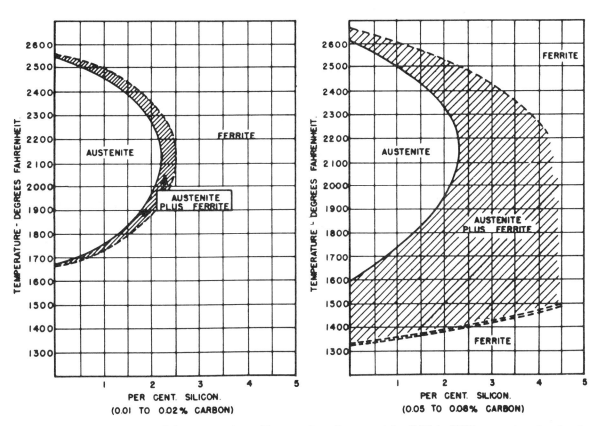

Figure 8-8: (Left) Section of the ternary iron-silicon-carbon diagram at the 0.01 to 0.02 per cent carbon level. (Right) Same at the 0.05 to 0.08 per cent carbon level.

Table 8-6
American Iron and Steel Institute Standard Type Numbers, Chemical Composition Limits and Ranges for Stainless Steels.

Type Number	Chemical Composition, Per Cent							
	C	Mn Max.	Si Max.	P Max.	S Max.	Cr	Ni	Other Elements
201	0.15 Max.	5.50-7.50	1.00	0.060	0.030	16.00-18.00	3.50-5.50	N: 0.25 Max.
202	0.15 Max.	7.50-10.00	1.00	0.060	0.030	17.00-19.00	4.00-6.00	N: 0.25 Max.
301	0.15 Max.	2.00	1.00	0.045	0.030	16.00-18.00	6.00-8.00	
302	0.15 Max.	2.00	1.00	0.045	0.030	17.00-19.00	8.00-10.00	
302B	0.15 Max.	2.00	2.00-3.00	0.045	0.030	17.00-19.00	8.00-10.00	
303	0.15 Max.	2.00	1.00	0.20	0.15 Min.	17.00-19.00	8.00-10.00	Mo: 0.60 Max.*
303 Se	0.15 Max.	2.00	1.00	0.20	0.06	17.00-19.00	8.00-10.00	Se: 0.15 Min.
304	0.08 Max.	2.00	1.00	0.045	0.030	18.00-20.00	8.00-12.00	
304L	0.03 Max.	2.00	1.00	0.045	0.030	18.00-20.00	8.00-12.00	
305	0.12 Max.	2.00	1.00	0.045	0.030	17.00-19.00	10.00-13.00	
308	0.08 Max.	2.00	1.00	0.045	0.030	19.00-21.00	10.00-12.00	
309	0.20 Max.	2.00	1.00	0.045	0.030	22.00-24.00	12.00-15.00	
309S	0.08 Max.	2.00	1.00	0.045	0.030	22.00-24.00	12.00-15.00	
310	0.25 Max.	2.00	1.50	0.045	0.030	24.00-26.00	19.00-22.00	
310S	0.08 Max.	2.00	1.50	0.045	0.030	24.00-26.00	19.00-22.00	
314	0.25 Max.	2.00	1.50-3.00	0.045	0.030	23.00-26.00	19.00-22.00	
316	0.08 Max.	2.00	1.00	0.045	0.030	16.00-18.00	10.00-14.00	Mo: 2.00-3.00
316L	0.03 Max.	2.00	1.00	0.045	0.030	16.00-18.00	10.00-14.00	Mo: 2.00-3.00
317	0.08 Max.	2.00	1.00	0.045	0.030	18.00-20.00	11.00-15.00	Mo: 3.00-4.00
321	0.08 Max.	2.00	1.00	0.045	0.030	17.00-19.00	9.00-12.00	Ti: 5 x C, Min.
347	0.08 Max.	2.00	1.00	0.045	0.030	17.00-19.00	9.00-13.00	Cb-Ta: 10 x C, Min.†
348	0.08 Max.	2.00	1.00	0.045	0.030	17.00-19.00	9.00-13.00	Cb-Ta: 10 x C, Min., Ta: 0.10 Max., Co: 0.20 Max.†
403	0.15 Max.	1.00	0.50	0.040	0.030	11.50-13.00		
405	0.08 Max.	1.00	1.00	0.040	0.030	11.50-14.50		Al: 0.10-0.30
410	0.15 Max.	1.00	1.00	0.040	0.030	11.50-13.50		
414	0.15 Max.	1.00	1.00	0.040	0.030	11.50-13.50	1.25-2.50	
416	0.15 Max.	1.25	1.00	0.06	0.15 Min.	12.00-14.00		Mo: 0.60 Max.*
416 Se	0.15 Max.	1.25	1.00	0.06	0.06	12.00-14.00		Se: 0.15 Min.
420	Over 0.15	1.00	1.00	0.040	0.030	12.00-14.00		
430	0.12 Max.	1.00	1.00	0.040	0.030	14.00-18.00		
430F	0.12 Max.	1.25	1.00	0.06	0.15 Min.	14.00-18.00		Mo: 0.60 Max.*
430F Se	0.12 Max.	1.25	1.00	0.06	0.06	14.00-18.00		Se: 0.15 Min.
431	0.20 Max.	1.00	1.00	0.040	0.030	15.00-17.00	1.25-2.50	
440A	0.60-0.75	1.00	1.00	0.040	0.030	16.00-18.00		Mo: 0.75 Max.
440B	0.75-0.95	1.00	1.00	0.040	0.030	16.00-18.00		Mo: 0.75 Max.
440C	0.95-1.20	1.00	1.00	0.040	0.030	16.00-18.00		Mo: 0.75 Max.
446	0.20 Max.	1.50	1.00	0.040	0.030	23.00-27.00		N: 0.25 Max.
501	Over 0.10	1.00	1.00	0.040	0.030	4.00-6.00		Mo: 0.40-0.65
502	0.10 Max.	1.00	1.00	0.040	0.030	4.00-6.00		Mo: 0.40-0.65

*At producer's option; reported only when intentionally added. †Columbium is also called niobium.

Table 8-7

Approximate Densities of Iron and Various Steels

Material (In Wrought Form)	Density (at 60° F)		
	Grams per cc	Lb. per Cu. In.	Lb. per Cu. Ft.
Pure Iron (99.9% Fe)	7.86	0.284	491
Soft Steel (0.06% C)	7.87	0.284	491
Carbon Steel (0.40% C)	7.84	0.283	489
Stainless Steel (18% Cr, 8% Ni)	8.03	0.29	501
Stainless Steel (17% Cr, 0.12% C)	7.75	0.28	484
Stainless Steel (27% Cr, 0.35% C)	7.47	0.27	467

8-4 The Processing of Steel Prior to Cold Rolling

Steel coils that are to be cold rolled have been subject to considerable prior deformation on hot strip mills. This hot processing, together with the subsequent pickling and oiling operations, can markedly affect the behavior of the steel during cold reduction as well as the properties and quality of the rolled product. For these reasons, it is appropriate to review the various processing operations that precede cold reduction.

Today, steel is made in oxygen steelmaking furnaces, open hearth furnaces, electric furnaces, and less frequently nowadays, Bessemer converters. Until recently, all steel produced was cast into ingot molds, but the 1960's saw the introduction of continuous casters, first for billets and then subsequently for slabs for conversion to hot rolled products.

When ingots are produced, these are stripped from the molds, "soaked" (i.e., reheated until they attain a reasonable uniformity of temperature), rolled on primary hot mills to slabs (or other semifinished products), reheated and then rolled on hot strip mills or plate mills. In the case of continuous casting, however, the cast is cut into slabs or other semifinished products, thus obviating the step of primary hot rolling. The slabs so produced are then processed in a manner essentially the same as that associated with the conventionally cast steel.

Before a slab is rolled into strip, it must be uniformly reheated to hot rolling temperature. The furnaces used for this purpose are constructed with hearths which permit the charge either to remain in a fixed position in the furnace or to be moved during the heating process. An example of the first type of hearth is the conventional batch type or "in-and-out" furnace. Examples of the second general type are found in roller hearth furnaces in which the material moves as the series of rollers that constitute the hearth rotate and in continuous furnaces in which a continuous line of material is pushed over skids. Batch type furnaces are especially suitable for heating blooms of mixed sizes and lengths in thicknesses over 8 inches; continuous furnaces are used for heating slabs or billets for large orders of uniform length and thickness, and car bottom furnaces are used for annealing miscellaneous shapes and sizes.

A counter-current fired continuous reheating furnace is shown in Figure 8-9. In this furnace the steel to be heated can be charged either from the end or through a side door. In either case, the steel is removed through the furnace by pushing the last piece charged with a pusher at the charging end. As each cold piece is pushed into the furnace against the continuous line of material, a heated piece is removed. The heated piece is discharged either through a door by gravity upon a roller table which feeds the mill, or it is pushed through a side door to the mill table by suitable manual or mechanical means.

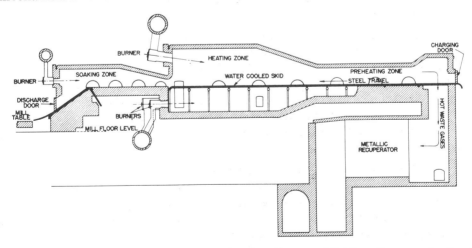

Figure 8-9: Schematic longitudinal section through a counter-current fired continuous reheating furnace.

A typical rolling train (see Figure 8-10) will consist of a roughing scale breaker, four, four-high roughing stands, a finishing scale breaker, and six, four-high finishing stands. Driven table rolls convey the steel from the furnace to the mill and also from stand to stand. If the mill is to produce strip, sheets or breakdowns of greater width than the maximum width of slab available, the first rougher or roughing stand is a broadside mill in which the width of the slab is increased in a single pass by cross rolling. In this case, turntables for manipulating the slab must precede and follow this stand. A slab squeeze also follows the broadside mill. The next three roughing stands usually are provided with integral vertical edgers in front of each stand. Separating the roughing train from the finishing train is a holding table, while the finishing end is a closely grouped tandem train composed of the finishing scalebreaker and six finishing stands.

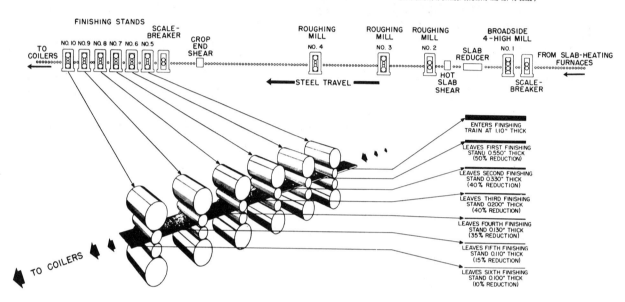

Figure 8-10: Typical reductions per pass in the finishing stands of a hot strip rolling mill.

High-pressure hydraulic sprays to remove scale from the hot slab are located after the two scalebreakers and perhaps at various roughing stands. Water is supplied to the spray nozzles by suitable high pressure pumps.

Following the last finishing stand there is usually a flying shear for cutting the rolled product into lengths, if so desired. As the steel proceeds from the mill, it is carried over a long table called the runout table, consisting of individually driven rollers. Two or more coilers are located in this table; they operate to coil the material when continuous long lengths are required. If short lengths are cut at the flying shears, the coilers are inoperative and the steel passes over them and onto a piler at the end of the table. Additional tables may be installed parallel to the central runout table, with suitable transfers for moving material to them; this equipment is used principally when the heavier gages are being rolled.

The finishing train must be operated with careful regulation to obtain a finished hot-rolled product of prime quality. Various automatic-control elements have been incorporated in mill design to assist the operators in producing strip of uniformly high quality. Surface, gage, width, finishing temperature and cross-sectional contour of the product, are all required to meet given standards depending upon the subsequent treatment or ultimate use of the material in question. As an example, metallurgical requirements may dictate a definite finishing temperature for a particular gage and width to be rolled. Time on the holding table prior to coiling, number of descaling sprays used during rolling in the finishing stands, speed of the finishing train, and method of drafting, all affect the finishing temperature and may be varied at the discretion of the operator to help meet

ALLOYS OF IRON

requirements. Defects in the surface of the rolled steel, if not evident in the rough slab, usually can be traced to defects in the surfaces of the work rolls on the finishing stands and are corrected readily by substituting newly surfaced work rolls. The principal factors affecting the overall dimensional accuracy of the finished hot rolled product include contour of the work rolls and backup rolls as installed, changes in the contour of the work rolls and backup rolls due to intermittent heating and cooling, method of drafting (i.e., amount of reduction in successive passes), and rolling sequence of various gages and widths. Also involved in the occurrence of gage (thickness) variations in hot rolled strip are: the difference in the speed of the strip leaving the last roughing stand and the entry speed of the strip entering the first finishing stand, which results in temperature variations along the length of the strip; and variations in the tension applied to the strip between stands. Many of these factors that affect product quality are amenable to control by automatic means to assist the mill operators in achieving the best results from the rolling operation by minimizing the number of elements that otherwise must be manually controlled.

During recent years, the practice of using a rolling lubricant on one or more of the stands of the finishing train has been introduced. This practice has certain decided advantages including:

1. Extended roll life (because of lesser roll wear).
2. Better surface quality of the rolled strip.
3. Increased reductions in the finishing train (and hence larger coils).
4. Lesser motor power requirements.
5. Decreased pickling costs (due to diminution of scaling resulting from the use of the lubricant).♦

However, the application of the lubricant must be carefully synchronized with the entry of the workpiece into the finishing train.■

Hot rolled coils, when cold, are covered with scale, or oxides, and they are therefore pickled and oiled prior to further processing, the oil being applied to prevent rusting, eliminate damage due to scuffing of the wraps on each other and to act as a lubricant at the first stand (and possibly all stands) in subsequent tandem cold mill processing.

A continuous pickling line for the processing of hot rolled carbon and electrical steels, generally capable of operating at speeds in the range of 600 to 1200 feet per minute, is illustrated in Figure 8-11.

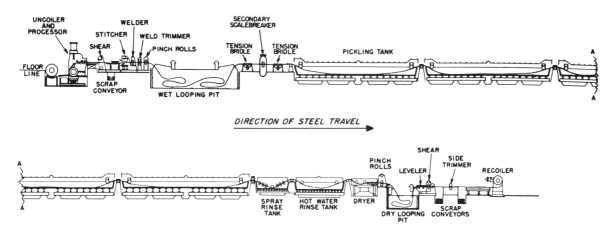

Figure 8-11: Schematic arrangement of the equipment comprising a continuous pickling line.

♦ M. R. Edmundson, "High Temperature Rolling Lubricant Aids Hot Rolling", Iron and Steel Engineer Year Book, 1970, pp. 522-525.
■ U. S. Patent 3,208,253, "Control of Rolling Mill Lubricant", Issued September 28, 1965.

While most of the pickling lines in the industry include stitchers for fastening coil ends for continuously processing hot rolled product, many also have installed flash butt welding as supplementary equipment. The main advantage of this method of joining coil ends is that it provides a joint which can be cold reduced, whereas the lapped and stitched joint cannot.

Following welding, the flash, or excess metal resulting from the upsetting action of the welder, is trimmed off by a cutter designed for the purpose. The looping pit is next in line and provides a continuous storage space for material to compensate for short delays at the charging end and to permit a uniform rate of travel through the acid tanks. Water is kept in the pit to minimize scratching and increase wetting action in the first pickling tank.

In some continuous pickling lines, an auxiliary or secondary scalebreaker follows the looping pit, to break the scale even further than was achieved in the processor at the entry end, and thus increase the speed at which the line can be operated and still produce satisfactorily pickled strip. The secondary scalebreaker may be a machine similar to the entry-end processor, or it may be a two-high temper mill preceded and followed by a tension bridle at the entry and exit sides of the mill.

The pickling zone consists of several individual acid-proof tanks located in a series, comprising an effective immersion length of about 250 to 300 feet. While most lines have from three to five tanks, each about 70 to 80 feet long, some modern lines have only one long tank, divided by weirs into four or five sections, thereby increasing effective immersion depth about 10 per cent to 15 per cent.

Following the acid tank are rinsing tanks consisting of a cold water spray rinse and a hot water tank. The hot water rinse is a tank with an effective product immersion length of 15 to 20 feet. This tank completes the rinsing and, by warming the steel, promotes flash drying prior to entering the succeeding set of pinch rolls. Situated between the final rinse tank and the pinch rolls are one, two or three banks of hot air dryers operating at low pressures. Pinch rolls at the exit end of the pickling tanks control the speed of product travel and, in conjunction with the pinch rolls which provide back tension at the entry end of the line, help to maintain the proper loops in the tanks.

The delivery end of the continuous pickling line has, in the order listed, a looping pit, pinch rolls, shear, oiler, recoiler and suitable supplementary equipment for conveying the finished product from the line. The pinch rolls preceding the shear are located so that the product delivery to the shear is facilitated. Stitches are removed at this point, as well as short sections which inspection has shown to be of inferior quality. Some lines are provided also with rotary side trimmers at the entry end or, more commonly, at the delivery end.

Prior to recoiling, the pickled steel passes between a set of oiling rolls which cover both surfaces with a small amount of oil. The type of oil used to lubricate the steel, and protect it from rusting during storage and from scratching during handling, is determined by the type of lubricating system on the cold-reduction mill unit. For example, palm oil diluted with light mineral oil, is applied to the steel at the pickling lines when a straight palm oil or a solution containing palm oil is used on the cold reduction mill. Finally, the pickled and oiled product is recoiled on a conventional coiler.

Hot rolled coils of stainless steels are processed as follows. In the case of the ferritic grades (400 series), the coils are (a) box annealed in air at a temperature in the range 790-820°C, (b) shot blasted and pickled, and (c) side trimmed and/or welded into larger coils (often with the addition of leader strips). Coils of the austenitic grades (300 series) are similarly processed except that the annealing operation requires a higher temperature (980-1150°C) but a much shorter annealing cycle. Accordingly, continuous annealing lines are generally used for this purpose.

8-5 Crystallographic Directions and Planes (Miller Indices)

In discussing slip directions and slip planes associated with plastic deformation, texture

ALLOYS OF IRON

and other topics related to the crystallographic nature of a metal, such as iron, it is desirable to use the concept of Miller indices.

A direction in a crystal is indicated by the indices enclosed in square brackets, e.g., [210], and, in cubic lattice systems, the direction is always perpendicular to the plane designated by the same indices. The direction [a, b, c] is that represented by a vector drawn from the origin to a point x=a, y=b and z=c, as shown in Figure 8-12, where a, b and c are expressed in terms of the lattice parameter. It should be noted that all directions parallel to this vector are represented by the same indices and that reciprocals are not involved in the specification of direction.

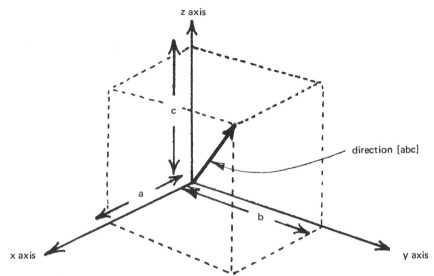

Figure 8-12: Sketch Illustrating Crystallographic Direction [abc].

In the cubic lattice structure, six directions, although specified differently in terms of Miller indices, are, in fact, equivalent. Thus the direction [100] is equivalent to the [$\bar{1}$00], [010], [0$\bar{1}$0], [001] and [00$\bar{1}$] directions where the bar above the digit represents the negative sign. These equivalent directions are designated a family and are represented by the single expression <100>, the angular brackets indicating that a family of directions is implied.

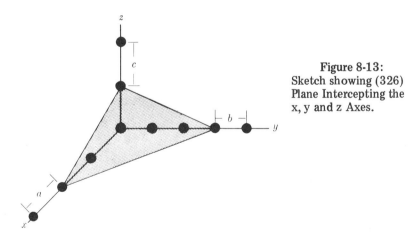

Figure 8-13: Sketch showing (326) Plane Intercepting the x, y and z Axes.

In determining the Miller indices of a crystallographic plane, the procedure is slightly more complicated. The first step involves the determination of the intercepts on the x, y and z axes made by the plane. In the example shown in Figure 8-13, these intercepts are 2, 3 and 1 lattice units on the x, y and z axes, respectively. The reciprocals of these intercepts are then taken and these are reduced to the smallest set of integers that are in the same ratio. In the example cited, the reciprocals are 1/2, 1/3 and 1, and the corresponding set of integers (of least value) are 3, 2, 6. Thus the plane illustrated has Miller indices (326), the indices being enclosed in curved brackets. Again, it should be remembered that all parallel planes are designated by the same set of indices and that the separation (d) of such planes is given by

$$d = \frac{a}{\sqrt{h^2 + k^2 + l^2}} \qquad (8-2)$$

where a is the lattice parameter and h, k and l are the Miller indices of the planes. As in the case of directions, families of equivalent planes are expressed by a single set of indices enclosed in doubly curved brackets, e.g. {326} .

Crystal lattices exhibit various kinds of symmetry, one of these being a rotational symmetry about various axes. A body is possessed of an n-fold rotational symmetry about an axis if a rotation of $360°/n$ brings it into self-coincidence. Thus a cube has a four-fold rotational axis normal to each face, a three-fold axis along each body diagonal and two-fold axes joining the centers of opposite edges. Some of these are shown in Figure 8-14 where the small plane figures (ellipse, triangle and square) designate the 2-, 3- and 4-fold axes, respectively.

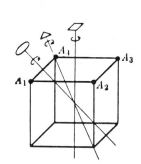

Figure 8-14:
Axes of Rotational Symmetry of Cubic Lattice.

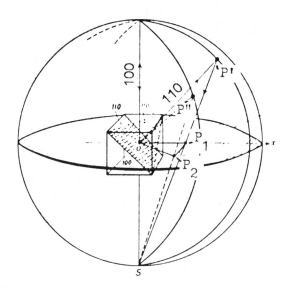

Figure 8-15: Sketch Illustrating Stereographic Projection.

For the convenient representation of crystallographic orientations, stereographic projections are used. Figure 8-15 illustrates how such a projection relates to the planes of a single crystal. The unit crystal is located at the center of a sphere and the normals from the center of each plane are drawn to the surface of the sphere which they intersect at points P', P", etc. Then one plane (e.g., the (100) plane) is chosen as the reference plane and the projection of its normal on the surface of the sphere becomes the pole S. Lines are then drawn from the pole S to the points P', P", etc., and the corresponding points of intersection (P_1, P_2, etc.) with the equatorial plane constitute the stereographic projections of the places. If this procedure is followed for all planes, the equatorial plane is covered with a symmetrical array of points as shown in Figure 8-16. Figure 8-17 shows the standard projections of cubic crystals with reference to the (001) and (011) planes, only the more important planes being illustrated, together with a square, triangle or ellipse surrounding each point to indicate its degree of rotational symmetry. For the cubic system, this array consists of 24 repeated triangles and it is possible to represent a unique axis, such as a tension or compression axis, by its position within a unit stereographic triangle as shown in Figure 8-18.

If a polycrystalline specimen were to be placed at the center of the sphere and the projection of only a certain plane (e.g., the (100) plane) for each crystallite is plotted, a stereographic representation similar to that shown in Figure 8-19(a) would be obtained for crystallites of random orientation. However, in rolling and other metal forming operations, the

ALLOYS OF IRON

crystallites develop a preferred orientation or texture so that the projections of the planes are no longer random but grouped as exemplified by the well-known cube texture shown in Figure 8-19(b). Such representations are known as pole figures.

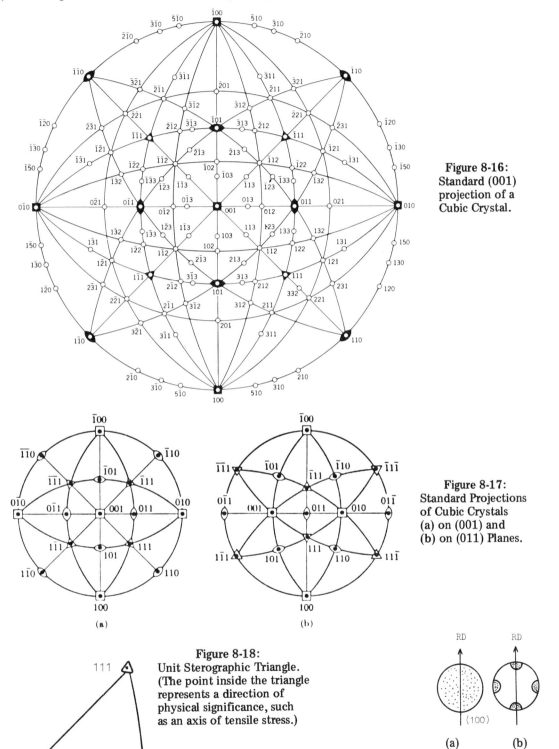

Figure 8-16: Standard (001) projection of a Cubic Crystal.

Figure 8-17: Standard Projections of Cubic Crystals (a) on (001) and (b) on (011) Planes.

Figure 8-18: Unit Sterographic Triangle. (The point inside the triangle represents a direction of physical significance, such as an axis of tensile stress.)

Figure 8-19: (100) Pole Figures (a) Random Orientation; (b) Preferred Orientation (RD = Rolling Direction).

8-6 Structural Defects in Crystal Lattices

Although metals, such as steels, exhibit crystallographic structures, the various types of atomic arrangements are not free from defects. Basically, there are three types of defects; those associated with single atoms or atomic positions (point defects); those identified with a line of atoms (dislocations) and those pertaining to surface areas of the lattice planes (partial dislocations).

Four types of point defects exist, as illustrated in Figure 8-20. In the case of a vacancy, an atom is missing from its normal place in the lattice whereas, in the case of an interstitialcy, an atom (like those constituting the rest of the crystal) is lodged between other atoms occupying their normal places in the lattice. A combination of a vacancy and an interstitialcy constitutes a Frenkel defect. At the same time, the introduction of a different type of atom into the lattice disturbs its perfection whether it occurs at a lattice point as a substitutional element or as an interstitial impurity.

From diffusion experiments, it is apparent that a significant number of lattice positions are vacant in all metals at all temperatures above absolute zero. The equilibrium fraction of vacancies (N) may be determined from the equation.

$$N = e^{-\Delta H/RT} \qquad (8\text{-}3)$$

where ΔH is the molar heat of reaction accompanying the formation of vacancies (about 20,000 cal./mol. for typical metals), R is the gas constant (approximately 2 cal./mol. deg. C) and T is the absolute temperature (°K).

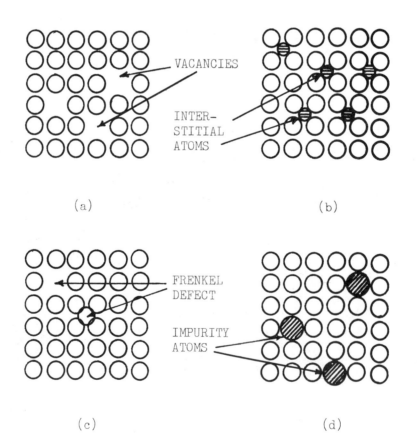

Figure 8-20: Four Types of Point Defects. (a) Vacancy, (b) Interstitialcy, (c) Frenkel Defect, (d) Impurity Atom.

ALLOYS OF IRON

Vacancies are important for a number of reasons. Under certain conditions, they may aggregate to form voids, particularly at grain boundaries, and hence may initiate cracks or failures of the metal. The energies of the conduction electrons in the metal are affected by the presence of vacancies with the result that the resistance of the metal is increased as is the magnetic susceptibility. Moreover, the internal friction of the metal appears to be affected by vacancy concentration. Most important, however, is the role that vacancies play in diffusion phenomena such as carburization and decarburization and various solid state reactions.

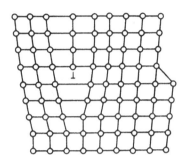

Figure 8-21: Diagrammatic Representation of Edge Dislocation.

An edge dislocation, one of the two types of line defects, is illustrated diagrammatically in Figure 8-21, and it may be imagined as being created by partially cutting the crystal vertically and inserting a half plane of atoms in the cut.◻ Alternately, it may be considered as created by the removal of a half plane of atoms below the line of the dislocation. In proximity to the dislocation, the crystal is stressed. Referring to Figure 8-21, immediately above the dislocation, the lattice structure is in compression, whereas below the defect it is in tension. These induced stresses are important inasmuch as they produce reactions with other defects tending to cause either aggregations of defects (such as the clustering of vacancies or impurity atoms around the dislocations), or internal stresses in the metal (making its deformation more difficult, as in the case of work hardening).

In the deformation of the crystal lattice, there may be a movement of the edge dislocation as indicated in Figure 8-22. The plane containing the path of the dislocation is known as the slip plane.

A quantitative description of a dislocation is given by its Burgers vector, $\underline{b}$. To obtain this vector, an atom-to-atom path is made in a defect-free portion of the crystal and a similar path is traced around the dislocation. The vector needed to close the loop (the vector from the end of the circuit to its starting point) is the Burgers vector describing the dislocation and is conveniently expressed in terms of the lattice parameters. In the example shown in Figure 8-23, the components of the vector are a, o, o with the vector being parallel to the x axis. A more concise representation of the same vector is 1[100] where the numbers in the brackets give the components of the vector as the number of lattice distances along each of the three coordinate axes and where the number in front of the first bracket indicates the length of the Burgers vector as a multiple of the vector inside the brackets.

It is important to note that an edge dislocation cannot end inside the crystal, but must terminate at the surface of the crystal, or a grain boundary or must change its characteristics to become a screw dislocation (see below). It is thus possible for edge and screw dislocations to join together in the crystal structure to form a dislocation loop.

The term screw dislocation is used to describe the other type of line defect because of the spiral surface formed by the atomic planes around the line of the dislocation. It may be considered as being formed by first partially cutting a crystal coincident with one of its planes and

---

◻ For simplicity, a cubic lattice is represented in Figure 8-21. In metals, however, bcc, fcc and close packed hexagonal structures are most common and the dislocation structure is more complicated than as shown in the figure.

then sliding the surfaces of the cut relative to each other as shown in Figure 8-24 giving an atomic arrangement for a cubic lattice as represented by Figure 8-25.

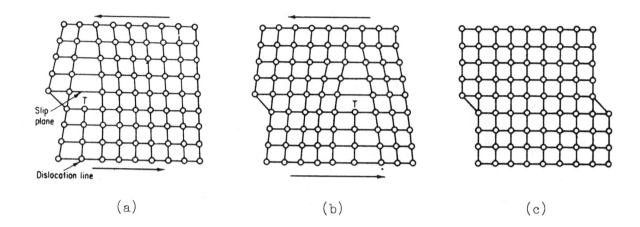

Figure 8-22:
The motion of an edge dislocation and the production of a unit step of slip at the surface of the crystal. (a) An edge dislocation in a crystal structure. (b) The dislocation has moved one lattice spacing under the action of a shearing force. (c) The dislocation has reached the edge of the crystal and produced unit slip.

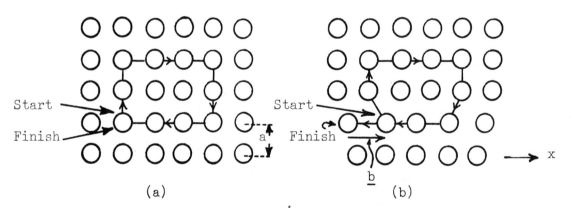

Figure 8-23:
Diagrams Illustrating the Meaning of the Burgers Vector. (a) A Burgers circuit in a dislocation-free material. (b) The same Burgers circuit passing through dislocation-free material, but encircling a dislocation of unit Burgers vector b.

ALLOYS OF IRON

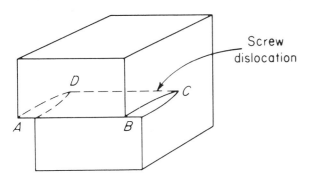

Figure 8-24:
Formation of a Screw Dislocation by Cutting in the Plane ABCD and Shearing.

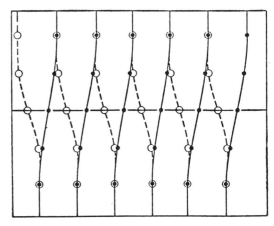

Figure 8-25:
Atomic Arrangement Around a Screw Dislocation. (The solid circles and Open Circles Represent Atoms in Adjacent Planes.)

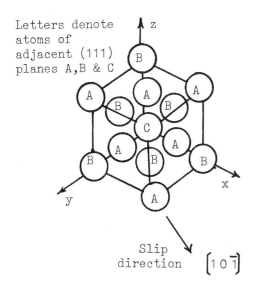

(a) (111) planes and $[10\bar{1}]$ slip direction.

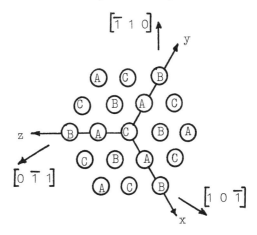

(b) The A, B and C planes viewed normally.

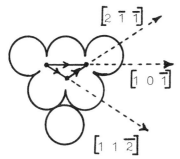

(c) Unit slip of an atom in the (111) plane via two paths.

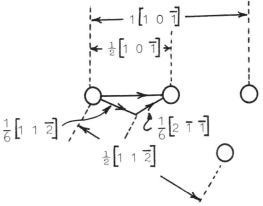

(d) The vectors involved in the slip in the (111) plane.

Figure 8-26:

The mechanism of the slip process in face-centered cubic metals.

In more complex metallic structures, dislocations may be split into partial dislocations, as indicated by the slip process in an fcc lattice illustrated in Figure 8-26. In this structure, the close packed slip planes are the {111} planes (see Figure 8-26(a)) with the slip occurring in the <10$\bar{1}$> directions (Figure 8-26(b)). In the latter figure, the fcc structure can be visualized as a similar alternation of three planes A, B and C that differ only in relative position.

When the slip in the [10$\bar{1}$] direction on the (111) plane is studied, it is seen that a direct path is not possible because of the obstruction of a "hump". However, a combination of two "valley" paths results in the same net displacement. These are partial dislocations in the <11$\bar{2}$> directions (Figure 8-26 (c) and (d)). If the 1[10$\bar{1}$] is the vector from the origin out to the point whose lattice coordinates are x=1, y=0 and z =—1 (in terms of the axial lengths), and if the 1[11$\bar{2}$] is an analogous vector, then, as shown in Figure 8-26 (d), the slip from one position to an adjoining equivalent position, given by the vector 1/2[101] actually occurs by the combination of 1/6[11$\bar{2}$] plus 1/6[2$\bar{1}\bar{1}$]. Thus the atomic movement is represented by

$$1/2[10\bar{1}] \rightarrow 1/6[11\bar{2}] + 1/6[2\bar{1}\bar{1}] \qquad (8-4)$$

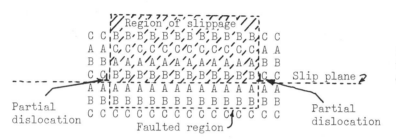

Figure 8-27: Stacking Fault containing two partial dislocations.

Two partial dislocations have their lowest energy when they are separated by about ten atomic distances.§ In the space between them, the lattice is not, strictly speaking, an fcc structure since the alternation of the atomic layers has been disturbed in the manner shown in Figure 8-27. This figure represents seven (111) planes in which the atoms are indicated by a letter (A, B or C). Starting at the left of the figure, the lattice is fcc up the first partial dislocation but, at this point, the positions of the atoms in the portion above the slip line are changed. Specifically, the layer immediately above the slip line changes from C positions to B positions since the A positions are excluded. When the second partial dislocation is reached, this layer of atoms changes in position again and returns to the C position. Thus the crystal to the right of the second partial dislocation resumes the symmetry of the fcc structure. The entire region that has slipped constitutes a stacking fault.

8-7 Grain Structure

Although solid metals are crystallized substances, they are generally polycrystalline in nature, i.e., they consist of an agglomerate of many small crystals designated as crystallites or grains. Each grain may consist of many thousands of unit cells all oriented in virtually the same manner. However, different grains have different lattice orientations and thus a mismatching of lattice structures occurs at the junctions of grains, or the grain boundaries.

In the simplest case where the orientation mismatch between adjacent grains is slight (corresponding to a slight tilting of the lattice to the extent of a few degrees), the tilt boundary so

ALLOYS OF IRON

formed consists of an array of edge dislocations as shown in Figure 8-28, the spacing between the dislocations is equal to b/θ where b is the spacing of the planes and θ is the angle of tilt (expressed in radians) between the two grains.

In other cases, however, a lattice structure of one crystallite can only be aligned with that of an adjacent crystallite by a rotation about an axis normal to the boundary. In such an instance, a twist boundary is said to exist consisting of two crossed arrays of screw dislocations as illustrated in Figure 8-29 where the top plane below the boundary is shown by the full lines and the twist plane above the boundary by the broken lines.

Because the energy of a crystal lattice is increased by the presence of a dislocation, grain boundaries represent regions of high energy analogous to the surface tension energy of a film. Accordingly, the system attempts to minimize this grain boundary energy by decreasing the areas of the boundaries. This is accomplished by grain growth which occurs at the high temperatures encountered in annealing operations.

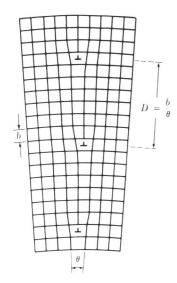

Figure 8-28:
A tilt boundary as a succession of added crystal planes. (The corresponding edge dislocations are indicated.)

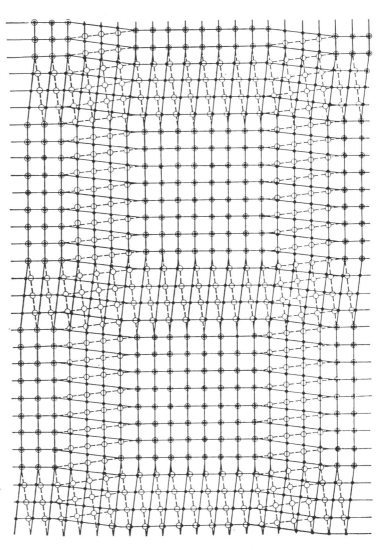

Figure 8-29:
Representation of a twist boundary as a network of screw dislocations.

Ideally, for minimum energy, the faces of grains should be flat and meet in threes with dihedral angles of 120° and the edges should meet at points with angles of 109° 28 min. (this being the angle between any two of four lines meeting symmetrically at a point). The tetrakaidecahedron, shown in Figure 8-30, most closely approaches these conditions, especially if slight curvatures are introduced in its faces. It should be remembered, however, that even if an idealized structure such as that shown in Figure 8-30 were to be cut, polished and examined, the grains would not be perfect hexagons because of the arbitrary choice of the section being studied. In practice, a nearly equilibrated grain structure is as shown in Figure 8-31, and in commercial metals and alloys, the grain sizes range from 10^{-1} to 10^{-4} cm. with typical values falling into the narrower range of 10^{-2} to 10^{-3} cms.

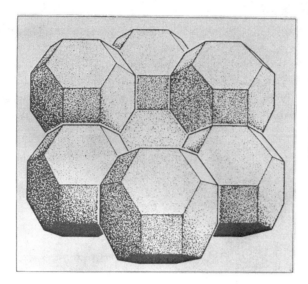

Figure 8-30:
Stack of tetrakaidecahedrons.
From Metal Interfaces, American Society for Metals, Cleveland, 1952, Figure 19, p. 90.

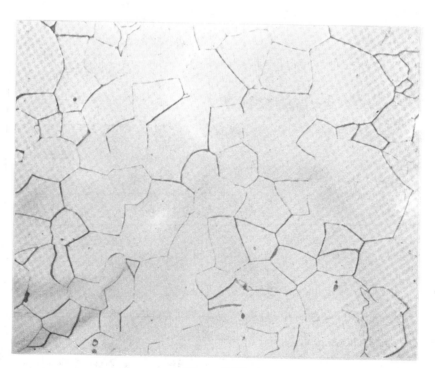

Figure 8-31:
Nearly equilibrated structure of polycrystalline specimen. Magnification (X780).

ALLOYS OF IRON

Grain and other interfacial boundaries have important effects on the properties of metals. They impede the movement of dislocations and hence make the material harder to deform. In the case of grain boundaries, the effect of grain size on the lower yield strength (σ_y) is described by the Hall-Petch relation

$$\sigma_y = \sigma_i + k_y d^{-1/2} \qquad (8\text{-}5)$$

where σ_i is the friction stress, k_y is thought to be a function of dislocation creation and d is the grain size.◈ Confirmation of this relationship is shown by the data of Figure 8-32 for Armco iron of various grain sizes tested at four temperatures at a strain rate of 4.2×10^{-4} seconds^{-1}.

In the case of grains containing particles of a dispersed phase, the relationship is often expressed in the form

$$\sigma_y = \sigma_i + h \ln(d) \qquad (8\text{-}6)$$

where σ_i and h are constants and where d is the mean free path in the matrix of the metal. In the absence of particles of a dispersed phase, d represents the average distance between grain boundaries (as, for example, the spacing of the platelets in lamellar pearlite).

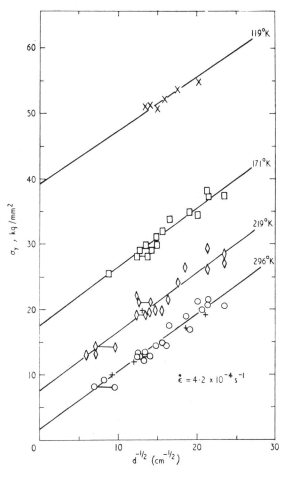

Figure 8-32: Lower Yield Stress σ_y of Armco Iron and Remelted Iron (+) at a Strain Rate of 4.2×10^{-4} S^{-1} as a Function of Grain Size at Various Temperatures.

During cold rolling, the microstructure of metals is severely distorted. Figure 12-3 shows the elongation of the ferrite and pearlite grains exhibited by full-hard, cold-reduced black plate. The number of point and line defects in this material has been increased by a factor of about 10^4 as a consequence of cold reduction, yet there is evidence that fragments of the crystal lattice are only moderately distorted as illustrated by Figure 12-4.

◈ E. Anderson and J. Spreadborough, "Yielding, Twinning and Fracture of Armco Iron and Alloys of Iron with Silicon, Chromium and Manganese", Journal of the Iron and Steel Institute, December, 1968, pp. 1223-1235.

8-8 Sub-Boundaries and Cells Within Grains

Within the confines of grain boundaries, the lattice structure is often far from perfect but features sub-boundaries or other cellular configurations. These, and a special type of grain boundary, known as a twin boundary, are discussed below.

During crystal or grain growth, slight misorientations (usually less than a degree) of the lattice structure occur with the consequent generation of sub-boundaries, generally similar in structure to the regular grain boundaries. It is possible that they always consist of a single array of screw dislocations as in a tilt boundary. This is confirmed to a certain extent by the fact that they can be made to move by the application of relatively small stresses exactly as would be predicted if each dislocation of the array were considered as moving independently in parallel slip planes.▲

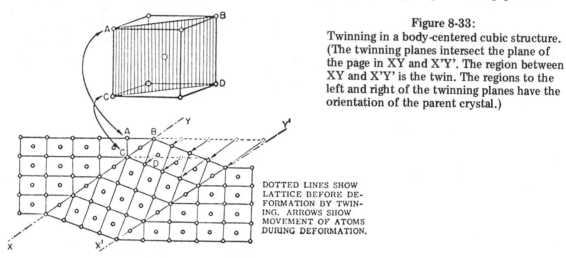

Figure 8-33:
Twinning in a body-centered cubic structure. (The twinning planes intersect the plane of the page in XY and X'Y'. The region between XY and X'Y' is the twin. The regions to the left and right of the twinning planes have the orientation of the parent crystal.)

DOTTED LINES SHOW LATTICE BEFORE DEFORMATION BY TWINING. ARROWS SHOW MOVEMENT OF ATOMS DURING DEFORMATION.

A special type of grain boundary in a pure metal is the twin boundary shown diagrammatically in Figure 8-33 and in a metallurgical specimen in Figure 8-34.◈ Here the two portions of the lattice separated by the boundary have a special orientation relationship that permits easy transition from one lattice to the other. This is a consequence of the fact that, at such a boundary, the crystals match plane for plane and are therefore coherent with each other. Such a boundary is common where the separated crystals have a twin or mirror-image relationship. This is exemplified by a stack of close packed planes, abcabc... as illustrated in Figure 8-35. A line of atoms

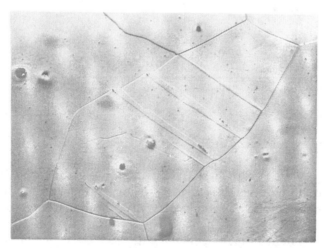

Figure 8-34:
Fe-1% Mn alloy which failed by ductile cleavage at 119° K: $d^{-1/2} = 11.1$ cm$^{-1/2}$. (The twins appear to have been initiated at the left-hand grain boundary; the shear at the "initiating" and "stopping" grain boundaries is evident.) X1000 (Anderson and Spreadborough).

▲ "Physical Metallurgy", by B. Chalmers, John Wiley and Sons, Inc., New York, 1959.
◈ E. Anderson and J. Spreadborough, "Yielding, Twinning and Fracture of Armco Iron and Alloys of Iron With Silicon, Chromium and Manganese", Journal of Iron and Steel Institute, December, 1968, pp. 1223-1235.

such as $A_1B_1C_1A_2B_2C_2$ in the figure, in which each successive atom is in the next layer upwards, lies in a <110> direction in a {111} plane of the crystal as indicated in Figure 8-36.

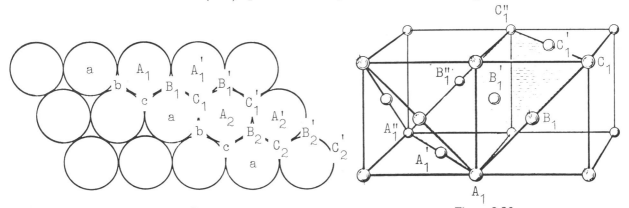

Figure 8-35:
Projection on (111) plane of atom sites in face-centered cubic crystal.

Figure 8-36:
Close-packed planes in face-centered cubic structure.

If a crystal defect occurs such that the stacking sequence (abcabc) is disturbed so that it reverses (abcabacbacb), then the points corresponding to $A_1B_1C_1$ (Figures 8-35 and 8-36) will no longer lie in a straight line but will change direction in such a manner that the new direction will be a mirror image of the former direction in the stacking plane (see Figure 8-37). However, it will be noted that each atom, including those actually in the twinning plane, still possess the same number of nearest neighbors at the same distance as originally. The second nearest neighbor relationships are modified, however, at the interface, where they are of the aba type characteristic of hexagonal close packing rather than the abc type of close packing encountered in bcc or fcc lattices. Because of this special configuration of the atoms, the energy associated with the coherent twin boundary is much lower than that of a typical grain boundary. In fcc and bcc lattices, the twinning planes are the {111} and {112} planes, respectively.

It will be noted that there is a similarity between a stacking fault and a twin boundary. In the former, the stacking sequence is abcababc whereas the latter it assumes the form abcbabc. Thus the stacking fault contains two layers with abnormal second nearest neighbors while the twin boundary contains only one. Moreover, whereas the stacking fault has a crystal of identical orientation each side of it, the twin boundary does not.

With respect to cell structure, Keh[°] found that whereas dislocations in alpha iron lightly strained at room temperature are usually joggy and nonuniformly distributed, under certain conditions of increased strain, they become arranged in a cellular structure (Figure 12-5). However, the cell size and the tendency to form cells decrease with decreasing deformation temperature.

8-9 Alloys of Iron

For commercial purposes, iron is almost always alloyed with other elements to form steels.[•] Steel is classified as carbon steel when no minimum content is specified or required for any element added to obtain a desired alloying effect. On the other hand, such materials are classified as alloy steels when the content of alloying element exceeds certain limits or in which a definite range of alloying elements is specified within the limits of the recognized commercial field of alloy steels.

[°] A. S. Keh, "Dislocation Arrangement in Alpha Iron During Deformation and Recovery" — portion of book, "Direct Observations of Imperfections in Crystals", edited by J. B. Newkirk and J. H. Wernick, Interscience Publishers, a division of John Wiley and Sons, New York, 1961.

[•] Steel is defined as an iron-base alloy, malleable in some temperature range as initially cast, containing manganese, usually carbon and other alloying elements. ("The Making of Steel", Second Edition, American Iron and Steel Institute, 150 East Forty-Second Street, New York, N.Y., 10017, 1964).

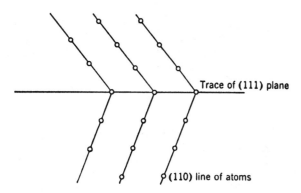

Figure 8-37: Twinned crystal in {110} plane.

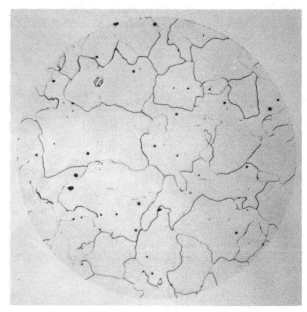

Figure 8-38:
The Microstructure of Ferrite. Magnification 100X (This sample has a coarse grain size).

Carbon steels at ambient temperatures contain two principal constituents, a relatively pure form of iron (ferrite) and iron carbide (cementite). The former is distinguishable in the microstructure as polyhedral grains as seen in Figure 8-38. In pure iron-carbon alloys, the ferrite consists of iron with a trace of carbon in solution but, as discussed below, in steels it may contain considerable quantities of alloying elements such as manganese, silicon or nickel dissolved in it. The cementite, on the other hand, is a much harder material corresponding to the chemical formula Fe_3C. When these two constituents precipitate together (as in a eutectoid steel) a characteristic lamellar structure known as pearlite is produced (Figure 8-39).

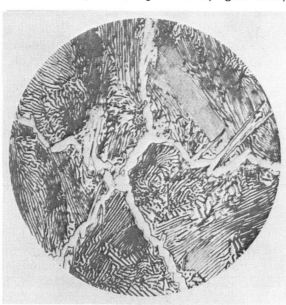

Figure 8-39:
The Microstructure of Slowly Cooled, High Carbon Steel Showing Pearlite with Cementite in the Grain Boundaries. Magnification 1000X.

Most of the common alloying elements are soluble to a certain extent in ferrite with the majority of them (including aluminum, copper, nickel, phosphorus and silicon) forming substitutional solid solutions. However, alloying elements with significantly smaller atomic radii (such as carbon, nitrogen and hydrogen) form interstitial solid solutions. Other elements, such as titanium, vanadium, molybdenum and tungsten, form carbide phases while others may be present in the cementite.

ALLOYS OF IRON

The general effects produced by the more common alloying elements, (manganese, nickel and chromium), are as follows. Manganese, (soluble in ferrite to 3 per cent and which is present in all commercial steels in the range 0.3 to 0.8 per cent), reduces oxides and counteracts the harmful influence of iron sulphide. When the manganese content is increased beyond this range, part is dissolved in the ferrite and part forms manganese carbide Mn_3C. It hardens the steel markedly and increases the hardenability moderately but inexpensively. Nickel is soluble in ferrite to 10 per cent, irrespective of the carbon content, and it strengthens unquenched or annealed steels. It also toughens pearlitic-ferritic steels, especially at low temperature and renders the high chromium iron alloys austenitic. Chromium, with unlimited solubility in ferrite, hardens the matrix and significantly increases its corrosion and oxidation resistance. It also increases hardenability, high temperature strength and (with high carbon) provides increased abrasion and wear resistance.

Since aluminum is generally used as a deoxidizing agent in steel making practices, it also appears as an alloying element in the solidified metal. It is highly soluble in ferrite (up to 36%), and it restricts grain growth by forming dispersed oxides or nitrides. It also exerts a hardening influence on the steel but only mildly increases the hardenability of the material.

Phosphorus also hardens strongly by solid solution and increases the resistance of steel to corrosion. The element is soluble in ferrite to 2.8% irrespective of carbon content.

Silicon, soluble to 18.5% in ferrite, hardens the steel with a loss of plasticity. It is often used as a general purpose deoxidizer, as a constituent for electrical steels (see Section 8-3) and for improving the oxidation resistance of steel.

8-10 The Iron-Carbon Equilibrium Diagram

To obtain a basic understanding of the properties and microstructures of iron-carbon alloys, reference should be made to the iron-carbon equilibrium diagram shown, in part, in Figure 8-40. It should be emphasized, however, that such a diagram relates to the compositions of phases which are present at various temperatures as a result of equilibrium cooling or heating of iron and carbon alloys at extremely small rates. The diagram indicates the ranges of compositions and temperatures within which the various phases are stable and the boundaries at which phase changes occur.

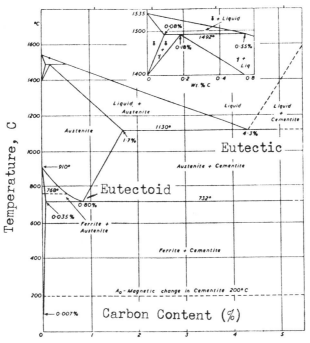

Figure 8-40: Equilibrium Diagram of Iron and Iron Carbide.

In the iron-carbon system, solid solutions of carbon in delta, gamma and alpha iron exist, namely, delta solid solution, austenite and ferrite. A eutectic (ledeburite) occurs at a composition of 4.3% carbon and consists of austenite (1.7% carbon) and cementite. There is also a eutectoid (pearlite), containing close to 0.8% carbon, consisting of ferrite and cementite.

When cooling curves (time-temperature curves) are obtained for alloys containing less than 1.7% carbon, "steps" or arrests are experienced when the transformations occur. The temperatures corresponding to these arrests are designated critical temperatures or critical points. The loci of such points separating the austenite and the austenite + ferrite region is designated the A_3 curve. The temperature corresponding to the magnetic transformation (the Curie point) is the A_2 temperature and that of the eutectoid as A_1 for alloys to the left of the eutectoid and by $A_{1,3}$ for alloys to the right of the eutectoid. The curve separating the austenite and the austenite + cementite regions is designated as A_{cm} and the interval between A_1 and A_2 is the critical range.

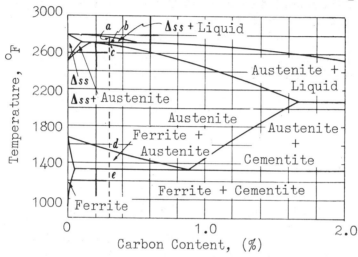

Figure 8-41: Portion of Equilibrium Diagram of Iron and Iron Carbide.

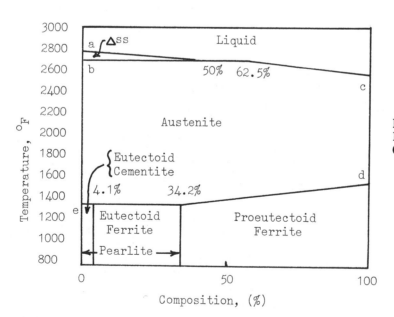

Figure 8-42: Phase Transformation Diagram of 0.3 Per Cent Carbon-Iron Alloy.

To illustrate the use of the equilibrium diagram, the slow cooling from 2800°F of a 0.3% carbon alloy may be considered. Referring to Figures 8-41 and 8-42, at the temperature a, solidification begins with the formation of crystals of delta solid solution (Δss), but with continued cooling, the temperature b is reached when the peritectic reaction occurs. On completion of this

ALLOYS OF IRON

reaction, the phases will consist of austenite and liquid. The percentage of austenite increases with decreasing temperature until, at the temperature corresponding to point c, all the solid is austenite. On further cooling, the point d on the A_3 transformation curve is reached and ferrite begins to form. (This ferrite is designated proto-eutectoid ferrite to distinguish it from that formed in the pearlite.) When the temperature reaches e, the remaining austenite transforms into pearlite (eutectoid ferrite and eutectoid cementite). With further cooling of this alloy, a small amount of cementite will precipitate from the ferrite because of the decreasing solubility of carbon in alpha iron with decreasing temperature.

To illustrate the cooling of a hypereutectoid steel with a carbon content in excess of 1.7 per cent, a 3 per cent carbon alloy may be considered. Referring to Figure 8-43, solidification of this alloy will begin to occur when the temperature has fallen to that corresponding to point a. Austenite (protoeutectic austenite) will continue to form until the temperature reaches b (2065°C). Below this temperature proto-eutectoid cementite will precipitate from the austenite until the temperature reaches 1333°F (point c). Below 1333°F, the remaining austenite converts to pearlite. Figure 8-44 illustrates the composition of the alloy with respect to the various phases as it cools from the liquid phase to a temperature of 800°F.

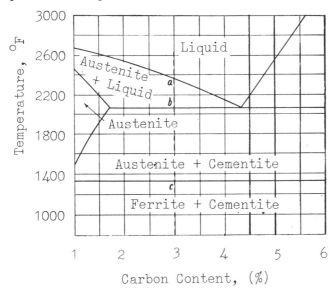

Figure 8-43: Portion of Equilibrium of Iron and Iron Carbide.

8-11 Austenite

Inasmuch as virtually all carbon steels have been in the austenitic form at some time in their existence, and their present microstructure results from a decomposition of this phase, it is appropriate to study the characteristics of the material. Basically, its atomic structure is that of gamma iron (face-centered cubic) with an atomic spacing that varies with carbon content. It is capable of dissolving up to 2% carbon and is stable at temperatures above those defined by the A_3 and A_{cm} curves, but below those fixed by the boundaries of the ($\gamma + \delta$) and the (γ + liquid) phases. (See Figure 8-40.)

Almost all heat treatments require that austenite be produced as the first step in heat treating operations, and to produce austenization within a reasonable time, the temperature of the steel must be raised somewhat above that corresponding to the A_3 or A_{cm} curve. Figure 8-45 shows the approximate times necessary for the isothermal formation of austenite in a normalized † eutectoid steel at various austenitizing temperatures, whereas Figure 8-46 shows the temperature ranges usually employed in various heat treatments.

---

† A normalized steel is one which has been heat treated to refine the grains and to obtain a carbide size and distribution which will be more favorable for carbide solution on subsequent heat treatment than the as-rolled material. (See Section 12-7.)

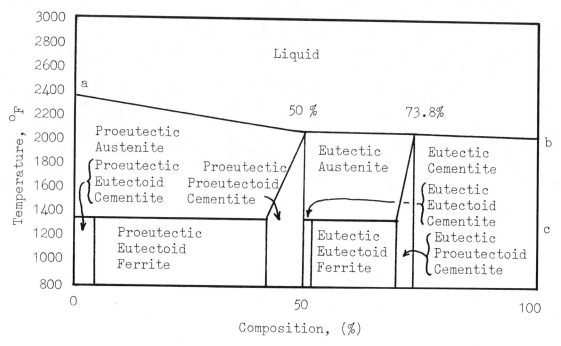

Figure 8-44: Phase Transformation Diagram for a 3 Per Cent Carbon-Iron Alloy.

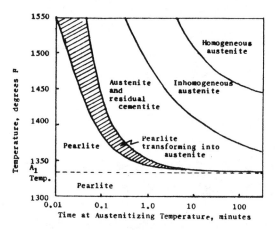

Figure 8-45:
Approximate times necessary for the isothermal formation of austenite in a normalized eutectoid steel at various austenitizing temperatures. (After Roberts and Mehl.)

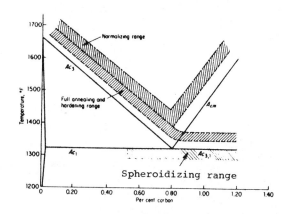

Figure 8-46:
A portion of the iron-carbon equilibrium diagram showing the temperature ranges usually employed in various heat treatments.

The first stage in the production of homogeneous austenite is the nucleation and growth of austenite from pearlite. However, even after the pearlite has been transformed, some carbide particles remain in the austenite and, even after these have dissolved, their trace remains for some time in the form of inhomogeneities of carbon concentration in the austenite. The rate of

ALLOYS OF IRON

austenization may be increased by increasing the temperature and by increasing the fineness of the initial carbide particles. The temperature used for hypoeutectoid and eutectoid steels is about 100°F above the A₃ curve and for the hypereutectoid steels about 1425°F. (Very low carbon steels must be heated to high temperatures to austenitize them completely, but as steels containing less than 0.2 per cent carbon respond poorly to hardening heat treatments, they are seldom heated for this purpose.)

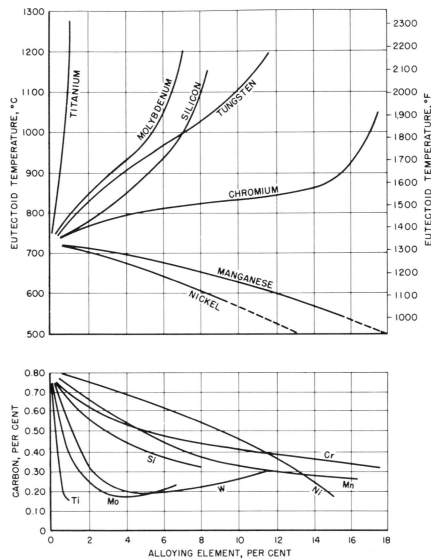

Figure 8-47: Eutectoid Composition and Temperature as Influenced by Several Alloying Elements (After Bain).

Alloying elements, such as manganese and nickel, lower the A_1 temperature but also shift the eutectoid point to a lower carbon content, thus facilitating austenitization. These effects, and those produced by other alloying elements, are illustrated in Figure 8-47. Those elements which enlarge the austenite field of the iron-carbon equilibrium diagram are descriptively known as austenite formers, whereas those elements, such as chromium, silicon, molybdenum, tungsten, vanadium, tin, columbium, phosphorus, aluminum and titanium are known as ferrite formers.

The grain size obtained in austenization is a function of the temperature above the critical to which the steel is heated as well as by the state of the metal prior to austenitization. Grain growth may be inhibited by carbides which dissolve slowly or remain undissolved in the austenite or by a suitable dispersion of non-metallic inclusions. Hot working refines the coarse grain size formed by heating to the relatively high temperatures used in hot rolling and the grain size of hot worked products is determined largely by the temperature at which the final stage of the hot working process is carried out.

8-12 The Decomposition of Austenite

Although the equilibrium diagram indicates that austenite is not stable at temperatures below those corresponding to the A_1 line, it is not informative with respect to the decomposition of that phase. To obtain more information with respect to the solid state reactions that ensue when austenite is cooled, recourse must be made to an appropriate isothermal transformation diagram (often called a TTT diagram, an S curve or a C curve), an example of which is shown in Figure 8-48. For each steel composition, there is a different diagram. Moreover, the presence of inclusions and other inhomogeneities can change the shape of the diagram for a given steel, but these complications are generally neglected.

In preparing such diagrams, a number of small specimens are austenitized and quickly transferred to a bath maintained at the desired temperature. After a given time in the bath, each specimen is removed and quenched in water. This causes the remaining austenite to be immediately changed to martensite which is distinguishable under the microscope. Thus the extent of the austenite decomposition can be readily ascertained and, if sufficient specimens are maintained at various bath temperatures, the isothermal transformation diagram may be constructed.

In the diagram for eutectoid steel illustrated in Figure 8-48, the austenite may decompose in two radically different ways. At high temperatures, the reaction products form with increasing time at constant temperatures. On the other hand, martensite forms only with decreasing temperature and has almost no tendency to continue forming at constant temperature.

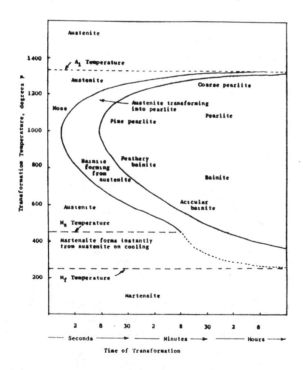

Figure 8-48:
Isothermal Transformation Diagram for the Decomposition of Austenite in a Eutectoid Carbon Steel.

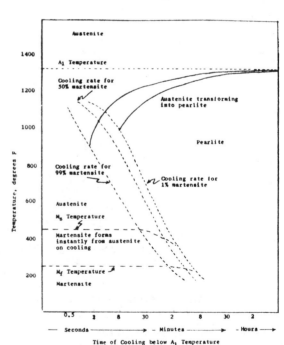

Figure 8-49:
An Approximate Continuous Cooling Transformation Diagram for Eutectoid Carbon Steel.

The products of the decomposition of austenite are spheroidite, pearlite, and bainite at high temperatures and martensite at low temperatures. It is possible to produce mixtures of two or

ALLOYS OF IRON

more of these constituents and martensite may be tempered to produce a second series of decomposition products.

In commercial heat treating operations, it is not usually practical to instantaneously drop the temperature of the steel to a given value and then maintain it constant as in the case of the specimens used in the preparation of isothermal transformation diagrams. Instead, the steel temperature continues to decrease with time so that a CCT (continuous cooling transformation) diagram would be more appropriate with respect to the decomposition of the austenite. Few such diagrams are available but that for the eutectoid carbon steel is shown in Figure 8-49. The most important rates on such a diagram are those that produce 99 percent and 50 percent martensite. The first of these, the critical cooling rate, establishes the position of the "nose" of the diagram, the time for which is about twice the time for the "nose" of the TTT diagram. The cooling rate that produces 50 per cent martensite is of use in heat treating practices since full hardening is usually defined in terms of 50% martensite at the center of a Jominy bar. (It is to be noted that no bainite forms in a eutectoid carbon steel during continuous cooling and even in alloy steels the bainite reaction may be slowed down by a factor of 10^3 as a result of changes that take place in the earlier stages of continuous cooling.)

8-13 Pearlite and Bainite

As indicated by the isothermal transformation diagram shown in Figure 8-48, both pearlite and bainite form isothermally from austenite, pearlite being the decomposition product at temperatures between A_1 and the "nose" of the diagram and bainite the product at lower temperatures.

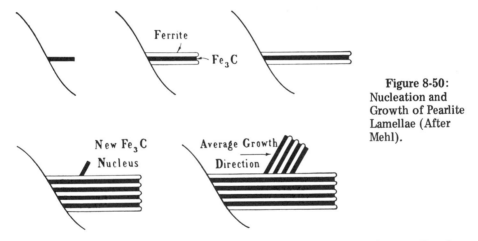

Figure 8-50: Nucleation and Growth of Pearlite Lamellae (After Mehl).

When austenite is first cooled to a temperature appropriate for pearlite formation, there is an initial period (corresponding to an incubation period) when no formation of pearlite can be detected. It is believed that the decomposition of the austenite begins with the formation of microscopic regions of cementite which act as nuclei for the pearlite nodules. This nucleation occurs preferentially at the grain boundaries of the austenite or at an inhomogeneity such as an inclusion as indicated in Figure 8-50.[◊] The nodule growth proceeds with the alternate formation of plates of ferrite and cementite with the simultaneous edgewise growth of these plates. With decreasing temperature, the rate of pearlite formation increases, the thicknesses of the ferrite and cementite plates decrease and their hardnesses increase.

In a temperature range near the "nose" of the isothermal transformation diagram, both pearlite and bainite may be formed, but below this range and above the M_s (martensite start) temperature, only bainite is produced. The nature of the bainite so formed depends upon the

---
[◊] R. F. Mehl, "The Structure and Rate of Formation of Pearlite", Trans. A.S.M. 29, 1941, pp. 813-862.

transformation temperature. In the upper part of the temperature range, feathery bainite (Figure 8-51(a)) is produced whereas acicular or needlelike bainite (Figure 8-51(b)) is obtained by decomposition of the austenite at lower temperatures. Although the process of bainite formation is different at different temperatures, in general, it is a ferrite-carbide aggregate that forms by growing from a ferrite nucleus.

Figure 8-51:
Bainite formed isothermally at two different temperatures in a eutectoid steel. (a) Feathery bainite formed at 925° F. Fine pearlite is also present in the martensite matrix. (b) Acicular bainite formed at 550° F. The matrix is martensite.

It should be noted that some isothermal transformation diagrams show two "noses", one of which is associated with the formation of pearlite and the other bainite. Under such circumstances, it may be possible to produce bainite by quenching but, in carbon steels, an isothermal reaction is necessary to develop this type of structure.

8-14 Martensite

The rapid quenching of austenite, which may contain as much as 1.4 per cent carbon, to a sufficiently low temperature produces martensite, which may be regarded as a ferrite (which has an equilibrium solubility limit for carbon of about 0.02 per cent) supersaturated with the element. The carbon atoms assume positions between the iron atoms in what would normally be a cubic lattice. Figure 8-52 illustrates the position of a carbon atom C between the two iron atoms A and B. Under such circumstances, there is an increase in the spacing of A and B and a decrease in spacing in the other two orthogonal directions. Thus tetragonality is introduced into the lattice structure. However, if such carbon atoms were randomly located at such sites throughout the material, there would be no tetragonality with respect to the lattice as a whole. If, however, the carbon atoms are relatively close together, they tend to cooperate to produce tetragonality that is constant in direction, as this results in a lower elastic strain. This long-range ordering or tetragonality probably occurs when the carbon content of the steel exceeds about 0.2% and the variation of the lattice constants (illustrated in Figure 8-53) with carbon content is as illustrated in Figure 8-54.

Individual crystals of martensite are platelike in form, the plates being elongated along one axis and thicker in the center than at the edges. The material itself is of high strength but low ductility, and has a density slightly less than ferrite (which is about 4% less than austenite). Because

ALLOYS OF IRON

of this density change during the martensitic transformation, large stresses may be induced in the transformed steel. To alleviate this situation, martensite is usually tempered.

In the cooling of austenite, a temperature (designated M_D) is reached below which martensite is more stable than austenite. However, there is a nucleation barrier that requires some supercooling below M_D so that the martensite start temperature M_S must be reached before transformation begins. The M_S temperature varies with carbon content as indicated in Figure 8-55. It should be pointed out, however, that in the M_D and M_S temperature range, transformation may be induced by plastic deformation.

The amount of martensite produced increases progressively as the quench temperature is lowered below the M_S temperature until the transformation is virtually complete at a temperature designated M_F. In some instances, M_F lies below room temperature so that a refrigeration treatment may cause additional martensite to form. Moreover, although the austenite-to-martensite reaction requires a finite time, in commercial steels, this is so short that the transformation may be regarded as instantaneous.

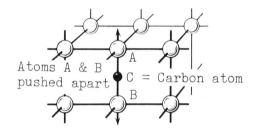

Figure 8-52:
Distortion of Ferrite by Interstitial Carbon.

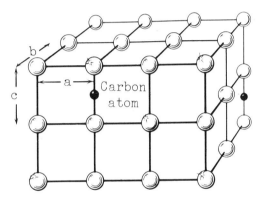

Figure 8-53: Carbon Sites in a Ferrite Crystal.

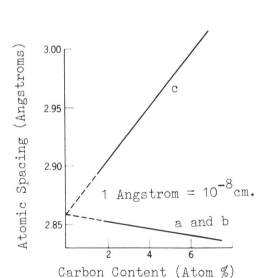

Figure 8-54:
Variation of Lattice Parameters of Martensite with Carbon Content.

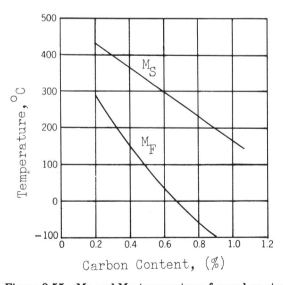

Figure 8-55: M_S and M_F temperatures for carbon steel.

The production of martensite is said to be a diffusionless transformation because the atoms of the existing phase (austenite) move into their new positions because of the strain energy resulting from the movement of neighboring atoms. The process is not thermally activated, but it takes place at a speed that is probably equal to the propagation of an elastic disturbance (namely, the speed of sound in the metal).

Because of the cooperative nature of this reaction, there must be coherent boundaries between the austenitic and martensitic phases. Although the coherency that exists during transformation may be subsequently lost by the plastic deformation of one or both phases, a definite relationship between the orientations of the two phases must persist and the interface (habit plane) must have a definite crystallographic significance. For iron-carbon alloys with carbon contents 0.5 — 1.4%, the $\{111\}$ planes of the austenitic fcc structure must be parallel to the $\{110\}$ planes of the martensitic (bc tetragonal) structure. Similarly, the $<110>$ directions in the former must be parallel with the $<111>$ directions of the latter.

8-15 Stainless Steels

As their name implies, such materials possess unusual resistance to attack by corrosive media at atmospheric and elevated temperatures, a property attributable to chromium used as an alloying element. The corrosion resistance of stainless steels increases with the chromium content, but 4% chromium is generally accepted as the dividing line between iron-chromium alloys and stainless steels.

All stainless and heat resisting steels fall into five general classifications according to their characteristics and alloy content.[8] These are as follows:

(1)   5% chromium, hardenable (martensitic) 500 series.

(2)   12% chromium, hardenable (martensitic) 400 series.

(3)   17% chromium, nonhardenable (ferrite) 400 series.

(4)   Chromium-nickel (austenitic) 300 series.

(5)   Chromium-nickel-manganese (austenitic) 200 series.

Some of the more commonly used types of steels in these categories are listed in Table 8-6. (See also Section 12·31.)

The general structural characteristics of stainless steels may perhaps best be understood by reference first to the iron-chromium alloys, then iron-chromium-carbon alloys followed by iron-nickel, iron-chromium-nickel and iron-chromium-nickel-carbon alloys.

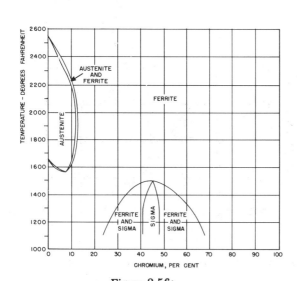

Figure 8-56:
——— Iron-Chromium Equilibrium Diagram.

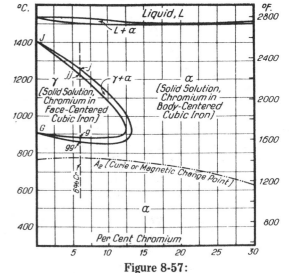

Figure 8-57:
Iron-Rich End of Iron-Chromium Equilibrium Diagram.

[8] "The Making of Steel", American Iron & Steel Institute, New York, Second Edition, 1964.

ALLOYS OF IRON

Figure 8-56 shows the iron-chromium equilibrium diagram with the iron-rich end of it shown enlarged in Figure 8-57. Increasing the chromium content significantly lowers the A_4 curve (the upper boundary of the gamma phase) and slightly decreases the A_3 curve (the lower boundary of the gamma phase). Above approximately 12.5% chromium, the gamma solid solution is nonexistent, and alloys corresponding to this region of the diagram consist of a solid solution of chromium dissolved in bcc iron. For chromium contents in the range 38-57%, a sigma phase is obtainable, this being a hard, brittle, intermetallic compound.

The effect of carbon on the iron-chromium equilibrium diagram is illustrated in Figure 8-58. These diagrams represent cross sections of a 3 dimensional composition-temperature diagram of the ternary system corresponding to levels of 8, 12, 15 and 20% chromium. Referring first to the 8% chromium level, it is seen that this addition of the alloying element has shifted the concentration of carbon in the eutectoid to a lower value (from about 0.8% to less than 0.2%). This same trend is continued in the next diagram for 12% chromium and it is to be noted that the delta solid solution-plus-austenite and the ferrite-plus-austenite regions are almost joined.

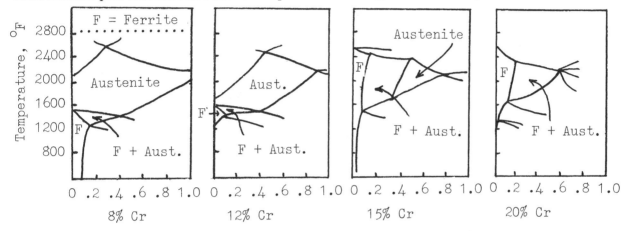

Figure 8-58: Sections of Iron-Chromium-Carbon Ternary Equilibrium Diagram for 8, 12, 15 and 20 Per Cent Chromium.

Alloys will respond to heat treatment if they exist in the form of austenite at high temperatures, and will transform to a mixture of chrome ferrite and chrome carbide on cooling. When the carbon content exceeds 0.35%, slow cooling from the austenite region results in the precipitation of excess chrome-iron carbide. Since many of the 12% chromium-iron alloys containing more than a small percentage of carbon undergo a phase change on cooling, they may be heat treated.

In the 15% chromium diagram of Figure 8-58, it is seen that for very low carbon contents, the austenite is stable down to room temperature (and hence such alloys cannot be heat treated). The austenite region of higher carbon contents is considerably restricted but hardening of such alloys by heat treatment can be achieved to a limited extent.

If the chromium content is increased to 20%, the ferrite-austenite region is greatly widened with complete elimination of the austenite region. High carbon steels of this type, when quenched from the austenite plus chrome iron region exhibit high wear resistance and hardness.

Figure 8-59 shows the equilibrium diagram for the iron-nickel alloys and it may be seen that increasing the nickel content decreases the temperature at which the gamma iron transforms to alpha iron upon cooling. However, the reverse transformation can only occur at much higher temperatures and, therefore, such alloys are designated "irreversible". The better known alloys of this type contain in excess of 35% nickel (such as Invar, with its very low coefficient of expansion and a similar alloy also containing chromium and tungsten, known as Elinvar, which has the additional property of retaining a constant modulus of elasticity over a certain temperature range). High nickel contents also impart high magnetic permeabilities to the alloys.

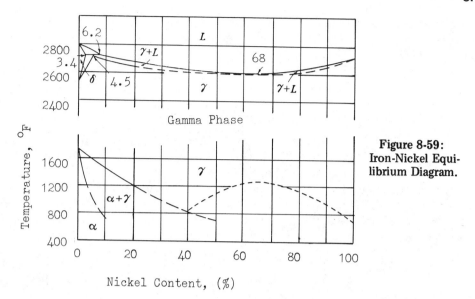

Figure 8-59: Iron-Nickel Equilibrium Diagram.

When chromium is added to the iron-nickel system, the alpha and alpha plus gamma regions are expanded and the gamma region restricted. These effects are illustrated in Figure 8-60. Decreasing amounts of nickel tend to extend the limits of the gamma solid solution and it is possible, with a normal rate of cooling, to obtain gamma solid solution in an alloy containing 20% nickel, 20% chromium and 60% iron even though equilibrium diagrams would indicate that the structure would consist of both alpha and gamma solid solutions.

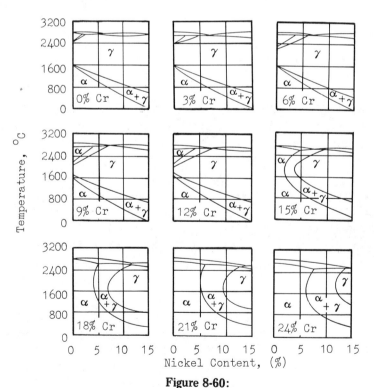

Figure 8-60:
Sections of Iron-Chromium-Nickel Ternary Equilibrium Diagram for 0, 3, 6, 9, 12, 15, 18, 21, and 24 Per Cent Chromium (Bain and Aborn, Metals Handbook, 1939).

ALLOYS OF IRON

An equilibrium diagram showing the phases in an 18% chromium — 8% nickel steel for carbon contents up to .4% is illustrated in Figure 8-61. At low carbon contents, austenite is stable down to room temperature and hence the designation of these alloys as austenitic stainless steel. Such alloys are not only very resistant to corrosion but also quite ductile. Since they cannot be hardened by heat treatment, increased strength is often attained by cold working. In the soft condition, the tensile strength is about 100,000 lbs./in.^{2} and the elongation about 65% in 2 ins. By cold working, the strength may be almost doubled, (see Figure 8-62).

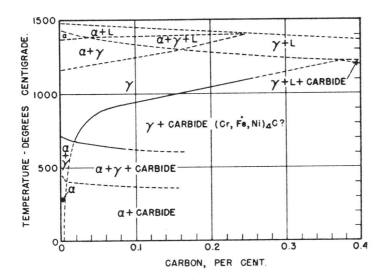

Figure 8-61: Phases in 18% Cr — 8% Ni Steel for Carbon Content Between 0 and .4% (Thum).

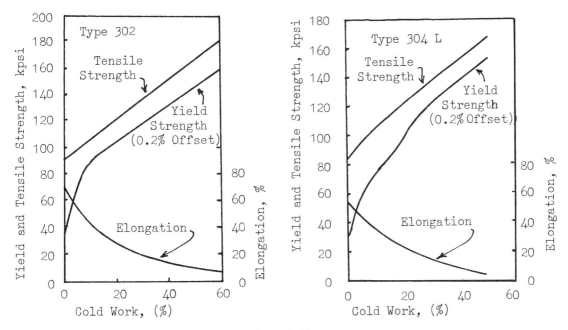

Figure 8-62:
Properties of Stainless Steel After Various Amounts of Cold Work

8-16 Magnetic Alloys

Because of the magnetic behavior of iron, various alloys, such as the silicon steels, are of considerable economic importance for use in the manufacture of electrical equipment, particularly motors and transformers. For these uses, such materials are generally cold rolled, often on small specialty mills to very light gages, and are specially processed to develop the desired magnetic properties. In this section, the magnetic behavior of relatively pure iron and some of the better known magnetic steels will be briefly surveyed.

Iron, nickel and cobalt all exhibit ferromagnetism, a property associated with the electrostatic interaction of adjacent atoms that affects the alignment of the resultant electron spins of the atoms.* Under ferromagnetic conditions, the atomic interaction is positive so that the spins are aligned parallel with one another within a given region. Such a region is called a domain (see Figure 8-63), and it is separated from an adjacent region by a boundary known as a Bloch wall (which may be of the order of 50 atoms wide) in which there is a gradual transition from one magnetic orientation to the other.

The exchange coupling between adjacent atoms of ferromagnetic materials that ensures parallelism of their magnetic moments is suddenly lost if the temperature is raised above a critical value known as the Curie point or temperature (1043°K for iron [the A_2 temperature]). Above this temperature, the material is paramagnetic.

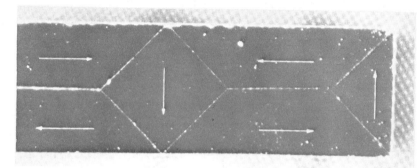

Figure 8-63: Domain Walls in an Unmagnetized Iron Whisker

The magnetic behavior of a steel is best understood with reference to the relationship that exists between the magnetization induced in the steel (the magnetic induction, B) when an external field, H, is applied to it. This relationship, when plotted, constitutes the B-H or magnetization curve, as illustrated in Figure 8-64. The significant features of this curve are the permeability μ (=B/H) and the saturation magnetization B_S.

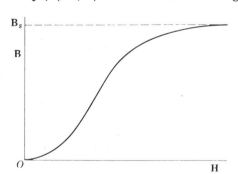

Figure 8-64: Magnetization Curve.

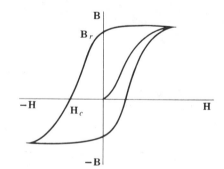

Figure 8-65: Hysteresis Loop.

When the magnetic field is first applied, taken to a maximum and then cycled, a B-H curve such as that shown in Figure 8-65 may be obtained. The three important features of this

---

* All orbital electrons display a quantity of magnetism or magnetic moment (defined as a Bohr magneton) as the result of "spin" of the electrons. The sign of this magneton may be positive or negative and a given quantum energy level can accommodate an equal number of electrons with spins of both types.

ALLOYS OF IRON

curve, from an engineering viewpoint, are the residual induction B_r (which is the magnetization that remains when the magnetizing field is removed); the coercive force H_c (which is the field required for demagnetization) and the area enclosed by the cyclical curve (representing the hysteresis loss in the material). Where the highest induction should be obtained with minimum hysteresis loss, as in the case of transformer cores, a "magnetically soft" material is required whereas for permanent magnets, magnetically hard materials (high B_r and H_c values) are desirable. (For further information relative to silicon steel, reference should be made to Section 12-30 which discusses the effect of orientation on the magnetic properties of a crystal.)

For most applications, the grain size of the material should be as large as possible. Figure 8-66 shows the effect of grain size on the hysteresis loss and Figure 8-67 its effect on coercive force. The effect of carbon content on the hysteresis loss of iron is shown in Figure 8-68 and the effects of various alloying elements on the saturation induction of iron in Figure 8-69.

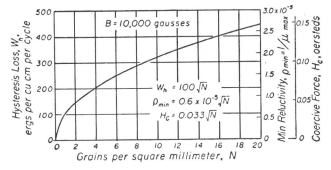

Figure 8-66:
Effect of Grain Size on the Hysteresis Loss of High Purity Iron.

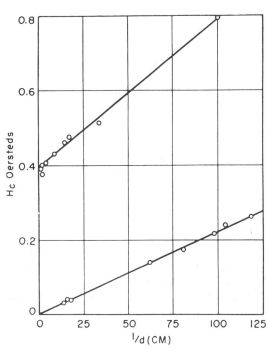

Figure 8-67:
Effect of Grain Size on Coercive Force.

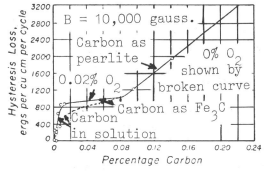

Figure 8-68:
Effect of Carbon Content on the Hysteresis Loss of Iron.

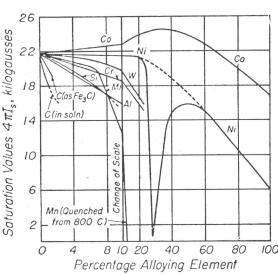

Figure 8-69:
Effect of Alloying Element on the Saturation Value of Iron.

Where magnetic materials are used in transformers and under other circumstances where eddy currents may be induced in them, it is desirable for the materials to exhibit a high resistivity so that such currents are minimized. Figure 8-70 shows the effect of alloying elements on the resistivity of iron and indicates the effectiveness of silicon in this respect.

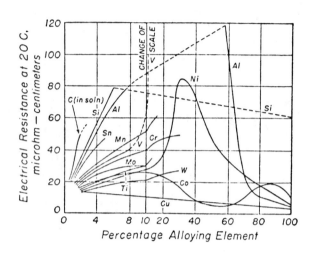

Figure 8-70:
Effect of Alloying Elements on
The Electrical Resistance of Iron.

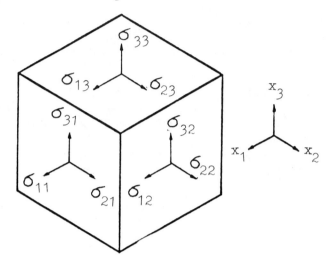

Figure 8-71:
The Stress Components That Describe a
General State of Stress at a Point.

8-17 Stresses

When one portion of a body exerts forces on neighboring portions, the body is said to be in a state of stress. In many instances, as, for example, in rolling, such stresses are induced by external or surface forces (as the work rolls acting on the strip) but in other cases internal stresses may exist as a result of centrifugal forces, thermal effects (or the consequences of thermal treatments) and non-uniform deformation.

The stress σ at any point across a small area ΔA can be defined by the limiting equation

$$\sigma = \lim_{\Delta A \to 0} \left(\frac{\Delta F}{\Delta A}\right) \qquad (8\text{-}7)$$

where ΔF is the internal force on the area ΔA surrounding the given point. It is said to be homogeneous if the forces acting on the surface of an element of fixed shape and orientation are independent of the position of the element in the body. For a complete description of a stress, it is necessary to specify its magnitude, direction, sense and the surface upon which it acts. Consequently, stress should be described by a second rank tensor because, in addition to its magnitude, direction and sense (which define a vector [a tensor of first rank]), it depends on another vector which represents the surface on which it acts.✢✢ Such a tensor represents a physical quantity that is, in actuality, independent of the choice of axes. When the axes are changed, only the method of representing the physical quantity is changed, not the quantity itself.

Referring to Figure 8-71 representing a small cube within an homogeneously stressed body, it will be assumed that all parts of the body are in static equilibrium and that there are no body forces or torques. The force transmitted across each face may be resolved into 3 components of stress. These are denoted by the general term σ_{ij} where the first subscript denotes the axis parallel to which the stress acts and the second subscript corresponds to the axis which is normal to the surface on which the stress acts. Hence σ_{11}, σ_{22} and σ_{33} are normal components of stress and

---

✢✢ "Applied Elasticity" by Chi-Teh Wang, McGraw Hill Book Company, Inc., New York, 1953.

ALLOYS OF IRON

σ_{12}, σ_{21}, etc., are shear components.‡ (Conventionally, normal stresses are regarded as positive if they are tensile stresses and negative if compressive stresses. Similarly, shear stresses are positive if they act in a positive direction with respect to the axis parallel to its direction when a tensile force on the same surface acts in a positive direction.) Because of the static equilibrium of the element

$$\sigma_{ij} = \sigma_{ji} \qquad (8\text{-}8)$$

The nine components of the stress tensor may be written in the form

$$\begin{bmatrix} \sigma_{11} & \sigma_{12} & \sigma_{13} \\ \sigma_{21} & \sigma_{22} & \sigma_{23} \\ \sigma_{31} & \sigma_{32} & \sigma_{33} \end{bmatrix}$$

However, since it is a symmetrical tensor (because $\sigma_{ij} = \sigma_{ji}$), it may be referred to the principal orthogonal axes, thus

$$\begin{bmatrix} \sigma_1 & 0 & 0 \\ 0 & \sigma_2 & 0 \\ 0 & 0 & \sigma_3 \end{bmatrix}$$

where σ_1, σ_2, and σ_3 are the principal stresses. Under these circumstances, it will be noted that the shear stresses disappear.

Special forms of the stress tensor exist. In the case of uniaxial stress (as, for example, a long vertical rod loaded by a weight at its bottom end), the tensor may be written

$$\begin{bmatrix} \sigma_1 & 0 & 0 \\ 0 & 0 & 0 \\ 0 & 0 & 0 \end{bmatrix}$$

In the case of biaxial stress (as, for example, a thin plate loaded by forces and couples applied to its edges), the tensor assumes the form

$$\begin{bmatrix} \sigma_1 & 0 & 0 \\ 0 & \sigma_2 & 0 \\ 0 & 0 & 0 \end{bmatrix}$$

For hydrostatic pressure, p, it may be written

$$\begin{bmatrix} -p & 0 & 0 \\ 0 & -p & 0 \\ 0 & 0 & -p \end{bmatrix}$$

and for pure shear stress

$$\begin{bmatrix} -\sigma & 0 & 0 \\ 0 & \sigma & 0 \\ 0 & 0 & 0 \end{bmatrix}$$

In the rolling operation, for those elements of the workpiece undergoing deformation in the roll bite, the principal axes are oriented as follows: (1) in a direction normal to the plane of the strip, (2) in a direction parallel to the axes of the mill rolls, and (3) in the rolling direction. The three principal stresses designated σ_1, σ_2, and σ_3 associated with these directions are illustrated in Figure 8-72. The stresses σ_1 and σ_2 are always compressive. The stress σ_3 may be tensile in proximity to the entry and exit ends of the roll bite and compressive throughout the remainder of the bite if strip tensions are used. Otherwise it, too, would be compressive throughout the roll bite.

The definition of stress given earlier in this section is that of true stress but, in certain instances, approximate stress values are also used. For example, in a tensile specimen with an initial

---

‡ Instead of numerical subscripts, x, y and z are frequently used and where they are identical, e.g. σ_{xx}, only one subscript is written, viz., σ_x.

cross section A_0, a tensile force F will produce a slight reduction of A_0. Thus, whereas the apparent or approximate stress is equal to F/A_0, in actual fact the true stress is slightly higher. Whereas the difference between true and approximate stress is usually inconsequential in elastic deformations, this is not usually the case in plastic deformation. Under such circumstances the stress should be calculated from actual measurements made of the cross section.

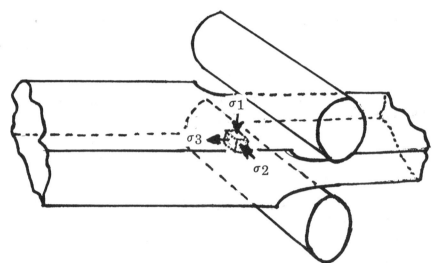

Figure 8-72: Sketch Showing the Principal Stresses in the Workpiece During Rolling.

8-18 Strains

When any physical body is stressed, the resulting deformation, whether elastic or plastic in nature, is generally spoken of as a strain. In a strict technical sense, strain is the direction and intensity of the deformation at any given point with respect to a specific plane passing through that point and a state of strain is a complete definition of the magnitude and direction of the deformation at a given point with respect to all planes passing through the point. Thus state of strain, like that of stress, should be represented by a tensor of the second rank.

Strains are generally expressed in terms of normal components ϵ and shear components γ, the normal strains referred to three orthogonal axes being defined as

$$\epsilon_1 = \lim_{x_1 \to 0} \frac{dx_1}{x_1}; \quad \epsilon_2 = \lim_{x_2 \to 0} \frac{dx_2}{x_2}; \quad \epsilon_3 = \lim_{x_3 \to 0} \frac{dx_3}{x_3} \qquad (8\text{-}9)$$

and the shear strains as, for example,

$$\gamma_{12} = \lim_{x_1 \to 0} \frac{dx_2}{x_1} = \tan \theta \simeq \theta \qquad (8\text{-}10)$$

where θ represents the deviation from an initial right angle.

The subscript notation is similar to that used for defining stresses. Thus γ_{12} is the strain resulting from taking adjacent planes perpendicular to the x_1 axis and displacing them relative to each other in the x_2 direction. Similarly, the sign convention follows directly from that of stresses in that a positive stress produces a positive strain and a negative stress produces a negative strain. Also, adopting a clockwise convention, γ_{21} would be negative and γ_{12} positive.

Whereas the strains ϵ_{11}, ϵ_{22} and ϵ_{33} are analogous to the stresses σ_{11}, σ_{22} and σ_{33}, respectively, it is to be noted that $\frac{\gamma_{12}}{2}$, $\frac{\gamma_{13}}{2}$ and $\frac{\gamma_{23}}{2}$ are analogous to τ_{12}, τ_{13} and τ_{23}, respectively.

Thus the symmetrical strain tensor (ϵ_{ij}) may be written

$$\begin{bmatrix} \epsilon_{11} & \epsilon_{12} & \epsilon_{13} \\ \epsilon_{21} & \epsilon_{22} & \epsilon_{23} \\ \epsilon_{31} & \epsilon_{32} & \epsilon_{33} \end{bmatrix}$$

with the diagonal components representing the stretches or tensile strains and the other components the shear strains. Because of its symmetry, this tensor may also be referred to mutually perpendicular principal axes and written

$$\begin{bmatrix} \epsilon_1 & 0 & 0 \\ 0 & \epsilon_2 & 0 \\ 0 & 0 & \epsilon_3 \end{bmatrix}$$

It is to be noted that the strain tensor is frequently written

$$\begin{bmatrix} \epsilon_x & \frac{\gamma_{xy}}{2} & \frac{\gamma_{zx}}{2} \\ \frac{\gamma_{xy}}{2} & \epsilon_y & \frac{\gamma_{yz}}{2} \\ \frac{\gamma_{zy}}{2} & \frac{\gamma_{yz}}{2} & \epsilon_z \end{bmatrix}$$

where $\gamma_{xy} = 2\epsilon_{12}$, $\gamma_{zx} = 2\epsilon_{13}$ and $\gamma_{zy} = 2\epsilon_{23}$

These gamma terms are engineering shear strains and are to be distinguished from the tensor shear strains denoted by the epsilon terms.

When one of the principal strains is zero, the body is said to be in plane strain, e.g.

$$\begin{bmatrix} \epsilon_1 & 0 & 0 \\ 0 & \epsilon_2 & 0 \\ 0 & 0 & 0 \end{bmatrix}$$

Pure shear is a special case of plane strain. For example, pure shear about the $0x_3$ axis may be written

$$\begin{bmatrix} 0 & \epsilon & 0 \\ \epsilon & 0 & 0 \\ 0 & 0 & 0 \end{bmatrix}$$

However, referred to the principal axes by rotation through 45°, this tensor becomes

$$\begin{bmatrix} -\epsilon & 0 & 0 \\ 0 & \epsilon & 0 \\ 0 & 0 & 0 \end{bmatrix}$$

Displacements, illustrated in Figure 8-73 and designated e_{11}, e_{22} and e_{33}, are related to the strains by the equations

$$e_{11} = \epsilon_1 l x_1; e_{22} = \epsilon_2 l x_2; e_{33} = \epsilon_3 l x_3 \tag{8-11}$$

Similarly, the displacements due to the shear strains are given by

$$e_{21} = \gamma_{21} l x_2; e_{32} = \gamma_{32} l x_3; e_{13} = \gamma_{13} l x_1 \tag{8-12}$$

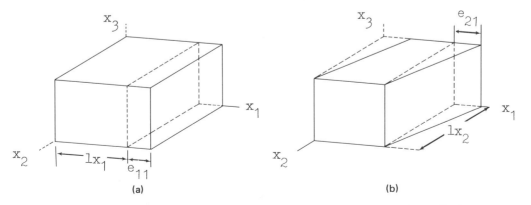

Figure 8-73: Sketches Illustrating Strains: (a) Normal Strain; (b) Shear Strain.

If the lengths lx_1, lx_2 and lx_3 are measured from the origin, so that x_1, x_2, and x_3 correspond to the coordinates of the infinitesimally small region being strained, then

$$e_1 = \epsilon_1 x_1 + \gamma_{21} x_2 + \gamma_{31} x_3 \quad (8\text{-}13)$$
$$e_2 = \gamma_{12} x_1 + \epsilon_2 x_2 + \gamma_{32} x_3 \quad (8\text{-}14)$$
$$e_3 = \gamma_{13} x_1 + \gamma_{23} x_2 + \epsilon_3 x_3 \quad (8\text{-}15)$$

Strains as discussed above are usually quite small (less than 0.2 per cent) in elastic deformations but in plastic deformations, considerably larger strains are involved. For this reason, the true strain (δ) of a bar of initial length (l_0) and instantaneous length (l) is defined as follows

$$\delta = \int d\epsilon = \int_{l_0}^{l} \frac{dl}{l} = \left[\ln l\right]_{l_0}^{l} = \ln l/l_0 \quad (8\text{-}16)$$

provided that the deformation is always uniform over the length considered.

8-19 Elastic Stress-Strain Relationships

The relationship between stress and strain under conditions of elastic deformation is best known in the form of Hooke's Law which states that, for sufficiently small stresses, the amount of strain developed is proportional to the magnitude of the applied stress.

Thus

$$\epsilon = S \sigma \quad (8\text{-}17)$$

where S is a constant of proportionality known as the elastic compliance constant or simply as the compliance. Alternately,

$$\sigma = C \epsilon \quad (8\text{-}18)$$

where C is the elastic stiffness constant, or simply the stiffness (but best known as Young's modulus E). However, in the full mathematical sense, since the stress is a second rank tensor (σ_{ij}) and the strain is a similar tensor (ϵ_{ij}), then Hooke's Law should be written

$$\epsilon_{ij} = S_{ijkl} \sigma_{kl} \quad (8\text{-}19)$$

where S_{ijkl} represent the compliances of the material and where k and l are dummy suffixes.

Thus an equation may be written for each of the 9 epsilon terms as exemplified by

$$\begin{aligned}\epsilon_{11} = &\ S_{1111}\sigma_{12} + S_{1112}\sigma_{12} + S_{1113}\sigma_{13} \\ &+ S_{1121}\sigma_{21} + S_{1122}\sigma_{22} + S_{1123}\sigma_{23} \\ &+ S_{1131}\sigma_{31} + S_{1132}\sigma_{32} + S_{1133}\sigma_{33}\end{aligned} \quad (8\text{-}20)$$

Thus there is a total of 81 S_{ijkl} coefficients.

Similarly, the stress-strain relationship may be written

$$\sigma_{ij} = C_{ijkl} \epsilon_{kl} \quad (8\text{-}21)$$

where C_{ijkl} represents the 81 stiffness coefficients.

Because of symmetry in the stress and strain tensors,

$$C_{ijkl} = C_{jikl} \quad (8\text{-}22)$$
$$\text{and} \quad C_{ijkl} = C_{ijlk} \quad (8\text{-}23)$$

Accordingly, the number of independent components is reduced from 81 to 36.

Thus, for a general state of 3-dimensional stress, Hooke's Law was generalized by Cauchy so that the six stress components were expressed as linear functions of all the strain

ALLOYS OF IRON

components. Using the subscripts x, y and z to denote the axes (instead of x_1, x_2, and x_3 used previously), the six stress components may be written

$$\sigma_x = C_{11}\epsilon_x + C_{12}\epsilon_y + C_{13}\epsilon_z + C_{14}\gamma_{xy} + C_{15}\gamma_{yz} + C_{16}\gamma_{zx} \quad (8\text{-}24)$$

$$\sigma_y = C_{21}\epsilon_x + C_{22}\epsilon_y + C_{23}\epsilon_z + C_{24}\gamma_{xy} + C_{25}\gamma_{yz} + C_{26}\gamma_{zx} \quad (8\text{-}25)$$

$$\sigma_z = C_{31}\epsilon_x + C_{32}\epsilon_y + C_{33}\epsilon_z + C_{34}\gamma_{xy} + C_{35}\gamma_{yz} + C_{36}\gamma_{zx} \quad (8\text{-}26)$$

$$\tau_{xy} = C_{41}\epsilon_x + C_{42}\epsilon_y + C_{43}\epsilon_z + C_{44}\gamma_{xy} + C_{45}\gamma_{yz} + C_{46}\gamma_{zx} \quad (8\text{-}27)$$

$$\tau_{yz} = C_{51}\epsilon_x + C_{52}\epsilon_y + C_{53}\epsilon_z + C_{54}\gamma_{xy} + C_{55}\gamma_{yz} + C_{56}\gamma_{zx} \quad (8\text{-}28)$$

$$\tau_{zx} = C_{61}\epsilon_x + C_{62}\epsilon_y + C_{63}\epsilon_z + C_{64}\gamma_{xy} + C_{65}\gamma_{yz} + C_{66}\gamma_{zx} \quad (8\text{-}29)$$

Fortunately, for isotropic elastic solids, only two of the 36 stiffness constants are independent, so that if C_{12} is expressed as λ (Lamé's constant) and C_{44} as G (the shear modulus or modulus of rigidity), then

$$\sigma_x = (2G + \lambda)\epsilon_x + \lambda(\epsilon_y + \epsilon_z) \quad (8\text{-}30)$$
$$\sigma_y = (2G + \lambda)\epsilon_y + \lambda(\epsilon_z + \epsilon_x) \quad (8\text{-}31)$$
$$\sigma_z = (2G + \lambda)\epsilon_z + \lambda(\epsilon_x + \epsilon_y) \quad (8\text{-}32)$$
$$\tau_{xy} = G\gamma_{xy} \quad (8\text{-}33)$$
$$\tau_{yz} = G\gamma_{yz} \quad (8\text{-}34)$$
$$\tau_{zx} = G\gamma_{zx} \quad (8\text{-}35)$$

These six equations may be rewritten using two elastic constants, E (Young's modulus) and ν (Poisson's ratio), as follows:

$$\epsilon_x = \frac{1}{E}[\sigma_x - \nu(\sigma_y + \sigma_z)] \quad (8\text{-}36)$$

$$\epsilon_y = \frac{1}{E}[\sigma_y - \nu(\sigma_z + \sigma_x)] \quad (8\text{-}37)$$

$$\epsilon_z = \frac{1}{E}[\sigma_z - \nu(\sigma_x + \sigma_y)] \quad (8\text{-}38)$$

$$\gamma_{xy} = \frac{2(1+\nu)}{E}\tau_{xy} = \frac{\tau_{xy}}{G} \quad (8\text{-}39)$$

$$\gamma_{yz} = \frac{2(1+\nu)}{E}\tau_{yz} = \frac{\tau_{yz}}{G} \quad (8\text{-}40)$$

$$\gamma_{zx} = \frac{2(1+\nu)}{E}\tau_{zx} = \frac{\tau_{zx}}{G} \quad (8\text{-}41)$$

These equations may be solved to obtain stress components as functions of strains, giving

$$\sigma_x = \frac{E}{(1+\nu)(1-2\nu)}[(1-\nu)\epsilon_x + \nu(\epsilon_y + \epsilon_z)] \quad (8\text{-}42)$$

$$\sigma_y = \frac{E}{(1+\nu)(1-2\nu)}[(1-\nu)\epsilon_y + \nu(\epsilon_z + \epsilon_x)] \quad (8\text{-}43)$$

$$\sigma_z = \frac{E}{(1+\nu)(1-2\nu)}[(1-\nu)\epsilon_z + \nu(\epsilon_x + \epsilon_y)] \quad (8\text{-}44)$$

$$\tau_{xy} = \frac{E}{2(1+\nu)}\gamma_{xy} = G\gamma_{xy} \quad (8\text{-}45)$$

$$\tau_{yz} = \frac{E}{2(1+\nu)}\gamma_{yz} = G\gamma_{yz} \quad (8\text{-}46)$$

$$\tau_{zx} = \frac{E}{2(1+\nu)}\gamma_{zx} = G\gamma_{zx} \quad (8\text{-}47)$$

For the special case in which the x, y, and z axes are coincident with principal axes 1, 2, and 3, these equations are simplified by virtue of all shear stresses and shear strains being equal to zero:

$$\epsilon_1 = \frac{1}{E}[\sigma_1 - \nu(\sigma_2 + \sigma_3)] \tag{8-48}$$

$$\epsilon_2 = \frac{1}{E}[\sigma_2 - \nu(\sigma_3 + \sigma_1)] \tag{8-49}$$

$$\epsilon_3 = \frac{1}{E}[\sigma_3 - \nu(\sigma_1 + \sigma_2)] \tag{8-50}$$

$$\sigma_1 = \frac{E}{(1+\nu)(1-2\nu)}[(1-\nu)\epsilon_1 + \nu(\epsilon_2 + \epsilon_3)] \tag{8-51}$$

$$\sigma_2 = \frac{E}{(1+\nu)(1-2\nu)}[(1-\nu)\epsilon_2 + \nu(\epsilon_3 + \epsilon_1)] \tag{8-52}$$

$$\sigma_3 = \frac{E}{(1+\nu)(1-2\nu)}[(1-\nu)\epsilon_3 + \nu(\epsilon_1 + \epsilon_2)] \tag{8-53}$$

For the commonly encountered biaxial-stress state, one of the principal stresses (say, σ_3) is zero, and the equations for the principal strains become

$$\epsilon_1 = \frac{1}{E}(\sigma_1 - \nu\sigma_2) \tag{8-54}$$

$$\epsilon_2 = \frac{1}{E}(\sigma_2 - \nu\sigma_1) \tag{8-55}$$

$$\epsilon_3 = -\frac{\nu}{E}(\sigma_1 + \sigma_2) \tag{8-56}$$

Referring to the equations for the principal stresses (σ_1, σ_2 and σ_3), if $\sigma_3 = 0$, then

$$\epsilon_3 = \frac{-\nu}{1-\nu}(\epsilon_1 + \epsilon_2) \tag{8-57}$$

Substitution of this expression into the other two equations yields

$$\sigma_1 = \frac{E}{1-\nu^2}(\epsilon_1 + \nu\epsilon_2) \tag{8-58}$$

$$\sigma_2 = \frac{E}{1-\nu^2}(\epsilon_2 + \nu\epsilon_1) \tag{8-59}$$

$$\sigma_3 = 0 \tag{8-60}$$

In the case of uniaxial stress, the equations for the principal strains and stresses reduce to

$$\epsilon_1 = \frac{1}{E}\sigma_1 \tag{8-61}$$

$$\epsilon_2 = \epsilon_3 = -\frac{\nu}{E}\sigma_1 \tag{8-62}$$

$$\sigma_1 = E\epsilon_1 \tag{8-63}$$

$$\sigma_2 = \sigma_3 = 0 \tag{8-64}$$

Bulk modulus or modulus of volume expansion (commonly designated by K) may be defined as the ratio between hydrostatic stress (in which $\sigma_1 = \sigma_2 = \sigma_3$) and volumetric strain (change in volume divided by initial volume).

For $\sigma_1 = \sigma_2 = \sigma_3 = \sigma$, then

$$\epsilon_1 = \epsilon_2 = \epsilon_3 = \epsilon = \sigma(1-2\nu)/E. \tag{8-65}$$

ALLOYS OF IRON

This state of uniform triaxial strain is characterized by the absence of shearing deformation; an elemental cube, for example, would change in size but remain a cube. The size of an elemental cube initially of unit dimension would change from 1^3 to $(1 + \epsilon)^3$ or to $1 + 3\epsilon + 3\epsilon^2 + \epsilon^3$. If we consider normal materials, ϵ is a quantity sufficiently small so that ϵ^2 and ϵ^3 are completely negligible, and the volumetric change is from 1 to $1 + 3\epsilon$. The volumetric strain is thus equal to 3ϵ or to $3\sigma(1 - 2\nu)/E$, and

$$K = \frac{E}{3(1 - 2\nu)} \tag{8-66}$$

For a material having a Poisson's ratio of $\frac{1}{3}$, values of E and K are identical.

For isotropic materials, the relationships between the elastic constants are as given in Table 8-8.

Table 8-8
Relations between elastic constants in isotropic materials

Constants involved	E, psi	ν, dimensionless	G, psi	K, psi
E, ν, G	$E = 2G(1+\nu)$	$\nu = \dfrac{E-2G}{2G}$	$G = \dfrac{E}{2(1+\nu)}$	...
E, ν, K	$E = 3K(1-2\nu)$	$\nu = \dfrac{3K-E}{6K}$	...	$K = \dfrac{E}{3(1-2\nu)}$
E, G, K	$E = \dfrac{9GK}{3K+G}$	...	$G = \dfrac{3EK}{9K-E}$	$K = \dfrac{GE}{3(3G-E)}$
ν, G, K	...	$\nu = \dfrac{3K-2G}{2(3K+G)}$	$G = \dfrac{3K(1-2\nu)}{2(1+\nu)}$	$K = \dfrac{2G(1+\nu)}{3(1-2\nu)}$

For steels, Young's modulus (E) is about 30×10^6 psi, Poisson's ratio (ν) about 0.3, the modulus of rigidity (G) close to 11.5×10^6 psi, and the bulk modulus 25×10^6 psi.

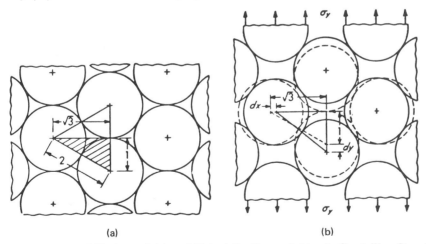

Figure 8-74: Diagram of Unstressed (a) and Uniaxially Stressed Atomic Crystalline Structure (b).

It is interesting to note that a simple physical model of a representative atomic crystalline structure predicts the value $\nu = 1/3$, which is a rough average of the experimentally determined values found for metals in general.※ Figure 8-74 (a) is a representation of the atomic

---

※ "Engineering Considerations of Stress, Strain and Strength", by R. C. Juvinall, McGraw-Hill Book Company, New York, 1967.

structure within a metal crystal. Regarding the circles as spheres of influence of the atoms, note that the mutual cohesive forces have brought the spheres uniformly close together at the minimum possible spacing. Figure 8-74 (b) shows the same structure under the action of vertical loading. Although the atom spacing has increased vertically, the cohesive forces keep the circles in contact along the diagonals. The figure also shows the displacement of typical circle centers to be dx and dy. Note that the hypotenuse of the right triangle remains 2 units long. Hence

$$(1 + dy)^2 + (\sqrt{3} - dx)^2 = 2^2 \qquad (8\text{-}67)$$

Since the differentials are extremely small in comparison with the unit spacing, we may neglect squared differentials with the result

$$1 + 2dy + 3 - 2\sqrt{3}dx = 4 \qquad (8\text{-}68)$$

$$dy - \sqrt{3}dx = 0 \qquad (8\text{-}69)$$

$$\frac{dy}{dx} = \sqrt{3} \qquad (8\text{-}70)$$

Lateral strain is $-dx/\sqrt{3}$, longitudinal tensile strain is $dy/1$, and the ratio ν is $dx/(dy\sqrt{3})$. From equation 8-70, therefore, ν exhibits a value of 1/3.

8-20 The Physical Mechanisms Involved in Plastic Deformation

Basically, the important factor in the deformation of a metal is considered to be the motion of dislocations within the lattice structure. However, the low dislocation density occurring in well-annealed metals ($\sim 10^8$ per cm^2) could not account for the macroscopic deformation such metals undergo in rolling and other metal working operations. To be effective in deformation, dislocations must have densities a thousand times greater (or be equally more effective) and considerable speculation has been made with respect to possible mechanisms which might cause dislocations to "multiply". Two such mechanisms that appear capable of creating large numbers of dislocation loops from a single dislocation have been proposed by Frank and Read.∥

The operation of one of these, the Frank-Read▣ spiral mechanism is illustrated in Figure 8-75 where a single dislocation ABC possesses a Burgers vector lying in the slip plane CDE. If the plane CDE is the only active slip plane, the portion AB of the dislocation (shown in Figure 8-75(a)) remains fixed but the portion BC is mobile and, when it moves it, causes a unit of slip in each part of the plane over which it sweeps (b). Thus when it has moved from BC to BD, the area BCD will have slipped one atom spacing while the rest of the plane remains unslipped. However, when it has made one complete revolution about B, the entire plane will have slipped by one spacing. The mobile part of the dislocation can then repeat its motion innumerable times so that a relatively large amount of slip can be produced on the plane.

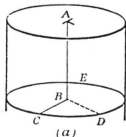
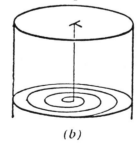

Figure 8-75: Sketch Showing How Unlimited Slip is Obtained from the Motion of a Single Dislocation Line.

The other concept, a double-ended Frank-Read source is illustrated in Figure 8-76. Here a straight portion BC of a dislocation lies in the slip plane while the ends of the dislocation are located outside of the plane. When a suitable stress is applied, the line BC will curve and it is

---

∥ "Dislocations and Plastic Flow in Crystals" by A. H. Cottrell, Oxford at the Clarendon Press, 1961.
▣ F. C. Frank and W. T. Read, "Multiplication Processes for Slow Moving Dislocations", Physical Review, 79, 722, (1950).

ALLOYS OF IRON

indicated in successive positions by the curves numbered 1, 2, 3, 4 and 5. The transition from 4 to 5 is made by the joining and mutual cancellation of the parts of the line which approach each other above BC. Continued application of a critical stress will produce a succession of closed dislocation loops.

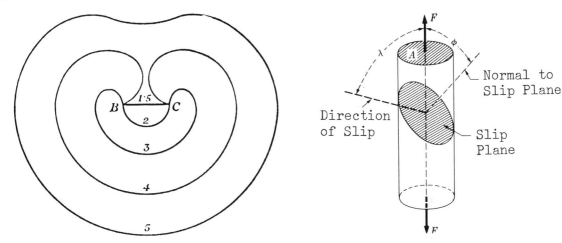

Figure 8-76:
Formation of a dislocation loop from a segment BC of a dislocation line (After Frank and Read).

Figure 8-77:
Sketch Illustrating the Angles λ and ϕ Relating to the Critical Shear Stress.

The slip mechanism associated with the Frank-Read sources appears to be the prevalent mode of deformation at ambient temperatures. Moreover, it accounts for the existence of slip planes, slip directions and, in the case of the double-ended source, the critical resolved shear stress required for deformation. The concept of the critical resolved shear stress may be understood with reference to Figure 8-77 showing how a force F acts upon the cross-sectional area A of a cylindrical crystal. However, only the component of force in the slip direction ($F \cos \lambda$) is effective in shearing the crystal and since the area of the slip plane is $A/\cos \lambda$, the corresponding shear stress τ is given by #[ϖ]

$$\tau = (F/A) \cos \lambda \cos \phi \qquad (8-71)$$

When "single slip" occurs, it does so on parallel sets of slip planes grouped to give slip bands. The spacing of slip lines within a band is about 200 Å whereas the separation of bands may be 100 times the distance. During single slip, the crystal lattice is not disturbed, except in the immediate neighborhood of a slip plane, and therefore the critical resolved shear stress remains nearly constant during the entire deformation by this process.

In polycrystalline metals, however, the obstructions resulting from grain boundaries requires that slip occur on several slip systems. Such systems for bcc and fcc lattices are discussed in detail in the next section. Where severe deformation occurs, the crystal lattice of each grain becomes quite disorganized giving rise to work or strain hardening discussed in detail in Section 8-32.

Another mechanism of plastic deformation that occurs under certain circumstances is twinning in which small regions are created with a mirror image lattice structure (as discussed in Section 8-8). In the case of bcc structures, twinning occurs only after some plastic deformation or when the stress is applied quickly. This is apparently associated with the higher stresses required to deform the metal at higher strain rates. In iron, the twinned areas are known as Neumann bands.

\# The factor $\cos \lambda \cos \phi$, which is used for converting tensile to resolved shear stress, is called the Schmid factor.

[ϖ] Equation 8-73 is frequently written in the form
$\tau = F/A \cos \lambda \sin \theta$
where θ is the angle between the slip plane and the direction of tension.

8-21 Slip Planes and Directions in BCC and FCC Structures

Examination of a lattice structure, as illustrated in Figure 8-22, shows that planes containing a low density of atoms lie close together whereas those with the maximum density have the maximum separation. These latter planes are important since, under deformation stresses, slip usually occurs parallel to them. In this connection, it should be noted that the slip plane and the direction of slip comprise a slip system specified in the form {hkl} <uvw>.

In fcc lattices, there are four planes known as octahedral planes which represent planes of closest packing of the atoms. These are the {111} planes and each plane has three directions of slip. Thus, for the (111) plane shown in Figure 8-78, the three directions of slip are [$\bar{1}$01], [0$\bar{1}$1] and [$\bar{1}$10] (which are the diagonals of three of the faces of the cube). Thus for fcc lattices there are 12 slip systems (4 planes x 3 directions).

In general, metals having a bcc lattice possess at least two types of planes of easiest slip, although these are not closely packed.◙ Generally speaking, slip occurs in the close-packed directions, <111>, but the slip planes are not well defined. It is believed that the {110} and {112} planes may all act as slip planes, the more prominent being the {110} planes. Some investigators believe slip also occurs on {123} planes but this appears doubtful.✦ However, it appears that 48 slip systems exist in iron but the slip lines are often forked, wavy and irregular.

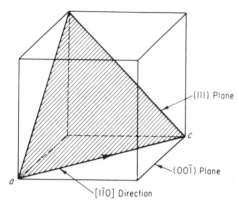

Figure 8-78:
The (111) Plane of a Cubic Crystal.

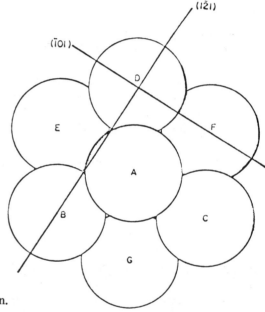

Figure 8-79:
The BCC lattice viewed along the [111] direction.

Referring to Figure 8-79, where the atomic arrangement is viewed along the [111] direction, an atom, such as that labelled A, must move in a vertical direction to the next equivalent crystallographic position. Now if A glides over D and F carrying E, C, B and G with it, slip has occurred on the ($\bar{1}$01) plane. This would involve a deviation from the [111] direction with the formation of partial dislocations according to the reaction ✕

$$1/2[111] \rightarrow 1/8[101] + 1/4[121] + 1/8[101] \qquad (8\text{-}72)$$

If atom A could glide over B and C, and then over G and F carrying E and D with it, such a movement would be equivalent to a dislocation with 1/2<111> Burgers vector splitting up

---

◙ "Metallurgy for Engineers", by E. C. Rollason, Edward Arnold, Ltd., Third Edition (Printed by Richard Clay and Co., Ltd., Bungay, Suffolk, England, 1963).

✦ C. M. Van der Walt, "Slip in the BCC Metals", Acta Metallurgica, Vol. 17, April, 1969, pp. 393-395.

✕ J. B. Cohen, R. Hinton, K. Lay and S. Sass, "Partial Dislocations on the {110} planes in the B.C.C. Lattice", Acta Met. 10, 1962, pp. 894-895.

ALLOYS OF IRON

into partials by the reaction

$$\frac{1}{2}\langle 111 \rangle \rightarrow \frac{1}{4}\langle 111 \rangle + \frac{1}{4}\langle 111 \rangle \qquad (8\text{-}73)$$

where the partials have different $\{112\}$ type slip planes defined by BC and GF. However, this configuration would be sessile (that is, unable to move) and therefore could not constitute a slip system. On the other hand, if A glided over D and then over E and F, a $\{112\}$ slip plane would be defined.

8-22 The Development of Rolling Textures in Steels

Although a polycrystalline metallic workpiece may have a more or less random orientation of grains prior to rolling, after emergence from the roll bite it will be characterized by a preferred orientation or texture, which may be simply defined as a condition in which the distribution of crystal orientations is nonrandom.[✠] In cold rolled sheet, most of the grains are oriented with a certain (hkl) plane parallel to the sheet surface and a certain direction [uvw] in that plane roughly parallel to the rolling direction. Accordingly, such textures are specified as (hkl) [uvw]. Such deformation textures, which are attributable to the tendency of the grains to rotate during plastic deformation, provide X-ray diffraction patterns (pole figures) of a unique type as discussed in Section 8-5.

For α-iron, the ideal orientations are stated to be (001) [1$\bar{1}$0], (112) [1$\bar{1}$0] and (111) [$\bar{1}\bar{1}$2].[♦] Lattice structures initially oriented so that the $\langle 112 \rangle$ direction lies between the rolling direction and the operative slip direction will generally assume the $\{111\} \langle 112 \rangle$ orientation. However, if slip occurs on $\{011\}$ planes in the $\langle 111 \rangle$ directions, the $\{111\} \langle 112 \rangle$ texture component will gradually be displaced toward the most stable $\{112\} \langle 110 \rangle$ texture. Where the $\langle 110 \rangle$ direction lies between the rolling and operative slip systems, the rolling plane may be $\{001\}$, $\{112\}$ or $\{111\}$ and the rolling direction $\langle 110 \rangle$. When double slip occurs on systems of the type $\{011\} \langle 111 \rangle$ and $\{112\} \langle 111 \rangle$, the two stable textures produced are $\{112\} \langle 110 \rangle$ and the $\{001\} \langle 110 \rangle$ orientations.

The textures developed during cold rolling are not entirely lost during the recrystallization that occurs in annealing, since an annealed or recrystallized texture develops that is related to the deformation texture. Generally speaking, some or all of the new crystals have orientations that belong to the deformation structure but in some cases the new and deformed lattices may be related by rotation about a certain crystallographic direction (e.g., the $\langle 111 \rangle$ direction in fcc metals). Figure 8-80 illustrates the recrystallized texture of rolled iron after annealing with reference to the (110) and (100) planes.

8-23 Stress-Strain Relationships in Plastic Deformation

Whereas, in the case of elastic deformation, strain occurs at any finite level of stress, under conditions of plastic deformation, no plastic strain occurs until the stress has reached a certain magnitude generally defined as the flow stress or yield strength σ_0. Thus, in the case of a simple tension test, the plastic deformation δ is specified by a pair of equations.

$$\delta = 0 \text{ for } \sigma < \sigma_0 \qquad (8\text{-}74)$$

$$\text{and} \quad \delta = f(\sigma) \text{ for } \sigma > \sigma_0 \qquad (8\text{-}75)$$

where σ is the applied stress.

Generally speaking, it is very difficult to ascertain precisely the stress at which plastic deformation begins. For engineering purposes, however, the yield strength is usually defined as that stress which produces a small, measurable plastic deformation. Usually this strain is 0.2 per cent but

---

[✠] "Elements of X-Ray Diffraction" by B. D. Cullity, Addison Wesley Publishing Company, Inc., Reading, Mass. 1956.

[♦] T. Kamijo, "A Consideration on the Development of Rolling Textures in δ Iron", Trans. JIM, 1969, Vol. 10, pp. 242-246.

0.1 per cent is often employed for steels and as much as 0.5 per cent for cast irons. The method of measuring this offset is discussed more fully in Section 8-26.

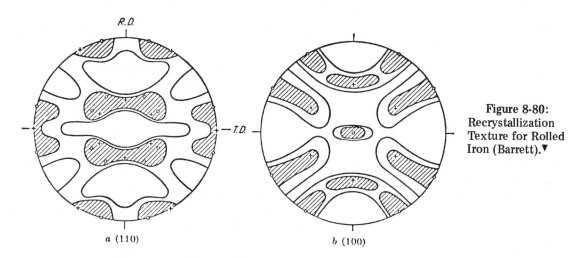

Figure 8-80: Recrystallization Texture for Rolled Iron (Barrett).▼

In practice, however, such as in cold rolling, the metal to be deformed is not subject merely to a uniaxial stress but to a combination of tensile and compressive stresses acting along orthogonal directions. It is therefore desirable to know the conditions for flow under combined stresses. Accordingly, from time to time, various yielding criteria have been developed but, to be valid, a criterion must not predict any plastic deformation in a material subjected to hydrostatic pressure, regardless of the magnitude of that pressure.⊞

Two yielding criteria that have found widespread acceptance are those of Tresca⊕ and Von Mises.♦ In Tresca's "Maximum Shear" theory, if the magnitude of the principal stresses σ_1, σ_2 and σ_3 are such that $|\sigma_1| > |\sigma_2| > |\sigma_3|$ the maximum yield stress in shear is equal to half the difference between σ_1 and σ_3 and yielding will occur when this maximum stress equals the yield stress of the material in shear. The latter can be shown to be equal to half the yield strength σ_T in pure tension so that yielding occurs when △

$$\sigma_1 - \sigma_3 = \sigma_T \quad (8\text{-}76)$$

The criterion of Von Mises, often referred to as the "Maximum Shear Strain Energy Theory", has perhaps been more widely used. It may be expressed by the relationship

$$(\sigma_1 - \sigma_2)^2 + (\sigma_2 - \sigma_3)^2 + (\sigma_3 - \sigma_1)^2 = 2\sigma_T^2 \quad (8\text{-}77)$$

In the case of a tensile test where σ_1 is the only principal stress with a finite value, it must reach a value

$$\sigma_1 = \sigma_T \quad (8\text{-}78)$$

before yielding occurs. In the case of biaxial stress, where a tensile stress σ_1 is applied in one direction and an equal compressive stress at right angles to it, for yielding to be initiated

$$\sigma_1 = 0.58\,\sigma_T \quad (8\text{-}79)$$

▼ "Structure of Metals", by C. S. Barrett. McGraw-Hill Book Company, Inc., 1952, Figure 14, p. 500.

⊞ "The Rolling of Metals, Theory and Experiment" by L. R. Underwood, Volume I, John Wiley and Sons, Inc., New York, 1950, p. 96.

⊕ H. Tresca, "Comptes Rendus Hebdomadaires des Seances de l'Academie des Sciences", 59 (1864), p. 756 and 64 (1867), p. 804.

♦ R. Von Mises, "Nachrichten der Akademie der Wissenchaften in Gottingen, II, Mathematisch — Physikalische Klasse", Berlin, 1913, p. 582, ("Mechanics of Solid Bodies Under Conditions of Plastic Flow").

△ "Mechanics of Materials for Engineers", by F. Charlton, John Wiley and Sons, Inc., New York, 1962, p. 129.

ALLOYS OF IRON

Since these criteria were developed, a number of investigators have attempted to establish their relative validity with respect to the yielding of steel,[□] but the inconclusive results of their work led Morrison[○] to undertake to attempt the task using tension, compression, combined torsion and tension and flexure tests. Despite elaborate precautions to ensure very reliable data, he was unable to show that either theory was correct. In examining Morrison's paper, Wood[¶] concluded that, in the absence of the ability to accurately control grain size, grain shape and crystallographic orientation in the specimens undergoing deformation, "attempts to establish criteria for yield solely by considerations of some critical set of external stress conditions could never lead to conclusions of universal application."

8-24 Slip Line Field Theory

When a body is stressed, it is possible to establish the location of lines along which the maximum shear stress occurs. Such lines correspond to the well-known Lüders bands frequently observed on the specimen in a tensile test on mild steel where they occur at about 45° to the axis of tension when the yield point has been reached.[⊞] However, after a few per cent elongation of the tensile specimen, the Lüders bands, or lines, lose their identity because the deformation has spread

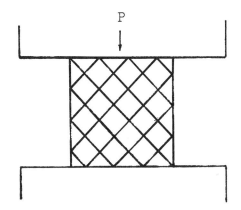

Figure 8-81:
Slip Line Field for Two-Dimensional (Plane Strain) Compression in the Absence of Friction.

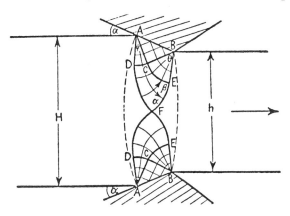

Figure 8-82:
Slip Line Field and Plastic Region for Drawing Through a Smooth Wedge-Shaped Die Giving a Moderate Reduction in Thickness.

[□] A. A. Scoble, "The Strength and Behavior of Ductile Materials under Combined Stress", Phil. Mag., Series 6, Vol. 12, 1906, p. 533.

W. Mason, "Mild Steel Tubes in Compression and Under Combined Stress", Proc. I. Mech. E., Dec. 1909, p. 1205.

C. A. M. Smith, "Compound Stress Experiments", Proc. I., Mech. E., Dec., 1909, p. 1237.

G. Cook and A. Robertson, "Strength of Thick Hollow Cylinders Under Internal Pressure", Engineer, Vol. 92, 1911, p. 786.

F. B. Seely, and W. J. Putman, "The Relation Between the Elastic Strengths of Steel in Tension, Compression and Shear", University of Illinois, Eng. Experiment, Bulletin No. 115, 1919.

M. Ros and A. Eichinger, "Experiments to Clear up the Danger of Fracture", Proc. 2nd International Congress of Applied Mechanics, Zurich, 1926, p. 315.

G. Cook, "The Yield Point and Initial Stages of Plastic Strain in Mild Steel Subjected to Uniform and nonuniform Stress Distributions", Phil. Trans. Roy. Soc. A., Vol. 230, 1931, p. 103.

W. Lode, "Yield of Metals", ZVDI, 1928, Forschungsarbeiten No. 303.

[○] J. L. M. Morrison, "The Yield of Mild Steel with Particular Reference to The Effect of Size of Specimen", Proc. I. Mech. E., Vol. 142, No. 3, January, 1940, pp. 193-223.

[¶] W. A. Wood, Discussion of the paper "The Yield of Mild Steel With Particular Reference to the Effect of Size of Specimen", by J. L. M. Morrison, Proc. I. Mech. E. (London), Vol. 144, No. 1, Nov., 1940, pp. 33-42.

[⊞] "Mechanical Treatment of Metals", by R. N. Parkins, American Elsevier Publishing Company, Inc., New York, 1968.

throughout the specimen. The plastically deformed region in its entirety may therefore be considered to be covered by two families of slip lines (designated α and β) each of which cuts the other orthogonally. Figure 8-81, for instance, shows the slip line field for the two dimensional (plane strain) compression of a long, rectangular workpiece of square cross section whereas Figure 8-82 shows the slip line field and plastic region for a strip drawn through a smooth, wedge-shaped die giving a moderate reduction in thickness.◄

The principal value of slip lines is that they define the shape and extent of the plastically deformed region subjected to a given stress system. The law governing the properties of slip-line families was first established by Hencky[1] and, for the case where the slip line has a curvature relative to a fixed direction measured by an angle Θ, it may be written

$$p + 2k\Theta = \text{constant along an } \alpha \text{ slip line} \qquad (8\text{-}80)$$

$$p - 2k\Theta = \text{constant along a } \beta \text{ slip line} \qquad (8\text{-}81)$$

with the value of the constant k, the yield stress in shear, being determined by the boundary conditions and where p represents the mean compressive stress. If the pattern of the slip line field for given geometrical conditions is known, then Hencky's law permits the variation of p along the α and β lines to be determined and the load necessary for the particular forming operation to be calculated.

Geiringer[○] developed analogous equations relating to the speed of deformation. These are

$$du - vd\Theta = 0 \text{ along an } \alpha \text{ slip line} \qquad (8\text{-}82)$$
$$\text{and} \quad dv - ud\Theta = 0 \text{ along a } \beta \text{ slip line} \qquad (8\text{-}83)$$

where u and v are the components of velocity of a particle along the α and β slip lines, respectively.

The application of slip line (shear plane) theory to the mathematical modelling of the rolling process is discussed more fully in Chapter 9.

8-25 The Constrained Yield Strength of the Workpiece

In the rolling operation, friction is necessary both to impart deformation energy from the rolls to the strip and to prevent the lateral spread of the strip as it is being rolled. Even so, it is still desirable to determine the pressure or stress that must be exerted by the work rolls on the strip to plastically deform the latter in the absence of frictional forces acting parallel to the rolling direction. Under these conditions, the frictional constraint in the transverse direction gives rise to the principal compressive stress (σ_2) that will permit no change in the width of the strip in the roll bite when the strip is subjected to the other two principal stresses (σ_1 and σ_3). Now the strain ϵ_2 of an elastic body in a direction coincident with that of the principal stress σ_2 is given by

$$\epsilon_2 = \frac{1}{E}\{\sigma_2 - \nu(\sigma_2 + \sigma_3)\} \qquad (8\text{-}84)$$

where E is Young's modulus and ν is Poisson's ratio. Since the strain ϵ_2 is zero because of the transverse frictional constraint exerted on the strip by the work rolls, the above equation yields

$$\sigma_2 = \nu(\sigma_1 + \sigma_3) \qquad (8\text{-}85)$$

In plastic deformations, Poisson's ratio ν is usually assigned a value of 0.5 so that

$$\sigma_2 = 0.5(\sigma_1 + \sigma_3) \qquad (8\text{-}86)$$

◄ "The Mathematical Theory of Plasticity" by R. Hill, Oxford, at the Clarendon Press, 1950.
[1] H. Hencky, Zeits. Ang. Math. Mech. 3 (1923), p. 241.
○ H. Geiringer, Proc. 3rd Int. Cong. App. Mech., Stockholm, 2 (1930), p. 185.

ALLOYS OF IRON

When this value for σ_2 is substituted in the expression representing the Von Mises' yielding criterion discussed in Section 8-23, it is seen that

$$\sigma_1 - \sigma_3 = \frac{2}{\sqrt{3}} \sigma_T = 1.155 \sigma_T = \sigma_C \quad (8\text{-}87)$$

where σ_C was termed by Nadai‡ the "constrained yield strength".

Thus, under "frictionless" conditions in the rolling direction, strip with a yield strength, σ_T, and with equal entry and exit tensile stresses, σ_3, would require a compressive stress, σ_1, to deform it given by

$$-\sigma_1 = -1.155 \sigma_T + \sigma_3 \quad (8\text{-}88)$$

or

$$\sigma_1 = 1.155 \sigma_T - \sigma_3 \quad (8\text{-}89)$$

Thus, the greater the magnitude of the strip tensions (represented by σ_3), the less the pressure σ_1 that must be exerted by the rolls to deform the strip.

8-26 The Measurement of Yield Stress at Low Strain Rates

From the preceding sections, it is seen that the material property of basic importance in the cold rolling process is the yield strength. Accordingly, consideration is given here in this section to its measurement at low strain rates and in Section 8-30 to its measurement at high strain rates comparable to those encountered under typical commercial rolling operations.

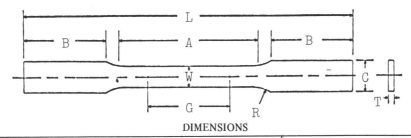

DIMENSIONS

G – Gage length	2.000 ± 0.005 in.
W – Width (Notes 1 and 2)	0.500 ± 0.010 in.
T – Thickness (Note 3)	thickness of material
R – Radius of fillet	1/2 in., min.
L – Overall length	8 in., min.
A – Length of reduced section	2-1/4 in., min.
B – Length of grip section (Note 4)	2 in., min.
C – Width of grip section (Notes 1 and 5)	3/4 in., approx.

Note 1. – When necessary a narrower specimen may be used. In such case the width should be as great as the width of the material being tested permits. If the width is 1/2-in. or less the sides may be parallel throughout the length of the specimen.

Note 2. – The ends of the reduced section shall not differ in width by more than 0.002 in. There may be a gradual taper in width from the ends to the center, but the width at either end shall not be more than 0.005-in. greater than the width at the center.

Note 3. – The dimension "T" is the thickness of the test specimen as provided for in the applicable material specifications.

Note 4. – It is desirable, if possible, to make the length of the grip section great enough to allow the specimen to extend into the grips a distance equal to two thirds or more of the length of the grips. If the thickness of the specimen is over 3/8-in., longer grips and correspondingly longer grip sections of the specimen may be necessary to prevent failure in the grip section.

Note 5. – The ends of the specimen shall be symmetrical with the center line of the reduced section within 0.01-in. However, for steel if the ends are symmetrical within 0.05-in. a specimen may be considered satisfactory for all but reference testing.

Figure 8-83: Standard Rectangular Tension Test Specimen with 2-Inch Gage Length.

---

‡ "Plasticity" by A. Nadai, McGraw-Hill Book Company, New York, 1943.

To measure the yield strength of a material in strip form, tensile specimens are usually prepared as illustrated in Figure 8-83. These are then pulled between grips shown in Figure 8-84 in a conventional testing machine such as that depicted in Figure 8-85, using a crosshead speed such that the rate of stressing, $\left(\frac{d\sigma_T}{dt}\right)$, maintains a desired value in terms of psi per second. The strain rate $\dot{\epsilon}$ during the elastic deformation is therefore given by

$$\dot{\epsilon} = \frac{1}{E}\frac{d\sigma_T}{dt} \qquad (8\text{-}90)$$

and, assuming the elastic modulus E has a value of 3×10^7 psi/inch/inch, the value of the strain rate $\dot{\epsilon}$ may be computed in terms of seconds^{-1}.

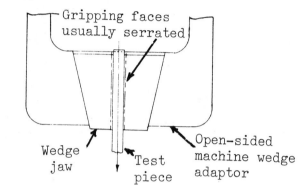
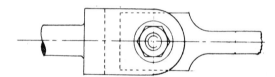
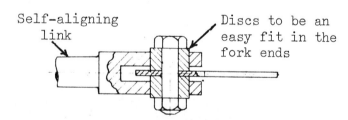

Figure 8-84: Gripping Devices for Sheet Specimens.

As the specimen is elongated under test, a graphical record of the force-elongation data similar to that shown in Figure 8-86 is continuously provided by the tensile-testing machine. In the case of low-carbon steel, the character of the curve depends on the previous history of the specimen. Fully annealed specimens will exhibit both an upper and a lower yield point as indicated in Figure 8-86 but only the lower yield point is regarded as being pertinent to rolling problems in general.

Strip that has been temper rolled after annealing provides a stress-strain curve exhibiting a smooth transition from the initial, linear portion of the curve (corresponding to elastic deformation) to the nonlinear portion corresponding to plastic deformation (see Figure 8-87). In the absence of well-defined yield point, the yield stress is that which, upon unloading, will produce a 0.1 or 0.2 per cent elongation, this stress level being referred to as the corresponding offset yield strength.

Compression tests have also been conducted by various investigators to establish the constrained yield strength using dies similar to those shown in Figure 8-88. In such tests, a lubricant must be used to minimize the restraint on the flow of the deforming metal due to friction between

ALLOYS OF IRON

the dies and the workpiece. Using calcium stearate as a lubricant in this method, Ford[a] found that for 0.2 per cent carbon steel, the ratio between the yield strength in compression to the yield strength in tension varied with reduction as follows

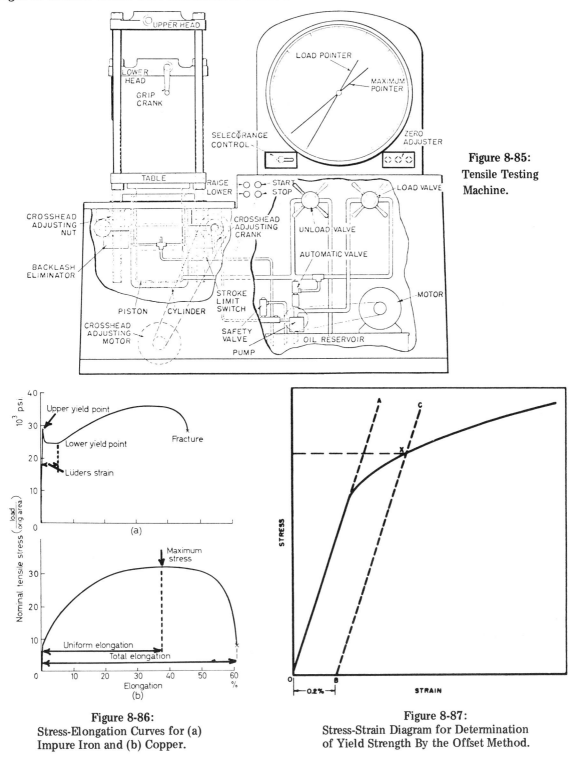

Figure 8-85: Tensile Testing Machine.

Figure 8-86: Stress-Elongation Curves for (a) Impure Iron and (b) Copper.

Figure 8-87: Stress-Strain Diagram for Determination of Yield Strength By the Offset Method.

---

[a] H. Ford, "Researches into the Deformation of Metals by Cold Rolling", "Researches on the Rolling of Strip", A symposium of Selected Papers, 1948–1958, The British Iron & Steel Research Association, Waterloo and Sons, Ltd., London.

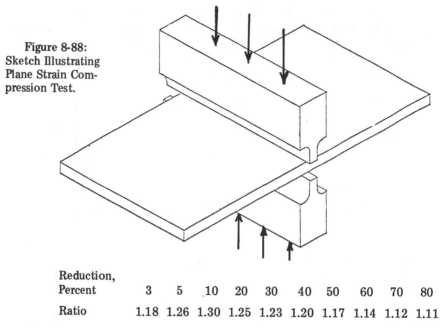

Figure 8-88: Sketch Illustrating Plane Strain Compression Test.

Reduction, Percent	3	5	10	20	30	40	50	60	70	80
Ratio	1.18	1.26	1.30	1.25	1.23	1.20	1.17	1.14	1.12	1.11

In making comparisons between compressive and tensile yield strengths, care should be taken to ensure that comparable rates of deformation are used.

Steel strip that has been annealed and subsequently rolled exhibits anisotropic characteristics. It is therefore necessary in referring to the yield strength of rolled strip to state how the axis of the tensile specimen relates to the rolling direction. Generally speaking, tensile specimens are prepared so that the yield strength is measured in the rolling direction and transverse to it.

It is interesting to note that the yield strength in the transverse direction is always higher than in the rolling direction and if the strip material were assumed to be isotropic, it would appear that the elongation in the transverse direction (which is actually zero) is about twice that in the rolling direction. In rolling theory, however, it is the general practice to utilize only the yield strength data in the longitudinal or rolling direction.

8-27 The Effect of Temperature on Yield Strength

Although the yield strength of a metal or alloy is frequently regarded as a physical characteristic of the material, this is not strictly correct since its value depends on the method and the circumstances of its measurement. One of the parameters influencing yield strength is temperature which has an appreciably larger effect in bcc structures than fcc or close-packed hexagonal lattices. Thus a 100°C rise in temperature will decrease the yield strength of mild steel about 4500 psi, whereas in the case of copper (with an fcc lattice) the change is only one-tenth of this amount. However, the temperature sensitivity of the yield strength of alloys may differ from that of the parent metal because of atomic rearrangements or phase transformations occurring as the temperature is raised. These effects in alloys, by influencing the interaction between dislocations and the solute "atmospheres", or by altering the amount or dispersion of a second phase will generally accentuate the reduction in yield strength resulting from a temperature increase but, in a few cases, the yield strength may not be significantly influenced or the change actually reversed in direction.

One of the earliest investigations of the effect of temperature on the yield stress for mild steel was made by Bach and Baumann▶ for temperatures ranging from 20 to 500°C. Their data,

---

▶ "Mechanical Properties and Physical Structure of Materials of Construction", C. Bach, and R. Baumann, Julius Springer, Berlin 1921, 2nd Edition.

ALLOYS OF IRON

reproduced in Figure 8-89, shows a marked decrease in the upper yield point with increasing temperature up to 300°C whereas for temperatures above this level, the yield point becomes poorly defined. On the other hand, the lower yield point rises slightly with temperature up to 100°C and then decreases, merging with the upper yield point about 300°C.

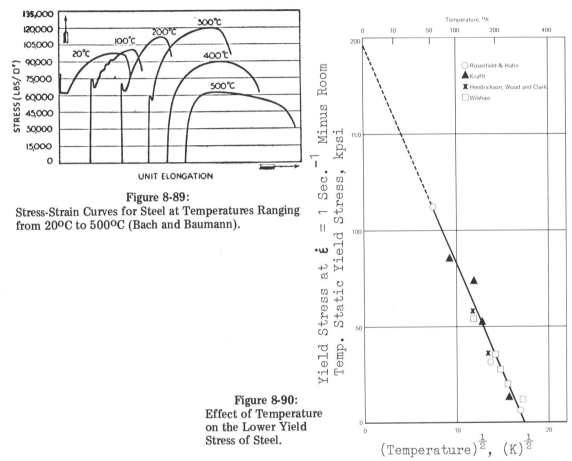

Figure 8-89:
Stress-Strain Curves for Steel at Temperatures Ranging from 20°C to 500°C (Bach and Baumann).

Figure 8-90:
Effect of Temperature on the Lower Yield Stress of Steel.

The effect of temperature on the lower yield stress of steel is also illustrated in Figure 8-90, showing data obtained by a number of investigators. In this case, the change in yield stress effected by raising the temperature above ambient is shown plotted against the square root of the absolute temperature.

Becker[♦] and Orowan[⊗], have suggested that the relationship between the applied stress, σ, and the absolute temperature, T, may be expressed as

$$\dot{\epsilon} = a \exp. - \left[(\sigma_0 - n\sigma)^2 \frac{V}{2aGT} \right] \qquad (8\text{-}91)$$

where $\dot{\epsilon}$ is the strain rate, G is the shear modulus, σ_0 is the yield stress at 0° K, V is a volume of the order of 200 Å^3, n is a stress concentration factor and a is a constant. Similarly, from a consideration of data published on several pearlitic steels, Hollomon and Zener[∅] concluded that, for many metals, the yield stress at a given strain rate ($\sigma_{\dot{\epsilon}}$) varies with temperature in accordance with the following relationship

$$\sigma_{\dot{\epsilon}} = b \exp. Q/RT \qquad (8\text{-}92)$$

[♦] R. Becker, Zeitschrift fuer Physik, 26, 1925, p. 919.
[⊗] E. Orowan, Zeitschrift fuer Physik, 89, 1934, pp. 605, 614, 634.
[∅] J. H. Hollomon and C. J. Zener, "Conditions of Fracture of Steel", Transactions, AIME, 158, 1944, pp. 283-297.

where T is the absolute temperature, R is the gas constant, b is a constant and Q is a parameter that depends partly on the physical properties of the metal.

8-28 The Influence of Microstructure on Yielding

Generally speaking, the smaller the grain size or the average distance between particles or discontinuities within the grains, the higher the yield strength of the material, as discussed in Section 8-7. For example, the properties of pearlitic steels are dependent on the interlamellar spacing of the pearlite, as well as the grain size, the coarser pearlite being softer than those with finer grained structures. This relationship is shown in Figure 8-91, the yield stress showing an excellent correlation with the average distance between the cementite particles (the mean free path in the ferrite), the shape of the particles having little or no influence on this relationship.

The dependence on the upper yield strength (σ_{uyp}) and the yield point drop (or the difference between σ_{uyp} and the lower yield point, σ_{lyp}) on grain size, d, has been investigated by Hutchison,[‡] Petch,[†] and Worthington[⊟] relative to Armco iron, mild steel and silicon steel, respectively. These researchers used center-annealed wires to avoid stress concentrations from specimen shoulders and to ensure uniformity of the applied stress." They observed that no yield point drop occurs in coarse-grained specimens, but as the grains become finer, the yield point drop increases inversely with the square root of the grain diameter. At the finer grain sizes, the yield point drop is given by

$$\sigma_{uyp} - \sigma_{lyp} \approx \Delta\sigma_0 \log_{10}(1/Nd^3) \qquad (8\text{-}93)$$

where N is the number of grains per unit volume undergoing plastic deformation and where $\Delta\sigma_0$ is the friction stress increase corresponding to a tenfold strain rate increase.

Christ, Smith and Burton[≡] found that, in the case of 0.050-inch diameter wire specimens made from zone-refined iron, a yield point drop occurred in specimens with grain diameters less than 0.1 mm but no drop occurred in specimens with a grain diameter greater than 0.1 mm, provided they were free of veining substructure. However, specimens with a grain diameter of 0.129 mm and which had been annealed in the austenitizing temperature range showed an upper yield point which was associated with the presence of a veining substructure.

Further studies relative to the effect of grain size on yield point behavior were made by Birkbeck and Douthwaite" using relatively pure iron (0.015%°C) in the form of center-annealed wires of three different diameters with grains free from veining. The specimens were tested in a tensile machine at a crosshead velocity of 3.4×10^{-4} in. per sec. Figure 8-92 shows the effect on the yield point drop of the number of grains across the cross section for the three specimen sizes used in the investigation. From this data, it is seen that the existence of a yield point drop does not depend uniquely on the number of grains across the cross section. When plotted as shown in Figure 8-93, the yield point drop is seen to be zero when

$$d^{-1/2} = 2 \text{ mm}^{-1/2} \qquad (8\text{-}94)$$

From these results, Birkbeck and Douthwaite concluded that grain size is the critical factor determining whether a yield point drop occurs.

[‡] M. M. Hutchison, "The Temperature Dependence of the Yield Stress of Polycrystalline Iron", Phil. Mag., 1963, Vol. 8, pp. 121-127.

[†] N. J. Petch, "The Upper Yield Stress of Polycrystalline Iron", Acta Met., 1964, Vol. 12, pp. 59-65.

[⊟] P. J. Worthington, "The Upper Yield Stress of Polycrystalline 3% Silicon Iron", Acta Met., 1967, Vol. 15, pp. 1795-1798.

" G. Birkbeck and R. M. Douthwaite, "The Influence of Grain Size and Specimen Size on the Upper Yield Stress of Iron". Trans. Met. Soc. AIME, Vol. 242, August, 1968, pp. 1595-1597.

[≡] B. W. Christ, G. V. Smith and M. S. Burton, "The Influence of Grain Boundaries and Veining Subgrain Boundaries on the Yield Phenomenon in Zone-Refined Iron", Trans. Met. Soc. — AIME, 1966, Vol. 236, pp. 9-13.

ALLOYS OF IRON

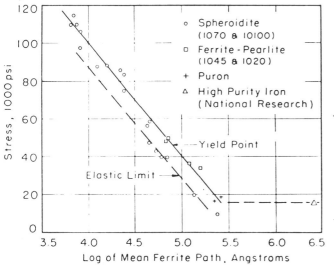

Figure 8-91:
Effect of the Distance Between Cementite Particles On the Yield Stress of a Ferrite Matrix.

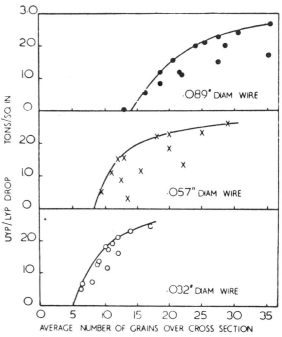

Figure 8-92:
UYP/LYP Drop in Pure Iron as a Function of Specimen Diameter and Average Number of Grains Over Cross Section.

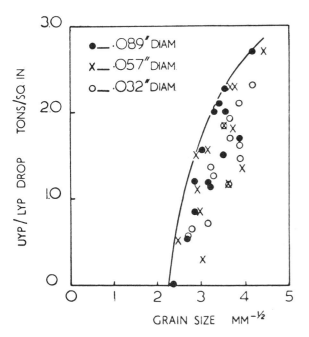

Figure 8-93:
UYP/LYP Drop in Pure Iron as a Function of Grain Size.

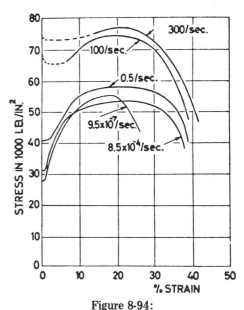

Figure 8-94:
Stress-Strain Curves of Mild Steel at Room Temperature for Various Rates of Strain (Manjoine).

8-29 The Effect of Strain Rate on Yield Strength

Experiments conducted by Nadai and Manjoine[1] about 1940 showed that, at room temperature (20°C), the yield stress of low carbon steel remains practically constant over a range of

---

[1] A. Nadai, and M. J. Manjoine, "High-Speed Tension Tests at Elevated Temperatures, Parts I and II", Proceedings ASTM 40, (1940), 822-37, Part III J. App. Mech. VI 8, 2 (1941) A77-A91.

strain rate from 10^{-5} to 10^{-1} seconds^{-1}. Perhaps for this reason most rolling theorists subsequently disregarded strain rate effects in their theoretical approach to cold rolling.[ø,Δ] Yet for strain rates in excess of 10^{-1} seconds^{-1}, the yield strength changes rapidly with strain rate, so much so that at strain rates commonly encountered in commercial cold rolling operations (10 to 1000 seconds^{-1}) the dynamic tensile yield strength may attain values about twice those obtained in conventional testing.

Figure 8-94 illustrates stress-strain data obtained later by Manjoine[Δ] for strain rates up to 300 seconds^{-1} and it will be noted that the increase in the yield stress with increasing strain (the workhardening effect) decreases with increasing strain rate. Similar data have been obtained by Green & Maiden[ø] of the G.M. Defense Research Laboratories at Santa Barbara, California, their data for a rimmed, commercial quality steel being shown in Figure 8-95.

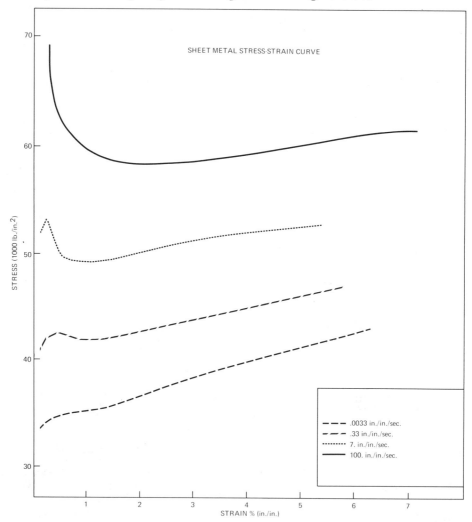

Figure 8-95: Flow Stress for Commercial Quality Rimmed Steel (Green and Maiden).

ø E. Orowan, "The Calculation of Roll Pressure in Hot and Cold Flat Rolling", "Research on the Rolling of Strip", A Symposium of Selected Papers 1948-1958, The British Iron and Steel Research Association, Waterloo and Sons, Ltd., London.

Δ M. J. Manjoine, "Influence of Rate of Strain and Temperature on Yield Stresses of Mild Steel", Trans. ASME, Vol. 66, 1944, pp. A211-A218.

ø S. J. Green and C. J. Maiden, "Strain Rate Effects in Sheet Steel and Their Application to Metal Forming", Final Report on P.O. MD 502079, Manufacturing Development, G. M. Defense Research Laboratories, Santa Barbara, California, Dec., 1965.

ALLOYS OF IRON

It is interesting to note that, in the case of rimmed steel, a very pronounced yield drop was developed as the strain rate was increased to 10 to 20 seconds^{-1}. In semikilled steels, the same trend was observed but not to such an extent. In this connection, Bodner[1] has stated that a general observation for metals is that strain rate sensitivity of the yield stress appears to be associated with the existence of a sharp yield point.

High-speed tensile tests show that the strain rate sensitivity of the yield strength decreases with increasing amounts of prior work as indicated in Figure 7-2. An isometric projection of the stress-strain rate surface for a mild steel is shown in Figure 8-96 and an empirical formula for the dynamic yield stress is presented in Section 7-2.

For annealed Type 301 stainless steel (with an austenitic fcc structure), the variations of the tensile strength and the yield strength with strain rate at various temperatures are illustrated in Figure 8-97.[2] It will be noted that, in this case, the strain rate sensitivity is much less pronounced.

Using equipment similar to that described in the next section, Ripperger[3] obtained dynamic stress-strain curves for Armco iron as illustrated in Figure 8-98. Under a hydrostatic pressure of 100,000 psi, curves such as those shown in Figure 8-99 were obtained. The strain rate effects for the two cases, with the excess strain ratio, $(\sigma - \sigma_0)/\sigma_0$, plotted against the strain rate are shown in Figures 8-100 and 8-101. (Here, σ_0 is the "static", or the value of the yield stress obtained at low strain rates.)

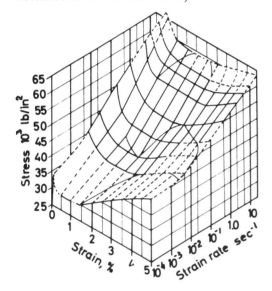

Figure 8-96:
Isometric Projection of Stress-Strain, Strain Rate Surface for a Mild Steel.

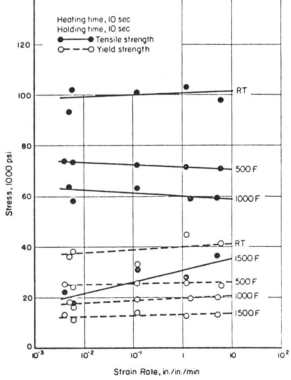

Figure 8-97:
Annealed Type 301 Stainless Steel Sheet, Longitudinal Specimens.

8-30 The Measurement of Yield Strength at High Strain Rates

Conventional tensile testing machines of the type shown in Figure 8-85, with hydraulic or screw-type actuation, usually produce a constant strain rate in the range 10^{-5} to 10^{-1}

---

[1] S. R. Bodner, "Strain Rate Effects in Dynamic Loading of Structures", Published in Behavior of Materials Under Dynamic Loading", ASME, New York, 1965.

[2] D. P. Moon and J. E. Campbell, "Effects of Moderately High Strain Rates on the Tensile Properties of Metals", DMIC Memorandum 142 (Battelle Memorial Institute), Dec. 18, 1961.

[3] E. A. Ripperger, "Dynamic Plastic Behavior of Aluminum, Copper and Iron", Paper published in "Behavior of Materials Under Dynamic Loading", Edited by N. J. Huffington, Jr., ASME, New York, 1965.

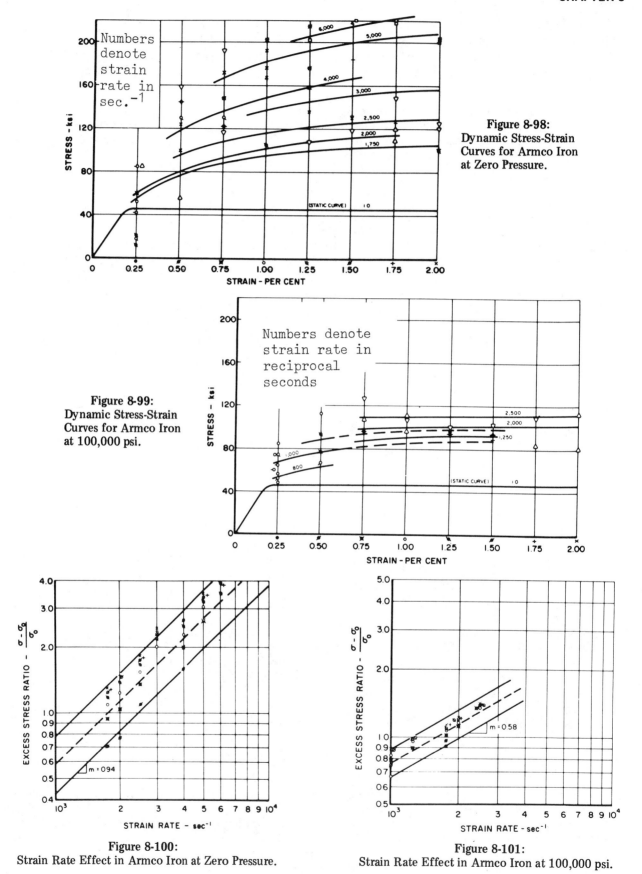

Figure 8-98: Dynamic Stress-Strain Curves for Armco Iron at Zero Pressure.

Figure 8-99: Dynamic Stress-Strain Curves for Armco Iron at 100,000 psi.

Figure 8-100: Strain Rate Effect in Armco Iron at Zero Pressure.

Figure 8-101: Strain Rate Effect in Armco Iron at 100,000 psi.

ALLOYS OF IRON

seconds^{-1} to provide so-called "static" stress-strain curves. Under such conditions, inertial forces in the measuring system may be neglected. In the approximate strain rate range 10^{-1} to 50 seconds^{-1}, fast acting hydraulic or pneumatic machines are used to provide "dynamic" stress-strain curves and for rates in the approximate range 50 to 10^4 seconds^{-1}, mechanical or explosive impact machines are used to obtain "impact" curves. In both these higher-rate ranges, inertial forces must generally be taken into account in interpreting the experimental data since they may not only produce additional loading of the specimen under test but also nonuniformity in the distribution of stress and strain over a finite gage length.[8]

Two methods, one direct and one indirect, of obtaining impact curves have been used. In the direct method, attempts have been made to measure both the load and deformation of the specimen independently as functions of time. The specimen, in the form of a cylinder, is loaded between two elastic bars and the load and displacement are measured only at the two faces of the cylinder. In the indirect method, only a deformation variable is measured (either strain or particle velocity) on the radial surfaces of slender rods through which a wave of plastic deformation is propagating.

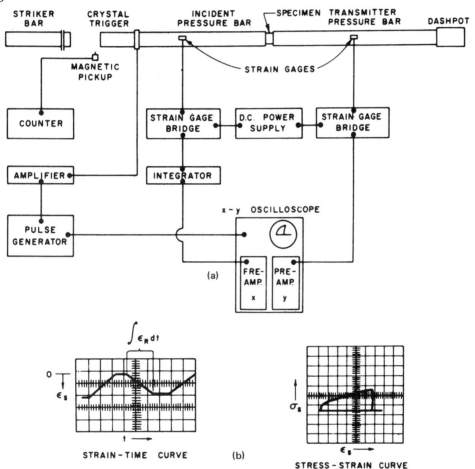

Figure 8-102: Schematic Diagram of Split Pressure Bar Apparatus (a) and Typical Oscillograms (b).

An example of the direct method has been described by Lindholm[8] and also by Ripperger.[∞] This technique, described as a split Hopkinson pressure bar method, was used to obtain

---

[8] U. S. Lindholm, "Dynamic Deformation of Metals", Paper Published in "Behavior of Materials Under Dynamic Loading", Edited by N. J. Huffington, Jr., ASME, New York, 1965.

[∞] E. A. Ripperger, "Dynamic Behavior of Aluminum, Copper and Iron", Paper Published in "Behavior of Materials Under Dynamic Loading", Edited by N. J. Huffington, Jr., ASME, New York, 1965.

data at rates in excess of 10^3 seconds^{-1}. The equipment is shown in diagrammatic form in Figure 8-102 and it is seen that the specimen is sandwiched between two long pressure bars which remain elastic throughout the impact. The loading is initiated by impacting the incident pressure bar with a striker bar producing a constant amplitude stress pulse whose duration (of the order of 200 microseconds) is established by the length of the striker bar.

Strain gages attached to the elastic bars record three separate strain pulses corresponding to the incident loading force (ϵ_L), the reflected pulse (ϵ_R) and the transmitted pulse (ϵ_T). It can be shown that, if the length of the specimen is l_0, the strain in the specimen (ϵ_S) is given by

$$\epsilon_S = \frac{C_0}{l_0}\int_0^t (\epsilon_L - \epsilon_R - \epsilon_T)dt \qquad (8\text{-}95)$$

and the stress (σ_S) by

$$\sigma_S = \frac{1}{2}E\left(A/A_S\right)(\epsilon_L - \epsilon_R + \epsilon_T) \qquad (8\text{-}96)$$

where C_0 and E are the elastic wave velocity and Young's modulus for the pressure bars, respectively, and (A/A_S) is the area ratio between the pressure bars and the specimen.[:] If the assumption is made that the stress is constant across the specimen, then

$$\epsilon_R = \epsilon_T - \epsilon_L \qquad (8\text{-}97)$$

and, as a consequence, the strain in the specimen is directly proportional to the time integral of the reflected pulse ϵ_R and the stress is directly proportional to the transmitted pulse ϵ_T. Since these two pulses may be made coincident in time by placing the two strain gages equidistant from the specimen, the stress-strain curve may be observed directly on an oscilloscope using the X and Y amplifiers (Figure 8-102(b)).

8-31 Theoretical Aspects of Dynamic Yielding

The experimental data discussed in the preceding sections indicate that the plastic deformation of steels is, in general, time dependent. Moreover, there also appears to be a degree of interdependence between the deformation rate sensitivity and the temperature sensitivity of the yield stress. Thus the dynamic yield stress (ϵ_D) appears to be a function of the strain rate ($\dot{\epsilon}$) and the temperature (T), thus

$$\sigma_D = \text{fn}(\epsilon, \dot{\epsilon}, T) \qquad (8\text{-}98)$$

Zener and Hollomon[▽] proposed a relationship of the form

$$\sigma_D = C(\dot{\epsilon})^n \qquad (8\text{-}99)$$

where C and n are material constants and where the temperature and the strain rate remain constant. They[⋔] later showed that the effects of temperature and strain rate are related (at least over limited ranges of the two variables) and, for conditions of constant strain, they derived a relationship of the form

$$\sigma_D = \text{fn}(\dot{\epsilon}\exp. Q/RT) \qquad (8\text{-}100)$$

where Q may also be a function of stress, R is the gas constant and T is the absolute temperature.

As Lindholm[8] has pointed out, of necessity, the general engineering approach has been to derive such relationships empirically from the results of mechanical tests, although hopefully a

---

[:] U. S. Lindholm, "Some Experiments with the Split Hopkinson Pressure Bar", J. Mech. and Phys. of Solids, Vol. 12, 1964, pp. 317-338.

[▽] C. Zener and J. H. Hollomon, "Effect of Strain Rate Upon Plastic Flow of Steel", Journal of Applied Physics, 15, 1944, pp. 22-32.

[⋔] C. Zener and J. H. Hollomon, "Problems in Non-Elastic Deformation of Metals", Journal of Applied Physics, 17, 1946, pp. 69-82.

[8] U. S. Lindholm, "Dynamic Deformation of Metals", Paper Published in "Behavior of Materials Under Dynamic Loading", Edited by N. J. Huffington, Jr., ASME, New York, 1965.

constitutive equation could eventually be derived on a theoretical basis. Such a theory would be based on the motion of dislocations and their interactions with other defects in the crystal lattice. Phenomena associated with creep and low strain rate behavior in metals can be described by the theory of thermally activated dislocation mechanisms which assumes that, as a dislocation moves through a crystal lattice, its motion is impeded by various obstacles, such as other dislocations. The moving dislocations may overcome these obstacles under conditions of increased applied stress or through the effects of random thermal fluctuations. As in other processes, where the rate is controlled by the overcoming of a potential barrier, the average strain rate $\dot{\epsilon}$ may be represented by an equation of the Arrhenius type, viz.,

$$\dot{\epsilon} = \dot{\epsilon}_0 \exp.-\left(\frac{H}{kT}\right) \quad (8\text{-}101)$$

where $\dot{\epsilon}_0$ is a frequency factor, H is the energy that must be supplied by the thermal fluctuations to overcome the obstacle, k is Boltzmann's constant and T is the absolute temperature.

Since the energy (H) that must be supplied thermally is predominantly a function of the stress (σ), it is possible to expand $H = H(\sigma)$ in a Taylor's series about the stress $\sigma*$ where $\sigma*$ is the magnitude of the long-range internal stress field opposing the motion of the dislocation. The expansion gives

$$H(\sigma) = H(\sigma*) + \frac{dH}{d\sigma}(\sigma*)[\sigma - \sigma*] + 1/2 \frac{d^2H}{d\sigma^2}(\sigma*)[\sigma - \sigma*]^2, \text{ etc.} \quad (8\text{-}102)$$

Truncating the series after two terms and representing $H(\sigma*)$ by H_0 and $\frac{-dH(\sigma*)}{d\sigma}$ by $v*$ (the activation volume, which may be regarded as the product of the Burgers vector, the length of the dislocation and the width of the barrier), it follows that

$$H(\sigma) = H_0 - v*(\sigma - \sigma*) \quad (8\text{-}103)$$

From the substitution of this expression for $H(\sigma)$ in the Arrhenius type equation, it can be shown that

$$\sigma = \sigma* + \frac{H_0}{v*} - \frac{kT}{v*} \ln \frac{\dot{\epsilon}_0}{\dot{\epsilon}} \quad (8\text{-}104)$$

The parameters $\dot{\epsilon}_0$, H_0, $v*$ and $\tau*$ (the shear stress in the plane of the dislocation due to the internal athermal stress field of the lattice) constitute the internal variables describing the deformed state of the material and σ, ϵ, $\dot{\epsilon}$ and T are the four external variables. By suitable measurements of the latter, it is possible to compute the former.

Experimentally, it has been found that, if a variation in the strain rate ($\Delta\dot{\epsilon}$) during plastic deformation is compensated by an appropriate temperature variation (ΔT), the stress for a given plastic strain may be kept at a constant value. However, in general, the relationship between $\Delta\dot{\epsilon}$ and ΔT will depend on the stress level σ. Sleeswyk[^] has shown that bcc metals are an exception to this rule for, in these metals, the functional relationship between $\Delta\dot{\epsilon}$ and ΔT is found to be independent of σ. Thus, $\dot{\epsilon}$ and T may be regarded as equivalent parameters. This is illustrated in Figure 8-103 showing the relationship between the yield stress (σ) and the temperature (T), for Armco ingot iron given a strain (ϵ) of 0.03 at four different strain rates.

From time to time, claims are made that the application of cyclical stresses has a beneficial effect in metalworking operations inasmuch as the yield strength of the metallic workpiece is believed somehow to be reduced. Careful experiments have shown that, whereas certain frictional effects may be reduced by audio and ultrasonic energy,[~] the plastic behavior of the metal is basically unaffected. This has been demonstrated by Winsper and Sansome[Θ] who

---

[^] A. W. Sleeswyk, "Equivalence of $\dot{\epsilon}$ and T in Plastic Deformation of BCC Metals", Scripta Metallurgica, Vol. 4, 1970, pp. 225-230.

[~] Such Reductions in Friction Could Usually be Obtained Much More Economically by the Application of Effective Lubricants.

[Θ] C. E. Winsper and D. H. Sansome, "The Influence of Oscillating Energy on the Stresses During Plastic Deformation", Journal of the Institute of Metals, Vol. 96, Part 12, December, 1968, pp. 353-357.

conducted tensile tests with the addition of a 100 cps cyclical stress on a modified fatigue machine (Figure 8-104). The analysis of their data showed that the apparent reduction in stress could be explained by the superposition of static and oscillatory stresses. A stress-strain curve for Duralumin is reproduced in Figure 8-105 in which the dashed curve represents the apparent yielding stress, but it can be seen that, when the cyclic stresses are superposed on it, the peak amplitude reaches the normal stress-strain curve (solid line up to "a" and the dot-dash projection). Confirmation of this superposition theory has been obtained in wire drawing experiments conducted with 100 cps vibratory energy applied to the system.

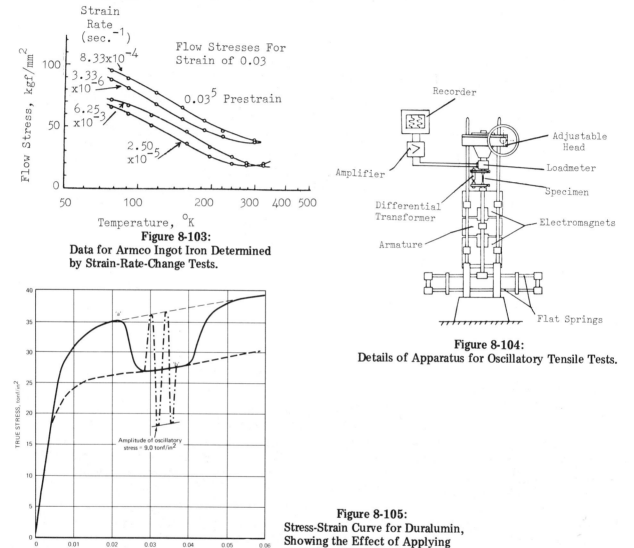

Figure 8-103:
Data for Armco Ingot Iron Determined by Strain-Rate-Change Tests.

Figure 8-104:
Details of Apparatus for Oscillatory Tensile Tests.

Figure 8-105:
Stress-Strain Curve for Duralumin, Showing the Effect of Applying Oscillatory Energy.

8-32 Stress-Strain (Workhardening) Curves

Where large reductions are to be given to a workpiece in a rolling operation, it is appropriate to develop stress-strain data relative to large ranges of strain. Such curves may be obtained, for ductile materials, from conventional tensile tests and, for materials of limited ductility, by first straining the material to various degrees and then measuring the corresponding yield strengths of the prestrained specimens.

When a typical tensile specimen in the form of a cylindrical bar (Figure 8-106) is pulled, an apparent stress-strain curve (in engineering units) is obtained as illustrated in Figure 8-107. From

ALLOYS OF IRON

measurements made on the specimen during its deformation, it is possible to derive the true stress-strain curve, as illustrated in Figure 8-108. To obtain the true stress, the tensile force applied to the specimen is divided by the cross-sectional area at the narrowest point of the necked region and the Bridgeman correction factor $[(1 + 4R/d) \ln (1 + d/4R)]$ where d is the minimum diameter of the neck and R is the radius of the neck profile (Figure 8-106(b)).

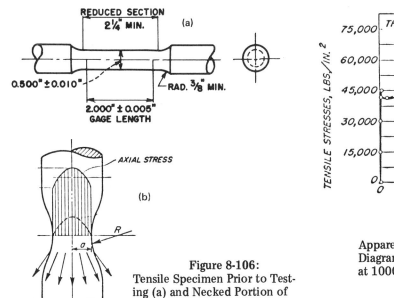

Figure 8-106: Tensile Specimen Prior to Testing (a) and Necked Portion of Strained Specimen (b).

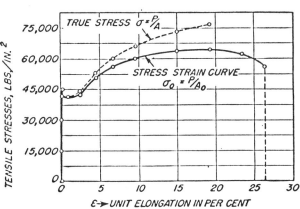

Figure 8-107: Apparent or Nominal Tensile Stress-Strain Diagram of Wrought Iron (Annealed 1 hour at 1000°C).

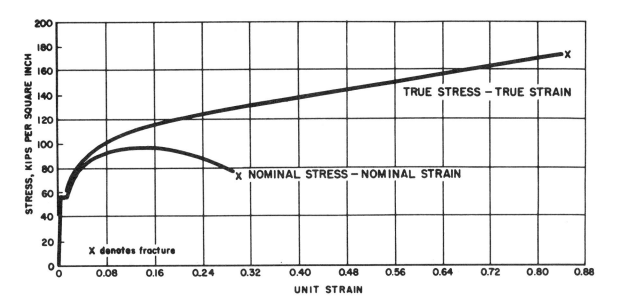

Figure 8-108: True Stress — True Strain Curve and Nominal Stress — Nominal Strain Curve.

Usually, the true stress (σ) and the true strain (δ) are related as indicated by the equation

$$\sigma = k \delta^n \qquad (8-105)$$

where k is defined as the strength coefficient and n is the strain-hardening exponent or coefficient. The latter is an important parameter with respect to metal forming since, in stretching operations, it is desirable for the metal to exhibit a relatively high value for n (~0.26). However, in operations where compressive stresses are applied, relatively low values for n would be more desirable. The strain hardening exponent in the case of iron and steels varies with yield stress as illustrated in Figure 8-109.

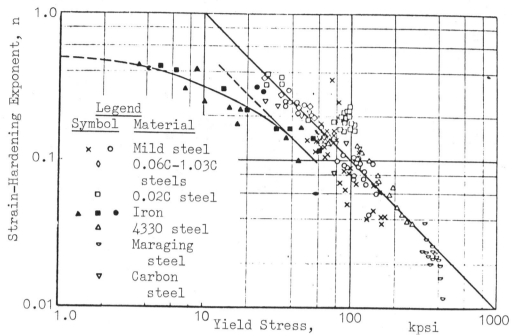

Figure 8-109: Effect of Yield Stress on Strain-Hardening Exponent.

Referring to Figure 8-107, the maximum value of the apparent stress-strain curve (where the slope, $d\sigma_E/d\epsilon$ is zero) corresponds to the ultimate tensile strength (σ_T) of the specimen. The true strain (δ_T) occurring at this stress level can be shown to be equal to the strain exponent, as follows. If the subscripts O and I denote initial and instantaneous values, respectively, and the symbols D, A, L and F denote, respectively, the diameter, area and length of the specimen and the force to which it is subjected, then, neglecting the Bridgeman correction factor, the true stress

$$\sigma = F_I/A_I \tag{8-106}$$

and the true strain

$$\delta = \int_{L_O}^{L_I} \frac{dL}{L} = \ln\left(L_I/L_O\right) \tag{8-107}$$

Assuming no volume change in the specimen

$$L_I/L_O = A_O/A_I = D_O^2/D_I^2 \tag{8-108}$$

Thus

$$\delta = 2 \ln\left(D_O/D_I\right) = \ln(1+\epsilon) \tag{8-109}$$

where ϵ is the engineering strain, and

$$\sigma = \sigma_E (1+\epsilon) \tag{8-110}$$

where σ_E is the engineering stress.

Since

$$\sigma = k\delta^n \tag{8-111}$$

$$\sigma_E(1+\epsilon) = k(\ln(1+\epsilon))^n \tag{8-112}$$

Differentiating

$$\frac{d\sigma_E}{d\epsilon} = k\left\{n[\ln(1+\epsilon)]^{n-1} - [\ln(1-\delta)]^n\right\} \quad (8\text{-}113)$$

Therefore when $d\sigma_E/d\epsilon$ is zero (at the stress level corresponding to the ultimate tensile strength)

$$n = \ln(1+\epsilon) = \delta \quad (8\text{-}114)$$

The exponential relationship of stress and strain was first proposed by Nadai [+] but it has the disadvantage of predicting a zero stress level corresponding to zero strain. A refinement of the equation was proposed by Swift [◁] and Rosenfield and Hahn [m] to ensure that the yield stress assumes a finite value at zero strain. Their suggested relationship may be written

$$\sigma = \frac{\sigma_{\dot{\epsilon}=0}^*}{B^m}(\delta + B)^m \quad (8\text{-}115)$$

where $\sigma^*_{\dot{\epsilon}=0}$ is the yield stress at zero strain rate, B a materials parameter (with values of approximately 0.005) which appears to be sensibly independent of strain rate and temperature. This relationship has been found to be valid for yield strengths up to about 300,000 psi.

In examining the workhardening behavior of a number of steels, Hollomon and Zener [T] concluded that:

(1) for any given steel, the product of the workhardening exponent and the yield stress is a constant with a value in the range 8000-16000 psi;

(2) this constant increases with increasing carbon content of the steel, viz.

$$\text{constant} = 6000[1 + 2(\text{wt \% C})] \quad (8\text{-}116)$$

(3) for a given carbon content, the constant for alloy steels was larger than for plain carbon steels.

Where a stress-strain curve is to be prepared from data obtained from a number of prestrained specimens, the curve is drawn as an envelope to the limited stress-strain curves for each specimen. Compressive or tensile data (as shown in Figures 8-110 and 8-111, respectively) may be used as required.[⊠] For low carbon steel, various empirical relationships have been developed relating tensile yield stress to the reduction r (expressed as a decimal fraction, equal to the draft divided by the initial strip thickness). One such relationship is

$$\sigma = \sigma_0 + ar^{1/2} \quad (8\text{-}117)$$

where σ_0 is the annealed yield strength of the steel and a is a constant generally in the range of 25,000 to 30,000 psi.[ϕ] Relative to tandem mill operation, polynomial expressions have been used as discussed in Section 7-2.

Although tensile tests conducted at low strain rates (such as those carried out to prepare Figures 8-107 and 8-108) show that workhardening occurs simultaneously with deformation, the analysis of cold rolling data indicates that, at high strain rates, low carbon steels do not appear to

---

[+] "Plasticity" by A. Nadai: McGraw-Hill Book Co., New York, 1943, p. 78.

[◁] H. W. Swift, "Plastic Instability Under Plane Stress", J. Mech. Phys. Solids, 1 (1952) 1.

[m] A. R. Rosenfield and G. T. Hahn, "Numerical Descriptions of the Ambient Low-Temperature, and High Strain Rate Flow and Fracture Behavior of Plain Carbon Steel", Transactions of the ASM, Vol. 59, 1966 pp. 962-980.

[T] J. H. Hollomon and C. J. Zener, "Conditions of Fracture of Steel", Trans. AIME, 158, (1944), pp. 283-297.

[⊠] H. Ford, "Researches into the Deformation of Metals by Cold Rolling", Paper Published in "Research on The Rolling of Strip", A Symposium of Selected Papers 1948-1958, The British Iron and Steel Research Association, Waterloo and Sons, Ltd., London.

[ϕ] W. L. Roberts, R. J. Bentz and D. C. Litz, "Cold Rolling Low-Carbon Steel to Minimum Gage", Iron and Steel Engineer Year Book, 1968, pp. 413-420.

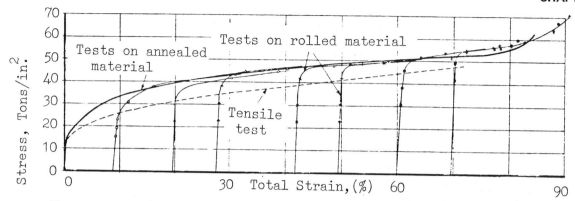

Figure 8-110: Yield Stress Curves from Compression Tests for 0.2 Per Cent Carbon Steel (Ford).

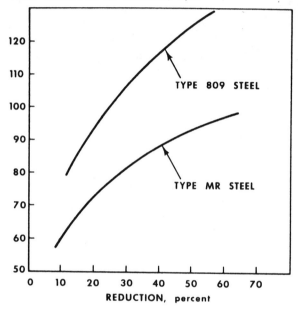

Figure 8-111:
Variation of Tensile Yield Strength With Reduction for Steel Types MR and 809.

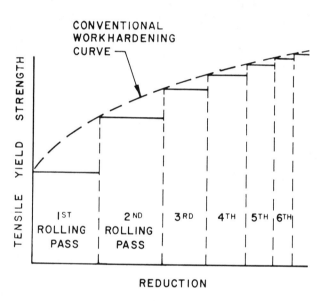

Figure 8-112:
Sketch Illustrating Changes in Yield Strength Effected by Successive Rolling Passes.

workharden while the metal is actually flowing.⁼ Thus, at high strain rates, the behavior of the steel is suggested by Bodner ⊣ to approach the elastic-perfectly plastic idealization but with a raised yield stress. Consequently, in recent years, many rolling theorists have taken strain-rate effects into consideration but have disregarded workhardening in the case of a single rolling pass.

After emergence from a roll bite, workhardening appears to take place with extreme rapidity so that, in the rolling of strip on high speed tandem mills, the strip is workhardened to the anticipated extent by the time it reaches the next mill stand. Thus, in calculating the rolling force and power requirements for a tandem six-stand cold mill, the strip should be considered to follow the stepped curve beneath the conventional workhardening curve as indicated in Figure 8-112.

⁼ R. J. Bentz and W. L. Roberts, "Speed Effects in the Second Cold Reduction of Steel Strip", Mechanical Working and Steel Processing V, Proceedings of the Ninth Mechanical Working and Steel Processing Conference, AIME, Gordon and Breach, New York, N. Y. 1968, pp. 193-222.

⊣ S. R. Bodner, "Strain Rate Effects in Dynamic Loading of Structures", published in "Behavior of Metals Under Dynamic Loading", Edited by N. J. Huffington, Jr., ASME, New York, 1965.

ALLOYS OF IRON 473

8-33 The Dependence of Shear Stress on Crystallographic Direction

Keh and Nakada[v] examined the plastic behavior of iron single crystals with respect to crystal orientation and temperature at low strain rates. Some of their data is reproduced in Figure 8-113 but, in this case, the shear stress is plotted against shear strain. (In this data, the flow stress value at a shear of 0.001 was taken as the critically resolved shear stress [CRSS].) Above 200°K, crystals of the different orientations have the same temperature dependence of the CRSS, as indicated in Figure 8-114. The relationship between shear stress and shear strain at 298°K for iron crystals is illustrated in Figure 8-115. Here the tensile axes of the crystal lie on the $[11\bar{1}] - [0\bar{1}\bar{1}]$ boundary of the stereographic triangle. Orientation B is a double-slip orientation. For the other orientations, the further the tensile axis is from B, the more pronounced is the change in workhardening behavior, this being attributable to the difference in the Schmid factor ratio of the major secondary slip system to the primary slip system. As the orientation moves away from B, this ratio decreases from 1 to a value 0.46 for orientation H. Not only does the orientation of the crystal affect the shape of the workhardening curve, but also the strain rate sensitivity of the flow stress. This is shown in Figure 8-116 with the change of shear stress $\Delta\tau$ due to a change in shear strain rate from $\dot{\gamma}_1 = 5.6 \times 10^{-4}$ seconds^{-1} to $\dot{\gamma}_2 = 2.8 \times 10^{-3}$ seconds^{-1} plotted against shear strain for the same orientations considered in Figure 8-115.

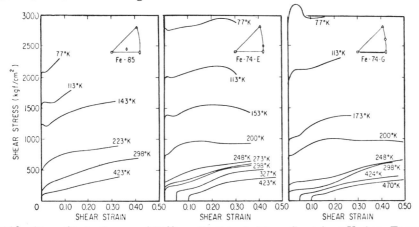

Figure 8-113: Stress-Strain Curves of Differently Oriented Iron Crystals at Various Temperatures.

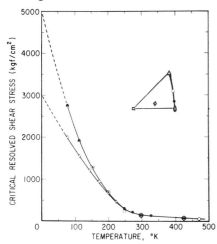

Figure 8-114:
Temperature Dependence of CRSS of Iron Crystals Oriented for (110) [111] and (211) [111] Slip.

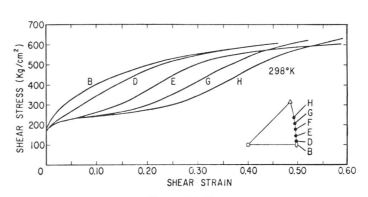

Figure 8-115:
Orientation Dependence of Stress-Strain Curves of Iron Crystals at 298°K.

---

[v] A. S. Keh and Y. Nakada, "Yielding, Plastic Flow and Dislocation Substructure in Iron Single Crystals", Supplement to Transactions of the Japan Institute Metals, Vol. 9, 1968.

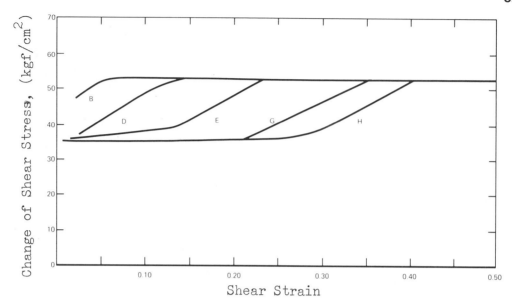

Figure 8-116:
Orientation Dependence of Strain-Rate Sensitivity of Flow Stress at 298°K, (γ_2/γ_1 = 5, ln (γ_2/γ_1 = 1.61).

8-34 Theoretical Aspects of Workhardening

In the deformation of polycrystalline metals, such as steels, slip occurs on many intersecting slip planes as discussed in Section 8-21. As the dislocations move on these planes, they not only increase considerably in numbers (Section 8-20), but considerable interactions or stresses develop between them. Thus, a progressive strengthening of the metal occurs as it is cold worked. Where dislocations cut across one another, a line of interstitialcies or vacancies may occur in the lattice if the dislocations are of the edge type. If they are of the screw type, however, part of the dislocation may be displaced into a slip plane parallel to the initial slip plane, thereby producing a discontinuity or a "jog" between the two portions of the dislocation. Yet, according to Friedel,[III] the hardening due to jogs should be small compared with that due to internal stresses.

A cold worked metal therefore contains a considerably increased number of dislocations, but also a very large concentration of point defects. Nevertheless, fragments of the original metal lattice survive the deformation and remain relatively intact. Such fragments, discussed in Section 12-3, have dimensions of about 10,000 Å and constitute a mosaic structure.

Apart from the mosaic structure, substructures can occur in grains as discussed in Section 8-8, and the low angle or tilt boundaries associated with them can have a marked effect on the plastic behavior of the material. As would be expected, the larger the subgrain size, the less the yield stress. This has been demonstrated by Warrington[⊢] and Ball[ω] and more recently by Baird[†] and MacKenzie.[8] Their data in part is reproduced in Figure 8-117, the increase in flow stress $\Delta\sigma$ due to the substructure being taken as the stress after straining (less an appropriate friction stress) and the dimensions of the subgrains being represented by t measured in microns. (The upper curve relates to nitrogenized specimens.) For subgrains about 2μ m in size, the increase in yield strength,

---

[III] "Dislocations", by J. Friedel, Addison-Wesley Publishing Co., Inc., Reading, Mass., 1964.
[⊢] D. H. Warrington, "The Flow Stress-Subgrain Size Relationship in Iron", JISI, 1963, 201, pp. 610-613.
[ω] C. J. Ball, "The Texture and Mechanical Properties of Iron", JISI, 1959, 191, pp. 232-236.
[†] J. D. Baird, "Effect of Subgrain Size on the Flow Stress of Iron", JISI, January, 1966, pp. 44-45.
[8] J. D. Baird and C. R. MacKenzie, "Effects of Nitrogen and Manganese on the Deformation Substructure of Iron Strained at 20°, 225° and 450°C", JISI, 1964, 202, pp. 427-436.

ALLOYS OF IRON

due to the substructure, is about 10 kg/mm² for the relatively pure iron carbon (0.036 per cent C) and about 20 kg/mm² for the nitrogenized (0.009 per cent N) specimens of the same alloy with approximately the same grain size.

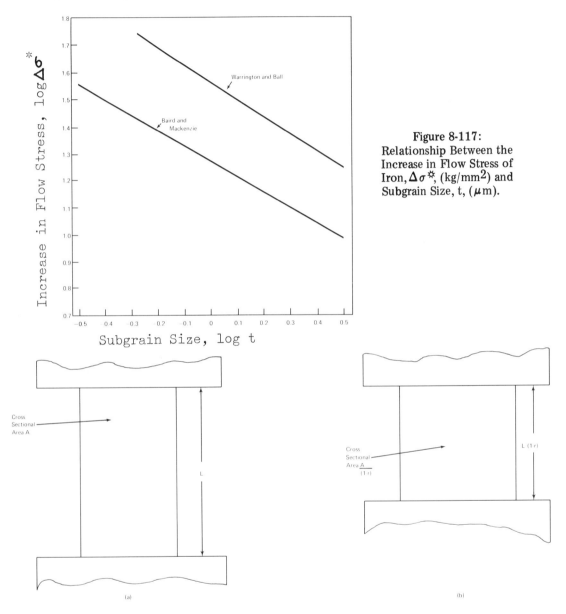

Figure 8-117: Relationship Between the Increase in Flow Stress of Iron, $\Delta \sigma^*$, (kg/mm²) and Subgrain Size, t, (μm).

Figure 8-118: Sketch Illustrating Deformation of Cylinder Between Platens.

8-35 The Energy of Deformation

Because of the large stresses required to plastically deform steels, even at low strain rates, considerable expenditures of energy are involved in the cold reduction of these metals. In deriving an expression for the energy of deformation, it is convenient to imagine a small right cylinder of cross-sectional area A and of length L placed between two frictionless platens as illustrated in Figure 8-118(a). After the platens have moved so that a reduction r has been obtained (as indicated in Figure 8-118(b), on the assumption that no volume change has occurred, the cross-sectional area will be A/(1-r). If the constrained yield stress is σ_c and if a further reduction (dr) occurs, then the force exerted on the platens is $\frac{A\sigma_c}{(1-r)}$ and the work performed in the incremental compression is $\frac{AL\sigma_c dr}{(1-r)}$. The total work W_T involved in compressing the cylinder from its initial length (L) to its final length (L(1-r)) is therefore given by

$$W_T = AL \int_0^r \frac{\sigma_c}{(1-r)} dr \qquad (8\text{-}118)$$

Since the volume of the cylinder is AL, the energy of deformation per unit volume, W, (the specific deformation energy) is given by

$$W = \int_0^r \frac{\sigma_c}{(1-r)} dr \qquad (8\text{-}119)$$

Because of the apparent lack of work hardening at rolling speeds during the flow of the metal (as discussed in Section 8-32), the value of σ_c may be regarded as a constant during a single pass reduction. Under these circumstances and assuming that no frictional constraints occur in the rolling direction

$$W = \sigma_c \int_0^r \frac{dr}{(1-r)}$$

$$= \sigma_c \ln\left(\frac{1}{1-r}\right) \qquad (8\text{-}120)$$

If σ_c is expressed in terms of psi, W is given in in.-lb./in.^{3}. Assuming the density of steel to be 0.283 lb./in.^{3}, the specific energy in terms of HP-hr./ton may be obtained by multiplying its value in in.-lb./in.^{3} by 2.98×10^{-4}. It should be remembered, however, that in the computation of specific energies, σ_c relates to the yield strength of the strip prior to its deformation corresponding to the rate of deformation occurring in the roll bite.

If it is assumed that the overall deformation occurs as a consequence of many small separate reductions, and that

$$\sigma_c = \sigma_0 + ar^{1/2} \qquad (8\text{-}121)$$

then

$$W = \int_0^r \frac{(\sigma_0 + ar^{1/2}) dr}{(1-r)}$$

$$= \sigma_c \ln \frac{1}{(1-r)} + 2a \left[\tanh^{-1} r - r\right] \qquad (8\text{-}122)$$

Measured values for the deformation of low carbon steel strip pulled through rotating dies is illustrated in Figure 8-119.[1]

8-36 Stored Energy and Its Release

Except for a small fraction (about 5 to 10 per cent), all the energy directly utilized in deforming the metal is dissipated as heat.[<] Since the thermal capacity of the steel remains essentially constant, it becomes possible to relate the specific deformation energy to a temperature

---

[1] D. G. Christopherson and B. Parsons, "The Effect of High Strain Rate in Strip Rolling", Proceedings of the Conference on "The Properties of Materials at High Rates of Strain", April 30 — May 2, 1957, Published by the Institution of Mechanical Engineers.

[<] A. L. Tichener and M. B. Bever, "The Stored Energy of Cold Work", Progress in Metal Physics, 7, Bruce Chalmers and R. King, Eds., Pergamon Press, London, 1958, pp. 247-338.

rise in the strip. (A specific deformation energy of 1 hp-hr/ton corresponds to a temperature rise of about 11.9°F in the steel, assuming complete conversion of the deformation energy to thermal energy.)

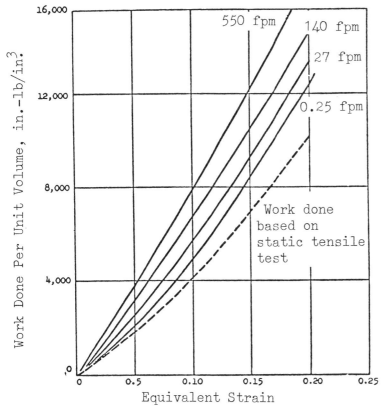

Figure 8-119: Energy Expended in Deforming Low Carbon Steel Strip by Drawing Through Rotating Dies. (Christopherson and Parsons)

The small quantity of deformation energy that is retained in the rolled strip (the so-called "stored energy") acts as a thermodynamic driving force which tends to return the metal to the properties it possessed in its undeformed state, provided the metal is at a temperature at which the reactions can occur at a reasonable rate.

The stored energy can be regarded as that of the dislocations introduced during the process of deformation. (0.2 cal./gram corresponds to the creation of 10^{13} dislocations/cm^2). In the deformation of iron, 0.66 cal./gram (or 8% of the deformation energy) has been measured as stored energy. The energy associated with vacancies and other point defects is largely released during recovery (see Section 12-6), but the energy associated with the other defects is released during recrystallization and grain growth (Section 12-7).

Chapter 9 Mathematical Models Relating to Rolling Force

9-1 Introduction

Mathematical expressions which theoretically relate the various rolling parameters to each other are generally referred to as models of the rolling process. They serve a number of purposes, the principal ones being the assistance they give to mill builders in the design of new rolling facilities and to operators in indicating how existing mills may be better or differently utilized. In addition, they are the basis of the proper computer control of cold mills.

Generally, the term "rolling model" is used to denote those sets of equations that relate rolling force and/or spindle torque to other factors such as the yield strength of the strip being used and the strip tensions. However, models of the process need not be limited to such kinetic considerations but may be concerned with such aspects as temperature distribution in the rolls and the strip (thermal models) or relate to the costs and profitability of the rolling operation (economic models). In this chapter, however, only those models relating to rolling force are discussed.

For a cold rolling model to be valid, it must be capable of elucidating all the various observations that have been made relative to the cold reduction process. For example, it must be able to explain, in quantitative terms, not only the more obvious relationships between the various parameters, like the increase in rolling force with increased draft, but the more subtle aspects of the process, such as the decrease in rolling force with increased strip tensions and the change in rolling force with increased mill speed.

The rolling process involves the interaction of three components, namely, the work rolls, the lubricant and the workpiece, which have been discussed in detail in Chapters 4, 6 and 8, respectively. The exact behavior of each is, generally speaking, too complex to be encompassed in any reasonable model so that rolling theorists must make simplifying assumptions with respect to each of these components. In the early sections of this chapter, the important characteristics of each are reviewed and the simplifications utilized for mathematical purposes are explained. In particular, the elastic deformation of the work rolls is discussed and the well known Hitchcock [§] relationship used for calculating the effective diameter of an elastically deformed work roll is presented.

A number of different models relating to rolling force have been developed during the last four decades, their diversity being attributable basically to the assumptions made. One of the better known models was developed by Von Karman, who expressed the pressure distribution along the arc of contact at the roll strip interface in the form of a differential equation. Attempts were then made by a few theorists to solve this equation, again using different simplifying assumptions as discussed in Sections 9-10 and 9-11. Others, including Orowan,[x] Jortner, et al.,[□] and Ekelund[◈] developed more sophisticated models, intending to establish mathematical equations of greater validity (see Sections 9-12 to 9-15).

Because such models were generally inconvenient to use, due to their mathematical complexity, attempts were made to develop simplified models more readily adaptable for engineering use and for the programming of control computers. Although the resulting models have not been claimed to be the most accurate from a predictive viewpoint, their convenience has considerably facilitated rolling research and mill design. For example, being basically of an algebraic form, they could be readily used for the calculation of any unknown in the rolling process, such as

---

[§] J. Hitchcock, "Roll Neck Bearings", ASME Research Publication, Appendix 1, 1935.

[x] E. Orowan, "The Calculation of Roll Pressure in Hot and Cold Flat Rolling", Proc. Inst. Mech. Eng., Vol. 150, No. 4, pp. 140-167.

[□] D. Jortner, J. F. Osterle and C. F. Zorowski, "An Analysis of the Mechanics of Cold Strip Rolling", Iron and Steel Engineer Year Book, 1959, pp. 403-411.

[◈] S. Ekelund, "The Analysis of Factors Influencing Rolling Pressure and Power Consumption in the Hot Rolling of Steel", Translated from Jernkontorets Ann., 1927, Vol. III, p. 39ff in Steel, 1933, Vol. 93(8), p. 27ff.

MATH MODELS RELATING TO ROLLING FORCE

the effective coefficient of friction in the roll bite, whereas, with the more sophisticated models, it was generally easier to assume values for the various parameters, develop theoretical curves and then compare these with actual mill data. Simplified models, capable of ready adaptation to digital computers are discussed in Sections 9-16 to 9-18.

The foregoing models relate to reasonably large reductions (generally in the range of 20-50 per cent) with the behavior of the lubricant and the workpiece being assumed to conform to conventional concepts. Section 9-19 presents a model for temper rolling with relatively small reductions (<20 per cent) and with dry, unlubricated surfaces in the roll bite exhibiting high coefficients of friction (in the general range 0.2 to 0.35).

In Section 9-20, a model is presented where purely hydrodynamic lubrication is assumed and, in Section 9-21, the behavior of the workpiece is postulated as that attributable to a viscous mass exhibiting sticking friction at the roll surfaces.

A different concept in modelling is described in Section 9-22. This approach, essentially an outgrowth of slip line theory (Section 8-24), develops an expression for the rolling force (and the spindle torques) from a consideration of a multiplicity of regions in the roll bite bounded by shear planes.

In Section 9-23, a few instances of the utilization of force models are presented and the predictive capabilities of the models examined.

Finally in Sections 9-24 and 9-25, the so-called minimum gage problem is discussed, its importance arising from the continuing trend towards the production of thinner and thinner strips for use in the packaging industry. The problem is associated with the minimum strip thickness that can be rolled on a given mill under a given set of circumstances. For many years, Stone's [1][2] formula was regarded as the cornerstone of the minimum gage theory, but in the last decade, it has been shown that it predicted minimum thicknesses that were too large by a factor of about 3. However, because of its importance in rolling theory, Stone's model and other minimum gage expressions are presented together with the more recent ideas on the subject.

9-2 Assumptions Relating to Rolling Lubrication

Critical to the rolling operation are the frictional conditions existing at the work roll — workpiece interfaces in the roll bite. Because of the difficulties encountered in trying to experimentally monitor them under high-speed rolling conditions, discrepancies between actual rolling forces and those predicted on the basis of various rolling models were attributed to the utilization of erroneous values of the coefficient assumed to exist at the interfaces. Thus, the coefficient assumed the role of a "fudge-factor", or a parameter that could be varied essentially at will to bring theoretical and experimental data into agreement. Though such a practice may be valid under certain circumstances, the values for other parameters used in the theoretical models were often in error and this led to the belief that abnormally high values of the coefficient existed in the roll bite even under conditions of excellent lubricity. As a consequence, sticking friction was believed to occur at least over part of the arc of contact. Although it seems highly improbable that such conditions exist in cold rolling, the concept has been used in the development of Kneschke's theory of rolling, presented in Section 9-21.

In the earlier models, Amanton's law was regarded as applicable to the friction at the roll-strip interface with the ratio of shear stress to normal stress assuming a constant value (equal to the coefficient of friction) along the entire length of the arc of contact. Such an assumption was made in both the case of dry rolling and where effective lubricants were used. Whereas the assumption is probably quite valid in the case of the former, it is to be seriously questioned with regard to the latter.

---

[1] M. D. Stone, "Rolling of Thin Strip — Part I", Iron and Steel Engineer Year Book, 1953, pp. 115-128.

[2] M. D. Stone, "Rolling of Thin Strip — Part II", Iron and Steel Engineer Year Book, 1956, pp. 981-1002.

Nadai [□] was one of the first theorists to consider the possibility of other frictional conditions existing in the bite. He derived solutions for Von Karmen's equation assuming (a) a constant coefficient of friction, (b) a constant shearing stress, and (c) a surface friction proportional to the velocity of slip. With respect to the second assumption, the coefficient of friction would be inversely proportional to the normal pressure exerted at the interface, whereas the third assumption relates to hydrodynamic lubrication based on the postulation of constant oil film thickness and viscosity.

Because of the "speed effect" which is frequently observed in cold rolling, wherein the rolling force decreases with increasing mill speed, the belief arose that hydrodynamic lubrication must at least play a part in roll-bite frictional phenomena. A number of models were subsequently developed on the basis of this type of lubricity, but the validity of such models is difficult to establish because of uncertainties with respect to the viscous behavior of the lubricant under the pressures and temperatures existing in the roll bite.

It became clear, however, that while hydrodynamic lubrication alone did not appear to be a satisfactory hypothesis in the case of wet rolling, neither was the postulation of a single coefficient of friction. Accordingly, a search was made for a method which would enable the effective coefficient of friction to be established for any given set of rolling conditions. Empirically, it was found that this could be accomplished by describing the behavior of the lubricant in terms of two frictional characteristics (discussed in detail in Section 6-31). Moreover, this approach had the added advantage of providing compensation for various roll and strip finishes and for metallic coatings on the strip surfaces(see Section 6-32).

9-3 The Strain Rate Occurring in Cold Reduction

In the 1960's it was realized that, from a theoretical standpoint, strain rate effects could not be ignored in cold rolling. Typically, strain rates in the range of 10 to 1000 seconds^{-1} are encountered in cold reduction operations and dynamic yield strengths measured in this range may be about double the yield strengths measured under conventional tensile testing conditions. It became desirable, therefore, to acquire more data with respect to the effect of strain rate ($\dot{\epsilon}$) on yield strength (σ_y) for various steels and also to derive an expression for the strain rate occurring during cold rolling.

Larke,[○] in connection with hot rolling, developed the following equation for the mean rate of deformation for slipping friction.

$$\dot{\epsilon} = \frac{V t_\alpha \cos \alpha}{t_1 t_2} \sqrt{\frac{2(t_1 - t_2)}{D}} \qquad (9\text{-}1)$$

where V is the rolling speed
    t_α is the thickness of the workpiece at the neutral plane,
    α is the angle subtended at the work roll center by the segment of the arc of contact lying between the neutral and exit planes,
    t_1 is the entry thickness of the workpiece,
    t_2 is the rolled thickness of the workpiece, and
    D is the work roll diameter.

In deriving this expression, it was assumed that the work rolls were rigid.

Alternately, the expression may be written

[□] A. Nadai, "The Forces Required for Rolling Steel Strip Under Tension", Journal of Applied Mechanics, ASME, June, 1939, pp. A54-A62.
[○] E. C. Larke, "The Rolling of Strip, Sheet and Plate", MacMillen Company, New York, 1957, pp. 258-273.

MATH MODELS RELATING TO ROLLING FORCE

$$\dot{\varepsilon} = \frac{Vt_\alpha \cos \alpha}{t_1^2 (1-r)} \sqrt{\frac{2t_1 r}{D}} \qquad (9\text{-}2)$$

where r represents the reduction (expressed as a decimal fraction). Assuming the neutral plane to be coincident with the exit plane (zero forward slip of the strip)

$$\dot{\varepsilon} = V \sqrt{\frac{2r}{Dt_1}} \qquad (9\text{-}3)$$

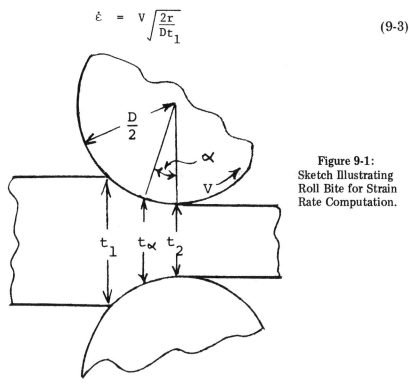

Figure 9-1: Sketch Illustrating Roll Bite for Strain Rate Computation.

This expression has been developed by other investigators [○] and, assuming the effective deformed diameter of the rolls to be 2.5 times their actual diameter, it follows that the strain rate $\dot{\varepsilon}$ may be calculated in terms of seconds^{-1} using the equation

$$\dot{\varepsilon} \simeq 0.179 \, V \sqrt{\frac{r}{Dt_1}} \qquad (9\text{-}4)$$

where V is expressed in terms of fpm and D and t are measured in inches.

It should be remembered, however, that the above expressions relate to the average strain rate occurring in the roll bite. In reality, the strain rate varies along the entire length of the arc of contact, decreasing continually from a maximum value occurring at the entry plane.

9-4 The Characteristics of the Deforming Workpiece

In the absence of strip tensions, the minimum pressure that must be exerted on the strip by the rolls must equal the constrained or compressive yield stress (σ_c) of the strip. As discussed in Section 8-25, this was shown by Nadai to be 1.155 times the yield strength in tension (σ_y). However, in the belief that the yield strength in tension is little influenced by the strain rate or the rate of deformation (which is true for the range of strain rates commonly encountered in conventional tensile testing), it was originally thought that the minimum deformation pressure in the roll bite was directly related to the value of the yield strength measured at very low strain rates.

[○] W. L. Roberts, R. J. Bentz and D. C. Litz, "Cold Rolling Low Carbon Steel Strip to Minimum Gage", Iron and Steel Engineer Year Book, 1970, pp. 413-420.

It was, however, realized by the early rolling theorists that the application of strip tensions lessened the pressures exerted by the rolls on the strip. At the entry end of the roll bite, the minimum roll pressure (σ_p) required to deform the strip would be given by

$$\sigma_p = \sigma_c - \sigma_1 = 1.155\,\sigma_y - \sigma_1 \qquad (9\text{-}5)$$

where σ_1 is the longitudinal stress in the strip on entry into the bite. Similarly it was believed that a stress existing in the strip on exit from the roll bite would lessen the roll pressure at the other end of the arc of contact.

To compensate for the effects of work hardening, the constrained or compressive yield stress of the workpiece was assumed to increase gradually from the entry to the exit of the roll bite in correspondence with the straining of the workpiece. Data relating to this work hardening was obtained by straining samples of annealed strip to various degrees (preferably by rolling) and then measuring the yield strength of each sample. The envelope of such curves was considered to be the work-hardening curve of the workpiece. For mathematical modeling purposes, the tensile stress required to yield a prestrained sample was considered to be equal to the stress corresponding to that strain in a sample being continuously deformed.

Recent studies have indicated, however, that contrary to earlier beliefs, annealed, low carbon steel does not work harden simultaneously with deformation, at least under conditions of large reductions at high speeds. Accordingly, models have been developed which assume that the constrained or compressive yield stress of an element of the strip at the entry of the roll bite, corrected for the rate of deformation (see Sections 7-2 and 8-29) remains constant as the element moves through the bite. At some time shortly after the cessation of the deformation, the yield strength of the strip increases rapidly to a value corresponding to the total deformation that has been given to the strip, in accordance with the conventional work-hardening curves. It should be pointed out, therefore, that where multi-pass rolling is undertaken, the constrained yield strength at the beginning of each pass corresponds to the total prior reduction given to the strip so that, though work-hardening effects are disregarded in each individual pass, they are considered with respect to successive reductions. Mathematical models based on these considerations are considerably simpler than some of the earlier rolling models but they have been found to be quite satisfactory for engineering purposes.

9-5 Elastic Flattening of the Work Rolls

In the earliest and simplest of the rolling theories, the work rolls were assumed to remain rigid, or to undergo no elastic deformation. In the heavy drafting of relatively soft material, the assumption of rigidity is reasonably valid but in the rolling of thin, hard strips, and in dry temper rolling, the rolls are significantly flattened so that the arc of contact is considerably longer than that which would exist with rigid rolls.

The elastic distortions of cylinders pressed against planar and other cylindrical surfaces were studied by Hertz,[△][+] Dinnik,[⊞] Huber,[◉] Fuchs[□] and others. They showed that when a cylinder exerts a given force on a planar surface, as illustrated in Figure 9-2, the pressure distribution along the region of contact is elliptical in nature. The length L of the region of contact is given by

---

[△] H. Hertz, J. Math (Crelle's J.), Vol. 92, 1881.
[+] Hertz, "Gesammelte Werke", Vol. 1, p. 155, Leipzig, 1895.
[⊞] A. N. Dinnik, Bull. Polytech. Inst., Kiew, 1909.
[◉] M. T. Huber, Ann. Physik., Vol. 14, 1904, p. 153.
[□] S. Fuchs, Physik. Z., Vol. 14, p. 1282, 1913.

MATH MODELS RELATING TO ROLLING FORCE

$$L = 1.6 \sqrt{fD\left[\frac{1-\nu_1^2}{E_1} + \frac{1-\nu_2^2}{E_2}\right]} \qquad (9\text{-}6)$$

where f is the force per unit length of the cylinder
    D is the diameter of the cylinder
    ν_1 and ν_2 are the values of Poisson's ratio for the cylinder and plate materials, respectively, and
    E_1 and E_2 are the corresponding values of Young's modulus.

If $E_1 = E_2 = E$ and $\nu_1 = \nu_2 = 0.3$

$$L = 2.15 \sqrt{\frac{fD}{E}} \qquad (9\text{-}7)$$

The maximum stress (σ_{max}) at the center of the region of contact is given by

$$\sigma_{max} = 0.798 \sqrt{\frac{f}{D\left[\frac{1-\nu_1^2}{E_1} + \frac{1-\nu_2^2}{E_2}\right]}} \qquad (9\text{-}8)$$

Again, assuming $E_1 = E_2 = E$ and $\nu_1 = \nu_2 = 0.3$

$$\sigma_{max} = 0.591 \sqrt{\frac{fE}{D}} \qquad (9\text{-}9)$$

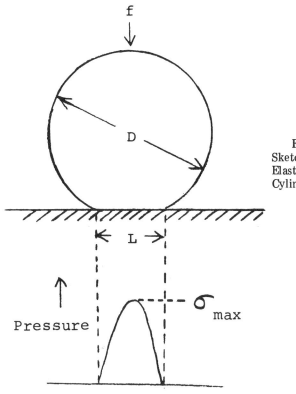

Figure 9-2: Sketch Illustrating Elastic Flattening of Cylinder on Plate.

In the simplified model discussed in Section 9-17, a portion of the theoretical length of the arc of contact is identified with this flattened interface between a cylindrical and a planar surface as expressed in the foregoing equation for L.

Whereas the region of contact between a cylinder and a plane, as discussed above, is generally regarded as flat, only in temper rolling does such a condition appear to occur. Under normal rolling conditions, the work roll appears to deform in such a manner that the arc of contact is generally regarded as possessing a curvature corresponding to a deformed work roll diameter (D') which is larger than the actual diameter (D). Under these circumstances, the length of the arc of contact L is given by

$$L = \sqrt{\frac{D'tr}{2}} \qquad (9\text{-}10)$$

whereas, for rigid rolls, the length is given by

$$L = \sqrt{\frac{Dtr}{2}} \qquad (9\text{-}11)$$

as illustrated in Figure 9-3. The relationship between D' and D and various other rolling parameters was first developed by Hitchcock[*] and, since the time of its publication, it has remained a cornerstone of rolling theory. For iron rolls, it expresses D' by the equation

$$D' = D\left(1 + 6.8 \times 10^{-4} \frac{f}{\Delta h}\right) \qquad (9\text{-}12)$$

where f is the specific rolling force (in pounds per inch) and Δh is the draft (in inches). Because of its importance, its derivation is presented in the next section.

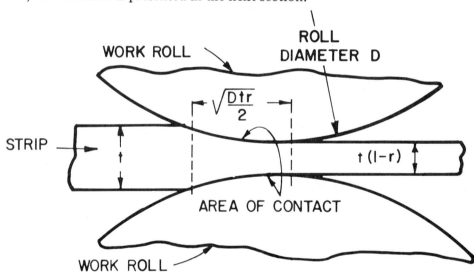

Figure 9-3: Sketch Illustrating Length of Arc of Contact Assuming Rigid Work Rolls.

Recent rolling research has, however, shown the Hitchcock relationship to give somewhat conservative values for D'. Empirically, the following modification of Hitchcock's equation has been found to be convenient in rolling force predictions.[◊]

---

[*] J. H. Hitchcock, "Elastic Deformation of Rolls During Cold Rolling", ASME Report of Special Research Committee On Roll Neck Bearings, June, 1935, pp. 33-41.

[◊] W. L. Roberts, "The Frictional Characteristics of Rolling Lubricants", Proceedings of the 11th Mechanical Working and Steel Processing Conference, Edited by G. A. McGrann, D. W. Murphy and F. E. Richardson, AIME, 1969, New York, N. Y., pp. 299-314.

MATH MODELS RELATING TO ROLLING FORCE

$$D' = D\left(1 + 2\sqrt{\frac{f}{Etr}} + 2\frac{f}{Etr}\right) \tag{9-13}$$

Photographic studies of the roll bite have also confirmed the conservative nature of Hitchcock's equation as is discussed in Section 9-7.

In dry temper rolling, it appears from studies of the forward slip of the strip that the interface between each work roll and the strip approximates to a planar surface. Under these conditions of high roll bite friction, the length of the arc of contact L may be represented by the expression [1]

$$L = \sqrt{\frac{Dtr}{2}\left(1 + \frac{\mu L}{t}\right)} \tag{9-14}$$

where μ is the effective coefficient of friction along the arcs of contact.

Jortner [□] was one of the first of the rolling theorists to establish the actual shape of the region of contact between the roll and strip surfaces and the pressure distribution along it, as discussed in Section 9-14. Under such circumstances, the length and shape of the so-called "arc" of contact is not related to a postulated deformed roll diameter.

9-6 Hitchcock's Expressions for the Length of the Arc of Contact and the Deformed Roll Diameter

Referring to Figure 9-4, the arc 206 represents a portion of the undeformed (or rigid) roll surface and the arc 345 the deformed surface such as exists when the roll is under compressive rolling stresses. Because of this elastic distortion, it should be noted that the length of the arc of contact is increased over that for the rigid roll and that the exit plane is shifted outwards from the line joining the roll centers. Assuming the point 0 to be the origin, the x and y directions as indicated and the length of the arc of contact to be small compared with the roll radius R, the equation for the undeformed roll surface is

$$y = \frac{x^2}{2R} \tag{9-15}$$

If it is further assumed that, at a position corresponding to x (subtending an angle α at the roll center), there is a radial deformation of the roll equal to z, such that the y coordinate has a value of y ,

then

$$y' = \frac{x^2}{2R} + z \tag{9-16}$$

Neglecting any elastic recovery of the strip after it leaves the roll bite, the top surface of the strip represented by the line 5-7 leaves the roll bite at the point 5 tangentially to the deformed roll surface. Assuming this point to have coordinates $-x_0, z_0$

$$\left[\frac{dy'}{dx}\right]_{-x_0} = 0 \tag{9-17}$$

Differentiation of equation (9-16) gives

---

[1] W. L. Roberts, "An Approximate Theory of Temper Rolling", Iron and Steel Engineer Year Book, 1972, pp. 530-542.

[□] D. Jortner, J. F. Osterle and C. F. Zorowski, "An Analysis of the Mechanics of Cold Strip Rolling", Iron and Steel Engineering Year Book, 1959, pp. 403-411.

$$\frac{dy'}{dx} = \frac{x}{R} + \frac{dz}{dx}$$

and hence, for the point represented by 5,

$$x_o = R\left[\frac{dz}{dx}\right]_{-x_o} \tag{9-18}$$

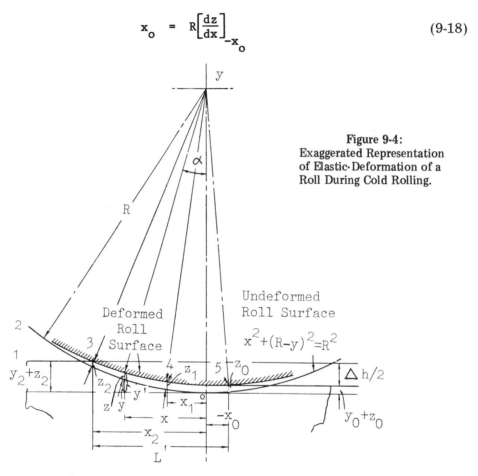

Figure 9-4:
Exaggerated Representation of Elastic Deformation of a Roll During Cold Rolling.

On the assumption that the pressure distribution along the arc of contact is elliptical, Prescott * has shown that

$$\frac{d^2z}{dx^2} = -\frac{4(1-\nu^2)f}{\pi E a^2} \tag{9-19}$$

where ν is Poisson's ratio, f is the specific force applied to the roll, E is Young's modulus and a is half the length of the arc of contact in the direction of rolling. For a given rolling situation, we may regard the right-hand side of equation 9-19 as a constant K,

such that

$$K = \frac{4(1-\nu^2)f}{\pi E a^2} \tag{9-20}$$

so that

$$\frac{d^2z}{dx^2} = -K \tag{9-21}$$

Integration of equation (9-21) gives

---

* J. Prescott, "Applied Elasticity", Longmans, Green and Co., London, 1924, p. 638.

MATH MODELS RELATING TO ROLLING FORCE

$$z = -\frac{Kx^2}{2} + C_1 x + C_2 \quad (9\text{-}22)$$

where C_1 and C_2 represent constants of integration. To determine the value of C_1, it is assumed that the greatest deformation of the roll occurs at the midpoint of the arc of contact (point 4 with an x coordinate x_1). Then

$$\left[\frac{dz}{dx}\right]_{x_1} = -kx_1 + C_1 = 0 \quad (9\text{-}23)$$

so that

$$C_1 = kx_1 \quad (9\text{-}24)$$

and

$$\frac{dz}{dx} = -k(x - x_1) \quad (9\text{-}25)$$

Referring back to equation (9-18), it is seen that

$$x_o = Rk(x_o + x_1) \quad (9\text{-}26)$$

From Figure 9-4, it is further observed that a, the semicontact length is equal to $(x_1 + x_o)$. Thus equation 9-26 may be written

$$x_o = R\left[\frac{4(1-\nu^2)f}{\pi E(x_o + x_1)}\right] \quad (9\text{-}27)$$

Rearrangement gives

$$x_1 = \frac{4R(1-\nu^2)f}{\pi E x_o} - x_o \quad (9\text{-}28)$$

Assuming that the point of contact of the strip at entry occurs at point 3 (with x coordinate x_2), and that the radial deformation of the roll surface is symmetrical about point 4, then

$$z_2 = z_o \quad (9\text{-}29)$$

and

$$y_2' - y_o' = \frac{\Delta h}{2} = \text{the semidraft} \quad (9\text{-}30)$$

From equation 9-16,

$$\left(\frac{x_2^2}{2R} + z_2\right) - \left(\frac{x_1^2}{2R} + z_o\right) = \frac{\Delta h}{2} \quad (9\text{-}31)$$

Since $z_2 = z_o$

$$x_2^2 - x_o^2 = \Delta h \cdot R \quad (9\text{-}32)$$

Because D is the midpoint of the arc of contact

$$x_2 + x_o = 2(x_1 + x_o) \quad (9\text{-}33)$$

or

$$x_o = x_2 - 2x_1 \quad (9\text{-}34)$$

Substitution in equation 9-34 of the value of x_1 given by equation 9-28, and writing, for simplicity,

yields
$$C = \frac{8R(1-\nu^2)}{\pi E} \qquad (9\text{-}35)$$

$$x_o(x_2 + x_o) = Cf \qquad (9\text{-}36)$$

and, from equation 9-32,

$$(x_2 - x_o)(x_2 + x_o) = \Delta h \cdot R \qquad (9\text{-}37)$$

Rearrangement of these equations gives

$$x_o = \frac{Cf}{(x_2 + x_o)} \qquad (9\text{-}38)$$

and

$$(x_2 - x_o) = \frac{\Delta h \cdot R}{(x_2 + x_o)} \qquad (9\text{-}39)$$

Doubling each side of equation 9-38 and adding it to equation 9-39 gives

$$(x_2 + x_o) = \frac{1}{(x_2 + x_o)}[2\,Cf + \Delta h \cdot R] \qquad (9\text{-}40)$$

or

$$(x_2 + x_o)^2 = 2\,Cf + \Delta h \cdot R \qquad (9\text{-}41)$$

Assuming that the specific rolling force is equal to the product of a mean pressure p and the length of the arc of contact $(x_2 + x_o)$, equation 9-41 may be written

$$(x_2 + x_o)^2 = 2\,C\bar{p}\,(x_2 + x_o) + \Delta h \cdot R \qquad (9\text{-}42)$$

Since this is a quadratic equation in terms of $(x_2 + x_o)$,

$$L' = (x_2 + x_o) = C\bar{p} \pm \sqrt{(C\bar{p})^2 + \Delta h \cdot R} \qquad (9\text{-}43)$$

Considering only the positive value of the square root term and substituting back for C,

$$L' = \frac{8R(1-\nu^2)\bar{p}}{\pi E} + \sqrt{\left[\frac{8R(1-\nu^2)\bar{p}}{\pi E}\right]^2 + \Delta h \cdot R} \qquad (9\text{-}44)$$

This expression is usually termed Hitchcock's equation.

It should be noted that if the rolls were rigid (and the modulus of elasticity becomes infinitely large), the length L' becomes equal to $\sqrt{\Delta h \cdot R}$. Moreover, the length of the arc of contact depends upon the modulus of elasticity, not upon roll hardness.

For most steel rolls, E may be taken as 3×10^7 psi and Poisson's ratio ν as 0.3, so that equation 9-44 reduces to

$$L' = \sqrt{(7.72 \times 10^8\,R\bar{p})^2 + \Delta h \cdot R} + 7.72 \times 10^{-8}\,R\bar{p} \qquad (9\text{-}45)$$

An expression for the effective radius R' of the deformed arc of contact can be easily obtained by equating $\sqrt{R'\Delta h}$ to the value of L' or $(x_2 + x_o)$. From equation 9-41

$$x_2 + x_o = \sqrt{2Cf + \Delta h \cdot R} \qquad (9\text{-}46)$$

Therefore

MATH MODELS RELATING TO ROLLING FORCE

$$\sqrt{R' \cdot \Delta h} = \sqrt{2Cf + \Delta h \cdot R} \qquad (9\text{-}47)$$

or

$$R'\Delta h = 2Cf + \Delta h \cdot R = \frac{16R(1-\nu^2)f}{\pi E} + \Delta h \cdot R \qquad (9\text{-}48)$$

Rearrangement gives

$$R' = R\left(1 + \frac{16(1-\nu^2)f}{\pi E \cdot \Delta h}\right) \qquad (9\text{-}49)$$

Underwood [B] points out that for steel rolls E = 13,400 tons [||] per square inch and ν = 0.3, so that

$$R' = R\left(1 + 3.458 \times 10^{-4} \frac{f}{\Delta h}\right) \qquad (9\text{-}50)$$

where f is expressed in tons/inch and Δh is the draft in inches.

For cast iron rolls, with a smaller elastic modulus

$$R' = R\left(1 + 6.8 \times 10^{-4} \frac{f}{\Delta h}\right) \qquad (9\text{-}51)$$

Figure 9-5 shows the relationship between the ratio of the deformed radius to the undeformed radius (R'/R) to the ratio of the specific rolling force to the draft for both cast iron and steel rolls. Underwood states that $f/\Delta h$ usually ranges between 300 and 10,000 tons-inch^2 units so that R'/R ranges from 1.1 to 4.46 for steel rolls and 1.2 to 7.8 for cast iron rolls.

Equation 9-49 is often written in terms of the deformed and the undeformed roll diameters (D' and D, respectively), the specific rolling force f (expressed in terms of lbs. per inch), the elastic modulus, E, and the draft expressed as the product of the initial thickness, t, and the reduction r.

Thus

$$D' = D\left(1 + \frac{4.63 \, f}{Etr}\right) \qquad (9\text{-}52)$$

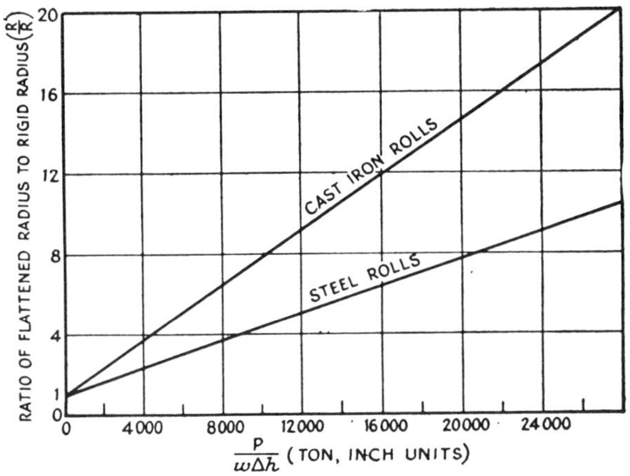

Figure 9-5:
Relation between R'/R and f/Δh According to Hitchcock's Equation for Cast Iron and Steel Rolls.

It is interesting to note that an equation very similar to 9-52 may be derived on the assumption that the arc of contact consists of three segments, one of length (L_R) corresponding to that obtained by a rigid roll of diameter (D) bounded at each end by a segment corresponding to

---

[B] L. R. Underwood, "The Rolling of Metals — Theory and Experiment", John Wiley and Sons, Inc., New York, N. Y., 1950.
[||] Long Tons (= 2,240 lbs.).

half the length of the region of contact (L_c) formed between a cylinder of diameter D elastically deformed against a plane surface. Thus

$$L_R = \sqrt{\frac{Dtr}{2}} \qquad (9\text{-}53)$$

and

$$L_c = 2.15\sqrt{\frac{fd}{E}} \qquad (9\text{-}54)$$

It is further assumed that, because of the roll bite geometry, the length of the resulting arc of contact L is equal to the square root of the sum of the squares of each segment, viz.

$$L = \sqrt{\left(\frac{L_c}{2}\right)^2 + \left(L_R\right)^2 + \left(\frac{L_c}{2}\right)^2}$$
$$= \sqrt{1.155\frac{fD}{E} + \frac{Dtr}{2} + 1.155\frac{fD}{E}} \qquad (9\text{-}55)$$
$$= \sqrt{2.31\frac{fD}{E} + \frac{Dtr}{2}}$$

Since the arc length L may be regarded as that corresponding to a rigid roll of diameter D' (the effective diameter of the elastically deformed roll of diameter D), then

$$\sqrt{\frac{D'tr}{2}} = \sqrt{2.31\frac{fD}{E} + \frac{Dtr}{2}} \qquad (9\text{-}56)$$

or

$$D' = D\left(1 + \frac{4.62\ f}{Etr}\right) \qquad (9\text{-}57)$$

9-7 The Validity of Hitchcock's Equation

Although the validity of Hitchcock's equation has often been questioned, the difficulty of measuring the actual length of the arc of contact under rolling conditions has left the question largely unresolved. In 1943, however, Orowan [x] published some experimental data which was claimed to throw some doubt on its validity. As has been seen in Section 9-6, Hitchcock assumed that the pressure distribution along the contact arc was elliptical and, under such circumstances, the shape of the arc of contact would be approximately circular. However, as may be seen in Section 9-9, the pressure distribution is believed to exhibit a "friction hill", with a peak pressure exceeding that of an equivalent elliptical pressure curve (see Figure 9-6). Orowan anticipated that this pressure peak would produce a "depression" in the rolls in the region of the pressure peak and he attempted to find whether such a depression did, in reality, exist.

To accomplish this purpose, Orowan used a two-high mill with 8-inch diameter work rolls to roll brass and steel strip about 3/4-inch to 1-1/2-inch wide. The mill was quickly stopped with the strip between the rolls, the rolls lifted and the shape of the arc of contact impressed into the partially rolled strip was determined by careful measurements. Some of the measured profiles are reproduced in Figure 9-7 with the dashed lines representing the undeformed roll surface. According to these tests, the shape of the arc of contact was not circular and, in the case of the brass strip, a noticeable "bump" (corresponding to the depression referred to above) was observed. However, while it is easy to establish the exact position of the beginning of the arc of contact on the partially rolled specimen (coincident with the entry plane), it is virtually impossible to establish the position of the exit plane. Thus Orowan's experiments must be regarded as inconclusive.

---

[x] E. Orowan, "The Calculation of Roll Pressure in Hot and Cold Flat Rolling", Proc. Inst. Mech. Eng., Vol. 150, 1943, No. 4, pp. 140-167.

MATH MODELS RELATING TO ROLLING FORCE

Foppl[*] and Hitchcock[§] showed that the effect of a peak on the shape of the arc of contact would be exceedingly small. For a parabolic distribution of pressure over the arc of contact, Foppl derived the following relation for the radial deformation at the center of the region of contact

$$z_1 = \frac{2(1-\nu^2)f}{\pi E}\left[1.026 + \log_e\left(\frac{R}{x_o + x_1}\right)\right] \quad (9\text{-}58)$$

For a rectangular or uniform pressure distribution, Hitchcock, using the same method, obtained the expression

$$z_1 = \frac{2(1-\nu^2)f}{\pi E}\left[0.694 + \log_e\left(\frac{R}{x_o + x_1}\right)\right] \quad (9\text{-}59)$$

Since the term $\log_e\left(\frac{R}{x_o + x_1}\right)$ in equations 9-58 and 9-59 is very much larger than unity, the deformation at the center of the arc of contact is virtually independent of the type of pressure distribution along the arc.

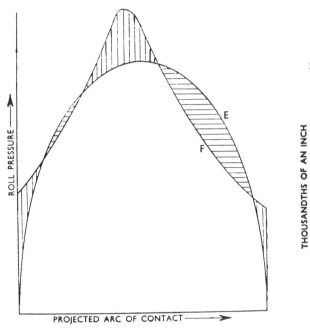

Figure 9-6:
Comparison of Elliptical and Probable Pressure Distribution in the Arc of Contact (Orowan).

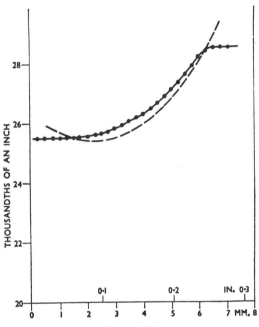

Figure 9-7:
Measurements on Steel Strip in Experiments on Roll Flattening (Orowan).

More recently, Pawelski and Schroeder,[*][•] measured the elastic changes in the radius of the work roll as they occurred during rolling. In particular, they measured the maximum value of y in the radial direction (see Figure 9-4) and the length of the arc of contact L, and compared these with values calculated from Hitchcock's equations. They found that the calculated values were always larger than the measured values. However, as the thickness of the strip increased, the difference between the calculated and measured values of L decreased. Moreover, they found that

---

[*] Foppl, "Vorlesungen uber technische mechanik", (Technical Mechanics), Vol. 5, p. 350.

[§] J. Hitchcock, "Roll Neck Bearings", ASME Research Publication, Appendix I, 1935.

[*] O. Pawelski and H. Schroeder, "Messung der Walzenabplattung beim Kaltwalzen in einem Vier-Walzen-Gerust", Arch. Eisenh., Vol. 39, 1968, pp. 747-754. (Address to the Max Planck Institute for Iron Research, Paper 1141, — Cold Rolling Committee 139.)

[•] O. Pawelski and H. Schroeder, "Effect of Roll Flattening on the Configuration of Roll Gap and Pressure Distribution During Cold Rolling", Arch. Eisenh., 1969, Vol. 40, pp. 867-873 (BISI Translation No. 8366).

the point of maximum flattening was not at the center of the arc of contact, as assumed by Hitchcock, but closer to the exit plane.

Figure 9-8 shows the length of the arc of contact calculated (1) on the basis of Hitchcock's relationship (ℓ'_d) (2) assuming rigid work rolls (ℓ_d) and (3) the length determined from actual measurement of roll deformation (designated the "actual" value ℓ_{dz}). For strip 0.2 mm thick, the "actual" lengths of the arc of contact were found to be midway between those corresponding to a rigid roll and those calculated from Hitchcock's relationship. As the strip thickness is increased, the "actual" values are very close to those derived from Hitchcock's model.

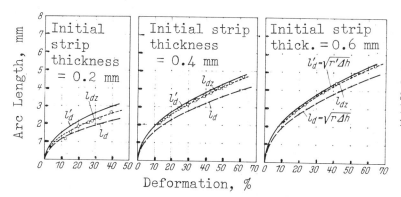

Figure 9-8:
Comparison of the Measured and Calculated Lengths of the Arc of Contact in the Rolling of Steel Strip of 3 Thicknesses (Using Rape-Seed Oil as the Rolling Lubricant).

Figure 9-9:
Experimental Setup of the Camera and Lights Used to Photograph the Roll Bite (Kobasa and Schultz).

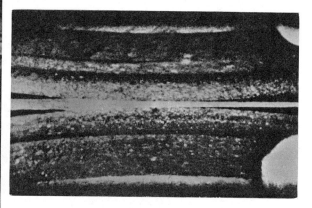

Figure 9-10:
Typical View of Roll Bite and Strip in Motion (Kobasa and Schultz).

In 1967, Kobasa and Schultz [■] studied the length of the arc of contact using a high-speed movie camera focused on the edge of prelubricated annealed black plate 0.015 inches thick as it underwent reduction in the roll bite. The experimental arrangement of the camera and lights is shown in Figure 9-9 and a typical frame taken of the strip during mill operation is reproduced in Figure 9-10. The lengths of the arcs of contact were measured on a motion analyzer and included not only the region of plastic deformation but also the regions of elastic compression and recovery as shown in Figure 9-11. Although the ends of the arc were difficult to establish

---

■ D. Kobasa and R. A. Schultz, "Experimental Determination of the Length of the Arc of Contact in Cold Rolling", Iron and Steel Engineer Year Book, 1968, pp. 283-288.

MATH MODELS RELATING TO ROLLING FORCE

visually, data satisfactory from a statistical viewpoint was acquired by 3 observers. This was compared with arc lengths theoretically derived on the basis of (a) Hitchcock's relationship, (b) rigid work rolls and (c) the expression

$$L = \left[\ln\left(\frac{f\left(\mu - \frac{T}{Df}\right)}{(\sigma_c - \sigma_1)\,t(1-r)} + 1\right)\right]\frac{t(1-r)}{\left(\mu - \frac{T}{Df}\right)} + \frac{1}{2}\left(2.15\sqrt{\frac{f_o D}{E}}\right) \qquad (9\text{-}60)$$

where:
- L = length of the arc of contact
- f = specific separating force corresponding to reduction r, lb. per in.
- f_o = specific separating force at incipient reduction, lb. per in.
- μ = average coefficient of friction
- T = specific torque, lb.-ft. per in.
- σ_c = compressive yield strength corrected for strain rate, psi
- σ_1 = average applied tension stress, psi and
- t = initial strip thickness, in.

The first term of equation 9-60, which is the effective arc length, is derived from a mathematical model of the cold rolling process and is discussed more fully in Section 9-18. The second term is half the actual elastic arc length at incipient reduction using the Hertz expression for a roll on a flat.

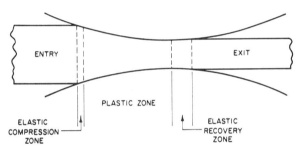

Figure 9-11:
Sketch Illustrating Three Regions of Deformation in the Roll Bite.

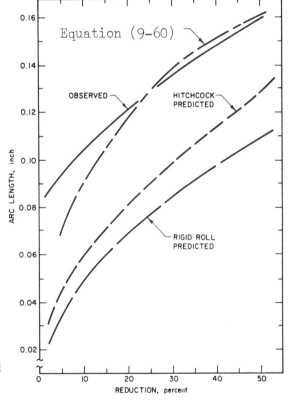

Figure 9-12:
Observed and Calculated Arc Lengths (Kobasa and Schultz).

The observed arcs of contact are noticeably larger than those predicted by Hitchcock (see Figure 9-12). This is explainable to some extent since the Hitchcock equation does not include the visible length of elastic strip flattening at the entry and exit ends of the roll bite. However, using the equation of Bland and Ford,[□] which adds elastic strip flattening to the Hitchcock arc length, the

---
[□] D. R. Bland and H. Ford, "An Approximate Treatment of the Elastic Compression of the Strip in Cold Rolling", Journal of Iron and Steel Institute (British), July, 1952.

total predicted arc length could be increased a maximum of 4 per cent at a 50 per cent reduction and only 1.5 per cent at a 2 per cent reduction. Thus, based on the experimentally observed arc lengths, the generally accepted Hitchcock equation, even considering elastic strip flattening, is not adequate to predict the length of the arc of contact between roll and strip. It should also be noted that the discrepancy between the experimental results and the Hitchcock prediction increases at the lower reductions.

The curve generated by the measured arc lengths is basically of equal slope to the one described by the equation for arc length obtained by use of a rigid roll (no elastic flattening) given by

$$L = \sqrt{\frac{Dtr}{2}} \qquad (9\text{-}61)$$

Although the slope of comparable points on both curves is similar, the observed arc lengths varied by a constant amount (approximately 0.054 in.) from the rigid roll predictions.

The best fit, particularly in the range of reductions from 20 to 50% occurs with the expression corresponding to item (c). In using this expression, the values for the coefficient of friction μ were calculated from slip, torque and force data as described in Section 6-23. It should be remembered again, however, that prelubrication of the strip and the low rolling speed (30 fpm) resulted in coefficients of friction about 0.12 which are much higher than those generally encountered on commercial mills with much larger diameter rolls operating at much higher speeds.

9-8 Assumptions Usually Made in Developing Rolling Models

To simplify the mathematics involved in developing models relating to rolling force, theorists have usually made certain very basic assumptions. These are listed below:

(1) The workpiece constitutes a continuous medium with essentially no change in volume (or density).

(2) The constrained yield stress of the metal undergoing deformation either remains constant or varies in a predictable manner along the length of each arc of contact.

(3) The workpiece is assumed to be homogenously compressed during its plastic deformation. In other words, if the strip entering the roll bite is assumed to consist of thin vertical segments perpendicular to the direction of rolling, these segments become shortened and expanded (in the rolling direction) but not curved.

(4) Lateral spreading of the strip does not occur. This condition is satisfied if the thickness of the strip is small compared with its width and it implies that the process is regarded essentially from a two-dimensional viewpoint.

(5) The lubrication in the roll bite is such that a uniform coefficient of friction occurs along the length of each arc of contact. The validity of this assumption appears highly doubtful since rolling lubrication appears to be attributable, in part, to hydrodynamic effects. For purely hydrodynamic lubrication, the frictional effects should be greatest at the entry into the roll bite where the greatest slipping speeds occur and decrease to zero at the neutral point where the roll and strip surface speeds match. On the other hand, the thickness of the lubricant film would be expected to gradually decrease from the exit to the entry plane so that the coefficient of friction attributable to boundary layer lubrication should increase.

(6) Elastic deformations in the sheet at the roll bite are negligible. This is reasonable if we realize that elastic deformation cannot exceed 0.1 to 0.2 per cent whereas cold rolling deformations are usually in the range of 20 to 50 per cent.

(7) In most of the theories, each arc of contact is assumed to have a constant radius of curvature corresponding to a certain roll diameter. In the early theories, the rolls

were assumed to be perfectly rigid but, in later modifications to the theories, the radius of curvature of the arcs of contact were assumed to correspond to a "deformed roll diameter" in consequence of the elastic distortions of the rolls. At the same time, the peripheral speed of the rolls is assumed to remain constant, i.e., there is no rotational acceleration or deceleration of the rolls.

(8) The contact angle, or the angle formed between the strip and the tangent to the roll surface at the entry plane is small. This is equivalent to assuming that the work roll diameter is quite large with respect to the thickness of the incoming strip.

(9) The acceleration of the strip in the roll bite is disregarded. Usually the energy expended for this purpose is negligible with respect to that utilized in the deformation of the strip.

(10) The principal stresses in the workpiece as it undergoes deformation in the roll bite occur in directions (a) normal to the plane at the center of the strip, (b) in the plane of the strip but normal to the direction of rolling, and (c) in the plane of the strip in the direction of rolling (see Figure 8-72).

(11) The tensile stresses in the strip reduce the compressive deformation stresses at the ends of the roll bite in accordance with the maximum shear theory of Tresca (see Section 8-23).

(12) Thermal effects are disregarded, even though it is known that temperatures of a few hundred degrees F may occur along the arcs of contact and that the temperature of the steel itself is increased by deformation. Such temperature changes could affect lubricity and the yielding behavior of the workpiece.

Without the simplifying assumptions listed above, however, the models developed would be very complex from both physical and mathematical viewpoints. Moreover, in some of the theories, other assumptions, such as sticking friction or lubricant films of constant thickness, had to be postulated to reduce mathematical complexities.

9-9 The Pressure Distribution Along the Arc of Contact (Von Karman's Equation)

As the first step in the development of an expression for rolling force, the pressure distribution along each arc of contact must be established on a theoretical basis.[8] From considerations of the restraint imposed upon the flow of the metal by the frictional effects at the roll-strip interfaces, the pressure distribution would be expected to take a general form as indicated in Figure 9-13.

Referring to Figure 9-14, the initial step in developing a rolling model is to consider the forces acting on a vertical element of the workpiece in the roll bite, the element having a height h, a thickness dx in the direction of rolling and to be located between the entry and neutral plane as shown. The radial pressure between the roll surface and the strip acting on the end of the element is p_r, the tangential frictional force is F, and the line joining the end of the element to the center of the roll makes an angle θ with the vertical.

If the workpiece is assumed to be of unit width and the rolls perfectly rigid, the normal force on one end of the element is $p_r \frac{dx}{\cos\theta}$, and the horizontal component of this force (which resists the entry of the strip into the bite) is

$$p_r \frac{dx}{\cos \theta} \sin \theta = p_r \tan \theta \, dx \qquad (9\text{-}62)$$

---

[8] L. R. Underwood, "The Rolling of Metals, Vol. I", John Wiley and Sons, Inc., New York, 1950.

Now the tangential frictional force F is equal to $\mu p_r \frac{dx}{\cos\theta}$ where μ is the coefficient of friction. The horizontal component of this frictional force tending to draw the element into the roll bite is

$$\mu p_r \frac{dx}{\cos\theta} \cdot \cos\theta = \mu p_r \, dx \qquad (9\text{-}63)$$

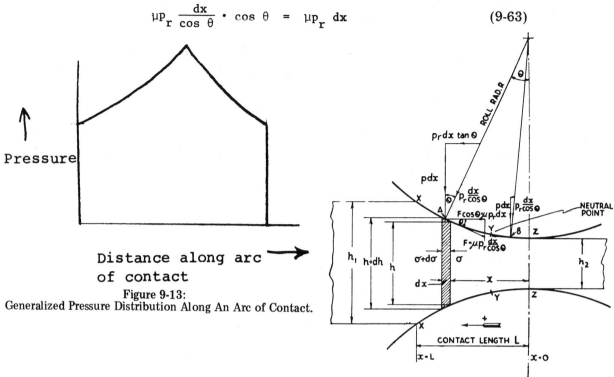

Figure 9-13:
Generalized Pressure Distribution Along An Arc of Contact.

Figure 9-14:
Forces and Stresses Acting Upon an Elemental Vertical Section of the Sheet Between the Rolls. (Forces Acting in the Direction of the Arrow are Regarded as Positive.)

The horizontal forces acting on the vertical faces of the elemental section produce compressive stresses assumed to be uniform over the height of the section. These are designated ($\sigma + d\sigma$) on the longer face of height (h + dh) and σ over the face of height h.

Resolving the horizontal forces on the element to give equilibrium yields

$$2p_r \tan\theta \, dx - 2\mu p_r \, dx = (h+dh)(\sigma+d\sigma) - h\sigma = d(h\sigma) \qquad (9\text{-}64)$$

If the vertical component of the force on the end of the element is considered to be pdx

$$p \, dx = p_r \frac{dx}{\cos\theta} \cdot \cos\theta = p_r \, dx \qquad (9\text{-}65)$$

and

$$p = p_r \text{ and } p(\tan\theta - \mu) = \frac{d(h\sigma)}{2} \qquad (9\text{-}66)$$

If a similar elemental section lying between the neutral plane and the exit plane is now considered, the corresponding equation is

$$p(\tan\theta + \mu) = \frac{d(h\sigma)}{2} \qquad (9\text{-}67)$$

The combination of equations 9-66 and 9-67 gives

$$p(\tan\theta \mp \mu) = \frac{d(h\sigma)}{dx} \qquad (9\text{-}68)$$

From geometrical considerations

$$\frac{1}{2}\frac{dh}{dx} = \tan\theta \tag{9-69}$$

so that equation 9-69 becomes

$$p\left(\frac{1}{2}\frac{dh}{dx} \mp \mu\right) = \frac{d\left(\frac{h\sigma}{2}\right)}{dx} \tag{9-70}$$

The total vertical force on the end of the element is $pdx + \mu F \sin\theta$ which is equal to $p(1 + \mu \tan\theta) dx$. Hence the vertical stress in the element is simply $p(1 + \mu \tan\theta)$. If it is assumed that this is the largest stress and the horizontal stress σ is the least, then by Tresca's yielding criterion

$$p(1 + \mu \tan\theta) = \sigma_c + \sigma \tag{9-71}$$

where σ_c is the constrained yield stress of the strip. Equations 9-70 and 9-71 can now be combined to give a differential equation in which p and x are the two variables since h can be expressed in terms of x. For mathematical simplicity, however, it is generally assumed that $\mu \tan\theta$ is very small compared with unity in the cold rolling of strip, this assumption being valid since the entry angles are small (generally less than 0.04 radians) and the values of μ generally much less than 0.1.

Thus on the basis that $\mu \tan\theta$ is negligibly small

$$\frac{p}{2}\frac{dh}{dx} \mp p\mu = \frac{d\left(\frac{h}{2}(p - \sigma_c)\right)}{dx} \tag{9-72}$$

The development of this equation, or one of its alternative forms, represents the initial step in determining the pressure distribution along the arc of contact. Since Von Karman first derived this differential equation in 1925, it is frequently referred to as Von Karman's equation.‡

9-10 Smith's and Tselikov's Solutions to Von Karman's Equation

In equation 9-69 of the preceding section, the left-hand side represents the slope of the arc of contact at a point located at a distance x from the exit plane. Smith ⁸ assumed a parabolic arc of contact (see Figure 9-15), the equation of which, referred to the longitudinal axis of the sheet and the intersection of this axis with the exit plane is

$$\frac{h}{2} = \frac{1}{2}\left[h_2 + m\left(\frac{x}{L}\right)^2\right] \tag{9-73}$$

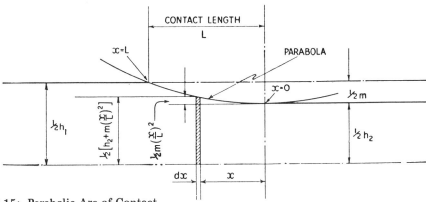

Figure 9-15: Parabolic Arc of Contact.

---

‡ T. Von Karman, "Beitrag zur Theorie des Walzvorganges", Zeitschrift fur angewandte, Mathematik und Mechanik, 1925, Vol. 5, p. 139.

⁸ L. R. Underwood, "The Rolling of Metals, Vol. I", John Wiley and Sons, New York, 1950.

where m is the draft ($h_1 - h_2$) and L is the projected length of the arc of contact. Differentiation of this expression gives

$$\frac{1}{2}\frac{dh}{dx} = \frac{mx}{L^2} \qquad (9\text{-}74)$$

and using this value for $\frac{1}{2}\frac{dh}{dx}$ in equation 9-70 gives

$$\frac{d\left(\frac{h\sigma}{2}\right)}{dx} = \frac{mpx}{L^2} \pm p\mu \qquad (9\text{-}75)$$

$$= p\left[\frac{mx}{L^2} \mp \mu\right]$$

If $\mu \tan \theta$ is assumed to be very much smaller than unity, then, from equation 9-71

$$\sigma = p - \sigma_c \qquad (9\text{-}76)$$

and since

$$h = h_2 + m\left(\frac{x}{L}\right)^2 \qquad (9\text{-}77)$$

equation 9-75 may be written

$$\frac{d\left[(p - \sigma_c)\left\{\frac{h_2}{2} + \frac{m}{2}\left(\frac{x}{L}\right)^2\right\}\right]}{dx} = p\left\{\frac{mx}{L^2} \mp \mu\right\} \qquad (9\text{-}78)$$

After differentiation, rearrangement of the terms and simplification, the equation 9-78 becomes

$$\frac{d\left(\frac{p}{\sigma_c}\right)}{d\left(\frac{x}{L}\right)} = \frac{\frac{2m}{h_2}\left(\frac{x}{L} \mp \frac{pL\mu}{\sigma\delta}\right)}{1 + \frac{mx^2}{t_2 L^2}} \qquad (9\text{-}79)$$

Since the equation cannot be directly integrated, Trinks♦ developed a graphical technique for its solution in the form of two sets of curves reproduced in Figure 9-16. From these curves, the average rolling pressure and the peak pressure, as given by equation 9-79, may be established for a wide range of rolling conditions. Each curve corresponds to a specific value of $\frac{L}{m}\tan \mu$ and, because of the assumed parabolic arc of contact, the ratio of draft to contact length $\left(\frac{m}{L}\right)$ is equal to the tangent of the entry angle.

Thus

$$\frac{L}{m}\tan \mu = \frac{\text{tangent of friction angle}}{\text{tangent of entry angle}} \qquad (9\text{-}80)$$

In plotting these curves, Trinks
(1) Selected various values of σ, h_1 and $\frac{L}{m}\tan \mu$.
(2) Calculated p from the equation 9-79 for various values of x from x = 0 to x = L.
(3) Drew up a diagram in which the vertical pressure p between the rolls and the strip was plotted over the corresponding points on the contact curve, and measured the average and peak pressures.

♦W. Trinks, "Pressures and Roll Flattening in Cold Rolling", Blast Furnace and Steel Plant", Vol. 25, 1937, pp. 617-19.

(4) Calculated the ratios of average roll pressure to the constrained yield stress σ_c and the peak pressure to the constrained yield stress σ_c.

(5) Plotted these ratios against corresponding values of m/h_1 as shown in Figure 9-16.

(6) Repeated the procedure until a sufficient number of points were established to draw the curves shown in the figure.

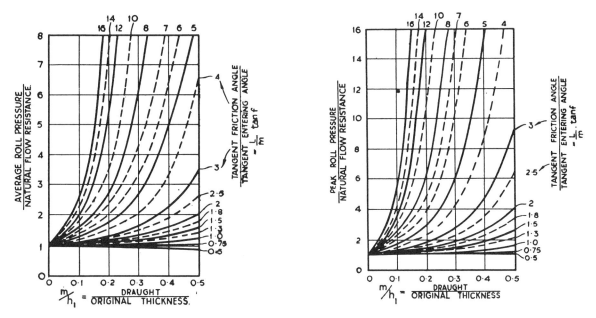

Figure 9-16: Trinks' Curves for the Graphical Solution of Von Karman's Equation.

Since the advent of the computer, however, such a solution as developed by Trinks is mainly of academic interest.

A different approach to the solution of Von Karman's equation was taken by Tselikov◇ who assumed the angle θ to be constant and equal to half the entry angle. Under these circumstances, the equation may be converted to a form capable of direct integration. Tselikov's solution for the pressure distribution assumed the form

$$p = \frac{\sigma_1}{1+\xi} \left[\left\{ \frac{p_1}{S_1}(1+\xi) - 1 \right\} \left(\frac{h_1}{h}\right)^{1+\xi} + 1 \right] \quad (9\text{-}81)$$

for the region between the entry and neutral planes

where

$$\xi = \frac{\tan(\tan^{-1}\mu - \theta)}{\tan \theta} \quad (9\text{-}82)$$

and

$$S_1 = \frac{S}{1 + \mu \tan \theta} \quad (9\text{-}83)$$

the symbol S denoting the constrained yield strength of the strip.

For the case of rolling without tension, equation 9-81 simplifies to

$$p = \frac{p_1}{1+\xi}\left[\xi\left(\frac{h_1}{h}\right)^{\xi-1} + 1\right] \quad (9\text{-}84)$$

◇ A. I. Tselikov, "Effect of External Friction and Tension on the Pressure of the Metal on the Rolls in Rolling", Metallurg No. 6, 1939, pp. 61-76.

where

$$p_1 = \frac{S}{1+\mu \tan \theta} \quad (9\text{-}85)$$

Similarly for the pressure distribution in the region between the neutral and exit planes

$$p = \frac{p_2}{\eta-1}\left[\eta\left(\frac{h}{h_2}\right)^{\eta-1} - 1\right] \quad (9\text{-}86)$$

where

$$\eta = \frac{\tan\left(\tan^{-1}\mu + \theta\right)}{\tan \theta} \quad (9\text{-}87)$$

and

$$p_2 = \frac{S}{1 - \mu \tan \theta} \quad (9\text{-}88)$$

It should be noted, however, that Tselikov's assumption of a constant angle θ is valid only for small contact angles occurring in the rolling of thin strip.

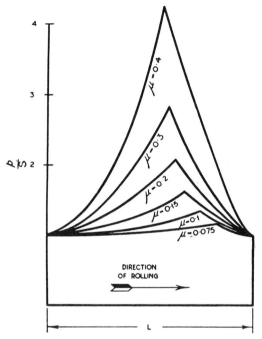

Figure 9-17:
Curves Showing Distribution of Pressure Over the Arc of Contact. According to Tselikov (Equations 9-84 and 9-86) in Rolling a Wide Strip for Different Coefficients of External Friction (μ), Other Conditions Being Equal. (Rigid Roll and No Work Hardening Assumed) Reduction, 30 Per Cent; Contact Angle, 5° 40′; h_2/D = 1.16 Per Cent.

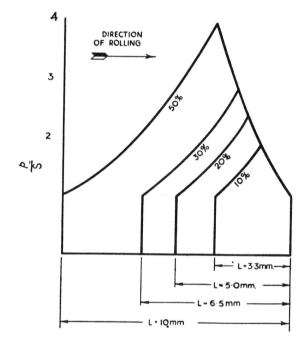

Figure 9-18:
Curves Showing the Distribution of Pressure Over the Arc of Contact, According to Tselikov (Equations 9-84 and 9-86) in Rolling Wide Strip with Different Amounts of Reduction (10, 20, 30 and 50 Per Cent), Other Conditions Being Equal in the Absence of Work Hardening and for a Rigid Roll.
h_2 = 1 mm.; D = 200 mm.; μ = 0.2

Tselikov published a series of diagrams illustrating the effect of various values of the coefficient of friction (μ), the percentage reduction ($\frac{m}{h_1} \times 100$) and the work roll diameter (D) on the pressure distribution along the arc of contact. These diagrams are reproduced in Figures 9-17 to 9-19, inclusive. The first shows the distribution of pressure over the arc of contact for a constant reduction of 30 per cent, an angle of contact 5° 40′ and for values of the coefficient of friction ranging from 0.075 to 0.4. The marked influence of the coefficient of friction μ on the peak pressure and the shift in the position of the peak away from the exit plane as friction increases should be noted. Figure 9-18 shows similar curves resulting from different degrees of reduction

MATH MODELS RELATING TO ROLLING FORCE

indicating that, as the arc length increases, so also does the peak pressure. Figure 9-19 indicates the effect of work roll diameter on pressure distribution and it is to be noted that the larger the work roll diameter the further the separation of the neutral and exit planes.

In deriving an expression for the mean specific roll pressure in the absence of strip tensions, Tselikov assumed uniformity in the specific rolling force (f) across the width of the strip. His expression for specific rolling force may be written

$$f = \int_{x=0}^{x=L} p(1 \mp \mu \tan \theta)dx \qquad (9\text{-}89)$$

Substituting values for p as given by equations 9-84 and 9-86 for the two regions of the roll bite (one on each side of the neutral point), writing

$$dx = \frac{dh_x}{2 \tan \theta} \qquad (9\text{-}90)$$

and integrating between the limits h_1 and h_χ for the region before the neutral point and between h_χ and h_2 for the other region, Tselikov obtained the following expression for the mean specific roll pressure p_s

$$p_s = \frac{2h_\chi \sigma_1}{(h_1 - h_2)(\delta - 1)}\left[\left(\frac{h_\chi}{h_2}\right)^\delta - 1\right] \qquad (9\text{-}91)$$

where

$$\delta = \frac{\mu}{\tan\left(\frac{\alpha}{2}\right)} \qquad (9\text{-}92)$$

(In this expression for δ, α is the contact angle and is equal to 2θ)

In deriving this expression, it was assumed

$$\xi_1 + 1 \simeq \eta - 1 \simeq \frac{\mu}{\tan\frac{\alpha}{2}} \qquad (9\text{-}93)$$

and that

$$\mu \tan\left(\frac{\alpha}{2}\right) \ll 1 \qquad (9\text{-}94)$$

Furthermore, work hardening of the strip was disregarded.

The thickness of the strip (h_χ) at the neutral point is determined from the intersection of the curves represented by equations 9-84 and 9-86. In this way, it may be shown that

$$\frac{h_\chi}{h_2} = \left\{\frac{1 + \sqrt{1 + (\delta^2 - 1)\left(\frac{h_1}{h_2}\right)^\delta}}{\delta + 1}\right\}^{\frac{1}{\delta}} \qquad (9\text{-}95)$$

If the material is rolled with entry and exit tensile stresses σ_1 and σ_2, respectively, then equation 9-95 becomes

$$\frac{h_\chi}{h_2} = \left\{\frac{1 + \sqrt{1 + \left(\frac{p_1 \delta}{\sigma} - 1\right)\left(\frac{p_2 \delta}{\sigma} + 1\right)\left(\frac{h_1}{h_2}\right)^\delta}}{\left(\frac{p_2 \delta}{\sigma} + 1\right)}\right\}^{\frac{1}{\delta}} \qquad (9\text{-}96)$$

where

$$p_1 = \sigma - \sigma_1 \tag{9-97}$$

and

$$p_2 = \sigma - \sigma_2 \tag{9-98}$$

Under these conditions, the mean specific roll pressure p_s is given by

$$p_s = \frac{\sigma h_\chi \tan\frac{\alpha}{2}}{(h_1-h_2)\mu} \left[\frac{1}{\delta-1}\left\{\left(1-\frac{\sigma_1}{\sigma}\right)\delta-1\right\}\left\{\left(\frac{h_1}{h_\chi}\right)^\delta - \left(\frac{h_1}{h_\chi}\right)\right\} + \left(\frac{h_1}{h_\chi} - 1\right) \right.$$
$$\left. + \frac{1}{\delta-1}\left\{\left(1-\frac{\sigma_2}{\sigma}\right)\delta+1\right\}\left\{\left(\frac{h_\chi}{h_2}\right)^\delta - \left(\frac{h_2}{h_\chi}\right)\right\} - \left(1-\frac{h_2}{h_\chi}\right) \right] \tag{9-99}$$

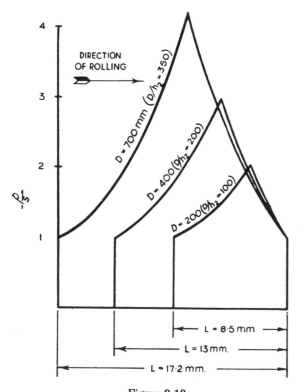
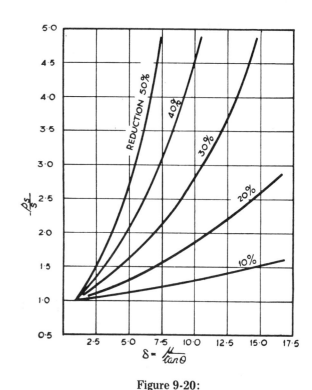

Figure 9-19:
Curves Showing the Distribution of the Pressure Over the Arc of Contact According to Tselikov (Equations 9-84 and 9-86) in the Rolling of Wide Strip with Rolls of Different Diameter (200, 400 and 700 mm.) with 30 Per Cent Reduction, Assuming Rigid Rolls and No Work Hardening.
h_2 = 2 mm.; h_1 = 2.86 mm.; μ = 0.3

Figure 9-20:
The Influence of the Coefficient of External Friction and the Angle of Contact (2θ) on the Mean Specific Pressure (p_s) for Percentage Reductions of 10, 20, 30, 40 and 50 Per Cent.
p_s Calculated from Tselikov's Equations.

Figure 9-20 shows curves developed from equation 9-91 which demonstrate the influence of the coefficient of friction μ and the angle of contact (2θ) on the mean specific pressure p_s for various reductions.

The relationship between the thickness of the strip at the neutral plane, $h\psi$ and the exit strip thickness, h_2, namely $h\psi/h_2$ and the parameter δ for various reductions is shown in Figure 9-21. These are instrumental in calculating p_s from equations 9-91 and 9-99.

MATH MODELS RELATING TO ROLLING FORCE

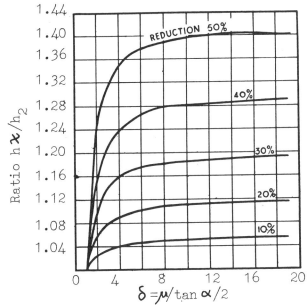

Figure 9-21:
Thickness (h X) of Section at Neutral Point in Terms of the Coefficient of External Friction (μ) and the Angle of Contact (α) for Percentage Reductions of 10, 20, 30, 40 and 50 Per Cent.

To verify the above model, Tselikov used certain published data relating to the rolling of steel with and without a lubricant.♦ Figure 9-22 shows a comparison of measured and calculated data with respect to the variation of p_s/σ, denoted by p_s/S, with h_2/D, with μ assumed to be 0.2 and with work-hardening effects neglected. The data of Rudbakh and Severenko⊕ is shown plotted in Figure 9-23 illustrating the variation of the mean specific rolling pressure with initial strip thickness with and without a lubricant.

As Underwood ⁸ has stated, although at first sight the agreement between experiment and theory looks reasonable, a closer consideration of the comparisons, however, is less reassuring. Such concern arises from (a) the disregard of work hardening (or perhaps, more reasonably, the disregard of strain rate effects), (b) the high coefficients of friction assumed (0.6 for dry steel and 0.3 with a lubricant) and (c) the assumption of rigid rolls.

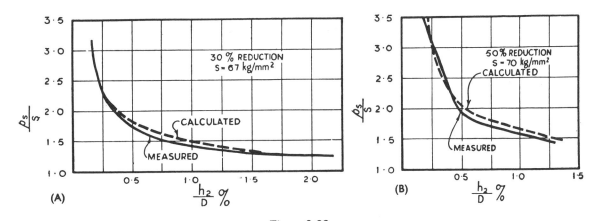

Figure 9-22:
Comparison of Results of Calculated Mean Specific Pressure (p_s) with S.K.F. Co. Experimental Data for the Cold Rolling of Steel with Different Ratios of Final Thickness (h_2) of the Strip to the Diameter (D) of the Rolls. For Calculations, μ is Assumed to be 0.2 and the Work-Hardening Effect is Neglected; p_s is Calculated from Tselikov's (Equations 9-91 and 9-95).

♦ Skefco Ball Bearing Co., Ltd., Luton, Bedfordshire, Booklet 1936, "Rolling Mill Data for Calculation and Design".
⊕ V. N. Rudbakh and V. P. Severdenko, ONTI, M-L, 1936, "Influence of External Friction on the Deformation of Metals in Rolling".
⁸ L. R. Underwood, "The Rolling of Metals", Vol. I, John Wiley and Sons, New York, 1950.

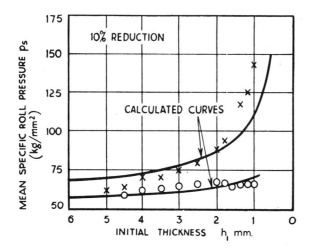

Figure 9-23:
Comparison of Values of Mean Specific Roll Pressure (p_s) Calculated from Tselikov's (Equations 9-91 and 9-95) with Tests of Rudbakh and Severdenko, 0.14 Per Cent. C Steel Strip, 10 Per Cent. Reduction, 150 mm. Roll Diameter. Crosses Show Experimental Results without Lubricant. Circles, with Linseed Oil as Lubricant. (Tselikov Assumed μ = 0.6 for Dry Rolling and 0.3 when Lubricated.)

9-11 Nadai's Solutions to Von Karman's Equation

In his treatment of Von Karman's equation, Nadai considered not only the use of strip tensions in the rolling operation, but also different modes of friction.[□] Assuming small contact angles, he derived the equation

$$(1 + z^2) \frac{dy}{dz} - \left(\frac{2}{\gamma S}\right)\tau = 2z \tag{9-100}$$

where

$$y = \frac{p_r}{S} \tag{9-101}$$

$$\gamma^2 = \frac{2h_2}{D} \tag{9-102}$$

$$z = \frac{2x}{\gamma D} \tag{9-103}$$

and

$$\tau = \pm \mu p_r \tag{9-104}$$

Nadai considered three frictional hypotheses; (1) a constant coefficient of friction, (2) a constant frictional shearing stress, and (3) a frictional resistance proportional to the relative velocity of slip between the strip and the roll. The solutions developed on the basis of these hypotheses are discussed separately below

(a) Constant Coefficient of Friction

Considering friction in the conventional sense with a constant coefficient μ, a constrained yield strength S, and with entry and exit strip tensile stresses of σ_1 and σ_2, respectively, Nadai established the two equations

$$p_r = S\left[\left\{y_1 + \frac{2(1-K\upsilon_1)}{K^2}\right\}e^{K(\upsilon_1-\upsilon)} - \frac{2(1 - K\upsilon)}{K^2}\right] \tag{9-105}$$

and

$$p_r = S\left[\left(y_2 + \frac{2}{K^2}\right)e^{K\upsilon} - \frac{2(1 + K\upsilon)}{K^2}\right] \tag{9-106}$$

[□] A. Nadai, "The Forces Required for Rolling Steel Strip Under Tension", Journal of Applied Mechanics, American Society of Mech. Engs., June, 1939, pp. A54-A62.

for the pressure distribution in the region before and the region after the neutral point respectively. In these equations

$$y_1 = \frac{S - \sigma_1}{S} \quad ; \quad y_2 = \frac{S - \sigma_2}{S} \qquad (9\text{-}107)$$

$$\upsilon = \tan^{-1}\left(\frac{2x}{\gamma D}\right) \qquad (9\text{-}108)$$

$$\upsilon_1 = \tan^{-1}\left(\frac{h_1 - h_2}{h_2}\right) \qquad (9\text{-}109)$$

and

$$K = \frac{2\mu}{\gamma} = \mu\sqrt{\frac{2D}{h_2}} \qquad (9\text{-}110)$$

Utilizing these equations, Nadai calculated the pressure curves for a number of rolling operations and his curves (some of which are shown in Figure 9-24) indicated that the application of front or back tension only, or of both together, reduces the rolling force and shifts the position of the neutral point. He also showed that under certain conditions the neutral plane may be shifted to coincide with either the entry or exit planes. Excessive exit tension tends to produce the former shift in which case the rolling operation becomes, in reality, a drawing operation.

Furthermore, Nadai showed that an increase in the coefficient also increased the rolling force and increased the slip (moved the neutral plane towards the entry end of the roll bite).

(b) Constant Shearing Stress

Equating K to $\frac{2\tau_o}{S}$, where τ_o represents a constant shearing stress, Nadai obtained the expression

$$\frac{dy}{dz} = \pm \frac{K}{1+z^2} + \frac{2z}{1+z^2} \qquad (9\text{-}111)$$

Upon integration, this equation gives

$$y = \frac{p_r}{S} = \pm K \tan^{-1} z + \log_e (1+z^2) + C \qquad (9\text{-}112)$$

the plus sign referring to the region of forward slip and the negative to that of backward slip and where C represents the constant of integration. This expression also yields the typical friction hill as shown in Figure 9-25 reproduced from Nadai's paper. In this case the frictional stress is assumed to be one tenth the constrained yield stress S, the ratio of roll diameter to entry strip thickness is 200 and the reduction given to the strip is 50 per cent.

(c) Surface Friction Proportional to Velocity of Slip

To investigate the hypothetical condition of hydrodynamic lubrication, Nadai postulates as the frictional law

$$\tau = \frac{\eta(V-v)}{\delta} = \frac{\eta V_2}{\delta}\left(\frac{1}{1+z^2} - \frac{1}{1-z_o^2}\right) \qquad (9\text{-}113)$$

where

η is the mean viscosity of the lubricant under the high rolling pressures
V is the variable velocity of the strip in the roll bite
v is the peripheral speed of the roll.

δ is the thickness of the oil film.
V_2 is the exit speed of the strip.
z is defined by Equation 9-103.
z_o is the value of z corresponding to the neutral point.

He derived the expression

$$y = \frac{p_r}{S} = 1 + \log_e (1+z^2) + \frac{K}{2}\left[\frac{z}{1+z^2} - \left(\frac{1-z_o^2}{1+z_o^2}\right)\tan^{-1}z\right] \qquad (9\text{-}114)$$

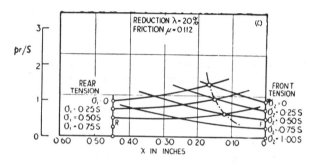

Figure 9-24:
Pressure Distribution Curves with an without Strip Tension According to the Equations of Nadai (Roll Diameter: 16 inches; Rolled Strip Thickness 0.10-inch).

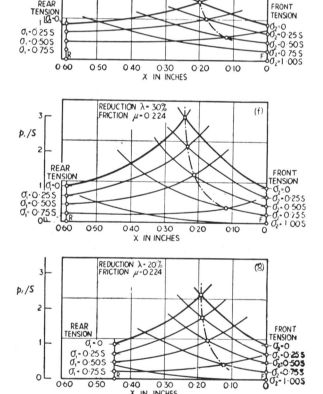

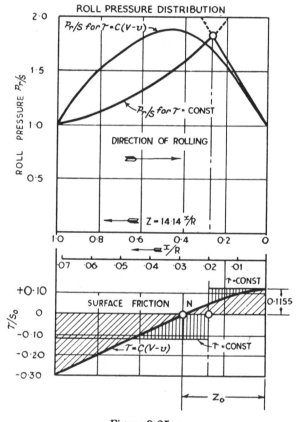

Figure 9-25:
Distribution of Rolling Pressure and Surface Friction Corresponding to Two Assumptions for Friction.

For the same conditions as in the preceding case, the pressure distribution for this expression is also shown in Figure 9-25. It will be observed that the sharp pressure peak has been replaced by a rounded peak and, consequently, the two distinct branches of the pressure curve have disappeared. Another interesting point is that the neutral plane does not coincide with the position of maximum pressure.

9-12 Orowan's Theory

As the result of his work at the Cavendish Laboratory in Cambridge, England, Orowan published a theory for the calculation of the pressure distribution along the arc of contact intended to be as general and accurate as possible. This theory avoids most of the mathematical approximations and assumptions, such as a constant coefficient of friction and a constant constrained yield strength, used by other rolling theorists, and permits any given variation in the yield stress and in the coefficient of friction along the arc of contact to be taken into account.

In deriving his theory, the usual assumption of homogeneous compression is not made. Instead, he considers the work of Prandtl[※] and Nadai[◇] relating to 2-dimensional plastic deformation and also takes into account the elastic deformation of the rolls. Orowan's theory does not assume that the material rolled slips on the rolls; instead, it provides criteria for determining in which areas of the arc of contact slipping occurs and in which areas sticking occurs. (In this treatment of Orowan's theory, however, only the theory as it relates to slipping friction will be discussed.) In addition, it enables spindle torques, power requirements and rolling efficiency to be calculated.

The compression of a plastic mass between two rough, parallel plates of infinite length (perpendicular to the plane of the drawing) is illustrated in Figure 9-26. On the assumption of no work hardening and the adherence of the plastic to the plates, Prandtl developed expressions for the vertical pressure q, the horizontal pressure t and the shear stress τ in vertical and horizontal planes, these being

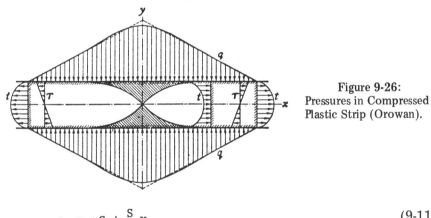

Figure 9-26:
Pressures in Compressed Plastic Strip (Orowan).

$$q = C + \frac{S}{h} x \qquad (9\text{-}115)$$

$$t = C + \frac{S}{h} x - S\sqrt{1 - 4y^2/h^2} \qquad (9\text{-}116)$$

and

$$\tau = -\frac{S}{h} y \qquad (9\text{-}117)$$

where   S is the constrained yield stress of the plastic material
h is the distance between the parallel plates
x and y are the coordinates in the directions as indicated in Figure 9-26, and
C is the constant of integration.

[※] L. Prandtl, Zeitschrift für angewandte Mathematik und Mechanik, Vol. 3, 1923, p. 401.
[◇] A. Nadai, "Plasticity" McGraw-Hill, New York and London, 1931.

If q = 0 at the edge of the plate where $x = \pm L$, then

$$c = \pm \frac{SL}{h} \qquad (9\text{-}118)$$

The resultant stress distribution is then as shown in Figure 9-26.

Nadai considered the stresses between two similar rough plates inclined to each other at a small angle 2θ. For flow towards the apex, he obtained the following expressions for the stresses (as illustrated in Figure 9-27).

$$t = q - S\sqrt{1 - \frac{\beta^2}{\theta^2}} \qquad (9\text{-}119)$$

and

$$\tau = -\frac{S\beta}{2\theta} \qquad (9\text{-}120)$$

where β is the angle subtended between the line joining the point under consideration to the apex and the plane of symmetry. These expressions apply only to flow towards the apex, but Orowan, in using them, assumed they were applicable for flow in the reverse direction.

Although the above equations apply to sticking friction, Orowan assumed that, for parallel plates, they would be valid for sliding friction provided a new value of h, namely h*, could be chosen so that the shear stress τ at the surface is equal to μq instead of $S/2$ as in the case of sticking. Under these circumstances

$$h^* = \frac{S}{2\mu q} h \qquad (9\text{-}121)$$

Similarly, for inclined plates, a semiangle $\theta^* (>\theta)$ was adopted so that the shear stress at the surface of the wedge of semiangle θ would be equal to μq ($<S/2$). The semiangle was given by

$$\theta^* = \frac{S\theta}{2\mu q} \qquad (9\text{-}122)$$

and hence, from equations 9-119 and 9-120

$$\tau = \frac{\mu q \beta}{\theta} \qquad (9\text{-}123)$$

and

$$t = q - S\sqrt{1 - \left(\frac{2\mu q}{S}\right)^2 \left(\frac{\beta}{\theta}\right)^2} \qquad (9\text{-}124)$$

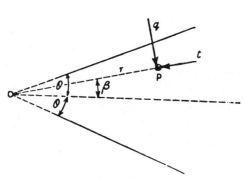

Figure 9-27:
Nonparallel Compression Plates (Orowan).

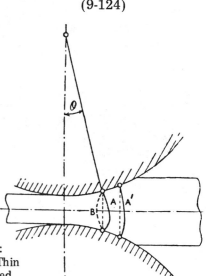

Figure 9-28:
Equilibrium of Thin Segment of Rolled Stock (Orowan).

MATH MODELS RELATING TO ROLLING FORCE

Thus, equations 9-123 and 9-124 apply to slipping conditions but if μq is made equal to $S/2$ they become the equations for sticking friction. These equations may now be utilized in the development of a generalized equation for rolling by considering the equilibrium of a segment of the workpiece in the roll bite.

Figure 9-28 shows a thin vertical section of the rolled material of arbitrary shape bounded by surfaces A and A' generated by straight lines moving always parallel to the roll axis and having their intersection with the rolls in a vertical plane. When the surface A subtends an angle θ at the roll center, the horizontal force acting across surface A is $f(\theta)$ and that across A' by $\left(f(\theta) + \frac{df}{d\theta} \cdot d\theta\right)$. The net difference is therefore $\frac{df}{d\theta} \cdot d\theta$ which is balanced by the frictional forces on the ends of the element and the horizontal component of the normal pressure.[+] The frictional forces (without assuming slipping or sticking) are given by $\pm 2\tau \cos\theta\, R\, d\theta$ (where the plus sign refers to the exit side of the neutral plane) and the normal force by $2q \sin\theta\, R\, d\theta$.

For equilibrium therefore

$$\frac{df}{d\theta} = Dq \sin\theta \pm D\tau \cos\theta \qquad (9\text{-}125)$$

This equation represents the variation of the horizontal force $f(\theta)$ along the arc of contact in terms of the normal pressure $q(\theta)$. Since the force $f(\theta)$ is the same for any surface configuration whose end points lie in the same vertical plane, it is convenient to choose cylindrical surfaces as illustrated in Figure 9-29. Considering unit width perpendicular to the plane of the figure, the area (dA) of the element of the surface AB subtending an angle $d\beta$ at point 0 is given by

$$dA = \frac{h}{2 \sin\theta} d\beta \qquad (9\text{-}126)$$

where h is the thickness of the strip or the distance between the points A and B.

The element of the horizontal force df acting across this element of the cylindrical surface is therefore given by (a) the contribution of the radial pressure t and (b) the contribution of the shear stress τ. The former is represented by

$$t \cos\beta\, dA = t \cos\beta \frac{h}{2 \sin\theta} d\beta \qquad (9\text{-}127)$$

Substituting for t from equation 9-124 and integrating between the limits $\beta = -\theta$ and $\beta = +\theta$, the contribution of the horizontal stress to $f(\theta)$ is

$$f_t(\theta) = hq - \frac{hS}{\sin\theta} \int_0^\theta \left[\sqrt{1 - \left(\frac{2\mu q}{S}\right)^2 \left(\frac{\beta}{\theta}\right)^2}\right] \cos\beta\, d\beta \qquad (9\text{-}128)$$

If

$$a = \frac{2\mu q}{S} \qquad (9\text{-}129)$$

and

$$w(\theta, a) = \frac{1}{\sin\theta} \int_0^\theta \left[\sqrt{1 - a^2 \frac{\beta^2}{\theta^2}}\right] \cos\beta\, d\beta \qquad (9\text{-}130)$$

then equation 9-128 may be written

$$f_t(\theta) = hq - hSw \qquad (9\text{-}131)$$

[+] In Orowan's original paper (q.v.) the notation $f(\theta)$, $w(\theta)$, $m+(\theta)$, etc., simply denotes the value the quantities have for a particular value of θ. Thus $f(\theta)$ denotes the value of the force f corresponding to a point on the arc of contact the radius to which makes an angle θ to the line connecting the roll centers. Orowan's notation has been retained in this discussion of his mathematical model.

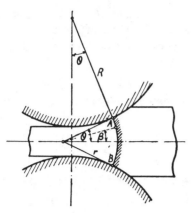

Figure 9-29: Diagram Relating to the Calculation of Horizontal Force in the Sheet Between the Rolls (Orowan).

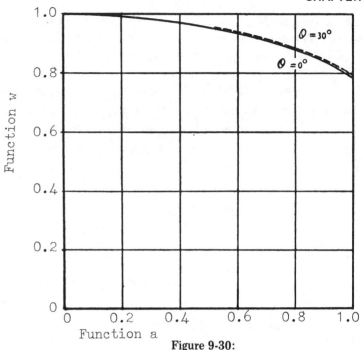

Figure 9-30: Graphical Representation of the Function $w(\phi, a)$.

From Figure 9-30, showing the variation of w with a, it is seen that w is practically independent of θ.

Considering now the contribution due to the shear stress τ

$$\tau \sin \beta \cdot dA = \frac{\mu q}{\theta} \beta \cdot \sin \beta \cdot \frac{h}{2 \sin \theta} \cdot d\beta \qquad (9\text{-}132)$$

the sign of the expression being positive for the exit side and negative for the entry side. Integration by parts yields

$$f_\tau(\theta) = \pm h \mu q \left(\frac{1}{\theta} - \frac{1}{\tan \theta} \right) \qquad (9\text{-}133)$$

The total horizontal force on the section AB is then

$$f(\theta) = f_t(\theta) + f_\tau(\theta) = h \left[q \left\{ 1 \pm \mu \left(\frac{1}{\theta} - \frac{1}{\tan \theta} \right) \right\} - Sw \right] \qquad (9\text{-}134)$$

Where slipping friction is concerned (as in cold rolling) $f_\tau(\theta)$ can be neglected and for values of $\mu < 0.2$ and values of $\theta < 10°$, Orowan states that $hq\mu\left(\frac{1}{\theta} - \frac{1}{\tan \theta}\right)$ is only about 1 per cent of hq. Thus for cold rolling, the following approximate formula may be used.

$$f(\theta) = h(q - Sw) \qquad (9\text{-}135)$$

or

$$q = \frac{f(\theta)}{h} + Sw \qquad (9\text{-}136)$$

Assuming $\tau = \mu q$ (slipping friction) and substituting the value of q as given by equation 9-136 in equation 9-125, the following expression is obtained

$$\frac{df}{d\theta} = f \cdot \frac{D}{h} (\sin \theta \pm \mu \cos \theta) + DSw (\sin \theta \pm \mu \cos \theta) \qquad (9\text{-}137)$$

MATH MODELS RELATING TO ROLLING FORCE

This equation may be solved for f by graphical integration techniques, details of which may be found in the published literature.[B,+,=] The normal pressure distribution (q) along the arc of contact is then calculated from equation 9-136.

The use of Orowan's model is tedious in that not only is the calculation of the pressure distribution laborious, but the integration of the pressure distribution curve must also be made if the specific rolling force is to be calculated.

Bland and Ford [†] utilized an approximation of Orowan's theory in developing the rolling model discussed in the following section and Jortner,[□] who took into account the elastic deformation of both the rolls and the strip, based his mathematical model on that of Orowan (see Section 9-15). Hocket,[※] in 1959, also utilized Orowan's concepts in conjunction with Hitchcock's formula for the length of the arc of contact.

9-13 The Approximate Theory of Bland and Ford

In their attempts to make practical calculations from various theories of cold rolling, Bland and Ford[†] developed a model of the cold reduction process which can be regarded as an approximation of Orowan's theory (as applied to homogeneous compression) discussed in the preceding section. Again, the rolling force is derived from an expression for the normal roll pressure at any point on the arc of contact between the roll and the strip.

Bland and Ford based their model on the differential equation developed by Orowan, viz.

$$\frac{df}{d\phi} = 2sR'(\sin \phi \pm \mu \cos \phi) \qquad (9\text{-}138)$$

(which is the same as equation 9-125, except for the fact that s is used to designate the normal roll pressure instead of q, the shearing stress at the roll surface is represented by μs instead of τ, ϕ replaces θ and the deformed roll radius R' is used instead of the deformed roll diameter D'.) Now the normal roll pressure s from the Huber-Mises condition of plasticity is related to the principal stress in the horizontal plane and the constrained compressive yield strength k by the equation

$$s = p + k = \frac{f}{h} + k \qquad (9\text{-}139)$$

or

$$f = h(s - k) \qquad (9\text{-}140)$$

Then substituting this value for f in equation 9-138

$$\frac{d}{d\phi}[h(s - k)] = 2sR'(\sin \phi \pm \mu \cos \phi)$$

or (9-141)

[B] L. R. Underwood, "The Rolling of Metals — Theory and Experiment", Vol. I, John Wiley and Sons, Inc., New York, 1950.

[+] "Research on the Rolling of Strip", A Symposium of Selected Papers, 1948-1958, The British Iron & Steel Research Association, London, England.

[=] M. Cook and E. C. Larke, "Computation of Rolling Load, Torque and Roll Face Pressure in Metal Strip Rolling", J. Inst. Met. 1945, pp. 557-579.

[†] D. R. Bland and H. Ford, "The Calculation of Roll Force and Torque in Cold Strip Rolling With Tensions", Proc. Inst. Mech. Eng., 1948, Vol. 159, pp. 144-153.

[□] D. Jortner, J. F. Osterle and C. F. Zorowski, "An Analysis of the Mechanics of Cold Strip Rolling", Iron and Steel Engineer Year Book, 1959, pp. 403-411.

[※] J. E. Hocket and R. K. Bates, "The Rolling Pressures of Uranium Sheet and Plate", Los Alamos Scientific Laboratory of the University of California, Los Alamos, New Mexico, LA-2233 Metallurgy and Ceramics, January, 1959.

$$\frac{d}{d\phi}\left[hk\left(\frac{s}{k} - 1\right)\right] = 2sR'(\sin \phi \pm \mu \cos \phi) \qquad (9\text{-}142)$$

Thus,
$$hk\frac{d}{d\phi}\left(\frac{s}{k}\right) + \left(\frac{s}{k} - 1\right)\frac{d}{d\phi}(hk) = 2sR'(\sin \phi \pm \mu \cos \phi) \qquad (9\text{-}143)$$

Assuming
$$\left(\frac{s}{k} - 1\right)\frac{d}{d\phi}(hk) \ll hk\frac{d}{d\phi}\left(\frac{s}{k}\right) \qquad (9\text{-}144)$$

$$hk\frac{d}{d\phi}\left(\frac{s}{k}\right) = 2sR'(\sin \phi \pm \mu \cos \phi) \qquad (9\text{-}145)$$

or
$$\frac{\frac{d}{d\phi}\left(\frac{s}{k}\right)}{\frac{s}{k}} = \frac{2R'}{h}(\sin \phi \pm \mu \cos \phi) \qquad (9\text{-}146)$$

and
$$\int \frac{\frac{d}{d\phi}\left(\frac{s}{k}\right)}{\frac{s}{k}} d\phi = \int \frac{2R'}{h}(\sin \phi \pm \mu \cos \phi)d\phi \qquad (9\text{-}147)$$

Since
$$h = h_o + 2R'(1 - \cos \phi) \simeq h_o + R'\phi^2 \qquad (9\text{-}148)$$

and
$$\sin \phi \pm \mu \cos \phi \simeq \phi \pm \mu \qquad (9\text{-}149)$$

$$\int \frac{\frac{d}{d\phi}\left(\frac{s}{k}\right)}{\frac{s}{k}} d\phi = \int \frac{2R'(\phi \pm \mu)}{h_o + R'\phi^2} d\phi \qquad (9\text{-}150)$$

Therefore
$$\log\left(\frac{s}{k}\right) = \log\frac{h}{R'} \pm 2\mu\sqrt{\frac{R'}{h_o}} \text{ arc tan}\left(\sqrt{\frac{R'}{h_o}}\phi\right) + \text{constant of integration} \qquad (9\text{-}151)$$

Thus
$$s = c \times k \times \frac{h}{R'} \times e^{\pm \mu H} \qquad (9\text{-}152)$$

where c is a constant and H is defined by the equation

$$H = 2\sqrt{\frac{R'}{h_o}} \text{ arc tan}\left(\sqrt{\frac{R'}{h_o}}\phi\right) \qquad (9\text{-}153)$$

At the exit end of the roll bite
$$p = -\sigma_o \qquad (9\text{-}154)$$

where σ_o is the exit tensile stress in the strip and H = 0.

Thus
$$s_o = k_o - \sigma_o \qquad (9\text{-}155)$$

so that, on the exit side,

MATH MODELS RELATING TO ROLLING FORCE

$$c = (k_o - \sigma_o)/(k_o \times h_o/R') \tag{9-156}$$

$$= \frac{R'}{h_o}\left(1 - \frac{\sigma_o}{k_o}\right) \tag{9-157}$$

Substituting back in equation 9-152 and denoting s on the exit side by s^+

$$s^+ = \frac{kh}{h_o}\left(1 - \frac{\sigma_o}{k_o}\right)e^{\mu H} \tag{9-158}$$

Similarly, for the entry side, where s is denoted by s^-

$$s^- = \frac{kh}{h_i}\left(1 - \frac{\sigma_i}{k_i}\right)e^{\mu(H_i - H)} \tag{9-159}$$

where h_i denotes the thickness at entry
k_i the constrained compressive yield stress at entry
σ_i is the entry strip tensile stress
H_i is the value of H at entry into the roll bite.

When no tensions are applied, and

$$\sigma_i = \sigma_o = 0 \tag{9-160}$$

$$s^+ = \frac{kh}{h_o}e^{\mu H} \tag{9-161}$$

and

$$s^- = \frac{kh}{h_i}e^{\mu(H_i - H)} \tag{9-162}$$

For computational purposes, it is desirable first to establish the neutral angle. Using the suffix n to denote this location

$$S_n^+ = S_n^- \tag{9-163}$$

and from equations 9-158 and 9-159

$$\frac{k_n h_n}{h_o}\left(1 - \frac{\sigma_o}{k_o}\right)e^{\mu H_n} = \frac{k_n h_n}{h_i}\left(1 - \frac{\sigma_i}{k_i}\right)e^{(\mu H_i - H_n)} \tag{9-164}$$

from which

$$H_n = \frac{H_i}{2} - \frac{1}{2\mu}\ln\left\{\frac{h_i}{h_o}\left(\frac{1 - \frac{\sigma_o}{k_o}}{1 - \frac{\sigma_i}{k_i}}\right)\right\} \tag{9-165}$$

or, in the absence of tensions

$$H_n = \frac{H_i}{2} - \frac{1}{2\mu}\ln\frac{h_i}{h_o} \tag{9-166}$$

From equation 9-153, the neutral angle ϕ_n is given by

$$\phi_n = \sqrt{\frac{h_o}{R'}}\tan\sqrt{\frac{h_o}{R'}} \times \frac{H_n}{2} \tag{9-167}$$

The specific rolling force, P, obtained by integration of the area under the pressure distribution curves is given by

$$P = \int_o^{\phi_i} q \times R'd\phi = R'\int_o^{\phi_i} q d\phi \tag{9-168}$$

where ϕ_i is the angular coordinate of the entry plane.

value $\underline{k}$ For simplicity, Bland and Ford assumed the constrained yield stress to assume a mean

where
$$\underline{k} = \frac{\int_o^{\phi_i} k d\phi}{\theta_i} \qquad (9\text{-}169)$$

Thus
$$P = R'\underline{k}\left[\int_o^{\phi_n} \frac{h}{h_o}e^{\mu H}d\phi + \int_{\phi_n}^{\phi_i} \frac{h}{h_i}e^{\mu(H_i-H)}d\phi\right] \qquad (9\text{-}170)$$

To facilitate computation, the following transformations were made
$$\psi = \phi/\mu \qquad (9\text{-}171)$$

$$a = \mu\sqrt{\frac{R'}{h_o}} \qquad (9\text{-}172)$$

and
$$r = \frac{h_i - h_o}{h_i} = 1 - \frac{h_o}{h_i} \qquad (9\text{-}173)$$

Thus
$$\frac{h}{h_o} = \frac{h_o + R'\phi^2}{h_o} = 1 + \frac{a^2}{\mu^2}\phi^2 \qquad (9\text{-}174)$$

$$= 1 + a^2\psi^2 \qquad (9\text{-}175)$$

$$\frac{h}{h_i} = \frac{h_o}{h_i}\frac{h}{h_o} = (1-r)(1+a^2\psi^2) \qquad (9\text{-}176)$$

and
$$\mu H = 2\mu\sqrt{\frac{R'}{h_o}} \text{ arc tan }\sqrt{\frac{R'}{h_o}}\phi = 2a \text{ arc tan } a\psi \qquad (9\text{-}177)$$

Also
$$\psi_i = \frac{\phi_i}{\mu} = \frac{1}{\mu}\sqrt{\frac{h_i-h_o}{R'}} = \frac{1}{\mu}\sqrt{\frac{h_o}{R'}}\sqrt{\frac{h_i-h_o}{h_o}} \qquad (9\text{-}178)$$

$$= \frac{1}{a}\sqrt{\frac{r}{1-r}} \qquad (9\text{-}179)$$

Hence
$$\mu H_i = 2a \text{ arc tan } a\psi_i = 2a \text{ arc tan }\sqrt{\frac{r}{1-r}} \qquad (9\text{-}180)$$

and
$$\mu H_n = \frac{\mu H_i}{2} - \frac{1}{2}\ln\frac{h_i}{h_o} = a \text{ arc tan }\sqrt{\frac{r}{1-r}} - \frac{1}{2}\ln\frac{1}{1-r} \qquad (9\text{-}181)$$

From equation 9-177
$$\psi_n = \frac{1}{a}\tan\frac{\mu H_n}{2a} \qquad (9\text{-}182)$$

whence
$$\psi_n = \frac{1}{a}\tan\left(\frac{1}{2}\text{ arc tan }\sqrt{\frac{r}{1-r}} - \frac{1}{4a}\ln\frac{1}{1-r}\right)$$

MATH MODELS RELATING TO ROLLING FORCE

Substituting equations 9-175, 9-176, 9-177 and 9-180 in equation 9-170 and changing the variable of integration from ϕ to ψ yields

$$\frac{P}{\mu R'\underline{k}} = \int_0^{\psi_n} (1 + a^2\psi^2)e^{2a \arctan a\psi} d\psi$$

$$+(1-r)e^{2a \arctan \sqrt{\frac{r}{1-r}}} \int_{\psi_n}^{\psi_1} (1 + a^2\psi^2)e^{-2a \arctan a\psi} d\psi \qquad (9\text{-}184)$$

It is seen from equation 9-184 that $\frac{P}{\mu R'\underline{k}}$ is a function of a, r, ψ_n and ψ_1. However, from equations 9-179 and 9-183, ψ_n and ψ_1 are functions of a and r, so that $\frac{P}{\mu R'\underline{k}}$ is a function of a and r only. Thus equation 9-184 may be written

$$P = \mu R'\underline{k} f_1(a,r) \qquad (9\text{-}185)$$

where the function $f_1(a,r)$ represents the right-hand side of equation 9-184.

However, another function $f_3(a,r)$ may be defined such that

$$f_3(a,r) = a\sqrt{\frac{1-r}{r}} f_1(a,r) \qquad (9\text{-}186)$$

so that

$$P = \underline{k} \times \sqrt{R'(h_i - h_o)} \times f_3(a,r) \qquad (9\text{-}187)$$

Thus the specific rolling force is seen as the product of a mean constrained compressive yield strength, the length of the arc of contact and a roll pressure function (to take account of the friction hill). Figure 9-31 shows the roll pressure function $f_3(a,r)$ as it relates to reduction for various values of a.

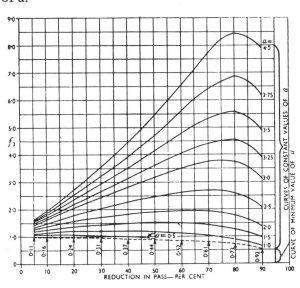

Figure 9-31:
Roll Pressure Function f_3 for Various Values of $a = \mu\sqrt{\frac{R'}{h_o}}$.

9-14 Jortner's Analysis

In his approach to the problem, Jortner▽ first developed expressions for the elastic deformation or flattening of the rolls with respect to external stresses applied over a very small

---

▽ D. Jortner, J. F. Osterle and C. F. Zorowski, "An Analysis of the Mechanics of Cold Strip Rolling", Iron and Steel Engineer Year Book, 1959, pp. 403-411.

portion of the circumference of the roll. Where a specific load P is applied at a single point on the circumference as illustrated in Figure 9-32 the radial deformation (u) of the cylinder is given at an angular position θ by

$$u = \frac{P}{\pi E}\left\{(1-\nu^2)\left(\cos\theta\,\log_e\left[\frac{1-\cos\theta}{1+\cos\theta}\right]+2\right)\right.$$

$$\left. - (1-\nu-2\nu^2)\sin\theta\left[\tan^{-1}\left(\frac{1+\cos\theta}{\sin\theta}\right)+\tan^{-1}\left(\frac{1-\cos\theta}{\sin\theta}\right)\right]\right\} \quad (9\text{-}188)$$

where ν represents Poisson's ratio and E the elastic modulus of the rolls.

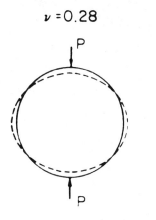

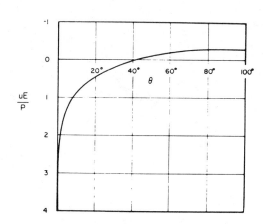

Figure 9-32:
Roll Deformed by Forces Applied at Two Diametrically Opposite Points.

Figure 9-33:
Variation of $\frac{uE}{P}$ with θ (Jortner).

The variation of the trigonometric expression within the outer brackets with θ is illustrated in Figure 9-33.

Replacing the single specific force P by a pressure p distributed over a small area (which subtends an angle 2α at the center of the cylinder of radius a), as shown in Figure 9-34, the expression assumes the form

$$u = \frac{pa}{\pi E}\left\{(1-\nu^2)\left[\sin\phi\,\log_e\left(\frac{1-\cos\phi}{1+\cos\phi}\right)\right]_{\theta-\alpha}^{\theta+\alpha}\right.$$

$$\left. + (1-\nu-2\nu^2)\cos\phi\left[\tan^{-1}\left(\frac{1-\cos\phi}{\sin\phi}\right)+\tan^{-1}\left(\frac{1-\cos\phi}{\sin\phi}\right)\right]_{\theta-\alpha}^{\theta+\alpha}\right\} \quad (9\text{-}189)$$

In Figure 9-35, dimensionless plots of equation 9-189 are shown for various sizes of the pressure angle α and compared to the result of equation 9-188 for the point load. For a given load P, the effect of the size of the pressure area on deformation decays rapidly outside the area but is quite important within it. In his computations, Jortner considered small pressure areas where $2\alpha = 0.1°$, thus allowing for 30 to 60 finite difference areas on the arc of contact. Figure 9-36 shows the actual form of the influence function used in the computation for the rolling of strip. The total deformation of a point on the cylinder is

$$u_{total} = \frac{a}{\pi E}\sum_{i=1}^{n}\left(\frac{\pi E u}{ap}\right)_{ij} p_i \quad (9\text{-}190)$$

MATH MODELS RELATING TO ROLLING FORCE

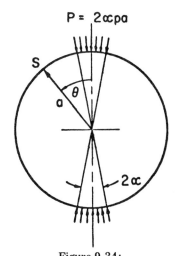

Figure 9-34:
Sketch Illustrating Small Zones of Uniform Pressure on the Work Roll.

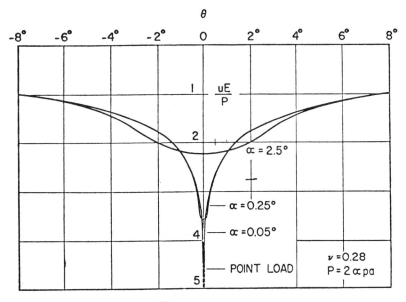

Figure 9-35: Variation of $\frac{uE}{P}$ with θ for Different Values of α (Jortner).

Jortner then analyzed the stress distribution in the elastically and plastically deformed portions of the strip in the roll bite (Figure 9-37) adapting Orowan's homogenous theory for the purpose.

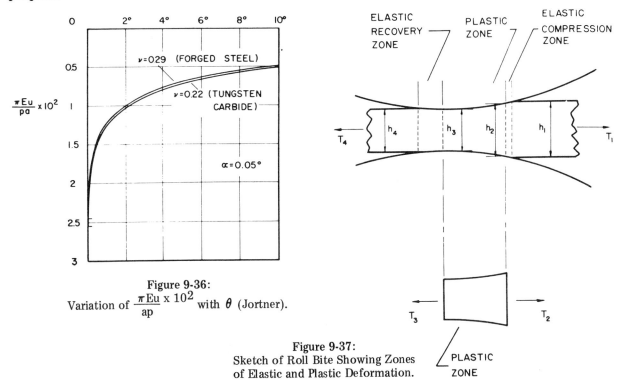

Figure 9-36:
Variation of $\frac{\pi E u}{ap} \times 10^2$ with θ (Jortner).

Figure 9-37:
Sketch of Roll Bite Showing Zones of Elastic and Plastic Deformation.

Referring to Figure 9-38, the vertical deformation V in the elastic region is given by

$$V = \left(\frac{1-\nu^2}{E}\right) \int_0^{h/2} \sigma_y dy - \frac{\nu(1+\nu)}{E} \int_0^{h/2} \sigma_x dy \qquad (9\text{-}191)$$

where σ_y and σ_x are the stresses in the y and x directions, respectively.

In the plastic region, (see Figure 9-39) the pressure on the nth segment of the bite is given by

$$[pr]_n = \frac{[hk]_n + [hp_r(1 \mp \mu \tan \beta) + p_r a \, \Delta\theta \, \cos \beta(\mu \pm \tan \beta) - hk]_{n-1}}{[h(1 \mp \mu \tan \beta) - a \, \Delta\theta \, \cos \beta(\mu \pm \tan \beta)]_n} \quad (9\text{-}192)$$

where β is defined as the angle between the strip surface and the horizontal, h is the strip thickness and k is the yield stress in plane compression.

In this relationship, the only unknown is the normal pressure. Starting with the known values at the plastic-elastic boundaries, each of the two forms of equation 9-192 are used to calculate the proper boundary into the plastic region. The neutral point is determined as that point at which both solutions give equal pressures.

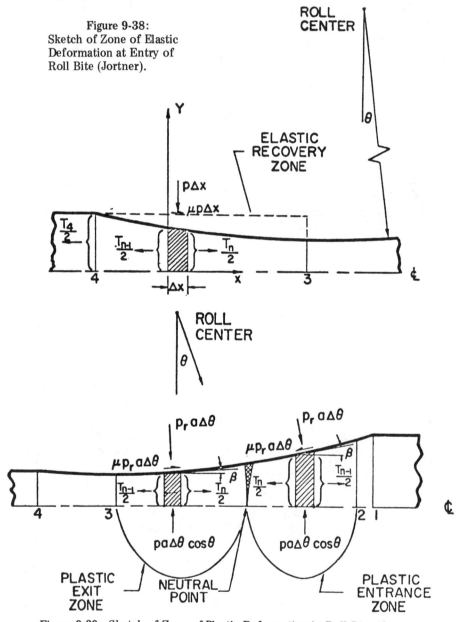

Figure 9-38: Sketch of Zone of Elastic Deformation at Entry of Roll Bite (Jortner).

Figure 9-39: Sketch of Zone of Plastic Deformation in Roll Bite (Jortner).

With the aid of a computer, Jortner then used the roll deformation and strip stress distribution equations alternately, continually adjusting his results to conform to boundary conditions. The shape of the roll was first approximated by its undeformed circular shape, and the distance between the rolls and the extent of the contact region were approximated from the boundary conditions relative to strip gage and the yield criterion. With the boundaries of the contact region thus determined, an approximate normal pressure distribution was established using the plastic and elastic strip solutions. Then another approximation to the roll shape was found by the roll deformation analysis using the previously calculated pressure distribution. Once again, the boundary conditions on height and the yield criterion determined the region of contact completely. This procedure was continued until the length of the contact region and the integral of the pressure distribution converged to the desired degree of accuracy.

To check his theory, Jortner used experimental results published in 1951 by Hessenberg and Sims[∞] relating to the rolling of steel strip on a 2-high mill with 10-inch diameter work rolls using a 10 per cent solution of soluble oil in water as a rolling solution. The average error in rolling force prediction was 4.3 per cent and in rolling torque 3.1 per cent.

Despite this good agreement, Jortner found that the analysis had a tendency to overestimate the effect of applied tensions in decreasing the rolling load.

9-15 A More Complex Model Based on Jortner's Analysis

In 1968, Cosse and Economopoulos[■] published a comprehensive mathematical model of the cold rolling process based in part on the prior work of Jortner. In this model, the fundamental assumptions made were restricted to a minimum and cognisance was taken of the elastic behavior and the work hardening of the workpiece, together with the dependence of its yield strength on strain rate. The elastoplastic straining of the workpiece was represented by the complete equation of Prandtl-Reuss and the profile of the arc of contact was calculated using influence functions (as undertaken by Jortner). This approach made it possible to consider frictional conditions other than those represented by a constant coefficient of friction.

With respect to the yielding of the strip, it was represented by the equation

$$\epsilon = \sigma/E + c\left(\sigma/\sigma_y\right)^n \quad (9\text{-}193)$$

where ϵ is the strain, σ is the true stress, E is the elastic modulus, c is a constant, σ_y is the yield strength defined in terms of a permanent strain (of say, 0.2%) and n is a work-hardening exponent. To allow for strain rate effects, the relationship proposed by Rosenfield and Hahn[▲] was used, this being

$$\sigma_y = \sigma_0 + a \log \dot{\epsilon} \quad (9\text{-}194)$$

where σ_0 is the static yield strength, a is a coefficient which depends only on temperature and $\dot{\epsilon}$ is the mean strain rate. The mean strain rate occurring in the rolling process was calculated from the following expression

$$\dot{\epsilon} = \frac{2}{\sqrt{3}} \left(1 - \frac{h_2}{h_1}\right) \cdot \frac{\pi N}{30(\phi_1 - \phi_4)} \quad (9\text{-}195)$$

where h_2 is the minimum separation of the work rolls
h_1 is the entry thickness of the strip
N is the rotational speed of the rolls
φ_1 and φ_4 are the angular positions of the entry plane and the position of the minimum roll separation (as illustrated in Figure 9-40)

[∞] W. C. F. Hessenberg and R. B. Sims, "The Effect of Tension on Torque and Roll Force in Cold Strip Rolling", Iron and Steel Institute Journal, Vol. 168, 1951, pp. 155-164.

[■] P. Cosse and M. Economopoulos, "Mathematical Study of Cold Rolling", C.N.R.M., No. 17, December, 1968, pp. 15-32.

[▲] A. R. Rosenfield and G. T. Hahn, Transactions of the ASM, 59, (1966), pp. 962-980.

The instantaneous strain rate $\dot{\varepsilon}_i$ which varies along the arc of contact, is given by

$$\dot{\varepsilon}_i = \frac{2}{\sqrt{3}} \frac{\dot{h}}{h} \qquad (9\text{-}196)$$

where $\dot{h}$ is the rate of change of thickness with time and h is the instantaneous thickness of the strip. This expression has a certain value at the entry of the bite and is zero where the thickness of the strip is minimum.

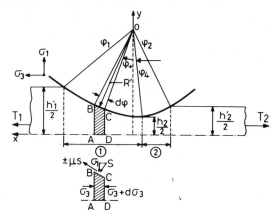

Figure 9-40:
Schematic Representation of Rolling.
1. Elasto-Plastic Zone;
2. Elastic Zone-Exit.

The derivation of the basic equations may be understood with reference to Figure 9-40. Assuming the inertial forces are negligible and that the tangential forces in the roll gap are proportional to the normal forces, the equations describing the equilibrium of an element ABCD in the roll bite are

$$\frac{d(\sigma_3 h)}{d\phi} = 2R'S(\sin \phi \pm \mu \cos \phi) \qquad (9\text{-}197)$$

and

$$\sigma_1 = S(1 \pm \mu \tan \phi) \qquad (9\text{-}198)$$

In these equations, μ is the coefficient of friction, the sign − or + to be used depends on whether the roll surface speed is higher or lower than the speed of the product. R′ is the distance from the center of the roll to its deformed surface and S is the stress normal to the roll surface.

At the angle ϕ, the thickness h of the strip is given by

$$\begin{aligned} h &= (2R + h_2) - 2R' \cos \phi + (h_2 - h_4) \\ &= 2h_2 - h_4 + 2R - 2R' \cos \phi \end{aligned} \qquad (9\text{-}199)$$

The term $(2R + h_2)$ represents the center-to-center spacing of the work rolls, $R' \cos \phi$ the projection of the line joining the roll center to the element in the roll bite and $(h_2 - h_4)$ represents the difference in thickness between the minimum value of the roll gap and the strip thickness at the plane of the roll centers.

Equating the pressure normal to the roll surface to σ_1, equation 9-197 becomes

$$\frac{d\sigma_3}{d\phi} = \frac{2}{h}\left[R'\sigma_1(\sin \phi \pm \mu \cos \phi) - \sigma_3\left(R' \sin \phi - \cos \phi \cdot \frac{dR'}{d\phi}\right)\right] \qquad (9\text{-}200)$$

The zone of elasto-plastic compression extends from the entry of the product between the rolls (angle φ_1) to the minimum height in the roll gap (angle φ_4) (Figure 9-40).

The most general equations relating stresses to strains are those of Prandtl-Reuss[♦]

[♦] R. Hill, "The Mathematical Theory of Plasticity", Clarendon Press, Oxford, 1950.

MATH MODELS RELATING TO ROLLING FORCE

$$\delta\varepsilon_1 = \frac{1}{E}\left[\delta\sigma_1 - \nu(\delta\sigma_2 + \delta\sigma_3)\right] + \frac{\delta\bar{\varepsilon}_p}{\bar{\sigma}}\left[\sigma_1 - \frac{1}{2}(\sigma_2 + \sigma_3)\right] \quad (9\text{-}201)$$

$$\delta\varepsilon_2 = \frac{1}{E}\left[\delta\sigma_2 - \nu(\delta\sigma_3 + \delta\sigma_1)\right] + \frac{\delta\bar{\varepsilon}_p}{\bar{\sigma}}\left[\sigma_2 - \frac{1}{2}(\sigma_3 + \sigma_1)\right] \quad (9\text{-}202)$$

$$\delta\varepsilon_3 = \frac{1}{E}\left[\delta\sigma_3 - \nu(\delta\sigma_1 + \delta\sigma_2)\right] + \frac{\delta\bar{\varepsilon}_p}{\bar{\sigma}}\left[\sigma_3 - \frac{1}{2}(\sigma_1 + \sigma_2)\right] \quad (9\text{-}203)$$

The variables with the p suffixes pertain to the plastic part of the total deformation and the first term in the right-hand side of the equation represents the elastic deformation.

In addition to the Prandtl-Reuss relationships, the solution of the problem calls for the use of a criterion of plasticity. The latter makes it possible to relate the behaviour of a material subjected to a triaxial system of stresses to its behaviour in a state of uniaxial stresses.

In this model, the von Mises criterion was selected, the effective stress of compression, $\bar{\sigma}$, being given by

$$\bar{\sigma} = \left[(\sigma_1 - \sigma_2)^2 + (\sigma_2 - \sigma_3)^2 + (\sigma_3 - \sigma_1)^2\right]^{1/2}/\sqrt{2} \quad (9\text{-}204)$$

The condition of plane strain is expressed by

$$\delta\varepsilon_2 = 0 \quad (9\text{-}205)$$

Making the substitutions

$$S_1 = \sigma_1 - \sigma_2 \quad (9\text{-}206)$$

$$S_2 = \sigma_2 \quad (9\text{-}207)$$

$$S_3 = \sigma_3 - \sigma_2 \quad (9\text{-}208)$$

$$S_2^2 = \bar{\sigma}^2 = S_1^2 + S_3^2 - S_1 S_3 \quad (9\text{-}209)$$

$$\delta\lambda = \delta\bar{\varepsilon}_p/\bar{\sigma} \quad (9\text{-}210)$$

$$G = (1 + \nu)/E \quad (9\text{-}211)$$

it can be shown from equations 9-201, 9-202, 9-203 and 9-205 that

$$\delta\varepsilon_1 = G\delta S_1 + (3\delta\lambda/2)S_1 \quad (9\text{-}212)$$

$$\delta\varepsilon_3 = G\delta S_3 + (3\delta\lambda/2)S_3 \quad (9\text{-}213)$$

Since $\bar{\varepsilon} = \bar{\varepsilon}_e + \bar{\varepsilon}_p$, (the mean elastic and plastic strains) then, from equation 9-193

$$\bar{\varepsilon}_p = |\bar{\sigma}|^{n-1} \bar{\sigma}/B^n \quad (9\text{-}214)$$

and

$$\delta\lambda = \frac{\delta\bar{\varepsilon}_p}{\bar{\sigma}} = \frac{n}{B^n}|\bar{\sigma}|^{n-1} \cdot \frac{d\bar{\sigma}}{\bar{\sigma}} \quad (9\text{-}215)$$

Taking Equations 9-206 to 9-211 into account, in the expression of $\delta\lambda$, equations 9-212 and 9-213 become

$$G\frac{dS_1}{d\varepsilon_1} + \frac{3n}{2B^n}\left(S_1^2 + S_3^2 - S_1 S_3\right)^{(n-2)/2} \cdot S_1 \frac{dS}{d\varepsilon_1} = 1 \quad (9\text{-}216)$$

$$G \frac{dS_3}{d\epsilon_3} + \frac{3n}{2B^n}\left(S_1^2 + S_3^2 - S_1 S_3\right)^{(n-2)/2} \cdot S_3 \cdot \frac{dS}{d\epsilon_3} = 1 \qquad (9\text{-}217)$$

Deriving $dS_1/d\epsilon_1$ and $dS_3/d\epsilon_3$ from equations 9-206 to 9-211, and if

$$F = \left(3n/4B^n\right)\left(S_1^2 + S_3^2 - S_1 S_3\right)^{(n-3)/2} \qquad (9\text{-}218)$$

equations 9-216 and 9-217 may be written as follows:

$$\left[G + FS_1(2S_1 - S_3)\right] dS_1/d\epsilon_1 + FS_1(2S_3 - S_1) \cdot dS_3/d\epsilon_1 = 1 \qquad (9\text{-}219)$$

$$\left[G + FS_3(2S_3 - S_1)\right] dS_3/d\epsilon_3 + FS_3(2S_1 - S_3) dS_1/d\epsilon_3 = 1 \qquad (9\text{-}220)$$

Summing up equations 9-201, 9-202 and 9-203, and taking equation 9-205 into account,

$$G_1 = (1 - 2\nu)/E \qquad (9\text{-}221)$$

$$\frac{d\epsilon_3}{d\epsilon_1} = -1 + G_1\left(\frac{dS_1}{d\epsilon_1} + 3\frac{dS_2}{d\epsilon_1} + \frac{dS_3}{d\epsilon_1}\right) \qquad (9\text{-}222)$$

Since $dS/d\epsilon_1 = (dS/d\epsilon_3)(d\epsilon_3/d\epsilon_1)$, equation 9-217 becomes

$$\frac{dS_3}{d\epsilon_1}\left[G + FS_3(2S_3 - S_1) - G_1\right] = \frac{dS_1}{d\epsilon_1}\left[G_1 - FS_3(2S_1 - S_3)\right] \\ + 3G_1 \frac{dS_2}{d\epsilon_1} - 1 \qquad (9\text{-}223)$$

Substituting the variable, ϕ for ϵ_1 using

$$\epsilon_1 = \ln(h/h_1) = \ln[(2h_2 - h_4 + 2R - 2R'\cos\phi)/h_1] \qquad (9\text{-}224)$$

noting that

$$dS/d\phi = (dS/d\epsilon_1) \cdot (d\epsilon_1/d\phi) \qquad (9\text{-}225)$$

and if

$$A = \frac{R'\sin\phi - (dR'/d\phi)\cos\phi}{h_2 + R - R'\cos\phi - 0.5 h_4} \qquad (9\text{-}226)$$

then equations 9-200, 9-219 and 9-223, (which form the system of equations to be solved in the elasto-plastic zone), can be written in the final form:

$$\frac{dS_1}{d\phi} = \frac{A}{G + FS_1(S_1 - S_3)} - \frac{FS_1(2S_3 - S_1)}{G + FS_1(2S_1 - S_3)} \cdot \frac{dS_3}{d\phi} \qquad (9\text{-}227)$$

$$\frac{dS_2}{d\phi} = \frac{A}{3G_1} - \frac{G_1 - FS_3(2S_1 - S_3)}{3G_1} \cdot \frac{dS_1}{d\phi} + \frac{G - G_1 + FS_3(2S_3 - S_1)}{3G_1} \cdot \frac{dS_3}{d\phi} \qquad (9\text{-}228)$$

$$\frac{dS_3}{d\phi} = \frac{R'(S_1 + S_2)(\sin\phi \pm \mu\cos\phi)}{h_2 + R - R'\cos\phi - 0.5 h_4} - A(S_2 + S_3) - \frac{dS_2}{d\phi} \qquad (9\text{-}229)$$

Thus, the solution of the elasto-plastic zone has the form of a system of three differential equations. The three unknown functions are the variables S_1, S_2 and S_3 which can be replaced by σ_1, σ_2 and σ_3, using equations 9-205 to 9-211; and the independent variable is the space variable φ.

MATH MODELS RELATING TO ROLLING FORCE

The negative sign in equation 9-229 is taken on the entry side of the neutral point, and the positive sign on the exit side. This condition gives rise to two systems of equations.

The elastic recovery zone extends from the φ_4 angle to the exit from the roll gap. In the elastic zone, the relationships between stresses and strains are expressed by Hooke's laws. Given a differential form, and written with respect to the three principal directions, they read:

$$d\varepsilon_1 = \left[d\sigma_1 - \nu(d\sigma_2 + d\sigma_3)\right]/E \tag{9-230}$$

$$d\varepsilon_2 = \left[d\sigma_2 - \nu(d\sigma_3 + d\sigma_1)\right]/E \tag{9-231}$$

$$d\varepsilon_3 = \left[d\sigma_3 - \nu(d\sigma_1 + d\sigma_2)\right]/E \tag{9-232}$$

Assuming the spread to be nil, $d\varepsilon_2 = 0$; hence

$$d\sigma_2 = \nu(d\sigma_3 + d\sigma_1) \tag{9-233}$$

Taking equation 9-233 into account, and effecting the change of variables:

$$d\sigma_1/d\phi = (d\sigma_1/d\varepsilon_1) \cdot (d\varepsilon_1/d\phi) \tag{9-234}$$

$$d\sigma_3/d\phi = (d\sigma_3/d\varepsilon_1) \cdot (d\varepsilon_1/d\phi) \tag{9-235}$$

Equation 9-230 gives

$$\frac{d\varepsilon_1}{d\phi} = \frac{1}{E}\left[(1 - \nu^2) \cdot \frac{d\sigma_1}{d\phi} - \nu(1 + \nu)\frac{d\sigma_3}{d\phi}\right] \tag{9-236}$$

Noting that $d\varepsilon_1/d\varphi = h^{-1} \cdot dh/d\varphi$, and taking equation 9-199 into account, equation 9-236 transforms to

$$\frac{d\sigma_1}{d\phi} = \frac{2E}{1 - \nu^2} \cdot \frac{R'\sin\phi - \cos\phi \, dR'/d\phi}{2h_2 - h_4 + 2R - 2R'\cos\phi} + \frac{\nu}{1 - \nu} \cdot \frac{d\sigma_3}{d\phi} \tag{9-237}$$

The equilibrium equation 9-200 and equation 9-237 form a system of two differential equations which represent the solution in the elastic zone at the exit from the roll gap. It is shown later that a second neutral point can exist in this zone. Depending on whether the point considered is upstream or downstream of this second neutral point, the sign to be taken in equation 9-200 is positive or negative. This condition gives rise to two systems of two equations in the elastic zone.

With respect to roll deformation, Jortner's approach was used. The radial elastic deformation at a point, S, of the surface (Figure 9-41) of a roll in a condition of plane deformation and subjected to two diametrically opposite concentrated forces is given by the following equation:

$$\Delta R = \left\{\frac{P}{\pi E}(1 - \nu^2)\left[\cos\theta \cdot \ln\left(\frac{1 - \cos\theta}{1 + \cos\theta}\right) + 2\right]\right.$$

$$- (1 - \nu - 2\nu^2)\sin\theta\left[\tan^{-1}\left(\frac{1 + \cos\theta}{\sin\theta}\right)\right.$$

$$\left.\left.+ \tan^{-1}\left(\frac{1 - \cos\theta}{\sin\theta}\right)\right]\right\} \tag{9-238}$$

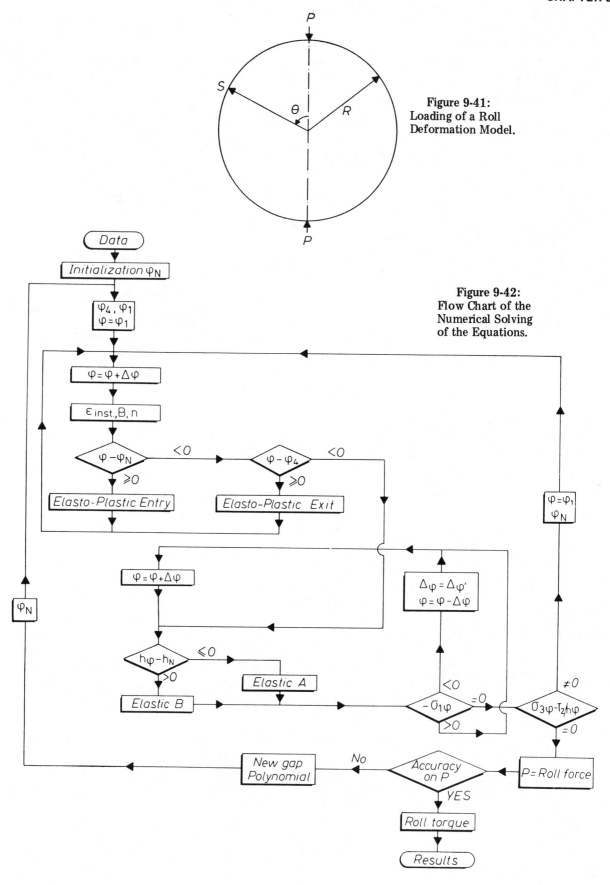

Figure 9-41: Loading of a Roll Deformation Model.

Figure 9-42: Flow Chart of the Numerical Solving of the Equations.

If the surface of the roll sustains a pressure distribution $\sigma_1 = F(\varphi)$, the unit force $\sigma_1 R d\varphi$ can be simulated by a concentrated force. Using equation 9-238 and the principle of superposition, the radial deformation of the roll due to a pressure distribution can be calculated point by point. For reasons of mathematical convenience, the deformed profile of the roll, obtained point by point, was transformed into a continuous function (polynominal) of the angle ϕ.

In solving the various equations, an iterative procedure was used with alternate computations being made relative to the deformation of the strip and the deformation of the rolls. The iterations were carried out according to the following scheme:

— The stress distribution within the strip was derived assuming that the rolls were deformable.

— The deformation of the rolls was calculated by applying this stress distribution to them,

— The stresses in the strip were recalculated so as to render them compatible with this new profile, and so on until the equilibrium condition was reached. The equilibrium condition was considered as being reached when the roll forces corresponding to two successive stress distributions were the same, to within a given percentage of accuracy.

The validity of the model was checked against various data published in the literature. For example, Figure 9-43 shows theoretical specific rolling force versus reduction curves developed using coefficients of friction of 0.045 and 0.057 together with experimental data published by Pawelski and Kading.◆

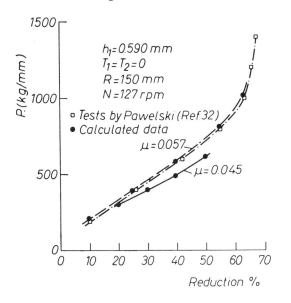

Figure 9-43:
Roll Force vs. Reduction Curve.

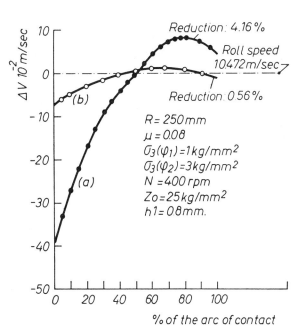

Figure 9-44:
Curves Showing Variation of Slip Along the Arc of Contact. Curve (a): One Neutral Point. Curve (b): Two Neutral Points.

Cosse and Economopoulos established the possibility of two neutral points in the case of light reductions. This occurrence is illustrated by Figure 9-44 showing, for two different reductions, the variation in the difference between the speeds of the product and of the rolls. It is

◆ O. Pawelski and G. Kading, "Untersuchugen über Grenzformanderung und Grenzdicke beim Kaltwalzen dunner Bänder, Stahl and Eisen, 22, (Nov. 1967), pp. 1340-1348.

seen that, in the case of the smaller reduction, the speed of the product at the exit is smaller than the roll surface speed.

9-16 Direct Methods for the Computation of Specific Rolling Force

Any method of determining the pressure distribution along the arc of contact can be used for computing the specific rolling force simply by integrating or measuring the area under the pressure distribution curve (the "friction hill"). However, such methods are tedious and time consuming and more direct approaches to the computation of specific rolling force are desirable. A number of these have been published and those of Ekelund, Cook and Parker, Bland and Ford and C. E. Davies are discussed briefly in this section.

    (a)   Ekelund's Equation

In 1927, Ekelund[◈] proposed the following equation for computing the specific rolling force, f. viz.

$$f = \bar{\sigma} \sqrt{\frac{D'd}{2}} \left(1 + \frac{1.6 \mu \sqrt{\frac{D'd}{2}} - 1.2d}{t_1 + t_2}\right) \qquad (9\text{-}239)$$

where

$\bar{\sigma}$ is the mean constrained yield strength
D' is the deformed roll diameter
d is the draft
μ is the coefficient of friction, and
t_1 and t_2 are the initial and final strip thicknesses, respectively.

In this expression, the factor contained within the parentheses represents a multiplier which takes friction into account.

The equation is stated to give remarkably good predictions of rolling loads over a wide range of rolling conditions. When strip tensions are used, the specific rolling force can be corrected by multiplying f by the factor $(1 - \sigma_A/\bar{\sigma})$ where σ_A represents the average of the entry and exit strip tensile stresses.

    (b)   The Cook and Parker Method

This method first involves the rolling of a number of annealed test strips on any single-stand mill with a pair of well-finished work rolls.[°] The rolling loads are measured and the results are plotted to show the variation of the ratio of the thickness of the annealed strip (t_b) to the work roll diameter (D) with the ratio of the square of the strip thickness $(t_b)^2$ to the specific rolling force (P_{ts}) for various reductions. (See Figure 9-45 relating to brass strip.) Then a first-pass curve can be plotted for any chosen thickness relating the specific rolling force to the reduction given to the strip (see Figure 9-46).

The method of establishing such a curve is most conveniently demonstrated by considering a particular thickness and roll diameter, and data will be derived, for example, for 24-inch wide x 0.2-inch thick 70/30 brass strip reduced on a mill fitted with 16-inch diameter steel rolls ground to a mean axial surface finish of about 11 microinch. The value of t_b/D corresponding to these conditions is 0.2/16 = 0.0125, and appropriate figures for the ratio of t_b^2/P_{ts} at single-pass reductions of 10-70% can be read from the family of

---

[◈] S. Ekelund, "The Analysis of Factors Influencing Rolling Pressure and Power Consumption in the Hot Rolling of Steel", Steel 93, Nos. 8-14, 1933 (Translated from Jernkontorets Ann., Feb. 1927, by B. Blomquist).

[°] M. Cook and R. J. Parker, "The Computation of Loads in Metal Strip Rolling by Methods Involving the Use of Dimensional Analysis", Journal of the Institute of Metals, 82, 1953, pp. 129-140.

MATH MODELS RELATING TO ROLLING FORCE

curves in Figure 9-45. Since $t_b = 0.2$-inch and $t_b^2 = 0.04$-inch^2, first-pass load data P_{ts} can be calculated from the equation:

$$P_{ts} = 0.04/t_b^2/P_{ts} \qquad (9\text{-}240)$$

and these values are given in column 3 of Table 9-1. As these figures refer to a t_b/w value of 0.0167, i.e. $w = 0.2/0.0167 = 12$ inches, the data for plotting the required first-pass curve, included in the last column of the table, are twice those recorded in the third column.

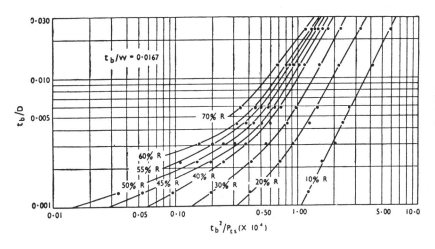

Figure 9-45: Data for Constructing First-Pass Curves for Annealed 70/30 Brass.

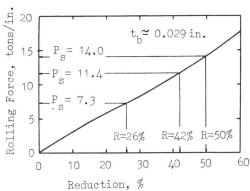

Figure 9-46: Annealed 70/30 Brass. First-Pass Curve for Conditions where $\frac{\text{initial thickness}}{\text{roll radius}} = 0.0125$.

Table 9-1
Method of Constructing First-Pass Curve
Annealed 70/30 brass, 24 in. wide x 0.2-in. thick; 16-in. dia. rolls

Reduction, %	t_b^2/P_{ts} (x 10^4) (from Fig. 9.45)	$0.04 \div t_b^2/P_{ts}$ = P_{ts}, tons (12 in. width)	Load Data for 24-in.-Wide Strip, tons
10	4.10	98	196
20	2.20	182	364
30	1.51	265	530
40	1.15	348	696
50	0.930	430	860
60	0.790	507	1014
70	0.675	593	1186

For multiple passes, Cook and Parker developed an expression for the energy expended in any intermediate pass. The sum of these energy expressions, for all the passes, is then equated to the energy which would be required to perform the total reduction in one pass assuming the mill to be capable of doing so. The specific force for this hypothetical total reduction in one pass is derived immediately from the first pass graph. The real first pass specific force is taken from this graph and the energy evaluated. A hypothetical combination of passes 1 and 2 then give the energy for pass 2 and hence the specific force for the second pass and so on for all passes.

This method can be applied to multiple-pass rolling schedules and involves only simple numerical calculations. Again, for cases where the measurements are made under appropriately comparable conditions, the predictions are stated to agree quite well with actual mill data.

(c) The Bland and Ford Graphical Solution

Under "frictionless" rolling conditions (if such were possible) the specific rolling force would be given by

$$f = \bar{\sigma}\sqrt{\frac{D'd}{2}} \qquad (9\text{-}241)$$

As discussed in Section 9-13, Bland and Ford [†] showed that the effect of friction can, however, be accounted for by multiplying the expression by a function of the reduction (r) and the dimensionless parameter a where

$$a = \mu\sqrt{\frac{D'}{2t_2}} \qquad (9\text{-}242)$$

Thus

$$f = \bar{\sigma}\sqrt{\frac{D'd}{2}} \cdot f_3(a,r) \qquad (9\text{-}243)$$

The value of the function $f_3(a,r)$ for various values of a and r may be derived from the graphs shown in Figure 9-31.

To calculate the specific rolling force from the initial and final thicknesses of the strip (t_1 and t_2, respectively), the yield stress curve, the undeformed roll radius (R), the elastic constant of the rolls (c) [*] and an assumed coefficient of friction μ, the reduction is first calculated from

$$r = \frac{t_1 - t_2}{t_1} \qquad (9\text{-}244)$$

A value for the mean yield stress $\bar{\sigma}$ is then determined using equation (9-169), a value of R′ (the deformed roll radius) is assumed and the term $\mu\sqrt{\frac{R'}{t_2}}$ calculated. The function $f_3(a,r)$ is conveniently and simply derived from Figure 9-31 and substituted in equation 9-243 to obtain the specific rolling force f. To ascertain whether or not the correct value of R′ has been assumed, $f/(t_1 - t_2)$ is calculated and, using this value and the appropriate value of c, R′/R is read off from the chart in Figure 9-47. If an incorrect value of R′ has been assumed, then the process may be repeated on an iterative basis or a graphical approach may be adopted as discussed by Bland and Ford [†] in their paper.

Graphical solutions for Bland and Ford's model have been developed by Sims [+] and Lianis and Ford. [•] Sims' method relates to the absence of tensions whereas that of Lianis and Ford is specifically intended for rolling with strip tensions.

[†] D. R. Bland and H. Ford, "The Calculation of Roll Force and Torque in Cold Strip Rolling with Tensions", Proc. Inst. Mech. Eng. 159, 1948, pp. 148-153.

[*] c represents the constant in Hitchcock's equation when expressed in the form

$$R' = R\left(1 + \frac{2cf}{t_1 - t_2}\right) \qquad (9\text{-}245)$$

where f is the specific rolling force. For steel rolls, $c = 1.67 \times 10^{-4}$ tons^{-1} in^2.

[+] R. B. Sims, "Calculation of Roll Force and Torque in Cold Rolling by Graphical and Experimental Methods", Portion of Book "Research on the Rolling of Strip", Published by BISRA, London, 1959.

[•] G. Lianis and H. Ford, "A Graphical Solution of the Cold Rolling Problem when Tensions are Applied to the Strip", Portion of Book "Research on the Rolling of Strip", Published by BISRA, London, 1959.

MATH MODELS RELATING TO ROLLING FORCE

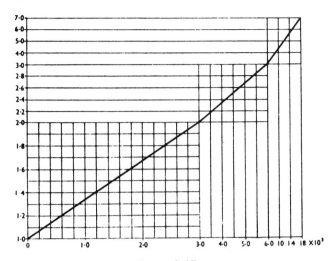

Figure 9-47:
Graphical Solution of Hitchcock's Formula for Flattened Roll Radius Steel Rolls $c = 1.67 \times 10^4$ Tons^{-1} in.^{2}.

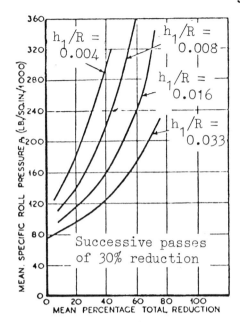

Figure 9-48:
Effect of Ratio $\frac{\text{initial thickness } h_1}{\text{roll radius R}}$ on mean specific roll pressure p_s for 0.15 per cent carbon steel.

(d) The C. E. Davies Method

In this simple method,[*] the mean rolling pressure p_s is calculated for a range of reductions by dividing the total rolling force by the area of contact between the strip and one roll assuming no roll deformation. The values of p_s are then plotted against the mean percentage total reduction for various values of h_1/R as shown in Figure 9-48, where h_1 represents the initial thickness of the strip. (The mean percentage total reduction is defined as the mean of the total reduction from the annealed state at the beginning and end of a pass under consideration.) This provides a chart for the particular metal rolled but in this form it relates only to first passes. To a first approximation, the chart may be scaled proportionally to allow for materials of different yield stresses, but because friction may also be changed, it is preferable to establish an individual chart for each metal.

This method is not too accurate but it does not require a knowledge of the yield stress or the coefficient of friction. It is therefore principally of use where the only equipment available is the rolling mill itself.

9-17 A Simplified Model Assuming Uniformity of Pressure Along the Arc of Contact

In a paper published in 1965,[**] the specific rolling force was considered to be simply the product of the average compressive yield strength σ_c and an "effective" (not the actual) length L_e of the arc of contact, namely

$$f = \sigma_c L_e \qquad (9\text{-}246)$$

It was further assumed that the length of the arc of contact was equal the sum of three contributing parts, the main one ℓ_p being associated with the actual plastic reduction of the strip,

---

[*] L. R. Underwood, "Rolling Mills, Methods of Roll Load and Power Calculation", Metal Ind., Feb. 27, 1948, pp. 166-169; March 5, 1948, pp. 187-190 and March 19, 1948, pp. 231-234.

[**] W. L. Roberts, "A Simplified Cold Rolling Model", Iron and Steel Engineer Year Book, 1965, pp. 925-937.

and the others ℓ_E and ℓ_F being associated with the flattening of the work rolls by the normal rolling force and the equivalent flattening due to frictional effects, respectively. From geometrical considerations it can be shown that ℓ_p (which would be the total length of the profile of contact if the rolls were perfectly rigid) is given by

$$\ell_p = \sqrt{\frac{Dtr}{2}} \qquad (9\text{-}247)$$

where D is the diameter of the work rolls and tr is the draft. (See Figure 9-49.)

Assuming rolling to be possible in the absence of friction in the roll bite, the length of the arc of contact was considered to be increased in actual length by $2.16\sqrt{\frac{fD}{E}}$ where E is the elastic modulus of the rolls and where Poisson's ratio for both the strip and roll materials is assumed to be 0.3 § (see also Section 9-5). Since half this addition is assumed to exist at each end of the profile of contact and over each portion, the rolling pressure is assumed to vary linearly from zero to a value corresponding to σ_c (see Figure 9-50), the effective length ℓ_E of these additions is given by

$$\ell_E = 1.08\sqrt{\frac{fD}{E}} \qquad (9\text{-}248)$$

Friction existing between the strip and each roll was assumed to produce similar extensions at each end of the arc of contact, that at the entry being equal to $\frac{\alpha^2 \mu Df}{Et}$ and that at the exit being equal to $\frac{\alpha^2 \mu Df}{Et(1-r)}$, where μ is the effective coefficient of friction existing between the strip and roll surfaces, t is the initial strip thickness and r is the reduction given to the strip. Since the pressure over these zones is not constant but is assumed to vary linearly as shown in Figure 9-50, the effective length of these zones ℓ_F is given by

$$\ell_F = \frac{\alpha^2 \mu Df(2-r)}{2Et(1-r)} \qquad (9\text{-}249)$$

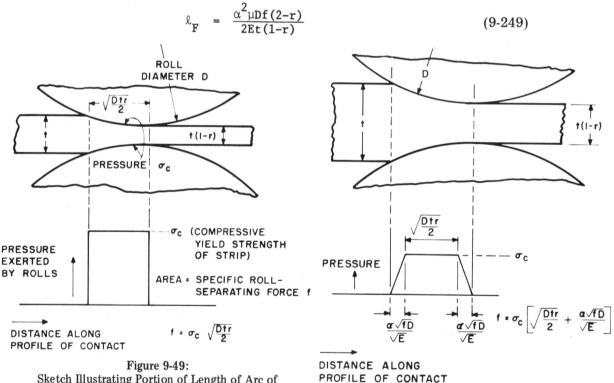

Figure 9-49:
Sketch Illustrating Portion of Length of Arc of Contact Associated with Rigid Work Rolls.

Figure 9-50:
Pressure Distribution in the Roll Bite for Elastic Rolls Under Frictionless Conditions.

---

§ R. J. Roark, "Formulas for Stress and Strain", McGraw-Hill Book Company, Inc., New York, 3rd Edition, 1954.

MATH MODELS RELATING TO ROLLING FORCE

The total effective length L_e of the arc of contact is therefore

$$L_e = \sqrt{\frac{Dtr}{2}} + \alpha\sqrt{\frac{fD}{E}} + \frac{\alpha^2 \mu Df(2-r)}{2Et(1-r)} \qquad (9\text{-}250)$$

and the specific roll-separating force f is

$$f = \sigma_c \left[\sqrt{\frac{Dtr}{2}} + \alpha\sqrt{\frac{fD}{E}} + \frac{\alpha^2 \mu Df(2-r)}{2Et(1-r)} \right] \qquad (9\text{-}251)$$

where

$$\sigma_c = 1.155\sigma_{y(r/2)} - \left(\frac{\sigma_1 + \sigma_2(1-r)}{(2-r)}\right) \qquad (9\text{-}252)$$

Equation 9-252 reflects the decrease in the constrained yield strength of the strip by an average tensile stress $(\sigma_1 + \sigma_2(1-r))/(2-r)$ the derivation of which is illustrated by Figure 9-52.

In equation 9-252, $\sigma_{y(r/2)}$ is the yield strength of the strip (as conventionally measured at low strain rates) corresponding to half the reduction (r) and σ_1 and σ_2 are, respectively, the entry and exit tensile stresses in the strip (see Figure 9-51).

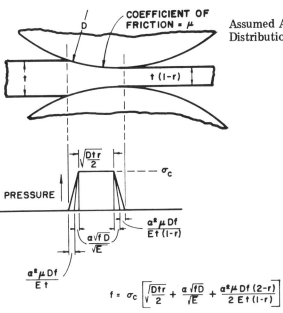

Figure 9-51:
Assumed Arc of Contact and Pressure Distribution for Simplified Model.

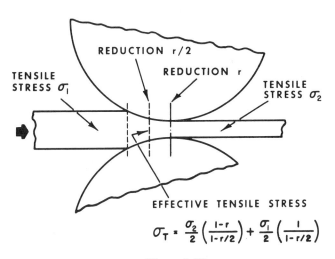

Figure 9-52:
Sketch Illustrating Derivation of Expression for Average Tensile Stress.

It is to be noted that f appears on both sides of equation 9-251, which is really a quadratic equation in terms of $\sqrt{f}$. Thus it may be rearranged as

$$f\left(\frac{1}{\sigma_c} - \frac{\alpha^2 \mu D(2-r)}{2Et(1-r)}\right) - \alpha\sqrt{\frac{D}{E}}\sqrt{f} - \sqrt{\frac{Dtr}{2}} = 0 \qquad (9\text{-}253)$$

and

$$f = \left[\frac{\alpha\sqrt{\frac{D}{E}} + \sqrt{\frac{\alpha^2 D}{E} + \sqrt{8Dtr}\left(\frac{1}{\sigma_c} - \frac{\alpha^2 \mu D(2-r)}{2Et(1-r)}\right)}}{2\left(\frac{1}{\sigma_c} - \frac{\alpha^2 \mu D(2-r)}{2Et(1-r)}\right)}\right]^2 \qquad (9\text{-}254)$$

In laboratory studies of cold rolling, this model was found to give predictions of rolling force as accurate as some of the more complex models considered and, consequently, in 1966, it was utilized in programming a computer used to control a 6-stand tandem cold mill. Moreover, it is of interest in that it provides expressions for (1) the effective diameter of the deformed work roll, (2) the minimum reducible strip thickness, and (3) the minimum attainable thickness in the rolled strip.

With respect to the deformed roll diameter, this may be obtained by equating the portions of the arc of contact corresponding to ℓ_p and ℓ_E with that provided by a rigid roll with a diameter equal to the deformed roll diameter, D'. Thus

$$\sqrt{\frac{D'tr}{2}} = \sqrt{\frac{Dtr}{2}} + 1.08\sqrt{\frac{fD}{E}}$$

$$= \sqrt{\frac{Dtr}{2}}\left\{1 + 1.08\sqrt{\frac{2f}{Etr}}\right\} \qquad (9\text{-}255)$$

Squaring each side of the equation yields

$$D' = D\left\{1 + 3.05\sqrt{\frac{f}{Etr}} + 2.33\frac{f}{Etr}\right\} \qquad (9\text{-}256)$$

This equation may be regarded as an alternate to Hitchcock's relationship discussed in Section 9-6. However, experience has indicated that equation 9-13 has greater validity.

In deriving an expression for the minimum reducible strip thickness (t_{mrt}), it is first convenient to rearrange equation (9-251) in the form

$$r = \frac{2}{Dt}\left\{f\left[\frac{1}{\sigma_c} - \frac{1.167\mu D(2-r)}{2Et(1-r)}\right] - 1.08\sqrt{\frac{fD}{E}}\right\}^2 \qquad (9\text{-}257)$$

By examination of this equation, it is seen that for r to be close to zero when f is very large,

$$\frac{1}{\sigma_c} - \frac{1.167\mu D}{Et_{mrt}} \simeq 0 \qquad (9\text{-}258)$$

rearrangement of which yields

$$t_{mrt} = \frac{1.167\mu D\sigma_c}{E} \qquad (9\text{-}259)$$

This expression is in general agreement with Stone's equation [Y] but the value of the constant of proportionality is approximately one third that developed by Stone (see Section 9-23).

It is also possible from equation (9-257) to establish an expression for the minimum attainable gage (t_{mat}). If r approaches unity when f is very large, and $t(1-r)$ is regarded as corresponding to t_{mat}, then

$$\frac{1}{\sigma_c} - \frac{1.167\mu D}{2Et_{mat}} \simeq 0 \qquad (9\text{-}260)$$

or

$$t_{mat} \simeq \frac{0.584\mu D\sigma_c}{E} \qquad (9\text{-}261)$$

or about half the value of t_{mrt}.

[Y] M. D. Stone, "Rolling of Thin Strip, Part I", Iron and Steel Engineer's Year Book, 1953, pp. 115-127.

MATH MODELS RELATING TO ROLLING FORCE

Figure 9-53 illustrates how the model relates the minimum rolled gage to the specific rolling force. In this case, the work roll diameter is 21 inches, the workpiece is annealed, low-carbon steel strip and the coefficient of friction assumed to be 0.04.

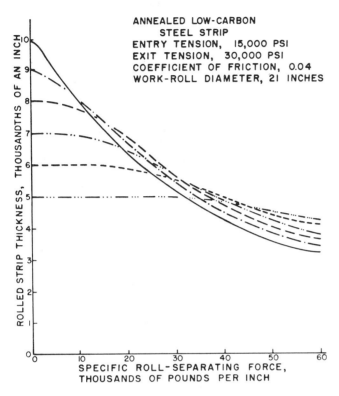

Figure 9-53: Family of Curves Relating the Rolled Thickness to Specific Roll Separating Force for Various Incoming Strip Thicknesses.

9-18 A Simplified, Friction-Hill Model

Subsequent to the publication of the simplified model described in the preceding section, it was established that

(1) strain rate effects in cold rolling were not insignificant as had been hithertofore supposed;▼ (see Section 9-4)

(2) the deformed roll diameter as postulated by Hitchcock underestimated the length of the arc of contact in normal rolling situations♦ (see Section 9-7), and

(3) that work-hardening effects could be disregarded during a single rolling pass.▼⊗

On the basis of these findings, a new simplified model of the rolling process was developed assuming a "friction hill" in the roll bite.● Although intended primarily for the computation of coefficients of friction from rolling data, the model is, of course, capable of the prediction of the rolling force if the frictional characteristics of the lubricant are known. The derivation of the mathematical model is described below.

▼ R. J. Bentz and W. L. Roberts, "Speed Effects in the Second Cold Reduction of Steel Strip", Mechanical Working & Steel Processing V, Proceedings of the Ninth Mechanical Working & Steel Processing Conference, AIME, Gordon & Breach, New York, New York, pp. 193-222.

♦ R. A. Schultz and D. Kobasa, "Experimental Determination of the Length of the Arc of Contact in Cold Rolling", Iron and Steel Engineer Year Book, 1968, pp. 283-288.

⊗ W. L. Roberts, "Thermal Considerations in Tandem Cold Rolling Operations", Iron and Steel Engineer Year Book, 1968, pp. 362-370.

● W. L. Roberts, "Computing the Coefficient of Friction in the Roll Bite from Mill Data", Blast Furnace and Steel Plant, 55, (1967), pp. 499-508.

Referring to Figure 9-54, an element of length dx on the arc of contact is considered to be at a distance x from the end of the region of plastic deformation in proximity to the exit plane. For simplicity, it is assumed that, between the neutral point (where the roll and strip surface speeds are equal) and the exit plane, the thickness of the strip is essentially constant and equal to the rolled gage of the strip.

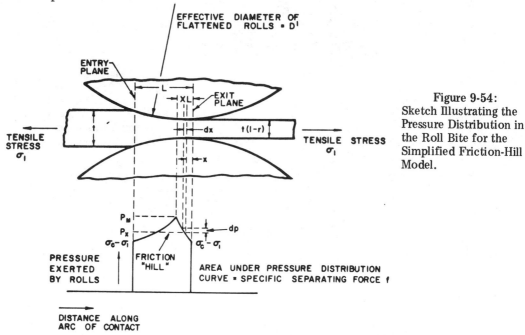

Figure 9-54: Sketch Illustrating the Pressure Distribution in the Roll Bite for the Simplified Friction-Hill Model.

Now the change in pressure dp across this element of length dx may be equated to the frictional shearing force developed on the element divided by half the rolled thickness of the strip. Thus

$$dp = \frac{2\mu p dx}{t(1-r)} \quad (9\text{-}262)$$

where p is the normal pressure on the strip, μ is the coefficient of friction along the arc of contact and $t(1-r)$ is the exit thickness of the strip. It is further assumed that the total length of the arc of contact associated with the plastic deformation is L, that the neutral point occurs at a distance XL from the exit plane, that no work hardening occurs during the deformation process and that the Von Mises' yielding criterion is valid. Then the integration of equation 9-262 gives

$$\int_{\sigma_c - \sigma_1}^{p_x} \frac{dp}{p} = \frac{2\mu}{t(1-r)} \int_0^x dx \quad (9\text{-}263)$$

where p_x is the pressure on the element dx at a distance x from the exit plane. From equation 9-263, it follows that

$$p_x = (\sigma_c - \sigma_1) e^{\frac{2\mu x}{t(1-r)}} \quad (9\text{-}264)$$

From this expression, the maximum pressure p_m at the peak of the friction hill is seen to be given by the relationship

$$p_m = (\sigma_c - \sigma_1) e^{\frac{2\mu XL}{t(1-r)}} \quad (9\text{-}265)$$

MATH MODELS RELATING TO ROLLING FORCE

Now the force exerted by the roll on the strip between the neutral point and the end of the plastic zone is given by

$$\int_0^{xL} P_x \, dx = \int_0^{xL} (\sigma_c - \sigma_1) e^{\frac{2\mu x}{t(1-r)}} dx \qquad (9\text{-}266)$$

$$= \frac{(\sigma_c - \sigma_1) \, t(1-r)}{2\mu} \left[e^{\frac{2\mu XL}{t(1-r)}} - 1 \right]$$

If it is now assumed that there is a geometrical similarity with respect to the pressure distribution on each side of the neutral point, then the specific rolling force f corresponding to the zone of plastic deformation is given by

$$f = \frac{(\sigma_c - \sigma_1) \, t(1-r)}{2\mu X} \left[e^{\frac{2\mu XL}{t(1-r)}} - 1 \right] \qquad (9\text{-}267)$$

Inasmuch as the elastic deformation is only about 0.1 per cent, it is believed that, for reductions in excess of 10 per cent, correction for the elastic entry and exit zones is not necessary.

In considering the horizontal forces acting on the rolls and the strip, it becomes necessary to consider the geometry of the roll bite. For simplicity, it is assumed that a linear change of strip thickness occurs in the roll bite. Under these conditions, the total force exerted on the strip to resist its entry into the roll bite (a "pinching-out" force) will be equal to the specific rolling force multiplied by the ratio of the draft to the arc length. There are, however, two other tensile forces to consider that affect the entry of the strip into the bite, namely the entry tensile force $t\sigma_1$ per unit width and the exit tensile force $t(1-r)\sigma_1$, per unit width, (it being assumed that equal entry and exit tensile stresses exist in the strip). The net tensile force per unit width is therefore approximately $tr\sigma_1$. If the "pinching out" and the net tensile force are equated to the frictional forces at the strip surfaces supplied by the total specific torque T, then

$$\frac{2T}{D} = \frac{(\sigma_c - \sigma_1) \, t(1-r)}{2\mu X} \left[\exp(2\mu XL/t(1-r)) - 1 \right] \frac{tr}{L} + tr\sigma_1$$

$$= tr \left\{ \frac{(\sigma_c - \sigma_1) \, t(1-r)}{2\mu XL} \left[\exp(2\mu XL/t(1-r)) - 1 \right] + \sigma_1 \right\} \qquad (9\text{-}268)$$

Because of the geometrical similarity of each side of the friction hill, the specific rolling force may be related to the specific total torque by the equation

$$\frac{T}{D} = \mu f (1-2X) \qquad (9\text{-}269)$$

Recapitulating, then, the five basic equations of the mathematical model are

$$D' = D \left(1 + 3.05 \sqrt{\frac{f}{Etr}} + 2.33 \frac{f}{Etr} \right) \qquad (9\text{-}270)$$

$$L = \sqrt{\frac{D'tr}{2}} \qquad (9\text{-}271)$$

$$f = \frac{(\sigma_c - \sigma_1) \, t(1-r)}{2\mu} \left[\exp(2\mu XL/t(1-r)) - 1 \right] \qquad (9\text{-}272)$$

$$\frac{2T}{D} = tr\left\{\frac{(\sigma_c - \sigma_1)\,t(1-r)}{2\mu XL}\left[\exp\left(2\mu XL/t(1-r)\right) - 1\right] + \sigma_1\right\} \quad (9\text{-}273)$$

$$\frac{T}{D} = \mu f(1-2X) \quad (9\text{-}274)$$

From the above equations, it is possible to show that

$$\frac{T}{D} = \frac{tr}{2}\left(\frac{f}{\sqrt{\frac{D'tr}{2}}} + \sigma_1\right) \quad (9\text{-}275)$$

Substituting this expression for T/D in equation 9-274,

$$X = \frac{1}{2}\left(1 - \frac{tr}{2\mu f}\left(f\sqrt{\frac{2}{D'tr}} + \sigma_1\right)\right) \quad (9\text{-}276)$$

Now if the exponent of equation 9-272 is small

$$f = (\sigma_c - \sigma_1)\,L\left[1 + \frac{\mu XL}{t(1-r)}\right]$$

$$= (\sigma_c - \sigma_1)\sqrt{\frac{D'tr}{2}}\left[1 + \frac{X\mu\sqrt{\frac{D'tr}{2}}}{t(1-r)}\right] \quad (9\text{-}277)$$

From equations 9-276 and 9-277,

$$f = (\sigma_c - \sigma_1)\sqrt{\frac{D'tr}{2}}\left[1 + \frac{1}{t(1-r)}\left\{\sqrt{\frac{D'tr}{2}}\left(\frac{\mu}{2} - \frac{tr\sigma_1}{4f}\right) - \frac{tr}{4}\right\}\right] \quad (9\text{-}278)$$

If the tensile stress σ_1 is kept small so that $\frac{tr\sigma_1}{4f}$ is considerably smaller than $\mu/2$, then the term $\frac{tr\sigma_1}{4f}$ may be neglected, and equation 9-278 becomes

$$f = \frac{(\sigma_c - \sigma_1)}{(1-r)}\sqrt{\frac{D'tr}{2}}\left[1 - \frac{5r}{4} + \frac{\mu}{2}\sqrt{\frac{D'r}{2t}}\right] \quad (9\text{-}279)$$

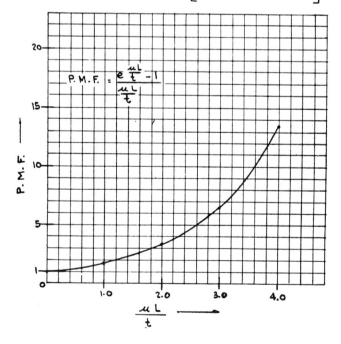

Figure 9-55: Pressure Multiplication Factor (PMF) is Shown as a Function of Parameter $\frac{\mu L}{t}$.

MATH MODELS RELATING TO ROLLING FORCE

To compute the specific rolling force, the simultaneous equations 9-270 and 9-279 must be solved. This can be accomplished graphically or by the use of a digital computer. A solution is predicated on an accurate knowledge of the coefficient of friction μ. However, as stated in Section 6-31, the coefficient of friction may be calculated for any given rolling situation provided the two frictional characteristics K_1 and K_2 are known.

It is interesting to note that, in the early 1950's, Stone◇ had developed a similar equation for specific rolling force. He considered the strip to be essentially flat, rather than wedge-shaped, in the roll bite and that the neutral plane occurred at the center of the arc of contact. Accordingly, Stone's formula could be derived from equation 9-265 by substituting an average thickness t for t (1 − r) and assigning a value of 0.5 to X. Thus

$$f = \frac{(\sigma_c - \sigma_1) t}{\mu} \left[\exp(\mu L/t) - 1 \right] \qquad (9\text{-}280)$$

If the mean pressure exerted by the rolls on the strip is $\bar{p}$, then

$$\bar{p} = \frac{f}{L} = (\sigma_c - \sigma_1) \left[\left(\exp(\mu L/t) - 1 \right) t/\mu L \right] \qquad (9\text{-}281)$$

The term inside the square parentheses was designated by Stone as the Pressure Multiplication Factor, or PMF, and this function of the term $\frac{\mu L}{t}$ is shown graphically in Figure 9-55. It should be noted, however, that Stone utilized Hitchcock's equation (equation 9-44) for establishing the length of the arc of contact and he assigned to the average thickness t, the arithmetic mean of the entry and exit strip thicknesses.

9-19 An Approximate Theory of Temper Rolling

Examination of the data obtained during the temper rolling of annealed, low-carbon steel strip on laboratory mills led to the belief that a uniquely different mathematical model was required to describe the temper-rolling process. This conclusion was reached from a study of the rolling force, spindle torque, and relative slip data which indicated that (a) the arc of contact at each roll-strip interface was more in the nature of a planar, rather than a cylindrical, surface; (b) the effective coefficient of friction in the roll bite was about an order of magnitude larger than in conventional cold rolling; (c) the neutral point was, therefore, in a relative sense, much closer to the center of the arc of contact; and (d) the efficiency of the temper-rolling process was appreciably reduced.

In an attempt to satisfactorily describe the temper rolling process in the light of the foregoing observations, a simplified theoretical model of the process was developed for single-pass processing, and experiments were conducted on laboratory type mills to establish its validity.[1] This model and its verification by further laboratory tests conducted on annealed black plate and galvanized strip are discussed in this section. It should be noted, however, that the range of reductions obtained in these laboratory experiments (up to approximately 15 per cent) was considerably larger than that pertaining to commercial temper rolling operations (up to approximately 2 per cent).

On the assumption that both the work rolls are driven so that a plane of symmetry exists coincident with the plane at the center of the strip, the four parameters of major importance involved in the model are:

  a. The average strain rate experienced by the strip as it undergoes deformation in the roll bite.

◇M. D. Stone, "Rolling of Thin Strip", Iron and Steel Engineer Year Book, 1953, pp. 115-128.
[1] W. L. Roberts, "An Approximate Theory of Temper Rolling", Iron and Steel Engineer Year Book, 1972, pp. 530-542.

b. The compressive stress normal to the plane of the strip required to plastically deform the steel at this strain rate.

c. The "average" strip tension in the roll bite.

d. The coefficient of friction existing at the roll bite interfaces.

The rolling model as developed below for a single-stand temper mill provides mathematical expressions for the following:

a. The length of the arc of contact.

b. The pressure distribution along each arc of contact and the specific rolling force.

c. The work dissipated in deforming the strip.

d. The frictional losses incurred at the interfaces of the work rolls and the strip.

e. The efficiency of the rolling process.

f. The spindle torques.

g. The maximum reduction attainable in temper rolling.

h. The relative slip between the emerging strip and roll surfaces.

Expressions for the length of the arc of contact and the specific rolling force (items (a) and (b) above) are discussed below, the remaining expressions being presented in Chapter 10.

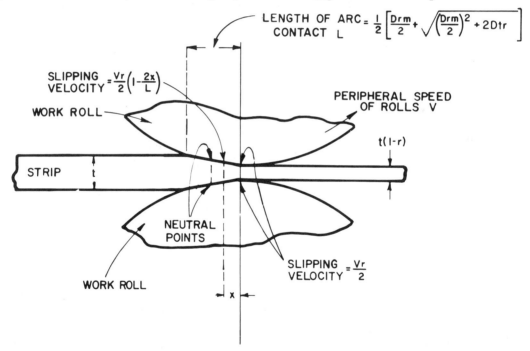

Figure 9-56: Sketch Illustrating Roll Bite Geometry Under Conditions of High Friction.

With respect to the length of the arc of contact (L) as illustrated in Figure 9-56, it has been found empirically that, for conditions of high friction, the most convenient expression for computing its value is

$$L = \sqrt{\frac{Dtr}{2}\left(1 + \frac{\mu L}{t}\right)} \qquad (9\text{-}282)$$

where D is the work roll diameter
t is the incoming strip thickness
r is the reduction (expressed as a decimal fraction) and
μ is the coefficient of friction.

This equation may be rearranged to give

$$L = \frac{1}{2}\left[\frac{Dr\mu}{2} + \sqrt{\left(\frac{Dr\mu}{2}\right)^2 + 2Dtr}\right] \quad (9\text{-}283)$$

The average strain rate ($\dot{e}$) associated with the temper rolling operation may, for the assumed roll bite geometry, be approximated by

$$\dot{e} \simeq \frac{2V}{D\mu} \quad (9\text{-}284)$$

where V is the rolling speed. Under these circumstances, it is to be noted that the strain rate does not depend on the reduction or the draft.

The minimum rolling pressure (σ_p) required to deform the strip is given by the expression

$$\sigma_p = 1.155\left[\sigma_t + a\,\log_{10} 1000\,\dot{e}\right] - \sigma_A \quad (9\text{-}285)$$

where σ_t is the yield strength as measured in tension at very low strain rates, a is the strain rate factor (the increase in tensile yield strength per tenfold change in strain rate) and σ_A is the average strip tension in the roll bite. (Equation 9-285 is discussed more fully in Section 8-29.)

The value of the coefficient of friction pertaining to the sliding of unlubricated, clean steel surfaces on each other appears to be in excess of 0.15 and can, under certain circumstances, be close to unity. Small changes in the condition of the surfaces may produce considerable variation in the value of the coefficient, and it is generally accepted that, as sliding continues, the coefficient decreases because of polishing of the surfaces. The coefficient also appears to be slightly speed sensitive, decreasing as the sliding speed increases. In studies of the coefficient of friction for cast iron brake shoes on steel-tired car wheels, it has been found that the coefficient of friction at very low sliding speeds was close to 0.3.[◄] In the temper rolling process, in spite of high rolling speeds, the average sliding speed of the roll surface relative to the strip surface is approximately rV/4 and is usually quite small. For example, in commercial, high-speed temper rolling, where the rolling speed may be as high as 5,000 feet per minute, the average sliding speed will be about 12.5 feet per minute per per cent reduction. Thus, assuming similar frictional conditions to occur in the bite of the mill, a coefficient of friction of about 0.3 would appear to be applicable to temper rolling. However, it is to be expected that roll and strip surface conditions and the rolling speed might modify the value to some extent.

The pressure distribution along the arc of contact would be expected to exhibit a typical friction hill and the specific rolling force f may be represented by the equation

$$f = \frac{\sigma_p\, t(1-r)}{\mu}\left[\exp\left(\mu L/t(1-r)\right) - 1\right] \quad (9\text{-}286)$$

as discussed in Section 9-18.

Thus equations 9-282 to 9-286 represent the temper rolling model as far as the specific rolling force is concerned. With respect to its verification, Figure 9-57 shows curves representing the theoretical variation of specific rolling force with reduction, together with the plots of experimental data for black plate rolled on a mill with 3.25-inch diameter work rolls. The experimental data agree reasonably well with the theoretical curve corresponding to a coefficient of friction of 0.25, but the data, contrary to expectation, indicated that the coefficient of friction tended to increase with increasing mill speed.

Data for slightly thinner black plate rolled on a mill with 6.5-inch diameter work rolls are shown in Figure 9-58. For a mill speed of 100 fpm, a value of about 0.3 for the coefficient

---

◄L. S. Marks, "Mechanical Engineers' Handbook", Fourth Edition, McGraw-Hill Book Company, Inc., New York, N. Y., 1941, p. 235.

appeared to be the most appropriate and, as in the case of the other black plate, the value for the coefficient tended to increase with increasing mill speed.

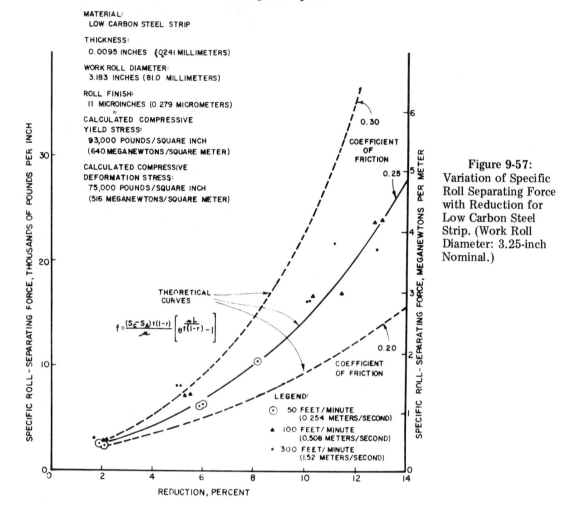

Figure 9-57: Variation of Specific Roll Separating Force with Reduction for Low Carbon Steel Strip. (Work Roll Diameter: 3.25-inch Nominal.)

Data pertaining to the rolling of galvanized strip is similarly illustrated in Figure 9-59. This indicates that the coefficient of friction lies in the range 0.20 to 0.25, the lower values being attributable presumably to the presence of zinc on the strip surfaces and the very smooth roll finishes used in the rolling experiment.

9-20 A Rolling Model Based on Plastohydrodynamic Lubrication

With the exception of Nadai's solution of Von Karman's equation assuming a hypothetical condition of hydrodynamic lubrication (as discussed in Section 9-11), all the mathematical models discussed in the preceding sections of this chapter have postulated the existence of a constant coefficient of friction or a constant shear stress (either constant or equal to half the yield strength of the workpiece) along the arc of contact. However, in 1966, Cheng[∞] presented a theory of rolling of thin sheet considering a full hydrodynamic film between the surfaces of the rolls and the workpiece. This theory provided numerical solutions for the pressure exerted by the rolls, the lubricant film thickness, lubricant temperature and surface friction along the arc of contact. Although the model recognized four regions in the roll bite, as illustrated in

---
[∞]H. S. Cheng, "Plastohydrodynamic Lubrication", Portion of Book, "Friction and Lubrication in Metal Processing", Edited by F. Ling, et al, ASME, New York, 1966.

MATH MODELS RELATING TO ROLLING FORCE

Figure 9-60, the regions of elastic deformation of the workpiece were disregarded and the work rolls considered to be perfectly rigid.

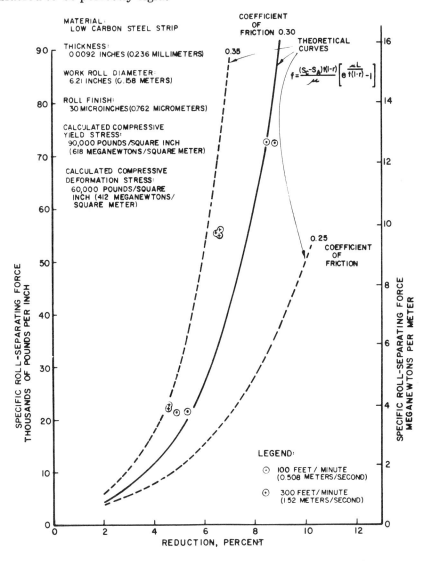

Figure 9-58: Variation of Specific Roll Separating Force with Reduction for Low Carbon Steel Strip. (Work Roll Diameter: 6.5-inch Nominal.)

The basic equations for the model were developed as discussed below. Referring to Figure 9-61, the equation describing the profile of velocity u across the film is

$$u = \frac{1}{2\eta}\frac{dp}{dx'}(y^2 - yh) + \left(\frac{u_2 - u_1}{h}\right)y + u_1 \qquad (9\text{-}287)$$

where η is the viscosity of the lubricant
  p is the pressure normal to the roll surface
  x' is the coordinate in the direction of lubricant flow
  y is the coordinate across the film
  h is the film thickness
  u_1 is the surface velocity of the strip, and
  u_2 is the surface velocity of the rolls.

Neglecting the Poisueille flow, the surface shear stress T_o is given by

$$T_o = \eta\left(\frac{u_2 - u_1}{h}\right) \qquad (9\text{-}288)$$

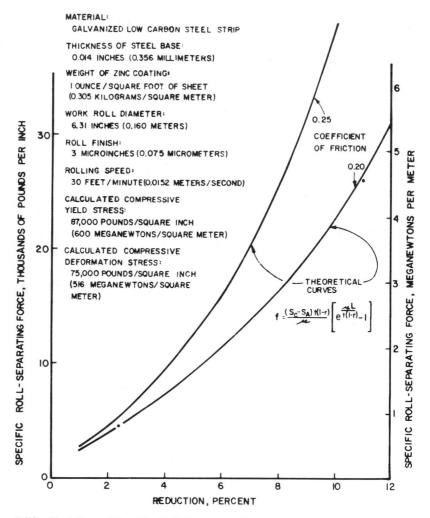

Figure 9-59: Variation of Specific Roll Separating Force with Reduction for Galvanized Strip.

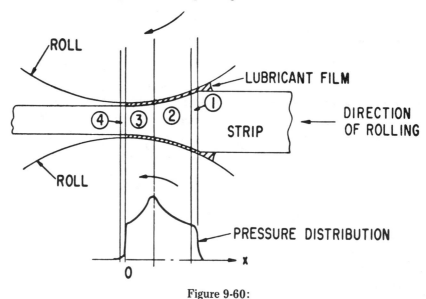

Figure 9-60:
Sketch Illustrating Roll Bite as Considered in the Plastohydrodynamic Lubrication Model (Cheng).
Region (1) — Elastic Entrance Region (3) — Plastic Exit
Region (2) — Plastic Entrance Region (4) — Elastic Exit

MATH MODELS RELATING TO ROLLING FORCE

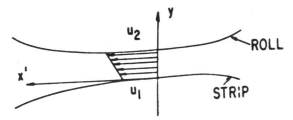

Figure 9-61:
Velocity Profile Across the Lubricant Film (Cheng).

and the total flow rate of the lubricant, q, by the equation

$$q = \left(\frac{u_2 - u_1}{2}\right)h \qquad (9\text{-}289)$$

Referring to Figure 9-62 showing the roll bite and Figure 9-63 depicting an element of the strip, the equilibrium of the element is represented by

$$\frac{d(\sigma_x t)}{dx} - 2\tau_o + \frac{2p}{\cos\phi}\sin\phi = 0 \qquad (9\text{-}290)$$

where σ_x is the stress in the strip in the direction of rolling,
t is the thickness of the element of strip, and
ϕ is the angular coordinate of the element.

Cheng assumed that σ_x and p were related by the yield criterion

$$\sigma_x + p = \sigma_o' = \frac{2}{\sqrt{3}}\sigma_o \qquad (9\text{-}291)$$

where σ_o' is the effective yield stress and σ_o is the yield stress as measured in a conventional tensile test.

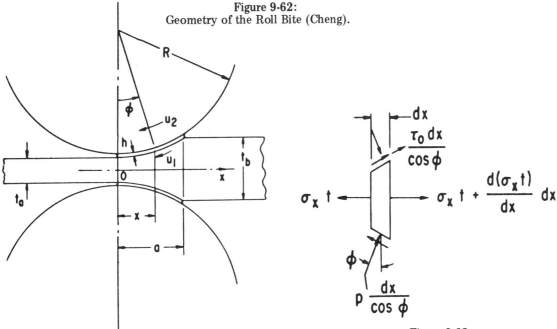

Figure 9-62:
Geometry of the Roll Bite (Cheng).

Figure 9-63:
Forces Acting on an Element of the Strip (Cheng).

From equations 9-290 and 9-291

$$\frac{dp}{dx} = \left(\sigma_o' - \frac{\tau_o}{\tan\phi}\right)\frac{1}{t}\frac{dt}{dx} \qquad (9\text{-}292)$$

Since the viscosity η is dependent upon the pressure p and the temperature T, the shear stress T_o may be expressed by

$$T_o = \eta(p,T) \left[\frac{u_2 - u_1(x)}{h(x)}\right] \qquad (9\text{-}293)$$

The dependence of the viscosity and temperature was assumed to be in accordance with the following equation

$$\eta = \eta_o \exp\left[\alpha_p - \beta(1/T_o - 1/T_m) + \gamma p/T_m\right] \qquad (9\text{-}294)$$

where η_o and T_o are the viscosity and the temperature of the lubricant at the entry of the roll bite, α, β, and γ are the pressure-viscosity, temperature-viscosity and pressure-temperature viscosity exponents of the lubricant, respectively, and T_m is the mean temperature of the lubricant film.

The strip velocity u_1 is governed by the continuity of plastic flow and may be represented by

$$u_1 = u_b t_b / t \qquad (9\text{-}295)$$

where u_b and t_b are the velocity and thickness of the strip at entry into the bite. Similarly, the film thickness, h, may be expressed

$$h = h_o \left(\frac{u_2 + u_b}{u_2 + u_1}\right) \qquad (9\text{-}296)$$

where h_o and u_b are the thickness and speed of the lubricant film at the entry of the bite.

From the two preceding equations

$$\frac{h}{h_o} = \frac{1 + \dfrac{u_b}{u_2}}{1 + \dfrac{u_b}{u_2} \cdot \dfrac{t_b}{t}} \qquad (9\text{-}297)$$

and substitution of the equation in equation 9-293 yields

$$T_o = \frac{\eta_o u_2}{h_o}\left(\frac{\eta}{\eta_o}\right) \frac{1 - \left(\dfrac{t_b u_b}{t u_2}\right)^2}{1 + \dfrac{u_b}{u_2}} \qquad (9\text{-}298)$$

In simplifying the mathematics, the following dimensionless parameters may be used

$$\bar{p} = \frac{p}{\sigma_o'} \qquad \alpha' = \alpha \sigma_o' \qquad (9\text{-}299)$$

$$B = \frac{\eta_o u_2}{\sigma_o' h_o} \qquad \beta' = \frac{\beta}{T_o} \qquad (9\text{-}300)$$

$$\bar{x} = \frac{x}{a} \qquad \gamma' = \frac{\gamma \sigma_o'}{T_o} \qquad (9\text{-}301)$$

$$\bar{t} = \frac{t}{t_b} \qquad \theta = \frac{T}{T_o} \qquad (9\text{-}302)$$

MATH MODELS RELATING TO ROLLING FORCE

where a is the length of the arc of contact. Equation 9-292 may then be written

$$\frac{d\bar{p}}{d\bar{x}} - \left\{ 1 - \frac{B}{\tan\phi}\left(\frac{\eta}{\eta_o}\right) \frac{\left[1 - \left(\frac{u_b}{u_2 \bar{t}}\right)^2\right]}{\left(1 + \frac{u_b}{u_2}\right)} \right\} \frac{1}{\bar{t}} \frac{d\bar{t}}{d\bar{x}} = 0 \qquad (9\text{-}303)$$

$$\text{where } \frac{\eta}{\eta_o} = \exp\left[\alpha'\bar{p} - \beta'\left(1 - \frac{1}{\theta_m}\right) + \frac{\gamma'\bar{p}}{\theta_m}\right] \qquad (9\text{-}304)$$

For a short arc of contact, the thickness of the strip $\bar{t}$ may be closely approximated by the expression

$$\bar{t} = \left(\frac{t_a}{t_b}\right) + \left(\frac{t_b - t_a}{t_b}\right)\bar{x}^2 \qquad (9\text{-}305)$$

where t_a is the final thickness of the strip. Since the reduction r is given by

$$r = \frac{t_b - t_a}{t_b} \qquad (9\text{-}306)$$

$$\bar{t} = \left((1-r) + r\bar{x}^2\right) \qquad (9\text{-}307)$$

and

$$\frac{d\bar{t}}{d\bar{x}} = 2r\bar{x} \qquad (9\text{-}308)$$

Furthermore

$$\tan\phi \simeq \left(\frac{a}{R}\right)\bar{x} = \frac{a}{\sqrt{t_b R}}\sqrt{\frac{t_b}{R}}\bar{x} = \sqrt{\frac{rt_b}{R}}\bar{x} \qquad (9\text{-}309)$$

From equation 9-303, 9-308 and 9-309, the final dimensionless pressure equation may be established, viz.,

$$\frac{d\bar{p}}{d\bar{x}} - \left\{ \sqrt{\frac{rt_b}{R}}\bar{x} - B\left(\frac{\eta}{\eta_o}\right) \frac{\left[1 - \left(\frac{u_b}{U_2}\right)\frac{1}{\bar{t}_2}\right]}{\left(1 - \frac{u_b}{u_2}\right)} \right\} \frac{2\sqrt{r}}{\bar{t}}\sqrt{\frac{R}{t_b}} = 0 \qquad (9\text{-}310)$$

On the basis of certain assumptions with respect to the thermal aspects of the lubricant film,

$$K_f \frac{d^2 T}{dy^2} = -\eta\left(\frac{\delta u}{\delta y}\right)^2 \qquad (9\text{-}311)$$

where K_f is the thermal conductivity of the lubricant. Since

$$\frac{\partial u}{\partial y} \simeq \left(\frac{u_2 - u_1}{h}\right) \qquad (9\text{-}312)$$

and

$$\theta = \frac{T}{T_o} \tag{9-313}$$

$$\frac{d^2\theta}{d\bar{y}^2} = -Q_m\left(\frac{\eta}{\eta_o}\right)\left(1 - \frac{u_1}{u_2}\right)^2 \tag{9-314}$$

where

$$Q_m = \frac{\eta_o u_2^2}{K_f T_o} \tag{9-315}$$

Assuming that all the energy required to deform the strip is converted to thermal energy

$$\theta_1(\bar{x}) = 1 + Q_s \ln\left(\frac{t_b}{t}\right) \tag{9-316}$$

where

$$Q_s = \frac{\sigma_o'}{\rho C_p T} \tag{9-317}$$

In the latter equation, ρ represents the density of the strip and C_p its specific heat.

Integrating equation 9-314 with respect to the proper boundary conditions yields

$$\theta = -Q_m\left(1 - \frac{u_1}{u_2}\right)^2 \frac{\eta}{\eta_o} \frac{1}{2}\left(\bar{y}^2 - \bar{y}\right) - \theta_1(\bar{y} - 1) + \bar{y} \tag{9-318}$$

from which the mean temperature θ_m can be derived as

$$\theta_m = \frac{Q_m}{12}\left(1 - \frac{u_1}{u_2}\right)^2 \exp\left[\alpha'\bar{p} - \beta\left(1 - \frac{1}{\theta_m}\right) + \gamma'\frac{\bar{p}}{\theta_m}\right] + \left(\frac{\theta_1 + 1}{2}\right) \tag{9-319}$$

By setting the initial value of $\bar{p}$ at the entry into the roll bite, and assuming a velocity ratio u_b/u_2 at the same location, the last two equations may be numerically integrated from the entry to the exit of the bite. The dimensionless total load $\bar{W}$ may then be calculated from

$$\bar{W} = \int_0^1 \left(\frac{P}{\sigma_o'}\right) d\bar{x} \tag{9-320}$$

and the dimensionless friction $\bar{F}$ from

$$\bar{F} = \int_0^1 \left(\frac{T_o}{\sigma_o'}\right) d\bar{x} \tag{9-321}$$

In substituting certain values for the input parameters, Cheng established pressure profiles as illustrated in Figures 9-64 and 9-65, these being smooth, continuous curves instead of the typically discontinuous friction hill as indicated in the latter figure. Whereas in the case of the conventional friction hill, the neutral point coincides with the location of the peak pressure, this is not generally the case under conditions of hydrodynamic lubrication since the maximum pressure occurs slightly to the entry side of the neutral point.

MATH MODELS RELATING TO ROLLING FORCE

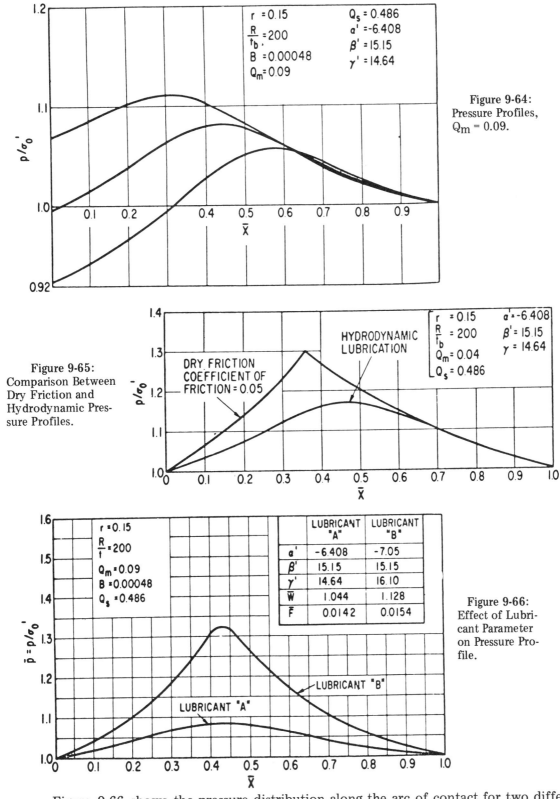

Figure 9-64: Pressure Profiles, $Q_m = 0.09$.

Figure 9-65: Comparison Between Dry Friction and Hydrodynamic Pressure Profiles.

Figure 9-66: Effect of Lubricant Parameter on Pressure Profile.

Figure 9-66 shows the pressure distribution along the arc of contact for two different lubricants, one (A) having a pressure viscosity exponent about 10 per cent less than the other (B). With increasing reduction, the pressures along the arc of contact are reduced with the neutral point shifting towards the exit plane as shown in Figure 9-67. Similarly, as the ratio of roll radius to initial strip thickness is decreased, the pressure along the arc of contact is also decreased, as seen in Figure 9-68.

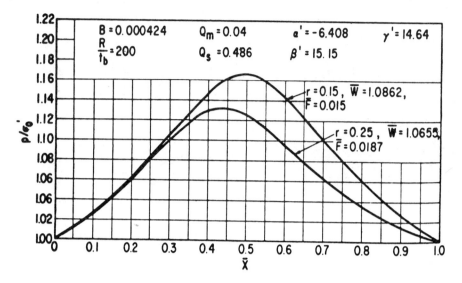

Figure 9-67: Effect of Reduction Ratio on Pressure Profile.

Figure 9-68: Effect of $\frac{R}{t_b}$ on Pressure Profile.

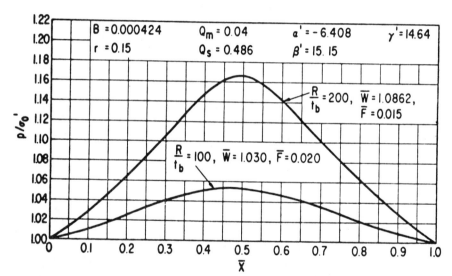

Unfortunately, this model predicts a neutral point much closer to the center of the arc of contact than is found in practice, since under rolling conditions, which are believed to most closely resemble hydrodynamic lubrication, the forward slip of the strip is close to zero. Moreover, the prediction of a decrease in slip with increasing reduction is also contrary to rolling experience.

9-21 The Workpiece Considered As a Viscous Mass in Motion

In 1957, Kneschke[‡,•] published a theory applicable to both hot and cold rolling in which the workpiece was regarded as a viscous mass in motion through the roll bite. The workpiece was assumed to be characterized by a static yield stress K_0 and a dynamic viscosity η. On the further assumption that the speeds of the rolls and strip surfaces are always identical, conclusions were reached which were claimed to be in general agreement with rolling experience.

---

‡ A. Kneschke: Frieberger Forschungsh. B. Met., 1957, 16, pp. 5-34.
• K. H. Weber, "Hydrodynamic Theory of Rolling", Journal of the Iron and Steel Institute, January, 1965, pp. 27-35.

MATH MODELS RELATING TO ROLLING FORCE

According to Kneschke's theory, the passage of the material through the roll bite is represented by the Navier-Stokes equations. Ignoring the inertial and gravitational forces, the metal flow is described by the equation

$$\eta \frac{\partial^2 \nu}{\partial y^2} = \frac{ds}{dx} \qquad (9\text{-}322)$$

where η is the dynamic viscosity of the rolled material
- ν is the linear velocity of a point within the material in the roll gap
- y is the distance on the ordinate of the coordinate system as measured from the center line of the rolled strip thickness in a vertical direction.
- s is the stress normal to the surface of the strip, and
- x is the distance on the abscissa of the coordinate system as measured from the exit plane of the roll bite towards the entry plane.

The corresponding system of forces acting on a work roll is illustrated in Figure 9-69.

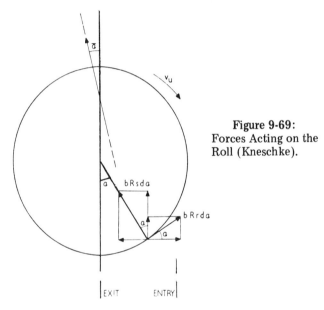

Figure 9-69: Forces Acting on the Roll (Kneschke).

Assuming that the outermost layers of the strip stick to the roll surface, Kneschke showed that the roll force and torque are dependent on the rolling speed and that the velocity distribution is parabolic at every section of the roll gap except at the neutral point where it is linear.

The basic solution for the specific rolling force f is

$$f = \bar{k}_o L + \int_o^\alpha pR d\alpha \qquad (9\text{-}323)$$

where $\bar{k}_o$ is the mean static yield stress (at a speed of deformation tending to zero)
- L is the length of the arc of contact
- p is the vertical stress
- R is the work roll radius, and
- α is the angular coordinate of the arc of contact.

The solution to this equation, in its final form, is

$$f = \bar{k}_o L + 3\eta v_u \left\{ 1 + \frac{R}{t_1} \cdot \frac{2}{(1-r)} \right\} \psi_r \qquad (9\text{-}324)$$

where
- v_u is the peripheral velocity of the work roll
- t_1 is the initial thickness of the strip prior to rolling
- r is the reduction, and
- Ψ_r is a function of the reduction

The values of Ψ_r have been tabulated by Kneschke and are reproduced in Table 9-2.

Table 9-2
Values for Ψ_r (Kneschke)

r	Ψ_r	r	Ψ_r
0	0	0.50	0.05739
0.025	0.0001057	0.55	0.07187
0.050	0.0004294	0.60	0.08856
0.10	0.001771	0.65	0.1077
0.15	0.004110	0.70	0.1294
0.20	0.007540	0.75	0.1541
0.25	0.01216	0.80	0.1819
0.30	0.01809	0.85	0.2132
0.35	0.02543	0.90	0.2484
0.40	0.03433	0.95	0.2881
0.45	0.04494	1.00	0.3333

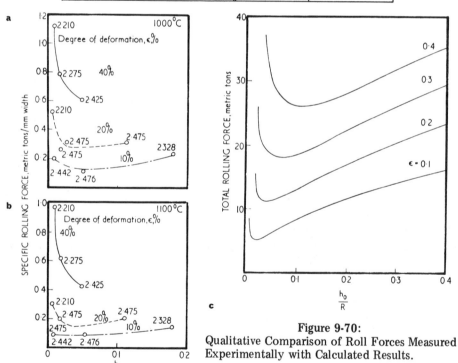

Figure 9-70:
Qualitative Comparison of Roll Forces Measured Experimentally with Calculated Results.

Pressure distributions as calculated on the basis of this model are said to be in general agreement with distributions experimentally established by Siebel and Lueg,[θ] Frolov and Golubev[+] and Korolev.[×]

[θ] E. Siebel and W. Lueg, Mitt. K. W. Inst., 1933, 15, pp. 1-14.
Stahl und Eisen, 1933, 53, pp. 346-352.
[+] E. J. Frolov and T. M. Golubev; Metallurgie, 1937, 7.
[×] A. A. Korolev, "Deformatsiya metallov pri prokatke", 1953, Moscow, Mashgiz.

MATH MODELS RELATING TO ROLLING FORCE

From the analysis of his results, Kneschke established the little known dependence of the rolling force on the t_1/R ratio and on the reduction r. Accordingly, at each particular value of the reduction and with the decrease of the t_1/R ratio, the magnitude of the roll force passes through a marked minimum and then rises again, as seen in Figure 9-70. This means that, for each given roll diameter, there exists an optimum value of the material thickness which, when increased or decreased, will produce an increase in the rolling force.

This curious property of the rolling force had, in the meantime, been experimentally confirmed by Dahl et al.,[1] and was evident in an investigation carried out by Emicke and Lucas.[2] A further proof of this property was also supplied from tests conducted by Wallquist.[3]

According to the hydrodynamic theory of rolling, the roll force (P) consists of a static (P_{STAT}) and a dynamic component (P_{DYM});

$$P = P_{stat.} + P_{dym.} \qquad (9\text{-}325)$$

The existence of these two components was experimentally demonstrated by Weber[4] who measured the force during the rolling of a bar and during a brief stoppage of the mill. The residual force after the rolling pass had been interrupted was believed to correspond to the static force value and the difference between the stationary and operational values of force corresponding to the dynamic component.

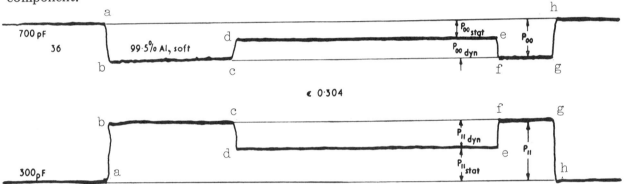

Figure 9-71: Oscillogram of Rolling Force Obtained from an Interrupted Cold Rolling Test (Weber).

Figure 9-71 shows an oscillogram of such an interrupted rolling pass. At the instant of the entry of the rolled bar into the roll gap, the roll force rises from zero to a value corresponding to the given conditions of rolling (see a-b), the force then remaining at this magnitude until the rolling process is interrupted (at point c). When the mill is at a standstill (see d-e), the roll force is reduced by the value of its dynamic component and, while the rolling speed is nil, only the static component remains. On the resumption of rolling the dynamic component is added to the static one (see e-f), and the roll force P reaches its original value (compare h-g with a-b), provided that the rolling conditions have remained unaltered.

The variation of the total roll force and its static component with reduction is seen in Figure 9-72. In this case, plain carbon steel was used as the rolling material. The ratio of the static component to the total roll force, and the ratio of the dynamic component to the total force are representative of the extent to which these two components participate in the building up of the roll force; these ratios, therefore, depend on the rolling conditions.

---

[1] W. Dahl, E. Wildschutz and J. Langer, "Messung von Walzkraft und Drehmoment beim Warmwalzen und Berechnung der Formanderungsfestigkeit", Arch. Eisenh., 32, (1961), pp. 213-219.

[2] O. Emicke and K. H. Lucas, "Das Walzen von Leichtmetallen zu Blechen und Bandern", 1944, Freiborg, Maukisch.

[3] G. Wallquist, "Investigation of the Influence of Different Factors on Roll Pressure, Energy Consumption, Spread and Forward Slip in Hot Rolling", Jernkont. Ann., 1962, 146, pp. 681-716.

[4] K. H. Weber, Freiberger Forschungsh. B. Met., 1963, 91, pp. 1-80.

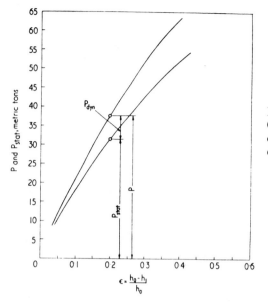

Figure 9-72:
Rolling Force P and its Static Component $P_{stat.}$ as Functions of the Relative Thickness Reduction (Weber).

The resolution of the expression for the roll force into the static and the dynamic component is not a characteristic unique to the hydrodynamic theory of rolling, since roll force equations derived by other investigators are also made up of static and dynamic terms. In particular, is the empirical formula proposed by Geleji [1]

$$f = k_f L \left\{ 1 + C\mu \frac{L}{t_1} \left(v_u \right)^{1/4} \right\} \qquad (9\text{-}326)$$

where k_f is the yield stress of the material taken at the appropriate temperature, C is a coefficient dependent on L/t_i and μ is the coefficient of friction in the roll bite.

9-22 A Shear Plane Theory of Flat Rolling

In 1962, a shear plane theory of the hot rolling of thick stock was published by Green and Wallace.[Δ] This was later extended to hot and cold flat rolling[m] and was found to be in accord with the hot rolling model of Sims [+] and the cold rolling model of Ford [Ω] (as discussed in Section 9-13). The development of this model, with respect to cold rolling, is presented in this Section.

Figure 9-73 shows a theoretical section of the upper half of the roll bite (the bottom half being disregarded because of symmetry about the center of the strip). It is assumed that the portion of the roll in contact with the metal is flat and that the arc of contact ends at the line joining the roll centers. The shear planes are chosen so as to give a simple analysis of the forces and are not necessarily planes on which the metal will yield. The nearest upper bound to the true deforming load is obtained simply by finding the configuration of shear planes which will give the lowest calculated load.

The following assumptions were also made in the model:

---

[1] A. Geleji, "Bildsame Formung der Metalle in Rechnung und Versuch", 1960, Berlin Akademie.

[Δ] J. W. Green and J. F. Wallace, "Estimation of Load and Torque in the Hot Rolling Process", J. Mech. Engng. Sci., 1962, 4, No. 2, pp. 136-142.

[m] J. W. Green, L. M. G. Sparling and J. F. Wallace, "Shear Plane Theories of Hot and Cold Flat Rolling", J. Mech. Engng. Sci., 1964, 6, No. 3, pp. 219-235.

[+] R. B. Sims, "Calculation of Roll Force and Torque in Hot Rolling Mills", Proc. Inst. Mech. Eng., London, 1954, Vol. 168, p. 191.

[Ω] H. Ford, "Researches into the Deformation of Metals by Cold Rolling", Proc. Inst. Mech. Eng., London, 1948, Vol. 159, pp. 115-143.

MATH MODELS RELATING TO ROLLING FORCE

(1) The mean yield stress is constant (equivalent to some "average" value throughout the roll bite,

(2) plane strain conditions exist because of the negligible spread in cold rolling,

(3) Coulomb (sliding) friction occurs along the entire length of the arc of contact and the coefficient is of constant value, and

(4) the strip under consideration is of unit width.

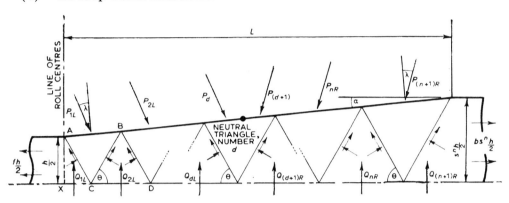

Figure 9-73: Forces Acting on Shear Planes in the Upper Half of a Wedge of Material Undergoing Cold Rolling.

Referring to Figure 9-73, the force vectors on the triangles (represented by the symbol P with the various subscripts) are seen to act at an angle to the normal of the flattened roll surface, the angle λ being dependent on the frictional force acting coincident with the roll surface. The Q forces act normal to the center line of the strip and, like the P forces, are taken to be the resultant of a constant pressure distribution over the planes on which they act. Front and back tensile stresses are taken to be uniformly distributed through the strip thickness. The diagram of forces acting on the various segments ACX, ABC, CBD, etc., are shown in Figures 9-74 and 9-75.

The following subscript notation is used: (1) Forces P_{mL} are those calculated from the exit force polygon shown in Figure 9-74, and they contribute to the combined force polygon where $m \leq d - 1$, d being the number of the neutral triangle [±] counted from the exit of the roll bite; (2) Forces P_{mR} are calculated from the entry force polygon except where $m \leq d + 2$; (3) P_D and $P_{(D+1)}$ are forces acting on the neutral triangle and are found by superposing the exit and entry force polygons (P_D acts on the exit side of the neutral point and $P_{(D+1)}$ on the entry side).

Referring to Figures 9-73 and 9-74, the force Q_{1L} (in units such that one unit equals the mean shear yield stress in plane strain K)[>] acting on triangle ACQ is given by

$$Q_{1L} = \frac{h}{2 \sin^2 \theta} - f \frac{h}{2} \cot \theta \qquad (9\text{-}327)$$

where

θ = the angle of inclination of the exit shear planes to the strip center line,
h = final strip thickness, and
f = exit tensile stress/shear yield stress.

The force P_{1L} is given, in similar units, by

$$P_{1L} = \left(2Q_{1L} + \frac{(S-1)h}{2 \sin^2 \theta}\right) \frac{\sin \theta}{\sin (\theta - \phi)} \qquad (9\text{-}328)$$

[±] This is the triangle containing the neutral point.
[>] The shear yield stress in plane strain is $1/\sqrt{3}$ times the yield stress in uniaxial compression.

where S = ratio of strip thickness before and after one pair of shear planes, and
ϕ = angle of ($\lambda + \alpha$), α being the angle of inclination of the flattened roll surface (the half angle of the wedge-shaped portion of strip in the roll bite) and λ being the friction angle (= $\tan^{-1} \mu$).

Referring to Figure 9-74, it may be shown in a similar manner that

$$Q_{(m+1)L} = P_{mL} \frac{\sin(\theta+\phi)}{\sin \theta} + \frac{h(S-1)S^{m-1}}{2 \sin^2 \theta} \qquad (9\text{-}329)$$

$$P_{(m+1)L} = \left(Q_{(m+1)L} + \frac{h(S-1)S^m}{2 \sin^2 \theta} \right) \frac{\sin \theta}{\sin(\theta+\phi)} \qquad (9\text{-}330)$$

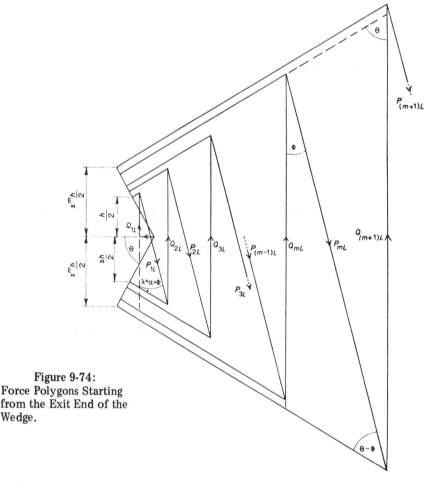

Figure 9-74: Force Polygons Starting from the Exit End of the Wedge.

for $1 \leq m \leq (n-1)$

Considering the entry end of the bite and referring to Figure 9-75

$$Q_{(n+1)R} = \frac{S^n h}{2 \sin^2 \theta} - bS^n \frac{h}{2} \cot \theta \qquad (9\text{-}331)$$

and

$$P_{(n+1)R} = \left(2Q_{(n+1)R} - \frac{h(S-1)S^{n-1}}{2 \sin^2 \theta} \right) \frac{\sin \theta}{\sin(\theta-\rho)} \qquad (9\text{-}332)$$

where $\rho = \lambda - \alpha$.

MATH MODELS RELATING TO ROLLING FORCE

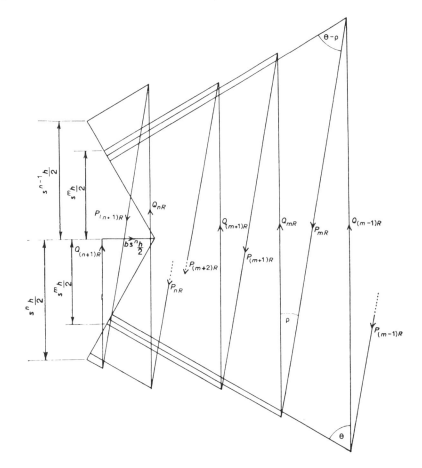

Figure 9-75: Force Polygons Starting from the Entry End of the Wedge.

In general terms

$$Q_{mR} = P_{(m+1)R} \frac{\sin (\theta+p)}{\sin \theta} - \frac{h(S-1)S^{m-1}}{2 \sin^2 \theta} \quad (9\text{-}333)$$

and

$$P_{mR} = \left(Q_{mR} - \frac{h(S-1)S^{m-2}}{2 \sin^2 \theta}\right) \frac{\sin \theta}{\sin (\theta-p)} \quad (9\text{-}334)$$

for $n \geq m \geq 2$.

The specific roll separating force, in terms of units of yield shear stress is given by

$$Q = \sum_{m=1}^{d} Q_{mL} + \sum_{m+d+1}^{n+1} Q_{mR} \quad (9\text{-}335)$$

However, to compute Q, the position of the neutral point (that is, the value of d) must be established. This may be accomplished from the relationship

$$\sum_{m=0}^{d} P_{mL} \sin \phi \geq \left[\sum_{m=d+2}^{n+1} P_{mR} \sin \rho\right] + \left(f - bS^n\right) \frac{h}{2} \quad (9\text{-}336)$$

Since Q is computed in units of the mean shear yield stress, the specific roll separating force, in conventional terms, is given by

$$f = Qk \tag{9-337}$$

As seen in Section 9-13, Bland and Ford expressed the specific rolling force f as follows

$$f = kL \times f_3(a,r) \tag{9-338}$$

where k is a "mean" yield stress, L is the length of the arc of contact and $f_3(a, r)$ is a function to compensate for the friction hill. Figure 9-76 shows this function plotted for constant values of $\mu\sqrt{\frac{R'}{h}}$ where R' is the effective radius of the deformed roll. The solid curves are those developed from the shear plane theory and the dashed ones by Bland and Ford.

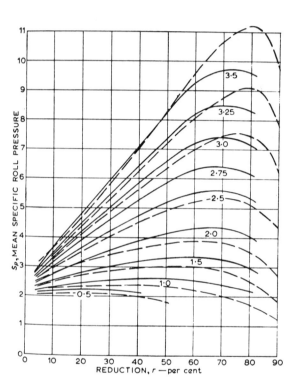

Figure 9-76:
Mean Specific Roll Pressure Against Reduction.

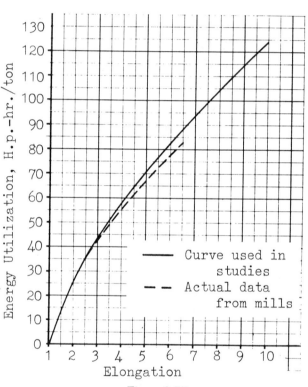

Figure 9-77:
Resistance to Compression Curves for Low Carbon Steels (0.06 to 0.10% C) (MacQueen).

9-23 Experiences in the Application of Rolling Force Models

The two parameters of paramount importance in the application of rolling force models are (1) the resistance to deformation of the material to be rolled and (2) the effective coefficient of friction in the roll bite. Unfortunately, sufficient data concerning these parameters has not, generally, been available or improper values have been assigned to one or the other. As a consequence, considerable difficulty has frequently been experienced in attempting to apply the various rolling force models to typical commercial rolling operations.

In the mid 1960's, MacQueen[⊕] investigated the applicability of rolling force models developed by Bland and Ford (Section 9-13) and Stone (Section 9-18) to the performance of an 80-inch cold reversing mill. The resistance to compression (constrained yield stress) data for the steels considered were measured in the laboratory and a typical curve for steels containing 0.06 to

---

[⊕] A. J. F. MacQueen, "Finding a Practical Method of Calculating Roll Force in Wide Reversing Cold Mills", Iron and Steel Engineer Year Book, 1967, pp. 425-440.

MATH MODELS RELATING TO ROLLING FORCE

0.10 per cent carbon is reproduced in Figure 9-77. Appropriate values for the coefficient of friction were derived from available rolling data (some of which is reproduced in Table 9-3) using mathematical expressions for spindle torque as developed by Bland and Ford (Section 10-5) and Stone (Section 10-3) and these values are presented in Table 9-4. The computed values of the coefficient of friction were then inserted into the corresponding force models (as well as the torque-arm and single-pass models as discussed in Sections 10-3 and 9-16, respectively), and the measured and predicted values of the rolling force were compared (Table 9-5).

MacQueen points out that the agreement in the values of the coefficient (Table 9-4) computed by the two methods was not too good in the earlier passes of a series but improved in the later passes. The predicted values of specific rolling force showed considerable variance, particularly at the later passes. However, the single pass method gave the best overall prediction of rolling force.

Table 9-3

Cold Rolling Test Data (MacQueen)

Width	Pass	Exit thickness, in.	Reduction, percent	Delivery speed, fpm	Drive hp	Back reel hp	Front reel hp	Back tension, psi	Front tension, psi	Roll force, lb.
35.25	1	0.068	32.0	790	1310	0	740	0	14,600	1,810,000
	2	0.048	29.4	1280	2700	770	1290	11,700	19,600	1,700,000
	3	0.034	29.2	1700	2700	1110	1400	18,000	22,700	1,590,000
	4	0.024	29.4	1500	1850	770	1280	20,000	33,300	1,480,000
	4	0.024	29.4	1500	2060	530	750	13,800	19,500	1,790,000
	5	0.017	29.1	1700	1920	880	820	28,500	26,600	1,570,000
	6	0.012	29.4	1800	1630	680	910	29,500	39,400	1,700,000
	7	0.0083	30.8	2000	1780	790	380	44,600	21,400	1,680,000
33.25	1	0.070	30.0	810	1400	0	690	0	12,100	1,700,000
	2	0.050	28.6	1310	2420	760	1210	11,500	18,300	1,660,000
	3	0.036	28.0	1700	2770	1310	1420	21,200	23,000	1,610,000
	4	0.026	27.8	1000	1230	580	880	22,100	33,600	1,640,000
	4	0.026	27.8	1750	2340	890	1460	19,400	31,800	1,660,000
	5	0.018	30.8	1000	1080	520	490	28,700	27,000	1,550,000
	5	0.018	30.8	1860	2270	940	990	27,900	29,300	1,700,000
	6	0.0125	30.5	1000	890	430	650	34,100	51,600	1,680,000
	6	0.0125	30.5	1800	1710	770	1080	34,000	47,600	1,720,000
	7	0.0088	29.6	2300	2200	1070	470	52,500	23,000	1,750,000
33.25	1	0.070	30.0	880	1200	0	920	0	14,800	1,660,000
	2	0.049	30.0	1600	2840	1010	1560	12,800	19,700	1,720,000
	3	0.035	28.6	1700	2240	1220	1640	20,400	27,400	1,610,000
	4	0.025	28.6	800	980	420	900	20,800	44,700	1,610,000
	4	0.025	28.6	1880	2420	1060	1870	22,400	39,500	1,610,000
	5	0.0175	30.0	800	820	420	440	29,800	31,200	1,410,000
	5	0.0175	30.0	1900	2200	950	960	28,400	28,600	1,660,000
	6	0.0125	28.5	750	610	310	460	32,800	48,700	1,660,000
	6	0.0125	28.5	2000	1780	920	1140	36,500	45,200	1,660,000
	7	0.0088	29.6	2300	2270	1070	410	52,500	20,100	1,700,000

Work roll, 12.1 in. diameter. Hot rolled strip thickness, 0.100-in.

Bryant and Osborn [1] developed a simpler, algebraic rolling force model based on the theories of Ford et al. [v] (Section 9-13) and Bland and Sims.[8] Their equation may be written

$$f = ((\bar{k} - \bar{\sigma}))\sqrt{R'\delta}\,(1 + 0.4a) + f_E \quad (9\text{-}339)$$

where f_E (the portion of the specific rolling force attributable to elastic recovery) is given by

---

[1] G. F. Bryant and R. Osborn, "Derivation and Assessment of Roll Force Models", Chap. 12 of Book "Automation of Tandem Mills", Edited by G. F. Bryant, Published by The Iron and Steel Institute, 1 Carlton House Terrace, London, SWIY, 5DB.

[v] H. Ford, F. Ellis and D. R. Bland, "Cold Rolling With Strip Tension, Part I, A New Approximate Method of Calculation and a Comparison with Other Methods", J. Iron & Steel Institute, May, 1951, 168, pp. 57-72.

[8] D. R. Bland and R. B. Sims, Proc. I.M.E., 1953, 167, p. 371ff.

$$f_E = (2/3)\sqrt{R't_2}\,(k_2 - \sigma_2)^{1.5}\sqrt{\frac{1-\nu_s}{E_s}} \qquad (9\text{-}340)$$

Table 9-4

Coefficient of Friction Values as Computed From Torque Data Using Models Developed by Stone and Bland and Ford (MacQueen)

Case	Required torque, lb.-ft.	Stone	Bland and Ford
1	27,600	*	0.0805
2	35,100	0.055	0.093
3	26,400	0.0455	0.045
4	20,500	0.054	0.080
5	22,900	0.0525	0.068
6	18,800	0.0735	0.0605
7	15,100	0.072	0.0755
8	14,800	0.061	0.052
9	28,800	*	0.102
10	30,800	0.035	0.083
11	27,100	0.076	0.066
12	20,500	0.0715	0.0915
13	22,300	0.080	0.103
14	18,000	0.063	0.046
15	20,300	0.077	0.0715
16	14,800	0.080	0.086
17	15,800	0.0815	0.084
18	15,900	0.070	0.063
19	22,700	*	0.087
20	29,600	0.019	0.053
21	21,900	0.031	0.0445
22	20,400	0.078	0.1205
23	21,400	0.0815	0.1095
24	17,100	0.068	0.057
25	19,300	0.077	0.0695
26	13,500	0.0785	0.085
27	14,800	0.084	0.084
28	16,400	0.069	0.063

*Not Possible.

Table 9-5

Comparison of Observed and Predicted Values of Specific Rolling Force (MacQueen)

	Roll Force, lb. per in.				
Pass	Observed	Stone	Bland and Ford	Torque arm	Single pass
1	51,350	*	41,300	49,650	51,770
2	48,230	39,500	44,000	51,490	50,280
3	45,110	36,620	33,000	43,400	48,690
4	41,990	35,210	40,200	47,380	48,470
5	50,780	40,280	41,600	50,500	52,660
6	44,540	43,660	33,200	44,400	48,100
7	48,230	46,040	45,300	56,740	49,810
8	47,660	61,870	32,500	45,960	51,690
9	51,130	*	43,100	58,200	50,560
10	49,930	36,080	41,500	46,470	49,760
11	48,420	40,400	35,100	44,660	46,490
12	49,320	37,790	41,200	48,720	45,940
13	49,930	41,890	47,800	55,790	47,090
14	46,620	39,020	29,000	40,150	48,880
15	51,130	45,480	38,200	49,770	48,770
16	50,530	43,450	46,200	58,020	50,250
17	51,730	47,780	46,400	61,200	50,930
18	52,630	*	36,000	53,080	48,080
19	49,930	*	40,200	44,660	50,130
20	51,440	34,000	36,000	42,710	50,200
21	48,420	31,500	31,000	38,350	46,600
22	48,420	37,630	52,700	55,790	45,570
23	48,420	39,950	48,300	54,740	45,900
24	42,410	38,840	30,900	42,110	46,380
25	49,930	45,880	36,900	48,420	48,240
26	49,930	42,390	45,500	56,090	46,260
27	49,930	48,420	43,800	58,350	45,630
28	51,130	*	37,000	52,480	53,660

*Not Possible.

In equations 9-339 and 9-340, the symbols have the following meanings

- $\bar{k}$ is the mean yield stress of the strip in the roll bite
- $\bar{\sigma}$ is the mean tensile stress in the strip
- R' is the deformed work roll radius
- δ is the draft
- k_2 is the yield stress of the strip on exit from the bite
- σ_2 is the tensile stress in the strip on exit from the bite
- ν_s is Poisson's ratio associated with the strip
- E_s is Young's modulus for the strip and
- a is given by

$$a = \sqrt{\frac{h_2}{h_1}}\, e^{\frac{\mu L}{(0.72 h_2 + 0.28 h_1)}} - 1 \qquad (9\text{-}341)$$

where h_1 and h_2 are the initial and final thicknesses of the strip and μ is the coefficient of friction.

The simplified equation, 9-340, was found to exhibit the following predictive accuracy when compared with the model of Ford et al., (from which values were computed which were designated "actual") in the rolling of strip 0.065 — 0.150-in. thick, with roll radii 9 to 12 inches and with assumed coefficients of friction 0.4 to 0.8. When the correct value of R' was utilized, the simplified model predicted a rolling force within 2,240 lbs./in. of the actual value in 81 per cent of the cases and within 5 per cent of the actual value in 77 per cent of the cases. Where R' had to be calculated by iterative techniques, the accuracies of the simplified model dropped to 78 and 73 per cent, respectively.

The validity of the simplified model discussed in Section 9-18 was studied with respect to data obtained during the operation of a 6-stand tandem mill." In this instance, the coefficients of friction for each stand were first calculated from frictional characteristics of the rolling lubricants used, the surface roughnesses of the rolls and strip and the drafts taken at the various stands. The strain rates associated with the reduction at each stand were calculated and the corresponding dynamic constrained yield strengths were computed for each roll bite taking into account the prior reductions achieved in the earlier stands. Typical interstand tensile stresses in the strip were assumed and the minimum rolling pressures required for deformation were worked out for each stand. The rolling forces and drive energy requirements were then calculated using the models described in Section 9-18 and Section 10-9, but with the deformed diameter of the work rolls calculated from equation 9-13.

Table 9-6 shows the comparison between predicted and actual values for two grades of steel used in tinplate manufacture. For the softer material, at the lower mill speed, the predicted and actual values of force agree to within 10 per cent and, at the higher mill speed, to within 5 per cent. For the harder material, the values agreed to within 10 per cent.

9-24 Stone's Formula for Minimum Gage

In rolling operations, it has long been noted that, regardless of the magnitude of the rolling force, or the number of passes made, a limiting, minimum thickness of the workpiece is reached when no further reduction appears to be possible. Whether or not this so-called "minimum gage" actually exists has been debated frequently during the last two decades, with some rolling theorists believing that effects, such as the work rolls pressing on each other beyond the edges of the workpiece, may actually be limiting the attainable reduction.∝ Others have believed that minimum gage is a physical reality and have developed formulae relating this rolled strip thickness to other process parameters.

Stone's formula for minimum gage is undoubtedly the best known and most frequently cited.◊ First published in 1953, its derivation is as briefly described below. The basic mathematical expressions used are those relating to the specific rolling force and the length of the arc of contact. Referring to the former, in Section 9-18, it was shown how Stone, making certain simplifying assumptions, developed the relationship for the average pressure $\bar{p}$ along the arc of contact

$$\bar{p} = (\sigma_c - \sigma_1) \left[\frac{e^{\frac{\mu L}{t}} - 1}{\frac{\mu L}{t}} \right] \quad (9\text{-}342)$$

This expression is shown plotted in Figure 9-78. If, for simplicity

$$\lambda = \frac{\mu L}{t} \quad (9\text{-}343)$$

" W. L. Roberts and R. J. Bentz, "Predicting Rolling Forces and Mill Power Requirements for Tandem Mills", Blast Furnace and Steel Plant, 58, 1970, pp. 559-568.

∝ O. Pawelski and G. Kading, "Investigation on the Limiting Degrees of Deformation and the Limiting Thickness in Cold Rolling of Thin Strip", Stahl und Eisen, 1967, 87, Nov. 2, pp. 1340-1348.

◊ M. D. Stone, "Rolling of Thin Strip", Iron and Steel Engineer Year Book, 1953, pp. 115-128.

Table 9-6
Predicted Rolling Forces In Operation of 6-Stand Cold Mill

<u>Material Rolled</u>: Low-Carbon Steel Strip (Soft Grade) 35.6 inches wide, 0.080-inch thick.

Coiling Speed, fpm	Stand No.	Reduction at Stand, Per Cent	Tensile Stress in Strip, kpsi		Calculated Coefficient of Friction in Bite	Rolling Force, Millions of Pounds	
			Entry into Bite	Exit from Bite		Predicted	Measured
1259	1	26.8	0.2	23.2	0.076	1.57	1.56
	2	28.2	23.2	20.2	0.031	1.74	1.77
	3	37.8	20.2	21.3	0.013	1.74	1.91
	4	33.5	21.3	22.5	0.017	1.76	1.85
	5	37.6	22.5	23.2	0.012	1.80	1.93
	6	15.0	23.2	3.3	0.0055	1.37	1.40
3236	1	24.8	0.2	23.4	0.089	1.65	1.67
	2	31.2	23.4	19.0	0.024	1.78	1.75
	3	37.0	19.0	22.3	0.0088	1.68	1.66
	4	34.5	22.3	24.7	0.011	1.64	1.69
	5	34.4	24.7	28.9	0.0083	1.58	1.65
	6	19.0	28.9	2.4	0.0049	1.43	1.47

<u>Material Rolled</u>: Low-Carbon Steel Strip (Hard Grade) 36.2 inches wide, 0.085-inch thick.

Coiling Speed, fpm	Stand No.	Reduction at Stand, Per Cent	Tensile Stress in Strip, kpsi		Calculated Coefficient of Friction in Bite	Rolling Force, Millions of Pounds	
			Entry into Bite	Exit from Bite		Predicted	Measured
4583	1	22.4	0.2	23.0	0.14	2.01	1.94
	2	25.8	19.0	26.0	0.016	1.63	1.66
	3	27.1	26.0	22.5	0.0065	1.62	1.75
	4	27.4	22.5	16.5	0.011	1.84	*
	5	25.1	16.5	11.0	0.0085	1.99	1.96
	6	17.0	11.0	2.0	0.0060	1.98	2.04

<u>Work Rolls</u>: 23 inches in Diameter; 35-Grit Finish in Stands 1 and 2; 80-Grit Finish in Stands 3, 4, 5 and 6.

<u>Rolling Lubricants</u>: Pickle, Oil and Water at Stand 1; Recirculated Emulsion of Palm Oil Substitute at Other Stands.

<u>Drive Motors (HP and Roll Speed at Rated Voltage)</u>: Stand 1—3,000 HP, 838 fpm; Stand 2—5,000 HP, 1,280 fpm; Stand 3—5,000 HP, 2,080 fpm; Stand 4—6,000 HP, 2,620 fpm; Stand 5—6,000 HP, 4,000 fpm; Stand 6—7,000 HP, 5,680 fpm.

then the slope of the curve for $\bar{p}$ is given by

$$\frac{d\bar{p}}{d\lambda} = (\sigma_c - \sigma_1)\frac{\lambda e^\lambda - (e^\lambda - 1)}{\lambda^2} \qquad (9\text{-}344)$$

Hitchcock's equation for the length of the arc of contact L (see Section 9-6) may be written in the form

$$L = \sqrt{\ell^2 + a^2\bar{p}^2} + a\bar{p} \qquad (9\text{-}345)$$

where ℓ is the length of the arc of contact that would be given by a rigid roll and a is a term associated with the roll flattening such that

$$a = \frac{8R(1-\nu^2)}{\pi E} \qquad (9\text{-}346)$$

where R is the work roll radius, ν is Poisson's ratio and E is Young's modulus (ν and E pertaining to the roll material).

MATH MODELS RELATING TO ROLLING FORCE

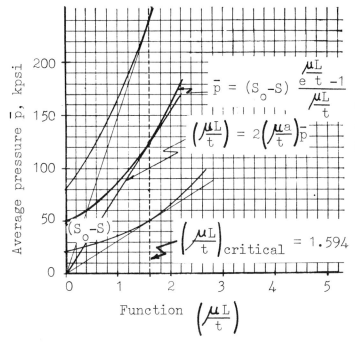

Figure 9-78: Curves Showing Relationship of $\bar{p}$ to $\left(\frac{\mu L}{t}\right)$ for minimum gage conditions.

Under conditions of limiting reduction,

$$\ell = 0 \tag{9-347}$$

and

$$L = 2a\bar{p} \tag{9-348}$$

so that

$$\frac{\mu L}{t} = 2\left(\frac{\mu a}{t}\right)\bar{p} \tag{9-349}$$

This last equation may be represented by the straight lines drawn in Figure 9-78.

Under conditions of limiting reduction, the second set of curves are tangential to the first set, as illustrated in the figure so that under these circumstances

$$\frac{d\bar{p}}{d\lambda} = (\sigma_c - \sigma_1)\frac{\lambda e^\lambda - (e^\lambda - 1)}{\lambda^2}$$
$$= \frac{\bar{p}}{\lambda} = \frac{(\sigma_c - \sigma_1)(e^\lambda - 1)}{\lambda^2} \tag{9-350}$$

Hence

$$\lambda \exp(\lambda) - 2(\exp(\lambda) - 1) = 0 \tag{9-351}$$

The solution to this equation is

$$\lambda = 1.594 \tag{9-352}$$

and therefore

$$\frac{\bar{p}}{\lambda} = 1.54(\sigma_c - \sigma_1) = \frac{t}{2\mu a} = \frac{t\pi E}{2\mu 8R(1-\nu^2)} \tag{9-353}$$

Then, if t is the limiting thickness, t_m,

$$t_m = \frac{3.58 \, D\mu(\sigma_c - \sigma_1)}{E} \qquad (9\text{-}354)$$

where D is the roll diameter and where σ_1 may be regarded as the average strip tension.

The value of the coefficient in the above formula was investigated by Takahashi, Nakajimi and Murata~ who rolled annealed, low carbon steel sheets without tension in single and multiple passes using a 2-high mill with 9.84-inch diameter rolls and a 4-high mill with 5.90-inch diameter rolls, using No. 60 spindle oil, cottonseed oil and palm oil as lubricants. From measurements of the rolling forces, the coefficients of friction were calculated from Stone's formula for specific rolling force and Hitchcock's equation for the length of the arc of contact. Figure 9-79 shows the variation of the mean specific roll pressure $\bar{p}$ with reduction for the three lubricants on one of the mills. The relationship between the calculated coefficient of friction and the variable DS_0/t_{mrt} for multipass rolling is shown in Figure 9-80 (S_0 corresponding to $(\sigma_c - \sigma_1)$ in equation 9-354). In this figure, the dotted curves represent Stone's minimum gage formula expressed in the form for the minimum reducible strip thickness (t_{mrt})

$$t_{mrt} = \alpha D \mu S_0 \qquad (9\text{-}355)$$

corresponding to the values of the coefficient α as indicated at the right-hand side of the figure. From the experimental data shown in Figure 9-80, it is seen that appropriate values for α lie in the range 13.3×10^{-5} to 27.3×10^{-5} mm^2/kg. Thus, if E is considered to be 2.1×10^4 kg/mm^2, the numerical constant of Stone's equation as usually written lies in the approximate range 2.8 to 5.7.

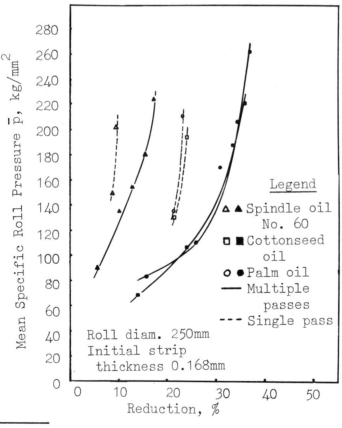

Figure 9-79: Relationship Between Mean Specific Roll Pressure and Reduction.

~K. Takahashi, K. Nakajima and K. Murata, "Experimental Study on the Minimum Thickness in Reduced Cold Rolling of Steel", portion of book, "Friction and Lubrication in Metal Processing", Edited by F. F. Ling, et al, A.S.M.E., New York, 1966.

MATH MODELS RELATING TO ROLLING FORCE

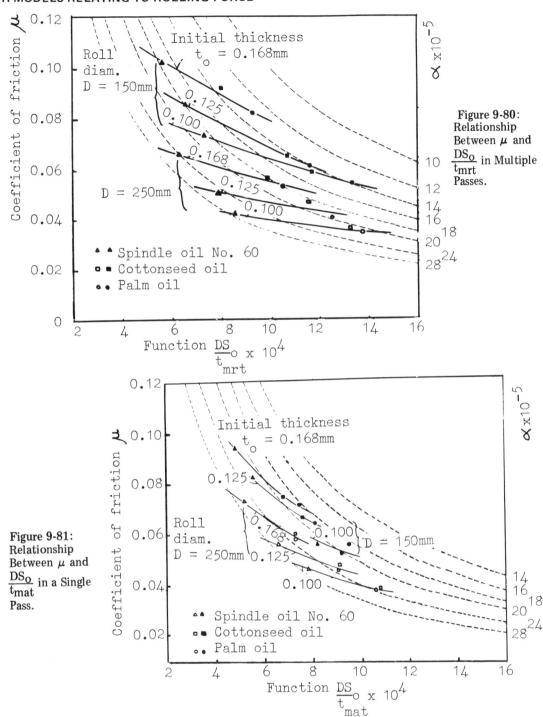

Figure 9-80: Relationship Between μ and $\frac{DS_o}{t_{mrt}}$ in Multiple Passes.

Figure 9-81: Relationship Between μ and $\frac{DS_o}{t_{mat}}$ in a Single Pass.

Figure 9-81 shows similar data for the coefficient of friction as a function of DS_o/t_{mat} for single pass reductions, t_{mat} being the minimum attainable thickness. Here the values for α lie in the range 18.9×10^{-5} to 27.0×10^{-5} mm^2/kg corresponding to a numerical coefficient for Stone's equation (in conventional units) lying in the range 4.0 to 5.7. The relationship of the computed values for the coefficient to a coefficient α for both multiple and single pass rolling is illustrated in Figures 9-82 and 9-83.

Takahashi and his coworkers believed that the coefficient α, because of its dependence on the coefficient of function μ, should be expressed as

$$\alpha = K_1 + K_2\mu \qquad (9\text{-}356)$$

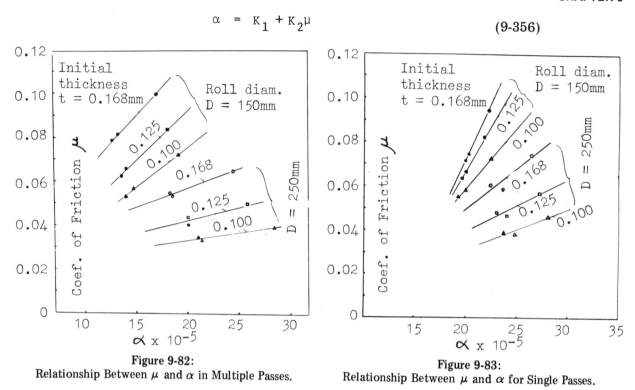

Figure 9-82:
Relationship Between μ and α in Multiple Passes.

Figure 9-83:
Relationship Between μ and α for Single Passes.

Figures 9-84 and 9-85 show the variation of K_1 and K_2 with the ratio of roll diameter to initial strip thickness for both single and multipass reductions.

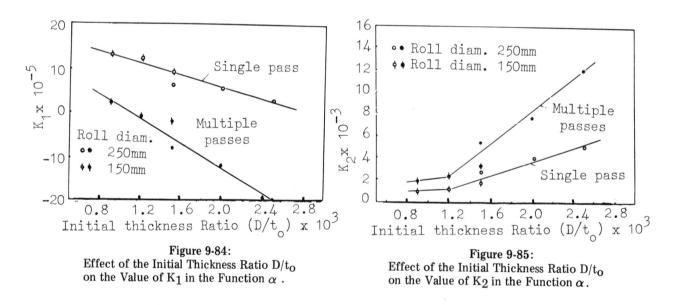

Figure 9-84:
Effect of the Initial Thickness Ratio D/t_o on the Value of K_1 in the Function α.

Figure 9-85:
Effect of the Initial Thickness Ratio D/t_o on the Value of K_2 in the Function α.

9-25 Other Formulae for Minimum Gage

Subsequent to the publication of Stone's formula, it gradually became apparent that thinner gages are obtainable than would be predicted by the formula. For example, tin plated, low carbon steel strip may be rolled as thin as 0.002-inch on a mill with 21-inch diameter work rolls whereas the formula would predict a minimum rolled thickness of about 0.006-inch. Accordingly, various other formulae have been developed concerning minimum gage and some of these are discussed in this section.

Tong and Sachs [θ] developed an expression for an "absolute minimum strip thickness... below which cold rolling is not possible even for very small reductions". This minimum reducible thickness, t_{mrt}, is given by the equation

where
$$t_{mrt} = \frac{0.78 \, \mu D k}{C} \qquad (9\text{-}357)$$

μ = coefficient of friction
k = the average flow stress of the material under rolling conditions
C = a rigidity constant of the roll given by

$$C = \frac{\pi E}{16(1 - \nu^2)} \qquad (9\text{-}358)$$

where E is the modulus of elasticity and
ν is the Poisson's ratio for the roll material.

Assuming Poisson's ratio ν (for the steel of the rolls and the strip) to have a value of 0.3, and assuming k = $(\sigma_0 - \sigma_1)$

where

σ_0 = the mean constrained yield stress (corresponding to half the reduction of the appropriate pass) of the strip, psi, and

σ_1 = the average tension stress in the strip, equal to 50 per cent of the entry tension stress plus 50 per cent of the delivery tension stress, psi,

the Tong and Sachs' expression may be simplified as

$$t_{mrt} = \frac{3.62 \, \mu D (\sigma_0 - \sigma_1)}{E} \qquad (9\text{-}359)$$

which is virtually identical to Stone's expression (equation 9-354).

For the rolling of steel strip with steel rolls, a more general expression for minimum thickness developed by Ford and Alexander [φ] can be written as

$$t_{min} = (4.53 + 7.11\mu)\frac{\mu D (1 - \nu^2)}{E}(1.155Y - \bar{t}) \qquad (9\text{-}360)$$

where

Y = the mean tensile yield strength of the strip
$\bar{t}$ = the mean value of the entry and exit tensile stresses in the strip being rolled

Considering $(1.155Y - \bar{t})$ to be equivalent to $(\sigma_0 - \sigma_1)$ and Poisson's ratio to be equal to 0.3, then equation 9-360 becomes

$$t_{min} = (4.13 + 6.49\mu)\frac{\mu D}{E}(\sigma_0 - \sigma_1) \qquad (9\text{-}361)$$

Considering this relation, Ford and Alexander believed that Stone's equation gave values for minimum gage that were too low.

On the basis of the simplified rolling model discussed in Section 9-17, it was seen that two minimum gages may be postulated; a minimum reducible thickness, t_{mrt} (or the maximum

---

[θ] K. Tong and G. Sachs, "Roll Separating Forces and Minimum Thicknesses of Cold Rolled Strip", J. Mech. & Phys. of Solids, Vol. 6, 1959, pp. 35-46.

[φ] H. Ford and J. M. Alexander, "Rolling Hard Materials in Thin Gages, Basic Considerations", J. Inst. Metals, Vol. 88, 1959, pp. 193-199.

thickness of strip that could not be reduced by rolling); and a minimum attainable thickness, t_{mat} (obtained by a single rolling pass from an optimum entry strip thickness). The expression for the former is given by

$$t_{mrt} = \frac{1.17 \, \mu D \sigma_c}{E} \quad (9\text{-}362)$$

where the constrained compressive flow stress of the strip in the roll bite, σ_c, is given by

$$\sigma_c = 1.15 \, \sigma_{Y(r/2)} - \left[\frac{\sigma_1 + \sigma_2(1-r)}{(2-r)}\right] \quad (9\text{-}363)$$

In equation (9-363), $\sigma_{Y(r/2)}$ was regarded as the tensile yield strength corresponding to half the reduction at the average strain rate encountered in the roll bite, σ_1 and σ_2 are the entry and exit tensile stresses in the strip, respectively, and the r is reduction given to the strip.

The minimum attainable thickness was shown to be about half the value of the minimum reducible thickness, t_{mrt}, namely,

$$t_{mat} = \frac{0.585 \, \mu D \sigma_c}{E} \quad (9\text{-}364)$$

In the model discussed in Section 9-18, the specific rolling force f is given by

$$f = \frac{(\sigma_c - \sigma)}{(1-r)} L \left(1 - \frac{5r}{4} + \frac{\mu L}{2t}\right) \quad (9\text{-}365)$$

where L is the length of the arc of contact, and
  t is the thickness of the strip prior to reduction.

Extensive use of this model has shown that a satisfactory equation for arc length appears to be

$$L = \frac{(\sigma_c - \sigma)D}{AE} + \sqrt{\frac{Dtr}{2}\left(1 + 4.63 \frac{f}{Etr}\right)} \quad (9\text{-}366)$$

where

  A = a constant with a value approximately unity

The term $(\sigma_c - \sigma)D/AE$ represents an effective arc length prior to the onset of the plastic flow of the strip, and the second term represents the length of the arc of contact as computed from Hitchcock's relationship for the deformed roll diameter.

$$D' = D\left(1 + 4.63 \frac{f}{Etr}\right) \quad (9\text{-}367)$$

A more valid complete expression for the specific rolling force is then given by the equation

$$f = \frac{(\sigma_c - \sigma)}{1 - r} \left[\frac{(\sigma_c - \sigma)D}{AE} + \sqrt{\frac{Dtr}{2}\left(1 + \frac{4.63f}{Etr}\right)}\right]$$
$$\times \left\{1 - \frac{5r}{4} + \frac{\mu}{2t}\left[\sqrt{\frac{Dtr}{2}\left(1 + \frac{4.63f}{Etr}\right)} + \frac{(\sigma_c - \sigma)D}{AE}\right]\right\} \quad (9\text{-}368)$$

From computations made with this model, it appears that, for constant values of the parameters, μ, D, $(\sigma_c - \sigma)$ and E, there is a certain minimum gage that can be attained regardless of the initial strip thickness and the reduction taken. This minimum value, t_{min}, is given by the approximate relationship.

MATH MODELS RELATING TO ROLLING FORCE

$$t_{min} \simeq \frac{1.25\mu D(\sigma_c - \sigma)}{E} \qquad (9\text{-}369)$$

The distinction between the minimum reducible and the minimum attainable thicknesses now depends primarily on the value of $(\sigma_c - \sigma)$. Where the minimum gage is to be produced by a single, large reduction (minimum attainable thickness), then, as will be seen later, the value of $(\sigma_c - \sigma)$ corresponds to that of the annealed strip. However, where minimum gage is to be produced by a number of relatively small reductions, the value of $(\sigma_c - \sigma)$ for the last pass corresponds to that of the work-hardened strip, which may be at least twice that of the annealed strip. Thus, in a practical case in which several passes are used to obtain minimum gage and in which the rolled strip is not annealed after every pass, the rolled thickness (corresponding to t_{mrt}) may be about twice the thickness (corresponding to t_{mat}) obtained by a single, large reduction.

It is important to realize that, in the use of these formulae for minimum gage that

(1) the coefficient of friction μ will depend on various processing parameters (as discussed in Section 6-31) and

(2) the flow stress term $(\sigma_0 - \sigma)$ should be computed with allowance for strain rate effects and, in the case of multiple passes, for work-hardening effects.

The formulae for minimum gage discussed so far have related to conditions where effective lubrication prevails. In temper, or dry rolling, where coefficients of friction may be in the range 0.25 to 0.35, different conditions exist. In the expression for specific rolling force, f, (equation 9-286) the value of f increases very rapidly with the reduction r once the exponent $\mu L/t(1-r)$ has attained a value of about 2.5. It may therefore be considered that the maximum attainable reduction (r_{max}) occurs when the length (L) of the arc of contact has acquired a value of approximately $2.5t/\mu$. From equation 9-283, therefore,

$$\frac{4t}{\mu} = \frac{1}{2}\left[\left(\frac{Dr_{max}\mu}{2}\right) + \sqrt{\frac{Dr_{max}\mu}{2} + Dtr_{max}}\right] \qquad (9\text{-}370)$$

and it can then be shown that

$$r_{max} \simeq \frac{3.6t}{\mu^2 D} \qquad (9\text{-}371)$$

That equation 9-371 is reasonably valid is shown by the data on Table 9-7 relating to the temper rolling of black plate and galvanized strip.

Table 9-7
Maximum Attainable Reductions

Material	Coefficient of Friction	Work Roll Diameter (inches)	Maximum Attainable Reduction (%)	
			Predicted	Actual
0.0095-Inch Black Plate	0.25	3.18	18.1	16.2
0.0092-Inch Black Plate	0.30	6.21	6.4	6.6
0.0140-Inch Galvanized Strip	0.20	6.31	14.2	15.0*

*Extrapolated from data obtained at relatively low-rolling forces.

Chapter 10 Torque Equations and Tandem Mill Control Models

10-1 Introduction

From the viewpoint of mill operation and design, the prediction of spindle torques is of the same importance as the prediction of rolling force. For example, a knowledge of torque requirements will indicate whether or not a certain rolling operation is possible with an existing mill drive system and, assuming the gear ratios are correct, the maximum possible rolling speed. Similarly, in the design of new facilities, a knowledge of rolling torque requirements is necessary in establishing motor sizes, the designs of gear trains, and minimum roll neck diameters. Furthermore, once the energy requirements are determined, it becomes possible to examine the thermal aspect of mill design, particularly with respect to the cooling of the rolls and the strip.

Frequently some confusion exists with respect to the exact meaning of rolling torques. In most rolling situations involving 2-high, 4-high and symmetrical clusters of rolls with two driving spindles, the symmetry inherent in the process ensures that the torques existing in both spindles are essentially the same. In such cases, the torque for each spindle is termed the "torque per spindle" which, divided by the width of the strip being rolled, becomes the "specific torque per spindle". When the sum of the two torques is implied, and for those cases where a single drive spindle is involved, the terminology "total torque" or, on the basis of unit width of strip, " specific total torque" is used.

Energy supplied by the mill drive system is utilized or dissipated in the rolling process as follows:

1. For the deformation of the workpiece in the roll bite.
2. In frictional energy dissipation at the roll-strip interfaces.
3. In frictional energy losses at the mill stand, particularly in the backup roll bearings and at the work roll — backup roll contacts.
4. In frictional energy losses in the pinions, reduction gears, etc.
5. As electrical resistance losses in electric motors, generators and other parts of the drive circuits.
6. In the acceleration of the rolls and the workpiece as the mill runs up to speed.

Frequently, rolling theorists are concerned only with the first two, inasmuch as they may be concerned with torques in the spindles with the assumption that the bearing losses in the roll chocks are negligible. However, the mill designer, in establishing power requirements for the mill, should take all expenditures of energy into consideration.

In the earlier sections of this chapter, mathematical expressions for the spindle torques associated with items (1) and (2) above are presented. The other energy requirements are not neglected, however, and are treated in sufficient detail to permit the mill designer to correctly specify spindle torques and take items (3) to (6) into account.

As will be seen in the following sections, a number of approaches to the determination of rolling torques have been developed. The simplest, designated the lever-arm method, derives the roll torque from the roll separating force and the distance separating the plane in which this force acts from the center of rotation of the rolls. Another method, involving the integration of the frictional stresses in the roll bite, utilizes a knowledge of the pressure distribution along the arcs of contact, the geometry of the roll bite and the effective coefficient of friction. A third method is based on the resolution of the horizontal forces on the strip and requires a knowledge of the draft and the compressive yield strength of the strip. Yet another method is based on an energy balance of the rolling process.

The forward slip of the strip as it emerges from the roll stand is discussed in Sections 10-14 and 10-15, and mathematical models permitting its computation are presented. An

TORQUE EQUATIONS AND TANDEM MILLS

understanding of this parameter is desirable in connection with tandem mill control equations treated in the last sections of the chapter.

The prediction of drive power requirements is the subject of Section 10-16 and the sizing and selection of drive motors for both mill stands and reels are considered in Sections 10-17 and 10-18.

The last sections of the chapter are devoted to mathematical models useful in the automatic control of tandem mills. Such models pertain to the interrelationships of the various parameters involved in tandem mill operation and are intended to predict all the effects that will result from a change in one or more of the operating parameters. For example, if the speed of one of the stands of a tandem mill is changed, an appropriate control model would permit the changes in exit strip gage, the rolling force and the spindle torque to be calculated for each stand. Needless to say, such models are finding increased use in the programming of computers for tandem mill control.

10-2 Empirical Methods for Predicting Mill Power Requirements

Because the spindle torques of a mill are so closely related to the energy of deformation of the workpiece, it is possible to compute reasonably accurate torque values from experimentally-established energy of deformation curves. Such curves are derived as follows.§ The electric power input to the main mill motor, the rolling speed and the reduction are noted for the various passes. The energy requirement in terms of horsepower-hours per ton of material rolled is then calculated for each pass and a chart is prepared in which the total horsepower-hours per ton up to and including that pass are plotted against the total elongation at the end of the pass. Figure 10-1 shows a curve obtained in this manner when rolling low-carbon steel on a three-stand tandem mill. Alternately, the data may be prepared in the form of a chart such as that shown in Figure 10-2 in

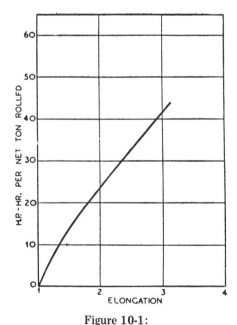

Figure 10-1:
Energy-Elongation Curve of Cold-Rolled Low-Carbon Steel.
(Average of tests on two different mills rolling 0.090 in., 0.093 in., 0.100 in., and 0.105 in. thick, 34.75 in. — 74 in. wide low-carbon steel strip to 0.035-0.042 in. thick in 3 passes, 380-390 fpm.)

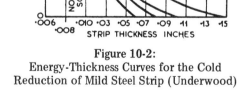

Figure 10-2:
Energy-Thickness Curves for the Cold Reduction of Mild Steel Strip (Underwood)

---

§ L. R. Underwood, "Rolling Mills — Methods of Roll Load and Power Calculation", Metal Industry, February 27, 1948, pp. 166-169; March 5, 1948, pp. 187-190 and March 19, 1948, pp. 231-234.

which the deformation energy per ton rolled is plotted against strip thickness. In this form, the data constitutes a family of curves in which more accurate interpolation may be made.

To determine the rolling horsepower required for a given pass from a curve such as that shown in Figure 10-1, the elongations at the beginning and end of a pass are calculated and the corresponding values of the horsepower-hours per ton are read off the chart. The rolling horsepower for the pass is then the difference between these two values multiplied by the weight of material in tons that would be rolled in one hour assuming the rolling to be continuous at the specified speed.

As Underwood has pointed out, however, in some published curves, the "horsepower-hours per ton rolled" relates to the mill motor input and, therefore, includes the power losses in the motor, reduction gears, pinions and roll neck bearings; whereas, in other curves, these losses have been allowed for. In using the former type of curve, care must be taken to allow for any change which will affect the power losses, such as the type of roll neck bearing or the inclusion or omission of a reduction gear. Thus, for example, curves obtained from mills with roller or fluid-film bearings would give power estimates which would be too low for a mill equipped with plain grease lubricated bearings.

In connection with energy curves, such as those shown in Figures 10-1 and 10-2, it is important to note whether or not strip tensions have been taken into account. In preparing such curves, the deformation energy should include the energy supplied by the coiler (or the tension energy supplied by the next mill stand). Conversely, the energy associated with the entry strip tension should be subtracted from the total deformation energy since this strip tension energy is theoretically recoverable.

This empirical method of mill power prediction has both advantages and limitations. Its chief merit lies in the fact that the energy curves are established on the basis of actual rolling mill tests and they automatically allow for such complications as those arising from roll flattening, strip lubrication and tensions. Thus they are reasonably valid when used for calculations relating to mills which are practically identical physically and which roll similar product under similar conditions. The principal limitation of the method results from the fact that, owing to the large number of variables in the rolling process, it is virtually impossible to develop curves for every combination of variables. Consequently, in practice, a relatively small number of curves representing only a limited range of roll sizes and rolling conditions has to suffice for a much wider range of variables. Thus, careful judgment must be exercised in using energy curves of this type and this judgment can only be developed by experience with the predicted and actual power requirements of a variety of mills. To overcome this difficulty, therefore, numerous investigators have studied the fundamentals of the rolling process with a view to deriving mathematical expressions embodying most of the variables but sometimes with arbitrary constants derived from rolling tests. Certain mathematical models of this type are discussed in succeeding sections of this chapter.

10-3 Calculation of Rolling Power by the Lever Arm Method

In its simplest form, this method assumes no roll flattening as indicated by the sketch of the roll bite shown in Figure 10-3. It assumes that the effect of the normal pressure along the arc of contact is equivalent to a single rolling force P acting parallel to but displaced from the line joining the roll centers by a distance a. This force, therefore, generates a moment Pa which must be counterbalanced by the roll torque. For both rolls, the total roll torque, therefore, is 2Pa.

If the rolls make one revolution, the total work done by the torques is $4\pi aP$. In one minute, rolls turning with a rotational speed of N revolutions per minute will perform $4\pi aPN$ units of work. If the distance a is measured in inches and the force P in pounds, the power required to generate the roll torques is

$$\frac{4\pi aPN}{12 \times 33,000} = 3.17 \times 10^{-5} \text{ aPN Horsepower} \qquad (10\text{-}1)$$

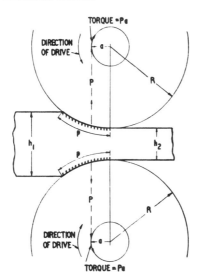

Figure 10-3:
Sketch Illustrating Lever Arm Method of Calculating Roll Torques.

For any given set of rolling conditions, the total rolling force P may be computed by multiplying the specific rolling force (derived by one of the methods discussed in Chapter 9) by the width of the strip. The question then arises as to the value of the length a of the lever arm to employ in the calculation of torque or horsepower. One method of choosing a value for a is to consider it to be a function λ of the theoretical length of the arc of contact assuming the rolls to be perfectly rigid, viz.,

$$a = \lambda \sqrt{Rtr} \qquad (10\text{-}2)$$

where R is the work roll radius and tr represents the draft. Ford and Bland■※* undertook extensive research on a well-instrumented mill to determine values of λ for various rolling conditions. Rolling force and torque values were obtained in rolling mild steel strip (carbon content 0.2%) 0.1 in. thick, rimmed carbon steel (carbon content 0.07%) 0.065 in. thick and carbon steel (carbon content 0.11%) 0.074 inch thick, all in the annealed condition. Reductions ranging from 10 to 70 per cent in single and multiple passes were given to all the steels.

During this research, the effect of roll finish was studied, one pair of rolls being ground to a mirror-like finish (final grinding being performed by a 400 grit wheel) while another pair, ground with a 180-grit wheel, had a smooth, matte surface providing a higher coefficient of friction in the roll bite. During most of the tests, the strips and rolls were lubricated with Vacuum 40A oil although some tests were conducted with dry rolls.

From the measured rolling force (P) and total torque (T_T) values, the length a of the lever arm was computed by means of the equation

$$a = \frac{T_T}{2P} \qquad (10\text{-}3)$$

and the corresponding value of λ being determined by the equation

$$\lambda = \frac{a}{\sqrt{Rtr}} \qquad (10\text{-}4)$$

■ H. Ford, "Researches into the Deformation of Metals by Cold Rolling", Proc. Inst. Mech. Eng. 158, 1948 pp. 115-143.

※ H. Ford, "Cold Rolling Technique", Sheet Metal Ind. 24, 1949, p. 734 ff.

* H. Ford, "Experimental Research into the Cold Rolling of Metals", J. West of Scotland Iron and Steel Inst., 52, 1944-45, p. 159 ff.

Some of the results obtained by Ford are shown in Table 10-1.‖

Table 10-1
Average Values of λ

Percent Carbon in Steel	Thickness (In.)	Roll Finish	Average Value of λ	Percentage of Values of λ Differing From The Average By Not More Than		
				5%	10%	20%
0.2	0.100	Mirror	0.40	59	86	100
0.2	0.100	Matte	0.32	42	61	97
0.2	0.100	Matte φ	0.33	11	53	100
0.11	0.074	Mirror	0.36	21	50	79
0.07	0.065	Mirror	0.35	58	84	100

φ Rolls dry. All other results refer to rolls lubricated by Vacuum 40A oil.

In an effort to assess the validity of this approach, Ford computed, for each set of results, the percentage of the number of tests where the ratio λ differed from the average by not more than 5, 10 and 20 per cent, this data being recorded in the last three columns of Table 10-1. It will be noted from these columns that the ratio λ is more consistent for the thicker than the thinner material and that the roll finish affects the value of the average ratio, the rougher the roll, the less the value of λ.

The values of λ discussed above assume that the work rolls are rigid. Ford ∎ developed an expression for computing a modified value of λ, namely λ', which is defined as the ratio of the length $(x_2 + x_0)$ in Figure 9-4 to the projected length l' of the deformed arc of contact, this expression being

$$\lambda' = \lambda\sqrt{\frac{R}{R'}} + \frac{1}{2}\left(1 - \frac{R}{R'}\right) \qquad (10\text{-}5)$$

where R' is the effective radius of the deformed roll.●

Equation 10-5 is derived as follows. Referring to Figure 9-4 in Section 9-6, it will be seen that x_2 represents the distance from the plane of entry of the roll bite to the plane of the roll centers and x_0 is the distance from the latter to the exit plane. From equations 9-32 and 9-38, it may be seen that

$$x_2 = \frac{\sqrt{(d^2 R^2 f^2) + R\delta(l')^2}}{l'} \qquad (10\text{-}6)$$

and

$$x_0 = \frac{dRf}{l'} \qquad (10\text{-}7)$$

‖ E. C. Larke, "The Rolling of Strip, Sheet and Plate", The MacMillan Company, New York, 1957.

∎ H. Ford, "Researches into the Deformation of Metals by Cold Rolling" Proc. Inst. Mech. Eng., 159, 1948, pp. 115-143.

● This approach to the calculation of spindle torques was also used by M. D. Stone in his paper "Rolling of Thin Strip", Iron and Steel Engineer Year Book, 1953, pp. 115-128.

TORQUE EQUATIONS AND TANDEM MILLS

where d is a constant of the material of the rolls (1.67×10^{-4} in.^{2}/ton (2,240 lb.)) for steel rolls, δ is the draft (= $t_1 r$), l' is the length of the deformed arc of contact and f is the specific rolling force, so that

$$l' = x_2 + x_o \qquad (10\text{-}8)$$

$$= \sqrt{R'\delta}$$

From equations 10-8 and 9-41,

$$R' = R\left(1 + \frac{2df}{\delta}\right) \qquad (10\text{-}9)$$

and

$$x_o = \frac{1}{2}\sqrt{R'\delta}\left(1 - \frac{R}{R'}\right) \qquad (10\text{-}10)$$

If the length of the lever arm, a, is regarded as equal to $\lambda\sqrt{R'\delta}$, then $a + x_o$, the distance from the point of application of the roll force to the exit plane is $\lambda'\sqrt{R'\delta}$. Thus,

$$\lambda'\sqrt{R'\delta} = \lambda\sqrt{R\delta} + \frac{1}{2}\sqrt{R'\delta}\left(1 - \frac{R}{R'}\right) \qquad (10\text{-}11)$$

so that

$$\lambda' = \lambda\sqrt{\frac{R}{R'}} + \frac{1}{2}\left(1 - \frac{R}{R'}\right) \qquad (10\text{-}12)$$

Using the values of λ previously developed (Table 10-1), Ford and Bland computed the values of λ' as listed in Table 10-2. It will be seen that the average values of λ' are more consistent than the corresponding values of λ. Based on these findings, Ford considered that, for all practical purposes, only two values of λ' are necessary; $\lambda' = 0.43$ for the rolls with a matte finish and $\lambda' = 0.48$ for rolls having a mirror finish.

Table 10-2
Average Values of λ'

Percent Carbon in Steel	Thickness (In.)	Roll Finish	Average Value of λ'	Percentage of Values of λ' Differing From The Mean By Not More Than		
				5%	10%	20%
0.2	0.100	Mirror	0.47	55	93	97
0.2	0.100	Matte	0.43	90	100	100
0.2	0.100	Matte*	0.44	79	100	100
0.11	0.074	Mirror	0.48	46	71	88
0.07	0.065	Mirror	0.47	53	95	100

* No Lubrication

It should be noted, however, that in using these preferred values of λ', the appropriate value of λ should be calculated from equation 10-12 using the rearranged expression

$$\lambda = 0.5\sqrt{\frac{R}{R'}} - (0.5 - \lambda')\sqrt{\frac{R'}{R}} \qquad (10\text{-}13)$$

With this value of λ, the horsepower Hp is now given by

$$H_p = 3.17 \times 10^{-5} \lambda\, PN\sqrt{Rtr} \qquad (10\text{-}14)$$

10-4 Torque Calculation from the Frictional Stresses Along the Arc of Contact

Once the pressure distribution along the arc of contact has been established (as, for example, by one of the models discussed in Chapter 9) and assuming a uniform coefficient of friction, integration of the frictional stresses becomes a method of computing the rolling torque.⊕

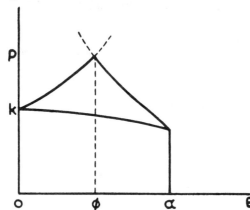

Figure 10-4:
Pressure Distribution Along the Arc of Contact (ϕ Represents Angular Position of the Neutral Point and α the Angular Position of the Entry Plane).

Referring to Figure 10-4, representing the pressure distribution along the arc of contact, it can be seen that, since there is a neutral point separating two regions of slip in opposite directions, the torque supplied by the rolls is equivalent to the difference of the frictional stresses integrated over the two regions multiplied by the radius of the undeformed roll. Thus, the specific torque per spindle G may be expressed by

$$G = R\left[\int_\phi^\alpha \mu p R'\, d\theta - \int_o^\phi \mu p R'\, d\theta\right] \qquad (10\text{-}15)$$

where

α is the angle corresponding to the entry of the roll bite
ϕ is the angular position of the neutral point
R' is the deformed roll radius
θ is the angular position of an element of strip in the roll bite
P is the pressure on the roll surface
μ is the coefficient of friction

If there are no tensile stresses applied to the strip at entry into or exit from the bite, then, resolving the forces in the strip in the roll bite

$$-\int_o^\alpha p \sin\theta\, R'\,d\theta + \int_\phi^\alpha \mu p \cos\theta\, R'\,d\theta - \int_o^\phi \mu p \cos\theta\, R'\,d\theta = 0 \qquad (10\text{-}16)$$

Since $\sin\theta \simeq \theta$ and $\cos\theta \simeq 1$,

⊕ W. C. F. Hessenberg, Introduction to "Research on the Rolling of Strip", The British Iron and Steel Research Association, 1959.

TORQUE EQUATIONS AND TANDEM MILLS

$$\int_0^\alpha p\theta R' d\theta = \int_\phi^\alpha \mu p R' d\theta - \int_0^\phi \mu p R' d\theta \qquad (10\text{-}17)$$

The combination of equations 10-15 and 10-17 gives

$$G = RR' \int_0^\alpha p\theta d\theta \qquad (10\text{-}18)$$

Where an entry stress σ_i and an exit stress σ_o are developed in the strip, the equation 10-18 must be modified, viz.

$$\begin{aligned} G &= RR' \int_0^\alpha p\theta d\theta + \left(\frac{t_i \sigma_i - t_o \sigma_o}{2}\right)R \\ &= RR'\left(\int_0^\alpha p\theta d\theta + \frac{t_i \sigma_i - t_o \sigma_o}{2R'}\right) \end{aligned} \qquad (10\text{-}19)$$

where t_i and t_o are the entry and exit strip thicknesses, respectively. As Hessenberg[⊕] has pointed out, equation 10-18 might seem to lack the physical significance of equation 10-17 because the coefficient of friction μ does not appear in it explicitly; it is, however, implicit in p. The main advantage of equation 10-18 is that it reduces errors in the computation of torque which may occur in the use of equation 10-15 involving the relatively small difference of two large numbers.

Solutions of equation 10-18 have been developed by a number of investigators. Some of these are discussed in the following sections.

10-5 The Solution of Bland and Ford

As described in Section 9-13, Bland and Ford[□] developed equations for the normal pressure distribution along the sides of the friction hill in the absence of front (exit) and back (entry) tensions, (see Figure 10-5), as follows

$$s^+ = \frac{k_r h \, e^{\mu H}}{h_o} \qquad (10\text{-}20)$$

and

$$s^- = \frac{k_r h \, e^{\mu(H_i - H)}}{h_1} \qquad (10\text{-}21)$$

where k_r is the constrained yield stress

        h is the thickness of the strip in the roll gap
        h_i is the entry thickness of the strip
        μ is the coefficient of friction
        h_o is the exit thickness of the strip and
        H is given by the relationship

[⊕] W. C. F. Hessenberg, Introduction to "Research on the Rolling of Strip", The British Iron and Steel Research Association, 1959.

[□] D. R. Bland and H. Ford, "The Calculation of Roll Force and Torque in Cold Strip Rolling with Tensions", Proc. Inst. Mech. Eng. 159, 1948, pp. 144-153.

$$H = 2\sqrt{\frac{R'}{h_o}} \arctan\left(\sqrt{\frac{R'}{h_o}}\,\phi\right) \qquad (10\text{-}22)$$

and H_i is the value of H at entry into the roll bite. (In equation 10-22, R' represents the deformed roll radius and ϕ is the angular position of the element of strip.)

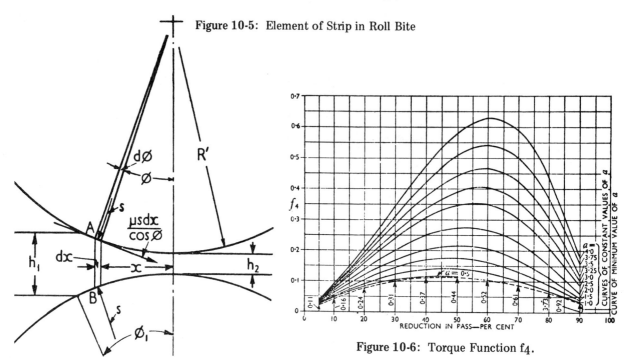

Figure 10-5: Element of Strip in Roll Bite

Figure 10-6: Torque Function f_4.

Bland and Ford used Simpson's rule to evaluate the integrals $\int_{\phi_i}^{\phi_n} \frac{s^-}{} d\phi$ and $\int_o^{\phi_i} s^+ d\phi$ separately for each side of the neutral point (ϕ_i corresponding to the entry plane of the roll bite and ϕ_n to the neutral point). These calculations proved to be tedious, however, and it was found convenient to express their final solution in the form

$$G = \bar{k} R \frac{h_i^2}{h_o} f_4(a,r) \qquad (10\text{-}23)$$

where $\bar{k}$ is a mean value of the yield stress and $f_4(a,r)$ is a torque function involving the coefficient of friction and the roll bite geometry. The torque function is shown graphically in Figure 10-6 plotted against reduction for various values of $a\left(=\mu\sqrt{\frac{R'}{h_o}}\right)$.

Where strip tensions must be considered, Ford and his coworkers[*] showed that, by making certain approximations to Orowan's general theory of rolling, it is possible to calculate roll torque by a method which is both rapid and easy to apply using the basic information of the yield stress characteristic of the material to be rolled, the coefficient of friction and the dimensions of the pass.

---

[*] H. Ford, F. Ellis and D. R. Bland, "Cold Rolling with Strip Tension, Part I — A New Approximate Method of Calculation and a Comparison with Other Methods", J. Iron and Steel Institute, May 1951, 168, pp. 57-72.

TORQUE EQUATIONS AND TANDEM MILLS

As an alternate form of equation 10-23, Ford, et al., developed the following expression for roll torque

$$G = R\bar{k}(h_i - h_o)\left(1 - \frac{\sigma_i}{\bar{k}}\right) f_5(a,r,b) \tag{10-24}$$

where the mean yield stress $\bar{k}$ is given by

$$\bar{k} = \frac{\int_0^{\phi_i} k\, d\phi}{\phi_i} \tag{10-25}$$

and where k represents the yield stress, ϕ represents the angular coordinate of the element in the roll bite, ϕ_i the angular coordinate of the entry plane, σ_i the entry tensile stress and $f_5(a,r,b)$ a nondimensional roll torque function such that

$$f_5(a,r,b) = \left[h_o \left\{ b \int_0^{x_n} (1+x^2) x e^{2a\,\tan^{-1} x} dx \right. \right. \\ \left. \left. + (1-r)e^{2a\,\tan^{-1} x_i} \int_{x_n}^{x_i} (1+x^2) x e^{-2a\,\tan^{-1} x} dx \right\} \right] \tag{10-26}$$

In this expression, x represents the distance along the roll bite in the direction of rolling, the subscripts n and i relating to the neutral and entry planes, respectively. For b see equation 10-33.

The expression $f_5(a,r,b)$ is derived as follows. Ford had shown that, if certain assumptions were made, the roll pressure in the presence of strip tensile stresses σ_1 and σ_2 at the entry and exit of the bite, respectively, could be expressed by modifications to equations 10-20 and 10-21.

$$s^+ = \frac{kh}{h_o}\left(1 - \frac{\sigma_2}{k_2}\right) e^{\mu H} \tag{10-27}$$

$$s^- = \frac{kh}{h_1}\left(1 - \frac{\sigma_1}{k_1}\right) e^{\mu(H_1 - H)} \tag{10-28}$$

where k_1 and k_2 represent the constrained yield stress at the entry and exit planes, respectively. Substitutions of these expressions in equation 10-19 (and the use of a slightly different notation) yields

$$G = RR'\bar{k}\left(1 - \frac{\sigma_1}{\bar{k}}\right) \left[\int_0^{\theta_n} \frac{h}{h_o}\left(\frac{1 - \frac{\sigma_2}{\bar{k}}}{1 - \frac{\sigma_1}{\bar{k}}}\right) e^{\mu H} \theta\, d\theta \right. \\ \left. + \int_{\theta_n}^{\theta_1} \frac{h}{h_1} e^{\mu(H_1 - H)} \theta\, d\theta \right] \tag{10-29}$$

578 CHAPTER 10

In this expression

$$\theta_n = \sqrt{\frac{h_o}{R'}} \tan\left(\sqrt{\frac{h_o}{R'}} \times \frac{H_n}{2}\right) \tag{10-30}$$

where

$$H_n = \frac{H_1}{2} - \frac{1}{2\mu} \log_e\left\{\frac{h_i}{h_o}\left(\frac{1 - \sigma_o/K_2}{1 - \sigma_1/K_1}\right)\right\} \tag{10-31}$$

Substituting

$$a = \mu\sqrt{\frac{R'}{h_o}} \tag{10-32}$$

$$b = \frac{1 - \sigma_o/\bar{k}}{1 - \sigma_1/\bar{k}} \tag{10-33}$$

$$r = \frac{h_1 - h_o}{h_1} \tag{10-34}$$

$$x = \sqrt{\frac{R'}{h_o}}\,\theta \tag{10-35}$$

$$dx = \sqrt{\frac{R'}{h_o}}\,d\theta \tag{10-36}$$

Then

$$\frac{h}{h_o} = \frac{h_o + R'\theta^2}{h_o} = 1 + x^2 \tag{10-37}$$

$$\frac{h}{h_1} = \frac{h_o}{h_1} \times \frac{h}{h_o} = (1 - r)(1 + x^2) \tag{10-38}$$

$$\mu H = 2a \tan^{-1} x \tag{10-39}$$

$$\mu H_1 = 2a \tan^{-1} x_1 \tag{10-40}$$

So that

$$G = R\bar{k}\left(1 - \frac{\sigma_1}{\bar{k}}\right)\left[h_o\left\{b \int_0^{x_n} (1+x^2)x\, e^{2a \tan^{-1} x}\, dx \right.\right.$$

$$\left.\left. + (1-r)e^{2a \tan^{-1} x_1} \int_{x_n}^{x_1} (1+x^2)x\, e^{2a \tan^{-1} x}\, dx\right]\right. \tag{10-41}$$

$$= R\bar{k}(h_1 - h_o)\left(1 - \frac{\sigma_1}{\bar{k}}\right) f_5(a,r,b)$$

TORQUE EQUATIONS AND TANDEM MILLS

For convenience in calculating the functions, the nondimensional parameter b (equation 10-33) was replaced by its Naperian logarithm B, i.e.

$$B = \log_e b = \log_e \left\{ \frac{1 - \frac{\sigma_2}{\bar{k}}}{1 - \frac{\sigma_1}{\bar{k}}} \right\} \qquad (10\text{-}42)$$

Charts for the determination of the function $f_5(a,r,B)$ are exemplified by those shown in Figure 10-7.

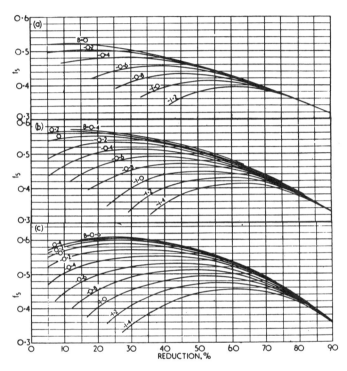

Figure 10-7:
Calculated Values of Function $f_5(a,r,B)$:
(a) a = 0.5; (b) a = 0.75; (c) a = 1.0

10-6 Hill's Solution

In 1948, Hill[*] proposed a model relating specific spindle torque to other rolling parameters. By consideration of various artificial distributions of pressure along the arc of contact, he showed that the ratio of the specific rolling force to the specific spindle torque depends only slightly on the precise shape of the friction hill.

Referring to Figure 10-8, O represents the point at which the axis of the roll of radius R intersects the section and O' the center of the circular arc of contact AB (corresponding to the radius R'). Since the pressure distribution is assumed to be symmetrical about the midpoint C of the arc of contact, O' must lie on an extension of the line CO. Consequently, the exit point B is some distance in advance of the line through the two roll centers.

Neglecting the elastic deformation of the strip, the roll surface is horizontal at the point of exit and O'B is vertical. Angle BO'O is $\alpha/2$, B is a distance $1/2 (R' - R) \alpha$ ahead of the roll center O; this being a fraction $1/2 (1 - R/R')$ of the arc of contact.

A small element of the roll surface of length $R\,d\theta$ exerting a pressure p on the workpiece is considered to be located at an angular displacement θ from the line O'B and ϕ is

---

[*] R. Hill, "Relations Between Roll Force, Torque and the Applied Tensions in Strip Rolling", Proc. Inst. Mech. Eng. 163, 1950, pp. 135-140.

regarded as the angular position of the neutral point. The normal force on the element is $pR'd\theta$ and the tangential force is $\mu pR'd\theta$ where μ is the coefficient of friction.

The lever arm of the force $pR'd\theta$ about the roll axis through O is approximately $\pm(R'-R)(1/2\alpha-\theta)$, the sign being dependent on the side of C on which the element lies. The lever arm of the force $\mu pR'd\theta$ is equal to R to the same order of approximation. Moments about O for the forces acting on the roll give

$$G = \int_\phi^\alpha \mu pRR' d\theta - \int_0^\phi \mu pRR' d\theta + \int_0^\alpha pR'(R'-R)\left(\theta - \frac{1}{2}\alpha\right)d\theta \qquad (10\text{-}43)$$

or

$$\frac{G}{RR'} = \mu\left(\int_\phi^\alpha pd\theta - \int_0^\phi pd\theta\right) + \left(\frac{R'}{R} - 1\right)\int_0^\alpha p\left(\theta - \frac{\alpha}{2}\right)d\theta \qquad (10\text{-}44)$$

Now the equation of equilibrium for the horizontal forces acting on the strip is,

$$T = 2\int_0^\alpha p\sin\theta\, R'd\theta - 2\left(\int_\phi^\alpha \mu p\cos\theta\, R'd\theta - \int_0^\phi \mu p\cos\theta\, R'd\theta\right) \qquad (10\text{-}45)$$

where T is the net tensile force per unit width of strip and is given by

$$T = T_f - T_b \qquad (10\text{-}46)$$

T_f and T_b representing the front and back tensions, respectively. For small values of θ, equation 10-45 may be approximated by

$$\frac{T}{2R'} = \int_0^\alpha p\theta d\theta - \mu\left(\int_\phi^\alpha pd\theta - \int_0^\phi pd\theta\right) \qquad (10\text{-}47)$$

Since the specific rolling force is given by

$$P = \int_0^\alpha p\cos\theta\, R'd\theta + \left(\int_\phi^\alpha \mu p\sin\theta\, R'd\theta - \int_0^\phi \mu p\sin\theta\, R'd\theta\right) \qquad (10\text{-}48)$$

it can be shown, by combining equations 10-44, 10-47, and 10-48 that

$$G = R'^2\int_0^\alpha p\theta d\theta - \frac{1}{2}RT - \frac{1}{2}(R' - R)\alpha P \qquad (10\text{-}49)$$

Hill examined the effect of applied tensions on the rolling process in general, including its influence on torque. Assuming a friction hill is represented by the solid curve of Figure 10-9 with a general pressure distribution p_0 acting to provide a neutral angle ϕ_0 in the absence of tensions, then the specific torque

TORQUE EQUATIONS AND TANDEM MILLS

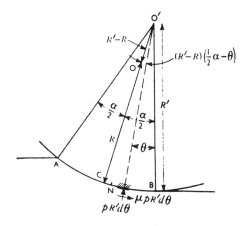

Figure 10-8:
Deformation of the Roll, According to Hitchcock.

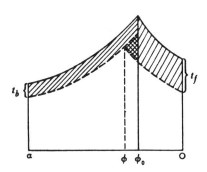

Figure 10-9:
Assumed Effect of Applied Tensions on the Pressure Distribution.

$$G_o = R'^2 \int_0^\alpha p_o \theta d\theta - \frac{1}{2} \alpha (R' - R) P_o \qquad (10\text{-}50)$$

where P_o is the specific rolling force. Assuming front and back tensions T_f and T_b per unit strip width, respectively, the pressure distribution is changed to that represented by the broken curve. The new value of the specific torque is

$$G = R'^2 \int_0^\alpha p \theta d\theta - \frac{1}{2} \alpha (R'-R) P - \frac{1}{2} R(T_f - T_b) \qquad (10\text{-}51)$$

Therefore

$$G_o - G = R'^2 \int_0^\alpha (p_o - p) \theta d\theta - \frac{1}{2} \alpha (R'-R)(P_o - P) + \frac{1}{2} R(T_f - T_b) \qquad (10\text{-}52)$$

This expression can be shown, by neglecting the forward slip and by making certain simplifying assumptions, to yield

$$G = G_o + \frac{1}{2} R \left[(1-r)T_b - T_f\right]$$
$$= G_o + \frac{1}{2} Rh(t_b - t_f) \qquad (10\text{-}53)$$

where h is the rolled strip thickness and t_b and t_f are the entry and exit tensile stresses, respectively. This relationship is confirmed by the experimental data shown in Figure 10-10 where G is shown plotted against $(t_b - t_f)$ for various reductions in the rolling of annealed, mild steel 0.0625 in. thick on an experimental mill.♦

♦ H. Ford, "The Sheffield Experimental Cold Rolling Mill", 1946, Iron and Steel Institute, Special Report No. 34, Section 3, page 69.

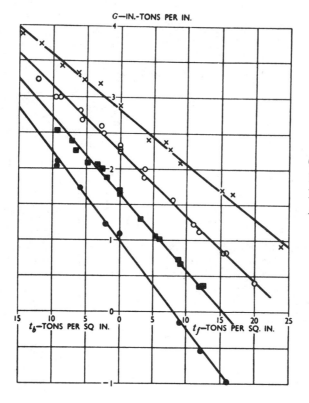

Figure 10-10:
Comparison of Theoretical Formula and Experimental Results for the Effect of Applied Tensions on Torque

× 50 per cent reduction.
○ 40 per cent reduction.
■ 30 per cent reduction.
● 20 per cent reduction.

10-7 Sims' Graphical Method of Solution

Sims‡ showed in 1953 that, if the yield stress/strain relationship for the material and the coefficient of friction in the roll gap are known, a relation may be derived between the roll force, torque and the geometry of the mill and strip which includes the effect of roll distortion. The form of the solution is complex but by graphical techniques, the rolling torques may be calculated in a few minutes. Sims utilized Ford's basic equation for the rolling torque per spindle in the absence of tensions, G_O (see Section 10-5), as follows

$$G_O = (RH^2/h)\bar{k}_G f_G(a,r) \tag{10-54}$$

where

  R is the radius of the undeformed roll
  H is the entry thickness of the pass
  h is the exit thickness of the strip
  $\bar{k}_G$ is the mean yield strength of the material in plane compression
$f_G(a,r)$ is an integral of a and r.

With respect to the term $f_G(a,r)$, r is the reduction and a is a dimensionless parameter given by

$$a = \mu\sqrt{\frac{R'}{h}} \tag{10-55}$$

where μ is the coefficient of friction, and R' is the deformed roll radius.

---

‡ R. B. Sims, "Calculation of Roll Force and Torque in Cold Rolling by Graphical and Experimental Methods", J. Iron and Steel Institute, 178, 1954, pp. 19-34.

TORQUE EQUATIONS AND TANDEM MILLS

Equation 10-54 is equivalent to

$$\frac{G_o}{\bar{k}_G H^2} = \frac{a^2}{\mu^2} f_G(a,r) \qquad (10\text{-}56)$$

and the quantity $G_o/\bar{k}_G H^2$ may be expressed as a function of the variables R/H, r, $\bar{k}_G$ and μ. This quantity is shown plotted against R/H for given values of $\bar{k}_G$, μ, and r in Figure 10-11.

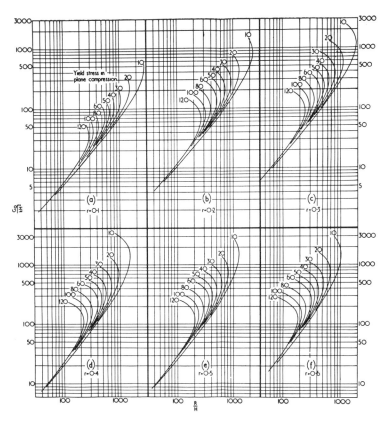

Figure 10-11:
Variation of the Parameter $G_o/\bar{k}_g H^2$ with R/H for Various Values of the Yield Stress and the Reduction ($\mu = 0.05$).

Equation (10-56) neglects the moment of the normal pressure about the roll axis, which may be appreciable when R'/R is very large. A correction to G_o may be made for this effect by means of the relation due to Hill[±], viz.

$$G'_o = G_o \left[1 - \left(\frac{R'}{R} - 1\right)\left(\frac{\alpha P_o R}{2G_o}\right) \right] \qquad (10\text{-}57)$$

It should be noted in the foregoing discussion that, for accuracy, the mean yield stress $\bar{k}_G$ should be derived from the expression

$$\bar{k}_G = \frac{2}{\alpha^2} \int_0^\alpha k\theta d\theta \qquad (10\text{-}58)$$

Where entry and exit strip tensile stresses, t_b and t_f, respectively, occur, then the specific torque per spindle G may be expressed by the approximate equation

---

[±] R. Hill, "The Mathematical Theory of Plasticity", 1950, Oxford, The Clarendon Press, p. 251.

$$G = G_o + \frac{Rh_2}{2}(t_b - t_f) \tag{10-59}$$

This equation is adequate except at very high values of the back tension, when the ratio of roll radius to entry thickness is of the order of 150:1 or more and the strip is hard.[⊗]

10-8 Determining Torque by Resolving the Forces on the Strip

A simple, approximate expression for the specific rolling torque may be developed by assuming a uniform pressure along the arc of contact.[°] As illustrated by Figure 10-12, it may be shown that, in the absence of strip tensions, the net force exerted by each roll on the strip in the direction of rolling must equal half the draft (tr) multiplied by the average dynamic constrained yield strength of the strip (σ_c). Thus

$$G = \frac{Dtr\sigma_c}{2} \tag{10-60}$$

where G is the specific total torque (for both spindles), and
D is the diameter of the work rolls.

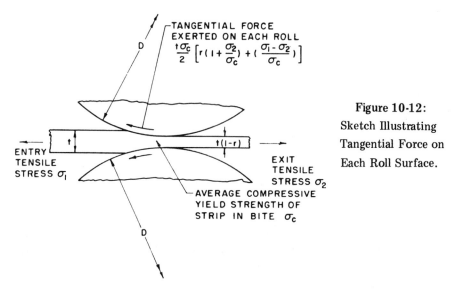

Figure 10-12: Sketch Illustrating Tangential Force on Each Roll Surface.

In the presence of strip tensile stresses σ_1 or σ_2 at the entry and exit ends of the bite, respectively, equation (10-60) may be modified to

$$G = \frac{Dt\sigma_c}{2}\left[r\left(1 + \frac{\sigma_2}{\sigma_c}\right) + \left(\frac{\sigma_1 - \sigma_2}{\sigma_c}\right)\right] \tag{10-61}$$

Equation (10-61) ignores the fact, however, that the strip is moving into the roll bite at a speed less than the peripheral speed of the rolls. Accordingly, from an energy balance viewpoint, the entry stress should, in effect, be decreased by the factor (1-r) so that equation 10-60 then becomes

---

[⊗] W. C. F. Hessenberg and R. B. Sims, "The Effect of Tension on Torque and Roll Force in Cold Rolling", J. Iron and Steel Institute, 168, 1951, pp. 155-164.
[°] W. L. Roberts, "A Simplified Cold Rolling Model", Iron and Steel Engineer Year Book, 1965, pp. 925-937.

TORQUE EQUATIONS AND TANDEM MILLS

$$G = \frac{Dt\sigma_c}{2}\left[r + \frac{(1-r)(\sigma_1-\sigma_2)}{\sigma_c}\right] \qquad (10\text{-}62)$$

an equation which, in slightly different form, was published by Stone.◆

In a more accurate model of the rolling process, involving the concept of a friction hill ▲ (as discussed in Section 9-18), the specific torque per spindle G was given by the relationship

$$G = \frac{Dtr}{4}\left(\frac{f}{\sqrt{\frac{D'tr}{2}}} + \sigma_1\right) \qquad (10\text{-}63)$$

where D' is the deformed roll diameter
  f is the specific rolling force, and
  σ_1 is the tensile stress in the strip (assumed equal on each end of the roll bite)

In connection with this model, the deformed roll diameter D' may be computed from the relationship

$$D' = D\left(1 + 2\sqrt{\frac{f}{Etr}} + \frac{2f}{Etr}\right) \qquad (10\text{-}64)$$

where E is the elastic modulus of the rolls. From tables or by the use of a computer, the values of f and D' may be computed for any rolling situation where the other parameters are known. The specific torque G may then be calculated from equation 10-63 using the values of f and D' so determined. It should be noted, however, that if the coefficient of friction μ is disregarded and σ_1 is very small compared with σ_c (the dynamic constrained yield stress corresponding to the strain rate of the rolling operation), then

$$\frac{f}{\sqrt{\frac{D'tr}{2}}} \simeq \sigma_c \qquad (10\text{-}65)$$

and

$$G = \frac{Dtr\sigma_c}{4} \qquad (10\text{-}66)$$

10-9 Calculation of Torque From an Energy Balance of the Rolling Process

If the inefficiencies of the rolling process are compensated for by a suitable correction in the constrained yield strength of the workpiece, and if the slip between the rolls and the strip is disregarded, the roll torque may be calculated from the theoretical energy required to deform the strip. Dahl ˟ has essentially proposed this method for computing rolling torques. On the other hand, consideration of the theoretical energy of deformation, together with the actual mill efficiency leads to expressions for the specific total torque identical with those discussed in Section 10-8.

If the modified, dynamic constrained yield strength of the strip is σ_c', then the energy required per unit volume to produce a reduction r is $\sigma_c' \log_e(1/(1-r))$. If a strip of unit width and thickness t is fed into the work rolls so that it emerges with a velocity V, then the volume rolled in unit time is $Vt(1-r)$ and the corresponding work expended is $Vt(1-r)\sigma_c' \log_e(1/(1-r))$. If each work roll has a diameter D and the specific torque per spindle is G, then the energy delivered by

---

◆ M. D. Stone, "Rolling of Thin Strip, Part II", Iron and Steel Engineer Year Book, 1956, pp. 981-1002.
▲ W. L. Roberts, "Computing the Coefficient of Friction in the Roll Bite from Mill Data", Blast Furnace and Steel Plant, 55, 1967, pp. 499-508.
˟ T. Dahl, "Die Ermittlung und Grasse der Reibungszahl beim Walzen", Stahl und Eisen, 57 (Pt I) 1937, pp. 205-209.

both rolls in unit time is 4GV/D. Equating this with the theoretical energy to deform the strip yields

$$G = \frac{Dt(1-r)\sigma_c'}{4} \log_e\left(1/(1-r)\right) \quad (10\text{-}67)$$

In using this method, however, it must be remembered that the value of σ_c' used compensates for the inefficiency of the mill which varies with the reduction given to the strip (see Section 10-13).

Instead of using a fictitious value of the constrained yield strength, it is perhaps more desirable to use the actual value σ_c of this parameter (corrected for the rate of deformation occurring in the roll bite) and take the mill efficiency directly into account. As shown in Section 7-3, the efficiency of the mill for a reduction r may be closely approximated by the fraction (1—r/2). Thus an actual energy balance, again assuming zero slip, would yield

$$G = \frac{Dt(1-r)\sigma_c}{4(1-r/2)} \log_e\left(1/(1-r)\right) \quad (10\text{-}68)$$

The logarithmic term may be expanded so as to give

$$\log_e\left(1/(1-r)\right) = -\log_e(1-r)$$

$$= r + r^2/2 + r^3/3 + r^4/4, \text{ etc.} \quad (10\text{-}69)$$

$$= r\left(1 + r/2 + r^2/3 + r^3/4, \text{ etc.}\right)$$

Hence

$$G = \frac{Dt(1-r)\sigma_c}{4(1-r/2)} r \left(1 + r/2 + r^2/3 + r^3/4, \text{ etc.}\right) \quad (10\text{-}70)$$

$$= \frac{Dt\sigma_c r}{4(1-r/2)} \left(1 - r/2 - r^2/6 - r^3/12, \text{ etc.}\right)$$

For reductions in the normal range of, say, 10 to 50 per cent, the terms involving r^2, r^3, etc., may be disregarded. Accordingly,

$$G \simeq \frac{Dtr\sigma_c}{4} \quad (10\text{-}71)$$

Thus, from equation 10-71, for values of r such that the above approximation is valid, the specific spindle torque should vary essentially linearly with reduction. Figures 10-13 and 10-14 verify this relationship and from the slopes (S) of such curves, the dynamic constrained yield strength σ_c may be determined from the relationship

$$\sigma_c = \frac{4S}{Dt} \text{ (psi)} \quad (10\text{-}72)$$

With a tensile stress σ_2 on the exit side of the mill, the energy supplied by the coiler or exit tension in unit time is $Vt(1-r)\sigma_2$. Similarly a theoretically recoverable energy utilized in establishing an entry tensile stress σ_1 on the entry side of the mill is $Vt(1-r)\sigma_1$. Thus the net tensional energy supplied is $Vt(1-r)(\sigma_2-\sigma_1)$. The energy balance thus yields

$$G = \frac{Dt}{4}\left\{\sigma_c''r - \frac{(1-r)(\sigma_2-\sigma_1)}{(1-r/2)}\right\} \quad (10\text{-}73)$$

TORQUE EQUATIONS AND TANDEM MILLS

where σ_c'' is the dynamic constrained yield strength modified by the presence of the tensile stresses σ_1 and σ_2[†].

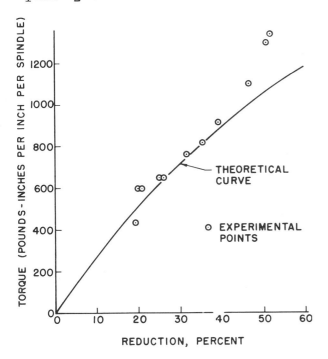

Figure 10-13:
Torque vs Reduction Curve for Full-Hard, Low-Carbon Steel. 0.015 in. Thick (Roll Diameter 3.25 in.: Minimum Strip Tensions) (Theoretical Curve, Corrected for Bearing Losses, Based on Equation 10-61)

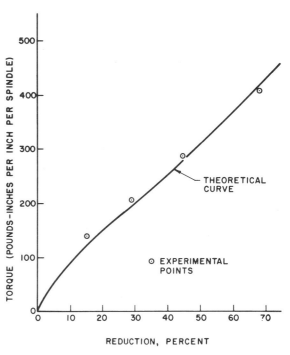

Figure 10-14:
Torque vs Reduction Curve for Annealed Black Plate 0.0065 In. Thick (Roll Diameter 3.25 In.; Entry Tensile Stress 17 ksi; Exit Tensile Stress 5 ksi) (Theoretical Curve, Corrected for Bearing Losses, Based on Equation 10-61)

10-10 The Derivation of Torque from the Shear Plane Theory

From the shear plane theory of flat rolling discussed in Section 9-22, it is possible to compute the specific spindle torque.[x] Referring to Figures 9-73 and 9-74, the roll torque is calculated by considering the forces acting externally and on the center line of the strip (rather than the forces acting directly on the roll), since this does not require an exact knowledge of the position of the neutral point. With respect to a triangular element of the strip, the line of action of the roll separating force Q is considered to act at a distance t from the line of the roll centers, thus the torque transmitted from the roll

$$C = Qt - (f - bs^n)R\frac{h}{2} \qquad (10\text{-}74)$$

where f is the exit tensile stress in the strip/shear yield stress of the strip
  b is the entry tensile stress in the strip/shear yield stress of the strip
  s is the ratio of stock thickness before and after one pair of shear planes
  n is the number of triangular sections of the strip in one half of the roll bite
  R is the natural roll radius and
  h is the final or rolled product thickness

[†] R. J. Bentz and W. L. Roberts, "Predicting Rolling Forces and Mill Power Requirements for Tandem Mills", Blast Furnace and Steel Plant, 58, 1970, pp. 559-568.

[x] J. W. Green, L. G. M. Sparling and J. F. Wallace, "Shear Plane Theories of Hot and Cold Flat Rolling", J. Mech. Eng. Sci., Vol. 6, No. 3, 1964, pp. 219-235.

The quantity Qt is given by the expression

$$Q_t = \frac{h}{2\tan\alpha}\left\{\frac{S-1}{2(S+1)}\left(Q_{1L} - S^n Q_{(n+1)R}\right) + \sum_{m=2}^{d} Q_{mL}(S^{m-1}-1) + \sum_{m=d+1}^{n+1} Q_{mr}(S^{m-1}-1)\right\} \quad (10\text{-}75)$$

where

α is the half angle subtended at the center of the roll by the arc of contact.

In the equations 10-74 and 10-75, the units are in terms of the mean yield shear stress so that the specific torque per spindle G is given by

$$G = Ck \quad (10\text{-}76)$$

where k is the shear yield stress in plane strain.

Green and his colleagues developed expressions for the specific force and spindle torque similar to those given by Ford and Bland[□] the equation for the specific spindle torque being in the form

$$G = 4\bar{k}_G R \frac{H^2}{h} S_G \quad (10\text{-}77)$$

where H is the initial strip thickness
$\bar{k}_G$ is the mean shear yield stress used for torque calculations and
S_G is a dimensionless function for the torque

The function S_G is analogous to the function $f_4(a,r)$ developed by Ford and Bland (See Section 10-5). Figure 10-15 shows a comparison of S_G with $f_4(a,r)$ for reductions in the range 0-90 per cent and for values of $a(=\mu\sqrt{R'/h})$ ranging from 0.5 to 3.5.

10-11 Torque in the Case of a Viscous Workpiece

The hydrodynamic theory of rolling, postulated by Kneschke,[♦] discussed by Weber[※] and presented in Section 9-21, is also capable of yielding an expression for torque. In this model, the workpiece is regarded as a viscous mass in motion characterized by a static yield stress k_0 and a dynamic viscosity η. Sticking friction between the workpiece and the rolls is assumed along the entire length of the arc of contact and, on this basis, the total torque M_d is given by the expression

$$M_d = \bar{k}_o l_d^2 b_m + 12 b_m R \eta \nu \sqrt{\frac{R}{h_o}} X_\epsilon \quad (10\text{-}78)$$

where $\bar{k}_o$ is the mean static yield stress (at a speed of deformation tending to zero),
l_d is the length of the arc of contact between the workpiece and the strip,
b_m is the width of the strip,
R is the roll radius,
ν is the peripheral speed of the roll,
h_o is the initial thickness of the strip prior to rolling, and
X_ϵ is a function of ϵ, the fractional reduction in thickness.

[□] D. R. Bland and H. Ford "Calculation of Roll Force and Torque in Cold Strip Rolling with Tensions", Proc. Inst. Mech. Eng., London 1948, Vol. 159, pp. 144-153.
[♦] A. Kneschke, Freiberger Forschungsh. B. Met., 1957, 16, pp. 5-34.
[※] K. H. Weber, "Hydrodynamic Theory of Rolling", J. Iron and Steel Institute, January 1965, pp. 27-35.

TORQUE EQUATIONS AND TANDEM MILLS

In equation 10-78, the first term on the right-hand side represents the contribution to the total torque of the static component of the yield stress whereas the second term relates to the viscous behavior of the workpiece.

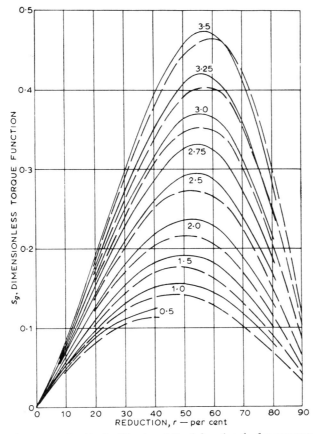

Figure 10-15: Dimensionless Torque Function S_G Shown Plotted Against Reduction.

S_g compared with the Bland and Ford function f_4, for constant values of $\mu \sqrt{R'/h}$.

——————— Shear plane.
— — — — Bland and Ford.

10-12 Temper Rolling Torques

In the case of temper rolling with light reductions and relatively high (~ 0.25 to 0.35) coefficients of friction, the torque is higher than would be predicted by the various equations presented in the earlier sections of this chapter. The formula for the specific spindle torque G is

$$G = \frac{Dtr(S_c - S_A)}{4}\left[1 + \frac{\mu L}{t}\right] \quad (10\text{-}79)$$

where D is the work roll diameter
  t is the initial strip thickness
  r is the reduction
  S_C is the constrained dynamic yield strength (corrected for strain rate)
  S_A is the average tensile stress in the strip
  μ is the coefficient of friction, and
  L is the length of the arc of contact given by the expression

$$L = \frac{1}{2}\left[\frac{Dr\mu}{2} + \sqrt{\left(\frac{Dr\mu}{2}\right)^2 + 2Dtr}\right] \quad (10\text{-}80)$$

Equation 10-79 is derived as outlined below.

At the temper rolling strain rates, the energy (E_W) in inch-pounds per cubic inch required to deform unit volume of the strip, assuming no redundant work, is given by

$$E_W = S_c \log_e\left(\frac{1}{1-r}\right) \qquad (10\text{-}81)$$
$$\simeq S_c r$$

since r is small compared to unity.

In deriving an expression for the frictional losses occurring in the roll bite, it is assumed that the neutral point is very close to the center of the arc of contact. Referring to Figure 9-56, the relative velocity of the rolls and strip at the exit end of the roll bite is approximately $Vr/2$. At a distance x from the exit end of the roll bite, the relative velocity is $\frac{Vr}{2}(1-\frac{2x}{L})$. In unit time, the frictional energy dissipation on the element of dimension dx in the rolling direction is

$$\mu\left(1 - \frac{2x}{L}\right)\frac{rV}{2}(S_c - S_A)\exp\left(\frac{2\mu x}{t(1-r)}\right) \times dx$$

Considering each side of the neutral point and both surfaces, the total frictional energy dissipation is

$$4\mu\frac{(S_c - S_A)}{2}rV\int_0^{L/2}\left[\exp\left(\frac{2\mu x}{t(1-r)}\right)\right]\left(1 - \frac{2x}{L}\right)dx$$

$$= 2\mu(S_c - S_A)rV\left[\int_0^{L/2} e^{2\mu x/t(1-r)}dx - \frac{2}{L}\int_0^{L/2} xe^{2\mu x/t(1-r)}dx\right] \qquad (10\text{-}82)$$

$$= (S_c - S_A)rVt(1-r)\left[\frac{t(1-r)}{\mu L}\left(e^{\frac{\mu L}{t(1-r)}} - 1\right) - 1\right]$$

Expressed in terms of unit volume of the strip, the frictional energy dissipation (E_f) is therefore

$$E_f = (S_c - S_A)r\left[\frac{t(1-r)}{\mu L}\left(e^{\frac{\mu L}{t(1-r)}} - 1\right) - 1\right] \qquad (10\text{-}83)$$

$$= r\left[\frac{f}{L} - (S_c - S_A)\right]$$

Therefore, the rate of frictional energy dissipation occurring in the roll bite during the reduction of strip of unit width is

$$(S_c - S_A)rVt(1-r)\left[\frac{t(1-r)}{\mu L}\left(\exp\left[\mu L/t(1-r)\right] - 1\right) - 1\right]$$

In the derivation of this expression, the "arc" of contact is assumed to be straight and the coefficients of friction on each side of the neutral point are considered identical. Since the volume of strip rolled in unit time is equal to $\frac{Vt(1-r)}{(1-r/2)}$, the frictional energy (E_f) given in inch-pounds per cubic inch unit volume rolled is given by

TORQUE EQUATIONS AND TANDEM MILLS

$$E_f = (S_c - S_A)r \left[\frac{t(1-r)}{\mu L} \left(\exp[\mu L/t(1-r)] - 1\right) - 1\right] \tag{10-84}$$

If the efficiency (j) of the process is defined as the fraction of the energy transferred from the rolls to the strip, then, from equations 10-79 and 10-81,

$$j = \frac{E_w}{E_w + E_f} = \frac{S_c r}{S_c r + (S_c - S_A)r \left[\frac{t(1-r)}{\mu L} \left(\exp[\mu L/t(1-r)] - 1\right) - 1\right]} \tag{10-85}$$

This expression may be simplified to

$$j = \frac{1}{1 + \frac{1}{S_c}\left[f/L - (S_c - S_A)\right]} \tag{10-86}$$

and if S_A is small compared with S_c, then

$$j \simeq \frac{S_c L}{f} \tag{10-87}$$

In establishing an expression for the specific total torque (T_T), it is desirable to consider an energy balance with respect to the quantity of strip rolled in unit time. The volume so rolled is $\frac{Vt(1-r)}{(1-r/2)}$ and the total deformation energy that must be supplied is, from equation 10-81, $\frac{Vt(1-r)rS_c}{(1-r/2)}$. Because the efficiency of the process is approximately $S_c L/f$ (equation 10-87), the total energy that must be supplied by the mill rolls in unit time is approximately $\frac{Vt(1-r)}{(1-r/2)} \frac{rf}{L}$. Moreover, since the peripheral speed of the work rolls is V, the roll radius D/2 and $r \ll 1$, then the specific total spindle torque (T_T) (exclusive of bearing losses) is given by

$$T_T = \frac{Dtrf}{2L} = \frac{Dtr(S_c - S_A)t(1-r)}{2L\mu}\left[\exp(\mu L/t[1-r]) - 1\right]$$

$$= \frac{Dtr(S_c - S_A)}{2}\left[1 + \frac{\mu L}{2t} + \frac{\mu^2 L^2}{6t^2}, \text{etc.}\right] \tag{10-88}$$

Since $\mu L/t$ is generally no more than about 3, the expression for the specific total torque may be approximated by

$$T_T = \frac{Dtr(S_c - S_A)}{2}\left[1 + \frac{\mu L}{t}\right] \tag{10-89}$$

To establish the general validity of the foregoing model, experiments were conducted in which annealed black plate (low-carbon-steel strip) of two thicknesses (0.0092 and 0.0095 inch) was rolled on two laboratory type mills. This material exhibited a yield strength in tension (as conventionally measured) of approximately 43,500 pounds/square inch and a surface roughness of about 5 microinches (0.13 micrometer).

The thicker strip was rolled on a four-high mill at speeds up to approximately 300 feet/minute using work rolls nominally 3.25 inches in diameter ground to a finish of 11 microinches. The thinner strip was rolled on another four-high mill using work rolls with a nominal diameter of 6.5 inches ground to a finish of about 30 microinches.

In addition to the rolling of black plate, 0.014 inch thick annealed low-carbon-steel strip (yield strength 46,000 pounds/square inch) with a hot-dip zinc coating weight of 1 ounce/square foot of sheet was rolled on the larger of the two mills at a speed of 30 feet/minute using rolls with a very smooth finish (approximately 3 microinches).

For the rolling of black plate on the smaller mill, the data relating to specific torque are shown in Figure 10-16. The theoretical curve, based on equation 10-89, in this instance involved a coefficient of friction of 0.25.

Similar data for the thinner black plate are shown in Figure 10-17, with the theoretical curve in this case using a value of 0.3 for the coefficient of friction.

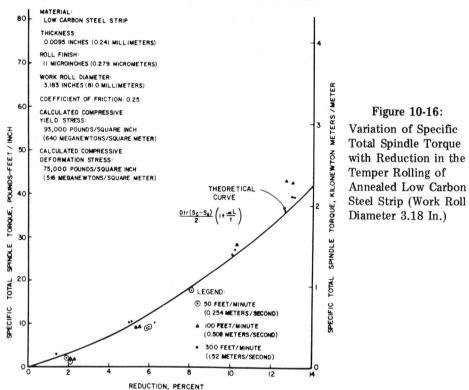

Figure 10-16: Variation of Specific Total Spindle Torque with Reduction in the Temper Rolling of Annealed Low Carbon Steel Strip (Work Roll Diameter 3.18 In.)

The corrected specific total torque data for the galvanized strip are illustrated by Figure 10-18 and are also seen to be in good agreement with the theoretical curve based on a value of 0.2 for the coefficient of friction.

For low carbon steel strip, the variation of theoretical horsepower with reduction for a range of strip thicknesses is illustrated by the series of curves shown in Figure 10-19. Such theoretical requirements would appear to be generally verified by the fact that temper rolling with single-stand mills using 18-inch diameter work rolls has been found to require energy as indicated in Figure 10-20.◊

10-13 The Efficiency of the Rolling Process

Of the energy delivered by the work rolls in the cold reduction process, a certain fraction is dissipated as frictional heat at the roll strip interfaces. The magnitude of this fraction is influenced principally by the reduction r given to the strip but also by the entry and exit tensile stresses σ_1 and σ_2, respectively. Thus the energy transfer from the rolls to the strip has a certain efficiency which we may define as the fraction of the roll energy that is used solely for the plastic deformation of the workpiece.

◊ J. F. Sellers, R. M. Peeples and A. C. Halter, "Temper Rolling", Iron and Steel Engineer Year Book, 1952, pp. 364-375.

TORQUE EQUATIONS AND TANDEM MILLS

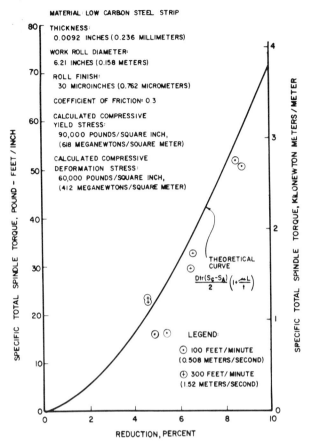

Figure 10-17:
Variation of Specific Total Spindle Torque with Reduction in the Temper Rolling of Annealed Low Carbon Steel Strip (Work Roll Diameter 6.2 In.)

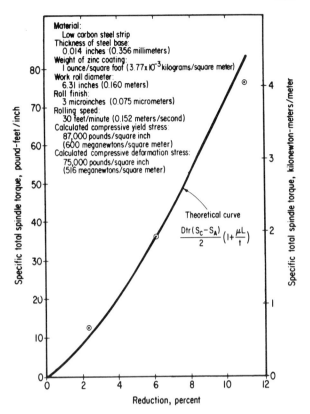

Figure 10-18:
Variation of Specific Total Spindle Torque with Reduction in Temper Rolling Galvanized Strip

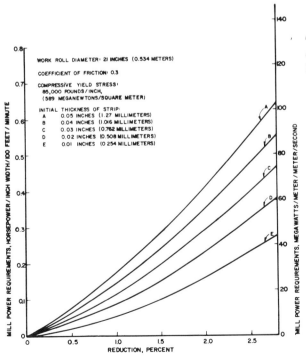

Figure 10-19:
Theoretical Curves Showing Variation of Mill Power Requirements in Temper Rolling Low-Carbon Steel Strip.

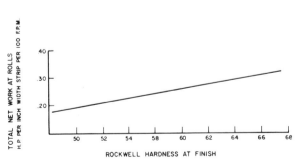

Figure 10-20:
Variation of Total Work on Strip with Hardness (Single-Stand Mills with 18 In. Diameter Work Rolls) (After Sellers)

Orowan[θ] computed the rolling efficiency η on the basis of the ratio of the theoretical specific energy of deformation W (see Section 8-35) to the actual work supplied by the rolls per unit volume of strip (the former being computed from the stress-strain curve of the material being rolled). From experiments conducted by Siebel and Lueg[✦], Orowan calculated an efficiency of 52 per cent with smooth rolls and 29 per cent for wet rough rolls, remarking that these efficiencies were rather low owing to the absence of lubrication and of front and back tension.

Ford[■] defined the efficiency η as follows

$$\eta = \frac{\int_{h_o}^{h_i} \frac{K}{h} dh}{\frac{2T}{Rh_n}} \qquad (10\text{-}90)$$

where K is the constrained yield stress of the workpiece
h_i is the entry thickness of the strip
h_o is the exit strip thickness
h_n is the strip thickness at the neutral point
T is the specific torque and
R is the roll radius

In computations, however, Ford utilized the exit thickness h_o instead of h_n. For single pass reductions ranging from 10 to 70 per cent, he experimentally established efficiencies close to 70 per cent and claimed that, in general, the efficiency falls as the material gets thinner and harder and as the pass reduction increases. He concluded that the cold rolling process is a fairly efficient one and that it is more efficient to make several small passes than one large pass. However, his data did not include the losses in the rolling mill drive, etc., and he felt the overall efficiency would be much lower.

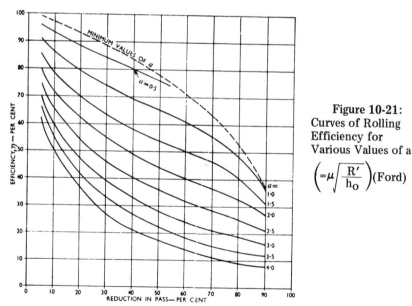

Figure 10-21: Curves of Rolling Efficiency for Various Values of a $\left(=\mu\sqrt{\frac{R'}{h_o}}\right)$ (Ford)

[θ] E. Orowan, "The Calculation of Roll Pressure in Hot and Cold Flat Rolling", Proc. Inst. Mech. Eng. 150, 1943, p. 152 ff.

[✦] E. Siebel and W. Lueg, 1933, Mitteilungen aus dem Kaiser Wilhelm Institut für Eisenforschung, Dusseldorf, Vol. 15, p. 1. (See also W. Lueg, Stahl und Eisen, 53, 1933, Part I, p. 159 ff.)

[■] H. Ford, "Researches in the Deformation of Metals by Cold Rolling", Proc. Inst. Mech. Eng. 159, 1948, pp. 115-143.

In a later paper, Bland and Ford[a] defined the efficiency as

$$\eta = \frac{E_c}{E_c + E'_c + E_f} \quad (10\text{-}91)$$

where E_c is the energy that would be involved in compression of the strip between frictionless parallel plates, E'_c is the frictional energy dissipated in the strip and E_F is the heat of external friction.

The overall efficiency of n consecutive passes is defined as

$$\eta_n = \frac{\sum E_c}{\sum (E_c + E'_c + E_f)} = \frac{\sum E_c}{\sum E_c + \sum E'_c + \sum E_F} \quad (10\text{-}92)$$

where Σ denotes summation from the first pass to the nth. It should be noted that η_n is not equal to $\frac{1}{n}(\eta_1 + \eta_2 \cdots \eta_n)$. For homogenous compression and no tensions, equation 10-91 reduces to

$$\eta = \frac{E_c}{E_T} \quad (10\text{-}93)$$

Curves showing the variation of η with reduction for various values of a $\left(=\mu\sqrt{\frac{R'}{h_o}}\right)$ are illustrated in Figure 10-21, it being assumed that the yield stress maintains a constant value.

In Section 7-3, the efficiency of energy transfer was shown, on the basis of certain simplifying assumptions, to be given by

$$\eta = \sqrt{\frac{(1-r)}{r}} \tan^{-1}\sqrt{\frac{r}{(1-r)}} \quad (10\text{-}94)$$

where r is the reduction given to the strip. This equation is represented by curve A in Figure 10-22, together with the expression

$$\eta = (1 - r/2) \quad (10\text{-}95)$$

(Curve B) which is seen to be a conservative approximation for the efficiency of energy transfer from the rolls to the strip.

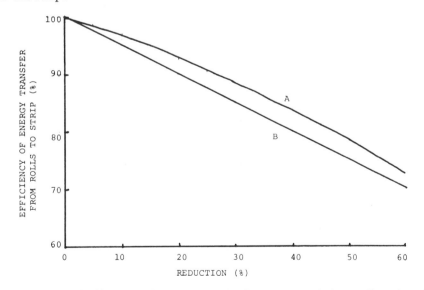

Figure 10-22: Variation of Efficiency of Energy Transfer from Rolls to Strip as a Function of Reduction.

---

[a] D. R. Bland and H. Ford, "The Calculation of Roll Force and Torque in Cold Strip Rolling with Tensions", Proc. Inst. Mech. Eng. 1948, Vol. 159, pp. 144-153.

10-14 The Forward Slip of the Strip

Because of the usual existence of a neutral point along the arc of contact, the strip generally emerges from the roll bite at a speed (V_s) greater than the peripheral speed of the rolls (V_R). This speed difference is designated the forward slip (f) defined by

$$f = \frac{V_s - V_R}{V_R} \tag{10-96}$$

Based on the mathematical model discussed in Section 9-13, Bland and Sims▼ developed expressions for calculating the slip. They used the following notation:

- R = radius of the undeformed roll, in.
- H = entry thickness of the strip in the pass, in.
- h = exit thickness of the strip in the pass, in.
- Y = thickness of strip at the neutral plane, in.
- δ = H − h, the draft in the pass, in.
- r = reduction in the pass = $\frac{\delta}{H}$ (written fractionally or as a percentage).
- α = angle of contact, radians.
- ϕ = angle of co-ordinate at the neutral plane, radians.
- μ = coefficient of friction between the rolls and strip.
- t_f = front tension stress, tons in.$^{-2}$.
- t_b = back tension stress, tons in.$^{-2}$.
- T_f = front tension force, in tons per inch width of strip.
- T_b = back tension force, in tons per inch width of strip.
- u = peripheral speed of rolls, ft. min.$^{-1}$.
- v = exit speed of strip, ft. min.$^{-1}$.
- f = forward slip (expressed fractionally or as a percentage).
- P = roll force in tons per inch width of strip.

The indices + and − denote quantities on the exit and entry side of the neutral plane, respectively; the suffix 0 denotes a no-tension measurement or quantity and 1 and 2 relates to the planes of entry and exit. At any point on the arc of contact in a plane perpendicular to the strip and parallel to the direction of rolling, the following notation is used:

- k = the yield stress of the material in plane homogeneous compression (tons in.$^{-2}$).
- R' = the deformed roll radius (in.) calculated by Hitchcock's Equation (See Section 9-6).
- θ = the angular co-ordinate (radians) with θ = 0 at the point of exit.
- y = thickness of strip at any point, in.
- s = the normal roll pressure (tons in.$^{-2}$).

The following expressions were derived for the pressure distribution along the arc of contact each side of the neutral point.

$$s^+ = k\left(\frac{y}{h} - \frac{t_f}{k}\right) \exp[\mu f(\theta)] \tag{10-97}$$

and

$$s^- = k\left(\frac{y}{H} - \frac{t_b}{k}\right) \exp[\mu[f(\alpha) - f(\theta)]] \tag{10-98}$$

where

$$f(\theta) = 2\sqrt{\frac{R'}{h}} \tan^{-1} \sqrt{\frac{R'}{h}}\, \theta \tag{10-99}$$

▼ R. B. Sims, "The Forward Slip in Cold Strip Rolling", Sheet Metal Industries, 29, 1952, pp. 869-877.

TORQUE EQUATIONS AND TANDEM MILLS

Starting from the plane of entry ($\theta = \alpha$) the roll pressure calculated from equation (10-98) is $k_1 \left(1 - \frac{t_b}{k_1}\right)$ and increases as θ decreases. At the plane of exit the roll pressure, from equation (10-97) is $k_2 \left(1 - \frac{t_f}{k_2}\right)$ and increases with θ. These solutions are continued independently until a value of $\theta = \phi$ is reached when the roll pressures are identical. Equations (10-97) and (10-98) may then be equated and

$$\left[\frac{\frac{Y}{h} - \frac{t_f}{k_\phi}}{\frac{Y}{H} - \frac{t_b}{k_\phi}}\right] \exp[2\mu f(\phi)] = \exp[\mu f(\alpha)] \tag{10-100}$$

where ϕ is the angular co-ordinate of the point of intersection and k_ϕ is the yield strength of the material in the plane compression at that point.

When the front and back tensions are removed from the strip equation (10-100) becomes

$$\frac{H}{h} \exp[2\mu f(\phi_o)] = \exp[\mu f(\alpha_o)] \tag{10-101}$$

On the assumption that the angle of entry is independent of tension and

$$\alpha = \alpha_o \tag{10-102}$$

equations 10-100 and 10-101 may be combined and by expanding $f(\phi)$ and $f(\phi_o)$ and writing

$$f = \frac{R'}{h} \phi^2 \tag{10-103}$$

the relationship between tension and forward slip is

$$\frac{1}{4a} \ln\left[\frac{\frac{Y}{h} - \frac{t_f}{k_\phi}}{\frac{Y}{h} - \frac{t_b}{(1-r)k_\phi}}\right] = \tan^{-1}\sqrt{f_o} - \tan^{-1}\sqrt{f} \tag{10-104}$$

where

$$a = \mu\sqrt{\frac{R'}{h}} \tag{10-105}$$

Alternatively, since $\sqrt{ff_o}$ is small in practice,

$$\sqrt{f} \simeq \sqrt{f_o} + \tan\left[\frac{1}{4a} \ln\left\{\frac{\frac{Y}{h} - \frac{t_b}{(1-r)k_\phi}}{\frac{Y}{h} - \frac{t_f}{k_\phi}}\right\}\right] \tag{10-106}$$

The no-tension slip may be found by expanding equation (10-101). It is

$$\tan^{-1}\sqrt{f_o} = \frac{1}{2}\left[\tan\sqrt{\frac{r}{1-r}} - \frac{1}{2a}\ln\left(\frac{1}{1-r}\right)\right] \tag{10-107}$$

The solution of equation (10-107) is given in Fig. 10-23 for values of forward slip against the reduction in the pass and the parameter

$$b = a\sqrt{1-r} = \mu\sqrt{\frac{R'}{H}} \qquad (10\text{-}108)$$

From it the value of the forward slip when rolling with applied tensions can be obtained from equation (10-106).

Sims stated, however, that the forward slip calculated from these equations would, at the best, be only an approximation since the mathematical model used was developed primarily for the calculation of rolling force and torque, quantities which are relatively insensitive to the form and position of the neutral plane.

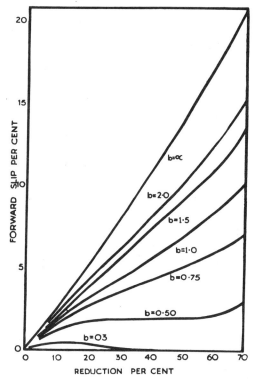

Figure 10-23:
Forward Slip as a Function of Reduction. (Sims) (Values of $b = \mu\sqrt{\frac{R'}{H}}$ marked on curves)

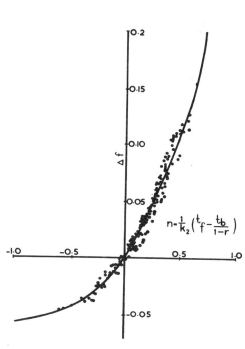

Figure 10-24:
The Relationship Between Δf and n for Experiments on Steel Coils.

The problem faced in the use of the foregoing equations is the difficulty in establishing appropriate values for the coefficient of friction μ for the dimensionless parameter a. Sims, therefore, investigated the relationship between the change in slip (Δf) and the difference in applied strip tensile stresses (t_b and t_f). A simple relationship between slip and applied tensions may be obtained from equation (10-106) when

$$\frac{Y}{h} \simeq 1 \qquad (10\text{-}109)$$

and the terms $\dfrac{t_b}{(1-r)k_\phi}$ and $\dfrac{t_f}{k_\phi}$ are small enough to neglect all but the first terms in the expansion of $\ln\left(1 - \dfrac{t_b}{(1-r)k_\phi}\right)$ and $\ln\left(1 - \dfrac{t_f}{k_\phi}\right)$. Then

$$\sqrt{f} - \sqrt{f_o} \simeq \tan\frac{1}{4ak_\phi}\left(t_f - \frac{t_b}{1-r}\right) = \tan\left[\frac{1}{4ahk_\phi}(T_f - T_b)\right] \qquad (10\text{-}110)$$

TORQUE EQUATIONS AND TANDEM MILLS

and equation (10-110) indicates that a general relationship may exist between Δf and the group $\frac{1}{k_\phi}\left(t_f - \frac{t_b}{1-r}\right)$.

The change in the forward slip, Δf, measured on a steel coil is plotted against the dimensionless group

$$n = \frac{1}{k_2}\left(t_f - \frac{t_b}{1-r}\right) \quad (10\text{-}111)$$

in Figure 10-24. (Here it should be noted that k_2 has been assumed to equal $k\phi$.)

In deriving Δf from the experimental results, values of the no-tension slip are required and these were obtained in the following way. A graph of load against tension was first drawn from the experimental data and in every case the relationship between these variables was very nearly linear.

A line was drawn through the points, and the no-tension load agreeing with this linear relationship obtained. In the majority of cases there was a measured load close to this value, and the corresponding slip was chosen as the no-tension slip in the calculation of Δf.

10-15 Forward Slip Under Conditions of Near-Minimum Friction

In the operation of modern high-speed tandem mills, effective lubricants provide values of the coefficient of friction reasonably close to the minimum values required for rolling operations. Accordingly, it is of interest to consider the magnitude of the forward slip under such conditions. This may be accomplished using a model of the rolling process as described below, this model assuming that the frictional effects along the arc of contact are due to boundary lubrication (where the shearing stresses are assumed constant and independent of the normal pressure along the arc of contact and the relative speed of the roll and strip surfaces). Thus the effective coefficient of friction μ_{eff} may be represented as

$$\mu_{eff} = S/p \quad (10\text{-}112)$$

where S represents the shearing stress and p the normal pressure. If, in the rolling operation, the normal pressure p is thought to be equal to the compressive yield strength σ_c of the workpiece undergoing deformation, then

$$\mu_{eff} = S/\sigma_c \quad (10\text{-}113)$$

The mathematical model developed below results from the resolution of longitudinal forces on the strip in the roll bite and from an energy balance with respect to the rolls and the strip. To simplify the mathematics, the following assumptions have been made:

1. The workpiece possesses a constant compressive yield strength, σ_c,
2. no tensions, either front or back, are used during rolling, and
3. the rolls are rigid (no elastic flattening of the work rolls occurs).

Initially, it is also assumed that the effective coefficient of friction is nearly the minimum permissible, that the frictional constraint on the workpiece is negligible, and that the pressure along the roll bite is equivalent to the compressive yield strength of the strip.

Figure 10-25 illustrates a section of the roll bite and it is assumed that the neutral point occurs at a distance x from the exit plane of the bite. Now the force on the strip resisting its entry into the bite is $tr\sigma_c$, where t is the entry thickness of the strip and r is the reduction expressed as a fraction. The net force tending to drag the strip into the bite is

$$2S\left(\sqrt{\frac{Dtr}{2}} - 2x\right)$$

where D is the work-roll diameter. Hence, since the net force on the strip under tensionless rolling is zero,

$$tr\sigma_c = 2S\left(\sqrt{\frac{Dtr}{2}} - 2x\right) \qquad (10\text{-}114)$$

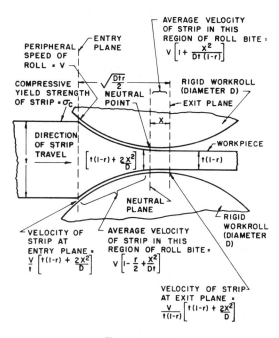

Figure 10-25:
Sketch Illustrating a Section of the Roll Bite.

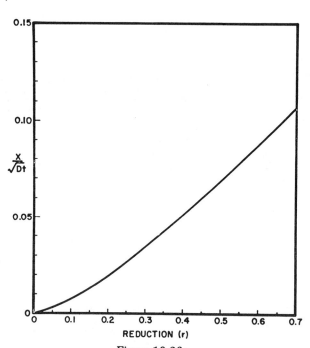

Figure 10-26:
The Relationship Between $X/\sqrt{Dt}$ and the Reduction (r) Under Conditions of Minimum Friction.

Before deriving an energy balance, however, it is first desirable to establish some expressions relating to conditions at the roll bite. If the surface velocity of each roll is V, then the velocity of the strip surface at the neutral point is also V. Now the thickness of the strip at the neutral point is

$$\left[t(1-r) + \frac{2x^2}{D}\right]$$

From this fact, it may be shown that the velocity of the strip at the entry of the bite (V_1) is given by

$$V_1 = \frac{V}{t}\left[t(1-r) + \frac{2x^2}{D}\right] \qquad (10\text{-}115)$$

and the velocity of the strip at the exit (V_2) is given by

$$V_2 = \frac{V}{t(1-r)}\left[t(1-r) + \frac{2x^2}{D}\right] \qquad (10\text{-}116)$$

To simplify the mathematics, an "average" velocity of the strip each side of the neutral point is considered. Such an average to the left of the neutral point is

$$\frac{1}{2}\left[V_1 + V\right] = V\left[1 - r/2 + x^2/Dt\right] \qquad (10\text{-}117)$$

TORQUE EQUATIONS AND TANDEM MILLS

Similarly, the average velocity of the strip on the right of the neutral point is

$$\tfrac{1}{2}[v + v_2] = v\left[1 + \frac{x^2}{Dt(1-r)}\right] \quad (10\text{-}118)$$

The energy imparted by each surface to the strip (assumed to be of unit width) to the left of the neutral point in unit time is then considered. This energy E_1 is given by

$$E_1 = S\left[\sqrt{\frac{Dtr}{2}} - x\right] v\left[1 - r/2 + x^2/Dt\right] \quad (10\text{-}119)$$

Similarly, the energy E_2 dissipated by the strip in rubbing against the roll to the right of the neutral point is given by

$$E_2 = SxV\left[1 + \frac{x^2}{Dt(1-r)}\right] \quad (10\text{-}120)$$

Now the energy E_D used in deforming the half of the strip associated with one surface is given by the volume of strip rolled in unit time multiplied by the energy expended per unit volume. Thus

$$E_D = \frac{V}{t(1-r)}\left[t(1-r) + \frac{2x^2}{D}\right]\frac{t(1-r)}{2}\,\sigma_c\,\log_e\left(\frac{1}{1-r}\right) \quad (10\text{-}121)$$

Therefore, from energy balance considerations, under conditions of minimum friction

$$E_1 - E_2 = E_D \quad (10\text{-}122)$$

or

$$SV\left\{\left[1 - r/2 + x^2/Dt\right]\left[\sqrt{\frac{Dtr}{2}} - x\right] - x\left[1 + \frac{x^2}{Dt(1-r)}\right]\right\}$$
$$= V\left[t(1-r) + \frac{2x^2}{D}\right]\frac{\sigma_c}{2}\log_e\left(\frac{1}{1-r}\right) \quad (10\text{-}123)$$

From equation 10-112, S may be expressed

$$S = \frac{tr\sigma_c}{2\left(\sqrt{\frac{Dtr}{2}} - 2x\right)} \quad (10\text{-}124)$$

Substituting for S in equation 10-123

$$\frac{tr}{\left(\sqrt{\frac{Dtr}{2}} - 2x\right)}\left\{\left[1 - r/2 + x^2/Dt\right]\left[\sqrt{\frac{Dtr}{2}} - x\right] - x\left[1 + \frac{x^2}{Dt(1-r)}\right]\right\}$$
$$= \left[t(1-r) + \frac{2x^2}{D}\right]\log_e\left(\frac{1}{1-r}\right) \quad (10\text{-}125)$$

This equation is a cubic in terms of x and the relationship between the value of $x/\sqrt{Dt}$ and the reduction r is given in Figure 10-26.

Under conditions of minimum permissible friction, the proximity of the neutral point to the exit plane is perhaps best understood either in terms of the fraction K, which represents the ratio of the distance between the neutral and exit planes to the length of the arc of contact, or in terms of the forward slip, ν, which is the ratio of the relative slipping speed between the strip and roll surfaces at the exit plane to the peripheral speed of the rolls.

Now the fraction K is equivalent to $\frac{x}{\sqrt{\frac{Dtr}{2}}}$ or $x\sqrt{\frac{2}{Dtr}}$. Since there is only a single rational solution for equation 10-125 in terms of $\frac{x}{\sqrt{Dt}}$ with respect to r, it will be realized that K is independent of roll diameter and is solely a function of the reduction r. This relationship between K and r is shown in Figure 10-27; and it will be seen that for reductions of less than 50 per cent, the distance between the neutral point and the exit plane represents less than 14 per cent of the arc of contact.

From geometrical considerations, it may be shown that the forward slip ν is given by the approximate relationship

$$\nu \simeq \frac{2x^2}{Dt(1-r)} \qquad (10\text{-}126)$$

This relationship between forward slip and reduction for conditions of minimum friction is shown in Figure 10-28, and it can be seen that, for reductions of less than 40 per cent, the disparity between coiling and roll peripheral speeds is less than 1 per cent. This disparity increases with increasing reduction, however, so that, at 70 per cent reduction, it is about 8 per cent.

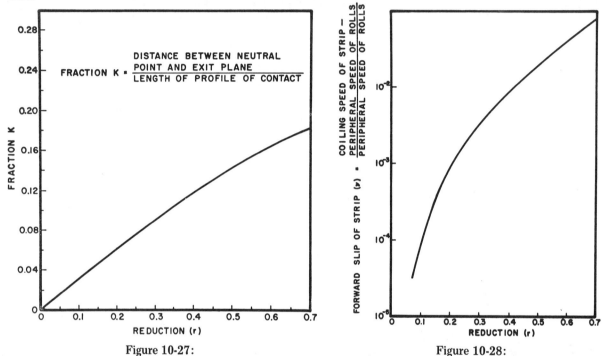

Figure 10-27:
The Relationship Between the Fraction K and the Reduction (r) Under Conditions of Minimum Friction.

Figure 10-28:
The Relationship Between Forward Slip (ν) and the Reduction (r) Under Conditions of Minimum Friction.

Where strip tensions exist, the slip may be corrected using the relationships presented in Section 10-14.

In temper or dry rolling, where a high coefficient of friction occurs and where the neutral point lies close to the center of the arc of contact, then, on the basis of a circular arc of contact

$$x \simeq \frac{1}{2} \sqrt{\frac{Dtr}{2}} \qquad (10\text{-}127)$$

and, therefore, equation 10-126 becomes

$$\nu \simeq \frac{r}{4(1-r)} \qquad (10\text{-}128)$$

This relationship appears to hold for the rolling of galvanized strip. (See Figure 10-29.) On the other hand, in the rolling of black plate, the relationship

TORQUE EQUATIONS AND TANDEM MILLS

$$\nu \simeq \frac{r}{2(1-r)} \qquad (10\text{-}129)$$

appears to be valid, indicating the roll bite is trapezoidal in shape.*

10-16 Prediction of Mill-Stand Drive Power Requirements

From a knowledge of the theoretical spindle torques (Sections 10-2 to 10-12) and the mill-stand losses (Section 7-4), it is possible to compute the power to be drawn from the mill-stand drive motor under steady-state conditions. However, it must be remembered that, in the operation of modern cold mills, a rapid acceleration of the mill is desirable if yield losses (through off-gage material) are to be held to a minimum. Accordingly, the mill-stand drive motors must be capable of delivering the high torques required for acceleration as discussed in Section 10-17.

Before mill-stand power requirements are studied, however, it is appropriate to examine the efficiencies of mill-drive motors. Under the most favorable conditions, such motors operate at about 95 per cent efficiency, but the efficiency varies appreciably with the loading of the motor and the speed at which it is operated.

Accurate computation of motor efficiency is a complicated procedure as there are a number of different losses associated with motor operations. A relatively simple approximate formula for the computation of efficiency ξ of mill motors of the conventional type is

$$\xi = \left[1 - 0.035 \left(\frac{S_R}{S_A} + \frac{S_A}{S_R} \cdot \frac{P_R}{P_s}\right)\right] \times 100\% \qquad (10\text{-}130)$$

where S_R is the rated speed of the motor
S_A is the actual speed (in the same units)
P_R is the rated horsepower of the motor, and
P_S is the horsepower delivered by the motor to the spindles.

Referring to Figure 10-29, if the mill speed is V (fpm), the roll diameter D (inches), the theoretical torque per spindle G (pounds-feet) and the width of the strip W (inches), then the theoretical horsepower P_L associated with the rolls is given by

$$P_L = \frac{48 \, VG}{33,000 \, D} = 1.454 \times 10^{-3} \, VG/D \qquad \text{(horsepower)} \qquad (10\text{-}131)$$

on the assumption that the forward slip is sufficiently small as to be negligible.

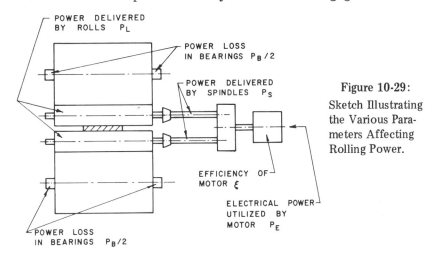

Figure 10-29:
Sketch Illustrating the Various Parameters Affecting Rolling Power.

---

*W. L. Roberts — "An Approximate Theory of Temper Rolling", Iron & Steel Engineer Year Book, 1972, pp. 530-542.

If the bearing losses, P_B, are taken into consideration, then the power, P_S, delivered by the spindles is

$$P_S = P_L + P_B \tag{10-132}$$

In many cases, the spindles are connected to the motor shafts by means of gear trains. Ideally, the efficiency of these units should also be taken into account in computing the electrical input power to the motor. However, the losses of gear trains are usually so low that their consideration is unnecessary. Therefore, if the motor efficiency is ξ, then the electrical power, P_E, (expressed in terms of horsepower) utilized by the motor is given by the equation

$$P_E = \frac{1}{\xi}[P_S] = \frac{1}{\xi}[P_L + B_B]$$
$$= \frac{1}{\xi}[1.454 \times 10^{-3} VG/D + P_B] \tag{10-133}$$

In the case of a mill stand utilizing effective rolling lubricants then, from equation 10-133,

$$P_E = \frac{1}{\xi}\left[\frac{tWV}{33,000}\left\{\sigma_c r + \frac{(\sigma_1 - \sigma_2)(1-r)}{(1-r/2)}\right\} + P_B\right] \tag{10-134}$$

where t is the thickness of the strip on entry into the roll bite,
  r is the reduction, and
  σ_c is the dynamic constrained yield stress of the strip in the presence of entry and exit strip tensile stresses, σ_1 and σ_2, respectively.

To verify equation 10-134, predictions were made with respect to the power requirements for the individual stands of a 6-stand tandem cold mill and these were compared with the corresponding measured data.[†] The tandem mill (similar to that illustrated by Figure 2-34) utilized work rolls with a nominal diameter of 23 inches and backup rolls with a nominal diameter of 55 inches, all with a face length of 54 inches. The rolls were conventional steel rolls ground to a 35-grit finish when used in the first two stands and an 80-grit finish when used in the remaining stands.

The successive stands of the mill were powered by motors of 3,000, 5,000, 5,000, 6,000, 6,000, and 7,000 HP, respectively. In the case of the first three stands, the spindles were attached to the work rolls and were connected to the motor by gears with step-down speed ratio of 1.67, 1.15 and 0.82, respectively. The last three stands featured direct backup-roll drive with no gear reducers. All stands featured twin drives, or drive systems that enabled the upper work roll and backup roll to be driven independently from the lower rolls, the twin drives being capable of being matched either on the basis of speed or load (torque).

The last five stands of the mill were equipped with a recirculating lubrication system capable of supplying high flow rates of the emulsion at each stand to cool both the rolls and the strip as well as to supply the necessary lubricity in the roll bite.

A proprietary brand of rolling lubricant was used in the recirculating system in emulsified form at a temperature of approximately 135 F. Laboratory tests showed that when used under the same conditions, the values of the frictional characteristics K_1 and K_2 provided by the lubricant were 1.5 and 1.5 x 10^{-3} (feet/minute)$^{-1}$, respectively.

The last stand work rolls of the hot mill on which these coils were rolled had been ground with a wheel of No. 35 grit. Since the first stand of the tandem mill had work rolls ground to a similar finish, it was assumed that the strip and these rolls had similar surface roughness in spite of the fact that the former had been pickled. Under these circumstances the product of the multipliers B_R and B_S (See Section 6-31) was assumed to be unity.

[†] R. J. Bentz and W. L. Roberts, "Predicting Rolling Forces and Mill Power Requirements for Tandem Mills", Blast Furnace and Steel Plant 58, 1970, pp. 559-568.

TORQUE EQUATIONS AND TANDEM MILLS

The pickle oil used to coat the strip was a proprietary brand lubricant found in laboratory tests to have K_1 and K_2 values of 2.18 and -8.37×10^{-4} (feet/minute)$^{-1}$ determined at ambient temperature for roll and strip surface roughnesses of 10 microinches. (It is to be noted that, inasmuch as the second frictional characteristic is negative, the coefficient of friction increased with mill speed if all other rolling conditions remained the same, at least up to rolling speeds of a few hundred feet per minute.)

Coils of "soft" and "hard" pickled and oiled hot band were rolled, the former being 0.080 inches thick and 35.6 inches wide, and the latter 0.085 inches thick and 36.2 inches wide. For the former, the yield strength in tension, in terms of pounds per square inch, may be related to the reduction (r), expressed as a percentage, by the empirical equation

$$\sigma = 40,000 + 1773r - 29.2r^2 + 0.195r^3 \quad (10\text{-}135)$$

For the latter, the equation would be

$$\sigma = 45,000 + 2090r - 44.4r^2 + 0.0377r^3 \quad (10\text{-}136)$$

The former coils were rolled to a thickness of 0.0093 inch and the latter to a thickness of 0.0161 inch, it being assumed that the effect of strain rate on dynamic constrained yield strength σ_c was, in both cases, represented by the equation

$$\sigma_c = 1.155 \, (\sigma + 4460 \log_{10} 1000 \, \dot{e}) \quad (10\text{-}137)$$

where $\dot{e}$ is the strain rate incurred by the deforming strip.

Table 10-3
Predicted Power Requirements For Six-Stand Mill

<u>Material Rolled</u>: Low-Carbon Steel Strip (Soft Grade) 35.6 inches wide, 0.080 inch thick.

Coiling Speed, fpm	Stand No.	Reduction at Stand, Per Cent	Tensile Stress in Strip, kpsi Entry into Bite	Tensile Stress in Strip, kpsi Exit from Bite	Calculated Coefficient of Friction in Bite	Predicted Power Reqmts., hp at Roll Bite	Predicted Power Reqmts., hp Bearing Losses	Total Spindle Power, hp	Calculated Drive Efficiency, Per Cent	Input Power to Motors, hp Predicted	Input Power to Motors, hp Measured
1259	1	26.8	0.2	23.2	0.076	−50	11	−39	—	—	48
	2	28.2	23.2	20.2	0.031	515	15	530	76.7	692	640
	3	37.8	20.2	21.3	0.013	779	26	806	79.0	1020	1220
	4	33.5	21.3	22.5	0.017	708	43	751	79.1	949	966
	5	37.6	22.5	23.2	0.012	958	78	1036	81.5	1271	1310
	6	15.0	23.2	3.3	0.0055	583	97	680	76.2	892	982
3236	1	24.8	0.2	23.4	0.089	−174	29	−145	—	—	182
	2	31.2	23.4	19.0	0.024	1556	46	1602	87.6	1828	1735
	3	37.0	19.0	22.3	0.0088	1861	83	1944	88.7	2193	2380
	4	34.5	22.3	24.7	0.011	1866	149	2014	87.8	2294	2275
	5	34.4	24.7	28.9	0.0083	2013	277	2290	88.6	2583	2770
	6	19.0	28.9	2.4	0.0049	1993	384	2377	88.0	2702	2570

<u>Material Rolled</u>: Low-Carbon Steel Strip (Hard Grade) 36.2 inches wide, 0.085 inch thick.

Coiling Speed, fpm	Stand No.	Reduction at Stand, Per Cent	Tensile Stress in Strip, kpsi Entry into Bite	Tensile Stress in Strip, kpsi Exit from Bite	Calculated Coefficient of Friction in Bite	Predicted Power Reqmts., hp at Roll Bite	Predicted Power Reqmts., hp Bearing Losses	Total Spindle Power, hp	Calculated Drive Efficiency, Per Cent	Input Power to Motors, hp Predicted	Input Power to Motors, hp Measured
4583	1	22.4	0.2	23.0	0.14	−553	134	−419	—	—	601
	2	25.8	19.0	26.0	0.016	2526	124	2650	89.2	2969	3065
	3	27.1	26.0	22.5	0.0065	3664	194	3858	92.0	4195	4221
	4	27.4	22.5	16.5	0.011	4262	315	4576	91.8	4986	5210
	5	25.1	16.5	11.0	0.0085	4193	497	4690	92.1	5094	5132
	6	17.0	11.0	2.0	0.0060	3297	675	3972	90.7	4380	4429

<u>Work Rolls</u>: 23 inches in Diameter; 35-Grit Finish in Stands 1 and 2; 80-Grit Finish in Stands 3, 4, 5 and 6.
<u>Rolling Lubricants</u>: Pickle Oil and Water at Stand 1; Recirculated Emulsion of Palm Oil Substitute at Other Stands.
<u>Drive Motors (HP and Roll Speed at Rated Voltage)</u>: Stand 1 — 3000 HP, 838 fpm; Stand 2 — 5000 HP, 1280 fpm; Stand 3 — 5000 HP, 2080 fpm; Stand 4 — 6000 HP, 2620 fpm; Stand 5 — 6000 HP, 4000 fpm; Stand 6 — 7000 HP, 5680 fpm.

The experimental and predicted data with respect to power requirements are listed in Table 10-3. It should be noted that the accuracy of the predicted values is affected by (1) the very low power utilized at the first stand (where the motors were almost acting as generators) and (2) the considerable imbalance of drive-motor loads occurring at the third stand. Excluding these two stands, it was then found that all values of predicted power agreed with the measured values with an accuracy of 10 per cent or better and that at the higher speed, agreement was within 7.5 per cent.

Because of the negligibly small entry tension and the high interstand tension between stands 1 and 2, most of the energy required to deform the strip at the first stand was supplied by the motors driving the second stand. As noted, the conditions at the first stand were such that the motors were almost acting as generators. From equation 10-134, it may be seen that such a condition would arise when $\left\{ \sigma_c r + \frac{(\sigma_1 - \sigma_2)(1-r)}{(1-r/2)} \right\}$ approximates zero. Since, for the conditions at the first stand, $\sigma_c = 72,500$ psi, $\sigma_1 = 0$ psi and $\sigma_2 = 23,000$ psi, the value of the reduction r corresponding to zero spindle torque would be close to 26 per cent. This theoretical value of reduction was close to that obtained in practice (23.6 to 26.8 per cent).

At the third stand the loading of the two motors, due to an unusual adjustment of the speed-balance control system, was in the ratio of about 1 to 4. Under these circumstances, the motor with the lighter load was operating with less efficiency than had been computed. Accordingly, the measured loads would be somewhat higher than had been predicted. However, at the higher speed, the predicted power was within 9 per cent of the measured value.

The data relative to the harder material are also shown in Table 10-3. It will be noted that the rolling power predictions were reasonably accurate for all stands, being within 5 per cent of the measured values.

10-17 The Selection of Mill-Stand Drive Motors

Since the selection of the main drive equipment powering a mill has a major influence on the future operation of the mill and the initial cost of the equipment, great care must be exercised in the selection. Limitations imposed by a marginal choice of equipment may result in lower expected yield per invested dollar and, in like manner, it would be just as foolhardy to overinvest in equipment which is not necessary or cannot be utilized. In making the selection, two basic criteria must be kept in mind; the first is to meet all of the rolling and operating requirements and the second is to obtain the most economical selection which will fulfill these requirements with the desired flexibility.✥

In the selection of mill-stand drive motors, the following requirements must be fulfilled.

1. Adequate drive power on each stand to roll the desired schedules for the rolling and tension loads.
2. Adequate drive power on stand No. 1 to thread with rolling load before tension is established between stands No. 1 and 2.
3. Adequate speed range on all stand motors to handle all desired reduction schedules.
4. Selection of motor base speeds which will utilize reasonable gear ratios.
5. Motor inertias low enough to maintain acceleration torques within the motor capabilities.
6. Duplication of motors where possible to reduce spare parts requirements.
7. Dimensions permitting installation in available interstand spaces.

✥ A. Jakimovich, "Power Requirements and Selection of Electrical Equipment for Tandem Cold Strip Mills", Iron & Steel Engineer Year Book 1963, pp. 963-977.

TORQUE EQUATIONS AND TANDEM MILLS

The horsepower requirements (items 1 and 2 above) may be calculated as discussed in Section 10-16 or from horsepower per ton charts similar to that shown in Figures 10-1 and 10-2 once the reduction schedules and the mill-stand speeds have been established. D.C. mill motors are designed with a certain base speed (in terms of revolutions per minute) but this may be increased by a factor about 2 or 3 by field weakening. Above the base speed, the delivered horsepower is maintained essentially constant so that the shaft torque is inversely proportional to the motor speed. Several economic factors must be borne in mind, however, in relationship to motor speeds. First is the fact that, ceteris paribus, the lower the base speed, the higher the cost, as indicated in Figure 10-30. Second, the wider the speed range, base to weak field, the higher the cost (Figure 10-31). Third, the

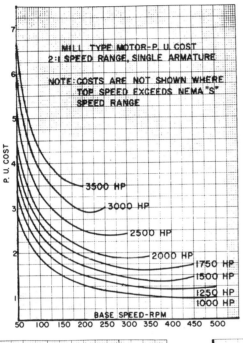

Figure 10-30:
Variation of Per Unit Cost of Mill Motors with Base Speed (Jakimovich)

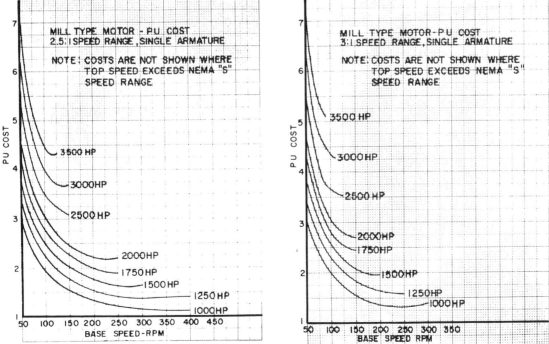

Figure 10-31: Per Unit Cost of D.C. Mill Motors of Two Different Speed Ranges (Jakimovich).

greater the number of armatures, the greater the cost. (However, as will be seen later, the number of armatures is important with respect to the rotational inertia of the drive system.)

Once the motor sizes (in terms of total horsepower) and gear ratios have been selected for each stand, it becomes possible to draw up a speed cone, as illustrated in Figure 10-32 for a four-stand mill.✢ The upper solid curve corresponds to the maximum possible speed at each stand and the lower solid curve to the base motor speeds. Also shown dotted is the curve corresponding to the reduction of 0.076 in. thick strip, 40 in. wide to a final thickness of 0.012 in.

With respect to the desired acceleration capabilities of the mill, the radius of gyration of the motor must be considered with respect to the motor current required for purposes of acceleration. The moment of inertia of the armature (WK^2) is, therefore, an important criterion. A general formula for this parameter, accurate to within 10 per cent is

$$WK^2 = (0.0044 + 0.00033L)D^{3.75} \qquad (10\text{-}138)$$

where L is the length of the armature core in inches, D is the diameter of the armature, and WK^2 is computed in lb.-ft.^{2}.

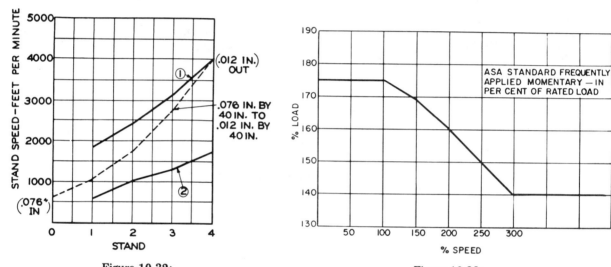

Figure 10-32:
Speed Cone for Four-Stand Tandem Mills.

Figure 10-33:
Load vs. Speed Curve for Frequently Repeated Overloads.

Steel mill main drive machines are normally rated 200 per cent load for one minute and 175 per cent load for frequently applied loads. (See Figure 10-33 for a motor with a 3:1 speed range.) The limit of overload is normally determined by commutation. Accelerating currents up to 75 per cent of full load rating represent good practice.

Since at any given rate of acceleration, WK^2 values and stored energies at top speed determine accelerating currents, the maximum rate of acceleration is determined by the intermittent overload rating of the machines. Accelerating currents at the minimum accelerating time must fall within the intermittent machine ratings and are determined by minimum WK^2 values consistent with a balanced design.

The accelerating current may be expressed as a percentage of the full load motor current. Under these circumstances

$$\% \, I_{acc} = \frac{100 \times WK^2 \times rpm^2}{1.613 \times 10^6 \times t_a \times hp} \qquad (10\text{-}139)$$

✢T. B. Montgomery and J. F. Sellers, "Inertia Studies for Modern Mill Drives", Iron and Steel Engineer Year Book, 1953, pp. 143-152.

TORQUE EQUATIONS AND TANDEM MILLS

where rpm corresponds to the maximum speed,
- t_a is the accelerating time in seconds and
- h_p is the rated power of the motor.

In addition to the inertia of the drive motor, the inertia of the mill rolls, spindles and other drive parts must also be taken into consideration. This will also place an acceleration demand on the motor. If the total acceleration load does not exceed 75 per cent of the full motor load (as is usually the case), then the overload capabilities of the motor will suffice during acceleration.

To ensure that the inertia of the motor is sufficiently low, it has become the general practice to utilize two or more motors in tandem. For example, Montgomery and Sellers † cite the case where a drive-power requirement of 4670 hp may be met with one motor of this size, two of 2333 hp, three of 1557 hp or four of 1167 hp, the total WK^2 values for each, respectively, being 220,000, 83,600, 50,400 and 32,500 lb.-ft.^{2}.

As will be seen in the later sections of this chapter, another motor characteristic of importance is its droop which relates to the decrease in the rotational speed of the motor as the torque increases. Thus, if a change ΔG in the torque produces a change ΔV in the roll speed, then the droop D is given by

$$D = \Delta V / \Delta G \qquad (10\text{-}140)$$

Relatively low values of D characterize a "stiff" drive system.

10-18 The Sizing and Selection of Reel Drive Motors

Today's tandem mills are designed with reels that can accommodate a coil build-up of approximately five to one. Consequently, the motors for these mills require a comparable speed range. However, a five-to-one speed range is approaching the upper practical limit of good d-c motor design in the sizes normally applied to a reel.※

Some additional coil build-up during acceleration is considered. In this case, the reel motor top speed can be geared at a speed corresponding to a coil build-up that would be reached after the end of the acceleration period. The reel motor current during this build-up is maintained by a higher voltage than normally required which in turn produces a higher tension during this portion of the coil build-up. Additional range can also be obtained by exceeding the motor range at the base speed or the maximum coil end. In this case, the reel motor current is maintained by a reduction in generator voltage producing a lighter tension.

However, this method of extending the coil build-up range is obtained at the expense of maintaining constant tension over the coil build-up, and should be avoided unless the limitations are acceptable to good operation. If additional range is necessary, it is desirable to obtain this at the maximum coil end rather than the empty mandrel end. The additional voltage required to maintain tension during acceleration has an adverse effect on the forcing and the counter emf regulator at its most critical operating point with the motor at weak field. This is not the case at the maximum coil end since the motor is at full field. In addition, the overtension limitations must be faced for every coil being rolled while the lower tension approach will affect only those coils that are coiled to the maximum diameter.

In the selection of the reel drive hp rating, two basic considerations must be satisfied. There must be adequate hp to hold the required tension, and there must be sufficient overload capabilities to enable the reel to accelerate in the required time and still maintain the desired tension. The reel hp required to hold the specified tension cannot be chosen from one schedule

---

† T. B. Montgomery and J. F. Sellers, "Inertia Studies for Modern Mill Drives", Iron and Steel Engineer Year Book, 1953, pp. 143-152.

※ A. Jakimovich, "Power Requirements and Selection of Electrical Equipment for Tandem Cold Strip Mills", Iron and Steel Engineer Year Book, 1964, pp. 963-977.

alone, but should be determined from the schedule which requires the maximum hp. A logical approximation to make without any specific data is to assume a maximum tensile stress of 10,000 psi.

To satisfy the acceleration requirements additional hp is required since it is usually impossible to increase the overload capabilities of a motor with a speed range that approaches or exceeds a five-to-one speed range. Without specific reel inertia data, a 50 per cent increase in the hp rating is usually sufficient to satisfy this requirement.

Where a wide range of coiling tensions are required, two reel motors may be employed. An example of such an arrangement on a foil mill is shown in Figure 2-50 where magnetic clutches are used to mechanically couple the motors when necessary.

10-19 The Interrelationships of Tandem Mill Operational Parameters

The rule that governs the steady state operation of a tandem mill is that the "output", namely, the volume of metal passing any point in unit time, is the same for all points. This output is the product of the gage, the width and the speed of the strip but, because of the virtually unchanged width of the strip in cold rolling operations, significant changes in output arise only from changes in gage and speed.

On the basis of this rule, a study was made by Hessenberg and Jenkins[∞] to determine the changes in interstand tension, mill speed and finish gage resulting from screw- and speed-setting changes made to the mill. In this study, one of the first of its kind and which is presented in this section, the following notation and terminology were used.

B	Width of strip, inches.	M	Rigidity of mill stand defined by M = $(\partial F/\partial h)_S$.
b	Backward slip.		
D	A disturbance to a stand, such as change of screw setting or speed.	P	Roll pressure.
E	Young's modulus.	Q	Defined by $Q = (\partial F/\partial h)_H$.
F	Roll force, tons.	S	Screw setting equal to the distance apart of the unloaded work rolls.
F_o	Roll force in absence of strip tension.	T_b	Back tension, tons.
f	Forward slip.	t_b	Back tension stress, tons per sq. in.
f_3	Bland and Ford's function. (See Section 9-13)	T_f	Front tension, tons.
		t_f	Front tension stress, tons per sq. in.
G	Total torque, tons in.	U	Volume output of mill per unit time, cu. in. per min.
H	Ingoing thickness of strip, inches.		
h	Outgoing thickness of strip, inches.	V	Ingoing speed of strip, in. per min.
$\bar{k}$	Mean yield stress of strip in roll gap.	v	Outgoing speed of strip, in. per min.
l	Distance between exit plane of one stand and entry plane of next.	ϕ ψ	Coefficients defined in text.

Suffixes.
1, 2 .... N   Number of stand.
12, 23, etc.   Interstand tensions.
n   Neutral point.

Assuming that the volume of metal flowing through all parts of the system in a given time is the same and that for strip being rolled through N stands,

$$v_1 h_1 = v_2 h_2 = \ldots = v_N h_N = \frac{U}{B} \qquad (10\text{-}141)$$

where v is the speed, h the gage, B the width of the strip, and U is the output, that is the volume of metal passing any point in unit time (Figure 10-34).

[∞] W. C. F. Hessenberg and W. N. Jenkins, "Effect of Screw and Speed-Setting Changes on Gauge, Speed and Tension in Tandem Mills", Proc. Inst. Mech. Eng., 169, 1955, pp. 1051-1058.

TORQUE EQUATIONS AND TANDEM MILLS

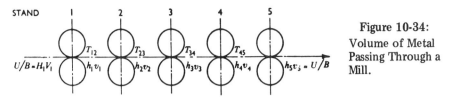

Figure 10-34: Volume of Metal Passing Through a Mill.

Where there are changes in v and h resulting in a new steady state,

$$(v_1 + \Delta v_1)(h_1 + \Delta h_1) = (v_2 + \Delta v_2)(h_2 + \Delta h_2) = \ldots = \frac{U + \Delta U}{B} \quad (10\text{-}142)$$

whence, for small deviations,

$$\left[\frac{\Delta v_1}{v_1} + \frac{\Delta h_1}{h_1}\right] = \left[\frac{\Delta v_2}{v_2} + \frac{\Delta h_2}{h_2}\right] = \ldots = \frac{\Delta U}{U} \quad (10\text{-}143)$$

Changes in v and h will occur when a stand is disturbed. The change in h will affect in turn the stands farther down the mill, while the change in v, by affecting the interstand tensions, will be felt by both preceding and following stands.

For any stand the components of $\Delta U/U$ are

$$\frac{\Delta h}{h} = \frac{1}{h}\left[\frac{\partial h}{\partial D}\Delta D + \frac{\partial h}{\partial T_b}\Delta T_b + \frac{\partial h}{\partial T_f}\Delta T_f + H\frac{\partial h}{\partial H}\frac{\Delta H}{H}\right] \quad (10\text{-}144)$$

$$\frac{\Delta v}{v} = \frac{1}{v}\left[\frac{\partial v}{\partial D}\Delta D + \frac{\partial v}{\partial T_b}\Delta T_b + \frac{\partial v}{\partial T_f}\Delta T_f + H\frac{\partial v}{\partial H}\frac{\Delta H}{H}\right] \quad (10\text{-}145)$$

where ΔD represents any disturbance to that stand such as an adjustment ΔS to the screw setting, ΔT_b and ΔT_f the changes in back and front tension, respectively, due to disturbances to earlier and later stands, and $\Delta H/H$ the fractional change in ingoing gage due to disturbances to earlier stands.

Before proceeding, it is desirable to simplify the notation, and equations 10-144 and 10-145 will be rewritten as follows:

$$\frac{\Delta h}{h} = \phi_D \Delta D + \phi_{Tb}\Delta T_b + \phi_{Tf}\Delta T_f + \phi_H \frac{\Delta H}{H} \quad (10\text{-}146)$$

$$\frac{\Delta v}{v} = \psi_D \Delta D + \psi_{Tb}\Delta T_b + \psi_{Tf}\Delta T_f + \psi_H \frac{\Delta H}{H} \quad (10\text{-}147)$$

where ϕ and ψ represent coefficients, the subscripts referring to the nature of the change.

By combining equations 10-143, 10-146 and 10-147, a set of simultaneous equations (perturbation equations) may be built up for each stand. In these mathematical models, changes in coiler tensions and gage variations in the unrolled strip are ignored. There are thus N simultaneous equations containing N unknowns, namely, the N—1 interstand tensions and the fractional change in output; they may therefore be solved for any known disturbance.

In order to make numerical calculations, the coefficients ϕ and ψ must be evaluated in terms of the mechanical and electrical characteristics of the mill and the conditions in the roll gap. This can be done with the aid of existing methods of calculating roll force and torque in single-stand mills as will be seen in the next section.

For simplicity small variations in the yield stress of the strip and the rolling friction are not considered, but the way in which these may be introduced will be apparent from existing theories of rolling.

10-20 Evaluating the Coefficients of the Perturbation Equations

Methods of evaluating the coefficients ϕ and ψ of the perturbation equations are discussed in detail below.

a. The Coefficients ϕ Relating to Gage Changes

Gage variations in single-stand mills have been discussed by Hessenberg and Sims[⊗][⊞] and they may be conveniently represented by curves showing the variation of rolling forces with strip thickness (Figure 10-35). The line $M = (\partial F/\partial h)_S$ exhibits the mechanical rigidity of the stand and the line $Q = (\partial F/\partial h)_h$ the effective resistance to deformation of the strip. The actual values of the roll force and strip gage are given by the point of intersection.

A change in screw setting shifts the M line parallel to itself by a distance ΔS. From the geometry of the diagram

$$\frac{\partial h}{\partial S} = \frac{M/Q}{1 + M/Q} \qquad (10\text{-}148)$$

The quantity M/Q measures the relative rigidity of the stand; this, for any particular stand, is larger when the strip is soft and thick and smaller when the strip is hard and thin. In the former case the screws will tend to 100 per cent effectiveness while in the latter this may be much reduced.

When the stand is disturbed by a screw-setting change, $\Delta D = \Delta S$, and the coefficient ϕ_D becomes

$$\phi_S = \frac{1}{h}\frac{\partial h}{\partial S} = \frac{1}{h}\frac{M/Q}{1 + M/Q} \qquad (10\text{-}149)$$

Similar reasoning about the effect of a change in ingoing gage, H, on the Q line in Figure 10-35 leads to

$$\frac{\partial h}{\partial H} = \frac{1}{1 + M/Q} \qquad (10\text{-}150)$$

whence

$$\phi_H = \frac{H}{h}\frac{\partial h}{\partial H} = \frac{H}{h}\frac{1}{1 + M/Q} \qquad (10\text{-}151)$$

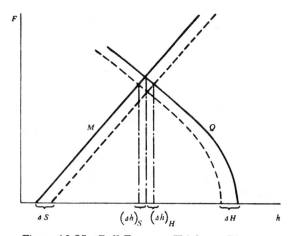

Figure 10-35: Roll Force vs. Thickness Diagram.

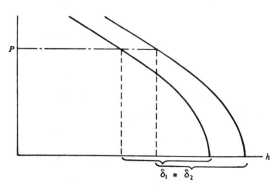

Figure 10-36: Relation Between Roll Pressure and Strip Thickness.

[⊗] W. C. F. Hessenberg and R. B. Sims, "The Effect of Tension on Torque and Roll Force in Cold Strip Rolling", J. Iron and Steel Institute, 168, pp. 155-164.

[⊞] W. C. F. Hessenberg and R. B. Sims, "Principles of Continuous Gauge Control in Sheet and Strip Rolling", Proc. Inst. Mech. Eng., 166, 1952, pp. 75-81.

TORQUE EQUATIONS AND TANDEM MILLS

This approximation is not sufficiently accurate for the numerical evaluation of ϕ_H since it implies that, for the same roll force per inch width, a change in H would involve no change in the draft δ (Figure 10-36).

Now according to Bland and Ford (See Section 9-13), the specific rolling force P is given by the equation

$$P = \bar{k}\sqrt{R'\delta}f_3 \qquad (10\text{-}152)$$

where the deformed roll radius R' is given by Hitchcock's relationship

$$R' = R\left(1 + \frac{2cP}{\delta}\right) \qquad (10\text{-}153)$$

(as discussed in Section 9-6). The combination of equations 10-152 and 10-153 gives

$$\delta = \frac{P}{R\bar{k}^2 f_3^2} - 2cP \qquad (10\text{-}154)$$

Changes in $\bar{k}$ with H are small except perhaps in a light, first pass on annealed material so that,

$$\frac{d\delta}{\delta} = -\frac{2P^2}{\delta R\bar{k}^2 f_3^2}\frac{df_3}{f_3} = -A\frac{df_3}{f_3} \qquad (10\text{-}155)$$

A ranges between 1 and 10. Inspection of Bland and Ford's curves for f_3 (Figure 9-31) show that the conditions under which df_3/f_3 is sufficiently small are rather restricted. For accurate evaluation, therefore, the Q curves must be calculated over the appropriate ranges. None the less, the approximate formula for ϕ_H is useful in indicating the role played by M; the mill with high relative rigidity, that is, $M>Q$, tends to "smooth out" or "attenuate" the fractional gage errors which it receives, and the relatively resilient mill, that is, $M<Q$, to "amplify" them. When M and Q and the reduction are such that $\phi_H = 1$, the errors are neither amplified nor attenuated.

The coefficients ϕ_T may be obtained from the relations developed by Hessenberg and Sims[⊗][⊞]

$$F = M(h-S) \qquad (10\text{-}156)$$

$$F = F_o\left(1 - \frac{t_f + t_b}{2\bar{k}}\right) \qquad (10\text{-}157)$$

which give,

$$\phi_{Tf} = \frac{1}{h}\frac{\partial h}{\partial T_f} = -\frac{F}{2\bar{k}MBh^2} \qquad (10\text{-}158)$$

$$\phi_{Tb} = \frac{1}{h}\frac{\partial h}{\partial T_b} = -\frac{F}{2\bar{k}MBHh} \qquad (10\text{-}159)$$

These are negative—an increase in either front or back tension reducing the gage—and inversely proportional to the rigidity of the stand. They are also independent of the width of the strip since F is directly proportional to it.

[⊗] W. C. F. Hessenberg and R. B. Sims, "The Effect of Tension on Torque and Roll Force in Cold Strip Rolling", J. Iron and Steel Institute, 168, 1951, pp. 155-164.

[⊞] W. C. F. Hessenberg and R. B. Sims, "Principles of Continuous Gauge Control in Sheet- and Strip-Rolling", Proc. Inst. Mech. Eng., 166, 1952, pp. 75-81.

A change in the speed setting, if large enough, will also affect the gage by means of changes in the friction between the rolls and the strip which, in turn, affects the roll force. This "speed effect" is well known and various means have been devised to mitigate its effects, the most usual being simply to reduce the accelerating and decelerating times of the mill as much as possible. The effect is most pronounced at the lower mill speeds and is negligible for small speed changes at full speed. In general, therefore, the effect of a speed-setting change under steady conditions on the thickness component of the output can be ignored.

b. The Coefficients ψ Relating to Strip Speed Changes

To evaluate the coefficients ψ, it is to be noted that changes in strip speed are composed of changes in neutral-point speed and forward slip, that is,

$$(dv/v) = (dv_n/V_n) + df/(1+f) \qquad (10\text{-}160)$$

Now the motor droop is defined as the percentage drop in speed at full-load torque and full speed. From this it is possible to determine the fractional change in neutral point speed per unit change in torque,

$$Z = (dv_n/V_n)/dG \qquad (10\text{-}161)$$

so that

$$(dv/v) = ZdG + df/(1+f) \qquad (10\text{-}162)$$

The problem of evaluating the coefficients ψ therefore involves estimating the appropriate changes in torque and in forward slip.

A change in screw setting merely affects the torque through the resulting change in outgoing thickness, so that

$$\frac{dG}{dS} = \frac{\partial G}{\partial h}\frac{dh}{dS} = h\frac{\partial G}{\partial h}\phi_s \qquad (10\text{-}163)$$

The forward slip change is

$$\frac{df}{dS} = \frac{\partial f}{\partial r}\frac{dr}{dS} = -\frac{h}{H}\frac{\partial f}{\partial r}\phi_s \qquad (10\text{-}164)$$

Thus

$$\psi_s = \frac{1}{v}\frac{dv}{dS} = Z\frac{dG}{dS} + \frac{1}{1+f}\frac{df}{dS} = h\phi_s\left[Z\frac{\partial G}{\partial h} - \frac{1}{H(1+f)}\frac{\partial f}{\partial r}\right] \qquad (10\text{-}165)$$

Usually the value of Z is such as to make the slip term of little consequence.

A change in ingoing thickness affects the torque directly as well as through the concomitant change in outgoing thickness. Similar reasoning to the above gives

$$\frac{dG}{dH} = \frac{\partial G}{\partial H} + \frac{\partial G}{\partial h}\frac{dh}{dH} = \frac{\partial G}{\partial H} + \frac{h}{H}\frac{\partial G}{\partial h}\phi_H \qquad (10\text{-}166)$$

and

$$\frac{df}{dH} = \frac{\partial f}{\partial H} + \frac{\partial f}{\partial r}\frac{dr}{dH} = \frac{\partial f}{\partial H} - \frac{h}{H^2}\frac{\partial f}{\partial r}\phi_H \qquad (10\text{-}167)$$

whence

$$\psi_H = Z\frac{\partial G}{\partial H} + \frac{1}{1+f}\frac{\partial f}{\partial H} + \frac{h}{H}\phi_H\left[Z\frac{\partial G}{\partial h} - \frac{1}{H(1+f)}\frac{\partial f}{\partial r}\right] \qquad (10\text{-}168)$$

The torque and slip terms may be evaluated as before.

TORQUE EQUATIONS AND TANDEM MILLS

The forward tension coefficient ψ_{Tf} is of great importance since, as will be seen later, its relation to ϕ_{Tf} vitally affects the behavior of the mill. It is the resultant of two opposing effects. There is first the fact that, other things being equal, an increase in front tension will reduce the amount of torque which must be supplied by the motor. At the same time the effect of an increase in front tension is a decrease in outgoing thickness which, in turn, involves an increase in torque.

The effect of front and back tension on the torque when there is no change in outgoing gage is given by Hill's equation (equation 10-53) expressed in a more accurate form

$$G = G_o + R(1-b)T_b - R(1+f)T_f \qquad (10\text{-}169)$$

so that

$$\frac{\partial G}{\partial T_b} = R\left(1 - b - T_b\frac{\partial b}{\partial T_b}\right) \qquad (10\text{-}170)$$

$$\frac{\partial G}{\partial T_f} = -R\left(1 + f + T_f\frac{\partial f}{\partial T_f}\right) \qquad (10\text{-}171)$$

To these must be added the changes in torque due to the fact that tension changes produce changes in the exit gage. These torque changes are,

$$\frac{\partial G}{\partial T_f} = \frac{\partial G}{\partial h}h\phi_{Tf} \qquad (10\text{-}172)$$

and

$$\frac{\partial G}{\partial T_b} = \frac{\partial G}{\partial h}h\phi_{Tb} \qquad (10\text{-}173)$$

Thus the neutral point components of ψ_{Tf} and ψ_{Tb} are

$$Z\left[\frac{\partial G}{\partial h}h\phi_{Tf} - R\left(1 + f + T_f\frac{\partial f}{\partial T_f}\right)\right]$$

and

$$Z\left[\frac{\partial G}{\partial h}h\phi_{Tb} + R\left(1 - b - T_b\frac{\partial b}{\partial T_b}\right)\right]$$

The forward slip changes are

$$\frac{df}{dT} = \frac{\partial f}{\partial T} + \frac{\partial f}{\partial r}\frac{dr}{dT} \qquad (10\text{-}174)$$

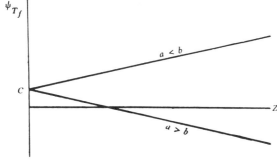

Figure 10-37: Relation Between Forward Tension and Speed.

ψ_{Tf} is therefore of the form $Z(a-b)+c$. Usually $a<b$, and Z being negative, ψ_{Tf} is positive. A is proportional to the width of the strip and inversely proportional to the rigidity of the stand. Calculations show that it is possible in the case of very wide mills with rather low rigidity for $a>b$; ψ_{Tf} then becomes negative (Figure

10-37) at high droops, which means that the stand will slow up under an increase in tension, a condition which tends to reduce the stability of the mill.

ψT_b is invariably negative.

Finally, if the disturbance is a speed-setting change

$$\Delta D \equiv \Delta v_n \qquad (10\text{-}175)$$

As far as is known, forward slip is not affected by speed changes except under those conditions where the speed affects the gage. Therefore,

$$\psi v_n = \frac{v_n}{v}\frac{dv}{dv_n} = 1 \qquad (10\text{-}176)$$

In all the foregoing expressions the roll force F, the torque G, and their derivatives, may be calculated with the aid of methods such as those discussed in Chapter 9 and the preceding sections of this chapter. The values of the forward slip terms, if required, may be computed using the mathematical expressions discussed in Sections 10-14 and 10-15.

10-21 Equations for a Two-Stand Mill

With the coefficients ϕ and ψ evaluated as discussed in Section 10-19, equations may now be developed for any given tandem mill. Since the output equations, although fundamentally simple, become superficially more complicated as the number of mill stands is increased, the simplest type of tandem mill, the two-stand facility, will be examined first.

In a two-stand tandem mill, an adjustment to either stand in the form of a change in the screw or speed setting brings about a change in interstand tension by creating a difference between the speed at which the strip leaves one stand and enters the next. The rate dT/dt at which the tension T tends to increase or decrease under the influence of such a speed difference is

$$\frac{dT}{dt} = \frac{BhE}{l}\left(v_2 - v_1\right) \qquad (10\text{-}177)$$

where
- B is the width of the strip
- h is the thickness of the strip emerging from the first stand
- E is the elastic modulus of the strip
- V_2 is the entry speed of the second stand
- v_1 is the exit speed of the first stand and
- l is the interstand distance between exit and entry planes.

An increase in the speed of stand 1 relative to stand 2 thus causes the tension to fall and vice versa.

If the reaction of the stands to the initial tension change is such that it tends to diminish the difference in their outputs created by the disturbance, the mill will be stable and a new equilibrium, incorporating the screw or speed-setting change which caused the original steady state to be disturbed, will be reached.

The output equations for a two-stand mill, on the assumptions of negligible entry tension and constant input gage, are therefore

Stand 1
$$\frac{\Delta h_1}{h_1} = \phi_{D1}\Delta D_1 + \phi_{Tf1}\Delta T_{12} \qquad (10\text{-}178)$$

$$\frac{\Delta v_1}{v_1} = \psi_{D1}\Delta D_1 + \psi_{Tf1}\Delta T_{12} \qquad (10\text{-}179)$$

TORQUE EQUATIONS AND TANDEM MILLS

Stand 2
$$\frac{\Delta h_2}{h_2} = \phi_{D2}\Delta D_2 + \phi_{Tb2}\Delta T_{12} + \phi_{H2}\left(\phi_{D1}\Delta D_1 + \phi_{Tf1}\Delta T_{12}\right) \quad (10\text{-}180)$$

$$\frac{\Delta v_2}{v_2} = \psi_{D2}\Delta D_2 + \psi_{Tb2}\Delta T_{12} + \psi_{H2}\left(\phi_{D1}\Delta D_1 + \phi_{Tf1}\Delta T_{12}\right) \quad (10\text{-}181)$$

The contribution to the output change on stand 2 of the thickness changes out of stand 1 should be noted. These equations may be solved for ΔT, $\Delta h/h$, and $\Delta v/v$ when the magnitude of the disturbances and the coefficients are known.

There are two special cases in which the sign of the tension and the finishing gage changes may be deduced without numerical calculation. These are when the fractional gage changes are either very large or very small compared with the fractional speed changes. The former case approximates to that of a mill with a very "stiff" drive to each stand (or a fixed-speed ratio between the stands) so that differences in output must be adjusted mainly by gage changes; the latter corresponds to a mill with stands driven by a very "soft" drive so that output differences are mainly absorbed by changes in the relative speeds of the stands.

In the first case, all the terms in ψ, and in the second case, those in ϕ_{Tf} and ϕ_{Tb} may be neglected. Under normal conditions $\phi_{Tf} > \phi_{Tb}$, $\psi_{Tf} > \psi_{Tb}$, and $\phi_H \doteq 1$, and these equations lead to the pattern of interstand tension and finishing-gage changes resulting from an increase in screw or speed setting as set forth in Table 10-4.

Table 10-4
Interstand Tension And Finishing Gauge Changes For A Two-Stand Mill

Stand	Positive Change	$\left(\frac{\Delta v}{v} \gg \frac{\Delta h}{h}\right)$		$\left(\frac{\Delta h}{h} \gg \frac{\Delta v}{v}\right)$	
		ΔT_{12}	$\Delta h_2/h_2$	ΔT_{12}	$\Delta h_2/h_2$
1	Speed	−	+	−	+
2	Speed	+	−	+	−
1	Screw	±	+	±	+
2	Screw	+	+	+	−

An increase is denoted by +, a decrease by − and a possibility of either by ±.

The results in the two cases are qualitatively the same except for a screw adjustment on stand 2.

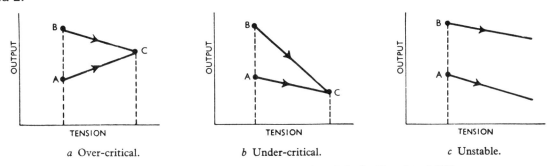

a Over-critical.   b Under-critical.   c Unstable.

Figure 10-38: Effect of Disturbance on Stand 2 of a Two-Stand Mill.

In Figure 10-38, output is plotted against tension. A represents the output of the two stands and the interstand tension before the screws on stand 2 are lifted whereas B is the output of stand 2 immediately after they have been lifted. Since this will have caused stand 2 to speed up, the tension will start to rise and continue to do so until the output of the two stands is once again the same at a value indicated by C. The slope of the lines AC and BC is given by $\phi_{Tf1} + \psi_{Tf1}$ and $\phi_{Tb2} + \psi_{Tf2} + (\phi_{H2} + \psi_{Tb2})\phi_{Tf1}$, respectively.

There are three possibilities corresponding to,

(a) $\phi_{Tf1} + \psi_{Tf1} > \phi_{Tb2} + \psi_{Tb2} + (\phi_{H2} + \psi_{H2})\phi_{Tf1}; \quad \phi_{Tf1} + \psi_{Tf1} > 0$ (10-182)

(b) $\phi_{Tf1} + \psi_{Tf1} > \phi_{Tb2} + \psi_{Tb2} + (\phi_{H2} + \psi_{H2})\phi_{Tf1}; \quad \phi_{Tf1} + \psi_{Tf1} < 0$ (10-183)

(c) $\phi_{Tf1} + \psi_{Tf1} < \phi_{Tb2} + \psi_{Tb2} + (\phi_{H2} + \psi_{H2})\phi_{Tf1}$ (10-184)

These relationships are illustrated in Figure 10-38, the diagram (c) representing a condition of instability. The fixed-drive mill will give the condition in Figure 10-38(b) because ϕ_{Tf1} is negative and ψ_{Tf1} negligibly small and, since the output change will be entirely in the form of a gage change, it is readily seen why raising the screws on the last stand of such a mill reduces the finishing gage. The behavior of the mill with a soft drive corresponds to Figure 10-38(a) and, since the tension changes mainly affect the speed of the strip, the gage increase originally caused by lifting the screws is hardly altered as the mill adjusts itself to equilibrium.

There is one condition of particular importance in tandem mills which would be represented in Figure 10-38 by the line AC being horizontal. This means that for stand 1

$$\phi_{Tf} + \psi_{Tf} = 0 \qquad (10\text{-}185)$$

The output of stand 1 is thus unaffected by any changes in the interstand tension or by any disturbance to stand 2, since the effects of this are only transmitted to stand 1 by way of changes in the tension. This is called the critical condition. Figure 10-38(a) and (b), respectively, display the over-critical and under-critical conditions.

The critical condition itself is of interest in relation to the effectiveness of the screws on stand 2. On a single-stand mill this is measured by $1/\left(1 + \frac{Q}{M}\right)$, so that when the strip becomes thin and hard and Q/M very large, it is difficult to make any gage adjustments by means of the screws. In a tandem mill the roll gap conditions are not the principal factor in determining screw effectiveness. Even when Q/M is large, the screws still have an appreciable effect on the torque; if the drive is stiff this will cause a tension change which will alter the gage. On the other hand, consider a mill whose first stand is adjusted to be critical, the drive to the last stand being stiff. The screws on the last stand will be practically ineffective, no matter what the condition in the roll gap, for a screw change will cause an output change

$$(\Delta U/U) = \{(\Delta h/h) + (\Delta v/v)\} = 0 \qquad (10\text{-}186)$$

The stiff drive will ensure that $\Delta v/v$ and therefore $\Delta h/h$ is very small. By varying the speed setting instead, substantial gage changes may be made.

10-22 Equations for a Three-Stand Mill

The behavior of a three-stand mill is, as would be expected, more complicated than that of a two-stand mill. There is, however, a simple graphical method which helps to visualize it as a whole; this is described later in this section. In the meantime, the results for disturbances to one stand at a time, together with finishing-gage changes, are summarized in Table 10-5.

TORQUE EQUATIONS AND TANDEM MILLS 619

Table 10-5

Interstand Tension And
Finishing Gauge Changes
For A Three-Stand Mill

Stand	Positive Change	$\left(\frac{\Delta v}{v} \gg \frac{\Delta h}{h}\right)$			$\left(\frac{\Delta h}{h} \gg \frac{\Delta v}{v}\right)$		
		ΔT_{12}	ΔT_{23}	Δh_3	ΔT_{12}	ΔT_{23}	Δh_3
1	Speed	−	−	+	−	±	+
2	Speed	+	−	±	+	−	−
3	Speed	+	+	−	−	+	±
1	Screw	±	±	+	±	±	+
2	Screw	+	±	+	+	±	−
3	Screw	+	+	+	−	+	+

A point of interest is that an increase in the speed of stand 3 may cause either an increase or a decrease in gage, depending on the values of the coefficients. By adjusting the rolling conditions it can be arranged that the back tension on the last stand may be varied freely to maintain the shape of the strip without affecting the finishing gage. A further adjustment makes it possible to vary the finishing gage by means of the screws on stand 2 without affecting the tension between stands 2 and 3. Completely independent control of finishing gage and final interstand tension thus appears possible.

This is of interest in connection with a method of gage control proposed by Shayne and Zeitlin.[*] A change in roll separating force on one stand is made to vary the screw setting on the preceding stand until the ingoing gage to the later stand is sufficiently changed to restore the roll force to its original value. Since the screws on this stand are not altered, this has the effect of keeping the outgoing gage from it constant. There are two drawbacks to this method which the present analysis brings out clearly. One is that unless the conditions are adjusted as described in the previous paragraph, there may be large tension changes. The second is that the gage change out of the first stand will take its time to reach the second, thus introducing an awkward time delay in the control system. Alternative methods proposed by Hessenberg and Sims[⊞] either use a tension change, which is much more rapidly transmitted than an ingoing gage change, to influence the roll force, or operate on the screws of the same stand from which the roll force signal is obtained; in both cases the time delay is obviated.

The intermediate stand through which the effects of disturbances to the first or third stands are passed forward or backward along the mill, respectively, is worth special consideration. The critical condition is particularly important in an intermediate stand. The output of a "critical" stand is not affected by changes in front tension, since the fractional change in gage is exactly compensated by the fractional change in strip speed, that is,

$$(\Delta U/U) = \{(\Delta h/h) + (\Delta v/v)\} = 0 \qquad (10\text{-}187)$$

But through the roll gap, the relation

$$(\Delta V/V) = \{(\Delta h/h) + (\Delta v/v)\} \qquad (10\text{-}188)$$

applies. Thus a change in front tension will have no effect on the ingoing speed; no difference will be created between the entry speed to this stand and the exit speed of the previous stand, so there will be no change in tension between the two. A critical stand thus acts as a barrier to the backward transmission of tension changes.

---

[*] A. Shayne and A. Zeitlin, 1944, U. S. Patent No. 2,339,359.
[⊞] W. C. F. Hessenberg and R. B. Sims, "Principles of Continuous Gage Control in Sheet and Strip Rolling", Proc. Inst. Mech. Eng., Vol. 166, 1952, pp. 75-81.

With an over-critical intermediate stand, an increase in front tension will cause an increase in entry speed and therefore an increase in back tension. It thus transmits tension changes backward without change of sign; through an under-critical intermediate stand the sign is changed.

The forward transmission of tension changes through a stand with a soft drive is governed by the coefficient ψ_{Tb} which, as has been shown in Section 10-20, is negative. Any increase in the back tension therefore causes a reduction in output in the form of a reduction in outgoing speed which, interacting with the next stand, will generate an increase in front tension. Such a stand thus transmits tension changes forward, as well as backward, without change of sign.

In a stand with a stiff drive, the initial reaction is determined by the coefficient ϕ_{Tb} which, like ψ_{Tb}, is negative. But the decrease in output is now in the form of a decrease in gage. This is transmitted to the next stand where, as a secondary disturbance, it may increase the entry speed of the strip sufficiently to reverse the sign of the initial change. Whether or not this reversal will take place depends, as may be seen at once from the geometrical display, in a rather complicated way upon the coefficients of the output equations, but it seems that whatever its sign the tension change will be relatively small.

So far, the effect of mill speed on gage, which occurs even on single-stand mills when no tensions are applied, has been ignored. As mentioned in Section 10-20, it is not serious for small speed adjustments at full rolling speed, but over the range of acceleration from threading to full speed, particularly the early part, the effect is very marked. In tandem mills it is complicated by the fact that it is not the same for each stand and that during acceleration the drive stiffens, so that the mill characteristic is changing all the time. The matter may be simplified by first imagining that the percentage speed-setting change during acceleration is made the same for each stand; then all gage and tension changes will be due to the "speed effect" alone. Since $\phi_H \cong 1$, these will be cumulative so that the later stands will be more affected than the earlier. This suggests the super-imposition of an unequal speed-setting change on each stand, so as to offset the output change caused by the "speed effect". Compensation could be accurately designed by first running the mill with equal speed-setting change on each stand, observing the cumulative effect of speed on tension and gage along the whole mill and then basing the additional speed-setting changes on this.

The mathematical representation of the behavior of a 3-stand mill has been published by Hessenberg and Jenkins[∞] and is as follows. The output equations for the mill may be written,

$$\Delta U/U = X_1 + a_1 \Delta T_{12} \tag{10-189}$$

$$\Delta U/U = X_2 + a_2 \Delta T_{12} + b_2 \Delta T_{23} \tag{10-190}$$

$$\Delta U/U = X_3 + a_3 \Delta T_{12} + b_3 \Delta T_{23} \tag{10-191}$$

where

$$a_1 = \phi_{Tf1} + \psi_{Tf1} \tag{10-192}$$

$$a_2 = (\phi_{H2} + \psi_{H2})\phi_{Tf1} + \phi_{Tb2} + \psi_{Tb2} \tag{10-193}$$

$$a_3 = (\phi_{H3} + \psi_{H3})(\phi_{H2}\phi_{Tf1} + \phi_{Tb2}) \tag{10-194}$$

$$b_2 = \phi_{Tf2} + \psi_{Tf2} \tag{10-195}$$

$$b_3 = (\phi_{H3} + \psi_{H3})\phi_{Tf2} + \phi_{Tb3} + \psi_{Tb3} \tag{10-196}$$

[∞] W. C. F. Hessenberg and W. N. Jenkins, "Effect of Screw and Speed-Setting Changes on Gauge, Speed and Tension in Tandem Mills", Proc. Inst. Mech. Eng., Vol. 169, 1955, pp. 1051-1058.

TORQUE EQUATIONS AND TANDEM MILLS

and

$$X_1 = \left(\phi_{D1} + \psi_{D1}\right)\Delta D_1 \tag{10-197}$$

$$X_2 = \left(\phi_{H2} + \psi_{H2}\right)\phi_{D1}\Delta D_1 + \left(\phi_{D2} + \psi_{D2}\right)\Delta D_2 \tag{10-198}$$

$$X_3 = \left(\phi_{H3} + \psi_{H3}\right)\phi_{H2}\phi_{D1}\Delta D_1 + \left(\phi_{H3} + \psi_{H3}\right)\phi_{D2}\Delta D_2 + \left(\phi_{D3} + \psi_{D3}\right)\Delta D_3 \tag{10-199}$$

In solving the simultaneous equations 10-189, 10-190 and 10-191, it can be seen that

$$\Delta T_{12} = \frac{\left(-b_3 X_1 + b_2 X_1 + b_3 X_2 - b_2 X_3\right)}{\left(a_1 b_3 - a_1 b_2 - b_3 a_2 + b_2 a_3\right)} \tag{10-200}$$

This expression may be written in the form of a determinant □

$$\Delta T_{12} = \frac{\begin{vmatrix} -1 & -X_1 & 0 \\ -1 & -X_2 & b_2 \\ -1 & -X_3 & b_3 \end{vmatrix}}{\begin{vmatrix} -1 & a_1 & 0 \\ -1 & a_2 & b_2 \\ -1 & a_3 & b_3 \end{vmatrix}} \tag{10-201}$$

Similarly, the expression for the change in tension between stands 2 and 3, namely ΔT_{23}, may be written

$$\Delta T_{23} = \frac{\begin{vmatrix} -1 & a_1 & -X_1 \\ -1 & a_2 & -X_2 \\ -1 & a_3 & -X_3 \end{vmatrix}}{\begin{vmatrix} -1 & a_1 & 0 \\ -1 & a_2 & b_2 \\ -1 & a_3 & b_3 \end{vmatrix}} \tag{10-202}$$

These determinants can be represented as triangles, whose area is half the value of the determinant. The denominator is represented, for example, by a triangle having co-ordinates $(a_1, 0)$ (a_2, b_2) (a_3, b_3). In a fixed-drive mill the elements of this determinant tend to

$$a_1 = \phi_{Tf_1} \tag{10-203}$$

$$a_2 = \phi_{H_2}\phi_{Tf_1} + \phi_{Tb_2} \tag{10-204}$$

$$a_3 = \phi_{H_3}\phi_{H_2}\phi_{Tf_1} + \phi_{Tb_2} \tag{10-205}$$

$$b_2 = \phi_{Tf_2} \tag{10-206}$$

$$b_3 = \phi_{H_3}\phi_{Tf_2} + \phi_{Tb_3} \tag{10-207}$$

All of these are negative. Furthermore

□Note: The reader not familiar with determinants may wish to consult a text dealing with the subject, as for example, "Survey of Applicable Mathematics" by K. Rektorys (Editor), The M.I.T. Press, Cambridge, Mass., 1969.

$$|a_1| < |a_2| \tag{10-208}$$

$$|b_2| < |b_3| \tag{10-209}$$

and

$$|a_2| \gtrless |a_3| \tag{10-210}$$

as

$$\phi_H \gtrless 1 \tag{10-211}$$

The denominator triangle, therefore appears as shown in Figure 10-39(a) for a mill with a stiff drive where

$$\phi_H > 1 \tag{10-212}$$

As the mill drive is made softer, the elements tend to

$$a_1 = \psi_{Tf_1} \tag{10-213}$$

$$a_2 = \psi_{Tb_2} \tag{10-214}$$

$$a_3 = 0 \tag{10-215}$$

$$b_2 = \psi_{Tf_2} \tag{10-216}$$

$$b_3 = \psi_{Tb_3} \tag{10-217}$$

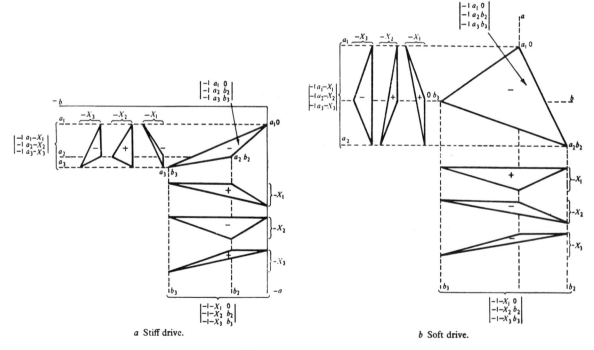

a Stiff drive.   *b* Soft drive.

Figure 10-39: Representation of Tension Changes in a Three-Stand Mill.

Under these circumstances, a_1 and b_2 are positive whereas a_2 and b_3 are negative. The shape of the denominator triangle changes to that shown in Figure 10-39(b). In both cases, the apices being in

TORQUE EQUATIONS AND TANDEM MILLS

clockwise sequence, the denominator is negative. Triangles corresponding to the numerators for different single disturbances are shown and, from the sequence of their apices in relation to those of the denominator triangle, the sign of the tension changes may be ascertained. This display is particularly useful when there are unequal disturbances on more than one stand; the reaction of the mill can be ascertained very quickly.

It also shows how the transition from "stiff" to "soft" conditions influences the behavior of the mill. It will have been noted that the area of the denominator triangle decreases as the mill drive is made stiffer. For a given disturbance the tension change with a stiff drive is larger than with a soft one. Should the denominator vanish altogether the mill will become unstable, any finite disturbance causing an unlimited change in tension. Stability in a three-stand mill therefore requires that the determinant

$$\begin{vmatrix} -1 & a_1 & 0 \\ -1 & a_2 & b_2 \\ -1 & a_3 & b_3 \end{vmatrix} \quad \text{should remain negative.}$$

10-23 Equations For a Five-Stand Mill

Hessenberg and Jenkins [∞] derived the equations for a five-stand mill as given in Table 10-6 (where the notation and terminology used is as set forth in Section 10-19). As they pointed out, there is no simple way

Table 10-6

The Output Equations for a Five-Stand Tandem Mill

$\frac{\Delta U}{U} = (\phi_{D1} + \psi_{D1})\Delta D_1$

$\frac{\Delta U}{U} = (\phi_{H2} + \psi_{H2})\phi_{D1}\Delta D_1 \quad + (\phi_{D2} + \psi_{D2})\Delta D_2$

$\frac{\Delta U}{U} = (\phi_{H3} + \psi_{H3})\phi_{H2}\phi_{D1}\Delta D_1 \quad + (\phi_{H3} + \psi_{H3})\phi_{D2}\Delta D_2 \quad + (\phi_{D3} + \psi_{D3})\Delta D_3$

$\frac{\Delta U}{U} = (\phi_{H4} + \psi_{H4})\phi_{H3}\phi_{H2}\phi_{D1}\Delta D_1 \quad + (\phi_{H4} + \psi_{H4})\phi_{H3}\phi_{D2}\Delta D_2 \quad + (\phi_{H4} + \psi_{H4})\phi_{D3}\Delta D_3 \quad + (\phi_{D4} + \psi_{D4})\Delta D_4$

$\frac{\Delta U}{U} = (\phi_{H5} + \psi_{H5})\phi_{H4}\phi_{H3}\phi_{H2}\phi_{D1}\Delta D_1 \quad + (\phi_{H5} + \psi_{H5})\phi_{H4}\phi_{H3}\phi_{D2}\Delta D_2 \quad + (\phi_{H5} + \psi_{H5})\phi_{H4}\phi_{D3}\Delta D_3 \quad + (\phi_{H5} + \psi_{H5})\phi_{D4}\Delta D_4 \quad + (\phi_{D5} + \psi_{D5})\Delta D_5$

$+ (\phi_{Tf1} + \psi_{Tf1})\Delta T_{12}$

$+ (\phi_{H2} + \psi_{H2})\phi_{Tf1}\Delta T_{12} \quad + (\phi_{Tb2} + \psi_{Tb2})\Delta T_{12} \quad + (\phi_{Tf2} + \psi_{Tf2})\Delta T_{23}$

$+ (\phi_{H3} + \psi_{H3})\phi_{H2}\phi_{Tf1}\Delta T_{12} \quad + (\phi_{H3} + \psi_{H3})\phi_{Tb2}\Delta T_{12} \quad + (\phi_{H3} + \psi_{H3})\phi_{Tf2}\Delta T_{23} \quad + (\phi_{Tb3} + \psi_{Tb3})\Delta T_{23} \quad + (\phi_{Tf3} + \psi_{Tf3})\Delta T_{34}$

$+ (\phi_{H4} + \psi_{H4})\phi_{H3}\phi_{H2}\phi_{Tf1}\Delta T_{12} \quad + (\phi_{H4} + \psi_{H4})\phi_{H3}\phi_{Tb2}\Delta T_{12} \quad + (\phi_{H4} + \psi_{H4})\phi_{H3}\phi_{Tf2}\Delta T_{23} \quad + (\phi_{H4} + \psi_{H4})\phi_{Tb3}\Delta T_{23} \quad + (\phi_{H4} + \psi_{H4})\phi_{Tf3}\Delta T_{34}$

$+ (\phi_{Tb4} + \psi_{Tb4})\Delta T_{34} \quad + (\phi_{Tf4} + \psi_{Tf4})\Delta T_{45} \quad + (\phi_{H5} + \psi_{H5})\phi_{H4}\phi_{H3}\phi_{H2}\phi_{Tf1}\Delta T_{12} \quad + (\phi_{H5} + \psi_{H5})\phi_{H4}\phi_{H3}\phi_{Tb2}\Delta T_{12} \quad + (\phi_{H5} + \psi_{H5})\phi_{H4}\phi_{H3}\phi_{Tf2}\Delta T_{23}$

$+ (\phi_{H5} + \psi_{H5})\phi_{H4}\phi_{Tb3}\Delta T_{23} \quad + (\phi_{H5} + \psi_{H5})\phi_{H4}\phi_{Tf3}\Delta T_{34} \quad + (\phi_{H5} + \psi_{H5})\phi_{Tb4}\Delta T_{34} \quad + (\phi_{H5} + \psi_{H5})\phi_{Tf4}\Delta T_{45} \quad + (\phi_{Tb5} + \psi_{Tb5})\Delta T_{45}$

of graphically representing such equations as in the case of the two-stand mill (Figure 10-38) or the three-stand mill (Figure 10-39). However, the general pattern of behavior may be deduced in part from the rules governing the transmission of output changes through the intermediate stands, as discussed in Section 10-19. The change in entry speed to a disturbed stand is of the same sign as that of the disturbance, and gives rise to a back tension change also of the same sign. By following the backward transmission rules appropriate to an under- or over-critical mill along to stand 1, the signs of the preceding interstand tension changes and of the change in output from stand 1 may be found. The change in output from the last stand will be the same as that from the first. In an under-critical mill the change in finishing gage will therefore also be the same. In an over-critical mill

---

[∞] W. C. F. Hessenberg and W. N. Jenkins, "Effect of Screw and Speed Setting Changes on Gauge, Speed and Tension in Tandem Mills", Proc. Inst. Mech. Eng., 169, 1955, pp. 1051-1058.

the output change from the last stand is mainly a speed change. The effects of speed and screw settings on interstand tensions and final strip thickness are presented in Table 10-7.

Table 10-7
Interstand Tension And Finishing Gauge Changes For A Five-Stand Mill

Stand	Positive Change	$\left(\dfrac{\Delta v}{v} \gg \dfrac{\Delta h}{h}\right)$					$\left(\dfrac{\Delta h}{h} \gg \dfrac{\Delta v}{v}\right)$				
		ΔT_{12}	ΔT_{23}	ΔT_{34}	ΔT_{45}	Δh_5	ΔT_{12}	ΔT_{23}	ΔT_{34}	ΔT_{45}	Δh_5
1	Speed	−	−	−	−	+	−	±	±	±	+
2	Speed	+	−	−	−	±	+	−	±	±	−
3	Speed	+	+	−	−	±	−	+	−	±	+
4	Speed	+	+	+	−	±	+	−	+	−	−
5	Speed	+	+	+	+	−	−	+	−	+	±
1	Screw	±	±	±	±	+	±	±	±	±	+
2	Screw	+	±	±	±	+	+	±	±	±	−
3	Screw	+	+	±	±	+	−	+	±	±	+
4	Screw	+	+	+	±	+	+	−	+	±	−
5	Screw	+	+	+	+	+	−	+	−	+	+

The sign of the tension changes following the disturbed stand are unambiguous only when speed changes alone are being propagated; as soon as gage changes are involved, numerical calculation is necessary to determine their sign.

It is possible, however, to prepare tables such as Table 10-7 which provide a partial answer to many of the questions posed concerning tandem mill operation. From the mathematical model presented in Section 10-19, the answer can be made complete for any particular case. Although the numerical solution of multistand problems is certain to prove very laborious, particularly if a range of working conditions is to be explored, simultaneous equations easily lend themselves to handling by digital computers.

A few additional properties of multistand mills may be deduced by a process of replacing two stands by one equivalent stand. The effective values of the output coefficients for the equivalent stand are derived in terms of those appertaining to each stand individually. The effective forward tension coefficient, for example, is derived as follows, using the notation of Section 10-19.

The output equations for a two-stand mill are,

$$\frac{\Delta U}{U} = X_1 + a_1 \Delta T_{12} \qquad (10\text{-}218)$$

$$\frac{\Delta U}{U} = X_2 + a_2 \Delta T_{12} \qquad (10\text{-}219)$$

where

$$a_1 = \phi_{Tf_1} + \psi_{Tf_1} \qquad (10\text{-}220)$$

and

$$a_2 = \left(\phi_{H2} + \psi_{H2}\right)\phi_{Tf1} + \phi_{Tb2} + \psi_{Tb2} \qquad (10\text{-}221)$$

If

$$X_1 = 0 \qquad (10\text{-}222)$$

TORQUE EQUATIONS AND TANDEM MILLS

and the disturbance to the second stand be an increase in forward tension so that

$$X_2 = \left(\phi_{Tf2} + \psi_{rf2}\right)\Delta T_{f2} \tag{10-223}$$

then the effective forward tension coefficient for the equivalent stand is,

$$\left(\phi_{Tf12} + \psi_{Tf12}\right) = \frac{\Delta U/U}{\Delta T_{f2}} = \frac{\left(\phi_{Tf2} + \psi_{Tf2}\right)a_1}{a_1 - a_2} \tag{10-224}$$

Now in an over-critical mill

$$0 < a_1/(a_1 - a_2) < 1 \tag{10-225}$$

so that the two stands taken together are effectively nearer to the critical condition than each stand alone. This process of "boxing" two stands may be repeated with a further decrease in the forward-tension coefficient each time another stand is added, so that a five-stand mill with over-critical stands must nevertheless be almost critical with respect to disturbances to the fifth stand, that is,

$$\left(\Delta h_5/h_5\right) + \left(\Delta v_5/v_5\right) = 0 \tag{10-226}$$

The speed of stand 5 can therefore freely be used to adjust the gage without affecting the tensions preceding stand 4. If, furthermore, the drive to stand 5 is kept soft, the consequent tension changes between stands 4 and 5 will be small.

In an under-critical mill, both $(\phi_{Tf2} + \psi_{Tf2})$ and $a_1/(a_1 - a_2)$ are negative; $(\phi_{Tf12} + \psi_{Tf12})$ is therefore positive, and a pair of "boxed" under-critical stands behaves like a single over-critical stand. The addition of another stand again reverses the sign of the forward tension coefficient. This is in full accord with the alternation of the sign of the tension changes in an under-critical mill.

To obtain the effective output coefficient for a change in ingoing gage $\Delta H_1/H_1$ put,

$$X_1 = \left(\phi_{H1} + \psi_{H1}\right)\frac{\Delta H_1}{H_1} \tag{10-227}$$

$$X_2 = \left(\phi_{H2} + \psi_{H2}\right)\phi_{H1}\frac{\Delta H_1}{H_1} \tag{10-228}$$

whence

$$\phi_{H12} + \psi_{H12} = \frac{\Delta U/U}{\Delta H_1/H_1} = \frac{\phi_{H1}\left(\phi_{H2} + \psi_{H2}\right)a_1 - \left(\phi_{H1} + \psi_{H1}\right)a_2}{a_1 - a_2} \tag{10-229}$$

In a stiff-drive mill, the speed coefficients are small compared with the gage coefficients and, if

$$\phi_{H1} \simeq \phi_{H2} \tag{10-230}$$

the above expression reduces to

$$\frac{\phi_{H12}}{\phi_{H1}} \simeq 1 + \frac{\left(\phi_{H1} - 1\right)a_1}{a_1 - a_2} \tag{10-231}$$

Now $a_1/(a_1 - a_2)$ is negative in the under-critical condition so that if

$$\phi_{H1} \gtrless 1 \tag{10-232}$$

then

$$\phi_{H12} \lessgtr \phi_{H1} \qquad (10\text{-}233)$$

This means that the two stands will amplify or attenuate fractional gage errors less when working in tandem than as independent stands.

In a soft-drive mill the gage changes due to tension changes will be small so that

$$\phi_{H12} \simeq \phi_{H1}\phi_{H2} \qquad (10\text{-}234)$$

Thus two such stands will amplify or attenuate more when in tandem. This suggests that a stiff drive is required for a mill with gage-error amplification and a soft drive for one that attenuates. But it has just been shown that if critical conditions are required between the last two stands of a multi-stand mill, a soft drive is necessary, in which case the stands should be such as to provide some degree of attenuation, that is, mechanically rigid.

In the foregoing analysis, the effects of coiler tensions were ignored. In a later model, developed by Courcoulas and Ham[◊], this omission was remedied but all the forward slips were assumed to be zero. As pointed out by Lianis and Ford[○], however, if full use is to be made of the analysis of Hessenberg and Jenkins, accurate means for calculating the differential coefficients in their equations are necessary, and it would also be preferable not to ignore any of the possible sources of disturbance. They, therefore, published a number of monographs and described their use in connection with tandem mill control equations.

Also in 1957, Phillips[δ] published an analytical study of tandem mill operation in which he assumed that, for a particular stand, the changes in roll separating force, rolling torque and neutral point thickness were expressed by linear equations involving changes of entry and exit strip tensions and strip thicknesses. He obtained the coefficient of the linear equations from the rolling theory of Ford, Ellis and Bland[■] (see Section 9-13) and using the results of his analytical study, he carried out an analog simulation of a vernier gage control system for a tandem mill.

10-24 Misaka's Control Equations for Tandem Cold Mills

In 1967 Misaka[+] published an account of his studies with respect to control equations for a five-stand mill, using a model and notation very similar to that of Hessenberg and Jenkins (see Section 10-19). Considering strip of unit width, the flow of metal through the mill is represented by

$$v_i h_i = U \qquad i = 1, 2, \ldots n \qquad (10\text{-}235)$$

where v is the velocity of the strip as it leaves a stand and h is its thickness. When a steady state has been reached after a disturbance

$$(v_i + \Delta v_i)(h_i + \Delta h_i) = U + \Delta U \qquad (10\text{-}236)$$

If the deviations Δv_i and Δh_i are small

[◊] J. H. Courcoulas and J. M. Ham, "Incremental Control Equations for Tandem Rolling Mills", Trans. Am. Inst. of Electrical Engineers, 1956, pp. 56-65.
[○] G. Lianis and H. Ford, "Control Equations of Multistand Cold Rolling Mills", Proc. Inst. Mech. Eng., 171, 1957, pp. 757 ff.
[δ] R. A. Phillips, Amer. Inst. Elec. Eng. (1957), pp. 355 ff.
[■] H. Ford, F. Ellis, and D. R. Bland, "Cold Rolling With Strip Tension, Part I — A New Approximate Method of Calculation and a Comparison with Other Methods", J. Iron and Steel Institute, 168, 1951, pp. 57-72.
[+] Y. Misaka, "Control Equations for Cold Tandem Mills", Transactions ISIJ, Vol. 8, 1968, pp. 86-96.

TORQUE EQUATIONS AND TANDEM MILLS

$$\left(\frac{\Delta v}{v}\right)_i + \left(\frac{\Delta h}{h}\right)_i = \frac{\Delta U}{U} \tag{10-237}$$

The speed of the outgoing strip at a given stand is expressed by

$$v_i = (1+f_i)(v_n)_i \tag{10-238}$$

where f is the forward slip and v_n is the peripheral speed of the work rolls. With respect to changes in roll speed, these may be attributed to changes in the set speed of the rolls and motor droop. Thus,

$$\left(\frac{\Delta v_n}{v_n}\right)_i = \frac{\Delta(N_i)}{N_i} + Z_i^* \cdot \Delta G_i \tag{10-239}$$

where N represents the set speed of the rolls, Z^* the damping coefficient (or the droop characteristic of the motor) and ΔG the total spindle torque (for the two work rolls). The relative change of speed of the outgoing strip at the ith stand may therefore be expressed

$$\left(\frac{\Delta v}{v}\right)_i = \left(\frac{\Delta N}{N}\right)_i + Z_i^* \cdot \Delta G_i + \frac{1}{(1+f_i)} \cdot \Delta f_i \tag{10-240}$$

Now the rolling torque G at any stand is a function of the thickness H_1 of the strip fed to the tandem mill, the thickness H of the strip fed to that stand, the rolled thickness h of the strip emerging from the stand, the back tensile stress t_b, the front or forward stress t_f, the coefficient of friction μ and a parameter l which corresponds to the resistance to deformation of the strip. Thus

$$G = G(H_1, H, h, t_b, t_f, \mu, l) \tag{10-241}$$

Similarly, the forward slip may be expressed

$$f = f(H_1, H, h, t_b, t_f, \mu, l) \tag{10-242}$$

Thus the derivatives may be written

$$\Delta G = \left(\frac{\partial G}{\partial H_1}\right) \cdot \Delta H_1 + \left(\frac{\partial G}{\partial H}\right) \cdot \Delta H + \left(\frac{\partial G}{\partial h}\right) \cdot \Delta h$$
$$+ \left(\frac{\partial G}{\partial t_b}\right) \cdot \Delta t_b + \left(\frac{\partial G}{\partial t_f}\right) \cdot \Delta t_f + \left(\frac{\partial G}{\partial \mu}\right) \cdot \Delta \mu \tag{10-243}$$
$$+ \left(\frac{\partial G}{\partial l}\right) \cdot \Delta l$$

and

$$\Delta f = \left(\frac{\partial f}{\partial H_1}\right) \cdot \Delta H_1 + \left(\frac{\partial f}{\partial H}\right) \cdot \Delta H + \left(\frac{\partial f}{\partial h}\right) \cdot \Delta h$$
$$+ \left(\frac{\partial f}{\partial t_b}\right) \cdot \Delta t_b + \left(\frac{\partial f}{\partial t_f}\right) \cdot \Delta t_f + \left(\frac{\partial f}{\partial \mu}\right) \cdot \Delta \mu \tag{10-244}$$
$$+ \left(\frac{\partial f}{\partial l}\right) \cdot \Delta l$$

The thickness of the outgoing strip at a given stand may be expressed by Sims' relationship (Section 5-14)

$$h = S + \frac{P}{K} \tag{10-245}$$

628   CHAPTER 10

where

    S represents the screw setting,
    P the total rolling force and
    K the mill spring.

The rolling force is also a function of the seven variables considered above.

$$P = P\left(H_1, H, h, t_b, t_f, \mu, l\right) \quad (10\text{-}246)$$

so that the change in thickness of the outgoing strip may be expressed

$$\Delta h = \frac{1}{K - \frac{\partial P}{\partial h}} \left\{ K \cdot \Delta S + \frac{\partial P}{\partial H_1} \cdot \Delta H_1 + \frac{\partial P}{\partial H} \cdot \Delta H \right. $$
$$\left. + \frac{\partial P}{\partial t_b} \cdot \Delta t_b + \frac{\partial P}{\partial t_f} \cdot \Delta t_f + \frac{\partial P}{\partial \mu} \cdot \Delta \mu + \frac{\partial P}{\partial l} \cdot \Delta l \right\} \quad (10\text{-}247)$$

With respect to a fractional change in horsepower $\frac{\Delta HP}{HP}$, this may be written as

$$\left(\frac{\Delta HP}{HP}\right)_i = \left(\frac{\Delta N}{N}\right)_i + \left(\frac{\Delta G}{G}\right)_i \quad (10\text{-}248)$$

Equations 10-237, 10-240, 10-243, 10-244, 10-247 and 10-248 may be combined to form a set of fundamental equations, as follows:

$$\sigma_{H1i} \cdot \left(\frac{\Delta H}{H}\right)_1 + \sigma_{Hi} \cdot \left(\frac{\Delta H}{H}\right)_i + (1 + \sigma_{hi}) \cdot \left(\frac{\Delta h}{h}\right)_i$$
$$+ \sigma_{tbi} \cdot \left(\frac{\Delta t_b}{t_b}\right)_i + \sigma_{tfi} \cdot \left(\frac{\Delta t_f}{t_f}\right)_i + \sigma_{\mu i}\left(\frac{\Delta \mu}{\mu}\right)_i$$
$$+ \sigma l_i \left(\frac{\Delta l}{l}\right) + \left(\frac{\Delta N}{N}\right)_i - \frac{\Delta U}{U} = 0$$

$$\nu_{H1i} \cdot \left(\frac{\Delta H}{H}\right)_1 + \nu_{Hi}\left(\frac{\Delta H}{H}\right)_i - \left(\frac{\Delta h}{h}\right)_i$$
$$+ \nu_{tbi} \cdot \left(\frac{\Delta t_b}{t_b}\right)_i + \nu_{tfi} \cdot \left(\frac{\Delta t_f}{t_f}\right)_i + \nu_{\mu i}\left(\frac{\Delta \mu}{\mu}\right)_i \quad (10\text{-}249)$$
$$+ \nu_{li}\left(\frac{\Delta l}{l}\right) + \nu_{si} \cdot \left(\frac{1}{h_i}\right) \cdot (\Delta S)_i = 0$$

$$\eta_{H1i}\left(\frac{\Delta H}{H}\right)_1 + \eta_{Hi}\left(\frac{\Delta H}{H}\right)_i + \eta_{hi} \cdot \left(\frac{\Delta h}{h}\right)_i$$
$$+ \eta_{tbi} \cdot \left(\frac{\Delta t_b}{t_b}\right)_i + \eta_{tfi} \cdot \left(\frac{\Delta t_f}{t_f}\right)_i + \eta_{\mu i}\left(\frac{\Delta \mu}{\mu}\right)_i$$
$$+ \eta l_i\left(\frac{\Delta l}{l}\right) + \left(\frac{\Delta N}{N}\right)_t - \left(\frac{\Delta HP}{HP}\right)_t = 0$$

TORQUE EQUATIONS AND TANDEM MILLS

$$\sigma_{H1} = Z^* \cdot H_1 \cdot \frac{\partial G}{\partial H_1} + \frac{1}{1+f} \cdot H_1 \cdot \frac{\partial f}{\partial H_1}$$

$$\nu_{t_f} = \frac{1}{K - \frac{\partial P}{\partial h}} \cdot \frac{t_f}{h} \cdot \frac{\partial P}{\partial t_f}$$

$$\sigma_H = Z^* \cdot H \cdot \frac{\partial G}{\partial H} + \frac{1}{1+f} \cdot H \cdot \frac{\partial f}{\partial H}$$

$$\nu_\mu = \frac{1}{K - \frac{\partial P}{\partial h}} \cdot \frac{\mu}{h} \cdot \frac{\partial P}{\partial \mu}$$

$$\sigma_h = Z^* \cdot h \cdot \frac{\partial G}{\partial h} + \frac{1}{1+f} \cdot h \cdot \frac{\partial f}{\partial h}$$

$$\nu l = \frac{1}{K - \frac{\partial P}{\partial h}} \cdot \frac{l}{h} \cdot \frac{\partial P}{\partial l}$$

$$\sigma_{t_b} = Z^* \cdot t_b \cdot \frac{\partial G}{\partial t_b} + \frac{1}{1+f} \cdot t_b \cdot \frac{\partial f}{\partial t_b}$$

$$\nu_s = \frac{K}{K - \frac{\partial P}{\partial h}}$$

$$\sigma_{t_f} = Z^* \cdot t_f \cdot \frac{\partial G}{\partial t_f} + \frac{1}{1+f} \cdot t_f \cdot \frac{\partial f}{\partial t_f}$$

$$\sigma_\mu = Z^* \cdot \mu \cdot \frac{\partial G}{\partial \mu} + \frac{1}{1+f} \cdot \mu \cdot \frac{\partial f}{\partial \mu}$$
$$\eta_{H_1} = \frac{H_1}{G} \cdot \frac{\partial G}{\partial H_1}$$

$$\sigma l = Z^* \cdot l \cdot \frac{\partial G}{\partial l} + \frac{1}{1+f} \cdot l \cdot \frac{\partial f}{\partial l}$$

$$\eta_H = \frac{H}{G} \cdot \frac{\partial G}{\partial H}$$

$$\nu_{H1} = \frac{1}{K - \frac{\partial P}{\partial h}} \cdot \frac{H_1}{h} \cdot \frac{\partial P}{\partial H_1}$$

$$\eta_h = \frac{h}{G} \cdot \frac{\partial G}{\partial h}$$

$$\nu_H = \frac{1}{K - \frac{\partial P}{\partial h}} \cdot \frac{H}{h} \cdot \frac{\partial P}{\partial H}$$

$$\eta_{t_b} = \frac{t_b}{G} \cdot \frac{\partial G}{\partial t_b}$$

$$\nu_{t_b} = \frac{1}{K - \frac{\partial P}{\partial h}} \cdot \frac{t_b}{h} \cdot \frac{\partial G}{\partial t_b}$$

$$\eta_{t_f} = \frac{t_f}{G} \cdot \frac{\partial G}{\partial t_f}$$

$$\eta_\mu = \frac{\mu}{G} \cdot \frac{\partial G}{\partial \mu}$$

$$\eta l = \frac{l}{G} \cdot \frac{\partial G}{\partial l}$$

(10-250)

In deriving the partial derivatives for the solutions of equations 10-249 and 10-250, Misaka used the following formulae:

$$P = b \cdot \varkappa \cdot \bar{K} \cdot \sqrt{R'(H-h)} \cdot D_R \quad (10\text{-}251)$$

where

$$\text{and } \bar{P} = P - b \cdot \varkappa \cdot \bar{K} \cdot \sqrt{R'(H-h)} \cdot D_R = 0$$

P is the rolling force
b is the strip width
$\varkappa$ is a factor defined below used to compensate for strip tensions
$\bar{K}$ is the mean yield stress in plane strain
R' is the deformed roll radius
H is the entry thickness of the strip
h is the exit thickness and
D_R is the approximate value of the f_3 function of Bland and Ford (See Section 9-13) and is given by

$$D_R = 1.08 + 1.79 \, r \cdot \sqrt{1-r} \cdot \mu \sqrt{\frac{R'}{h}} - 1.02r \quad (10\text{-}252)$$

where r is the reduction.

The factor $\varkappa$ is expressed as

$$\varkappa = 1 - \frac{(\alpha-1)t_b + t_f}{\alpha \bar{K}} \qquad (10\text{-}253)$$

where α is the reciprocal of the reduction
t_b is the back tensile stress and
t_f is the forward tensile stress in the strip.

The total torque G is given by

$$G = G_o + R \cdot b \cdot (Ht_b - ht_f) \qquad (10\text{-}254)$$

where R is the work roll radius
and G_o is the total torque in the absence of tensions expressed by

$$G_o = b \cdot \varkappa' \cdot \bar{K} \cdot R_o(H - h)D_G \qquad (10\text{-}255)$$

In this expression, D_G is the approximate value of twice the factor f_5 used by Bland and Ford (Section 10-5) for equal front and back tensions, where

$$D_G = 1.05 + (0.07 + 1.32r)\sqrt{1-r} \cdot \mu\sqrt{\frac{R'}{h}} - 0.85r \qquad (10\text{-}256)$$

The factor $\varkappa'$ may be calculated from

$$\varkappa' = 1 - \frac{(\alpha' - 1)t_b + t_f}{\alpha' \cdot \bar{K}} \qquad (10\text{-}257)$$

where α' is the reciprocal of the reduction.

The slip f is represented by

$$f = \phi_n^2 \frac{R'}{h} \qquad (10\text{-}258)$$

where the neutral angle ϕ_n is given by

$$\phi_n = \sqrt{\frac{h}{R'}} \cdot \tan\left(\sqrt{\frac{h}{R'}} \cdot \frac{\bar{H}_n}{2}\right) \qquad (10\text{-}259)$$

such that the dimensionless factor $\bar{H}_n$ is represented by

$$\bar{H}_n = \frac{\bar{H}_b}{2\mu} - \frac{1}{2} \cdot \ln\left(\frac{H}{h} \cdot \frac{1 - \frac{t_f}{S_f}}{1 - \frac{t_b}{S_b}}\right) \qquad (10\text{-}260)$$

where S_f and S_b are the constrained yield strengths of the strip on exit from and on entry into the roll bite and where

$$\bar{H}_b = 2\sqrt{\frac{R'}{h}} \cdot \tan^{-1}\left(\sqrt{\frac{R'}{h}} \cdot \phi_b\right) \qquad (10\text{-}261)$$

In this case ϕ_b is the angle of the arc of contact such that

$$\phi_b = \sqrt{\frac{H-h}{R'}} \qquad (10\text{-}262)$$

and the deformed roll radius R' is given by the Hitchcock relationship

$$R' = R \cdot \left\{1 + c \cdot \frac{P}{b(H-h)}\right\} \qquad (10\text{-}263)$$

TORQUE EQUATIONS AND TANDEM MILLS

where

$$C = 2.14 \times 10^{-4} \text{ mm/Kg}^2 \qquad (10\text{-}264)$$

Misaka assumed that the plastic behavior of the workpiece could be expressed by the relationship

$$s = l\,(r+m)^n \qquad (10\text{-}265)$$

where s = yield stress in plane strain
l, m and n = constants, and
r = total reduction given to the strip.

The mean resistance to deformation, $\bar{K}$ was derived from

$$\bar{K} = l\,(\bar{r}+m)^n \qquad (10\text{-}266)$$

where

$$\bar{r} = 0.4 r_b + 0.6 r_f \qquad (10\text{-}267)$$

$$r_b = 1 - \frac{H}{H_1} \qquad (10\text{-}268)$$

and

$$r_f = 1 - \frac{h}{H_1} \qquad (10\text{-}269)$$

All the partial derivatives used in equation 10-226 were derived from the rolling theory equations cited above and are listed below

1. The Partial Derivatives of Rolling Force

$$\frac{\partial P}{\partial \lambda} = -\left(\frac{\partial \bar{P}}{\partial \lambda}\right) \Big/ \left(\frac{\partial \bar{P}}{\partial P}\right) \qquad (10\text{-}270)$$

λ represents H, h, t_b, t_f, μ, l, H_1.

$$\frac{\partial \bar{P}}{\partial P} = 1 - \frac{R'-R}{2R'} \cdot \left(1 + \frac{1.79\, r\mu}{D_p} \cdot \sqrt{\frac{R'}{H}}\right) \qquad (10\text{-}271)$$

$$\frac{\partial \bar{P}}{\partial H} = -\frac{P}{H} \cdot \left\{ -\frac{0.4n}{\bar{r}+m} \cdot \frac{H}{H_1} \cdot \frac{1}{\kappa} + \frac{1}{2r} \cdot \frac{R}{R'} + \frac{1.79\mu}{2D_p} \right.$$
$$\left. \times \sqrt{\frac{R'}{H}} \cdot \left(1 - 3r + \frac{R}{R'}\right) - \frac{1.02}{D_p} \cdot (1-r) \right\} \qquad (10\text{-}272)$$

In the case of stand 1, the first term in the doubly-curved parentheses is unnecessary.

$$\frac{\partial \bar{P}}{\partial h} = -\frac{P}{H} \cdot \left\{ \frac{0.6n}{\bar{r}+m} \cdot \frac{H}{H_1} \cdot \frac{1}{\kappa} - \frac{1}{2r} \cdot \frac{R}{R'} \right.$$
$$\left. -\frac{1.79\mu}{2D_p} \cdot \sqrt{\frac{R'}{H}} \cdot \left(1 + \frac{R}{R'}\right) + \frac{1.02}{D_p} \right\} \qquad (10\text{-}273)$$

$$\frac{\partial \bar{P}}{\partial t_b} = \frac{P}{\bar{k}} \cdot \frac{1}{\kappa} \cdot \frac{\alpha-1}{\alpha} \qquad (10\text{-}274)$$

$$\frac{\partial \bar{P}}{\partial t_f} = \frac{P}{\bar{k}} \cdot \frac{1}{\kappa} \cdot \frac{1}{\alpha} \tag{10-275}$$

$$\frac{\partial \bar{P}}{\partial \mu} = \frac{P}{\mu} \cdot \left(1 + \frac{1.02r - 1.08}{D_P}\right) \tag{10-276}$$

$$\frac{\partial \bar{P}}{\partial l} = -\frac{P}{l} \cdot \frac{1}{\kappa} \tag{10-277}$$

$$\frac{\partial \bar{P}}{\partial H_1} = -\frac{P}{H_1} \cdot \frac{1}{\kappa} \cdot \frac{1-\bar{r}}{\bar{r}+m} \cdot n \tag{10-278}$$

In the case of stand 1, $\partial \bar{P}/\partial H_1$ is as follows.

$$\left(\frac{\partial \bar{P}}{\partial H_1}\right)_1 = -\frac{P}{H_1} \cdot \frac{1}{\kappa} \cdot \frac{n}{\bar{r}+m} \cdot \frac{0.6h}{H_1} \tag{10-279}$$

2. The Partial Derivatives of Rolling Torque

$$\frac{\partial G}{\partial H} = G_o\left(\frac{1}{\bar{k}} \cdot \frac{\partial \bar{k}}{\partial H} + \frac{1}{\kappa'} \cdot \frac{\partial \kappa'}{\partial H} + \frac{1}{H-h} + \frac{1}{D_G} \cdot \frac{\partial D_G}{\partial H}\right) + R \cdot b \cdot t_b \tag{10-280}$$

$$\frac{\partial G}{\partial h} = G_o\left(\frac{1}{\bar{k}} \cdot \frac{\partial \bar{k}}{\partial h} + \frac{1}{\kappa'} \cdot \frac{\partial \kappa'}{\partial h} - \frac{1}{H-h} + \frac{1}{D_G} \cdot \frac{\partial D_G}{\partial h}\right) - R \cdot b \cdot t_f \tag{10-281}$$

$$\frac{\partial G}{\partial t_b} = G_o \cdot \frac{1}{\kappa'} \cdot \frac{\partial \kappa'}{\partial t_b} + R \cdot H \cdot b \tag{10-282}$$

$$\frac{\partial G}{\partial t_f} = G_o \cdot \frac{1}{\kappa'} \cdot \frac{\partial \kappa'}{\partial t_f} - R \cdot h \cdot b \tag{10-283}$$

$$\frac{\partial G}{\partial \mu} = G_o \cdot \frac{1}{D_G} \cdot \frac{\partial D_G}{\partial \mu} \tag{10-284}$$

$$\frac{\partial G}{\partial l} = G_o\left(\frac{1}{\bar{k}} \cdot \frac{\partial \bar{k}'}{\partial l} + \frac{1}{\kappa'} \cdot \frac{\partial \kappa'}{\partial l}\right) \tag{10-285}$$

$$\frac{\partial G}{\partial H_1} = G_o\left(\frac{1}{\bar{k}} \cdot \frac{\partial \bar{k}}{\partial H_1} + \frac{1}{\kappa'} \cdot \frac{\partial \kappa'}{\partial H_1} + \frac{1}{D_G} \cdot \frac{\partial D_G}{\partial H_1}\right) \tag{10-286}$$

$$\frac{\partial \bar{k}}{\partial H} = -\frac{n}{\bar{r}+m} \cdot \frac{0.4}{H_1} \cdot \bar{k} \tag{10-287}$$

In the case of stand 1,

$$\partial \bar{k}/\partial H = 0 \tag{10-288}$$

$$\frac{\partial \bar{k}}{\partial h} = -\frac{n}{\bar{r}+m} \cdot \frac{0.6}{H_1} \cdot \bar{k} \tag{10-289}$$

$$\frac{\partial \bar{k}}{\partial l} = \frac{\bar{k}}{l} \tag{10-290}$$

$$\frac{\partial \bar{k}}{\partial H_1} = \frac{n}{\bar{r}+m} \cdot \frac{1-\bar{r}}{H_1} \cdot \bar{k} \tag{10-291}$$

In the case of stand 1, $\partial \bar{k}/\partial H_1$ is as follows.

TORQUE EQUATIONS AND TANDEM MILLS

$$\left(\frac{\partial \bar{k}}{\partial H_1}\right)_1 = \frac{n}{\bar{r}+m} \cdot \frac{0.6(1-r_f)}{H_1} \cdot \bar{k} \tag{10-292}$$

$$\frac{\partial \kappa'}{\partial H} = -\frac{n}{\bar{r}+m} \cdot \frac{0.4}{H_1} \cdot (1-\kappa') \tag{10-293}$$

In the case of stand 1,

$$\partial k'/\partial H = 0 \tag{10-294}$$

$$\frac{\partial \kappa'}{\partial h} = -\frac{1}{\bar{r}+m} \cdot \frac{0.6}{H_1} \cdot (1-\kappa') \tag{10-295}$$

$$\frac{\partial \kappa'}{\partial t_b} = -\frac{\alpha-1}{\alpha} \cdot \frac{1}{\bar{k}} \tag{10-296}$$

$$\frac{\partial \kappa'}{\partial t_f} = -\frac{1}{\alpha} \cdot \frac{1}{\bar{k}} \tag{10-297}$$

$$\frac{\partial \kappa'}{\partial l} = \frac{1}{l} \cdot (1-\kappa') \tag{10-298}$$

$$\frac{\partial \kappa'}{\partial H_1} = \frac{n}{\bar{r}+m} \cdot \frac{1-\bar{r}}{H_1} \cdot (1-\kappa') \tag{10-299}$$

In the case of stand 1, $\partial \kappa'/\partial H_1$ is as follows.

$$\left(\frac{\partial \kappa'}{\partial H_1}\right)_1 = \frac{n}{\bar{r}+m} \cdot \frac{0.6(1-r_f)}{H_1} \cdot (1-\kappa') \tag{10-300}$$

$$\frac{\partial D_G}{\partial H} = 1.32 \cdot \frac{h}{H^2} \cdot \mu \cdot \sqrt{\frac{R'}{H}} + \frac{(0.07 + 1.32r)}{2} \cdot \mu$$
$$\times \frac{1}{\sqrt{\frac{R'}{H}}} \cdot \frac{1}{H} \cdot \left(\frac{\partial R'}{\partial H} - \frac{R'}{H}\right) - 0.85 \cdot \frac{h}{H^2} \tag{10-301}$$

$$\frac{\partial D_G}{\partial h} = -1.32 \cdot \frac{1}{H} \cdot \mu \sqrt{\frac{R'}{H}} + \frac{(0.07 + 1.32r)}{2} \cdot \mu$$
$$\times \frac{1}{\sqrt{\frac{R'}{H}}} \cdot \frac{1}{H} \cdot \frac{\partial R'}{\partial h} + 0.85 \cdot \frac{1}{H} \tag{10-302}$$

$$\frac{\partial D_G}{\partial \mu} = (0.07 + 1.32r) \cdot \sqrt{\frac{R'}{H}} + \frac{(0.07 + 1.32r)}{2} \cdot \mu$$
$$\times \frac{1}{\sqrt{\frac{R'}{H}}} \cdot \frac{1}{H} \cdot \frac{\partial R'}{\partial \mu} \tag{10-303}$$

$$\frac{\partial D_G}{\partial H_1} = \frac{(0.07 + 1.32r)}{2} \cdot \mu \cdot \frac{1}{\sqrt{\frac{R'}{H}}} \cdot \frac{1}{H} \cdot \frac{\partial R'}{\partial H_1} \tag{10-304}$$

In the case of stand 1,

$$\partial D_G / \partial H_1 = 0 \tag{10-305}$$

3. The Partial Derivatives of Forward Slip

$$\frac{\partial f}{\partial \lambda} = 2\phi_n \frac{R'}{h} \cdot \frac{\partial \phi_n}{\partial \lambda} + \frac{\phi_n^2}{h} \cdot \frac{\partial R'}{\delta \lambda} \tag{10-306}$$

λ represents H, t_b, t_f, μ, l, H_1.

$$\frac{\partial f}{\partial h} = 2 \cdot \phi_n \cdot \frac{R'}{h} \cdot \frac{\partial \phi_n}{\partial h} \cdot \frac{\phi_n^2}{h} \cdot \left(\frac{\partial R'}{\partial h} - \frac{R'}{h} \right) \tag{10-307}$$

$$\frac{\partial \phi_n}{\partial \lambda} = -\frac{\phi_n}{2} \cdot \frac{1}{R'} \cdot \frac{\partial R'}{\partial \lambda} + \frac{\frac{h}{R'}}{\cos^2\left(\sqrt{\frac{h}{R'}} \cdot \frac{H_n}{2}\right)}$$
$$\times \left(-\frac{H_n}{4R'} \cdot \frac{\partial R'}{\partial \lambda} + \frac{1}{2} \cdot \frac{\partial H_n}{\partial \lambda} \right) \tag{10-308}$$

λ represents H, t_b, t_f, μ, l, H_1.

$$\frac{\partial \phi_n}{\partial h} = -\frac{\phi_n}{2} \cdot \frac{1}{R'} \cdot \left(\frac{\partial R'}{\partial h} - \frac{R'}{h} \right) + \frac{\frac{h}{R'}}{\cos^2\left(\sqrt{\frac{h}{R'}} \cdot \frac{H_n}{2}\right)}$$
$$\times \left\{ -\frac{H_n}{4R'} \left(\frac{\partial R'}{\partial h} - \frac{R'}{h} \right) + \frac{1}{2} \cdot \frac{\partial H_n}{\partial h} \right\} \tag{10-309}$$

$$\frac{\partial H_n}{\partial H} = \frac{1}{2} \cdot \frac{\partial H_b}{\partial H} - \frac{1}{2\mu} \cdot \frac{1}{H} - \frac{1}{2\mu}$$
$$\times \left(\frac{t_f}{s_f} \cdot \frac{1}{s_f - t_f} \cdot \frac{\partial s_f}{\partial H} - \frac{t_b}{s_b} \cdot \frac{1}{s_b - t_b} \cdot \frac{\partial s_b}{\partial H} \right) \tag{10-310}$$

$$\frac{\partial H_n}{\partial h} = \frac{1}{2} \cdot \frac{\partial H_b}{\partial h} + \frac{1}{2\mu} \cdot \frac{1}{h} - \frac{1}{2\mu}$$
$$\times \left(\frac{t_f}{s_f} \cdot \frac{1}{s_f - t_f} \cdot \frac{\partial s_f}{\partial h} - \frac{t_b}{s_b} \cdot \frac{1}{s_b - t_b} \cdot \frac{\partial s_b}{\partial h} \right) \tag{10-311}$$

$$\frac{\partial H_n}{\partial t_b} = \frac{1}{2} \cdot \frac{\partial H_b}{\partial t_b} - \frac{1}{2\mu} \cdot \frac{1}{s_b - t_b} \tag{10-312}$$

$$\frac{\partial H_n}{\partial t_f} = \frac{1}{2} \cdot \frac{\partial H_b}{\partial t_f} + \frac{1}{2\mu} \cdot \frac{1}{s_f - t_f} \tag{10-313}$$

$$\frac{\partial H_n}{\partial \mu} = \frac{1}{2} \cdot \frac{\partial H_b}{\partial \mu} + \frac{1}{\partial \mu} \cdot (H_b - 2 \cdot H_n) \tag{10-314}$$

TORQUE EQUATIONS AND TANDEM MILLS

$$\frac{\partial H_n}{\partial l} = \frac{1}{2} \cdot \frac{\partial H_b}{\partial l} - \frac{1}{2\mu}$$
$$\times \left(\frac{t_f}{s_f} \cdot \frac{1}{s_f - t_f} \cdot \frac{\partial s_f}{\partial l} - \frac{t_b}{s_b} \cdot \frac{1}{s_b - t_b} \cdot \frac{\partial s_b}{\partial l} \right)$$

$$\frac{\partial H_n}{\partial H_1} = \frac{1}{2} \cdot \frac{\partial H_b}{\partial H_1} - \frac{1}{2\mu} \quad (10\text{-}\circ$$
$$\times \left(\frac{t_f}{s_f} \cdot \frac{1}{s_f - t_f} \cdot \frac{\partial s_f}{\partial H_1} - \frac{t_b}{s_b} \cdot \frac{1}{s_b - t_b} \cdot \frac{\partial s_b}{\partial H_1} \right)$$

$$\frac{\partial H_b}{\partial \lambda} = \frac{H_b}{2R'} \cdot \frac{\partial R'}{\partial \lambda} + (1-r) \cdot \left(\frac{\phi_b}{h} \cdot \frac{\partial R'}{\partial \lambda} + \frac{2R'}{h} \cdot \frac{\partial \phi_b}{\partial \lambda} \right) \quad (10\text{-}317)$$

λ represents H, t_b, t_f, μ, l, H_1.

$$\frac{\partial H_b}{\partial h} = \frac{H_b}{2R'} \cdot \left(\frac{\partial R'}{\partial h} - \frac{R'}{h} \right) + (1-r)$$
$$\times \left\{ \frac{\phi_b}{h} \cdot \left(\frac{\partial R'}{\partial h} - \frac{R'}{h} \right) + \frac{2R'}{h} \cdot \frac{\partial \phi_b}{\partial h} \right\} \quad (10\text{-}318)$$

$$\frac{\partial \phi_b}{\partial \lambda} = - \frac{\phi_b}{2R'} \cdot \frac{\partial R'}{\partial \lambda} \quad (10\text{-}319)$$

λ represents t_b, t_f, μ, l, H_1.

$$\frac{\partial \phi_b}{\partial H} = - \frac{\phi_b}{2R'} \cdot \left(\frac{\partial R'}{\partial H} - \frac{1}{\phi_b^2} \right) \quad (10\text{-}320)$$

$$\frac{\partial \phi_b}{\partial h} = - \frac{\phi_b}{2R'} \cdot \left(\frac{\partial R'}{\partial h} + \frac{1}{\phi_b^2} \right) \quad (10\text{-}321)$$

$$\frac{\partial s_b}{\partial H} = - \frac{n}{r_b + m} \cdot \frac{1}{H_1} \cdot s_b \quad (10\text{-}322)$$

In case of stand 1,

$$\partial s_b / \partial H = 0 \quad (10\text{-}323)$$

$$\frac{\partial s_b}{\partial h} = 0 \quad (10\text{-}324)$$

$$\frac{\partial s_b}{\partial l} = \frac{s_b}{l} \quad (10\text{-}325)$$

$$\frac{\partial s_b}{\partial H_1} = \frac{n}{r_b + m} \cdot \frac{1 - r_b}{H_1} \cdot s_b \quad (10\text{-}326)$$

In case of stand 1,

$$\partial s_b / \partial H_1 = 0 \qquad (10\text{-}327)$$

$$\frac{\partial s_f}{\partial H} = 0 \qquad (10\text{-}328)$$

$$\frac{\partial s_f}{\partial h} = -\frac{n}{r_f + m} \cdot \frac{1}{H_1} \cdot s_f \qquad (10\text{-}329)$$

$$\frac{\partial s_f}{\partial l} = \frac{s_f}{l} \qquad (10\text{-}330)$$

$$\frac{\partial s_f}{\partial H_1} = \frac{n}{r_f + m} \cdot \frac{1 - r_f}{H_1} \cdot s_f \qquad (10\text{-}331)$$

4. The Partial Derivatives of Deformed Roll Radius

$$\frac{\partial R'}{\partial \lambda} = (R' - R) \cdot \frac{1}{P} \cdot \frac{\partial P}{\partial \lambda} \qquad (10\text{-}332)$$

λ represents t_b, t_f, μ, l, H_1.

$$\frac{\partial R'}{\partial H} = (R' - R) \cdot \left(\frac{1}{P} \cdot \frac{\partial P}{\partial H} - \frac{1}{H-h} \right) \qquad (10\text{-}333)$$

$$\frac{\partial R'}{\partial h} = (R' - R) \cdot \left(\frac{1}{P} \cdot \frac{\partial P}{\partial h} + \frac{1}{H-h} \right) \qquad (10\text{-}334)$$

With the partial derivatives computed, it becomes possible to calculate the effects of small changes in screw position, roll speed setting, coefficient of friction or the thickness and resistance to deformation of the strip entering the mill from the equations

$$\left(\frac{\Delta h}{h}\right)_i = \sum_{J=1}^{5} A_{ij} \cdot (\Delta S)_j + \sum_{J=1}^{5} B_{ij} \cdot \left(\frac{\Delta N}{N}\right) + \sum_{J=1}^{5} C_{ij} \cdot \left(\frac{\Delta \mu}{\mu}\right)$$
$$+ D_i \left(\frac{\Delta H}{H}\right)_1 + E_i \left(\frac{\Delta l}{l}\right), \quad i=1\ldots 5 \qquad (10\text{-}335)$$

$$\left(\frac{\Delta t}{t}\right)_i = \sum_{j=1}^{5} A'_{ij} \cdot (\Delta S)_j + \sum_{j=1}^{5} B'_{ij} \cdot \left(\frac{\Delta N}{N}\right) + \sum_{j=1}^{5} C'_{ij} \cdot \left(\frac{\Delta \mu}{\mu}\right)$$
$$+ D'_i \left(\frac{\Delta H}{H}\right)_1 + E'_i \left(\frac{\Delta l}{l}\right) \quad i=1\ldots 4 \qquad (10\text{-}336)$$

For the mill constants as given in Table 10-8, the transfer coefficients ABDE, A'B'D' and E' were calculated for the mill data presented in Table 10-8 and the three rolling schedules presented in Table 10-9.

Table 10-8
Mill Data

	Stand 1	Stand 2	Stand 3	Stand 4	Stand 5
Roll radius (mm)	273	273	292	292	292
Elastic constant of stand (kg/mm)	470×10^3	470×10^3	470×10^3	470×10^3	470×10^3
Motor droop (1/(kg·mm))	0	0	0	0	0
Coefficient of friction (--)	0.07	0.07	0.07	0.07	0.07

TORQUE EQUATIONS AND TANDEM MILLS

Table 10-9
Rolling Schedule Data

Schedule No.	Thickness and tension	Stand 1	Stand 2	Stand 3	Stand 4	Stand 5
I	H_i(mm)	3.20	2.64	2.10	1.67	1.34
	h_i(mm)	2.64	2.10	1.67	1.34	1.20
	t_{b_i}(kg/mm^2)	0	10.2	12.8	16.1	16.1
	t_{f_i}(kg/mm^2)	10.2	12.8	16.1	16.1	4.5
II	H_i(mm)	2.60	2.12	1.57	1.17	0.89
	h_i(mm)	2.12	1.57	1.17	0.89	0.80
	t_{b_i}(kg/mm^2)	0	12.7	17.1	23.0	24.1
	t_{f_i}(kg/mm^2)	12.7	17.1	23.0	24.1	6.7
III	H_i(mm)	2.00	1.50	0.99	0.67	0.46
	h_i(mm)	1.50	0.99	0.67	0.46	0.40
	t_{b_i}(kg/mm^2)	0	17.9	21.7	24.1	23.4
	t_{f_i}(kg/mm^2)	17.9	21.7	24.1	23.4	10.8

Remarks:
(i) Strip width is 930 mm.
(ii) l, m, and n in Eq.(10-265) were determined by stress-strain curve of low-carbon steel (C=0.08%) as follows: l=84.6, m=0.00817, and n=0.3.
(iii) a (in Eq. (10-253))=3.0, a' (in Eq. (10-257))=10.0.

The transfer coefficients so calculated are listed in Tables 10-10 to 10-17.

Table 10-10

Transfer coefficient relating the change of thickness of outgoing strip at i stand to the change of screw position at j stand A_{ij}.

Schedule No.	j \ i	1	2	3	4	5
I	1	0.143	0.107	0.107	0.111	0.106
	2	−0.023	0.085	0.040	0.038	0.036
	3	−0.000	−0.041	0.038	0.003	0.001
	4	0.000	0.001	−0.040	0.029	0.001
	5	−0.000	−0.000	0.004	−0.040	0.026
II	1	0.160	0.123	0.124	0.128	0.123
	2	−0.026	0.084	0.039	0.038	0.036
	3	0.000	−0.046	0.038	−0.000	−0.001
	4	−0.000	0.003	−0.042	0.029	0.000
	5	0.000	−0.000	0.007	−0.044	0.024
III	1	0.195	0.161	0.162	0.164	0.158
	2	−0.038	0.074	0.027	0.028	0.027
	3	0.004	−0.042	0.027	−0.005	−0.002
	4	−0.000	0.006	−0.022	0.017	−0.001
	5	0.000	−0.001	0.006	−0.020	0.012

Table 10-11

Transfer coefficient relating the change of thickness of outgoing strip at i stand to the change of roll setting speed at j stand, B_{ij}.

Schedule No.	j	1	2	3	4	5
I	1	0.150	0.647	0.734	0.769	0.733
	2	−0.146	−0.385	0.178	0.236	0.228
	3	−0.003	−0.269	−0.661	−0.031	0.004
	4	0.000	0.009	−0.288	−0.647	−0.031
	5	−0.000	−0.001	0.036	−0.327	−0.934
II	1	0.140	0.705	0.790	0.821	0.789
	2	−0.145	−0.458	0.164	0.206	0.199
	3	0.004	−0.267	−0.727	−0.023	−0.006
	4	−0.000	0.025	−0.285	−0.672	−0.005
	5	0.000	−0.005	0.058	−0.332	−0.975
III	1	0.159	0.830	0.895	0.908	0.873
	2	−0.181	−0.620	0.134	0.133	0.128
	3	0.026	−0.253	−0.874	0.029	−0.024
	4	−0.006	0.066	−0.239	−0.816	0.103
	5	0.002	−0.023	0.085	−0.253	−1.080

Table 10-12

Transfer coefficient relating the change of front tension at i stand to the change of screw position at j stand, A'_{ij}.

Schedule No.	j	1	2	3	4
I	1	−0.716	−0.534	−0.323	−0.352
	2	1.636	0.149	−0.083	−0.117
	3	0.041	1.613	0.267	0.026
	4	−0.001	−0.050	1.311	0.406
	5	0.000	0.005	−0.146	1.582
II	1	−0.556	−0.343	−0.183	−0.207
	2	1.379	0.120	−0.044	−0.061
	3	−0.043	1.180	0.162	0.010
	4	0.003	−0.097	0.867	0.265
	5	−0.000	0.017	−0.159	1.066
III	1	−0.371	−0.240	−0.176	−0.247
	2	1.012	0.094	−0.030	−0.042
	3	−0.116	0.778	0.090	−0.008
	4	0.017	−0.115	0.510	0.130
	5	−0.005	0.034	−0.154	0.607

Table 10-13

Transfer coefficient relating the change of front tension at i stand to the change of roll setting speed at j stand, B'_{ij}.

Schedule No.	j	1	2	3	4
I	1	−10.572	−4.429	−2.298	−2.445
	2	10.306	−5.836	−1.077	−0.800
	3	0.273	10.581	−4.776	−0.537
	4	−0.009	−0.361	9.330	−8.917
	5	0.001	0.045	−1.177	12.701
II	1	−7.244	−2.607	−1.198	−1.327
	2	7.471	−3.636	−0.446	−0.342
	3	−0.246	6.755	−2.940	−0.140
	4	0.023	−0.647	5.767	−6.091
	5	−0.004	0.132	−1.182	7.902
III	1	−4.192	−1.495	−0.974	−1.362
	2	4.774	−2.387	−0.135	−0.201
	3	−0.699	4.672	−2.392	0.412
	4	0.183	−1.224	5.432	−6.459
	5	−0.065	0.435	−1.930	7.610

Table 10-14

Transfer coefficient relating the change of thickness of outgoing strip at i stand to the change of thickness of a strip entering the mill, D_i.

Schedule No.	1	2	3	4	5
I	0.721	0.684	0.688	0.700	0.677
II	0.752	0.733	0.737	0.750	0.727
III	0.799	0.799	0.804	0.807	0.781

Table 10-15

Transfer coefficient relating the change of the thickness of outgoing strip at i stand to the change of resistance to deformation of a strip entering the mill E_i.

Schedule No.	1	2	3	4	5
I	0.161	0.209	0.222	0.211	0.273
II	0.173	0.231	0.245	0.223	0.284
III	0.255	0.319	0.337	0.302	0.363

Table 10-16

Transfer coefficient relating the change of front tension at i stand to the change of thickness of a strip entering the mill, D'_i.

Schedule No.	1	2	3	4
I	−1.390	−1.204	−0.822	−1.154
II	−0.862	−0.669	−0.425	−0.660
III	−0.460	−0.432	−0.486	−0.863

Table 10-17

Transfer coefficient relating the change of front tension at i stand to the change of resistance to deformation of a strip entering the mill E'_i.

Schedule No.	1	2	3	4
I	2.984	3.812	3.589	4.483
II	2.771	3.211	2.703	3.338
III	2.333	2.845	2.591	3.751

From the foregoing computations, it was found that

1. The effect of screw position change on the finished gage is the largest at No. 1 stand and comparatively small at other stands.

   The effect of roll setting speed change on the finished gage is conspicuous at Nos. 1 and 5 stands and comparatively small at other stands.

2. Raising the screws of No. 1 stand decreases the strip tension at every interstand position. Raising the screws at each of Nos. 2 to 5 stands increases the back tension of the stand. Speed-up of No. 1 stand decreases every interstand tension. Speed-up at each of Nos. 2 to 5 stands increases the back tension of the stand and decreases the front tension.

3. An increase in the thickness of the strip entering the mill increases the finished gage as shown in Figure 10-40.

4. An increase in the resistance to deformation of the strip entering the mill makes the finished gage increase as indicated in Figure 10-41.

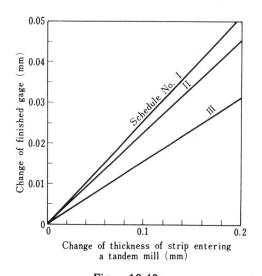

Figure 10-40:
The Effect of Change of Thickness of Strip Entering a Tandem Mill on the Change of Final Gage.

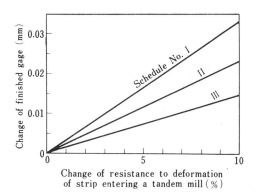

Figure 10-41:
The Effect of Change of Resistance to Deformation of Strip Entering a Tandem Mill on the Change of Final Gage.

Since it was difficult to establish all the transfer coefficients A to E and A' to E', Misaka experimentally obtained only the transfer coefficients A'_{ij} in equation 10-306 using the five-stand cold mill at the Wakayama Works of Sumitomo Metal Industries, Ltd. The work roll radii and the stiffness of the mill stands were as shown in Table 10-8, and the rolling schedule for experimental purposes as shown in Table 10-18. After the acceleration of the mill had been terminated and

TORQUE EQUATIONS AND TANDEM MILLS

steady-state conditions reached, the AGC system was deactivated and the screw position of a certain stand was altered slightly by the operator. Four interstand total tension

Table 10-18
Rolling Schedule Data

Thickness and Tension	Stand 1	Stand 2	Stand 3	Stand 4	Stand 5
Ingoing strip thickness (mm)	3.10	2.50	1.90	1.44	1.11
Outgoing strip thickness (mm)	2.50	1.90	1.44	1.11	1.00
Total back tension (t)	0	27.5	27.5	22.5	22.5
Back tension stress (kg/mm^2)	0	11.9	15.6	20.6	21.8
Total front tension (t)	27.5	27.5	22.5	22.5	5.0
Front tension stress (kg/mm^2)	11.9	15.6	20.6	21.8	5.4

Remarks:
(i) Strip width is 930 mm.
(ii) l, m, and n in Eq.(10-265) are as follows: l=84.6, m=0.00817, and n=0.3.
(iii) a, and a' in Eqs. (10-253) and (10-257) are as follows:
a=3.0, a'=10.0.

Table 10-19

Transfer coefficient relating the change of total front tension at i stand to the change of screw position at j stand, $(A_{ij} + A'_{ij})$.

j \ i	1	2	3	4
1	−0.449	−0.279	−0.103	−0.121
2	1.380	0.196	−0.014	−0.037
3	−0.004	1.243	0.225	0.017
4	0.000	−0.072	0.942	0.335
5	0.000	0.011	−0.140	1.147

values were then recorded by an oscillograph, as illustrated in Figure 10-42.

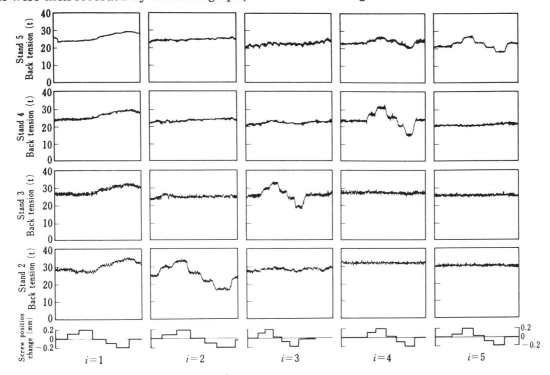

Figure 10-42: The Effect of the ith Stand Screw Position Change on Interstand Tension.

It is clearly shown in Figure 10-42 that when the screw position of a certain stand is changed, the back tension of that stand changes significantly. Transfer coefficients ($A_{ij} + A'_{ij}$) were calculated for the conditions stated in Tables 10-8 and 10-18 and are listed in Table 10-19. The results of the experiment are shown in Figures 10-43 and 10-44 which depict the effect of the change of screw position at four stands on the change of total back tension at the same stands.

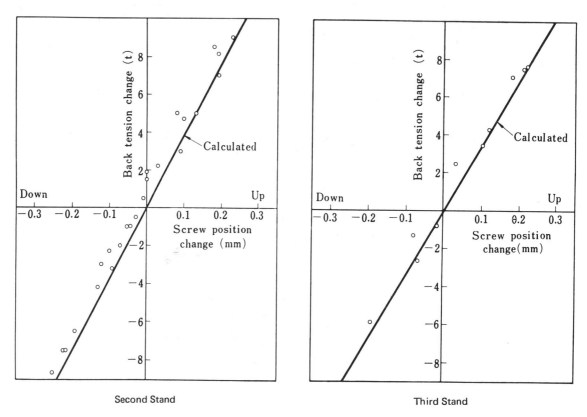

Figure 10-43: The Effect of Screw Position Change on Back Tension.

On the other hand, total back tension change ΔT_{Bi} due to the change of screw position ΔS_i at the ith stand is calculated by

$$\Delta T_{Bi} = T_{Bi} \cdot \left(A_{i-1,i} + A'_{i-1,i} \right) \cdot \Delta S_i \qquad (10\text{-}337)$$

Calculated correlations between total back tension change and screw position change for the last four stands are shown in Figures 10-43 and 10-44. As is seen, the agreement between the measured and theoretical data is reasonably good.

10-25 The Analysis of the Dynamic Behavior of Tandem Mills

The control equations discussed in Sections 10-19 to 10-24 are concerned essentially with the "static" or steady-state behavior of tandem mills. Such models are valuable in that they predict the effect of a change in one variable (such as the initial strip thickness) on another parameter (such as the final rolled thickness of the strip). However, they do not indicate the time required for the transition from one steady state condition to another.

In high speed mill operation, the dynamic response of the mill is of considerable importance for it determines the length of rolled product affected by process changes. It also controls the degree of stability with which the mill operates. The dynamic characteristics are not only dependent upon the mechanical aspects of the workpiece and the mill, but also on the effectiveness of the rolling lubricant and the electrical control system.

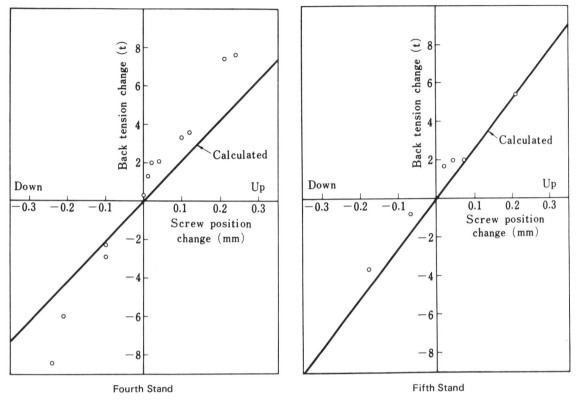

Figure 10-44: The Effect of Screw Position Change on Back Tension.

A study of the dynamic behavior of a mill is best undertaken with the aid of a high speed computer connected to a data plotter. Such an investigation relative to the simulation of a five-stand mill on a digital computer was made by Arimura et al.,[⊞] for the purpose of improving the accuracy of an existing automatic gage control (AGC) system and of developing a "gage alteration in rolling" (GAIR) system. The nomenclature used was as follows:

> H; Thickness of ingoing strip.
> h; Thickness of outgoing strip, or gage.
> S_r; Roll opening, or screw setting.
> K; Rigidity of mill stand.
> P; Roll force per unit width.
> G; Roll torque per unit width.
> $\mathcal{G}$; Torque generating function.
> t_f; Forward tension per unit area.
> t_b; Backward tension per unit area.
> f; Forward slip.
> ϵ; Backward slip.
> Vout; Speed of outgoing strip.
> Vin; Speed of ingoing strip.
> $V_{i, i+1}$; Speed of strip between the ith and the $(i+1)^{th}$ stands.
> V; Peripheral speed of roll, or speed of stand.
> L; Distance between stands.

[⊞] T. Arimura, M. Kamata and M. Saito, "An Analysis Into the Dynamic Behavior of Tandem Cold Mills by a Digital Computer and Its Application", Proc. International Conference on the Science and Technology of Iron & Steel, Supp. to Trans. Iron & Steel Institute of Japan, Vol. 11, 1971, pp. 777-781.

L_x; Distance between a stand and an x-ray gage meter.
E; Young's modulus of steel.
D; Torque characteristic of main motor.
i; Suffix for number of stand.
Δ; Symbol which denotes the small deviation.

The thickness of strip leaving a roll bite was determined by Sims' equation (Section 5-14), viz.

$$h = S_r + P/K \tag{10-338}$$

At each stand, the effect of small changes in the entry and exit thicknesses of the strip and the entry and exit tensions on the exit gage may be represented by

$$\Delta h = \Delta S_r \cdot K \Big/ \left(K - \frac{\partial P}{\partial h}\right) + \Delta H \cdot \left(\frac{\partial P}{\partial h}\right) \Big/ \left(K - \frac{\partial P}{\partial h}\right)$$
$$+ \Delta t_f \cdot \left(\frac{\partial P}{\partial t_f}\right) \Big/ \left(K - \frac{\partial P}{\partial h}\right) + \Delta t_b \cdot \left(\frac{\partial P}{\partial t_b}\right) \Big/ \left(K - \frac{\partial P}{\partial h}\right) \tag{10-339}$$

The speed of ingoing strip is slower and that of outgoing strip is faster than the peripheral speed of roll. This relationship is expressed as

$$V_{out} = (1 + f) V \tag{10-340}$$

$$V_{in} = (1 + \epsilon) V \tag{10-341}$$

Taking the thicknesses of the ingoing and outgoing strip, and the forward and backward tensions as the variables which influence the forward and the backward slips, equations (10-340) and (10-341) may be linearized as follows:

$$\Delta V_{out} = V \left\{ \left(\frac{\partial f}{\partial H}\right) \Delta H + \left(\frac{\partial f}{\partial h}\right) \Delta h + \left(\frac{\partial f}{\partial t_b}\right) \Delta t_b + \left(\frac{\partial f}{\partial t_f}\right) \Delta t_f \right\} + (1+f) \Delta V, \tag{10-342}$$

$$\Delta V_{in} = V \left\{ \left(\frac{\partial \epsilon}{\partial H}\right) \Delta H + \left(\frac{\partial \epsilon}{\partial h}\right) \Delta h + \left(\frac{\partial \epsilon}{\partial t_b}\right) \Delta t_b + \left(\frac{\partial \epsilon}{\partial t_f}\right) \Delta t_f \right\} + (1 + \epsilon) \Delta V \tag{10-343}$$

Since the rotational speed of the roll varies with the torque, this characteristic of the motor may be expressed as

$$\Delta V = D \Delta G \tag{10-344}$$

The variation of the torque is related to the variations of the thicknesses of the ingoing and outgoing strip, and the forward and backward tensions. Therefore, the deviation of the torque is expressed as

$$\Delta G = G\left(\Delta H \cdot \Delta h, \ \Delta t_f, \ \Delta t_b\right)$$
$$= \left(\frac{\partial G}{\partial H}\right) \Delta H + \left(\frac{\partial G}{\partial h}\right) \Delta h + \left(\frac{\partial G}{\partial t_f}\right) \Delta t_f + \left(\frac{\partial G}{\partial t_b}\right) \Delta t_b \tag{10-345}$$

Tension is applied to the strip between stands, and deforms it elastically. The deviation of the interstand tension is determined by the speed of the outgoing strip at one stand and that of the ingoing strip at the next stand as follows:

TORQUE EQUATIONS AND TANDEM MILLS

$$\Delta t_{fi} = \frac{E}{L} \int_0^t (\Delta V_{in,\,i+1} - \Delta V_{out,\,i}) dt \qquad (10\text{-}346)$$

The relationship between the thicknesses of the outgoing strip and the ingoing strip is expressed by using the Laplacian operator as

$$H_{i+1} = h_i \cdot e^{-s(L/V_{i,i+1})} \qquad (10\text{-}347)$$

The partial derivatives for the mathematical model were calculated from the theories of Hill[▶] and Ford, Ellis and Bland[†]. The simulation of the mill is represented in block diagrammatic form in Figure 10-45, the blocks labelled D_i representing the automatic speed regulation system and the blocks, G_i, the torque generating function.

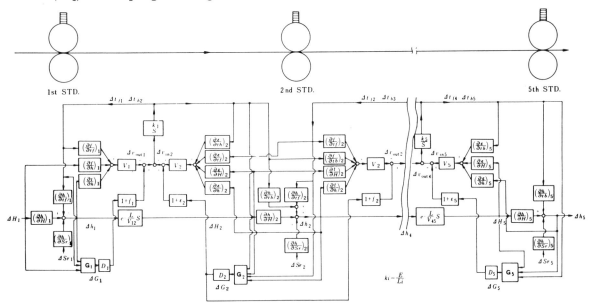

Figure 10-45: Block Diagram of the 5-Stand Tandem Mill Simulation.

Step changes in the hot band gage, the peripheral speed of the rolls and the roll openings were introduced into the system and the corresponding responses were recorded for a rolling schedule as represented in Table 10-20.

Table 10-20
Drafting Schedule Used On Five-Stand Mill

Hot band gauge; 3.2mm Radius of work roll; 273mm
Distance btw. stands; 4600mm Coefficient of friction; 0.07
Rigidity of mill; 470 ton/mm Width of strip; 930mm

Std.	Strip Gauge (mm)	Forward Tension (Kg/mm^2)	Peripheral Speed of Roll (m/s)
1	2.64	10.2	9.62
2	2.10	12.8	12.35
3	1.67	16.1	15.44
4	1.34	16.1	19.30
5	1.20	4.5	22.00

[▶] R. Hill, The mathematical theory of Plasticity, Clarendon Press, 1950.
[†] H. Ford, F. Ellis, & D. R. Bland, "Cold Rolling with Strip Tension, Part I — A New Approximate Method of Calculation and a Comparison with Other Methods", J. Iron and Steel Inst., 168 (1951), pp. 57-72.

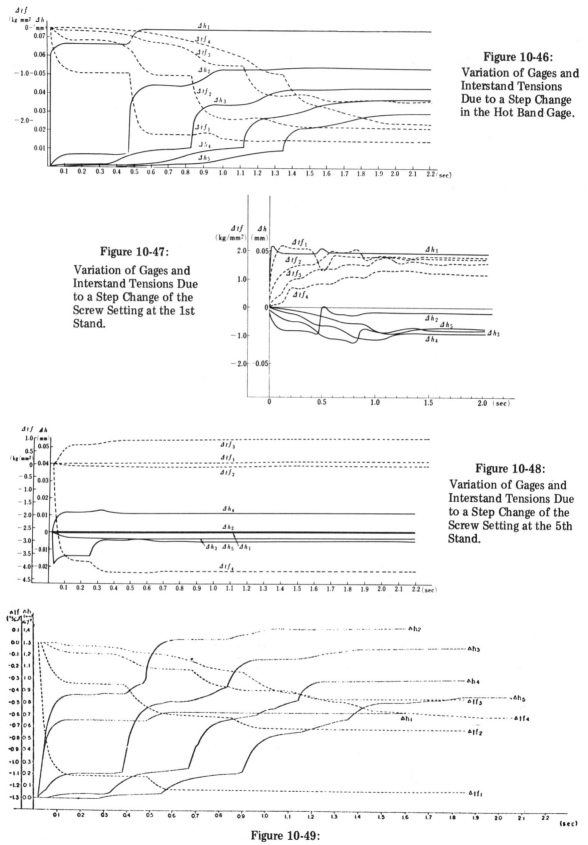

Figure 10-46: Variation of Gages and Interstand Tensions Due to a Step Change in the Hot Band Gage.

Figure 10-47: Variation of Gages and Interstand Tensions Due to a Step Change of the Screw Setting at the 1st Stand.

Figure 10-48: Variation of Gages and Interstand Tensions Due to a Step Change of the Screw Setting at the 5th Stand.

Figure 10-49:
Variation of Gages and Interstand Tensions Due to a Step Change in the Rolling Speed of the 1st Stand.

TORQUE EQUATIONS AND TANDEM MILLS

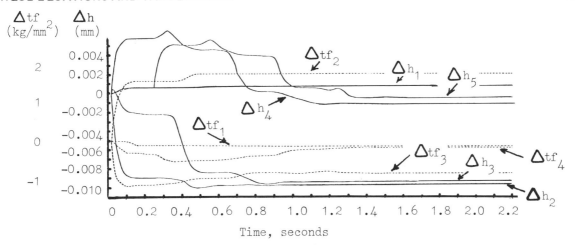

Figure 10-50:
Variation of Gages and Interstand Tensions Due to a Step Change in the Rolling Speed of the 3rd Stand.

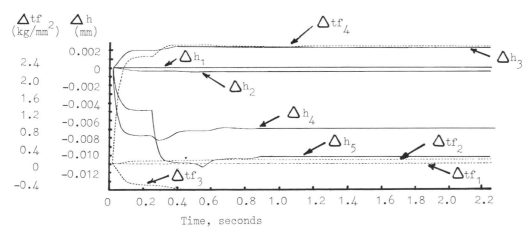

Figure 10-51:
Variation of Gages and Interstand Tensions Due to a Step Change in the Rolling Speed of the 5th Stand.

The results of the study, represented by Figures 10-46 to 10-51, may be summarized as follows:

1. An abrupt increase of the hot band gage decreased the interstand tensions. When the gage deviation arrived at one stand, the forward tension decreased, and it decreased further when the deviation arrived at the next one.

2. The outgoing strip gage increased the most by an increase in the ingoing strip gage. However, it was further increased by the decrease of the tension between one stand and the next stand caused by the arrival of the thickness increase at the latter stand.

3. When the roll opening at the first stand was decreased, the gage at all stands and all the interstand tensions decreased.

4. A decrease in the roll opening at the fifth stand decreased the finished gage. However, at the same time, the tension between the fourth and the fifth stands decreased to a large extent, causing an increase in the gage at the fourth stand. Yet, after certain elapsed time, only a small deviation was observed in the finished gage.

5. An increase in the speed of the first stand resulted in a decrease in all the interstand tensions and all the strip gages.

6. With respect to the intermediate stands, an increase in speed caused a decrease in the exit tension and an increase in the entry tension, with the finished gage finally changing slightly.

7. An increase of the speed at the fifth stand increased the tension between the fourth and the fifth stands, decreased the gage at the fourth stand and decreased the finished gage.

8. The "backward" effect of the rolling conditions of one stand conveyed through interstand tensions was limited to the preceding stand and affected the stand before the preceding one only to negligibly small extent.

9. Accordingly, the factors which influenced the finished gage were the change in the speed of the first stand, the roll opening of the first stand, and the hot band gage. The other factors changed the finished gage to a significant extent only on a transient basis.

Using the simulation discussed above, Arimura and his coworkers showed that an increase by a factor of 3 in the speed of the screwdown motor at the first stand would considerably reduce the time required by the mill stand to respond to a change in hot band gage (Figure 10-52). These results were confirmed by experiment in which the screwdown motor speed of stand 1 was increased by a factor of 1.44 to 1, as illustrated in Figure 10-53 for comparable changes in hot band gage.

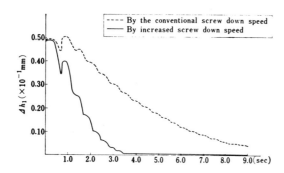

Figure 10-52:
Variation in the Gage Out of Stand 1 Caused by a 0.1 mm Change in Hot Band Gage.

Figure 10-53:
Response of the AGC System with Conventional Screwdown Speed and with the Speed Increased by a Factor of 1.44.

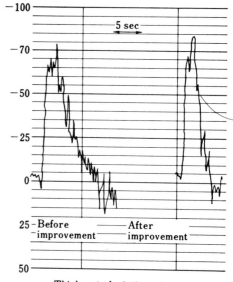

A similar study concerning AGC systems on hydraulic mills was made by Watanabe and Takahashi using an analog computer.[‡] In their mathematical model, based on very small changes in the rolling parameters, the following notation was used:

H_i — entering gage of strip

h_i — exit gage of strip

V_i — entering velocity of strip

v_i — exit velocity of strip

v_{mi} — velocity of rolls

t_{fi} — front tension (stress)

[‡] H. Watanabe and N. Takahashi, "Control Equations for Dynamic Characteristics of Cold Rolling Tandem Mills", Iron & Steel Engineer, February 1974, pp. 59-64.

TORQUE EQUATIONS AND TANDEM MILLS

T_{fi} — front tension (force)
T_{bi} — back tension (force)
E — modulus of elasticity of strip
H_i^o — effect of gage entering first stand on the yield stress of strip at i^{th} stand
l_i — strip length between i^{th} and $(i+1)^{th}$ stand
r_i — reference input to speed control
s_i — reference input to screwdown
e_i — roll eccentricity
f_i — forward slip
Q_i — rolling torque
P_i — rolling force
L_{hi} — time delay due to transport delay of strip between i^{th} and $(i+1)^{th}$ stand
L_{mi} — time delay due to transport delay of strip between i^{th} stand and x-ray gage monitoring just after i^{th} stand
K_{mi} — mill modulus, without mill modulus control
K_{ei} — mill modulus, when mill modulus is controlled
s — Laplace operator
ω — angular frequency

Transfer functions are:

$W_{hsi}(s)$ — from reference input of screwdown to gage
$W_{hei}(s)$ — from roll eccentricity to gage
$W_{hpi}(s)$ — from disturbance of rolling force to gage
$W_{mri}(s)$ — from reference input of speed control to roll velocity
$W_{mqi}(s)$ — from rolling torque to roll velocity

As the time taken by the strip to travel between stands introduces transport delays, the equation of thickness between adjacent stands is:

$$\frac{\Delta H_{i+1}}{H_{i+1}} = \frac{\Delta h_i}{h_i} \exp(-L_{hi} s) \qquad (10\text{-}348)$$

There also exists some transport delay between the change in incoming strip thickness and its effect on the yield stress change of the strip at the ith stand.

$$\frac{\Delta H^o_{i+1}}{H^o_{i+1}} = \frac{\Delta H_1}{H_1} \exp\left\{-(L_{h_1} + L_{h_2} + \ldots + L_{h_i})s\right\} \qquad (10\text{-}349)$$

Disturbances, such as changes to the screwdown or sometimes to the hydraulic device, roll eccentricity, and change of interstand tensions or the ingoing gage will change the output gage at the ith stand. Thus, using Sims' relationship between gage, screw setting, rolling force and mill spring.

$$\frac{\Delta h_i}{h_i} = W_{hsi}(S) \cdot S_i + W_{hei}(S) \cdot e_i$$

$$+ W_{hpi}(S)\left(H_i^o \frac{\partial P_i}{\partial H_i^o} \frac{\Delta H_i^o}{H_i} + H_i \frac{\partial P_i}{\partial H_i} \frac{\Delta H_i}{H_i} + \right. \qquad (10\text{-}350)$$

$$\left. T_{bi} \frac{\partial P_i}{\partial T_{bi}} \frac{\Delta T_{bi}}{T_{bi}} + T_{fi} \frac{\partial P_i}{\partial T_{fi}} \frac{\Delta T_{fi}}{T_{fi}} \right)$$

The strip speed after each stand is changed by the roll speed and the slip.

$$\frac{\Delta v_i}{v_i} = \frac{\Delta v_{mi}}{v_{mi}} + \frac{1}{1+f_i}$$

$$\times \left(H_i^\circ \frac{\partial f_i}{\partial H_i^\circ} \frac{\Delta H_i^\circ}{H_i^\circ} + H_i \frac{\partial f_i}{\partial H_i} \frac{\Delta H_i}{H_i} + h_i \frac{\partial f_i}{\partial h_i} \frac{\Delta h_i}{h_i} + \right. \quad (10\text{-}351)$$

$$\left. T_{bi} \frac{\partial f_i}{\partial T_{bi}} \frac{\Delta T_{bi}}{T_{bi}} + T_{fi} \frac{\partial f_i}{\partial T_{fi}} \frac{\Delta T_{fi}}{T_{fi}} \right)$$

Neglecting the change of width of strip and the longitudinal length of contact between the strip and the roll,

$$\frac{\Delta H_i}{H_i} + \frac{\Delta V_i}{V_i} = \frac{\Delta h_i}{h_i} + \frac{\Delta v_i}{v_i} \quad (10\text{-}352)$$

When the deviation of gage H_i and h_i, and the change of outgoing strip speed v_i is given, the change of ingoing speed of strip can be derived from equation 10-352. When the strip speed change is known, change of interstand tension can be calculated from

$$\frac{\Delta T_{fi}}{T_{fi}} = \frac{\Delta T_{bi+1}}{T_{bi+1}} = \frac{E v_i}{l_i t_{fi}} \left(\frac{\Delta v_{i+1}}{\Delta v_{i+1}} - \frac{\Delta v_i}{v_i} \right) \frac{1}{s} \quad (10\text{-}353)$$

In this equation, tension appears as force and not as stress. Although the change of thickness of strip between the stands also causes a change in tension, the amount is small compared with the effect of change of stress. Therefore, in equation 10-353 the effect of gage change is neglected.

The peripheral velocity of a roll is changed by the speed reference change and rolling torque change. Thus,

$$\frac{\Delta v_{mi}}{v_{mi}} = W_{mri}(s) r_i + W_{mqi}(s) \left(H_i^\circ \frac{\partial Q_i}{\partial H_i^\circ} \frac{\Delta H_i^\circ}{H_i^\circ} + \right.$$

$$\left. H_i \frac{\partial Q_i}{\partial H_i} \frac{\Delta H_i}{H_i} + h_i \frac{\partial Q_i}{\partial h_i} \frac{\Delta h_i}{h_i} + T_{bi} \frac{\partial Q_i}{\partial T_{bi}} \frac{\Delta T_{bi}}{T_{bi}} + T_{fi} \frac{\partial Q_i}{\partial T_{fi}} \frac{\Delta T_{fi}}{T_{fi}} \right) \quad (10\text{-}354)$$

Neglecting the response time of the load cell signal, the rolling force change detected is expressed as:

$$\Delta P_i = h_i \frac{\partial P_i}{\partial h_i} \frac{\Delta h_i}{h_i} + \frac{K_{mi}}{K_{mi} - \frac{\partial P_i}{\partial h_i}}$$

$$\left(H_i^\circ \frac{\partial P_i}{\partial H_i^\circ} \frac{\Delta H_i^\circ}{H_i^\circ} + H_i \frac{\partial P_i}{\partial H_i} \frac{\Delta H_i}{H_i} + \right. \quad (10\text{-}355)$$

$$\left. T_{bi} \frac{\partial P_i}{\partial T_{bi}} \frac{\Delta T_{bi}}{T_{bi}} + T_{fi} \frac{\partial P_i}{\partial T_{fi}} \frac{\Delta T_{fi}}{T_{fi}} \right)$$

From this equation, the variation of gage detected by the gagemeter is

TORQUE EQUATIONS AND TANDEM MILLS

$$\left(\frac{\Delta h_i}{h_i}\right)_g = \frac{1}{h_i}\frac{\Delta P_i}{K_{mi}} \tag{10-356}$$

This load cell output can be fed back to controllers to correct the deviation. In equations 10-348 to 10-355, only the local feedback (gagemeter screwdown or motor speed control) are considered: feedback from x-ray gage monitoring or interstand tension controls is not included. To complete the AGC system, more equations need to be derived.

The deviation of the gage, detected by x-ray gage monitoring, is

$$\left(\frac{\Delta h_i}{h_i}\right)_m = \frac{\Delta h_i}{h_i}\exp(-L_{mi}s) \tag{10-357}$$

The error signal from tension meters or x-ray gage monitoring, can be fed back to any stand, so the reference to the roll actuating cylinder system and motor speed control is:

$$s_i = \sum \left(W_k(s)\cdot X_k\right) \tag{10-358}$$

and

$$r_i = \sum \left(W_l(s)\cdot X_l\right) \tag{10-359}$$

In equations 10-358 and 10-359 $W_k(s)$ and $W_l(s)$ are transfer functions of detectors and controllers appearing in feedback paths and X_k and X_l are the controlled variables.

Rearranging the equations and substituting $j\omega$ for (s) in the transfer function, control equations for the dynamic characteristics or the frequency response of cold rolling mills may be written as follows:

$$\frac{\Delta H_{i+1}}{H_{i+1}} = \frac{\Delta h_i}{h_i}\exp\left(-j\omega L_{hi}\right) \tag{10-360}$$

$$\frac{\Delta H^\circ_{i+1}}{H^\circ_{i+1}} = \frac{\Delta H_i}{H_i}\exp\left\{-j\omega(L_{h1} + L_{h2} + \ldots + L_{hi})\right\} \tag{10-361}$$

$$\frac{\Delta h_i}{h_i} = W_{hsi}(j\omega)\cdot\sum\left\{W_{ki}\cdot(j\omega)\cdot X_{ki}\right\} + W_{hei}(j\omega)\cdot e_i$$

$$+ W_{hpi}(j\omega)\left(H^\circ_i\frac{\partial P_i}{\partial H^\circ_i}\frac{\Delta H^\circ_i}{H^\circ_i} + H_i\frac{\partial P_i}{\partial H_i}\frac{\Delta H_i}{H_i} + \right. \tag{10-362}$$

$$\left. T_{bi}\frac{\partial P_i}{\partial T_{bi}}\frac{\Delta T_{bi}}{T_{bi}} + T_{fi}\frac{\partial P_i}{\partial T_{fi}}\frac{\Delta T_{fi}}{T_{fi}}\right)$$

$$\frac{\Delta v_i}{v_i} = W_{mri}(j\omega)\sum\left\{W_{li}(j\omega)\cdot X_{li}\right\} + \frac{\Delta H^\circ_i}{H^\circ_i}Y_i(H^\circ_i) + \frac{\Delta H_i}{H_i}Y_i(H_i)$$

$$+ \frac{\Delta h_i}{h_i}Y_i(h_i) + \frac{\Delta T_{bi}}{T_{bi}}Y_i(T_{bi}) + \frac{\Delta T_{fi}}{T_{fi}}Y_i(T_{fi}) \tag{10-363}$$

where

$$Y_i(\alpha) = \alpha \cdot \left(W_{mqi}(j\omega) \frac{\partial Q_i}{\partial \alpha} + \frac{1}{1+f_i} \frac{\partial f_i}{\partial \alpha} \right) \quad (10\text{-}364)$$

$$\frac{\Delta V_i}{V_i} = \frac{\Delta v_i}{v_i} + \frac{\Delta h_i}{h_i} - \frac{\Delta H_i}{H_i} \quad (10\text{-}365)$$

$$\left(\frac{\Delta h_i}{h_i} \right)_m = \frac{\Delta h_i}{h_i} \exp(-j\omega L_{mi}) \quad (10\text{-}366)$$

$$\frac{\Delta T_{fi}}{T_{fi}} = \frac{\Delta T_{bi+1}}{T_{bi+1}} = \frac{Ev_i}{l_i T_{fi}} \left(\frac{\Delta V_{i+1}}{V_{i+1}} - \frac{\Delta v_i}{v_i} \right) \frac{1}{j\omega} \quad (10\text{-}367)$$

$$\left(\frac{\Delta h_i}{h_i} \right)_g = \frac{1}{K_{mi} h_i} \left\{ h_i \frac{\partial P_i}{\partial h_i} \frac{\Delta h_i}{h_i} + \frac{K_{mi}}{K_{mi} - \frac{\partial P_i}{\partial h_i}} \right.$$

$$\left. \times \left(H_i^\circ \frac{\partial P_i}{\partial H_i^\circ} \frac{\Delta H_i^\circ}{H_i^\circ} + H_i \frac{\partial P_i}{\partial H_i} \frac{\Delta H_i}{H_i} + T_{bi} \frac{\partial P_i}{\partial T_{bi}} \frac{\Delta T_{bi}}{T_{bi}} + T_{fi} \frac{\partial P_i}{\partial T_{fi}} \frac{\Delta T_{fi}}{T_{fi}} \right) \right\} \quad (10\text{-}368)$$

When the frequency ω of the disturbances (such as an entry thickness change) is assigned, the coefficients of variables in equations 10-361 to 10-367 may be expressed as complex constant numbers and by combining a set of simultaneous equations for each stand, control equations for tandem cold rolling mills may be developed. In solving these equations, transfer coefficients for dynamic characteristics of tandem cold rolling mills are obtained. The solution is given as complex numbers from which both the amplitude and phase shift of transfer coefficients can be determined. Thus the effect of certain disturbances to other variables are known. By assigning different values to ω and solving the equations, a Bode diagram for tandem cold rolling mills results.

Use of the foregoing model was made in connection with a 6-stand tandem mill featuring hydraulic stands each with a controllable mill modulus and a roll eccentricity compensating system as shown in Figure 10-54. In the compensating system, the rolling force signal detected by the load cell is sent to the analyzer and from the rolling force signal, only the component which changes simultaneously with the rotation of backup rolls is separated and fed back to the roll force cylinder instrumentation to compensate for the effect of roll eccentricity. The rotation of the backup roll is detected by a pulse generator installed at the end of the roll; the signal is transmitted to the analyzer. When an abrupt change of rolling force is detected, the analyzer ignores the signal. This control system requires no additional detectors to be installed on the stand except the pulse generator.

The AGC system is shown diagrammatically in Figure 10-55. Since the disturbances that occur are assumed to be small, controls with "dead-band" features are not utilized.

Figure 10-56 shows the calculated behavior of a screwdown type mill and a hydraulic mill, both using AGC systems. Shown is the ratio of the deviation of finished gage to the incoming gage at the first stand as a function of the frequency of the disturbance. At low frequencies, there is no significant differences between the two screwdown systems. However, when the frequency of

TORQUE EQUATIONS AND TANDEM MILLS

the disturbance increases, the hydraulic mill combined with gagemeter control gives a smaller deviation of finished gage.

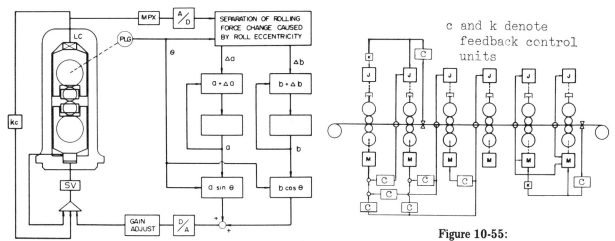

Figure 10-54:
Roll Eccentricity Compensation System.

Figure 10-55:
Model of Automatic Gage Control System. "M" Represents the Drive Motors and "J", the Hydraulic Actuating Instrumentation. (Note X-Ray Gages Following 2nd and 6th Stands.)

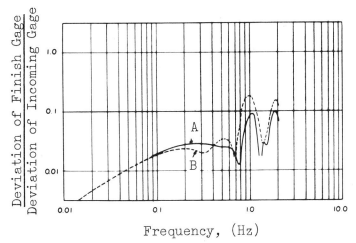

Figure 10-56:
Comparison of Characteristics of Tandem Cold Rolling Mill with an AGC System. "A" Represents a Hydraulic Mill and "B" a Mill with Conventional Screwdowns.

Figure 10-57:
Characteristics of Hydraulic Mills with AGC System and Variable Mill Modulus. "A" Represents AGC Turned Off, "B" AGC on, and "C" AGC on with Increased Mill Modulus.

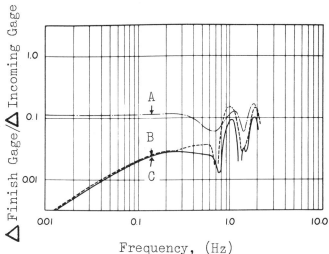

In Figure 10-57, the solid line represents the characteristics of the hydraulic mill when the stiffnesses of the first and the second stands are increased to 76,290 tons/in. and the last stand is decreased to 5,086 tons/in. The dotted line shows the characteristic of the mill when the stiffnesses of stands are not increased. When the frequency of the disturbance is increased, the stiffer mill results in smaller deviations of the finished gage after the sixth stand. The dash-dot line shows the characteristic of the mill when the x-ray gages are not used. When the output gage is not controlled by the x-ray gages, deviation of gage persists and is much greater at lower frequencies.

By selecting a stiffer mill modulus for each stand, the effect of the gage deviation of the strip entering the first stand to the finished gage after the last stand can be eliminated. In this case, a stiffer mill modulus will result in a better finished gage. But calculations according to the control equations for static or dynamic characteristics show that there is no need to select stiffer mill moduli for all stands. Most of the effect of ingoing strip gage deviation can be eliminated effectively by the screwdowns of the first and the second stand. Moreover, the soft mill modulus may be preferable for the last stand, in order to obtain good strip shape.

Consequently, there is some need for selecting stiff mill moduli for the first and the second stands and a soft mill modulus for the last stand. By selecting stiffer mill moduli for both the first and second stand, the effect of the roll eccentricity on the first stand can be eliminated to a certain extent, because the strip gage deviation at the first stand may be eliminated by the second stand. But the gage deviation caused by the roll eccentricity on the second stand cannot be eliminated by the next stand.

To predict the amount of gage deviation in the outgoing strip caused by the roll eccentricities on each stand, calculations were made assuming the mill moduli of the first and second stands to be increased to 76,290 tons/in. and the last stand to be decreased to 5086 tons/in. The result is shown in Table 10-21. The roll eccentricity on the second stand would be expected to give the most significant effect to the finished gage. At the same time, the roll eccentricity on the third stand should give a substantial amount of gage deviation to the outgoing strip. This was not expected. The calculated result indicates that, if the stiffness of the third stand cannot be reduced, the installation of roll eccentricity compensating instruments on this stand is preferable.

Table 10-21
Predicted deviation of gage after sixth stand caused by the roll eccentricity on each stand

Stand	Gage, In.	K_e, Tons/In.	Δh_6, In. $\times 10^{-3}$	K_e, Tons/In.	Δh_6, In. $\times 10^{-3}$	K_e, Tons/In.	Δh_6, In. $\times 10^{-3}$
1	0.070	76,290	0.047	76,290	0.0484	15,258	0.0299
2	0.048	76,290	0.093	76,290	0.0881	15,258	0.035
3	0.032	15,258	0.0495	5,086	0.0197	15,258	0.0326
4	0.020	15,258	0.00354	15,258	0.0236	15,258	0.00236
5	0.0134	15,258	0.00275	15,258	0.00275	15,258	0.00236
6	0.00985	5,086	0.00079	5,086	0.00079	15,258	0.00079

Note: Amplitude of roll eccentricity on each stand is assumed to be 0.787×10^{-3} in. K_e = mill modulus and Δh_6 = deviation of gage at delivery of sixth stand.

As mentioned before, incoming strip gage deviation can be eliminated by the first and the second stand. And even if the stiffnesses of other stands are decreased, a rapidly changing incoming gage deviation can be effectively eliminated. As indicated in Table 10-21, by reducing the stiffness of the third stand to 5,086 tons/in., the effect of the roll eccentricity in the third stand can be reduced significantly.

Finally, it was decided that the roll eccentricity compensating instruments should be installed only on the first and the second stand and, if necessary, the effect of roll eccentricities on other stands should be reduced by decreasing the mill moduli.

Chapter 11 Strip Shape: Its Measurement and Control

11-1 Shape — Its Meaning and Importance

The term "shape" as applied to rolled strip is rather ambiguous in that it may refer to the cross-sectional geometry of the strip or to the ability of the strip to lie flat on a horizontal, planar surface. The former, particularly in the case of plates and sheets, is commonly designated "crown", especially when it is intended to denote the difference in thickness between the center of a strip and its edges.□ The latter, on the other hand, relates to the presence or absence of defects in the unrestrained workpiece which tend to distort its geometrical shape, good shape being ascribed to strip or sheet product which is essentially free from such distortion.● The emphasis of this chapter is upon this second and more generally used meaning of shape.

In the commercial utilization of sheet and strip product, it is desirable that the gage of the material should be as uniform as possible and the material devoid of all types of shape defects. For example, in the fabrication of rocket engine casings and other aerospace hardware, the use of sheet of nonuniform gage could lead to serious discrepancies between the actual and computed values for component weights and their centers of mass.★ Moreover, shape defects could result in undesirable stresses and the subsequent distortion of such components. Furthermore, their fabrication, by such methods as automatic welding, could be rendered appreciably more difficult if the sheets to be welded together were to be of different gage.

In some cases, sheet product is slit and, as narrow strip, is further reduced by cold rolling. Where a coil with excessive crown is slit, the edge portions would possess a very slightly wedge-shaped cross section, and the strip, if rolled to uniform thickness, would exhibit a camber or sweep.† On the other hand, if the roll gap were adjusted to avoid the occurrence of camber, the rolled workpiece might be difficult to satisfactorily coil on a mandril.

Flatness of cold rolled products is especially important where they find application in such items as curtain walls, appliances, furniture, trailer bodies and mobile homes. Departures from flatness in large exposed areas become immediately apparent and objectionable from an aesthetic viewpoint. Moreover, in such applications, the sheet product must also exhibit certain desirable surface features as will be discussed more fully in Chapter 12.

Shape defects may be enhanced by subsequent shearing or slitting of a workpiece. A rolled product that may appear to be of acceptable shape but possessing a slight fullness just within an edge may develop an unacceptable wavy edge when sidetrimmed. Strip, that may appear satisfactory in long lengths uncoiled on a flat surface, may sometimes be sheared into sheets which have a tendency to curl. This effect is decidedly disadvantageous where sheet or strip is to be sheared and stacked for further processing, as in can manufacture. Under these circumstances, shape defects, such as coil set, twist and crossbow, make it difficult for the sheared pieces to be successfully fed to punch presses and other machines in a sequential manner without jamming the machines.

11-2 Standard Thickness, Camber and Flatness Tolerances

The American Iron and Steel Institute (AISI) has established certain thickness, camber

---

□ G. R. Christoph and J. F. Griffin, "Influence of Hot Strip Profile on Subsequent Operations" Flat Rolled Products I, Interscience, New York, 1959, pp. 47-55.

● M. Kotyk and J. W. Stewart, "The Influence of Surfaces and Shape on the Formability of Carbon-Steel Sheet and Strip" — Paper Presented at the Material Forming Seminar Sponsored by the American Society of Tool and Manufacturing Engineers, January 20, 1969.

★ N. H. Polakowski, D. M. Reddy and H. N. Schmeissing, "Principles of Self Control of Product Flatness in Strip Rolling Mills" — Transactions of the ASME, Journal of Engineering for Industry, Paper No. 68-WA/Prod-18.

† J. D. Shelesnow et al., "Effect of Strip Tension on the Quality of Cold Strip and Control of Strip Tension During the Rolling Process" Neue Hütte, 1968, 13 Sept., pp. 531-534.

and flatness tolerances for cold rolled steel products.■ Thickness tolerances for sheets over 12 inches in width, either in coil form or cut lengths, are given in Table 11-1 below.

Table 11-1
THICKNESS TOLERANCES
Cold Rolled Sheet
Coils and Cut Lengths Over 12 Inches (300 mm) In Width

Specified Width Inches (mm)	Thickness Tolerances Over, Inch (mm) No Tolerance Under						
	Specified Minimum Thickness, Inch (mm)						
	Over .098 (2.49) to .142 (3.61) incl.	Over .071 (1.80) to .098 (2.49) incl.	Over .057 (1.45) to .071 (1.80) incl.	Over .039 (.99) to .057 (1.45) incl.	Over .019 (.48) to .039 (.99) incl.	Over .014 (.36) to .019 (.48) incl.
Over 12 (300) to 15 (380) incl.	.010 (.25)	.010 (.25)	.010 (.25)	.008 (.20)	.006 (.15)	.004 (.10)
Over 15 (380) to 72 (1830) incl.	.012 (.30)	.010 (.25)	.010 (.25)	.008 (.20)	.006 (.15)	.004 (.10)
Over 72 (1830)	.014 (.36)	.012 (.30)	.010 (.25)	.008 (.20)	.006 (.15)	—

Note 1. Thickness is measured at any point across the width not less than 3/8 in. (9.5 mm) from a side edge.
Note 2. The specified thickness range captions noted above also apply when sheet is specified to a nominal thickness, and the above tolerances are divided equally, over and under (based upon ASTM A568).

Camber is defined as the greatest deviation of a side edge from a straight line, the measurement being taken on the concave side with a straight edge. The camber tolerance for sheets in coil form is one inch in any 20 feet. For sheets in cut lengths, not resquared, the camber tolerances are as given in Table 12-2 below.

Table 11-2
CAMBER TOLERANCES
Cold Rolled Sheet
Over 12 Inches (300 mm) Wide
(ASTM A568)

Camber tolerances for cut lengths, not resquared, are as follows:

Cut Length, feet (m)	Camber Tolerances, Inches (mm)
To 4 (1.2) incl.	1/8 (3.2)
Over 4 (1.2) to 6 (1.8) incl.	3/16 (4.8)
Over 6 (1.8) to 8 (2.4) incl.	1/4 (6.4)
Over 8 (2.4) to 10 (3.0) incl.	5/16 (7.9)
Over 10 (3.0) to 12 (3.7) incl.	3/8 (9.5)
Over 12 (3.7) to 14 (4.3) incl.	1/2 (12.7)
Over 14 (4.3) to 16 (4.9) incl.	5/8 (16)
Over 16 (4.9) to 18 (5.5) incl.	3/4 (19)
Over 18 (5.5) to 20 (6.1) incl.	7/8 (22)
Over 20 (6.1) to 30 (9.1) incl.	1-1/4 (32)
Over 30 (9.1) to 40 (12.2) incl.	1-1/2 (38)

The camber tolerance for coils is one inch (25 mm) in any 20 feet (6.1 m).

Flatness tolerances are published by the AISI for cold rolled sheets based on whether or not the sheets have been specified to a stretcher-leveled standard of flatness. Tolerances for sheets not so specified are given in Table 11-3 whereas the more stringent requirements for the other sheets are given in Table 11-4.

For sheets coated with zinc or aluminum, the flatness tolerances are similar to those given in Tables 11-3 and 11-4. However, the reader is referred to the AISI Steel Products Manual — Carbon Steel Sheets for exact specifications.

■ Steel Products Manual — Carbon Steel Sheets, American Iron and Steel Institute, 1000 16th Street, N.W., Washington, D.C. 20036, April 1974.

Table 11-3

FLATNESS TOLERANCES
Cold Rolled Sheet
Cut Lengths Over 12 Inches (300 mm) in Width
Not Specified to Stretcher Level Standard
of Flatness
(ASTM A568)

Specified Minimum Thickness Inch (mm)	Specified Width Inches (mm)	Flatness Tolerances (maximum deviation from a horizontal flat surface), Inch (mm)
.044 (1.12) and thinner	To 36 (910) incl.	3/8 (9.5)
	Over 36 (910) to 60 (1520) incl.	5/8 (16)
	Over 60 (1520)	7/8 (22)
Over .044 (1.12)	To 36 (910) incl.	1/4 (6.4)
	Over 46 (910) to 60 (1520) incl.	3/8 (9.5)
	Over 60 (1520) to 72 (1830) incl.	5/8 (16)
	Over 72 (1830)	7/8 (22)

Note 1. The above table also applies to lengths cut from coils by the consumer when adequate flattening measures are performed.
Note 2. The above table does not apply when product is ordered:
(a) Full Hard or to a hardness range.
(b) Class 2.

Table 11-4

FLATNESS TOLERANCES
Cold Rolled Sheet
Cut Lengths Over 12 Inches (300 mm) in Width
Specified to Stretcher Level
Standard of Flatness
(ASTM A568)

Specified Minimum Thickness, Inch (mm)	Specified Width, Inches (mm)	Specified Length, Inches (mm)	Flatness Tolerances (maximum deviation from a horizontal flat surface), Inch (mm)
Over .015 (.38) to .028 (.71) incl.	To 36 (910) incl.	To 120 (3050) incl.	1/4 (6.4)
	Wider or Longer		3/8 (9.5)
Over .028 (.71)	To 48 (1220) incl.	To 120 (3050) incl.	1/8 (3.2)
	Wider or Longer		1/4 (6.4)

11-3 Hot-Band Shape

Since the sideways spread of thin workpieces during cold reduction is severely restricted due to transverse frictional effects in the roll bite, the profile existing in the hot-rolled strip prior to its reduction is repeated with considerable exactness in the profile of the cold reduced strip.[♦][□] It is

---

[♦] G. P. Bernsmann, "Lateral Material Flow During Cold Rolling of Strip", Iron and Steel Engineer Year Book, 1972, pp. 162-166.

[□] H. W. O'Connor and A. S. Weinstein, "Shape Flatness in Thin Strip Rolling", ASME Paper 71-WA/Prod-13.

therefore found that the crown and contour features decrease in height from hot-rolled to cold-rolled strip in just about the same ratio as the overall thickness of the strip is reduced.

A typical hot-band profile such as that shown in Figure 11-1 is drawn from very accurate thickness measurements made at one-inch intervals across the width of the strip made at a location along its length where the profile is believed to represent its average contour. As illustrated in the figure, the gage falls sharply from each "shoulder" to the corresponding edge of the strip, the shoulders being located about 2 inches in from the edges of the strip. (For statistical purposes, Sibakin et al.,[*] found a distance of 1-1/2 inches from the edge as being the most satisfactory location for the shoulder.)

Hot strip profiles may be classified to three basic types: convex, flat and concave. They may be symmetrical, asymmetrical (if they are associated with a wedge condition) or irregular (if associated with high spots).[◘] These basic types of profiles are illustrated in Figure 11-2. A crown, or convex profile is one which has a fairly uniform increase in gage of not less than 0.0015 in. from each shoulder to the center of the strip (as illustrated in Figure 11-1). A flat profile shows little or no variation in gage between shoulders, and a concave profile exhibits a center thickness less than that at the shoulders. A wedge profile is one in which the thickness at one shoulder exceeds that at the other by 0.001 inch or more.[□] While most contours fall into these general shape patterns, sharp irregularities or peaks may occur in the profiles. These persist through cold reduction and give rise to "build up" as discussed in the next section.

Figure 11-1: Hot Rolled Strip Profile (Crown Contour).

Of the different contours, the crown or convex one is the most desirable from the viewpoint of cold reduction inasmuch as it is believed it provides easier tracking of the strip through a tandem cold mill and other processing lines. Even so, hot band with a flat profile has been processed into cold-rolled strip with good shape on many occasions. However, as Christoph and Griffin point out, whatever type of hot-rolled strip is being supplied to the cold mill, it is of extreme importance that the strip contours be as consistent as possible from coil to coil, so that continual adjustments to the mills are unnecessary. In practice, a cold mill essentially starts the rolling of each strip on the premise that its shape is the same as the last similar strip processed. If this proves not to be the case, several hundred feet of strip may be rolled with improper shape before adjustments can be made to the mill. Wedge and concave profiles in the hot band are always

---

[*] J. G. Sibakin, J. S. Ride and W. Sherwood "Factors Affecting Strip Profile in Cold and Hot Strip Mill", Flat Rolled Products I, Interscience, New York 1959, pp. 3-45.

[◘] W. F. Gilbertson, "Hot Rolled Strip Abnormalities Leading to Ridge in Cold Reduced Sheets and Tinplate", J. Iron and Steel Institute, June 1965, pp. 553-561.

[□] G. R. Christoph and J. F. Griffin "Influence of Hot Strip Profile on Subsequent Operations", Flat Rolled Products I, Interscience, New York 1959, pp. 47-65.

STRIP SHAPE: ITS MEASUREMENT AND CONTROL

undesirable, for not only will the strip gage variations near the shoulders persist through cold reduction but major shape irregularities, such as build-up, wavy edge and buckles will occur. □

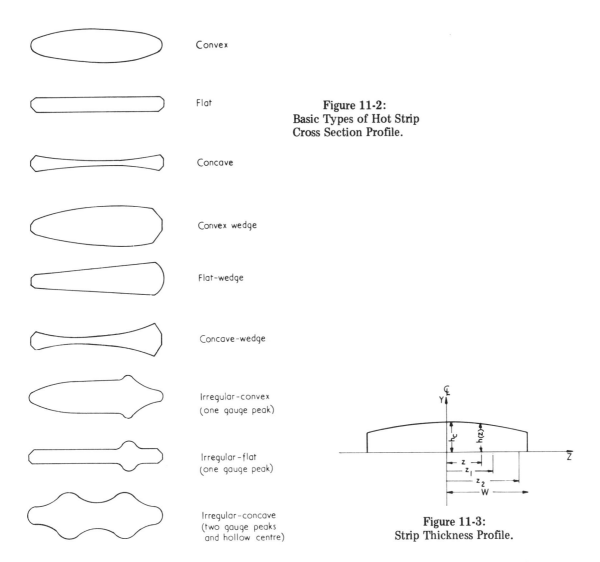

Figure 11-2:
Basic Types of Hot Strip Cross Section Profile.

Figure 11-3:
Strip Thickness Profile.

In computer studies involving the profile of hot rolled strips, it is desirable to express the thickness of the strip h(z) at a distance z from the centerline in the form of a mathematical expression. As discussed more fully in Section 11-23, Sabatini and Yeomans[§] found the following equation satisfactory for their purpose

$$h(z) = g + k_1 z^2 + k_2 z^4 \ldots k_n z^{2n} \tag{11-1}$$

where g is the thickness of the strip at the centerline and k_1, k_2, etc., are constants. (See Figure 11-3.) Together with g, these constants are referred to as "profile parameters". No significant improvements in accuracy were found by utilizing higher powers of z than the squared term, but utilization of only a parabolic approximation led to considerable errors.

□ G. R. Christoph and J. F. Griffin, "Influence of Hot Strip Profile on Subsequent Operations", Flat Rolled Products I, Interscience, New York 1959, pp. 47-65.

§ B. Sabatini and K. A. Yeomans, "An Algebra of Strip Shape and its Application to Mill Scheduling", J. Iron and Steel Institute, December 1968, pp. 1207-1213.

11-4 Build-Up

Strip profile irregularities in the hot band produce defects known as build-up or ridges in the cold-reduced coil. Such ridges, even though only very slightly thicker than the bulk of the strip, can account for a build-up of a quarter of an inch or more in the hundreds of wraps of the coiled strip. They may contribute to abrasion damage in the subsequent handling of the coil, or cause wraps to stick together during box annealing. Furthermore, the trend towards the utilization of coils of greater weight has aggravated the build-up problem so that hot strip mill practice must be carefully supervised to prevent its occurrence.

Such irregularities in the hot-band profile may result from one or more conditions which lead to very slight variations in the roll gaps in the finishing train of the hot strip mill. The most obvious cause is work-roll wear which may not be uniform. Moreover, in the first two stands of finishing trains, there is a tendency for the work rolls to "build-up" by the accretion of material transferred to the rolls from the strip. Where the build-up occurs, it may not be uniform and, even where it does not result in a net increase in roll diameter, it may prevent the roll from wearing uniformly. In addition to the wear and build-up phenomena, nonuniform cooling of the mill rolls, due to misaligned or plugged spray nozzles, may produce irregularities of the roll gap. Lastly, nonuniformity in the frictional conditions in the roll bites, (due to variations in the thickness of the scale on the workpiece across its width, or the improper application of hot-rolling lubricants) may also give rise to abrupt changes in the cross-sectional profile of the hot-rolled strip.

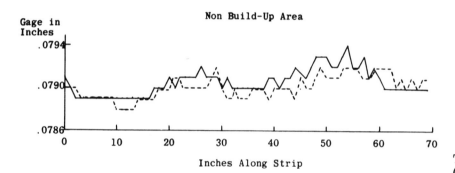
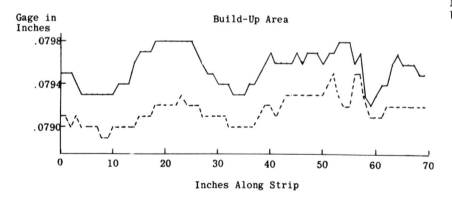

Figure 11-4: Thickness Measurements (1 in. apart) Along 70 in. of Hot-Rolled Strip in Non-Build-Up and Build-Up Areas.

Figure 11-4 shows the gage variation along 70 in. of hot-rolled strip measured at one-inch intervals. In the area free of build-up, the solid line and the dotted line represent gage measurement on parallel lines one inch apart along the length of the strip. As can be seen, the variation between the two is small, averaging about 0.0001 in. and not exceeding a value of about 0.00025 in. Similar measurements made in an area of build-up show appreciably wider variations in gage between the two measurements, the average differential being about 0.0004 in. and the maximum value about 0.0006 in. The thickness difference between the ridge and the adjacent strip is not always consistent and, in some instances, it has been found that the build-up area is

STRIP SHAPE: ITS MEASUREMENT AND CONTROL

characterized by a series of uniformily spaced lumps or welts with an apparently normal profile between the lumps. □

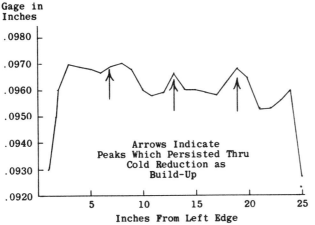

Figure 11-5:
Hot-Rolled Strip Profile Exhibiting Wedge Contour With Peaks.

Strips with wedge and concave profiles appear to be particularly subject to abrupt gage variations. Figure 11-5 is a wedge profile of a hot-rolled band showing three such ridges and, as would be expected, the cold-reduced coil from this strip showed build-up at the corresponding locations. Figure 11-6 illustrates the profile of a concave strip with three ridges which also persisted after cold reduction. Whether or not build-up would occur on strip with wedge or concave profiles in the absence of abrupt gage changes has not been established, but the problem is really an academic one inasmuch as such profiles should be avoided in hot mill practice. Convex or crown profiles are the most desirable from a cold-reduction viewpoint but ridges, and subsequent build-up, can still occur even with these profiles.

11-5 Residual Stresses in Cold Rolled Products and Their Effect on Shape

Because of the nature of the cold rolling process, residual stresses exist in the rolled strip even when it is not subject to any external forces. This has been established by Baldwin,■ Baker, Ricksecker,◊ Ostermann,✠ Davidenkow, Bugakow,◆ and others who have been concerned about the distribution of these stresses and their effect on the mechanical properties of the product.

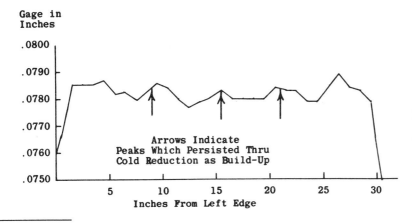

Figure 11-6:
Hot-Rolled Strip Profile Displaying Concave Contour With Peaks.

□ G. R. Christoph and J. F. Griffin, "Influence of Hot Strip Profile on Subsequent Operations", Flat Rolled Products I, Interscience, New York 1959, pp. 47-65.

■ W. M. Baldwin, "Residual Stresses in Metals", Proc. ASTM, 49 (1949), pp. 538-583.

◊ R. Baker, R. Ricksecker and W. Baldwin, "Development of Residual Stresses in Strip Rolling", Trans. AIME 175 (1948), pp. 337-354.

✠ F. Ostermann, Metallwirtsch 10 (1931), pp. 329-336.

◆ N. Davidenkow and W. Bugakow, Metallwirtsch 10 (1932), pp. 317-324.

The residual stress distribution in the rolled strip is as illustrated in Figure 11-7 showing a maximum residual tensile stress in the plane of each surface and a maximum residual compressive stress in the plane of symmetry. The exact shape of the distribution curve varies with the reduction given to the strip and it would be expected that some degree of nonuniformity of stress distribution would exist in the transverse direction.

Where the stress distribution is symmetrical with the central plane of the strip, the rolled strip should not exhibit any tendency to curl. However, curling (coilset) may result from an imbalance of these residual stresses due to one or more of the following:

  a. different diameters for the upper and lower work rolls,
  b. different surface speeds for the two work rolls,
  c. different frictional conditions along the two arcs of contact,
  d. the plane containing the work rolls not being perpendicular to the plane of the strip, and
  e. the strip being deflected from its symmetrical pass line at entry and/or exit from the roll bite.

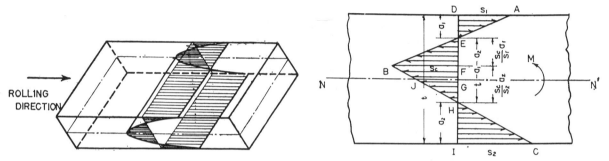

Figure 11-7:
Residual Stress Distress Distribution in a Cold-Rolled Strip.

Figure 11-8:
Schematic Residual Stress Distribution in Longitudinal Cross Section of Sheet.

Assuming an asymmetrical linear stress distribution on the longitudinal cross section of the strip as illustrated in Figure 11-8 and uniformity of stresses in the transverse direction, Tanaka et al.,[♦] derived an expression for the curling curvature in terms of the strip thickness and the surface stress.

Designating the residual tensile stresses in the upper and lower surfaces as S_1 and S_2, respectively, the maximum residual compressive stress as S_c and the strip thickness as t, the distance between the upper surface and the nearer stress-free plane as a_1 and the distance between the lower surface and its corresponding stress-free plane as a_2, then, assuming that the maximum compressive residual stress exists in a plane which divides the strip thickness in the ratio of the maximum tensile stresses at the surfaces

$$S_c = \frac{S_1^2 + S_2^2}{S_1 + S_2} \tag{11-2}$$

$$a_1 = \frac{S_1^2}{2S_1^2 + S_1 S_2 + S_2^2} \cdot t \tag{11-3}$$

♦ E. Tanaka, K. Tsunokawa and T. Fukuda, "Curling and Bowing of Rolled Strips", Trans. J.I.M. Vol. 4, 1963, pp. 124-133.

$$a_2 = \frac{S_1^2}{S_1^2 + S_1 S_2 + 2S_2^2} \cdot t \tag{11-4}$$

Referring to Figure 11-8, the bending moment M tending to curl the strip is given by

$$M = M_{CIH} - M_{JGH} + M_{BJGF} - M_{ADE} = \frac{S_1 - S_2}{6} bt^2 \tag{11-5}$$

where b is the width of the strip. The radius of curvature ρ of the rolled strip is then given by

$$\rho = \frac{Et}{2(S_2 - S_1)} \tag{11-6}$$

where E is the modulus of elasticity of the strip.

The magnitude of a surface stress S may be derived from an empirical formula developed by Baker, Ricksecker and Baldwin,◊ viz.

$$S = CK_f (t/L)^2 \tag{11-7}$$

where C is a constant of proportionality or coefficient of residual stress (about 0.3 for mild steel) ♦

K_f is the mean resistance to deformation

t is the strip thickness after rolling, and

L is the length of a chord joining the ends of the arc of contact between the roll and strip surfaces (assuming the rolls to be perfectly rigid).

The application of these formulae by Tanaka et al., is discussed more fully in Section 11-8.

Residual stress patterns may cause modes of distortion other than coil set. Nonuniformity of longitudinal stresses may cause camber in a side-trimmed or slit strip as discussed in Section 11-3. Moreover, transverse residual stresses can cause "crossbow", and a mixture of transverse and longitudinal stresses may give rise to "twist". These two types of shape defects are examined in greater detail in Sections 11-9 and 11-10.

Wistreich ⊕ has pointed out that while residual stresses may, under certain circumstances, not create any geometrical distortion of the workpiece, if the workpiece is slit, they may cause the cut portions to curve or curl to an objectionable extent. For this reason he proposed a broader connotation to bad shape so that it comprises both its "manifest" and "latent" states. Wistreich considered the theory of buckling of thin plates in which the condition for buckling is given by

$$S = \frac{K\pi^2 E}{12(1 + \nu)} (h/b)^2 \tag{11-8}$$

where S is the critical stress
E is Young's modulus
ν is Poisson's ratio
h is the plate thickness, and
b is the plate width.

◊ R. Baker, R. Ricksecker and W. Baldwin, "Development of Residual Stresses in Strip Rolling", Trans AIME 175 (1948), pp. 337-354.

♦ E. Tanaka, K. Tsunokawa and T. Fukuda, "Curling and Bowing of Rolled Strips", Trans. J.I.M. Vol. 4, 1963, pp. 124-133.

⊕ J. G. Wistreich, "Control of Strip Shape during Cold Rolling", J. Iron & Steel Institute, December 1968, pp. 1203-1206.

The constant K is a function of the stress distribution and the conditions of support at the plate edges. It may range in value from 1/2 (when a uniform compressive stress acts across the width of the plate, one longitudinal edge of which is free to deflect) or it may be as high as 25 (if half the plate width is in tension). The direction of the buckles is dependent on the nature of the stresses. They are parallel to the plate width in the case of compressive stresses and assume a diagonal orientation in the case of shear stresses.

Wistreich notes, however, that the formula is not strictly applicable to wide rolled strip on account of membrane effects and different boundary conditions. Yet it is useful for a qualitative assessment of the effects of stresses and it illustrates the importance of strip thickness to width ratio in the stressed regions. He also suggested that, by considering idealized patterns of stress distribution, it is possible to gain some idea of their connection with the various forms of shape defects encountered.

11-6 Nonuniform Straining and Its Effect on Shape

While strip may emerge from the roll bite and, in the absence of internal stresses, would be of ideal shape, conditions frequently arise in rolling where the percentage reduction given to the strip is not uniform across the width of the strip. Some portions of the strip emerge from the roll bite at greater speeds than others with the result that all portions of the strip are not of the same length. Thus, some are in tension, others in compression and localized buckling of the strip occurs. Where such buckles occur, the strip is said to be over-rolled at that location. Where the reduction given to the strip gradually varies from one side of the strip to the other, the strip may be flat but exhibit a lateral curvature known as camber or sweep (and sometimes as bowing). If the nonuniformity of plastic straining is not severe, subjecting the strip to high tension may put all elements of the strip under tension and give the strip a flat appearance devoid of camber.

Nonuniform straining may be attributable to one or more of a number of causes. Incoming strip may have a profile not exactly matching the roll gap, as for example, with the occurrence of ridges. The roll gap of the mill may not be uniform due to nonuniform crowning of the rolls or there may be nonuniform lubrication across the width of the strip in the roll bite. Furthermore, the metallurgical condition of the hot band may not be uniform but may contain striations of coarse-grained material which exhibits a different resistance to deformation than the bulk of the workpiece.♦

11-7 Various Types of Shape Defects

Through the years, the various shape defects have acquired a considerable number of names. Accordingly, since the same defects carry different appellations in different mills, some degree of confusion exists with respect to their identification. Given below, however, is a listing of the principal defects encountered in cold rolling operations:

    a. coil set, or curl (Figure 11-9 [a])

    b. crossbow or "guttering" (Figure 11-9 [b])

    c. twist (Figure 11-9 [c])

    d. full edge(s), also known as "loose edges", "long edges", "over-rolled edges", "floppy edges" and "piecrust edges" (see Figure 11-10 [a])

    e. full center, also known as a "loose", or "over-rolled center", "long middle", and sometimes identified as "center buckles", or "oil-can type areas". (See Figure 11-10 [b])

    f. herringbone, bananas, chevroning or fluting represented by a series of elongated buckles running diagonally. (Figure 11-10 [c])

♦ H. Kuntz, "Causes of the Generation of Local Warpings of Cold Rolling of Thin Wide Strip" (British Iron & Steel Research Association Translation BISI 7276.)

STRIP SHAPE: ITS MEASUREMENT AND CONTROL

g. quarter buckles or a fullness occurring about a quarter of the way across the width of the strip (Figure 11-10 [d]) and pockets, which may be a localized fullness occurring at any position on the rolled strip

h. camber or sweep (see Figure 11-11).

j. Lines in the form of narrow ridges or valleys, and

k. "orange peel" (sometimes designated "waves" or "slack bands").

These defects are discussed in greater detail in the succeeding sections of this chapter.

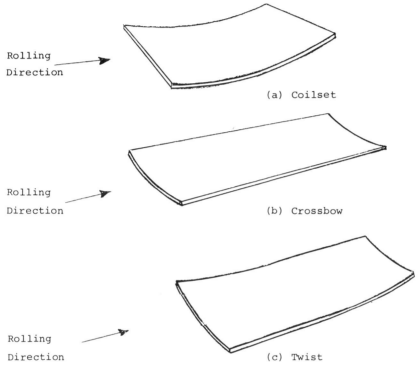

Figure 11-9: Sketches Illustrating Coil Set, Crossbow and Twist.

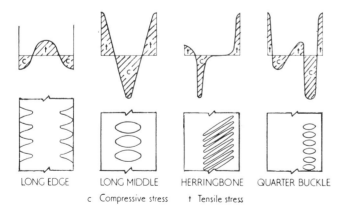

Figure 11-10:
Sketches Illustrating Certain Shape Defects Resulting From Localized Over Rolling.

11-8 Coilset or Curl

As its name implies, strip characterized by this defect exhibits a tendency to curl as shown in Figure 11-9 [a]. The defect results from a condition of asymmetry with respect to the normal pass line of the strip in the rolling operation and, as stated in Section 11-5, may involve asymmetrical conditions involving the positions, diameters, surface conditions, or speeds of the work rolls or the strip may be deflected from its normal pass line at the entry and/or exit of the roll bite.

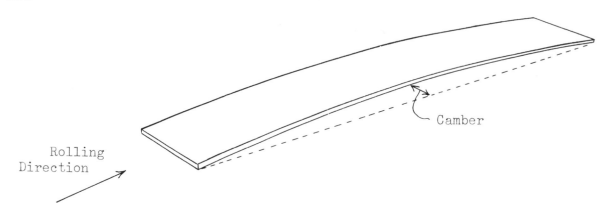

Figure 11-11: Sketch Illustrating Camber or Sweep.

Tanaka et al.,♦ investigated the development of coil set by (a) offset rolls, (b) rolls of different diameters, (c) an inclined pass line, and (d) a combination of (b) and (c), establishing the radius of curvature ρ of the defective strip in terms of other process parameters, using equations 11-2 to 11-5. In the case of the offset rolls (Figure 11-12), the radius of curvature was found to be given by the following expression

$$\rho = \frac{2ER^2}{CK_f t \left(\dfrac{1}{\sin^2\left(\dfrac{\theta + \gamma_2 - \phi}{2}\right)} - \dfrac{1}{\sin^2\left(\dfrac{\theta - \gamma_1 - \phi}{2}\right)} \right)} \quad (11-9)$$

the terms in the expression being as illustrated in the figure. From this expression, it was shown that the strip curls to the side on which the work roll has a slight offset in the rolling direction.

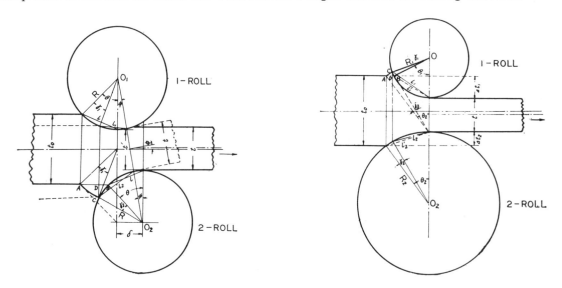

Figure 11-12: Offset Work Rolls. Figure 11-13: Work Rolls of Different Radii.

♦ E. Tanaka, K. Tsunokawa and T. Fukuda, "Curling and Bowing of Rolled Strips" Trans. J.I.M. Vol. 4, 1963, pp. 124-133.

STRIP SHAPE: ITS MEASUREMENT AND CONTROL

For the rolling situation where the two work rolls have different radii R_1 and R_2 (Figure 11-13), the equation for the radius of curvature was

$$\rho = \frac{ER_1R_2\left[R_1(\cos\theta_1-1)+R_2(\cos\theta_2-1)\right]\left[R_1(\cos\theta_1+1)+R_2(\cos\theta_2-1)\right]}{2CK_ft(R_2-R_1)(R_1\cos\theta_1+R_2\cos\theta_2)} \quad (11\text{-}10)$$

For this situation, it was shown that the strip tends to curl to the side with the smaller diameter work roll.

Inclining the pass line on the exit side of the roll bite, as shown in Figure 11-14, led to the equation

$$\rho = \frac{2ER^2(t_o - t)}{CK_ft(t_o - t')\left[\dfrac{1}{\sin^2\left(\dfrac{\theta - \psi}{2}\right)} - \dfrac{1}{\sin^2\left(\dfrac{\theta + \psi}{2}\right)}\right]} \quad (11\text{-}11)$$

where

$$t' = (2R + t)\sec\psi - 2R \quad (11\text{-}12)$$

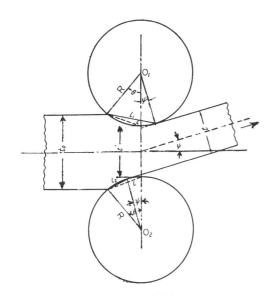

Figure 11-14:
Inclined Pass Line.

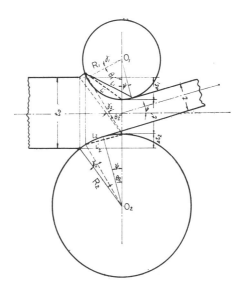

Figure 11-15:
Different Roll Radii and Inclined Pass Line.

In this case, the strip tends to curl towards the same side in which the pass line was inclined. When the two work rolls have different radii and the exit pass line is inclined (See Figure 11-15) the radius of curvature is given by

$$\rho = \frac{2E(t_o-t)}{CK_ft(t_o-t')\left\{\dfrac{1}{R_2^2\sin^2\left(\dfrac{\theta_2-\gamma_2-\psi}{2}\right)} - \dfrac{1}{R_1^2\sin^2\left(\dfrac{\theta_1+\gamma_1+\psi}{2}\right)}\right\}} \quad (11\text{-}13)$$

where
$$t' = (R_1 + R_2 + t) \sec \psi - (R_1 + R_2) \quad (11\text{-}14)$$

Accordingly, the radius of curvature in the direction of the side with the smaller roll will be decreased, since both the smaller roll and the inclined pass line both tend to curl the strip in this same direction.

Tanaka and his coworkers verified the general validity of their equations by rolling annealed mild steel strip (0.15 mm x 100 mm wide) to a thickness of 0.064mm on a 20-high Sendzimir mill with the wipers adjusted to give the desired roll configurations.

The fact that differential lubrication can provide coilset has also been proven by experimental work.[▲] If the coefficient of friction on the upper surface is higher than that on the lower surface, because of the relative positions of the neutral points on the two arcs of contact, the upper surface will try to leave the roll bite faster than the lower surface. As a consequence, after the strip has left the roll bite, it will tend to curl so that its lower surface becomes concave. Figure 11-16 illustrates this effect with samples of low carbon steel rolled on a laboratory mill with 6-1/2 in. diameter work rolls.[⊞] In practice, differential conditions may arise from (a) differences in the work roll finishes, (b) differences in the finishes of the strip surfaces, and (c) different rolling oils or differing concentrations of the rolling solution applied to the two surfaces of the strip on entry into the roll bite. To obviate the occurrence of coilset, it is desirable to achieve complete symmetry in the rolling operation with respect to the pass line of the strip. If this is not possible, due to the incoming strip having different surface finishes, then compensation should be attempted by using a slightly poorer lubricant on the strip surface with the rougher finish.

Strip With Coil Set (Concave Downwards)	Flat Strip	Strip With Coil Set (Concave Upwards)
Calculated Coefficient of Friction — Upper Surface		
0.130	0.046	0.046
Calculated Coefficient of Friction — Lower Surface		
0.046	0.046	0.130

Figure 11-16: Coil Set Produced by Differential Lubrication.

In connection with coilset, however, it should be remembered that it can be induced in strip by pulling it over a roll with too small a diameter. To ensure that this does not occur, it is desirable that all rolls associated with the deflection of strip have diameters about 1,000 times the thickness of the strip they engage.

11-9 Crossbow and Hooked Edges

Although the literature contains few references to these defects, they are perhaps the most commonly encountered of those types due solely to residual stresses. Crossbow is usually a symmetrical distortion of the strip so that the cross section of the strip assumes an arcuate form as

---

[▲] W. L. Roberts, "The Influence of the Rolling Lubricant on Sheet and Strip Quality", Tribology in Iron & Steel Works — ISI Publication 125, The Iron & Steel Institute, 1970.

[⊞] D. C. Litz, "Rolling of Light Gage Strip on a Laboratory Mill", Blast Furnace and Steel Plant, 55, 1967, pp. 1027-1035.

STRIP SHAPE: ITS MEASUREMENT AND CONTROL

illustrated in Figure 11-9 [b]. In an asymmetrical form, where only half the strip or less is affected (Figure 11-17), it is usually termed hooked edge.

Such defects are believed attributable to the horizontal bending of the work rolls to different degrees and tend to become more prevalent with less rigid rolls. This horizontal bending may result from excessive crowns on the work rolls so that, under the influence of the rolling force exerted on them by the backup rolls, one work roll tends to deflect more towards the exit side of the mill than the other. Under these conditions, when an element of the strip is about to emerge from the roll bite at the cessation of its plastic deformation, the surface layers of one face are subject to a different stress level in a transverse direction than the other face. Thus, when the strip is freed from any external restraining forces, the strip assumes a curved cross section with the surface that was under the higher tensile or lesser compressive stress facing the center of curvature. Figure 11-18 illustrates the manner in which crossbow may be developed.

Figure 11-17: Sketch Illustrating a Hooked Edge.

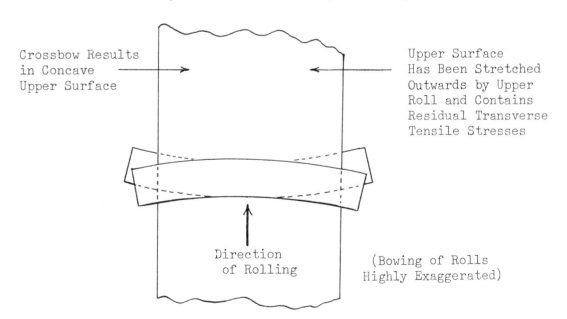

Figure 11-18: Sketch Illustrating Development of Crossbow.

Hooked edges may be regarded as "half crossbow", and this type of defect may result from a combination of the horizontal bowing and misalignment of the work rolls.

To minimize the problems associated with crossbow and hooked edges, the crowning of the work rolls should be held to a minimum to lessen the tendency for the rolls to be deflected horizontally in opposite directions. Reducing the rolling force through the use of more effective lubricants or higher strip tensions will alleviate the problem. It is interesting to note, however, that on small laboratory mills with work rolls having small face-to-diameter ratios, it is difficult, if not virtually impossible, to produce strip with crossbow.

11-10 Twist

Residual surface stress of a different magnitude in the two surfaces acting in a direction intermediate between the rolling and transverse directions give rise to a defect known as twist. Every sheet cut from a strip exhibiting this defect will tend to curl in an identical manner as illustrated in Figure 11-9. The defect is believed attributable to the misalignment of the work rolls, not only with respect to each other, but also with respect to a tension roll on the exit side of the mill. To obviate this type of defect, an accurate alignment of mill rolls is necessary and the use of lesser exit strip tensions is advisable where possible.

11-11 Full Center

This type of defect is the most prevalent in cold rolling operations and a slight fullness in the center of the rolled strip is sometimes deliberately produced in the belief that better tracking of the strip in subsequent processing lines, such as continuous annealing lines, is afforded by this characteristic. Moreover, it has been shown in Section 11-3 that the preferred profile for the hot band is that with a crown. Such a profile can more readily give rise to a full-centered strip on rolling than any other.

From geometrical considerations, it can be shown that, if the peak-to-peak amplitude of the center-buckle waves in the strip is A and their wave length λ, then the ratio of the full or actual length of the strip to its foreshortened length is $(1+[A\pi/2\lambda]^2)$ so long as the amplitude A is very much smaller than the wave length λ.

Full center results from an over-rolling of the center portion of the strip (nonuniform straining) due generally to excessive crowning of the cold mill rolls. If strip of satisfactory shape has been produced earlier by essentially the same rolling conditions, the defect may be attributable to (a) an increase in the crown of the incoming strip, (b) an increase in the effective cold mill roll crowns, or (c) a decrease in the rolling force. Increased roll crowns may result from the development of a nonuniform temperature distribution in the mill rolls or, where roll bending systems are used, a change in the amount of roll bending employed. A decrease in the rolling force, on the other hand, may result from (a) incoming strip that is narrower and/or softer than normal, (b) a lesser draft being given to the strip, or (c) improved rolling lubrication. In turn, better lubricity may result from (a) chemical or physical changes in the rolling lubricant, (b) rougher surfaces on the incoming strip, (c) smoother work roll surfaces, and (d) a change in rolling speed.

To alleviate the problem of full center, the effective crown of the rolls must be decreased as, for example, by substituting rolls with lesser crowns, changing the temperature distribution in the rolls by adjustment of the cooling sprays, adjustment of the roll bending system if such is used or increasing the rolling force by the use of a less effective rolling lubricant. In the last stand of a tandem mill, strip shape may be improved by increasing the reduction at that stand and by correspondingly decreasing the overall reduction taken at the preceding stands.

11-12 Full Edges, Quarter Buckles and Pockets

In a sense, full edges are the converse of a full center and they result from an over rolling of the strip at its edges. As would be expected, therefore, this defect occurs as a consequence of inadequate roll crowns. If encountered unexpectedly in cold rolling operations, it may be attributable to incoming strip that is harder and/or wider or requires a larger draft. Decreased rolling lubricity may also give rise to full edges.

This type of defect may be alleviated by increasing the roll crowns, suitably bending either the work rolls or the backup rolls, or by decreasing the rolling force. The latter may be accomplished, entirely or in part, by using improved rolling lubricity, smoother rolls and higher

STRIP SHAPE: ITS MEASUREMENT AND CONTROL

strip tensions. In the case of tandem mills, a rescheduling of the reductions taken at the various stands (with a lesser reduction at the last stand) could also ameliorate the situation.

A tendency to fulness in strip edges is often regarded as advantageous in that the tension exerted on the strip on various processing lines will then result in lower tensile stresses at the strip edges and lessen the possibility of strip breaks.

The occurrence of over rolling at portions of the strip other than the center or edges gives rise to quarter buckles and pockets. However, since the other portions of the strip exhibit satisfactory flatness, the roll crowns and other rolling parameters must be generally satisfactory and should not be changed.

These defects may be attributable to one or more of the following causes: (1) a localized "barrelling" of one or more of the mill rolls; (2) a localized region of improved rolling lubricity; and (3) a region in the incoming strip that is softer and/or thicker than the remainder of the strip. The first may be due to a misalignment or improper operation of the roll cooling sprays so that a portion of one or more of the mill rolls is not being cooled as effectively as the remainder of their surfaces. A more intensive cooling of this area of the roll surfaces would alleviate the problem regardless of whether or not this is the primary cause of the defect.

A region of improved lubricity at the roll bite may result from the drippage of contaminating oils on the strip or the rolls. Figure 11-19 illustrates the effect of nonuniform lubrication as observed on a laboratory mill rolling light-gage steel strip.▲ Usually, if this is the cause, it may be eliminated by appropriate maintenance work. Where the incoming strip is at fault, however, the problem is best solved by remedying the hot mill practice to ensure temperature uniformity in the hot band when it is rolled and coiled and thereby prevent build-up in the cold-rolled strip as discussed in Section 11-4.

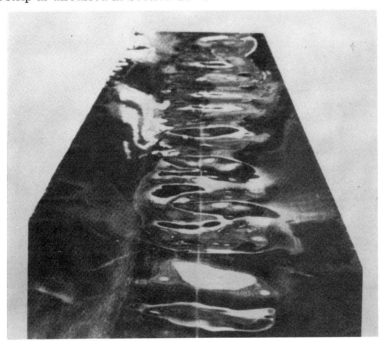

Figure 11-19:
Defective Strip
Produced by
Nonuniform
Rolling Lubricity.

11-13 Camber or Sweep

This defect, sometimes referred to as "bowing" and manifested by a lateral curvature of the strip as illustrated in Figure 11-11, may result from the rolling of strip with a wedge-shaped

---

▲ W. L. Roberts, "The Influence of the Rolling Lubricant on Sheet and Strip Quality" — Tribology in Iron and Steel Works — ISI Publication 125, The Iron and Steel Institute 1970.

section on a mill with a uniform roll gap, or the rolling of strip of acceptable shape on a mill with a nonuniform gap. In either case, an adjustment of the roll gap to provide proper tracking of the strip will usually eliminate the problem.

Alternately, the slitting of edge-rippled or center-buckled strips may also lead to cambered strip. Tanaka et al.,♦ have investigated cambering attributable to both these shape defects as a consequence of elastic and plastic straining, and have established the dependence of the radius of curvature of the camber in terms of the strip width, the position of the slit, the strip profile and various other parameters.

In doing so, these investigators made two assumptions. The first involved the profile of the strip which they assumed to be given by the equation

$$t = t_m - \frac{4(t_m - t_n) y^2}{b^2} \qquad (11\text{-}15)$$

where, with respect to the strip,
  t is its thickness at a distance y from its center line,
  t_m is its thickness at its center line
  t_n is its thickness at its edge, and
  b is its width.

Furthermore, an idealized compressive stress-strain curve of the form shown in Figure 11-20 was considered to be applicable.

For strips which are edge-buckled due to elastic strains, slitting produces a radius of curvature ρ given by the formula

$$\rho = \frac{b^2 t_o (E-P) \{5t_m - (2m^2+m+2)(t_m-t_n)\}}{2v(a+b)(t_m-t_n)\left[2\{5t_m-(2m^2+m+2)(t_m-t_n)\}P - 5t_o\{\varepsilon_o(E-P)+P\}\right]} \qquad (11\text{-}16)$$

where t_o is the initial strip thickness
  E is the elastic modulus
  P is the slope of the stress-strain curve in the plastic region
  a is twice the distance between the slit and the center line
  b is the width of the strip
  E_o is the elastic strain at yielding, and
  m equals a/b

For center buckled strip, the radius is given by

$$\rho = \frac{b^2 t_o (E-P) \{5t_m + (2m^2+m+2)(t_n-t_m)\}}{2v(a+b)(t_n-t_m)\left[2\{5t_m+2(2m^2+m+2)(t_n-t_m)\}P - 5t_o\{\varepsilon_o(E-P)+P\}\right]} \qquad (11\text{-}17)$$

where the symbols are as previously defined.

Where plastic strains are involved and the cross-sectional profile is as shown in Figure 11-21, slitting of the edge rippled strips results in a camber curvature ρ given by

---

♦ E. Tanaka, K. Tsunokawa and T. Fukuda, "Curling and Bowing of Rolled Strips" Trans. J.I.M. 1963, Vol. 4, pp. 124-133.

STRIP SHAPE: ITS MEASUREMENT AND CONTROL

$$\rho = \frac{(b^2-a^2)t_m + (a^2+b^2)t_n}{4(a+b)(t_m-t_n)} \qquad (11\text{-}18)$$

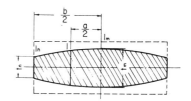

Figure 11-21:
Cross Section of Edge-Rippled Strip.

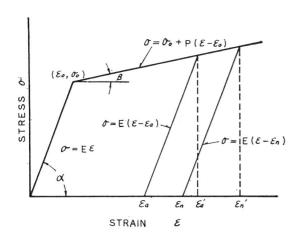

Figure 11-20:
Idealized Stress-Strain Curve $E = \tan \alpha$, $P = \tan \beta$

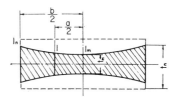

Figure 11-22:
Cross Section of Center-Buckled Strip.

Similarly, when slitting strips center-buckled by plastic strains, (Figure 11-22) the curvature ρ may be derived from the expression

$$\dot{\rho} = \frac{(b^2-a^2)t_m + (a^2+b^2)t_n}{4(a+b)(t_n-t_m)} \qquad (11\text{-}19)$$

11-14 A Quantitative Evaluation of Strip Fullness

Perhaps because fullness in the rolled strip is generally regarded as an undesirable characteristic which should be avoided, or at least minimized, little attention has been given to its quantitative measurement. However, a method for evaluating strip shape defects (involving over-rolled portions of the strip) on a quantitative basis has been developed by Pearson.[°] He considered a narrow rectangular, longitudinal element cut from the strip with a long edge, as illustrated in Figure 11-23. On releasing the buckling stresses, the element will lie flat but will exhibit a lateral curvature (camber) assumed to be circular in form. From Figure 11-24, it can be seen that an

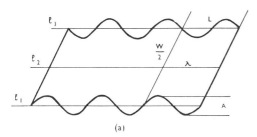

Figure 11-23:
Diagram of Full-Edged Sheet With Sinusoidal Waves.

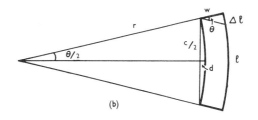

Figure 11-24:
Curvature of Strip Element After Relief of Buckling Stresses When Cut From Sheet Shown in Figure 11-23.

° W. J. K. Pearson "Shape Measurement and Control" J. Institute of Metals, Vol. 93, 1964-65, pp. 169-178.

annular element of length l on the shorter side and $l + \Delta l$ on its longer side subtends an angle θ at a radius r. The curvature is due to the difference in length Δl on opposite sides of the element of width w.

$$w \ll r \qquad (11\text{-}20)$$

$$\ell = r\theta \qquad (11\text{-}21)$$

and

$$\ell + \Delta \ell = (r+w)\theta \qquad (11\text{-}22)$$

Thus

$$\frac{\Delta \ell}{\ell} = \frac{w}{r} \qquad (11\text{-}23)$$

Defining shape Σ as the fractional difference in length per unit width,

$$\Sigma = \frac{\Delta \ell}{\ell w} \qquad (11\text{-}24)$$

it can be seen that Σ is equal to the curvature $1/r$ of the equivalent flat annular element. Since this quantity is usually small, Pearson found it convenient to adopt units designated mons, a mon being defined as the shape which would develop a lateral curvature of 10^4 cms radius in an element 1 cm wide when removed from the strip and laid flat. On this basis, it has been found that strip with a shape value of 0.02 mons/cm displays extremely good flatness. One higher order of shape at 0.2 mons/cm still possesses excellent flatness and is stated to be roughly representative of the best shape achievable by direct rolling without a subsequent flattening operation.

Pearson's method of evaluating fullness, as for example, with respect to the over-rolled edge of a strip, is informative with respect to the very small variation in rolled-strip length with transverse distance from the edge of the strip. It does not, however, predict the peak-to-peak amplitude or the wavelength of the ripples characterizing the edge (Section 11-11). Yet, as is seen in Section 11-2, flatness tolerances specify the maximum deviation from a horizontal flat surface which is closely related to the peak-to-peak amplitude of the wavy portion of the strip.

In the U. S., the percentage length difference between the longest and the shortest ribbon across the width of the strip has been used as a measure of shape. The Aluminum Co. of Canada (ALCAN) also defines the shape or flatness as the relative length difference between the longest and the shortest ribbon across the width of the strip, but has introduced an "I-unit" equal to 10^{-5} instead of percentage to obtain more convenient figures.[*][+]

Table 11-5 shows typical commercial tolerances for aluminum strip, defined in I-units.

Table 11-5
Typical Commercial Tolerances For Aluminum Strip

Commodity	I-units
Mill products	100
Leveled products	25
Stretched products	10

---

[*] O. G. Sivilotti, W. E. Davies, M. Henze and O. Dahle, "ASEA-ALCAN AFC System For Cold Rolling Flat Strip", Iron and Steel Engineer Year Book 1973, pp. 263-270.

[+] O. G. Sivilotti, B. Johansson, "Measurement and Control of Strip Flatness in Cold Rolling Mills", IEEE Conference Record of 1969, Fourth Annual Meeting of the IEEE Industry and General Applications Group, pp. 251-259.

STRIP SHAPE: ITS MEASUREMENT AND CONTROL

11-15 Monitoring Shape Defects

Basically, the methods for the detection of shape defects fall into two categories. In the one, the strip is examined in a relatively stress-free condition; in the other it is subject to high tensile stresses.

The former is exemplified by the practice of mill operators in slowing down the mill, relaxing the tension in the strip and then visually inspecting its shape. Similarly, cold-rolled coils are frequently placed on other lines, such as recoil lines, where the strip shape may be conveniently examined at any position along the length of the coil. Samples of the rolled product are generally examined visually in the absence of externally applied stresses sometimes being suspended from one edge for the purpose of detecting curl, twist or crossbow.

An instrument to monitor the shape of strip in a stress-relaxed condition is described in Section 11-16. This unit measures the relative lengths of the "wavy" and the "tight" regions of the strip and provides a quantitative evaluation of strip shape.

When the strip is held taut and flat under relatively high longitudinal stresses as, for example, on a cold mill between the last stand and the tension bridle, strip of poor shape may be pulled into a condition of flatness. Under these circumstances shape can only be assessed by a determination of the transverse distribution of longitudinal stress.[*] Operators frequently check this distribution in a crude manner by hitting the strip with a broom handle. However, a number of instruments to perform this function on a continuous basis have recently been developed. The simplest instrument of this type is a tensiometer roll utilizing a load cell at each bearing. A more complex variation of this approach involves the use of a segmented roll with the loading on each segment being measured by a load cell. Some of these stress monitoring systems are discussed more fully in the later sections of this chapter.

A noncontacting method of measuring the longitudinal stress distribution in the strip has also been under development. This utilizes a number of magnetic sensors placed at suitable intervals across the width of the strip as discussed more fully in Section 11-21.

The detection of certain shape defects resulting from residual stresses, such as crossbow, hooked edges, coilset and twist is much more difficult in strip that lies flat under tension. To date, no instruments have been developed for this purpose.

11-16 Pearson's Method of Shape Measurement

Since the taut and wavy portions of the strip with respective lengths l and $l + \Delta l$ emerge from the roll bite during the same time period, it is possible to determine the ratio $\Delta l/l$ from the speeds V_1 and V_2 with which the two portions of the strip leave the mill. Thus, shape may be established in terms of $(V_1-V_2)/V_2$ where the two speeds are measured at the center and at one edge of the strip by means of identical rollers. An instrument capable of making these measurements and utilizing them to give a metered indication of shape was built by Pearson[°] and its block schematic diagram is shown in Figure 11-25. One roller is connected to a Synchro transmitter (CX) and the other to a Synchro differential transmitter (CDX), the output signal from these two units being proportional to the velocity difference (V_1-V_2) of the two rollers. This signal is applied to the stator of a Synchro Control Transformer (CT) the rotor of which is driven through a gear box N_2 by a d.c. servomotor (M). When the tracking error signal, induced in the rotor winding and amplified by the amplifier G_1, is fed back negatively to M, this motor drives the rotor CT exactly at the velocity $N_1(V_1-V_2)$. This positional servo is stablized by feedback from the d-c tachogenerator (T) through a blocking capacitor and the velocity signal is amplified.

The electrical analog of $(V_1 - V_2)/V_2$ is computed by multiplication in a tachogenerator (XT) where the armature is driven at velocity $N_1N_2(V_1-V_2)$ and the field excited

---

[*] J. P. Barreto and M. J. Hillier, "Shape Control in Cold Strip Rolling", Sheet Metal Industries, October 1968, pp. 707-722.

[°] W. J. K. Pearson, "Shape Measurement and Control", J. Institute of Metals, Vol. 93, 1964-65, pp. 169-178.

by a voltage proportional to $1/V_2$, which is obtained from a tachogenerator T driven by a pass line roll and inverted by a reciprocal function generator. The signal given by the (XT) armature is then the required shape signal.

The meter performance during extended trials on a production mill met the design requirements and no difficulty was experienced with roller skidding, thermal effects, or damage to precision components.

With very poor shape in thick strip, and at high speeds, a limit to successful tracking must eventually be reached and roller skidding would ensue. This condition should not be arrived at in normal mill operation but if it were, retraction of the head could be automatic, after which the operator could solve his problem by visual shape monitoring. Such bad shape would be rejectable but could be measured, if need be, by running the strip more slowly on a coil-processing line. This condition was not found with the prototype instrument on any strip from 0.005 to 0.050 in. thick and with speeds from 50 to 1000 ft./min., except that obvious trouble sometimes occurred during setting up a mill for a new rolling operation.

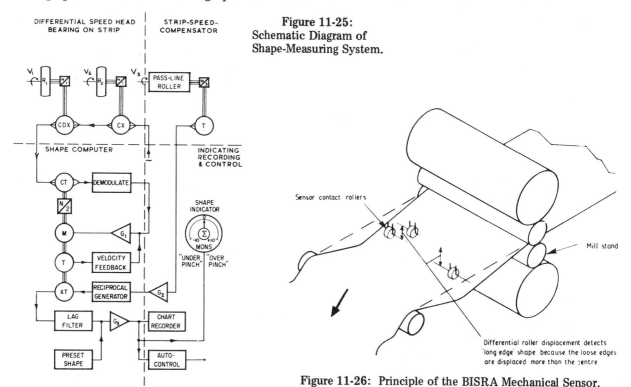

Figure 11-25: Schematic Diagram of Shape-Measuring System.

Figure 11-26: Principle of the BISRA Mechanical Sensor.

11-17 The BISRA Mechanical Shape Sensor

Subsequent to the development of Pearson's method of shape measurement, the British Iron and Steel Research Association (BISRA) constructed a shape sensor consisting of 5 independently-mounted rollers disposed across the strip width, all pressed against the strip with equal force.‡▼ The lateral distribution of the tensile stress in the strip is inferred from the measurement of the displacement of the rollers in the direction normal to the direction of strip movement (See Figure 11-26).

---

‡ J. G. Wistreich, "Measurement and Control of Strip Shape in Cold Rolling", Proceedings ICSTIS, Section 4. Suppl. Trans. ISIJ, Vol. 11, 1971, pp. 674-679.

▼ N. A. Townsend and J. C. E. Winch, "Development of a Contact Shape Sensor", Sheet Metal Industries, May 1971, pp. 365-386.

Roller forces in this type of sensor must be large enough to straighten out the buckles in the strip but must not be large enough to mark its surface.

Tests on a laboratory mill suggested that a suitable roller would be 76 mm. diameter with a generous meridional curvature and made of Delrin. (Delrin is a dimensionally stable thermo-plastic with high creep resistance and elastic modulus.) With this design no marking occurred with a 130N force even when the roller was purposely made to skid, and this load is far greater than is necessary to suppress buckles in the thickest strip encountered in practice.

With long edge/middle shape, the closer the outer rollers are to the strip edges, the larger the output signal from the sensor. However, the practical consideration of imprecise tracking of the strip through the mill, and the need to avoid spurious readings from "frilly edges", meant that the outer rollers could not be too near the strip edges. Moreover, as the differential roller displacement is to be a measure of shape, then ideally, with zero shape, there should be zero differential displacement, and this requirement really governs the position of the outer rollers. In other words, the deflection of all rollers should be identical for zero shapes.

It was anticipated that if the outer rollers are too near the edges of the strip, the unsupported boundary of the strip would cause their displacements to be larger than that of the center roller. To find the appropriate location for the outer rollers, a series of tests was conducted on flat strip of different widths and thicknesses, and at different levels of roller force. It was concluded that, with a three-probe unit, the outer rollers should be approximately one-sixth of the strip width from the edge.

With the upper limit of the roller force established, it was necessary to find what displacements would result so as to select suitable displacement transducers. Experiments on strip with gross shape implied that the differential displacements would be large enough to enable relatively unsophisticated transducers to be used. More important, however, was what differential displacements would exist when the shape was on the threshold of acceptability. A useful instrument would need to measure an elongation difference of the order of 0.01 per cent across the roll gap, which means for steel being able to resolve to a stress differential of approximately 20 kN/m^2. From the above-mentioned experiments it was concluded that, if the sensor has to resolve a stress differential of this magnitude at, say, a mean tension of 80 kN/m^2 (representative of the mean tension between the last stand of a tandem cold mill and its coiler and well above that found in a temper mill), then differential displacements have to be resolved to approximately 0.4 mm, which can be readily achieved by commercially available transducers.

A sensing probe is shown in Figure 11-27. The contact roller which runs on a needle-roller bearing, is supported by a fabricated light-alloy bridge piece and attached to the bridge piece at either end are posts (hidden by the gaiters in Figure 11-27) that permit the bridge to move only vertically. Each bridge is propelled vertically by a hydraulic jack situated midway between the bridge posts. The lower trunnion mounting of the jack and the linear ball bearings that guide the posts are incorporated in a fabricated steel housing.

A beam — which in practice would be mounted underneath the mill exit table — with V-slides, carries the roller assemblies, thus allowing the distance between the rollers to be varied to deal with different strip widths. The distance between the roller assemblies is altered by means of a motorized load screw engaging with captive nuts in the outer roller housing.

The electronic system of a prototype unit is shown in Figure 11-28. In this unit, only three displacement transducers were used and these were supplied with a 9-volt supply from batteries, B_1, the supply being stabilized by a Zener diode, D_1; the potentiometer, P_1, and shunt, S_1, are used to adjust the current supply to the required level of 15mA, which is displayed on meter M_1. The sensitivities of the left-hand and right-hand transducers are matched to that of the middle transducers by potentiometers P_2 and P_3.

The output signals from the left-hand and right-hand transducers are zeroed with reference to output from the middle transducers by potentiometers P_4 and P_5. The difference of output voltages from the left-hand and middle transducers is amplified by A_1 and fed to a U-V recorder; similarly, the difference between the right-hand and middle transducer outputs is fed to the U-V recorder after being amplified in A_2. Potentiometers P_6 and P_7 provide a means of gain adjustment for the output signal at the recorder, and DS_1 is a decade switch which enables a stepwise calibration of the output signal to be made.

Figure 11-27:
Details of Sensing Probe.
(1) Contact Roller.
(2) Displacement Transducer.
(3) Hydraulic Cylinder.
(4) Adjusting Tappet For Displacement Transducer.

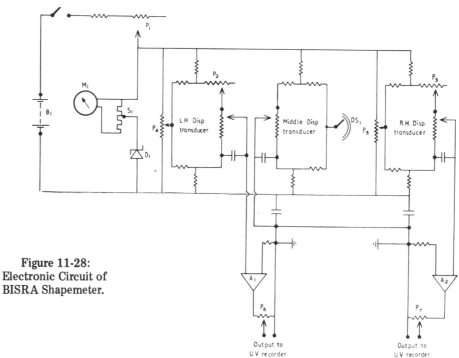

Figure 11-28:
Electronic Circuit of
BISRA Shapemeter.

STRIP SHAPE: ITS MEASUREMENT AND CONTROL

The prototype BISRA contact shapemeter using three sensors was installed on the exit side of a temper mill and it was found that, contrary to expectations, the commonest shape defect at this mill is not the symmetrical long edge/middle modes, but wide, asymmetrically located bands of slackness or "collars". For this reason, it was decided to utilize five sensors and a visual output (Figure 11-29) for an improved version of the shapemeter.

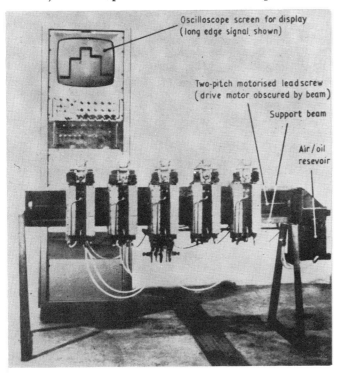

Figure 11-29:
Five-Probe Sensor and Associated Electronics and Display System of Improved BISRA Mechanical Shape Sensor.

11-18 An Optical Shape Sensing System

A system for monitoring the shape of rolled strip under conditions of relaxed tension has been described by Santo et al.[x] The sensor and the associated shape control circuits are illustrated in Figure 11-30. A fluorescent light source is placed near one edge of the strip and its reflection in the strip is viewed by a television camera mounted beyond the other edge as indicated in Figure 11-31. If the corrected video signal outputs corresponding to the strip edges are Q_A and Q_C and the output corresponding to the strip center is Q_B, then the relationship

$$\frac{1}{2}\left(Q_A + Q_C\right) = Q_B \tag{11-25}$$

is valid for flat strip,

$$\frac{1}{2}\left(Q_A + Q_C\right) > Q_B \tag{11-26}$$

corresponds to an edge wave, and

$$\frac{1}{2}\left(Q_A + Q_C\right) < Q_B \tag{11-27}$$

represents a center fullness in the strip. Figure 11-31(a) shows how the video outputs (denoted only by their subscripts) varied with the force applied to the work roll for bending purposes in a skin-pass rolling operation. Figure 11-31(b) shows the corresponding values of the shape of the

---

[x] T. Santo, T. Sakaki, T. Matsushima, H. Shibata, T. Emori and M. Arimuza, "Automatic Shape Control", Proceedings ICSTIS Suppl. Trans. ISIJ, Vol. II, 1971, pp. 698-702.

sheared product in the absence of strip tensions (measured in terms of the "inclination" or the peak-to-peak amplitude of the wave divided by the wavelength).

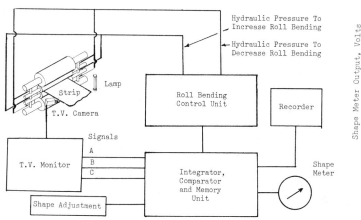

Figure 11-30:
Block Diagram of Shape Control System. (Santo, et al.)

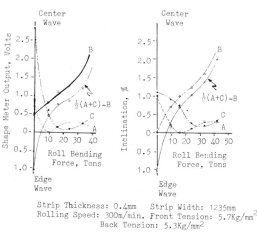

Figure 11-31:
Strip Shapes With Roll-Bending Force Varied During a Skin Pass (Santo et al.).

11-19 Monitoring Shape by Direct Measurement of Tensile Stresses Using Rollers

The use of a roll or a plurality of rollers pressing against the strip for the purpose of measuring the distribution of the longitudinal tensile stresses in the strip was patented by Misaka et al.[±] The method is illustrated in Figure 11-30. All the rollers in contact with the strip are in the same relative position and the force exerted by the strip on each is measured by means of strain gages.

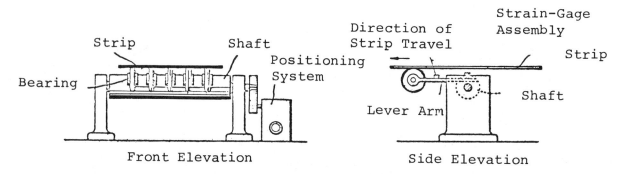

Figure 11-32: Misaka's Shape Monitoring Method.

To provide an automatic system for producing proper shape during rolling, Misaka and his co-inventors proposed the use of a work-roll bending system (See Section 11-24), with the outputs from the strain gages controlling the hydraulic pressure in the roll separating jacks.

A similar system for monitoring shape has been proposed and patented by T. Kajiwara.[▣] This is shown in Figure 11-33 and it utilizes the output signals from the rollers to

---

[±] Yoshisuke Misaka et al., "Controlling Method and Measuring Instrument for the Flatness of Strips", U. S. Patent 3,442,104, issued May 6, 1969.

[▣] Toshiyuki Kajiwara, "Control Apparatus and System for Strip Rolling", U. S. Patent 3,475,935, issued Nov. 4, 1969.

STRIP SHAPE: ITS MEASUREMENT AND CONTROL

automatically rotate and elevate a pair of tensioning rolls so as to provide for greater uniformity of tension in the strip.✤

This two-roller type of shape monitor was installed on the five-stand tandem cold mill in the Mizushima Works of Kawasaki Steel Corp., Japan. It utilized the outputs of four-load cells, supporting the two rollers (See Figure 11-33), as discussed below.

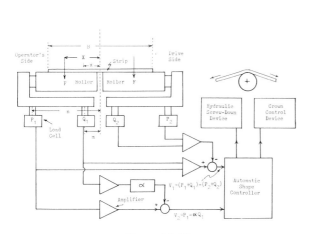

Shape	(a) Flat	(b) Op. side or dr. side wavy edge	(c) Wavy edge	(d) Full strip
Tension distribution	▮▮▮▮▮▮	▁▂▃▅▇	▅▇▅	▇▃▇
V_1	≈ 0	$\neq 0$	≈ 0	≈ 0
V_2	≈ 0	> 0	< 0	> 0
Center of tension distribution	$X = \dfrac{B}{4}$	$X > \dfrac{B}{4}$	$X < \dfrac{B}{4}$	$X > \dfrac{B}{4}$

Figure 11-33:
Principle of Shape Detector. (Kawamata, Hotta and Kajiwara)

Figure 11-34:
Chart Showing Relation Between Strip Shape and Tension Distribution.

If the outputs of the two outermost load cells are P_1 and P_2 and of the two corresponding center load cells Q_1 and Q_2, then the difference between the total load for the two rollers V_1 may be written

$$V_1 = (P_1 + Q_1) - (P_2 + Q_2) \tag{11-28}$$

If $V_1 = 0$, the tensile stress distribution in the strip is symmetrical but if V_1 does not equal 0, then a wavy condition exists on one edge of the strip or the other (See Figure 11-34). For conditions of equilibrium

$$P_1 = \frac{F}{n - m}(X - m) \tag{11-29}$$

$$Q_1 = \frac{F}{n - m}(n - X) \tag{11-30}$$

where, F = the total load of the vertical components of the tension distribution.

X = the distance from the center of the pass line to the center of the load of the vertical components of the tension distribution.

m, n = distance from the center of the pass line to each load cell.

For flat strip (Column (a) of Figure 11-34)

$$X = B/4 \tag{11-31}$$

where B is the full width of the strip. If under such circumstances, the values of P_1 and Q_1 are P_{10} and Q_{10}, respectively, and the ratio of P_{10} to Q_{10} is defined as α, (the width correcting coefficient), then

---

✤T. Kawamata, M. Hotta and T. Kajiwara, "Automatic Shape Control in Cold Rolling", Proceedings ICSTIS, Suppl. Trans. ISIJ, Vol. 11, 1971, pp. 702-704.

$$\alpha = \frac{B - 4m}{4n - B} \tag{11-32}$$

Defining a force parameter V_2 such that

$$V_2 = P_1 - \alpha Q_1 \tag{11-33}$$

then

$$V_2 = \frac{F}{n - \frac{B}{4}} \left(X - \frac{B}{4} \right) \tag{11-34}$$

Thus, for a symmetrical tensile stress distribution in the strip where

$$V_1 = 0 \tag{11-35}$$

(1) In case of flat strip, $X = B/4$, so $V_2 = 0$
(2) In case of wavy edge, $X < B/4$, so $V_2 < 0$
(3) In case of full centered strip, $X > B/4$, so $V_2 > 0$

Therefore, it is possible to detect the shape of strip by measuring the load cell outputs P_1, Q_1, P_2, and Q_2 and computing the magnitude and polarity of V_1 and V_2.

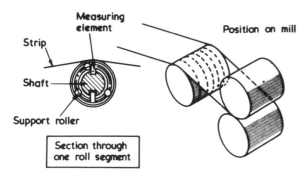

Figure 11-35:
The Principle of Operation of the Max Planck Institute Shapemeter.

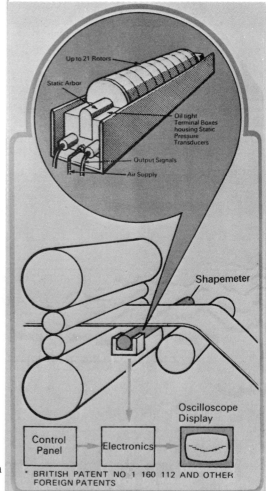

Figure 11-36:
The Loewy-Robertson "Vidimon" Visual Shapemeter.

STRIP SHAPE: ITS MEASUREMENT AND CONTROL

Yet another shape monitoring system of this type was under development at the Max Planck Institute in Düsseldorf.[●] It is shown diagrammatically in Figure 11-35 and it utilizes a billy roll divided axially into a number of segments, the strip load on each segment being measured by the use of strain gages on its supporting pillar. To date, this device does not appear to have been used in practice outside of the research laboratory, although the test results achieved there were good.[❏][§]

Built along similar lines is the "Vidimon" visual-reading shape meter marketed by Loewy-Robertson. This system is illustrated in Figure 11-36 and it utilizes air bearings for the individual rollers.

11-20 The Stressometer

To avoid the necessity of a plurality of rollers, Sivilotti, et al.,[*] patented a single roll housing a number of pressure sensitive transducers spaced axially along the roll surface as illustrated in Figure 11-37. The electrical outputs from these sensors (designated collectively as a "Stressometer") are utilized to provide a visual indication of strip shape for the convenience of the mill operator.

This shapemeter resulted from more than six years of development and field testing. The first commercial unit has now been in operation since 1967 at ALCAN's Kingston works in Ontario. The shapemeter is installed in an 84-in. single-stand, nonreversing 4-high cold mill. The measuring equipment consists of the measuring roll (which replaces the normal billy roll), a slip-ring device, photocell device, an electronic unit and a display unit for the multi-channel presentation of the tensile stress profile.

The measuring roll is divided into 25 measuring zones across the roll and the display unit has the same number of indicating instruments.

Each zone measures the stress in the corresponding part of the strip, independent of adjacent zones. A condition for this independence is that the whole roll assembly and the individual measuring zones are very much stiffer than the curved part of the strip.

A true shape indication on the instruments can naturally be obtained if the stress distribution in the strip is exclusively the result of the shape. To obtain the best possible approximation of this ideal state, it is obviously necessary to make sure that the roll gap, measuring roll and coiler are parallel. Another condition for a stress distribution proportional to shape is naturally that the coil builds up in a true cylindrical form. Any deviation from this will introduce a false stress pattern superimposed on a shape-induced pattern.

The measuring roll has a diameter of 313 mm (12.33 in.), and a total measuring length of 2,130 mm. (84 in.) divided into 25 zones, each with a length of 85 mm. (3.36 in.). Four axial slots are equally spaced along the roll periphery. The width of the slots partly determines the length of the output pulses from sensors placed in the slots and actuated as the roll rotates under strip load. The slot width of 50 mm. (2 in.) has been chosen as a suitable value corresponding to the smallest normally used deflection angle of the strip.

Each slot contains 25 sensors, which have a slightly larger height than the slot depth during mounting and are subsequently ground together with the roll core to the right dimension.

[●] N. Davis, "Measurement of Strip Shape, Speed and Length", Iron and Steel, August 1972, pp. 391-394.

[❏] O. Pawelski, V. Schuler and B. Berger, "Control of Flatness in Cold Strip Rolling", International Eisen Huttentagung, Düsseldorf, 1970.

[§] O. Pawelski, V. Schuler and B. Berger, "Development of a Shape Control System in Cold Strip Rolling", Proceedings ICSTIS, Section 4, Suppl. Trans. ISIJ, Vol. II, 1971, pp. 692-697.

[*] O. G. Sivilotti, W. E. Davies, M. Henze and O. Dahle, "ASEA-ALCAN AFC System for Cold Rolling Flat Strip", Iron & Steel Engineer Year Book, 1973, pp. 263-270.

Each zone is covered by a 10 mm. (0.39 in.) thick sleeve shrunk on the roll while standing vertically so that the gap between the sleeves becomes negligible (0.02 to 0.04 mm.) and disappears after grinding.

If the surface should be damaged, the roll can be reground until the diameter has been reduced by 6 mm. (0.24 in.) without substantially changing the strength of the measuring properties.

The roll is fitted with roller bearings, and the 28-pole slip-ring device and photocell device are mounted on one of the shaft ends.

The sensors are modified magnetoelastic Pressductor force transducers. To obtain a roll with a low moment of inertia and high flexural rigidity, a compact sensor had to be developed. (Figure 11-38)

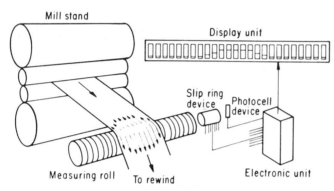

Figure 11-37:
Schematic Diagram of Shapemeter,
Developed by Sivilotti et al.

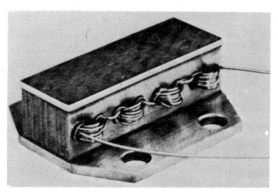

Figure 11-38:
Magnetoelastic Transducer
for the Measuring Roll.

The normal Pressductor transducers have two windings arranged crosswise so that the output signal is zero in the unloaded state and increases linearly with load. As implied, this normal design would give too high a sensor for this application. Instead, two of the compact sensors with an output signal that is largest at no-load and decreases with load are connected with their secondary windings in series opposition and are placed in adjacent slots. Another sensor pair, similarly connected, is placed in the two other slots, and the pairs are connected in series so that the signals from diametrically opposite sensors are added. This arrangement provides good compensation for the constant force from the shrunk-on sleeve, variable forces caused by centrifugal force or temperature variations, bending forces due to roll weight and strip tension, and for disturbances from external magnetic fields.

The primary windings of all the sensors are connected in series and magnetized at a frequency of 2000 Hz through two of the slip rings. The signals from the secondary windings of the zones are taken out individually through the other slip rings. One output lead from each zone is connected to a common slip ring.

The output signal from each zone is amplitude modulated with pulses whose sign changes each time a sensor is loaded by the strip. The amplitude of modulation is proportional to the load on the zone and hence to the stress in the part of the strip covering the zone in question.

Pulses from the photocell device make the electronic equipment change the sign of every other pulse in the demodulated signal. Filtering of the resultant signal gives a direct voltage that is a measure of the force on the sensors in the individual zone.

The basic functions of the electronic equipment are described with reference to the block diagram in Figure 11-39, where the following notation is used:

F_x = force on measuring zone x

σ_x = mechanical stress in the strip across zone x

STRIP SHAPE: ITS MEASUREMENT AND CONTROL

F_m = mean value of all forces acting on zones covered by the strip

σ_o = mean value of the mechanical stress

T = strip tension

B = strip width

t = strip thickness

n = number of zones

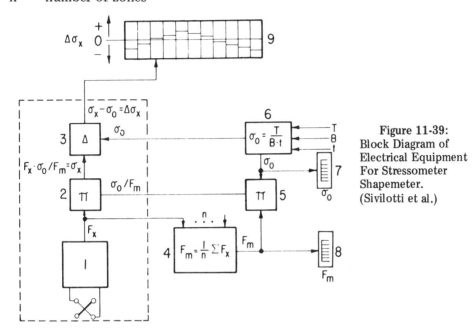

Figure 11-39: Block Diagram of Electrical Equipment For Stressometer Shapemeter. (Sivilotti et al.)

The figure shows the signal processing for one measuring zone (the blocks within the broken line) as well as the common part for the entire equipment.

The transducer signal from zone x is rectified and filtered in block 1. The output signal is a d.c. voltage proportional to the force F_x. In block 4 the mean value F_m from all zones covered by the strip is formed. The mechanical mean stress σ_o is calculated in block 6 with the aid of data for the strip.

The quotient σ_o/F_m is formed in block 5 and multiplied by F_x in block 2. The output signal $F_x \sigma_o/F_m$ is equal to the local mechanical stress σ_x in the strip.

Finally σ_o is subtracted in block 3 and the difference $\Delta\sigma_x = \sigma_x - \sigma_o$ constitutes the output signal and is displayed on a separate instrument for each measuring zone.

Factors influencing the sensitivity of the equipment, such as angle of wrap and changing areas of contact between the strip and the roll with gage and alloy are compensated for by an automatic scaling unit in the electronic equipment. The electronic equipment also incorporates an automatic gain regulation system to prevent amplifier saturation, which would otherwise result from a large maximum to minimum tension range.

Figures 11-40 and 11-41 show the correlation between the instrument's output signal and the experimentally measured strip shape for two strips of different widths. From this data, Sivilotti and his co-workers concluded that the Stressometer system exhibited a measuring accuracy within ± 5 I-units for strip thicknesses down to 0.25 mm (0.01 in.).

11-21 Magnetic Strip-Shape Monitors

Contact type stress measuring instruments have the advantage of making a direct measurement and hence being independent of any intermediate material properties. They are therefore applicable to nonferromagnetic materials. The magnetic instruments, on the other hand, use the properties of the strip itself and so have to be calibrated for each material grade and gage. However, an important aspect of shape control is the control of locked-in stresses which do not reveal themselves as nonplanarity or in-plane distortion until the material is slit or stamped. In the production of tin plate and electrical steels, the rolled products are frequently sheared. Accordingly, there is an advantage in magnetic monitors in that they can sense residual internal stresses in a workpiece that would normally be judged to be of good shape.

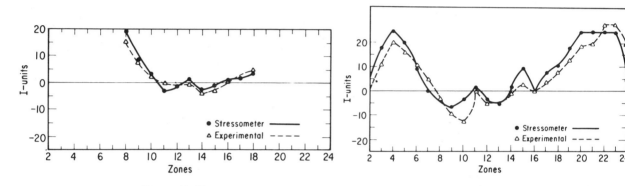

Figure 11-40:
Correlation of Stressometer Results and Experimental Results, Narrow Strip.

Figure 11-41:
Correlation of Stressometer Results and Experimental Results, Wide Strip.

For the reasons outlined above and also with a view to minimizing the possibility of strip surface damage, several noncontact (principally magnetic) shapemeters have been developed. These rely for their operation on the fact that the magnetic properties (mainly induction) of a ferromagnetic material are changed by the application of stress. The magnitude of this change is dependent on a number of factors, in particular, the crystal orientations present in the material.

J. D. Shelesnow et al.,[◊] have described a "magnetic elastic transmitter", developed at the Moscow Steel Institute for measuring the specific tension in strip as it is being cold rolled. The transmitter consists of two electrical circuits — a primary and a secondary — and is intended to be mounted in a noncontacting position above the strip. A voltage proportional to the stress is induced in the secondary circuit by reason of the anisotropy of the strip.

The basic unit was mounted in a cylindrical brass case 50 mm. in diameter and 70 mm. long and filled with resin. It was found to be satisfactory for rolling speeds up to about 5,000 feet/minute with its output monitored by a highly sensitive thermionic voltmeter and an oscilloscope.

To permit the simultaneous use of several such transmitters without their mutual interference, they had to be carefully oriented. The outputs of the instruments were found to be dependent on their separation from the strip surface, but this dependence could be reduced by correctly proportioning the sizes of the primary and secondary circuits.

Prototypes of the sensor utilizing three detecting units developed by Shelesnow were used experimentally: (1) between the third and fourth stands of the 1700 mm. wide strip mill at Tscherepowets, Russia; (2) on the 1700 mm. skin pass mill at Shdanow, Russia, and (3) on an experimental four-high mill (80 mm. and 300 mm. x 250 mm.) at the Research Institute for Non-Ferrous Metals, Freiburg, Germany.[◄] These tests demonstrated the feasibility of using such devices to improve the quality of cold rolled strip.

[◊] J. D. Shelesnow et al., "Effect of Strip Tension on the Quality of Cold Strip and Control of Strip Tension During the Rolling Process", Neue Hütte, 1968, 13 September, pp. 531-534.

[◄] K. Oppermann, et al., "Works Tests With Contactless Strip Tension Detectors", Neue Hütte, 15, 1970, pp. 224-228. (British Iron & Steel Industry Translation Service, BISI 8913).

STRIP SHAPE: ITS MEASUREMENT AND CONTROL

The British Iron & Steel Research Association (BISRA) also developed a magnetic monitor.[‡] In the BISRA unit, alternating flux (125Hz) is generated by an exciter core on one side of the strip, and is detected by a detector core on the other side. The construction of a single probe is shown in Figure 11-42. The amount of flux reaching the detector is dependent upon the amount which uses the strip as part of its circuit, and hence on the stress in the strip. As can be seen from Figure 11-42, the measurement is made in two directions: parallel to the strip tension direction and transverse to it. The output signal from the probe is the difference between these two measurements. Making the measurements in the two directions compensates for the different crystalline anisotropy encountered in different steels. Typical outputs from the transducers are illustrated in Figure 11-43 where the V_L and V_T curves represent the outputs in the longitudinal (rolling) and transverse directions, respectively.

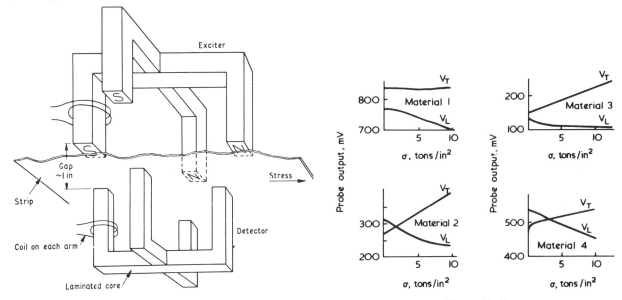

Figure 11-42:
Arrangement of Single Probe of BISRA Shapemeter.

Figure 11-43:
Typical Probe Outputs of the BISRA Shapemeter.

The complete instrument consists of several stationary probes, limited to five in number, disposed across the strip width, mounted between mill stand and bridle roll in the manner shown in Figure 11-44.

This figure also shows the form of display adopted for the experimental instrument, which is based on an electronic sample-and-hold system with a one-second total scanning time for all five probes. Full height of the indication represents a tensile stress of 22,000 psi.

Figure 11-45 shows the extent of correlation between the electromagnetic scan of a strip length under tension and its visual appearance when laid out freely on a plane surface. The latter was quantified by projecting a shadow grid onto the strip and measuring the distortion of the longitudinal lines associated with the departure from flatness. Further confirmation of this has been obtained by measurement of the stresses in stationary sheets by means of electric resistance strain (ERS) gages, when these have been tensioned in a specially constructed stretching frame. An example is shown in Figure 11-46.

Following laboratory investigations, the instrument was installed in a four-high reversing mill in the Orb Works of the British Steel Corporation to undergo extensive trials. (Figure 11-47) These trials demonstrated the importance of the correct alignment of mill components. For example, a one-minute tilting of the strip plane through misalignment of the bridle roll relative to

---

[‡] J. G. Wistreich, "Measurement and Control of Strip Shape in Cold Rolling", Proceedings, ICSTIS, Suppl. Trans. ISIJ, Vol. II, 1971, pp. 674-679.

the mill rolls caused a stress difference of about 2,000 psi across a 39-in. wide strip. Equally, a 0.010" camber on the bridle roll produced a 1,000 psi difference between center and edge stress.

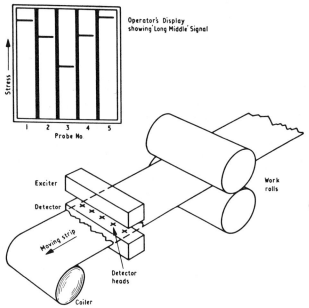

Figure 11-44:
Arrangement of BISRA Magnetic Shapemeter.

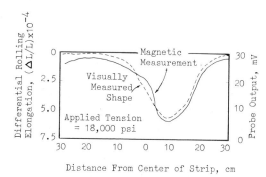

Figure 11-45:
Comparison of Visual Shape With Magnetic Measurement.

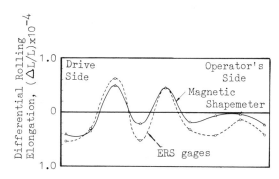

Figure 11-46:
Comparison of ERS Gage and Magnetic Measurements.

Figure 11-47:
Shapemeter Installed at Orb Works.

Figure 11-48: Secondary Cold Reduction Mill.

In 1968, the Graham Research Laboratory of the Jones and Laughlin Steel Corporation began the development of a magnetic shape sensor, similar in many respects to that produced by

STRIP SHAPE: ITS MEASUREMENT AND CONTROL

BISRA.[‖][‡] In this case, however, the sensor was to be located only on one side of the strip which necessitates a more constant pass line than the BISRA unit.

The first prototype unit was installed at the exit end of a secondary cold reduction mill, between the last stand and the tension reel, as shown in Figure 11-48. This mill is a 48-in., three-stand mill which reduces annealed strip from 0.0094 inches to 0.0061 inches at speeds up to 5,000 fpm. A close-up of the sensor is shown in Figure 11-49. The strip comes out of the last stand, over the 8-in. billy roll, over the sensor, under the 10-in. billy roll, and onto the tension reel. Note that there are nine sensing units equally spaced across the width of the strip.

Figure 11-49: Shape Sensor on 3-Stand Mill.

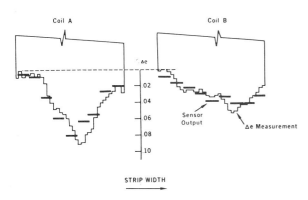

Figure 11-50: Comparison of Sensor With Δe Measurement.

Figure 11-51: Traversing Sensor on 3-Stand Mill.

The output of the instrument was fed to an oscilloscope with the horizontal sweep divided into 9 segments, each segment corresponding to a sensing head.

Figure 11-50 illustrates the correlation obtained between the sensor and the flatness or Δe measurements. Two coils are shown. Coil A on the left has a very full center, while coil B on the

---

[‖] J. R. Dahm, "Measurement of Shape on Double Reduced Tin Plate", Proceedings of the Twentieth International ISA Iron and Steel Instrumentation Symposium, March 23-25, 1970, Pittsburgh, Pa., pp. 1-9.

[‡] S. E. Renner and J. R. Dahm, "Progress in Measuring and Controlling Tin Plate Shape", AISI Regional Meetings 1970, pp. 202-222.

right has a very light center. While this model of the sensor was able to reliably indicate center fullness, it did not consistently indicate edge waves. This is attributed to the fact that tension was sensed at only nine points across the strip, and only one point on the outer six inches, where most edge waves occur. Better resolution of the tension distribution was therefore needed. To achieve this, a new model of the sensor was developed. In this, shown in Figure 11-51, a single sensing element was moved back and forth across the strip width, giving a continuous tension distribution from edge to edge.

11-22 The Control of Strip Shape in Cold Rolling

It has been shown in Section 11-6 that defects such as full center and full edges are directly related to the mismatch of the cross-sectional profiles of the incoming strip and of the stressed roll gap and the resulting transverse variation in the elongation of the strip.

Referring to equation 11-1, it will be realized that, to control gage and shape at the same time, the three most important profile parameters, g, k_1 and k_2, must be simultaneously controlled in one integrated scheme.◈ In this connection, it is first desirable to examine the dependence of g, k_1 and k_2 upon the available control means.

As will be seen in later sections, several different methods of shape control exist. In this instance, however, work roll bending will be assumed to be the method used.

The functional relationships between the profile parameters g, k_1 and k_2 and the controlling factors, assumed to be the screw position S, rolling tension T and roll bending force J, can be indicated as follows:

$$g = g(S,T,J) \tag{11-36}$$

$$k_1 = k_1(S,T,J) \tag{11-37}$$

$$k_2 = k_2(S.T.J) \tag{11-38}$$

These functions cannot be expressed in algebraic form but their numerical values can be computed for any set of control factors S, T, J under specified rolling conditions.

Derived from equation 11-36 to 11-38, small changes in g, k_1, k_2 around specified operating conditions, can be expressed by:

$$\Delta g = \left(\frac{\partial g}{\partial S}\right)_o \cdot \Delta S + \left(\frac{\partial g}{\partial T}\right)_o \cdot \Delta T + \left(\frac{\partial g}{\partial J}\right)_o \cdot \Delta J \tag{11-39}$$

$$\Delta k_1 = \left(\frac{\partial K_1}{\partial S}\right)_o \cdot \Delta S + \left(\frac{\partial K_1}{\partial T}\right)_o \cdot \Delta T + \left(\frac{\partial k_1}{\partial J}\right)_o \cdot \Delta J \tag{11-40}$$

$$\Delta k_2 = \left(\frac{\partial k_2}{\partial S}\right)_o \cdot \Delta S + \left(\frac{\partial k_2}{\partial T}\right)_o \cdot \Delta T + \left(\frac{\partial k_2}{\partial J}\right)_o \cdot \Delta J \tag{11-41}$$

In the above equations the partial derivatives represent the sensitivity of each of the output quantities g, k_1 and k_2 to each of the controlling factors S, T, J around the initial operating point. For example $\left(\frac{\partial g}{\partial S}\right)_o$ measures the sensitivity of gage to a change in screw setting (known as "roll gap transfer function"). Similarly, $\left(\frac{\partial k_1}{\partial T}\right)_o$ is the sensitivity of the profile parameter k_1 to a change in rolling tension. The partial derivatives in equations 11-39 to 11-41 describe the behavior

---

◈ B. Sabatini and M. Tarokh, "Control of Gauge and Shape in Flat Rolling", BISRA Open Report, PE/B/28/69.

STRIP SHAPE: ITS MEASUREMENT AND CONTROL

of the mill with respect to control factors and are here referred to as "plant parameters". Clearly, a knowledge of the plant parameters enables one to determine the set of adjustments ΔS, ΔT, ΔJ capable of correcting measured errors Δg, Δk_1, Δk_2.

In practice, gage errors Δg can be directly measured, for example, by an X-ray gage meter. Shape errors Δk_1 and Δk_2 can be measured indirectly through the tension distribution across the width. The plant parameters can be found by deliberately changing the control factors one at a time and measuring the corresponding changes in output quantities.

However, equations 11-39 to 11-41 represent only a static model for gage and shape regulation. This would solve the control problem if the set of controlling actions ΔS, ΔT and ΔJ could be applied to the plant instantaneously. In practice this is not so because each individual control system (for screw position, rolling tension and roll bending force) responds with finite time to changes in demanded values. Following a disturbance in shape and/or gage there will be a transient period during which the interaction between gage and shape may cause instability. It will be shown that this problem can be overcome by the design of an integrated scheme for combined shape and gage control.

The proposed scheme is shown in Figure 11-52 in schematic form. Deviations from the desired values of g, k_1, k_2 are measured by sensing devices of the type described above. Such deviations will, in general, arise as a consequence of disturbances occurring in the rolling process such as changes in input gage and/or shape, lubrication conditions, etc. The errors are fed to a suitably designed "controller" which, taking into account the interaction and the time response of the measuring devices, determines appropriate corrective actions for the individual control systems of screw position, rolling tension and roll bending force. The control systems respond to the demanded changes according to their individual characteristics and act on the mill with their outputs.

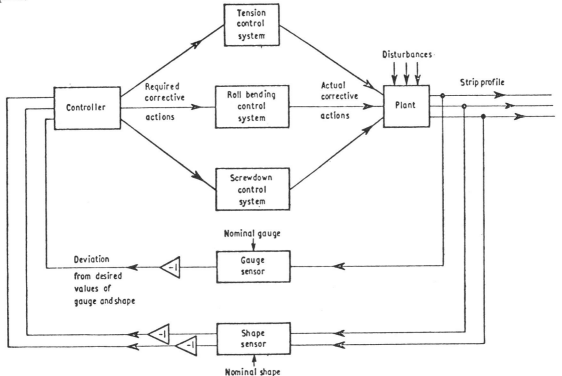

Figure 11-52: Shape and Gage Control Scheme. (Sabatini and Tarokh)

From a mathematical analysis of the control scheme shown in Figure 11-52, Sabatini and Tarokh showed that the stability of the whole system is not ensured by the stability of the individual gage and shape control systems but also depends on the values of the plant parameters and on the transport delays associated with the measuring devices. Consequently, separately controlled gage and shape are likely to pose a stability problem.

These results lead to the conclusion that in designing new rolling plants, it is desirable to devise a combined control of gage and shape in one integrated scheme.

In most of the existing plants, however, a gage control system is already installed. For such plants, it is of interest to examine under which conditions separate control of gage and shape is satisfactory; that is, a shape control system can be simply "added" to the mill, without altering the existing gage control. In the following section this problem is analyzed.

11-23 Separate Gage and Shape Control

In the automatic gage control (AGC) systems commonly in use today, screw position and strip tensions are varied to maintain constant gage. Furthermore, in many instances, only roll bending is available for shape control and it therefore follows that a one-parameter definition of shape should be used. This implies the existence of rolling conditions where strip profiles h(z) can be adequately expressed in terms of two profile parameters only; g for gage and k for shape.

For simplicity gage should be assumed to be controlled solely by screw position. The system under study, therefore, reduces to a system having two outputs, gage g and shape k and two controlling factors, S and J. The set of static control equations 11-39 to 11-41 in this case becomes:

$$\Delta g = \left(\frac{\partial g}{\partial S}\right)_o \cdot \Delta S + \left(\frac{\partial g}{\partial J}\right)_o \cdot \Delta J \qquad (11\text{-}42)$$

$$\Delta k = \left(\frac{\partial k}{\partial S}\right)_o \cdot \Delta S + \left(\frac{\partial k}{\partial J}\right)_o \cdot \Delta J \qquad (11\text{-}43)$$

By mathematical analysis, Sabatini and Tarokh[◊] showed that control of gage and shape is satisfactory provided that, in equations 11-42 and 11-43 above:

$$\left(\frac{\partial g}{\partial S}\right)_o \cdot \left(\frac{\partial k}{\partial J}\right)_o \gg \left(\frac{\partial g}{\partial J}\right)_o \cdot \left(\frac{\partial k}{\partial S}\right)_o \qquad (11\text{-}44)$$

This "non-interaction condition" requires that one or other of the following conditions is satisfied:

(1) the gage sensitivity to changes in screw position is much greater than shape sensitivity to these same screw changes. The physical significance of this is that the flexural rigidity of the rolls is high compared with the stiffness of the mill stands. For this the roll diameters must be large.

(2) The jack force influences shape much more than gage, i.e., the shape controlling means should affect the roll contour much more than the mill stretch. For this the mill modulus must be large.

Investigations of the "non-interaction condition" have indicated that if the left-hand side of equation 11-44 is at least one order of magnitude greater than the right-hand side, then separate control of gage and shape is satisfactory.

Sabatini and Tarokh carried out experiments on a cold mill to determine the plant parameters required for equations 11-42 and 11-43 and the feasibility of adding a shape control loop to an existing AGC system. The plant parameters were established by deliberately introducing step changes in the screw position(S) and the roll bending force (J) and measuring the corresponding changes in the gage and shape (Δg and Δk, respectively). In this case, shape was measured through

---

[◊] B. Sabatini and M. Tarokh, "Control of Gauge and Shape in Flat Rolling", BISRA Open Report PE/B/28/69.

STRIP SHAPE: ITS MEASUREMENT AND CONTROL

tension distribution using a noncontact magnetic sensor with five probes equally spaced across the width. The outputs of the five sensing heads were used to fit a parabolic tension distribution thus determining a parameter proportional to k_1. Gage was measured by an X-ray gage meter on the output of the stand. Screw position and roll bending force were measured using suitable transducers.

The values established for the plant parameters were

$$\left(\frac{\partial g}{\partial S}\right)_o = 0.85 \times 10^{-1} \quad mm/mm \tag{11-45}$$

$$\left(\frac{\partial g}{\partial J}\right)_o = 0.26 \times 10^{-3} \quad mm/KN \tag{11-46}$$

$$\left(\frac{\partial k}{\partial S}\right)_o = 0.64 \times 10^{-2} \quad KN/mm^3 \tag{11-47}$$

$$\left(\frac{\partial k}{\partial J}\right)_o = 0.72 \times 10^{-3} \quad KN/mm^2/KN \tag{11-48}$$

and, using these values, it may be shown that the left-hand side of equation 11-44 is approximately 35 times the right-hand side. Under these circumstances, a shape control system could be added to the existing AGC system, as represented by Figure 11-53.

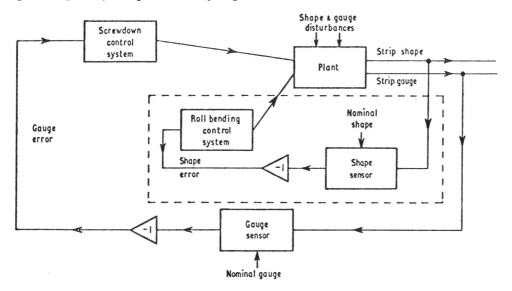

Figure 11-53:
Separate Control of Gage and Shape. (Sabatini and Tarokh)
A Shape Control System, Encircled in Dotted Lines, is Added to an Existing Gage Control System.

Roll bending has been considered as the method of strip shape correction in this and the preceding section, and this topic is discussed in greater detail in the next two sections.

11-24 Work-Roll Bending

As its name implies, this method of shape control involves the bending of one or both work rolls around the corresponding backup roll which is ground with an exaggerated crown. The basic elements of the system are illustrated in Figures 11-54 and 11-55 in which it may be seen that hydraulic cylinders are used to apply separating forces between the work-roll chocks and the

694   CHAPTER 11

corresponding backup-roll chock (System A) or between the roll chocks themselves (System B).⊗ It was patented by T. A. Fox• in 1966 and has found extensive use in the steel industry in the USA.

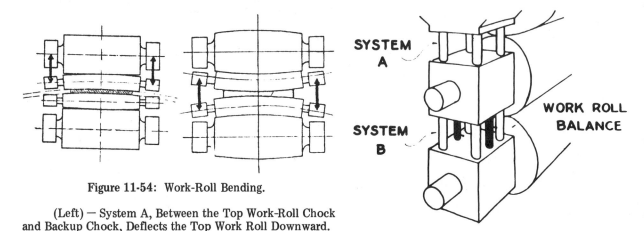

Figure 11-54: Work-Roll Bending.

(Left) — System A, Between the Top Work-Roll Chock and Backup Chock, Deflects the Top Work Roll Downward.

(Right) — System B Consists of Hydraulic Cylinders Located in the Top of Each Bottom Work-Roll Chock. This Force Deflects the Top Work Roll Upward and the Bottom Work Roll Downward.

Figure 11-55:
The Two Hydraulic Roll Bending Systems.

The shaping force exerted by each cylinder or set of cylinders is small compared with the rolling forces used in reducing the strip. With System A, when the bending forces are increased, the effective roll crown is diminished. Conversely, when the bending forces are decreased, the effective crown is accentuated. Of considerable advantage is the fact that these forces can be controlled during rolling to provide a rapidly acting shape-control system.

This system has the advantage that it can be installed conveniently on nearly all existing four-high mills as well as on new mills of the same type. The maximum shaping force used is determined by the work-roll bearing capacity and the maximum permissible stresses in the work roll.► An electrical control station may be provided for each stand equipped with three spring-centered joy-stick type switches for pressure control. These switches permit control of the hydraulic pressure on either the drive side, operator side or both sides of the mill. On continuous hot-strip mills, a manual-off-automatic switch is provided. When this switch is on automatic, the main mill drive motors actuate valves through load sensitive relays and timers or with heat sensitive devices. This is to ensure that the shape-control cylinders are energized only when the strip is between the work rolls.

A mathematical analysis of work-roll bending has been carried out by Stone and Gray.× Referring to Figure 11-56, showing a four-high mill with crowned work rolls in both the unloaded and loaded conditions, the crown is correctly selected so that with a rolling force P a uniform roll gap is provided that will exert a uniform pressure on the strip. The diametral work-roll crown Δ corresponds to the bending deflection of each backup roll. At the same time, the deflection of each work roll is $\Delta/2$. The forces acting on the lower work roll are shown in Figure 11-57a wherein the specific rolling force P' is uniform and is given by P/l. The equation for the deflection curve of the top of the bottom roll where it contacts the strip is $y = 0$. The work roll (Figure 11-57b) deforms under the additional action of the end forces F and it is the magnitude and shape of this

---

⊗R. A. Somerville, "Hydraulic Shaping in Cold Rolling", Iron & Steel Engineer Year Book, 1965, pp. 667-674.
•T. A. Fox, "Strip Rolling", U. S. Patent 3,228,219, Issued January 11, 1966.
►T. A. Fox, "Strip Mill Shape and Contour Control", Iron and Steel Engineer Year Book, 1966, pp. 519-522.
×M. D. Stone and R. Gray, "Theory and Practical Aspects in Crown Control", Iron and Steel Engineer Year Book, 1965, pp. 657-667.

STRIP SHAPE: ITS MEASUREMENT AND CONTROL

deformation curve that must be evaluated in order to establish the nature of the crown control attainable by this work-roll bending system.

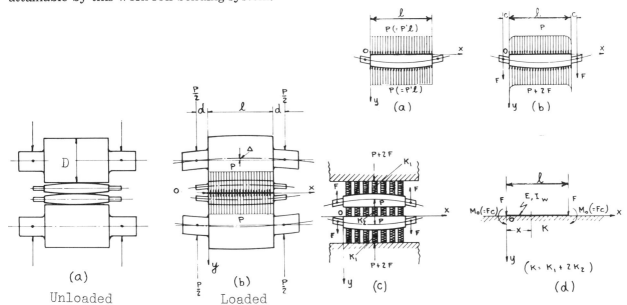

Figure 11-56:
Sketches of Roll Crowns and Rolling Force Distribution.

Figure 11-57:
Forces and Pressure Distribution on Work Roll and Beam on an Elastic Foundation.

Stone and Gray[x] approached the analysis from the viewpoint of a beam on an elastic foundation. Referring to Figure 11-57c, K_1 is the elastic beam constant representing the elastic contact conditions between the backup roll and its contacting work roll, and K_2 is the elastic beam constant for the two contacting work rolls including the effect of the workpiece. The equivalent constant is shown in Figure 11-57d, along with the forces, pressure distributions and the elastic foundation conditions.

From their analysis, they showed that the crown shape due to work-roll bending is equal to

$$\frac{4F\beta}{K(\sinh\beta l + \sin\beta l)} \left[\{\cosh\beta x \cos\beta(l-x) + \cosh\beta(l-x)\cos\beta x\} + c\beta\{\sinh\beta x \cos\beta(l-x) - \cosh\beta x \sin\beta(l-x) + \sinh\beta(l-x)\cos\beta x - \cosh\beta(l-x)\sin\beta x\} \right] + \frac{F(d+e)}{EI_b}[x(l-x)] \quad (11\text{-}49)$$

where

F = work-roll bending force at each end of the mill roll, (lb.)

K = beam on elastic foundation constant, (lb. per in. per in.)

$$\beta = \sqrt[4]{\frac{K}{4EI_w}} \cdot (\text{in}^{-1}) \quad (11\text{-}50)$$

I_w = moment of inertia of work-roll body, (in.^{4})

c = moment arm of forces F, to end of roll body, (in.)

---

[x] M. D. Stone and R. Gray, "Theory and Practical Aspects in Crown Control", Iron & Steel Engineer Year Book, 1965, pp. 657-667.

l = body length of work roll, (in.)

x = distance to any position along body of work roll from end of body, (in.)

I_b = moment of inertia of the backup-roll body, (in.^{4})

(d+e) = moment arm of forces F, acting on the backup roll, d representing the distance between the centerline of the backup roll-neck bearings and the end of the roll body, in. and e is the distance from the end of the roll body to the resultant line of action of force F, acting between the work roll and the backup roll, (in.)

The mill crown defined by the equation above is not a true parabola and under certain conditions assumes the profile shown in Figure 11-58.[x] Under these circumstances, there is a tendency to over roll the strip at the quarter points.

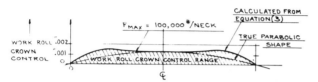

Figure 11-58: Effective Work-Roll Crown.

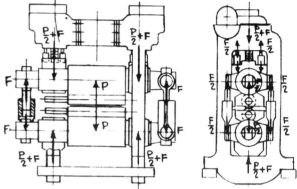

Figure 11-59: B.U.R.B. System Applicable to Existing Mill.

Regardless of whether System A or System B is used (work rolls pushed together or pushed apart), the same amount of crown control is obtainable. However, an operating difficulty is introduced with System A since the bending forces now tend to bang the work rolls together when the strip leaves the mill. Thus, for mills not operating below face, the hydraulic pressure must be shut off automatically at the time of strip delivery and turned on again at the time of strip entry.

11-25 Backup-Roll Bending

Inasmuch as backup rolls are utilized to provide a rigidity that cannot be achieved solely by the smaller diameter work rolls, the thought of deliberately bending the backup rolls may seem rather paradoxical. Yet backup-roll bending has been used on hot-strip and plate mills for the rolling of both steel and aluminum and it appears that it may soon find acceptance with respect to the cold rolling of steel. Conceived by Stone,[++] the backup-roll bending (BURB) method has the added advantages of controlling the crown of the thicker workpieces where sideways spread of the workpiece is possible and of controlling the gage of thinner workpieces.[□]

A number of methods of bending the backup rolls are possible, each having its own unique design advantages. Figure 11-59 shows one of the earliest installations where hydraulic cylinders in pairs apply bending moments to extended backup-roll necks by pulling the roll necks together. Work rolls are changed in the normal manner by moving them through the spacing between the two cylinders on the operator's side of the mill stand. Figure 11-60 illustrates an improved version of the BURB system, particularly from the viewpoint of roll changing. In this construction, open-ended bending cylinders are pivoted off the housing posts, each cylinder having

---

[x] M. D. Stone and R. Gray, "Theory and Practical Aspects in Crown Control", Iron and Steel Engineer Year Book, 1965, pp. 657-667.

[++] M. D. Stone, "Method and Apparatus for Rolling Flat Strip", U. S. Patent No. 3,459,019, Issued August 5, 1969.

[□] M. D. Stone, "Backup-Roll Bending — For Crown and Gage Control", Iron and Steel Engineer Year Book, 1969, pp. 669-688.

STRIP SHAPE: ITS MEASUREMENT AND CONTROL

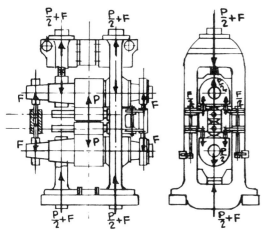

Figure 11-60: B.U.R.B. System.

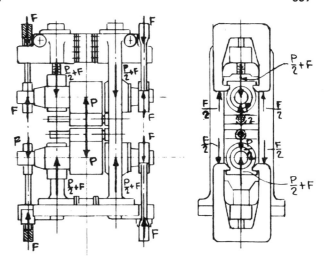

Figure 11-61: B.U.R.B. System for New Mill

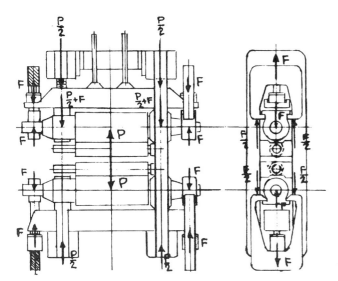

Figure 11-62: B.U.R.B. System with Reaction Beams.

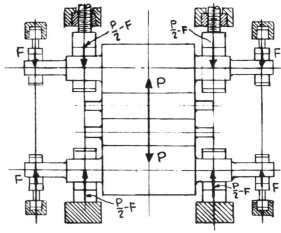

Figure 11-63:
B.U.R.B. System Well Adapted for Tight Mills Such as Foil Mills.

a central post and dual plungers, one to push up and against the top backup-roll bending bearing chocks and the other to push downward against the bottom backup-roll bending chocks. When either work rolls or backup rolls are to be changed, unloading of the bending cylinders automatically permits the plungers to recede slightly and the entire assemblies are swung clear of the housing windows by power cylinders and the rolls changed in a convenient manner. Figure 11-61 shows another design variation best used in a new mill design. Here the backup rolls are bent by cylinders mounted on projections at the top and bottom of the mill housing. To prevent these offset forces from objectionably bending the mill housings, adequately large and correctly designed separators are provided between the mill housings. Figure 11-62 shows a method of installing BURB using reaction beams and this method has two significant advantages. First, it is a self-contained system in that the BURB force reactions are taken by the reaction beams and, hence, the BURB forces do not affect the screws or the mill housings. Second, the stroke of the BURB cylinder is small (of the order of 1/2 to 1 inch as compared to 10 to 20 inches for the previously discussed designs). These two features are most desirable when using BURB for automatic gage control purposes. Figure 11-63 illustrates the use of BURB to decrease the crowns of the backup rolls and to decrease the load on the screws and housings. However, unless this design is used in foil mills or temper mills, which roll "below face", it is necessary to unload the bending cylinders prior to the workpiece leaving the roll bite. Figure 11-64 shows another type of self-contained BURB system where, by the use of additional bending bearings, the loads on the mill housings and main bearings are determined only by the rolling forces, being unaffected by the BURB forces. From a roll changing viewpoint, this method does not appear as attractive. In a somewhat similar system patented by Diolot [*] and illustrated in Figure 11-65, the bending moments are applied directly to the backup roll chocks.

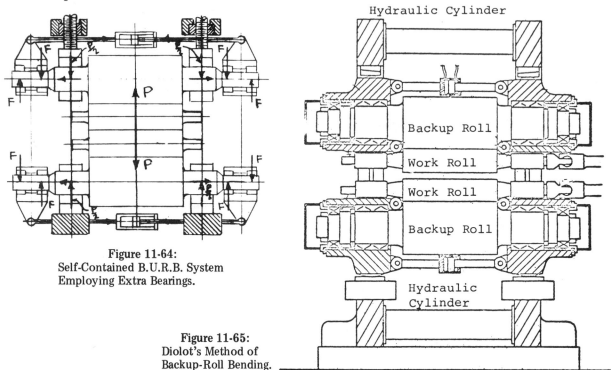

Figure 11-64:
Self-Contained B.U.R.B. System Employing Extra Bearings.

Figure 11-65:
Diolot's Method of Backup-Roll Bending.

Stone [x] has shown that the mill roll bending deflection due to rolling is

$$2 \times \frac{5P'W^4}{384EI_b}\left[\left(1 + \frac{24}{5}\frac{h}{W} + 2\left(\frac{D}{W}\right)^2\right) - 8\left(\frac{3}{5} + \frac{12}{5}\frac{h}{W} + \left(\frac{D}{W}\right)^2\right)\left(\frac{x}{W}\right)^2 + \frac{16}{5}\left(\frac{x}{W}\right)^4\right] \quad (11\text{-}51)$$

where

P' = specific rolling force (lb./in.)

W = width of the strip (in.)

I_b = moment of inertia of backup rolls $=\left(\frac{\pi}{64}D^4 (\text{in.}^4)\right)$

D = diameter of backup rolls (in.)

h = distance from the edge of the strip to the centerline of the backup-roll bearings (in.) and

x = distance from centerline of the mill to any point along the body from 0 to W/2 (in.)

The crown shape attainable by backup roll bending is

$$\frac{(Fa)W^2}{4EI_b}\left[1-4\left(\frac{x}{W}\right)^2\right] \quad (11\text{-}52)$$

---

[*] L. Diolot — "Rolling Mills", U. S. Patent No. 3,442,109, Issued May 6, 1969.

[x] M. D. Stone, "Theory and Practical Aspects in Crown Control", Iron and Steel Engineer Year Book 1965, pp. 657-674.

STRIP SHAPE: ITS MEASUREMENT AND CONTROL

where

F = backup-roll bending forces (lb.) and
a = moment arm of force F to centerline of the backup-roll bearing (in.)

The bending deflection curves under rolling conditions and under the influence of the proper external bending moments are very similar, the maximum deviation between the two being only of the order of 2 percent.

Letting $x = o$ in the above expressions, it may be seen that the total mill roll bending (due to rolling) is

$$\frac{5P'w^4}{192EI_b}\left[1 + \frac{24}{5}\left(\frac{h}{w}\right) + 2\left(\frac{D}{w}\right)^2\right] \tag{11-53}$$

and that due to the BURB system is $\frac{(Fa)w^2}{4EI_b}$.

With this magnitude of crown control possible with the BURB system, it is practical in many cases to provide all the desired crown by backup-roll bending only and therefore to use flat rolls. When this is the case, the axes of the backup rolls operate in the undeflected position, contrary to conventional backup rolls in 4-high mills, which necessarily rotate about a deflected axis. This not only eliminates all angularity in the main mill bearings (contributing to longer bearing life), but also eliminates the usual condition of microslippage along the line of contact between the backup and the work rolls, thereby affording longer roll life. Under such conditions, the roll-neck fillet stresses and, in fact, all roll-body stresses are much lower than for conventional 4-high mills. This results from the fact that the bending moments applied by the BURB system act to oppose the rolling pressure bending moments and, for all points between the roll-neck fillets, the resultant bending moments are greatly reduced.

To understand the use of the BURB system as a means of automatic gage control, reference should be made to the design of very stiff mills as discussed in Section 5-29. The no-load roll-gap S_o of such mills remains constant in spite of changes in the entry gage of the strip or changes in its resistance to deformation. The outgoing gage t_2 of a mill of this type is given by the equation

$$t_2 = S_o + \frac{P}{M_{mill}} - \frac{F}{M_{RB}} \tag{11-54}$$

where

P = rolling force (lbs.)
M_{mill} = modulus of the mill housing (lb./in.)
F = BURB forces
M_{RB} = modulus of the mill backup-roll bending mechanical system

Differentiating

$$\Delta t_2 = \Delta S_o + \frac{\Delta P}{M_{mill}} - \frac{\Delta F}{M_{RB}} \tag{11-55}$$

However, for an infinitely stiff mill $\Delta S_o = 0$ so that

$$\frac{\Delta P}{M_{mill}} - \frac{\Delta F}{M_{RB}} = 0 \tag{11-56}$$

constitutes a definitive condition whereby constant delivery gage is obtained. Thus, if the ratio of the change in the bending forces ΔF to the change in the rolling force ΔP is maintained constant and equal to the ratio of the moduli (M_{RB}/M_{mill}) then the BURB system will not only provide proper shape but an AGC system that will respond 10 to 15 times faster than those associated with the conventional mill screws.

11-26 A Comparison of the Effectiveness of Work-Roll and Backup-Roll Bending

From a computer study of the deflections of cold-mill rolls under a variety of operational conditions, Turley[a] reached some interesting conclusions relative to the merits of work and backup-roll bending. He verified the findings of Saxl[o] that the secondary deflection of the work roll (due to the fact that the backup roll contacts the work roll across its full face length and the workpiece contacts it over a lesser extent thereby creating bending moments in the work roll) has a greater effect on roll force distribution than the backup roll deflection. Moreover, this secondary deflection increases in importance as the strip width is reduced. Turley found that the camber produced under these circumstances is not generally parabolic.

The effect of camber of the backup rolls, whether by mechanical or thermal crowning or by bending, is attenuated at the strip under all cold rolling conditions. This is due to the combined effect of flattening between work rolls and backup rolls, and bending resistance of the work rolls. One interesting result of this is that a given diametral crown on the work rolls is generally much more effective than a radial crown of the same magnitude on the backup rolls.

Turley also established that excellent shape can be produced even when there is a nonuniform roll-force distribution provided that front tension is employed. Front tension is capable of giving a reduction of two orders in the magnitude of the shape error.

Back tension on the other hand has been shown to be detrimental to the production of good shape and has such a great effect that it can cause a shape reversal in a given pass. High back tensions (desirable, of course, for maximizing drafts) can only be tolerated if adequate front tensions are used to maintain the shape. In practice, a reasonable balance between front and back tensions is required for other reasons, so a condition of high back tension and no front tension seldom, if ever, occurs.

Because the length-to-diameter ratios are a function of face width for practical rolling-mill configurations, Turley established some guidelines for the use of roll-bending techniques. Work-roll bending he believed to be generally unsatisfactory for mills more than 100 in. wide as its effect becomes localized at the strip edges. Backup-roll bending is considered to be only satisfactory on mills more than 80 in. wide, as only these mills have high enough length to diameter ratios to give a deflection under the action of the backup-roll bending device which is sufficient to affect the strip shape significantly. However, this flexibility of the backup rolls restricts the capability of wide 4-high mills to produce strip to close gage tolerances, for it results in a very low mill modulus. For this reason, Turley states that the cluster mill will become of greater importance in the future in the rolling of very wide strip. At the present time, designs are available for a Sendzimir mill capable of rolling 150 in. wide strip at a roll separating force of 60,000 lb./in., with a mill modulus comparable with the narrower mills. These designs offer shape control devices, namely, "AS-U-ROLL" crown control to compensate primarily for housing deflection, and lateral adjustment of the specially crowned first intermediate rolls to compensate primarily for secondary deflection of the work rolls, and furthermore, the designs are available with a range of work-roll sizes (from 3.5 to 10 in. diameter) to give optimum performance for a wide variety of materials.

11-27 Flatness Control with a Flexible Work-Roll and Backup Bearings

A different approach to shape control during rolling was suggested by Ungerer[†] who conceived of a mill with a thin, relatively flexible work roll cradled between two parallel rows of staggered, hydraulically loaded rollers as illustrated in Figure 11-66. Such a device, Ungerer believed, would be limited in application to light reductions such as are encountered in temper rolling.

---

[a] J. W. Turley, "Extracts From Behavior of Rolls in Four-High Rolling Mills", Iron and Steel Engineer Year Book, 1973, pp. 430-434.

[o] K. Saxl, "Transverse Gauge Variations in Strip and Sheet Rolling", Proc. Instn. Mech. Eng. 172, 1958, No. 22, p. 727 ff.

[†] K. F. Ungerer, "Finishing Mill for Thin Sheet Metal", U. S. Patent No. 2,828,654.

STRIP SHAPE: ITS MEASUREMENT AND CONTROL

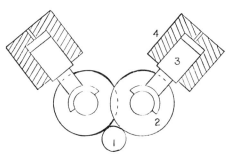

Figure 11-66:
Ungerer's Machine, 1: Flexible Work Roll;
2: Support Rollers; 3 and 4: Hydraulic Cylinder.

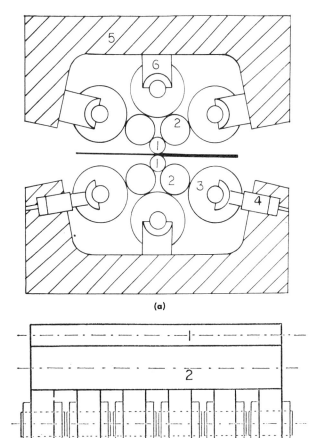

Figure 11-67:
Modified Flexible Roll Mill.

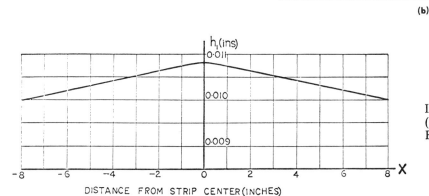

Figure 11-68:
Initial Strip Profile (Underside Assumed Flat).

Polakowski and his coworkers* have analyzed Ungerer's concept and have found it desirable to modify the design by the insertion of intermediate rolls between the work rolls and the backup bearings as shown in Figure 11-67. They assumed that the entering strip exhibited two levels of yield stress (40,000 and 80,000 psi), had a profile as indicated in Figure 11-68 and was as wide as

---

* N. H. Polakowski, D. M. Reddy and H. N. Schmeissing, "Principles of Self Control of Product Flatness in Strip Rolling Mills", Journal of Engineering for Industry, Paper No. 68-WA/Prod 18.

the face length of the 1-1/2 inch diameter work rolls. Using a computer, they calculated the optimum intermediate roll diameters that would give the most uniform percent reduction to the strip at all points across its width for three reductions, 15, 25 and 35 percent. In a similar manner, they calculated the optimum diameter for backup rolls used in a 4-high configuration with 1-1/2 inch diameter work rolls. Their data are shown in Table 11-6 from which it may be seen that the continuously supported mill provides a more uniform reduction across the width of the strip than the conventional 4-high mill.

Table 11-6

Comparison of Two Methods of Roll Support

Example No.	Roll Optimized for Pass Parameters				Continuously Supported Mill Diff. in % Redn. Between Center & Edge (%)	Conventional Mill Diff. in % Redn. Between Center & Edge (%)
	Yield Stress (psi)	Draft (percent)	Yield Stress (psi)	Draft (percent)		
1	40,000	25	40,000	15	0.4	+2.1
2	40,000	25	40,000	50	−0.80	−4.8
3	80,000	25	40,000	25	+0.82	+2.8
4	80,000	25	92,000	25	−0.32	−1.05

However, Polakowski et al., reported that strips with a width less than the face length of the rolls would pose complications, especially in connection with edge conditions.

11-28 Flatness Control by the Horizontal Deflection of a Work Roll

In a recently developed mill, known as the Taylor Mill,[+][*] rolled strip shape is controlled by the horizontal flexing of a slender work roll. Two configurations of this mill, a 5-high version and a 6-high unit, are illustrated in Figure 11-69.

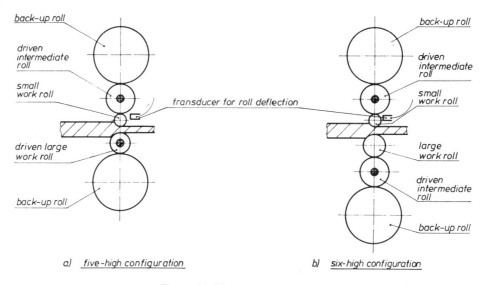

Figure 11-69: The Taylor Mill.

[+] K. Bosenberg and G. Hanel, "New Process for Rolling Thin and Flat Strip", Sheet Metal Industries, July 1973, pp. 397-399.

[*] Pamphlet 4-67 published by The Youngstown Research and Development Company, 1200 Slambaugh Building, Youngstown, Ohio 44501.

STRIP SHAPE: ITS MEASUREMENT AND CONTROL

In the Taylor Mill, the roll gap is formed by two work rolls of different diameter. Both work rolls are independently driven by intermediate rolls of larger diameter which are supported so that a five-high or six-high stand is formed.

The special feature of the Taylor Mill is that one of the work rolls is intentionally kept small in diameter in order to use its deflection for control. This roll is supported vertically by the intermediate roll and the backup roll which largely prevent a deflection in this direction. The roll, which can only deflect in the horizontal direction under the influence of the force components, is held at the roll necks in the chock bearings as shown in Figure 11-70.

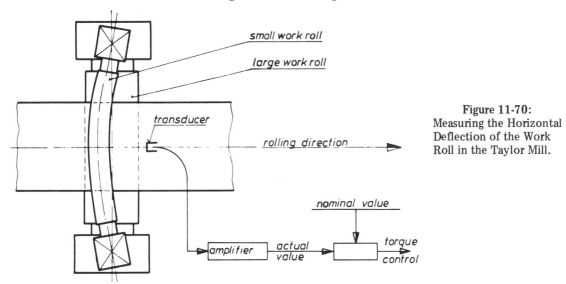

Figure 11-70: Measuring the Horizontal Deflection of the Work Roll in the Taylor Mill.

Deflection of the smaller work roll is measured by a non-contact transducer. As shown in Figure 11-70, the transducer is located in a device situated behind the top work roll halfway between the bearing chocks, a point at which the greatest deflection occurs. The measuring principle is based on the determination of the eddy-current losses which result between the two arms of an electrical bridge circuit and the work roll. When the work roll position changes, the electromagnetic field, and therefore the eddy-current losses, are altered. These losses are transformed into a voltage which is proportional to the spacing between the work roll and the transducer.

Horizontal deflection control is effected through the drive motor torques. By amplifying or reducing the torques, the acting forces, and through them the deflection of the work roll, can be altered. Since, in this case, the strip deformation work has to be performed at the same time, a condition may be achieved in which the sum of the torques remains constant:

$$\Sigma M = M_1 + M_2 = \text{const.} \tag{11-57}$$

where M_1 is the torque of the smaller work roll and M_2 is the torque of the larger work roll.

The transducer measures the deflection of the work roll and compares it with the assumed nominal value. If this nominal value is not reached, the torque M_1 is varied by means of the electrical control until the deflection of the top work roll corresponds to the assumed value. At the same time, the torque of the bottom work roll M_2 is altered in the opposite direction to the top work roll so that the sum of the two torques remains constant.

In contrast to certain cluster mills, the Taylor Mill is operated deliberately at low strip tensions so that the roll may deflect freely according to the torque. In addition, the risk of strip cracks is reduced and the operator is given the opportunity to examine the flatness of the strip.

The first commercial version of the Taylor Mill has been in operation in the U.S. for some time and mainly rolls low carbon steel with a final thickness of 0.10 mm for use in television aperture masks which place high demands on the flatness of the material. The six-high stand has the following technical characteristics:

Work roll 1	100 mm in diameter
Work roll 2	178 mm in diameter
Intermediate rolls	252 mm in diameter
Backup rolls	710 mm in diameter
Barrel length	710 mm
Rolling Speed	0-300-900 m/min.

Material 670 mm wide and up to 50 micrometer thick is rolled on this stand. Stainless steel 301 and 304 has been rolled also. During the last passes on the material, the stand is controlled by the operator in such a way that a slightly long edge develops and strip cracks are avoided. He is assisted in examining the strip flatness by a magnetic strip shape sensor.

11-29 Other Methods of Shape Control

Apart from the techniques for bending the work or backup rolls in the vertical or horizontal direction as discussed in Sections 11-24 to 11-27, other methods have been used to control the shape of the rolled strip. These are outlined below.

    a. Variable Rolling Lubricity

The effective crowns of mill rolls may be increased by reducing the rolling force necessary to attain a given reduction. This may be possible by improving the lubricity of the rolling lubricant. For example, in the case of a direct application lubrication system, if a blend of a mineral oil and a fatty material is used as the lubricant, decreasing the percentage of the mineral oil will usually decrease the effective coefficient of friction in the roll bite thereby providing a lesser rolling force, increasing the effectiveness of the roll crowns and alleviating the condition of over rolled edges. Conversely, for strip that is being rolled with a full center, the increase in rolling force necessary to eliminate this defect may be created by increasing the relative amount of mineral oil in the lubricant. This effect is illustrated in Figure 11-71 showing strips with full edges, good shape and full center rolled to the same gage from annealed black plate using different blends of lubricants.▲ It should be noted, however, that as the coefficient of friction provided by the lubricant is increased, the brightness of the strip is also increased as evidenced by the markedly darker-appearing, full-edged strip at the left of the figure.

Equipment suitable for lubricity control is illustrated in Figure 11-72. Here, a relatively good and a relatively poor lubricant are applied separately to the strip on the entry side of the roll bite through two header systems. Such a direct application system admits of separate control of the two types of lubricant and affords a convenient method of rolling force control. It is particularly useful as a means of compensating for the speed effect in cold rolling whereby the lubricity improves with increasing mill speed.✦ Under such conditions, as the mill speed is increased, the ratio of mineral oil to fatty materials applied to the strip is increased. This system, disclosed in a U. S. patent,△ may also be conveniently coupled to a shape monitor by a feedback control system. Alternately, it may be

---

▲ W. L. Roberts, "The Influence of the Rolling Lubricant on Sheet and Strip Quality", Tribology in Iron & Steel Works, ISI Publication 125, Iron and Steel Institute, 1970.

✦ R. J. Bentz and W. L. Roberts, "Speed Effects in the Second Cold Reduction of Steel Strip", Mechanical Working and Steel Processing V, Proceedings of the Ninth Mechanical Working and Steel Processing Conference, AIME, Gordon and Breach, New York, New York, pp. 193-222.

△ W. L. Roberts, U. S. Patent 3,150,458, "Method and Apparatus for Controlling the Thickness of Rolled Strip", issued September 29, 1964.

coupled to load cells monitoring the rolling force by a control system maintaining a constant rolling force.

Increasing Fullness of Center →

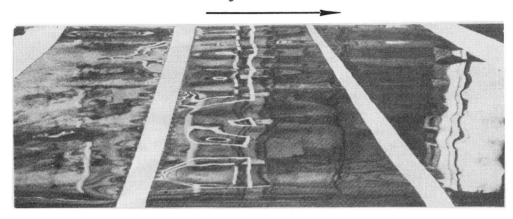

Specific Rolling Force
75,200 lb/in   64,600 lb/in   40,500 lb/in

Computed Coefficient of Friction
0.134   0.101   0.075

Figure 11-71: The Effect of Rolling Lubricity on Shape.

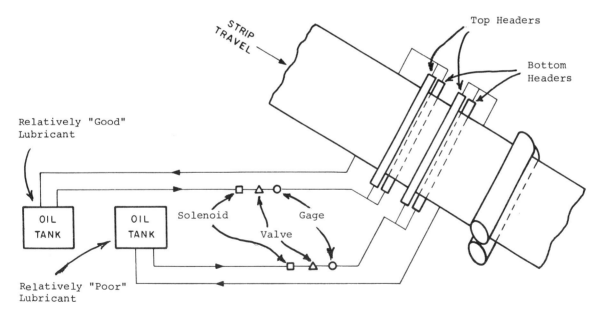

Figure 11-72: System for Rolling-Lubricity Control.

b. Changing the Strip Tensions

As has been shown in Chapter 9, the rolling force is, to a first approximation, proportional to the difference between the dynamic yield strength of the strip at the rolling strain rate and an "average" strip tensile stress occurring in the roll bite. At the strain rates usually encountered in cold rolling, low carbon steel strip exhibits a dynamic yield strength generally in the range of 60,000 to 150,000 psi depending on the prior work hardening experienced by the strip. Strip tensile stresses utilized in such reductions are usually in the range of 5,000 to 30,000 psi depending on the state of the strip and the features of the cold mill. An increase of 5,000 psi in both the entry and exit tensile stresses could therefore produce changes in the rolling force of at least a few percent and thereby reduce the edge strains (or the edge fullness). ⊞

The effect of tensional changes is, however, greater than would be expected by this first approximation. This is due to the fact that the rolling force is also dependent on the length of the arc of contact (which is assumed proportional to the square root of the deformed roll diameter and which is also a function of the rolling force) and the effective coefficient of friction.

Clearly, however, the extent of shape control by variation of strip tension is limited, inasmuch as there is only a finite range of strip tensile stresses permissible in the rolling operation. Tensions of too low a magnitude may introduce tracking difficulties with the attendant danger of the strip cobbling, whereas the use of excessive tensions involves the risk of strip breakage. Moreover, the use of tension variations to affect an independent control of the strip shape is not possible where the automatic adjustment of tension is used to control the outgoing gage of the strip as discussed in Section 5-30.

c. Localized Heating of the Mill Rolls

Where no rolling oils or coolants are employed in the rolling process, as for example, in temper rolling, backup roll crowns have been varied by playing open flames on the rolls or exposing them to radiant heating devices. On wet mills, where rolling lubricants and coolants are used, adjustment of the sprays to provide nonuniform cooling of the rolls is often undertaken by mill operators as a method of crown control.

These methods are slow, however, due to the large thermal inertia of the rolls (See Section 7-14) and, in the case of large mills, such as wide sheet mills, it may take as long as half an hour to create the desired changes in the crowns. Moreover, such methods do not lend themselves to precise crown control. Such objections also apply to other similar methods, such as the application of jets of warm or cold air and the generation of frictional heat in the rolls by motor driven buffing wheels pressed against the rolls.

d. Misalignment of the Work Rolls

Stone [×] has also discussed the attempted use of slightly crossed work rolls as a means of controlling shape. In 1932, a patent [○] was issued relating to this method of control, which was a consequence of the experience of sheet rollers of a few decades ago that, as the mill rolls lost their crowns as the rolling campaign drew to an end, extended use of the rolls could be satisfactorily made by driving wedges between the roll chocks and the housing windows at opposite corners.

It can be shown mathematically that, by crossing the work rolls or by crossing the backup rolls with respect to the work rolls, an equivalent increase in crown can be obtained. The theoretical crown shape that results from crossing a pair of rolls has been shown by Carrier[♦] to be equal to $4(\alpha^2 L^2)/(D_1 + D_2)$ where α is the angular displacement of each roll (radians), L is the body length of the rolls (in.), and D_1

⊞ H. W. O'Connor and A. S. Weinstein, "Shape Flatness in Thin Strip Rolling", ASME Paper No. 71-WA/Prod-13.

× M. D. Stone and R. Gray, "Theory and Practical Aspects in Crown Control", Iron and Steel Engineer Year Book, 1965, pp. 657-667.

○ A. T. Keller, U. S. Patent No. 1,860,931.

♦ G. F. Carrier, "The Use of Skewed Rolls in Calendering Operations", ASME Journal of Applied Mechanics, December 1950.

and D_2 are the roll diameters (in.). It is to be noted that this theoretical crown shape is parabolic.

Singer and O'Brien [1] found by experimentation with a small, 4-high laboratory mill that the actual crowns were less than one fourth of the theoretically calculated values. This was probably due to the flattening of the work rolls under load. Moreover, when the rolls are misaligned to the extent necessary to make significant changes in the effective crowns, large axial forces may be set up which can damage work-roll bearings, holding bolts and latches. As a consequence, this method of shape control has never achieved success in commercial metal-rolling operations.

e. Temper Rolling

Although the suppression of the yield point in annealed strip is the basic purpose of the temper rolling operation, the process can effect some shape correction. This correction is enhanced by increasing the number of stands in the temper mill and using high interstand tensions. Accordingly, temper rolling may be regarded as the most widely used, if not the most efficient shape correction process. ◊

11-30 Automatic Systems for Shape Control

As a consequence of the development of strip shape sensors, automatic shape control systems have been engineered. At the present time, these are being tested on various production facilities and it is likely that, within a decade, some of them may find widespread use in cold reduction operations.

Systems with a relatively limited resolution that can, in effect, differentiate between full center and full edges generally utilize work-roll bending. For example, the optical monitoring unit developed by Santo et al.× (Section 11-18) was connected to the hydraulic roll-bending cylinders as illustrated in Figure 11-30. Under automatic control, the shape of strip rolled on a skin pass mill (as measured by "inclination" or the ratio of peak-to-peak amplitude of the waves divided by their wave length) could be maintained under about 0.45%. Moreover, the system permitted the strip to be finished in any shape between a center fullness and edge waves.

In a similar manner, the two-roll sensor developed by Kajiwara and his co-workers ❅ (Section 11-19) was utilized to adjust the work-roll bending system (as well as the hydraulic screwdowns) on the fifth stand of a five-stand cold mill. This arrangement was reported to provide "almost satisfactory control" except in the vicinity of strip welds.

The Stressometer described in Section 11-20 has also been utilized to bend the work rolls of a four-high mill. This equipment, designated the ALCAN-ASEA Automatic Roll Deflection Control (ARDC) is shown in Figure 11-73.

The ARDC system is designed to control the force between two jack systems, the hydraulic jacks located between the back-up and work roll chocks (contour jacks) and those located between the two work-roll chocks (balance jacks). The pressure in each jack system is measured by a transducer and the difference constitutes the feedback in a closed-loop pressure-control system. References for the pressure difference are:

---

[1] A. R. E. Singer and J. J. O'Brien, "Control of the Shape of Metal Strip on an Experimental Variable Camber Rolling Mill", J. Iron and Steel Institute, December 1962, pp. 1003-1010.

◊ T. Sheppard and J. M. Roberts, "Shape Control and Correction in Strip and Sheet", International Metallurgical Reviews, Vol. 18, 1973, pp. 1-18.

× T. Santo, T. Sakaki, T. Matsushima, H. Shibata, T. Emori and M. Arimuza, "Automatic Shape Control", Proceedings ICSTIS Suppl. Trans. ISIJ, Vol. II, 1971, pp. 698-702.

❅ T. Kawamata, M. Hotta and T. Kajiwara, "Automatic Shape Control in Cold Rolling", Proceedings ICSTIS, Suppl. Trans. ISIJ, Vol. II, 1971, pp. 702-704.

1. a manually-controlled potentiometer (Shape) which allows continuous adjustment of the roll bending forces between maximum positive and negative,
2. total roll-separating force corrected by strip width to maintain the level of flatness irrespective of any changes in roll-separating force (Load Compensation),
3. correction signal from STRESSOMETER to control crown changes during rolling.

The main components of this equipment are as shown in Figure 11-73 and they make use of electronic and hydraulic devices to achieve the flexibility and high force levels required.

When used with a Sendzimir mill, the Stressometer effects crown control by the automatic adjustment of the eccentrics in the saddles of the two upper central back-up rolls by means of a hydraulic drive which is on-off controlled by solenoid valves actuated by signals from the crown control system.

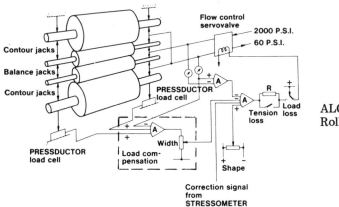

Figure 11-73:
ALCAN-ASEA Automatic Roll Deflection Control.

Figure 11-74 shows that each crown is controlled by means of a superimposed closed-loop position-control system. Reference for each crown position is one potentiometer (Shape) which allows continuous adjustment of the basic roll camber.

The use of a Stressometer to automatically adjust the thermal crowning of the mill rolls by selective control of the roll cooling sprays is illustrated in Figure 11-75. This mode of operation is designated the ALCAN-ASEA Automatic Thermal Crown Control (ATCC), and its use permits localized changes in the thermal crowns of the rolls to be made automatically.

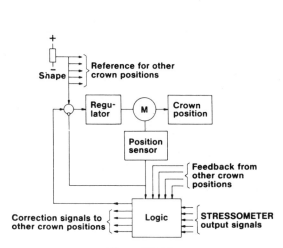

Figure 11-74:
Use of Stressometer to Control Shape in Sendzimir Mill Operation.

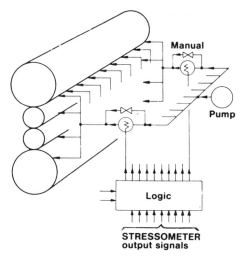

Figure 11-75:
The ALCAN-ASEA Automatic Thermal Crown Control (ATCC).

11-31 Obtaining Flatness After Rolling

In spite of the constant attempt to obtain satisfactory flatness in commercial rolling operations, considerable quantities of sheet and strip are rolled with improper shape. To convert such material to saleable product, it is usually flattened or levelled in subsequent operations as described more fully in the following sections of this chapter.

In the case of blackplate used for the manufacture of tinplate, the annealing of the strip under tension may result in a considerable improvement in shape. For example, strip with a full center entering a continuous annealing (CA) line will usually emerge with a very acceptable degree of flatness. However, the shape of the strip entering the furnace should be such that tracking difficulties do not occur on the line. Even if the strip is not as flat as might be desired from the CA line, its processing on the temper mill will usually enhance its shape to commercial quality.

Other cold rolled products must be flattened or levelled by rerolling or on processing lines specifically designed for the purpose. On such lines, the strip is subject to high tensile stresses, which produce plastic strains in the material and flatness is achieved on account of the nonuniform yielding of the strip.

A number of different machines have been devised for the levelling of strip. All of these develop the necessary deformation stresses with bending, tension or a combination of tension and bending. Simple roller levelling, as discussed more fully in Section 11-32 utilizes essentially simple bending stresses, whereas stretcher levelling described in Section 11-33 is achieved by the use of two sets of bridle rolls applying tension to the strip. In stretch-bend levellers a number of strip deflector rolls are located between the bridles so as to introduce a combination of bending and tensile stresses into the strip.

11-32 Roller Levelling

When strip is bent around a roll as illustrated in Figure 11-76, compressive stresses are established in layers of the strip surface adjacent to the roll and tensile stresses on the opposite surface. So long as the roll-diameter-to-strip-thickness (D/t) ratio is sufficiently large, these stresses produce only elastic strains which disappear when the bent portion of the strip is moved past the roll. However, as the roll diameter to strip thickness ratio is decreased, a point is reached where the stresses are such that plastic deformation occurs and a "coil set" will be given to the strip. The minimum ratio of roll diameter to strip thickness that can be used without plastic deformation is given by

$$D/t = E/\sigma_{YS} \qquad (11\text{-}58)$$

where E is Young's modulus and σ_{YS} is the yield strength of the strip material. Since for annealed low carbon steel $D \simeq 3 \times 10^7$ psi/in./in. and σ_{YS} may be as low as 30,000 psi, the D/t ratio should not be less than 1000:1. Obviously, rolls that are to be used for roller levelling must have D/t ratios appreciably less than this value and may range from a value of about 12 in the case of a cold plate leveller to about 230 in the case of a cold strip leveller. ‖

A roller levelling machine basically consists of a number of rolls nested as shown in Figure 11-77. One row has an even number of rolls and the other an odd number, the total number usually ranging from 11 to 27. These rolls are usually all driven by individual spindles from the same drive system, the peripheral speed of the rolls being normally less than 1,000 feet per minute. It is the general practice also to utilize rows of backup rolls as shown in Figure 11-78 to minimize the bowing of the rolls under load. Lead-bronze or phosphor-bronze sleeve bearings are commonly employed for the work-roll bearings with roller bearings for the back-up rolls.

‖ R. Jamieson, "Maintenance of Levellers for Plate and Sheet", J. Iron & Steel Institute, November 1965, pp. 1129-1132.

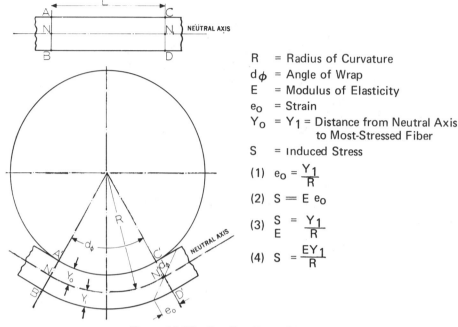

Figure 11-76: Bending Stress Diagram.

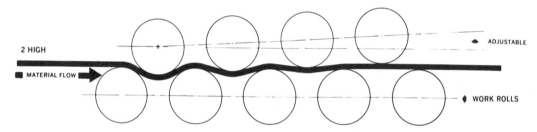

Figure 11-77: Staggered Arrangement of Rolls For Flattening Sheet Metal.

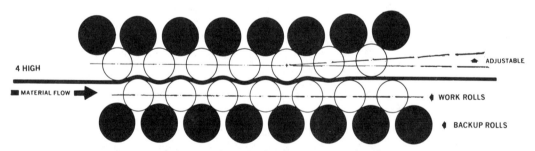

Figure 11-78: Arrangement of Back-Up Rolls in Leveller.

In practice, the center-to-center spacing of one set of work rolls is about 5 to 11 percent larger than the roll diameter.[•] This close spacing limits the size of the bearings and, hence, the speed at which these machines will continuously operate.

The total driving power P_L required for a roller leveller may be developed from the formula

$$P_L = \left[M_R + \Sigma P \left(f + \frac{\mu d}{2}\right)\right] \frac{2V}{D} \times \frac{1}{75\eta} \text{ (horsepower)} \qquad (11\text{-}59)$$

[•] A. I. Tselikov and V. V. Smirnov, "Rolling Mills", Translated from the Russian by M. H. T. Alford — edited by W. J. McG. Tegart, Pergamon Press, London, 1965.

where

M_R is the straightening moment applied to the rollers which is used for the plastic deformation of the metal (kg-cms)

ΣP is the total force on the rollers (Kg)

f is the coefficient of rolling friction (m) (In the case of flat products, $f \simeq 0.8$ mm)

μ is the coefficient of friction in the bearings of the roller (for roller bearings $\mu \simeq 0.005$, for plain bearings $\mu = 0.05 - 0.07$)

d is the diameter of the journal or bearing (m)

D is the diameter of the barrel of the rolls (m)

V is the speed of straightening (m/sec), and

η is the efficiency of the transmission

The straightening moment M_R may be derived from the equation

$$M_R = D/2 \times 1.5 \times \sigma_{ys} W\left(\frac{1}{r_o} + \frac{1}{r_2} + 2\sum_3^{n-2} \frac{1}{r_i}\right) \tag{11-60}$$

where

W is the moment of resistance with elastic bending

$1/r_o$ is the initial curvature of the workpiece

$1/r_2$ is the curvature at the second roll

$1/r_i$ is the curvature at the ith roll, and

σ_{ys} is the yield strength of the workpiece

The various designs of roller levellers are manifold. In some machines, each top roll is independently adjustable with respect to its position, whereas in other machines, the rolls are adjustable in groups. However, to achieve flattening the axes of the upper and lower rolls are closer together at the feed or insertion end of the machine than at the side of emergence. When the sheet is first introduced, therefore, it is thus more severely flexed than when it leaves the machine; only the last few rolls actually straightening it.

Roller-levelling can be effective as a tempering process; indeed, before the introduction of temper rolling, it was a common method of removing the yield-point elongation. The elongation obtained is limited because after 4-5% elongation, fluting occurs which spoils the appearance of the product. However, elongations of this magnitude are not easily obtained and roller-levelling is still used to eliminate fluting tendencies in strip.

In brief, roller-levelling is a well established process which can successfully flatten plate and strip down to thicknesses close to 0.5 mm (0.020 in.).◊

11-33 Continuous Stretcher Levelling

Developed in 1960, the continuous stretcher levelling process flattens the strip as it is stressed between two bridles driven at speeds differing by the extent of the elongation desired in the strip.• The entire cross section of the strip is stressed beyond its elastic limit and, after processing, the strip is essentially free of residual stresses and shows less directionality and greater ductility than its rolled counterpart.

Stretcher levellers usually have the capability of extending the strip up to three percent. This is accomplished by the use of a differential gear box connecting the inputs of the entry and

---

◊ T. Sheppard and J. M. Roberts, "Shape Control and Correction in Strip and Sheet", International Metallurgical Reviews, Vol. 18, 1973, pp. 1-18.

• R. J. Bell and G. R. Vassily, "The Continuous Strip Stretcher Levelling Process", Iron & Steel Engineer Year Book, 1967, pp. 355-362.

exit stretch bridles. Generally, however, provision is made for speed differentials up to seven percent to be attained thereby allowing for the mismatching of roll diameters resulting from the grinding of worn bridle rolls.

The driving motor is usually selected for rating by assuming that the power required is about one-fifth of that obtained by multiplying the maximum strip tension (rating of the machine) and its corresponding line speed. For example, a line with a stretch force capability of 50,000 pounds at 300 fpm would be powered by a 100-hp motor.

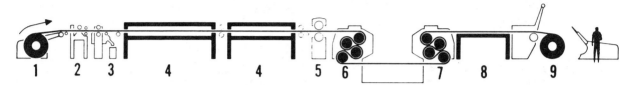

Figure 11-79: Schematic Diagram of Typical Stretcher Leveller Line.

1. Payoff Reel
2. Strip Splicer
3. Edge Trimmer
4. Cleaning and Rinsing Sections
5. Embossing Unit
6.&7. Entry and Exit Stretch Bridles
8. Inspection Table
9. Rewind Station

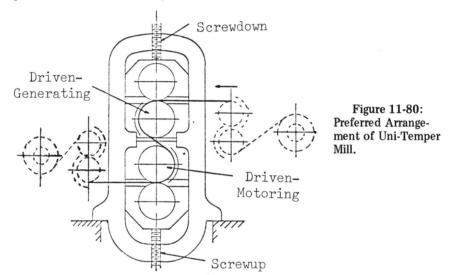

Figure 11-80: Preferred Arrangement of Uni-Temper Mill.

A schematic diagram of a stretcher-leveller line is shown in Figure 11-79. The payoff stand is usually equipped with a drag-generator and edge detection equipment that will provide for proper tracking of the strip. This unit is followed by a shear and strip joining unit (welder or stitcher) from which the strip passes to an edge trimmer and side scrap choppers. If necessary, strip cleaning and embossing units may be installed at the next positions prior to the flattening of the strip by the tension bridles. For examining the strip after welding, an inspection table may be provided. Finally, the rewind station may include an exit shear, a strip coiler and a belt wrapper.

Bell and Vassily[*] have reported other recent design trends in stretch levellers. The use of polyurethane as a lagging material for the stretch bridle rolls has been found to be superior to rubber in extending roll life. The older practice of heating the strip to temperatures in the range 200 to 300 F has been abandoned, but better instrumentation has been provided to measure both the elongation of, and the stretch force applied to, the strip.

An earlier levelling device that found successful application was the "Uni-temper Mill" developed by Stone.[✦] This is essentially two 2-high mills located in the same housing between

---

[*] R. J. Bell and G. R. Vassily, "The Continuous Strip Stretcher Levelling Process", Iron & Steel Engineer Year Book, 1967, pp. 355-362.

[✦] M. D. Stone, "The Uni-Temper Mill and Process", Iron & Steel Engineer Year Book, 1945, pp. 12-23.

STRIP SHAPE: ITS MEASUREMENT AND CONTROL

which the strip is stretched as shown in Figure 11-80. The two-high mill at the top provides a cold reduction of about a quarter of a percent and the two-high mill at the bottom takes another one percent after the stretching operation. The purpose of the first slight reduction is to prevent the formation of Lueder's lines and that of the second is to restore the bright surface which is slightly dulled by the stretching operation.❖ Provision is made for stretch elongations up to 12 percent that may be controlled to within 1/10 of one percent. The top two rolls are powered by a double armature 250 hp drag generator and the lower two rolls by a 1500 hp motor.

11-34 Tension Levelling

A tension levelling line differs from a stretcher leveller in that it incorporates a leveller roll assembly between the two sets of tension bridles as illustrated in Figure 11-81. The leveller assembly is similar to those used in the roller levellers, except that in this case the rolls are undriven and forces may be applied at various locations along the lengths of certain of these rolls to provide far improved shape correction. For example, where the entering strip has full edges, center pressure is applied; for material with a full center, edge pressure is used.

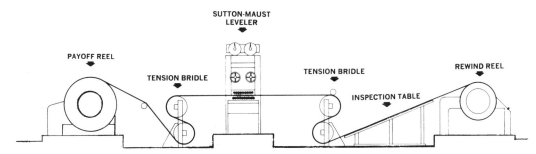

Figure 11-81: Tension Levelling Line.

According to Bland and Alters,[B] approximately 85 percent of the strip thickness should be stressed to the yield point to obtain good levelling results. Knowing the portion of the strip not required to be stressed to the yield point, the theoretical roll diameter D can be established from the equation

$$D = \frac{0.15 \, t \, E}{S_{ys}} \tag{11-61}$$

where t is the strip thickness, E is Youngs' Modulus for the strip material and S_{ys} is the yield strength of the strip. However, for economic reasons, a range of workpiece thicknesses must be processed using rolls of a given diameter. The work-roll diameter selected must be small enough to effectively work the thin material and yet large enough to permit the use of back-up roll bearings of sufficient capacity to withstand the load produced by deflecting the thicker strip. To provide ample back-up roll bearing capacity in the leveller, it normally becomes necessary to use rolls of larger diameter than that determined by the theoretical calculations for the minimum strip thickness. Under these circumstances, the stress S_I induced by bending is given by

$$S_I = 0.15 \, tE/D \tag{11-62}$$

and the tensile stress S_T that must supplement the bending stress is given by

$$S_T = S_{ys} - S_I = S_{ys} - 0.15 \, tE/D \tag{11-63}$$

❖ M. D. Stone, discussion after paper by R. J. Bell and G. R. Vassily, "The Continuous Strip Stretcher Levelling Process", Iron & Steel Engineer Year Book, 1967, pp. 355-362.

[B] R. A. Bland and M. F. Alters, "Tension Levelling of Ferrous and Non-Ferrous Strip", Iron and Steel Engineer Year Book, 1967, pp. 613-623.

Tension levellers are of several basic designs. A two-row configuration of rolls is known as a "flattener" and when support rolls are used they are known as backed-up levellers. When adjustments may be made to the back-up rolls and closer tolerances are used in their design and manufacture, the term "precision, backed-up leveller" is used. When full length intermediate rolls are added, the unit becomes a "6-high leveller".

Machines are available in both the original 17-roll double-tilt design, the modified 14-roll exit wing-tilt, four-high design (See Figure 11-82) with either symmetrical or staggered backup roll arrangement, and many other designs.

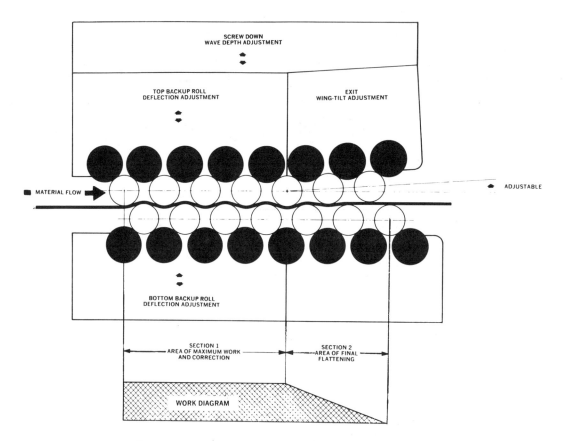

Figure 11-82: 14-Roll, Exit Wing-Tilt Four-High Design.

The effectiveness of a tension leveller in correcting the shape of thin steel strip has been demonstrated by Kusakabe and Hirasawa.[b] They showed it to be capable of removing defects impossible to correct by a conventional leveller and to provide the same elongation as a stretcher leveller but with only 1/3 to 1/4 of the tensile stress.

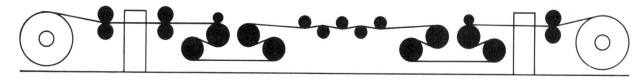

Figure 11-83: Diagram of Herr Continuous Tension Levelling System.

---

[b] T. Kusakabe and T. Hirasawa, "Shape Improvement of Thin Steel Strip by Using the Roller Stretcher", 33 Magazine, December 1967, pp. 93-106.

11-35 Other Levelling Devices

A number of other levelling machines have been developed in recent years which are presently being used on an experimental basis in commercial operations.

The Herr leveller,[1] shown in Figure 11-83, is somewhat analogous to the conventional tension leveller except that (1) the bridles provide an accurately controlled extension to the strip, (2) each bridle roll can have its own electrical power unit, thus permitting the proper proportioning of torque distribution in the bridle, (3) the rolls used to deflect the strip are larger than normal, (e.g., nonbacked-up 6-inch diameter rolls may be used to process aluminum 0.010 to 0.125 inch thick) and, hence, may use large bearings and thus be rotated at much higher speeds, and (4) the strip is elongated between the deflector rolls and not as it is bent around them.

Polakowski[2] and his coworkers[3] have investigated the capabilities of a tension levelling machine design (known as the Floatrol Tension Levelling System) utilizing very small diameter work rolls as illustrated in Figure 11-84. As shown in the drawing, the work roll A is kept in position solely by the taut strip which traps it. Otherwise, it is free to perform small movements along the direction of strip travel. Owing to the separating forces exerted by the tension components normal to the pass line, these rollers must be supported against deflection by the use of outer rows of roll bearings E with thick outer races like those of a Sendzimir mill.

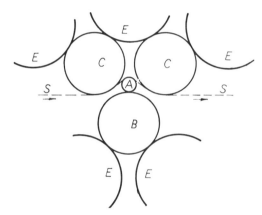

A—Work Rolls, B—Support Rolls, C—Deflector Rolls, E—Backup Roller Bearings, S—Strip.

Figure 11-84:
Roll Arrangement in the "Floatrol" Tension Levelling System.

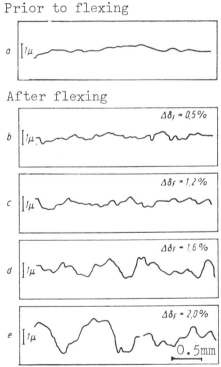

Figure 11-85:
The Change in Strip Surface Roughness with Increasing Elongation Achieved by Flexing. (Panknin et al.)

11-36 The Influence of Flattening Techniques on the Surface Condition and Width of the Strip

Where compressive deforming stresses are induced into the strip to improve its flatness, as in temper rolling, the surface finish of the strip after rolling bears a close correlation with the surface finish of the rolls. Under circumstances where the strip is flattened by tension and/or

---

[1] "Continuous Tension Levelling of Metal Strip", Sheet Metal Industries, May 1969, pp. 397-399.

[2] N. H. Polakowski, "The 'Floatrol' Tension Levelling System — Present Status and Perspectives", Journal of Metals, May 1969, pp. 36-40.

[3] J. J. Buchinski, P. Dion and N. H. Polakowski, "Technological Principles of Floating Work Roll Tension Levellers for Metal Strip", J. Iron & Steel Institute, December 1965, pp. 1244-1249.

flexing, the surface condition of the strip may be changed.[∞] This effect has been confirmed by Panknin et al.[∧] from the results of flexing trials carried out on steel strip. Figure 11-85 shows the longitudinal surface roughness after various elongations achieved by flexing the strip.

It will be noted that the greater the elongation, the greater the roughness of the strip. Longitudinal and transverse surface roughnesses corresponding to a 2 percent elongation are compared in Figure 11-86.

This change in surface condition is created by the fact that when materials with a defined yield point are bent, closely-spaced kinks are produced transversely to the bending direction. Although these are subsequently flattened during levelling, they nevertheless create a deterioration in surface finish regardless of the strip thickness.

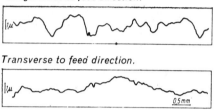

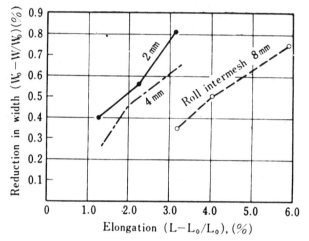

Figure 11-86:
Longitudinal and Transverse Surface Roughnesses Corresponding to an Elongation of 2%. (Panknin et al.)

Figure 11-87:
Relation Between Elongation and Reduction in Width. (Kusakabe and Hirasawa)

The influence of flexing on strip surface can limit the application of this technique for the elimination of the yield point because changes in surface structure are generally undesirable. In accordance with Figure 11-85, changes in surface profile when creating elongations up to approximately one percent are still slight so that flexing can perhaps be used where only slight reductions are necessary to suppress the yield point. Thus, flexing does not represent an alternative to temper rolling annealed strip. However, it is probably an efficient method for eliminating the slight aging effect which occurs after the temper rolling of strips of unkilled steels if these are stored for any length of time prior to further conversion.

As would be expected, the use of large tensile stresses in flattening operations may result in a slight decrease in the width of the strip. Using a tension leveller, Kusakabe and Hirasawa [♭] found that the relationship between the reduction in width to elongation depends on the degree of intermesh of the rolls, the larger the intermesh, the less the reduction in width as indicated in Figure 11-87.

[∞] M. Verduzco and N. H. Polakowski, "Control of Lüders Markings on Mild Steel Strip by Roller-Flexing Under Tension", J. Iron and Steel Institute, 204, 1966, pp. 1027-1033.

[∧] W. Panknin, H. Thiele and W. Zeigler, "Research into the Levelling, Straightening and Flexing of Coiled Strip Material and Its Effect on Surface Finish", Sheet Metal Industries, October 1973, pp. 578-586.

[♭] T. Kusakabe and T. Hirasawa, "Shape Improvement of Thin Steel Strip by Using the Roller Stretcher", 33 Magazine, December 1967, pp. 93-106.

Chapter 12   The Rolled Strip —
Its Properties and Further Processing

12-1 The Effect of Cold Rolling on the Properties and Subsequent Processing of the Strip

Although cold rolling is used primarily to obtain the desired physical dimensions of the workpiece, the process cannot achieve that purpose without affecting the physical condition of the surfaces and the metallurgical condition of the metal undergoing deformation. Thus, in rolling strip, its surfaces, usually covered with a residue of the rolling solution, are dirty and sometimes contain defects directly attributable to processing on the cold mill. Of even greater importance is the work hardening or the increased hardness and the loss of ductility that occurs as a consequence of the deformation.

Inasmuch as the further processing or utilization of sheet or strip, such as electrotinning, demands real cleanliness of its surfaces, roll oil residues must be completely removed from the surfaces. At the same time, heat treatments are necessary to restore the desired metallurgical properties to the strip. As will be seen, however, the kinetics of these heat treating processes are affected by the reduction given to the strip and, even after such treatments, the effects of cold reduction are still apparent in the anisotropic characteristics of the strip and its texture.

In this chapter, the reader is briefly introduced to those manufacturing steps associated with the processing of sheet and strip subsequent to the primary cold rolling of the hot band. These include cleaning, heat treating, temper rolling, secondary cold rolling, foil rolling, shearing, side trimming and slitting, metallic and organic coating, embossing, roll forming and oiling. The metallurgical aspects of some of these processes are discussed and sections are included in the chapter briefly describing the processing of electrical and stainless steels.

12-2 Rolling Lubricant Residues and Their Removal

Although rolling lubricant residues are useful in providing the rolled strip with a certain degree of corrosion resistance, such residues must be completely removed prior to, or during the further processing of the strip. Generally speaking, cleaning of the strip is accomplished by one of the following methods.

1. The application of a detergent at the last stand of the cold reduction mill;
2. electrolytic cleaning of the strip either on a separate line or in tandem with another processing line, such as a continuous annealing line;
3. the vapor degreasing or chemical cleaning of the strip on a separate line; and
4. the burnoff of the lubricant in a box annealing furnace.

A number of proprietary brand detergents are commercially available for mill application and they are generally used in the form of aqueous solutions. Headers and sprays are designed to provide high-pressure impingement of the detergent solution on the strip and rolls and, as a consequence of this treatment, the oil residue on the surface of the strip is considerably reduced.

A typical electrolytic cleaning line for tinplate products is shown in Figure 12-1. It utilizes alkaline detergent solutions, such as caustic soda, sodium orthosilicate and trisodium phosphate, but sodium metasilicate and sesquisilicate may also be used to clean certain types of product. The electrolytic action, while not universally employed, is believed to have merit, and the total electrical power utilization is in the range 8 to 10 KW-hr. per ton. Although the operating practices for such cleaning lines vary considerably, the strip is usually made cathodic and current densities of the order of 60 to 75 amps per square foot are used.⊕ After emerging from the

---

⊕ W. E. Hoare and E. S. Hedges, "Tinplate", Edward Arnold & Co., London.

electrolytic cell, the strip, in some cases, is dipped into a short rinse tank and then passed through water sprays and revolving brushes to remove the last traces of alkali and smut. A dip in a hot water rinse table and the hot air drier enables the strip to dry rapidly in air before reaching the pinch rolls and takeup reel.

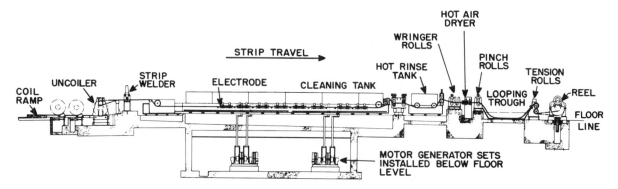

Figure 12-1: Schematic arrangement of a typical electrolytic cleaning line for processing cold reduced steel strip.

It is important to note, however, that where the strip is to be further cold reduced, no cleaner residues should be left on the strip surface. Residual films of materials such as silicates do not volatilize from the strip surface during annealing and the resultant contamination may adversely affect the behavior of the cold rolling lubricants in secondary cold rolling operations.

Cleaning lines utilizing tanks of organic solvents may also be used to clean the strip by immersion or by vapor degreasing. In such cases, chlorinated hydrocarbons, such as trichlorethylene, may be used as the cleaner.

In the case of sheet products, the roll oil is usually removed by "burnoff" in the box annealing furnaces (discussed more fully in Section 12-8). Ideally, the residual oil is completely vaporized from the strip surface. However, if the oil residues contain iron fines and other sludge-like materials, the oil will not "burn off" cleanly but will leave objectionable deposits on the strip known as "smut", "annealed, stained edge", "snakey edge", or "carbon edge".♦ It is easily recognizable as a wavy band, varying in degrees of darkness, found near the edge of the strip or sheets after annealing. It is characterized by a black deposit with a sharp outer limit gradually diffusing into the steel background as it penetrates towards the center of the sheet or strip. The extreme edge of the steel so affected is always very clean and bright. The sharp transition from bright to black may occur at any distance up to approximately 1/2-inch from the coil edge while the width of the black region may vary from a pencil line thickness to 2 inches or more. As stated by Strefford and McCallum,♦ this mark is lyophobic and, hence, it is difficult to wet with paints, lithographic materials or liquid metals. These investigators found that snakey-edge arises during annealing due to the catalytic breakdown of carbon-containing gases which can arise from the oil carryover. They also found that the major practical factor influencing snakey edge was the extent of the reduction given to the strip on the cold mill. (See Figure 12-2.) From these data they inferred (1) that as the amount of cold reduction increases, the amount of surface energy stored in fine particles and available for gas catalysis increases, (2) that oil constitution can affect the incidence of snakey edge by changing the frictional characteristics during rolling and (3) that a rolling oil with superior load-bearing characteristics would minimize the prevalence of snakey edge.

Lillie & Levinson§ found that carbon can be deposited during annealing at certain temperatures if the protective atmosphere has an unfavorable CO/CO_2 ratio.

♦ R. Strefford and N. McCallum, "Influence of Cold Mill Processing Conditions on the Formation of Snakey-Edge on Cold Reduced Steel", Tribology in Iron and Steel Works, ISI Publication 125, Iron and Steel Institute, 1970, pp. 313-316.

§ C. R. Lillie and D. W. Levinson, "Surface Staining of Cold Rolled Sheet, Part I: Types of Defects and Their Occurrence", Iron and Steel Engineer Year Book, 1957, pp. 377-389.

PROPERTIES OF THE ROLLED STRIP

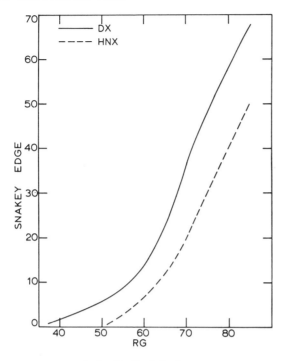

Figure 12-2:
Relationship Between Snakey-Edge Index and RG Factor in DX and HNX Gases
(After Strefford and McCallum.)

(RG represents % cold reduction for low carbon rimming steel, % cold reduction +10 for high carbon rimming steel and % cold reduction −10 for stabilized steel.)

12-3 The Microstructure of the Rolled Strip

In the cold-rolling process, the original grains of the workpiece become highly distorted and elongated as exemplified by the microstructure of full-hard black plate (85 percent reduction) illustrated in Figure 12-3.± The elongation of the ferrite and pearlite grains and the generally distorted microstructure are characteristic, the elongation of the grains coinciding with the elongation of the strip.

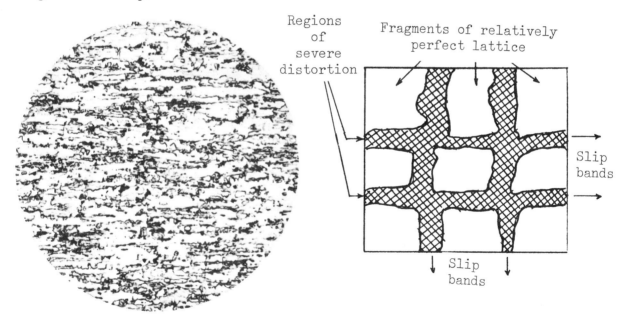

Figure 12-3:
Microstructure of full-hard cold-reduced black plate (85 percent reduction). Nital etch; magnification 200X.

Figure 12-4:
The structure of severely cold-worked metal.

± The Making, Shaping and Treating of Steel, Edited by H. E. McGannon, Eighth Edition, U. S. Steel, Pittsburgh, Pa., 1964.

Although it might be thought that all elements of the strip would experience similar distortions, such is not the case. From X-ray diffraction studies, it has long been known that very small elements of a workpiece survive the severest deformation without appreciable change. These elements, with dimensions of a few microns, constitute a "mosaic" and are surrounded by intense dislocation networks as illustrated in Figure 12-4.

12-4 The Rolled Crystallographic Texture

In the deformation of bcc lattices, it is possible for slip to occur in four directions on each of 12 planes. This leads to the formation of considerable dislocation tangles and the formation of a mosaic structure as discussed in the preceding section. Observations with the electron microscope have shown that, as rolling proceeds, the cells of the mosaic tend to elongate in the

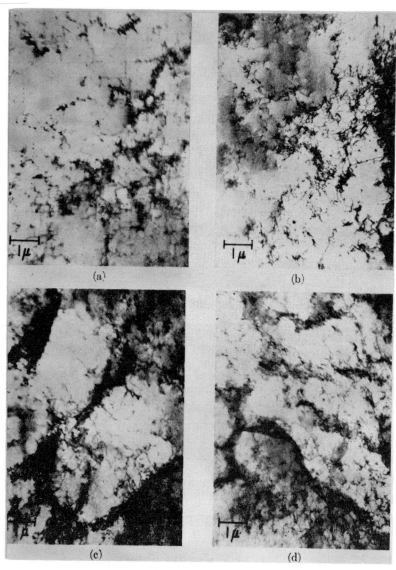

Figure 12-5:
Dislocation arrangements in iron deformed at room temperature. (a) 5%, (b) 9%, (c) 16%, and (d) 30%. (After A. S. Keh.)

PROPERTIES OF THE ROLLED STRIP

rolling direction and the cell walls become thicker as deformation continues▲*♦ (Figure 12-5). When the metal is severely deformed (50-90% reduction), the dislocation density is of the order of 10^{10} to 10^{12}/sq. cm. and the minor axes of the cells have a dimension of about 1 micron or less.

Texture develops in a way that most easily accommodates slip □ and the texture becomes more sharply defined as deformation is increased. Transmission pole figure data indicate that the main component of the deformation is the {100}<110>(see Figure 12-26) with a spread about this orientation that includes the {112 <110>, {111}<110> and, in some cases, the {111} <112> *

12-5 Work Hardening of the Strip

Due to the cold working of the steel, certain changes in its physical properties occur which constitute the phenomenon of work- or strain-hardening. (See Section 8-32.) These include

(a) increased yield and ultimate tensile strengths
(b) increased hardness, and
(c) decreased ductility

and result from the interaction of the high density dislocations created during deformation. For low-carbon sheet and strip, the variation of yield strength is shown in Figures 8-110 and 8-111, of tensile strength in Figure 12-6, and ductility in Figure 12-7.

Because of these changes in the physical properties, heavily reduced steel strip finds little commercial application. For this reason, heat treatments are usually given to the rolled product to restore its ductility and formability. However, strip given a 30 to 40 percent reduction after annealing is used in tinplate manufacture (double-reduced product) where the high strength is attained at the cost of some loss of ductility.

In Section 8-32, the mathematical relationship between the yield stress σ and the true strain is shown to be

$$\sigma = k \delta^n \qquad (12\text{-}1)$$

where k is the strength coefficient and n is the strain-hardening exponent or coefficient. The latter is important with respect to formability (particularly stretching operations) where a relatively high value (in the range of 0.28 to 0.40) is desirable inasmuch as it tends to promote uniformity of straining. However, for other types of forming operations, lower values are to be preferred.

Gensamer ■ has shown that alloy additions decrease the strain hardening of ferrite which is free of carbon and nitrogen. Recent work by Schunk, cited by Blickwede,□ is reproduced in Table 12-1.

Usually the only alloying element of significant quantity in commercial sheet steels is manganese in the range 0.28 to 0.35 percent which, from the above table, would reduce n by about 0.01 relative to pure iron. However, since some of the manganese is chemically combined with other elements, its effect is probably much reduced.

▲ A. S. Keh and S. Weissman, Conference on the Impact of Transmission Electron Microscopy on Theories of the Strength of Crystals, Berkeley, Cal., 1961.

* W. C. Leslie, J. T. Michalak and F. W. Aul, "The Annealing of Cold Worked Iron and its Dilute Solid Solutions", John Wiley & Sons, New York, 1963, p. 119.

♦ A. S. Keh, "Dislocation Arrangement in Alpha Iron During Deformation and Recovery", Proc. of AIME Tech. Conf. held in St. Louis, Missouri, March 1-2, 1961, Interscience Publishers, New York.

□ D. J. Blickwede, "Sheet Steel — Micrometallurgy by the Millions", ASM Trans. Quart., Vol. 61, 1968, pp. 653-679.

* I. L. Dillamore and W. T. Roberts, "Preferred Orientation in Wrought and Annealed Metals", Met. Rev. (1965), 39, No. 10, pp. 271-380.

■ M. Gensamer, "Strength & Ductility", Trans. ASM 36 (1946), pp. 30-60 (1945 Campbell Memorial Lecture).

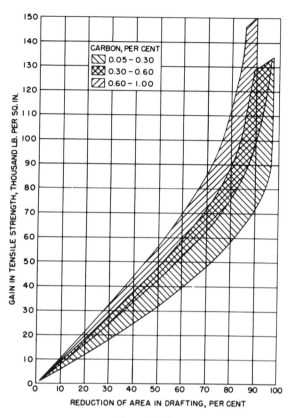

Figure 12-6:
Increase of tensile strength of plain carbon steel with increasing amounts of cold working.

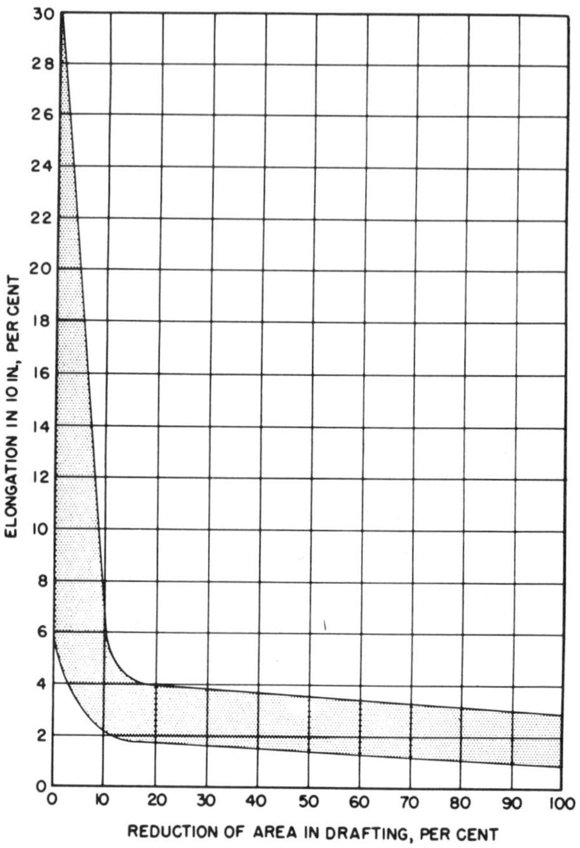

Figure 12-7:
Effect of cold working on the ductility of plain carbon steel.

Table 12-1

The Effect of Substitutional Alloying Elements on the Strain Hardening Exponent of Ferrite

Alloying Element	Average change in n per wt. percent [1]	Approximately Solubility in Ferrite at Room Temperature, wt. %
Cu	−0.06	1
Si	−0.06	15
Mo	−0.05	3
Mn	−0.04	3
Ni	−0.04	10
Co	−0.04	75
Cr	−0.02	100[2]

1) Data corrected for variation in grain size
2) Assumes no sigma phase

Increasing grain size also increases the strain-hardening exponent. Morrison* found that the following empirical relationship holds true for mill-produced steels

$$n = \frac{5}{10 + d^{-1/2}} \quad (12\text{-}2)$$

where d is the average grain diameter in millimeters.

---

* W. B. Morrison, "Effect of Grain Size on the Stress-Strain Relationship in Low Carbon Steel", ASM Trans. Quart., 59, December, 1966, pp. 824-846.

In hot rolled or annealed sheet steels, the grain size is usually in the range ASTM 6 to 8 (0.045 to 0.022 mm diameter) and the variation in n due to grain size is about 0.04. Although it may be thought desirable to increase the grain size so as to increase the exponent, it must be remembered that grain sizes larger than ASTM 6 are usually detrimental to the surface appearance of the steel after it has been formed.

12-6 Recovery

Recovery may be defined as property changes produced in cold-worked metals by heat treating times and temperatures that do not cause appreciable changes in microstructure.° Under these circumstances, only the more mobile imperfections can undergo rearrangement in the lattice structure. It is believed that vacancies and interstitialcies existing in the slip bands are eliminated first and that some dislocations of opposite signs are annihilated. However, the majority of the dislocations and the bulk of the strain energy are not removed by the usual recovery treatments.

In the early stages of recovery, there is a change in the electrical characteristics of the metal but no observable change in its mechanical properties.♦▼ Thus, the electrical resistance decreases, presumably due to vacancy migration.● As recovery proceeds, there is a decrease in dislocation density, dislocation networks of subgrains with low energy boundaries are formed and the yield strength of the material decreases. During the later stage of recovery, subgrain growth takes place with the development of higher-angle, higher-energy boundaries. The subgrains slowly increase in size until they have mobile, high-angle boundaries and can act as the nuclei of recrystallized grains.

12-7 Annealing and Normalizing

To render the flat-rolled products suitable for their intended end uses, sheet and strip are usually given heat treatments to effect changes in their mechanical properties. However, other heat treating objectives include (a) the solution of chromium carbides to attain maximum corrosion resistance of austenitic stainless steels, (b) development of optimum magnetic properties and the formation of an insulating oxide film on silicon-bearing electrical steels and (c) the dispersion or spheroidization of carbides to influence later heat treating characteristics of alloy and high-carbon sheets.

Annealing and normalizing involve the recrystallization of the work hardened strip, the drastic reduction in dislocation density and the development of a new crystallographic texture. Recrystallization starts when the high-angle boundary subgrains begin to grow at the expense of their neighbors to form new grains that are relatively free of substructure. The kinetics of the annealing process are illustrated in Figure 12-8 showing the variation of mechanical properties of strip annealed continuously at different speeds through a pilot line furnace at various temperatures.⊗□

The fraction of the material recrystallized may be plotted against time at annealing temperature as indicated by the solid curves of Figure 12-9. A characteristic feature of these curves

---

° A. G. Guy, "Elements of Physical Metallurgy", Sec. Edition, Addison-Wesley Pub. Co., Inc., Reading, Mass., 1960, p. 422.

♦ A. Seeger, "Recent Advances in the Theory of Defects in Crystals", Phys. Stat. Solidi 1, 1961, pp. 669-698.

▼ A. Seeger and H. Kronmüller, "Stored Energy and Recovery of Deformed F.C.C. Metals", Phil. Mag. 7, 1962, pp. 897-913.

● J. T. Michalak and H. W. Paxton, "Some Recovery Characteristics of Zone Melted Iron", Trans. AIME 221, 1961, pp. 850-857.

⊗ W. A. Johnson and R. F. Mehl, "Reaction Kinetics in Processes of Nucleation and Growth", Trans. AIME 135, 1939, pp. 416-458.

□ J. W. Cahn, "The Kinetics of Grain Boundary Nucleated Reactions", Acta Met. 4, 1956, pp. 449-459.

is the incubation or nucleation period that precedes the first visible recrystallization. For this reason, they are often referred to as "nucleation and growth" curves.

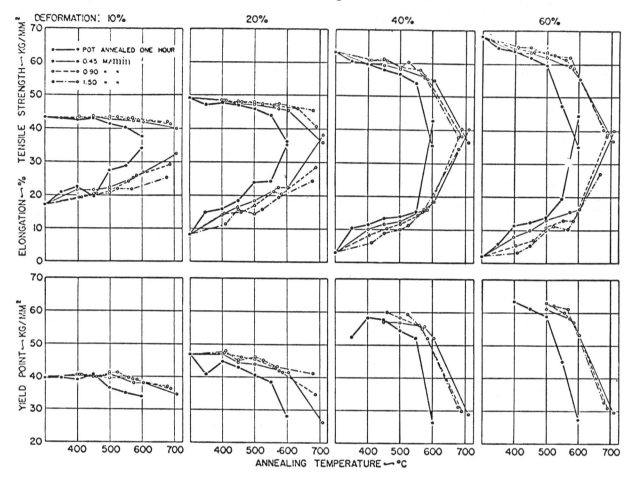

Figure 12-8: Tensile Properties of Cold-Rolled Low-Carbon Steel after Continuous Annealing at the Indicated Strip Speeds. Rimmed Open-Hearth Heat (0.04% C) (Heating section of furnace 14.8 feet long: cooling section of furnace 26.2 feet long).

A number of factors besides time and temperature influence the recrystallization process, however. The most important of these are (1) alloying elements, (2) the degree of cold work,° (3) grain size, and (4) the rate at which the temperature is raised.

Low metalloid steels (enameling irons) recrystallize in a very sluggish manner presumably due to their lower carbon and manganese contents and the interactions of these elements with oxygen.[*,+,‡,♦] According to Leslie[x] and his coworkers, the manganese content of the steel must be in the 0.02 to 0.15 percent range for sluggish recrystallization to occur. Witner[x] found

---

° A. G. Guy, Elements of Physical Metallurgy, 2nd Ed., Addison-Wesley Publishing Co., Inc., Reading, Mass., 1960.

* W. C. Leslie, J. T. Michalak and F. W. Aul, "The Annealing of Cold Worked Iron and Its Dilute Solid Solutions", John Wiley & Sons, New York, 1963, p. 119.

+ R. L. Kenyon, "Recrystallization and Grain Control in Ferrous Metals, Grain Control in Industrial Metallurgy", ASM, Metals Park, Ohio, 1949, p. 74.

‡ J. D. Baird and J. M. Arrowsmith, "Recrystallization Behavior of Some High Purity Irons", J. Iron and Steel Inst. 204, 1966, pp. 240-247.

♦ W. C. Leslie, J. T. Michalak, A. S. Key and R. J. Sober, "The Effects of a Manganese-Oxygen Interaction on the Recrystallization of Iron", ASM Trans. Quart. 58, December, 1965, pp. 645-657.

x D. A. Witner, "Effect of Thermal History on the Recrystallization Behavior of Low Carbon Irons Containing Manganese, Oxygen and Sulfur", MS Thesis — Lehigh University, Bethlehem, Pa., 1968.

PROPERTIES OF THE ROLLED STRIP

that the sluggish recrystallization behavior of one alloy could be eliminated by heating in the range 1500 to 1600° F prior to cold reduction as illustrated in Figure 12-9. Heating above or below this range resulted in varying degrees of retarded recrystallization.

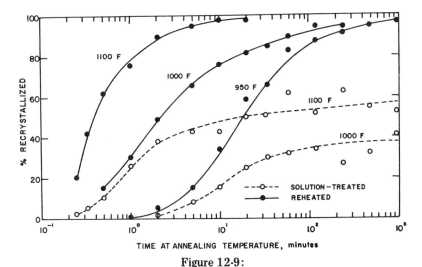

Figure 12-9:
Isothermal recrystallization curves for steel containing 0.003% C, 0.03% Mn, 0.074% O. Samples were pretreated by (a) solution treating 2100° F for 30 min. and air cooling (ASTM No. 5.1), or (b) solution treating, then reheating 1500° F for 2 hrs. and air cooling (ASTM No. 5.3). (After Blickwede)

The effect of cold working on the kinetics of recrystallization of low-carbon steel, electrolytic iron and 70-30 brass is illustrated in Figure 12-10. Here the recrystallization temperature is defined as the lowest temperature at which equiaxed, stress-free grains appear in the structure of a previously plastically deformed metal. It is found that significantly increased temperatures are necessary, in the cases of the iron and the steel, for reductions less than 20%. However, the capacity for recrystallization produced by a given reduction is also influenced by the grain size of the metal, the finer the grain size, the faster recrystallization occurs.

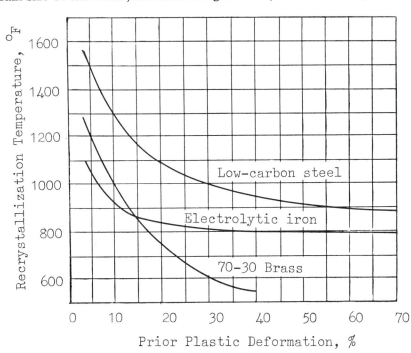

Figure 12-10:
Effect of Amount of Plastic Deformation on the Recrystallization of Iron, Steel and Brass.

The more rapidly deep-drawing sheet steel is increased in temperature during annealing, the finer the grain size produced. Blickwede ✥ shows that the material can be recrystallized in less than 10 seconds producing a grain size of about 850 grains per square inch. Even after subsequently soaking for 20 hours at 1300° F, no significant coarsening of grain size occurred. However, in batch annealing, it may require in excess of 1,000 minutes to reach final anneal temperature. Under these circumstances, a much coarser grain size of 200 grains per square inch is developed.

In normalizing operations, the strip is heated to temperatures around 1800° F (above the upper critical temperature) and cooled at a rate which permits the formation of ferrite with the desired grain size.

A number of investigators have found that changes in sheet surface occur during annealing resulting from the migration of alloying elements to the surface. Goldberg & Horton ☐ found that carbon and manganese migrate to the surface within a depth of 2 to 10 microns, and that their concentration becomes 2 to 3 times that of the base metal. This effect occurs not only during normal annealing practices but also during vacuum annealing.

Ainslie † and his co-workers have suggested that this effect results from the movements of dislocations to the strip surface during annealing and that these dislocations carry the impurity atoms with them. However, another possibility is that a greater density of dislocations may be produced at the sheet surface during rolling and that these would attract and hold the impurity atoms.

12-8 Facilities for Annealing and Normalizing

Heat treatments may be classified as the "batch type", as exemplified by box annealing and open-coil annealing, and the "continuous type", such as continuous annealing, strand annealing and normalizing. In the former, stationary, lightly-wound or open-wound coils are subjected to a long heat-treating cycle by varying the temperature of the furnace that surrounds them, whereas in the latter, a single thickness or a few thicknesses of the cold reduced strip are passed through a furnace in a relatively short time and are subjected to a heat-treating cycle determined largely by the temperature distribution in the furnace and the dimensions and rate of travel of the strip.

Box annealing equipment, such as that illustrated in Figure 12-11, consists basically of bases on which to place the coils, inner covers which fit over the coils and contain the protective atmosphere that prevents oxidation of the steel, furnaces that can be lowered over the coils to apply the heat and fans to improve the heat transfer from the inside of the inner cover to the core of the stack of coils.✦ For low carbon steels, temperatures in the range 1275 to 1350° F are used with firing times ranging from 30 to 90 hours. When no steel decarburization occurs, as is the case with the proper atmosphere, the gross differences in time at temperature throughout the annealing load have little effect on sheet properties so long as all portions have been adequately annealed (and so approach the inherent limit of grain growth and steel softening).

Such time at temperature differences are more critical when continuously annealing black plate for tinplate manufacture since well defined higher hardness levels are desired. However, the heavy cold reductions given to the strip provide a relatively fine grain size and, therefore, a relatively hard structure regardless of any annealing temperature employed so long as it is below the transformation temperature. Also, the heavy reductions promote full recrystallization in slightly lower temperature ranges than would be necessary were a light reduction to have been taken.

✥ D. J. Blickwede, "Continuous Annealing of Deep Drawing Sheet", Flat Rolled Products; Rolling and Treatment, AIME Met. Soc. Conf., Vol. 1, Interscience, New York, 1959.
☐ D. J. Blickwede, "Sheet Steel — Micrometallurgy by the Millions", ASM Trans. Quart., Vol. 61, 1968, pp. 653-679.
† N. G. Ainslie, R. E. Hoffman and A. V. Seybolt, "Sulfur Segregation at α— Iron Grain Boundaries — I and II", Acta Met. 9, 1960, pp. 523-528.
✦ G. Gordon, "Conventional Batch Annealing", Iron and Steel Engineer Year Book, 1961.

Figure 12-11:
A 500-ton capacity annealing furnace for coils. (The furnace is being lowered by crane over the base loaded with eight stacks of coils.)

For products other than low-carbon strip, box-annealing practices can vary considerably. Temperatures as low as 1000° F may be used for stress relieving (instead of recrystallizing) on certain products and temperatures as high as 2000° F on other products.

With open coil annealing, shorter heating times are possible due to the improved transfer of heat to all parts of the coil and the open coil technique may be used to deliberately change the chemical composition of the sheet, or its surface.♦ ◘ + However, these advantages are to some degree offset by the additional "tight-to-loose" and "loose-to-tight" coiling operations and by the larger size of the furnaces.

A schematic diagram of the equipment designed to open sheet-gage coils before annealing is shown in Figure 12-12.

In the process as originally developed, the separations between wraps were obtained by means of a nylon string that was withdrawn from between the wraps before annealing of the open coil. The nylon string is still satisfactory for many kinds of annealing. However, when the annealing is done to change the composition of the steel by what has been referred to as "gas alloying", it is often advantageous to use a more positive method for maintaining uniform spacings between wraps. In these instances, the nylon string can be replaced, for example, by 1/4-inch wide corrugated strip that remains in the open coil during annealing and that insures positive wrap separation without restricting the flow of the hot gas through the coil.

Commercial sheet units are designed to operate at speeds up to 1,200 fpm during winding of an open coil before annealing and at speeds up to 800 fpm during winding of a tight coil after annealing. Commercial tinplate units are designed to operate at 1,500 fpm during the winding of both open and tight coils.

♦ J. A. Bauscher, "Use of Open Coil Process to Change Composition and Improve Sheet Steels", Iron and Steel Engineer Year Book, 1961, pp. 401-410.

◘ J. Arnold, "Open Coil Process Shows Way to Gas Alloying", Iron and Steel Engineer Year Book, 1960, pp. 561-581.

+ "Unitized Open Coil Sheet Steel Annealing System", Sheet Metal Industries, March, 1969, pp. 227-230.

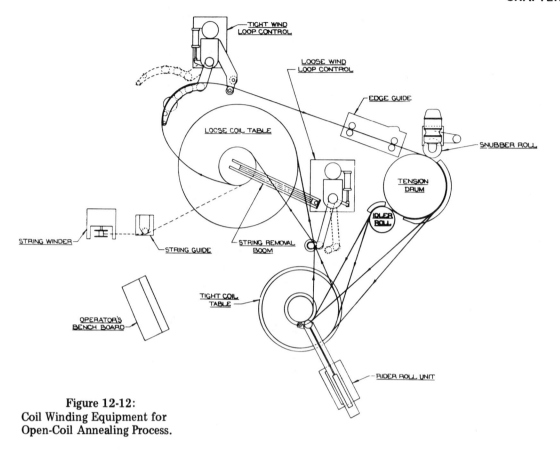

Figure 12-12:
Coil Winding Equipment for
Open-Coil Annealing Process.

The annealing of the open coils is carried out in specially designed single-stack equipment with: (1) bases, inner-covers and furnaces capable of annealing opened coils with OD of 108 and 114 inches; (2) large base fans that circulate the hot annealing gas at 25,000 cfm; and (3) specially designed plenums and diffusers for circulating the gas up past the inner-cover and down through the opened coil. A schematic diagram showing this equipment, and the flow path of the atmosphere gas is presented in Figure 12-13.

Normalizing furnaces are designed to heat sheets singly or in thin packs of two, three or four sheets. They usually have three sections (preheating, soaking and cooling zones), with the sheets moved through the zones on disc rollers. They are built up to 100 inches in width, varying from 120 to 200 feet in length, utilize fuel at the rate of 2.0 to 4.5 million Btu/ton and have throughputs in the range of 3 to 12 tons per hour. An example of such a furnace is shown in Figure 12-14. Such furnaces utilize gas or oil and do not employ protective atmospheres. Thus the workpieces scale during the heat treatment.

Normalizing furnaces have been built so as to provide a free loop or catenary of the sheet or strip in the furnace as it is unwound from coils. Furnaces of this type with steam or water quenching facilities and pickling or other descaling equipment at the exit end are widely used for the heat treatment of stainless steels at temperatures in the range of 1900 to 2200° F.

Continuous annealing furnaces process light gage cold reduced steel strip in a deoxidizing atmosphere at relatively high speeds (up to 1,500 feet per minute).[❈] In such a furnace, the strip is brought to a temperature just above the lower critical in a very short time. Recrystallization is almost instantaneous and after passage through a cooling zone, the strip emerges into the air at a temperature low enough to prevent oxidation. The annealed strip, however, is

---

[❈] M. D. Stone and E. A. Randich, "High Speed Continuous Annealing of Tin Plate", Iron and Steel Engineer Year Book, 1953, pp. 673-682.

PROPERTIES OF THE ROLLED STRIP

generally slightly harder than that produced by box annealing but has improved uniformity of hardness[○][◇] (Figure 12-15).

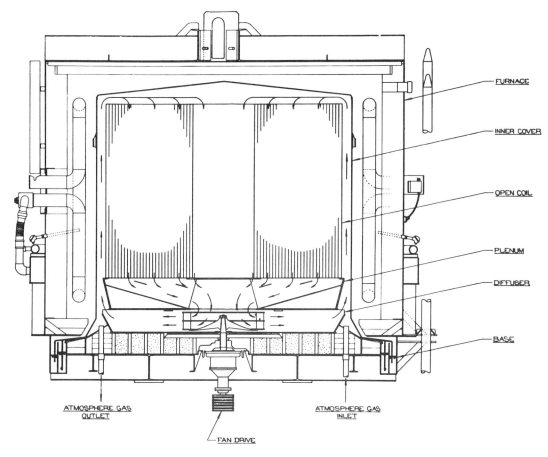

Figure 12-13: Furnace for Open-Coil Annealing.

Figure 12-14: Continuous Disc-Roller-Hearth Type of Normalizing Furnace.

[○]W. H. Swisshelm, "Quality Comparison of Box vs. Strip Annealing of Tin Mill Products", Iron and Steel Engineer Year Book, 1955, pp. 176-178.

[◇]"Symposium on Continuous Annealing of Steel Strip", Iron and Steel Engineer Year Book, 1957, pp. 109-134.

A diagrammatic illustration of a continuous annealing line is shown in Figure 12-16. It is to be noted that furnaces of this general type, providing temperatures up to 1900° F, are used for the processing of high silicon electrical steels.◈

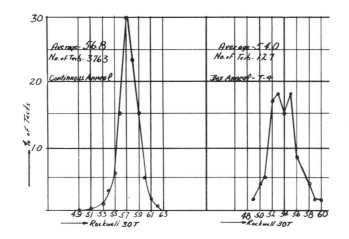

Figure 12-15:
A comparison of hardness uniformity for continuously annealed product on the left and box-annealed T-4 on the right.

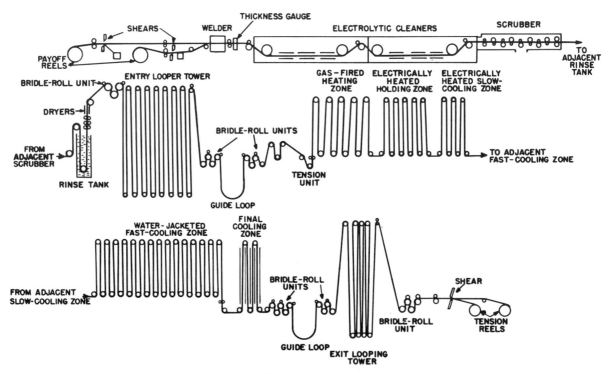

Figure 12-16:
Schematic diagram (not to scale) of a continuous-annealing line designed to operate at a strip speed of 1,500 feet per minute.

In 1972, the first continuous annealing and processing line for sheet product was put into operation at Nippon Steel's Kimitsu Works.⸸ This is illustrated schematically in Figure 12-17, and it is seen that electrolytic cleaning, annealing, cooling, temper rolling and inspection are all incorporated in a single line. Moreover, the end product is deep-drawing quality strip. The new line

---

◈ F. L. Prentiss, "Electrical Strip — A New Product", Iron Age, 1933.
⸸ K. Toda, B. Kawasaki and T. Saiki, "World's First Continuous Annealing and Processing Line for Cold Rolled Strip", Iron and Steel Engineer Year Book, 1973, pp. 415-418.

PROPERTIES OF THE ROLLED STRIP

has a production capacity of 34,000 metric tons a month, and accommodates cold-rolled strip in widths from 750 to 1240 mm (30 to 49 inches) and in thicknesses from 0.4 to 1.2 mm (0.01575 to 0.0472-inch). Coils with a maximum weight of 45 metric tons and a maximum diameter of 2540 mm (100 inches) can be handled. Line speed is 200 mpm (660 fpm) maximum through the furnace, and 260 mpm (855 fpm) max. on the entry and exit ends.

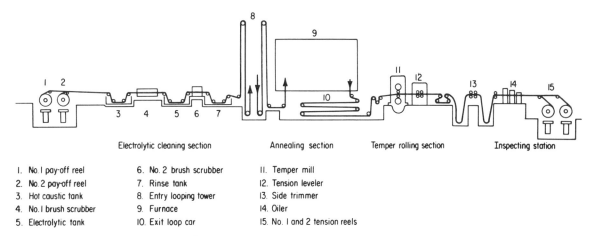

Figure 12-17: Layout of the continuous annealing and processing line for sheet product.

The furnace section, through which the strip passes vertically, has a heating capacity of 60 metric tons/hr. Strip accumulating equipment, at the entry and delivery end of the furnace, permits the furnace to continue operating at a constant speed, maintaining uniform quality of annealing. Overall length of the line is 291 m (955 ft.) and the total drive power is 2500 kw.

It will be seen from Figure 12-17 that the line incorporates a 4-high temper mill. This mill, driven only by its bottom backup roll, is equipped with 380 mm and 890 mm by 1500 mm (15 and 34.7 by 59 inches) rolls and it features a hydraulic screwdown system to control the elongation to within ± 0.1 percent. In addition to the variable mill modulus mode, constant roll pressure and constant gap control systems are available. A work-roll shape-control system permits operation with only one work-roll crown for the full product range.

A push-out type rapid work-roll-changing device enables the replacement of work rolls in 3 minutes without the strip being cut.

A tension leveller with 40 mm (1.57-inch) diameter work rolls is located after the temper mill. A flying shear is installed at the delivery end of the line which permits the changeover of the two tension reels without stopping the line. When the strip is coiled to a predetermined weight, the changeover is accomplished automatically.

Typical properties of the continuously annealed products rolled from low carbon capped steel are presented in Figure 12-18.

12-9 Quench Aging

Quench aging is generally assumed to be caused by the disintegration of a supersaturated solid solution and, in the case of low carbon steels, by the precipitation of iron carbide from a supersaturated α — solid solution.[±] Referring to Figure 12-19(A) showing a portion of the iron-carbon equilibrium diagram, it can be seen that a solubility of carbon in the ferrite increases to a value close to 0.05 percent as the temperature is increased to 1333° F. If, for compositions and temperatures above the curve corresponding to the maximum solubility under equilibrium

---

± The Making, Shaping and Treating of Steel, Edited by H. E. McGannon, Eighth Edition, U. S. Steel, Pittsburgh, Pa., 1964.

conditions, the temperature is suddenly decreased, the carbon will be retained in super-saturated solid solution but will tend to precipitate out as iron carbide on standing. Moreover, as the temperature of the quenched steel is raised, the aging effect will proceed at a faster pace. However, if at any given temperature, aging is allowed to proceed too far, coagulation or coarsening of the particles occurs (i.e., the small ones tend to redissolve and the large ones to grow still larger) with the numerous finely dispersed small particles gradually replaced by a smaller number of more widely dispersed, coarse particles. In this state, the material becomes softer and is said to be in an overaged condition.*

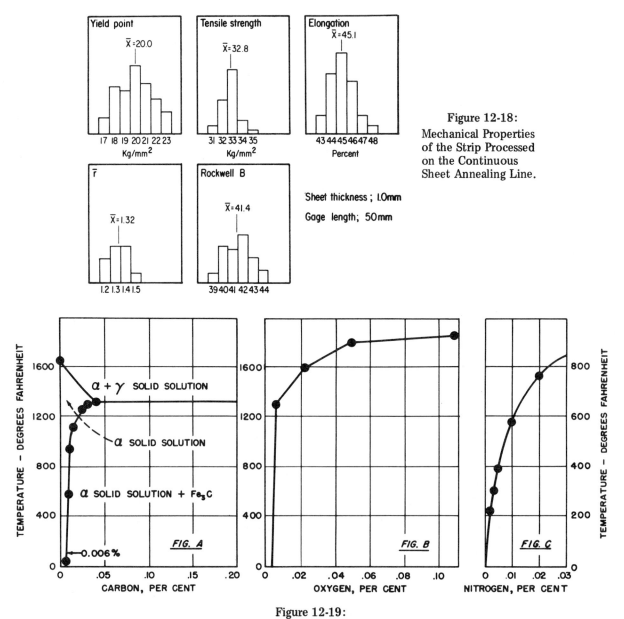

Figure 12-18: Mechanical Properties of the Strip Processed on the Continuous Sheet Annealing Line.

Figure 12-19:
Effect of temperature upon the solubilities of carbon, oxygen, and nitrogen in ferrite.
(From "Metals Handbook", 1948 Edition; American Society for Metals.)

Although carbon appears to be mainly responsible for aging, oxygen and nitrogen also have similar solubility curves with respect to ferrite as shown in Figure 12-19 (B and C) and

---

*R. E. Smallman, "Modern Physical Metallurgy", Butterworths, London, 1963, p. 293.

probably all three of these elements play a part in the aging of steel, but it is very difficult to isolate their individual effects.[±]

The changes in hardness with time of 0.06 percent carbon steel quenched from 1325° F after aging at various temperatures are illustrated in Figure 12-20.[○] From these curves it will be seen that the hardening sets in much more quickly as the temperature is raised, but at the same time, it will be noted that there is a decided tendency for the maximum to become less for the higher temperatures.[⊠] Killed steels show a lesser tendency to quench age than rimmed steels as indicated in Figure 12-21.

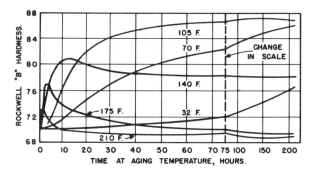

Figure 12-20:
Changes in hardness of 0.06 per cent carbon steel quenched from 1325° F after aging at indicated temperatures. (From "Metals Handbook", 1948 Edition; American Society for Metals.)

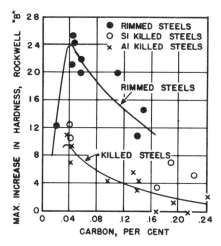

Figure 12-21:
Effect of deoxidation practice on quench aging characteristics of carbon steels. (From "Metals Handbook", 1948 Edition: (American Society for Metals.)

12-10 The Crystallographic Texture After Annealing

During the recrystallization of the steel that occurs during annealing, the crystallographic texture of the sheet or strip changes. In rimmed steel, the annealed textures have about the same proportions of $\{111\}\{332\}$ and $\{211\}$ as the corresponding deformation texture but less $\{100\}$ except at very high reductions.[◊] In the case of killed steel, however, the change is similar but more pronounced.

Although the amount of cold reduction given to the strip prior to annealing affects its post-annealed texture, the texture of the material prior to its cold reduction also has an effect. Whitely & Wise found that sheet hot-rolled at a low finishing temperature often exhibits a strong $\{110\}$ component.[✦✦] Matsudo et al,[‖] showed that, where the rolling and annealing steps were repeated, the final texture is dependent on the distribution of the reduction between the two stages of rolling rather than the total reduction. Moreover, the amount of cold reduction in the last stage has a greater effect on texture than the initial reduction and the results of this work indicate that

---

[±] The Making, Shaping and Treating of Steel, Edited by H. E. McGannon, Eighth Edition, U. S. Steel, Pittsburgh, Pa., 1964, p. 1076.

[○] E. S. Davenport and E. C. Bain, "The Aging of Steel", Trans. Amer. Soc. Met., 1935, Vol. 23, pp. 1047-1106.

[⊠] C. A. Edwards, "The Structures and Properties of Mild Steel", ASM, Cleveland, Ohio, 1953.

[◊] P. C. Hancock and W. T. Roberts, "Influence of Rolling Temperature on Textures in Rimming Steel", J. Iron and Steel Inst., 205, 1967, pp. 547-550.

[✦✦] R. L. Whiteley and D. E. Wise, "Relationship Among Texture, Hot Mill Practice & Deep Drawability of Sheet", Flat Rolled Products III, AIME, New York, 1962, pp. 47-63.

[‖] K. Matsudo, T. Shimomura and Y. Hashimoto, "Effect of Two Stage Cold Rolling-Annealing Process on Deep Drawability of Low Carbon Rimmed Steel Sheet". Report of Tech. Res. Inst., Nippon Kokan, Ltd., 1966.

two-stage rolling with an intermediate anneal does not produce as strong a texture as a one-stage rolling operation which provides the same overall reduction.

Three theories have been advanced relative to the development of annealed textures. One holds that the texture is established by the formation of nuclei with certain orientations and that these nuclei then proceed to grow without competition from nuclei of appreciably different orientation. Another theory postulates that nuclei of all orientations are found but some nuclei grow faster than others depending on their orientation. A third combines the ideas of oriented nucleation and oriented growth.[◊][♭] Blickwede[□] reports that Dillamore, Smith and Watson at the University of Birmingham and Higgins and Dunn at Liverpool University have shown by electron diffraction techniques that the relative stored energy in oriented cold-rolled grains of iron decreases in the order {111}{112}{100}. They have shown that, because the {111} oriented grains have the highest stored energy, they are the first to recrystallize. Furthermore, the recrystallized orientation is also {111} and is usually nucleated within the grain. This takes place by forming the full {111} orientation around the normal to the plane. At some stage after the {111} grains have started to recrystallize, the remaining orientations recrystallize by the nucleation of new grains mainly at the as-rolled grain boundaries. These boundary nucleated grains have a wide range of orientations and grow by absorbing the remaining unrecrystallized orientations. Thus, the intensity of the {111} component remains approximately constant whereas that of the {100} decreases. Kubotera and Nakaoka [§] also found that deformed {111} grains tend to form subgrains more readily than the deformed {100} grains. As a consequence, the recrystallized {100} grains are smaller than the {111} grains and are consumed by the {111} grains during grain growth. However, as Blickwede[□] points out, this mechanism breaks down with very large cold reductions.

12-11 The Effect of Alloying Elements on the Annealed Crystallographic Texture

From the studies of a number of researchers, it now appears that aluminum, in combination with nitrogen, affects the final stage of recovery and the subsequent nucleation stage of recrystallization during which the subgrains grow and coalesce.[●][=][❏][✦] On annealing, the {111} component of aluminum-killed steels increases and the {100} decreases to almost zero. Rickett and Leslie[⊠] found that copper addition produced a similar effect in rimmed steels, but the amount of copper required was large (0.4 to 0.8%). Phosphorus, titanium and oxygen also affect the textures[⊞][▶] with the amount of oxygen present affecting the efficiency of the titanium addition. Oxygen contents less than 0.015% result in very strong {111} textures and very weak {100} textures but, above this concentration, the effectiveness of the titanium falls off appreciably. Grain growth has been found to further develop the texture established at the beginning of recrystallization.

◊ C. G. Dunn, "Cold Rolled and Primary Recrystallization Textures in Cold Rolled Single Crystals of Silicon Iron", Acta Met. 2, 1954, pp. 173-183.

♭ J. E. Burke, "Origin of Recrystallization Textures", Trans. AIME, 194, 1952, pp. 263-264.

□ D. J. Blickwede, "Sheet Steel — Micrometallurgy by the Millions", ASM Trans. Quart. Vol. 61, 1968, pp. 653-679.

§ H. Kubotera and N. Nakaoka, Paper presented at the 1967 AIME Metal Working Conference.

● F. A. Hultgren, "The Reversion and Re-precipitation of Aluminum Nitride in Aluminum-Killed Drawing Quality Steel", Blast Furnace and Steel Plant 56, February, 1968, pp. 149-156.

= R. H. Godenow, "Recrystallization and Grain Structure in Rimmed and Aluminum Killed Low Carbon Steel", ASM Trans. Quart. 59, December, 1966, pp. 804-823.

❏ W. Jolley, "Effect of Heating Rate on Recrystallization and Anisotropy in Aluminum Killed & Rimmed Sheet Steels", J. Iron and Steel Inst. 205, 1967, pp. 321-328.

✦ I. L. Dillamore and S. R. Fletcher, "Recrystallization, Grain Growth & Textures", ASM Metals Park, Ohio, 1966, p. 448.

⊠ R. L. Rickett and W. C. Leslie, "Recrystallization, Structure and Hardness of Low Carbon Steels Containing up to 1% Copper", Trans. ASM 51, 1959, pp. 310-334.

⊞ E. H. Mayer and D. E. Wise, "Deep Drawing Steel & Method of Manufacture", U. S. Patent No. 3,244,565, April 5, 1966.

▶ M. Simizu, K. Matsuda, Y. Sadamura, N. Takahaski and M. Kawhardanda, "Process for Manufacturing Cold Rolled Steel Sheets with Excellent Disposition for Press Work", French Patent, No. 1,511,529, February 16, 1967.

12-12 The Plastic Strain Ratio and Its Measurement

In drawing operations, it is desirable for the steel sheet being formed not only to possess a low flow strength in all directions in the plane of the sheet, but to resist deformation in a direction normal to the plane of the sheet. The ratio of strengths in the plane and thickness directions can be obtained from the ratio of true strains in the width and thickness directions as measured in a simple tension test.[1] This parameter is known as the plastic strain ratio r (or the Lankford r value) expressed as

$$r = \frac{\bar{\epsilon} \text{ width}}{\bar{\epsilon} \text{ thickness}} \tag{12-3}$$

Due to anisotropy in the sheet, the plastic strain ratio assumes different values when measured in different directions. Therefore, it is desirable to use the average of the strain ratios ($\bar{r}$) measured parallel to, transverse to and 45° to the rolling direction, as expressed by

$$\bar{r} = \frac{r_{\text{rolling direction}} + 2r_{45°} + r_{\text{transverse direction}}}{4} \tag{12-4}$$

If $\bar{r}$ has a value of unity, the flow stresses in all directions in the plane of the sheet and in the thickness direction are equal. However, if the strength in the thickness direction is greater than the average strength in the plane of the sheet, $\bar{r} > 1$ and the material is resistant to thinning. The depth of drawing possible in the formation of a cup has been shown by Whiteley to be directly related to the $\bar{r}$ value[Δ,Θ] as indicated in Figure 12-22.

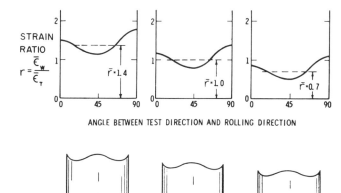

Figure 12-22: Upper curves show the typical manner in which the strain ratio r varies with test direction in low-carbon steel. (After Blickwede)

The average strain ratio is a partial measure of the "normal" anisotropy of the sheet since it gives the ratio of the flow strength normal to the plane of the sheet to the average flow stress in the plane of the sheet. The "planar" anisotropy, Δr, or the variation of the strain ratio in different directions of the sheets is expressed

$$\Delta r = \frac{r_{\text{Rolling Direction}} + r_{\text{Transverse Direction}} - 2r_{45°}}{2} \tag{12-5}$$

It is to be noted that a completely isotropic material would yield $\bar{r} = 1$ and $\Delta r = 0$.

---

[1] W. T. Lankford, S. C. Snyder and J. A. Bauscher, "New Criteria for Predicting the Press Performance of Deep Drawing Sheets", Trans. ASM 42, 1950, pp. 1197-1232.

[Δ] R. L. Whiteley, "The Importance of Directionality in Drawing-Quality Sheet Steel", Trans. ASM 52, 1960, pp. 154-169 (1960 Grossman Award).

[Θ] R. L. Whiteley, D. E. Wise and D. J. Blickwede, "Anistropy as an Asset for Good Drawability, Sheet Metal Industries" 38, 1961, pp. 349-353 and 358.

[O] J. Woodthorpe and R. Pearce, "The Effect of r and n Upon the Forming Limit Diagrams of Sheet Steel", Sheet Metal Industries, December, 1969, pp. 1061-1067.

As previously indicated, the plastic strain ratio may be computed from the width and thickness strains measured in a simple tension test. This, however, is a time-consuming operation so that in recent years, efforts have been directed towards the development of a simple, rapid method for measuring this characteristic of the material.

One approach to the automatic measurement of this parameter has been described by L. J. Owen and his co-workers.θ A tensile specimen is inserted in the machine, the initial and final widths are measured using an integrating pneumatic width gage and the initial and final lengths are monitored using a linear displacement transducer. All the measurements are electronically stored and the r value is calculated by an analog computer.

The plastic strain ratio may now be measured conveniently by an instrument known as the "MODUL-r" manufactured by Control Products Co., Inc., Carnegie, Pa., under license from U. S. Steel Corporation.⋒ This unit is illustrated in Figure 12-23.

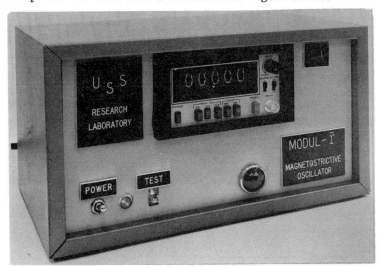

Figure 12-23: The "MODUL-r" unit for the rapid measurement of the plastic strain ratio.

12-13 The Relationship Between Crystallographic Texture and the Plastic Strain Ratio

In the bcc crystal structure of iron, the greatest strength lies in the direction of the cube diagonal or <111> direction. It is weaker in the direction of the face diagonal <110> and weakest along the cube edge <100> as indicated in Figure 12-24. Any deviation from a perfectly random orientation of the many crystals in the sheet will therefore result in an anistropy with respect to the mechanical properties of the material.

A cube-on-corner texture, i.e., {111} with the strongest crystal direction oriented normal to the sheet is favorable to the development of high strain ratios. (See Figure 12-25.) Conversely, the cube-on-face texture 100} with the weakest crystal direction oriented normal to the sheet is unfavorable to the development of high strain ratios.⁖^ The relationship between strong {111} texture and higher r̄ values is evidenced by X-ray pole figures. This is illustrated in Figure 12-26 showing the pole figure for a killed steel with an r̄ value of about 1.9 beside the weaker texture of a rimmed steel with an r̄ value of 1.3. Whiteley and Wise~ have shown that increasing the reduction

---

θ L. J. Owen, M. P. Sidey and M. Atkinson, "Automatic Measurement of r Values", Sheet Metal Industries, June, 1971, pp. 452-459 and 482.

⋒ P. R. Mould and T. E. Johnson, "Rapid Assessment of Drawability of Cold-Rolled, Low-Carbon Steel Sheets", Sheet Metal Industries, June, 1973, pp. 328-332 and 348.

⁖ R. S. Burns and R. H. Heger, "Orientation and Anistropy in Low-Carbon Steel Sheets", Steel Metal Industries 35, 1958, pp. 261-275.

^ W. F. Hosford and W. A. Backofen, "Strength & Plasticity of Textured Metals", Proc. Ninth Sagamore Conference, AMRA, 1962.

~ R. L. Whiteley and D. E. Wise, "Relationship Among Texture, Hot Mill Practice and Drawability of Sheet", Flat Rolled Products III, AIME, New York, 1962, pp. 47-63.

PROPERTIES OF THE ROLLED STRIP

given to the strip prior to annealing increases the intensity of the {111} or cube-on-corner texture up to reductions of about 75%. For increasingly larger reductions, the r value falls even though the intensity of the cube-on-corner texture continues to increase with reduction, up to a reduction of about 90%. This may be attributable to an increase in the {100} or cube-on-face texture for reductions above 75%.[^][B]

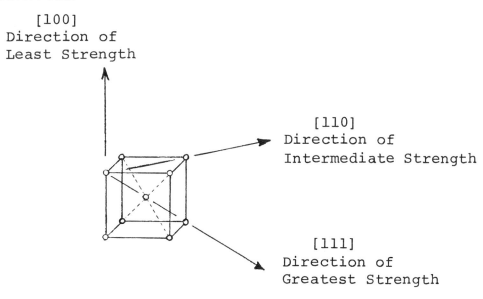

Figure 12-24: Effect of Orientation on the Strength of a Crystal.

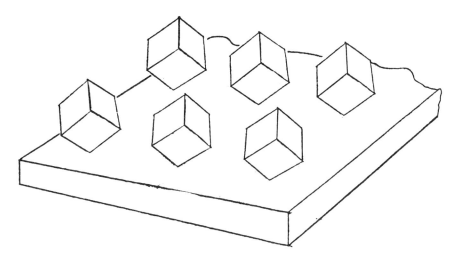

Figure 12-25: "Cube-on-Corner" Orientation (Cube diagonal normal to plane of strip surface).

The extent of the earing experienced in deep drawing is dependent on the variation of strain ratio with test direction Δr. Whiteley and Wise[~] showed that variation in strain ratio Δr varied with cold reduction as shown in Figure 12-27 and that nonearing behavior occurred when the reductions taken prior to annealing were about 25 and 90% when $\Delta r \approx 0$. Steels given reductions

---

[^] W. F. Hosford and W. A. Backofen, "Strength & Plasticity of Textured Metals", Proc. Ninth Sagamore Conference, AMRA, 1962.

[B] H. Piehler, "Plastic Anisotropy of Body Centered Cubic Metals", Second Thesis, MIT, August, 1967.

[~] R. L. Whiteley and D. E. Wise, "Relationship Among Texture, Hot Mill Practice and Drawability of Sheet", Flat Rolled Products III, AIME, New York, 1962, pp. 47-63.

between these limits develop ears at 0° and 90° to the rolling direction, whereas steels given very light (<25%) or very heavy (>90%) reductions develop ears at 45° to the rolling direction.

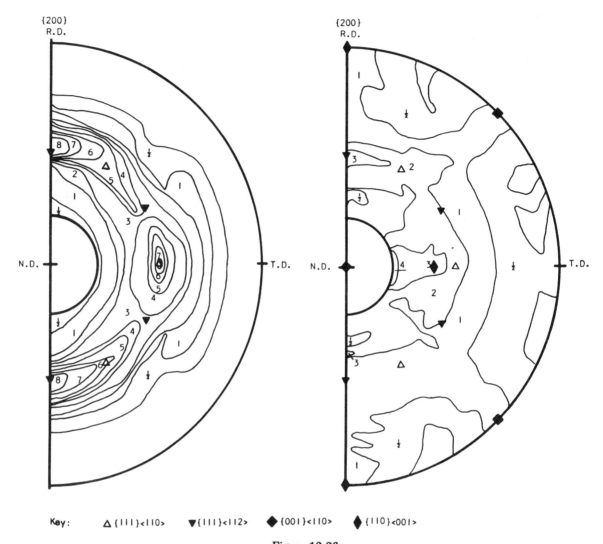

Key: △ {111}<110>   ▼ {111}<112>   ◆ {001}<110>   ◆ {110}<001>

Figure 12-26:
Effect of steel composition (Al-killed vs. rimmed) on texture produced by cold rolling 70% and box annealing 24 hr. at 1300° F. (Left) Al-killed steel, r = 1.89; (right) rimmed steel, r = 1.27.

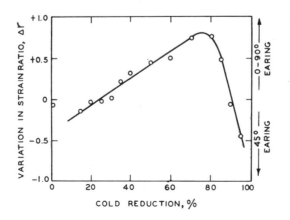

Figure 12-27:
Effect of cold reduction prior to annealing on the earing behavior of a rimmed steel as indicated by Δr. (After Whiteley & Wise)

PROPERTIES OF THE ROLLED STRIP

12-14 Temper Rolling

The purpose of temper rolling depends upon the type of products processed. In some sheet products, the main purpose is to develop the proper stiffness or temper by cold working the steel to a controlled extent. This is also true with respect to the temper rolling of black plate for tinplate manufacture. In addition, however, temper rolling may be used to improve the flatness of annealed strip, to develop desired mechanical properties and to impart a desired finish to the strip surface.±

For tinplate, the temper designations are as listed in Table 12-2, the superficial Rockwell hardness scale being used as the basis for temper classification.

Table 12-2
Temper Designations

Designation	Expected Average Control Hardness Rockwell 30-T*	Lower Yield Point 1000 psi	Tensile Strength 1000 psi	Elongation in 2 in. (percent)	Characteristic	Example of Usage
T-1	46 - 52	25 - 45	50	28 - 40	Soft for drawing	Drawn requirements, nozzles, spouts, closures
T-2	50 - 56	35 - 50	55	25 - 32	Moderate drawing where some stiffness is required	Rings and plugs, pie pans, closures, shallow drawn and specialized can parts
T-3	54 - 60	45 - 55	60	20 - 28	Shallow drawing, general purpose with fair degree of stiffness to minimize fluting	Can ends and bodies large diameter closured, crown caps
T-4	58 - 64	48 - 58	62	18 - 24	General purpose where increased stiffness desired	Can ends and bodies, crown caps
T-5	62 - 68	56 - 68	70	16 - 22	Stiffness, rephosphorized steel used for hardness to resist buckling	Can ends and bodies
T-5A TU	62 - 68	56 - 66	60 - 70	14 - 20	Stiffness, rephosphorized steel used for hardness to resist buckling	Can ends and bodies
T-6	67 - 73	68 - 80	75 - 85	12 - 20	Rephosphorized steel for great stiffness	Beer can ends

*These values are based on the use of the diamond anvil.
For material 0.007-inch thick and less, the 15-T scale should be used.

The effect of temper mill extension during temper rolling at 70°F on the hardness of a typical sample of temper-rolled and aged⊢ type T-U black plate is typified by the data shown in Figure 12-28. The corresponding change in elongation with reduction in the longitudinal direction for the same material is shown in Figure 12-29 and Figure 12-30 presents the corresponding change in the transverse direction.

Various work-roll finishes are used in temper rolling; for tinplate, the No. 7 (with a 10-12 R.M.S. microinch surface finish) and the No. 5 (45 to 50 microinches) are the most commonly used finishes.

Sheet temper mills are used only when it is necessary to process the steel in sheet form, otherwise coil temper mills are used. The changes in the mechanical properties given to the sheet or strip depend on the reduction which is usually referred to in temper rolling as extension. Sheets intended for deep drawing applications receive 0.25 to 1 percent extension, whereas sheets having lesser ductility requirements are given 1.0 to 1.5 percent extension. Sheet temper mills are of both two-high and four-high types. Coil temper mills for sheet product usually consist of single-stand, four-high units while two-stand tandem coil temper mills are used to produce the harder tempers required for tinplate product.

± The Making, Shaping and Treating of Steel, Edited by H. E. McGannon, Eighth Edition, U. S. Steel, Pittsburgh, Pa., 1964.

⊢ Aging was accomplished by holding the samples at 420° F for 30 minutes.

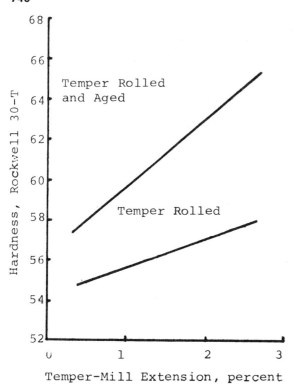

Figure 12-28:
The Effect of Temper Mill Extension on the Hardness of Continuously Annealed Black Plate and Tinplate.

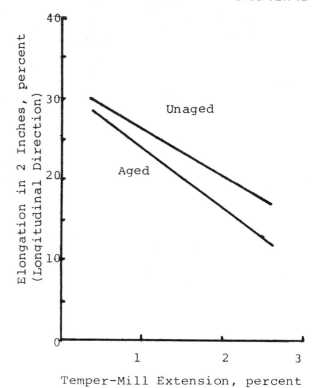

Figure 12-29:
The Effect of Temper Mill Extension on the Longitudinal Elongation of Continuously Annealed Tinplate.

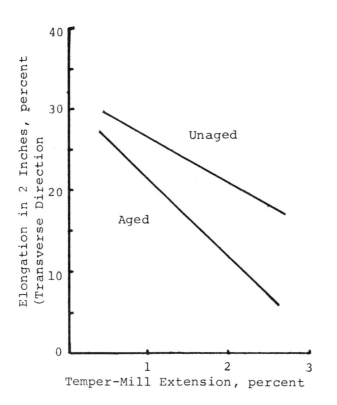

Figure 12-30:
The Effect of Temper Mill Extension on the Transverse Elongation of Continuously Annealed Tinplate.

PROPERTIES OF THE ROLLED STRIP

The finish of the rolled product is controlled by the finish of the work rolls used in the mill. Roll finishes for this purpose range from ground and polished surfaces (to provide a bright finish) to shot-blasted rolls (that produce a dull velvety finish on the steel surface).

Although temper-rolled sheets appear to have been deformed homogeneously, close examination of the surface reveals a fine surface rippling transverse to the rolling direction indicative of a nonhomogeneous mode of deformation.[✦][≡] (See Figure 12-31.)

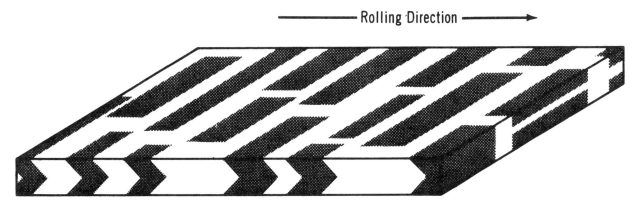

Figure 12-31:
Schematic Diagram Showing the Distribution of Deformed and
Undeformed Regions Formed During a Light Temper Rolling Pass.

Such strain markings may become objectionable when the steel sheet has been lightly drawn and painted. Investigation shows that a complex arrangement of Luder's bands propagates into the sheet from the area of roll contact at an angle of approximately 45° to the sheet surface. Further deformation which involves stretching the sheet can occur at low stresses by the propagation of these Luder's fronts. The number and distribution of the bands depend on many variables both in the sheet itself and in the temper rolling process. Although the exact effects of all these variables on the Luder's band distribution are not known, it is believed that discontinuous yielding occurs to a greater extent in rimmed steels than in aluminum-killed steels.[≡]

12-15 Strain Aging

One of the effects of temper rolling, as has been seen in Section 12-14, is to remove the sharp yield point present in most commercial annealing steels. Yet, if rimmed or capped steels are maintained at room temperatures for prolonged periods, the sharp yield point reappears. (See Figure 12-32.) Moreover, other properties that affect drawability are impaired, particularly the strain hardening exponent, which increases initially and then decreases slowly and the yield strength which undergoes similar changes.[□]

Strain aging is attributable to the segregation of carbon and nitrogen atoms at the dislocations formed during temper rolling.[✤] These so-called atmospheres pin the dislocations restraining their movement and hence recreating the sharp yield point. Four stages in this process

---

✦ B. B. Hundy, "Inhomogeneous Deformation During The Temper Rolling of Annealed Mild Steel", J. Iron & Steel Institute 181, 1955, pp. 313-315.

≡ E. J. Paliwoda and I. I. Bessen, "Temper Rolling and Its Effect on Stretcher-Strain Sensitivity", Flat Rolled Products II, AIME, 1960, pp. 63-81.

□ D. J. Blickwede, "Sheet Steel — Micrometallurgy by the Millions", Campbell Memorial Lecture, Trans. of the ASM, Vol. 61, 1968, pp. 653-679.

✤ A. H. Cottrell and B. A. Bilby, "Dislocation Theory of Yielding and Strain Aging of Iron", Proc. Phys. Soc. 62A, 1949, pp. 49-62.

have been identified by Wilson and Russell[◈] and this sequence has been confirmed by others.[T,¢,<,∧,∀] Table 12-3 lists the four stages and the corresponding changes in mechanical properties.[∀]

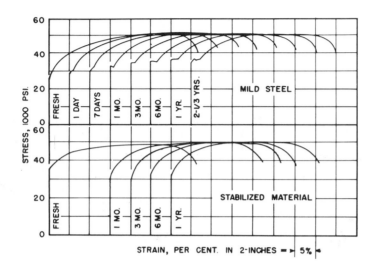

Figure 12-32: Effects of strain aging on the characteristics of the stress-strain curves obtained by tension testing of plain carbon steels. (From "Metals Handbook", 1948 Edition; American Society for Metals.)

Table 12-3
Physical Changes Occurring During Strain Aging

Stage	Mechanical Property	Change in Property
I	Lower Yield Point	Increases
	Yield Elongation	Increases
	Flow Stress	Constant
	Tensile Strength	Constant
II	Lower Yield Point	Increases
	Yield Elongation	Constant
	Flow Stress	Increases
	Tensile Strength	Constant
III	Lower Yield Point	Increases Slowly
	Yield Elongation	Constant
	Flow Stress	Increases Slowly
	Tensile Strength	Increases
	Strain Hardening Exponent	Increases
	Elongation, total and uniform	Decreases
IV	Lower Yield Point	Decreases Slowly
	Yield Elongation	May increase
	Flow Stress	May decrease
	Tensile Strength	Decreases slowly
	Strain-hardening exponent	Decreases slowly
	Elongation, total and uniform	Increases slowly

[◈] D. V. Wilson and B. Russell, "The Contribution of Atmosphere Locking to the Strain Aging of Low-Carbon Steels", Acta Met. 8, 1960, pp. 36-45.

[T] J. W. Edington, T. C. Lindley and R. E. Smallman, "Strain Aging of Vanadium", Acta Met. 12, 1964, pp. 1025-1031.

[¢] J. D. Baird, "Strain Aging of Steel — A Critical Review", Iron and Steel, 36, 1963, pp. 186-326,368,400.

[<] A. S. Keh and W. C. Leslie, "Recent Observations on Quench Aging and Strain Aging of Iron & Steel", Materials Science Research, Vol. I, Plenum Press, 1963, p. 208.

[∧] S. Bergh, "Influence of Carbon Content on Strain Aging in Annealed, Rimming, Deep-Drawing Steels at Different Low Nitrogen Contents", Jorn. Ann. 150, 1966, p. 250.

[∀] A. Josefsson and B. Backstrom, "Strain Aging in Carbon Saturated Ferrite with Various Nitrogen Contents", Iron & Steel, 39, 1966, p. 43-49.

The amount of carbon and nitrogen dissolved in the ferrite and the type of prestrain are the most important factors affecting strain aging. Killed, rimmed and capped annealed sheet steel contain very little dissolved carbon but the rimmed and capped do possess significant amounts of nitrogen in solution (in the range 8 to 30 parts per million). Since the killed steels have very little dissolved nitrogen and practically no tendency to strain aging, it is believed that dissolved nitrogen is mainly responsible for the aging effect.

The suppression of strain aging by the addition of aluminum is attributable to the formation of stable aluminum nitrides in the annealed material, leaving no significant amount of nitrogen in solution. Other elements such as vanadium, columbium, titanium and boron also reduce strain aging by the formation of nitrides. However, as pointed out by Blickwede,[□] these elements are not normally used in place of aluminum simply for the purpose of preventing the strain aging since, not only are they costly, they also affect the rimming action.

Less strain is required to eliminate the yield point by temper rolling than by stretching the steel under tension. According to some researchers, the temper rolling gives rise to very narrow alternating bands of deformed and undeformed metal running transverse to the rolling direction.[✦,⌖,▽,≡] On a larger scale, these bands produce a relative uniformity of yielding in the steel or strip. Under conditions of stretching, however, the strain to produce uniform yielding must exceed the yield elongation.

It is interesting to note that the aging effect is more rapid in tension-strained sheets than in temper rolled material.[∿] The direction of tensile testing with respect to the rolling or straining direction also reveals some interesting effects. Specimens of temper-rolled material pulled in a direction parallel to the rolling direction exhibit slower aging effects than when pulled in the transverse direction. However, for tension-strained sheets the reverse is true,[ʀ] the effects being presumably attributable to the heterogeneous patterns of residual microstresses.[⋔]

The temperature at which the steel is temper rolled also affects the rate at which the yield point returns.[⊠] Karlin found that, for specimens of rimmed steel given a 5 percent reduction at room temperature, a yield elongation of 1 percent developed after a period of 48 hours. However, those rolled at liquid nitrogen temperatures were found to have an elongation of only 0.5 percent even after aging 1,400 hours.

12-16 Specifications for Sheet Products

On March 3, 1892, the United States Standard Gage for Sheet and Plate Iron and Steel (Table 12-4) was established by an Act of Congress as the only standard gage for these materials after July 1, 1893. This gage is a weight gage based upon weights per square foot in pounds

---

[□] D. J. Blickwede, "Sheet Steel — Micrometallurgy by the Millions", Campbell Memorial Lecture, Trans. of the ASM, Vol. 61, 1968, pp. 653-679.

[✦] B. B. Hundy, "Inhomogeneous Deformation During the Temper Rolling of Annealed Mild Steel", Jour. Iron & Steel Inst. 181, 1955, pp. 313-315.

[⌖] J. F. Butler, "Inhomogeneous Deformation and Its Relationship to Strain Markings on Annealed and Temper-Rolled Sheets", Flat Rolled Products — III, Met. Soc. Conf. Vol. 16, Interscience, New York, 1962, pp. 65-84.

[▽] R. D. Butler and D. V. Wilson, "The Mechanical Behaviour of Temper Rolled Sheets", J. Iron & Steel Inst. 201, 1963, pp. 16-33.

[≡] E. J. Paliwoda and I. I. Bessen, "Temper Rolling and Its Effect on Stretcher Strain Sensitivity", Flat Rolled Products — II, Met. Soc. Conf., Vol. 6, Interscience, New York, 1960, pp. 63-81.

[∿] B. B. Hundy, "Elimination of Stretcher Strains in Mild Steel Pressing", J. Iron & Steel Inst., 178, 1954, pp. 127-138.

[ʀ] H. P. Tardif and C. S. Ball, "The Effect of Temper Rolling on the Strain Aging of Low-Carbon Steel", J. Iron & Steel Inst. 182, 1956, pp. 9-19.

[⋔] J. D. Baird, R. D. Butler and D. V. Wilson, "Discussion of Paper by J. D. Baird, Strain Aging of Steel — A Critical Review", J. Iron & Steel Inst., 201, 1963, pp. 9-19.

[⊠] R. G. B. Yeo, British Patent No. 1,051,307, December 14, 1966.

avoirdupois. The gage table as established by Congress began with 20 pounds per square foot, No. 7/0's gage, and ended with 0.25 pound per square foot, No. 38 gage, but the light side of the table has been extended by custom to 0.1875 pound per square foot, or No. 44 gage. In this country the gage is standard for all uncoated iron and steel sheet and plate, and is also used for tinplate in the lighter gages.

However, the manufacturers of steel sheets in the U. S. have adopted a new gage, known as the Manufacturers' Standard Gage for Sheet Steel (Table 12-5). The gage numbers and corresponding weights in this gage are identical to those contained in the U. S. Standard Gage, but the equivalent thicknesses are less since they are based on the density of steel, not that of wrought iron. The conversion factor used in determining these thicknesses is actually greater than the density of steel by an amount necessary to allow for the fact that sheet weights are calculated on the basis of ordered width and length with shearing tolerances on the overside, and that sheets are thicker in the center than they are at the edges where thickness is commonly and most conveniently measured.

Table 12-4
United States Standard Gage

Name of Gage	United States Standard Gage, U.S.S.G.		Name of Gage	United States Standard Gage, U.S.S.G.	
Principal Use	Uncoated Carbon Steel Sheets and Light Plates		Principal Use	Uncoated Carbon Steel Sheets and Light Plates	
Gage No.	Equivalent Thickness, Inch	Lb. per Sq. Ft.	Gage No.	Equivalent Thickness, Inch	Lb. per Sq. Ft.
7/0's	0.4902	20.0000	20	0.0368	1.5000
6/0's	0.4596	18.7500	21	0.0337	1.3750
5/0's	0.4289	17.5000	22	0.0306	1.2500
4/0's	0.3983	16.2500	23	0.0276	1.1250
3/0's	0.3676	15.0000	24	0.0245	1.0000
2/0's	0.3370	13.7500	25	0.0214	0.8750
0	0.3064	12.5000	26	0.0184	0.7500
1	0.2757	11.2500	27	0.0169	0.6875
2	0.2604	10.6250	28	0.0153	0.6250
3	0.2451	10.0000	29	0.0138	0.5625
4	0.2298	9.3750	30	0.0123	0.5000
5	0.2145	8.7500	31	0.0107	0.4375
6	0.1991	8.1250	32	0.0100	0.4062
7	0.1838	7.5000	33	0.0092	0.3750
8	0.1685	6.8750	34	0.0084	0.3437
9	0.1532	6.2500	35	0.0077	0.3125
10	0.1379	5.6250	36	0.0069	0.2812
11	0.1225	5.0000	37	0.0065	0.2656
12	0.1072	4.3750	38	0.0061	0.2500
13	0.0919	3.7500	39	0.0057	0.2344
14	0.0766	3.1250	40	0.0054	0.2187
15	0.0689	2.8125	41	0.0052	0.2109
16	0.0613	2.5000	42	0.0050	0.2031
17	0.0551	2.2500	43	0.0048	0.1953
18	0.0490	2.0000	44	0.0046	0.1875
19	0.0429	1.7500			

Note: The table above is based on the theoretical weight, which makes the weight of a plate one foot square and one inch thick 40.8 pounds. Sheets and light plates are gaged on the edge, and the spring in the rolls causes the centers to be slightly thicker than the edges. To have the estimated weights of sheets and light plates equal the actual weight, the average weight of a square foot one inch thick is taken as 41.82 pounds.

PROPERTIES OF THE ROLLED STRIP

The factor commonly used in converting from weight to thickness of steel sheets is 41.82 pounds per square foot per inch thick (see Footnote, Table 12-4).

In the foregoing discussion, the density of steel was assumed to be 489.6 pounds per cubic foot or 40.8 pounds per square foot per inch of thickness, which figure has been adopted as the standard density of steel of the grades and kinds generally used in plates. The actual density of steel varies slightly with composition and treatment, and thus may be at variance with the adopted standard density.

Table 12-5
Manufacturers' Standard Gage for Sheet Steel

Gage thickness equivalents are based on 0.0014945-inch per oz. per sq. ft.; 0.023912-in. per lb. per sq. ft. (reciprocal of 41.820 lb. per sq. ft. per in. thick); 3.443329-inch per lb. per sq. in.

Manufacturers' Standard Gage No.	Ounces per Square Foot	Pounds per Square Inch	Pounds per Square Foot	Inch Equivalent for Steel Sheet Thickness
3	160	0.069444	10.0000	0.2391
4	150	.065404	9.3750	.2242
5	140	.060764	8.7500	.2092
6	130	.056424	8.1250	.1943
7	120	.052083	7.5000	.1793
8	110	.017743	6.8750	.1644
9	100	.043403	6.2500	.1495
10	90	.039062	5.6250	.1345
11	80	.034722	5.0000	.1196
12	70	.030382	4.3750	.1046
13	60	.026042	3.7500	.0897
14	50	.021701	3.1250	.0747
15	45	.049531	2.8125	.0673
16	40	.017361	2.5000	.0598
17	36	.015625	2.2500	.0538
18	32	.013889	2.0000	.0478
19	28	.012153	1.7500	.0418
20	24	.010417	1.5000	.0359
21	22	.0095486	1.3750	.0329
22	20	.0086806	1.2500	.0299
23	18	.0078125	1.1250	.0269
24	16	.0069444	1.0000	.0239
25	14	.0060764	0.87500	.0209
26	12	.0052083	.75000	.0179
27	11	.0047743	.68750	.0164
28	10	.0043403	.62500	.0149
29	9	.0039062	.56250	.0135
30	8	.0034722	.50000	.0120
31	7	.0030382	.43750	.0105
32	6.5	.0028212	.40625	.0097
33	6	.0026042	.37500	.0090
34	5.5	.0023872	.34375	.0082
35	5	.0021701	.31250	.0075
36	4.5	.0019531	.28125	.0067
37	4.25	.0018446	.26562	.0064
38	4	.0017361	.25000	.0060

12-17 Surface Defects on Strip Products

As in the case of shape, defects are known by a large number of names and, as a consequence, considerable confusion exists with respect to their identification. In 1964, however, the "Surface Treatment of Hot-Rolled and Cold-Rolled Strip" Sub-Committee of the Cold-Rolling Committee of the "Verein Deutscher Eisenhuttenleute" set up a special Working Group to undertake a detailed consideration of surface defects.[3] This group made a fundamental study of numerous defect phenomena in seven research establishments of the steel industry, and defined the various types of defects encountered. The definitions and descriptions of the various defects discussed below are based on the information published by this Working Group.

a. Shell — consists of irregular, flaky overlapping material permeated with nonmetallic inclusions and/or scale inclusions. Usually the overlapping material is separated from the base material by scale or nonmetallic materials. The defect is believed to originate most frequently during casting[3] although it may be caused during rolling by the exposure of fairly large clusters of nonmetallic inclusions immediately beneath the surface of the strip.

b. Scale — This defect is characterized by irregular, flake-like surface discontinuities which occur as numerous minute particles of shell. They are usually elongated in the direction of rolling, are generally still connected to the base metal at certain points and are more prevalent near the edges of the strip.

c. Rokes (or Seams) — These are longitudinal marks on the strip surface parallel to the direction of rolling. They may be associated with shell and contain nonmetallic inclusions and/or scale at the surface. Such defects are very long with respect to their width and range from being wide open cracks to extremely narrow bands. They are caused by nonmetallic inclusions close to the surface being exposed during rolling or by minute longitudinal defects being doubled over in rolling.

d. Holes — Holes are discontinuities in the strip extending from surface to surface and range in size from pinholes to large jagged holes. The larger holes may be characterized by tears extending in the transverse direction of the strip. These defects may arise as a result of pipe, blowholes, coarse inclusions or rolled-in materials, or by severe mechanical damage to the strip prior to rolling.

In can manufacture, holes constitute a serious problem. Accordingly, tinplate and black plate is usually examined for the presence of such defects by means of a pinhole detector. Such detectors utilize a strong light source on one side of the strip and a sensitive light detector on the other, the penetration of light through the holes being immediately sensed by the photocells.

e. Doublings — Cavities in the rolled product are the causes of these defects, the cavities being prevented from welding shut by the presence of nonmetallic inclusions. During hot and cold rolling they are usually not apparent, becoming so only during the commercial utilization of the steel strip. They can occur in various forms. Usually they are greatly elongated in the direction of rolling (as a consequence of pipe) or they may be localized in nature resulting from "stringers".[V]

f. Rolled-in Extraneous Matter — Defects of this type result from nonmetallic materials such as scale, dirt, annealing dust, pickling and rolling emulsion residuals adhering to the surface and being embedded during rolling. Such defects exhibit no uniform shape and vary from speck-like or small streaking defects to relatively

---

[3] H. G. Grunhofer, et al, "Surface Defects on Cold Rolled Strip and Sheet", Edited by the Verein Deutscher Eisenhüttenleute, Verlag Stahleisen, M. B. H. Dusseldorf, 1967.

[3] C. J. McLean, "Control of Defects — Flat Rolled Products", Blast Furnace and Steel Plant, 54, 1966, No. 3, pp. 231-240.

[V] M. de Sars, "Les defauts de l'ingot d'acier", Rev. Metallurg. No. 43, 1946, No. 5/6, pp. 137-155.

large defective regions. They can arise in both hot and cold rolling, the most common being

(1) rolled-in scale attributable to inadequate spraying of the strip during hot rolling,
(2) scale present on the strip due to inadequate scale breaking and pickling and subsequently cold rolled into the surface,[**]
(3) rolled-in metal particles abraded from the strip during processing and subsequently rolled into it,
(4) dirt and residues present in the cold rolling solution and
(5) rolled-in dust originating in box annealing furnaces.

g. Pitting — Pitting is characterized by open, small depressions in the surface generally distributed irregularly over the surface of the strip. The defect may be attributable to pickling in spent pickle liquors, when pickling is carried out for too long a time or when insufficient inhibitors are used. However, it may also be caused by local corrosion such as, for instance, rust formation and by nonmetallic or metallic inclusions which break out from the surface.

h. Roll Marks — These are surface markings, elevations or depressions, of various shapes, occurring in either hot or cold rolling and attributable to defective roll or roller surfaces. They may arise from cracks on the roll surfaces or metallic pieces which have welded to the peripheries of the rolls. They occur periodically on the strip surface and the source of the defects may usually be established by measuring the interval between the defects and comparing this with the circumferences of the rolls or rollers.

i. Fire Crack Transfer Marks — Such defects are lines or patterns of a characteristic nature attributable to cracks, known as fire cracks, occurring in the surfaces of the work rolls of the hot strip mill. They are usually small elevations on the strip surface and may be easily rolled over in subsequent deformation, oftentimes with the inclusion of scale. Heavy reductions in hot rolling are greatly conducive to the occurrence of fire cracking of the rolls[>] but the use of hot rolling lubricants has considerably diminished the frequency of this defect.

j. Hot Strip Scratches — All scratches occurring during hot rolling and subsequent processing up to the entry of the first pass in a cold reduction mill are known as hot strip scratch. As separate scratches, they can occur in any direction, but if in groups, they are usually in the direction of rolling. During the actual scratching, a chip is frequently formed which is pushed sideways or to the end of the scratch.

If the strip is so damaged during hot rolling, the scratch may be scaled over and rolled up with the result that scale inclusions are found in the microsection. Gouges commonly occur in pickling lines, particularly in wet looping pits where numbers of loops of the strip are thrown on top of each other. Alternately, they may be caused by relative movements of the strip during coiling and uncoiling.[3]

k. Odd Strip Scratches — These are grooves, scratches, or gouges of varying dimensions occurring during or after cold rolling usually parallel and at right angles to the direction of rolling, all scratches relating to the same defect source being parallel to each other. They are open or closed and are free from nonmetallic and/or scale inclusions. One of the main causes of this defect is scoring by sharp

---

[**] J. P. Morgan and D. J. Shellenberger, "Hot Band Pickle-Patch: Its Cause and Elimination", J. Metals, Vol. 17, 1965, No. 10, pp. 1121-1125.

[>] N. R. Arant, "Fire Cracking in Rolling Mill Rolls", Iron and Steel Engineer Year Book, 1959, pp. 881-887.

[3] C. J. McLean, "Control of Defects — Flat Products", Blast Furnace & Steel Plant, Vol. 54, 1966, No. 3, pp. 231-240.

corners or objects and machine parts or by particles of dirt being trapped and held stationary in guides, etc.

l. <u>Annealed Stained Edges</u> — This defect is discussed in detail in Section 12-2.

m. <u>Stretcher Strains</u> — Stretcher strains are lines, the shapes of which are not always clearly defined, which occur during processing as the result of local strains. They usually occur at an angle of about 45 degrees with respect to the direction of the deforming force and, in certain cases, may look like "orange peel" (see later). When this defect occurs, the sheet or strip exhibits a pronounced yield-point elongation in the stress-strain diagram. The inconsistencies in the stress-strain curve and, hence, the stretcher strains, may be suppressed by roller levelling and the anchoring effect of the carbon and nitrogen atoms against the onset of plastic flow is thereby overcome. However, if the material is stocked for a fairly long period, strain aging once more causes a pronounced yield-point elongation because of the renewed anchoring effect. As a consequence, stretcher strains reoccur. Steel killed by aluminum undergoes scarcely any aging and consequently exhibits no tendency to form stretcher strains even after long periods.[∞]

n. <u>Cross Breaks</u> — Cross breaks occur at regular or irregular intervals and run transversely across the width of the strip normal to the rolling direction. As in the case of stretcher strains, they are attributable to local straining in the strip during uncoiling either on the mill or during subsequent processing.[⊖] High yield-point elongations, particularly when combined with low yield-point values, tend to increase the formation of cross breaks. The tendency is reduced by increasing the speed of uncoiling.

o. <u>Shadow Lines</u> — These defects are cross breaks which have been made invisible during temper rolling but which reoccur during deep drawing. This effect results from the fact that the material in the crossbreaks is work hardened to a greater extent than the rest of the material and, hence, offers a higher resistance to deformation in the final processing.

p. <u>Feathering (Feather)</u> — Feathering is evidenced by lines which occur during temper rolling and are caused by nonuniform local straining. They may occur locally or over the entire width of the strip and are particularly prevalent when either poor shape is being corrected or produced on the mill.

q. <u>Sticker Breaks</u> — As the name implies, these defects result from the sticking of adjacent wraps of the coil during annealing. They are characterized by transverse, slightly crescent-shaped markings. Frequent causes of underlap sticking are unfavorable strip profiles due to eccentric coiling causing bulging or coil tension which is too high during coiling prior to annealing.[?]

r. <u>Coil Breaks</u> — These defects occur transverse to the direction of deformation when sheets are being bent. They appear at more or less regular intervals so that instead of being cylindrical, the bent sheet assumes the appearance of a multisided pillar. This effect results from the presence of a yield-point elongation in the stress-strain diagram, attributable either to insufficient temper rolling or as a result of aging.

s. <u>Orange Peel</u> — This defect is characterized by a rough, pebbly surface and is produced by processing (deforming) coarse-grained material. A coarse-grained

---

[∞] B. B. Hundy, "Recent BISRA Work on the Elimination of Stretcher Strains in Mild Steel Pressings", Sheet Met. Ind., Vol. 31, 1954, No. 331, pp. 909-920.

[⊖] R. S. Burns and U. H. Latour, "Limiting Factors for Prevention of Coil Breaks", Blast Furnace & Steel Plant, Vol. 49, 1961, No. 12, pp. 1182-1188.

[?] A. A. Zdanov, V. P. Volegov, V. Ju, and V. Silken, "Weldability of Cold Rolled Strip During Annealing", Metallurgist, (1964), No. 7/8, pp. 424-425.

PROPERTIES OF THE ROLLED STRIP

structure can arise because of recrystallization after deformation in the critical range or because of excessive annealing temperatures. Inasmuch as the tendency to produce an "orange peel" type of surface cannot be recognized in the rolled strip, it is generally advisable to detect such a tendency by performing a cupping test.

t. Rust — As its name implies, this defect is a layer of the products of corrosion adhering to the surface of the strip. It varies in appearance from reddish-yellow to black patches on the surface. Its formation is promoted by temperature fluctuations, high humidity in the atmosphere and prolonged warehousing. The tendency to rust is enhanced by surface inclusions, dirt particles and acid residues and is sometimes difficult to obviate even with ideal processing conditions.

12-18 Secondary Cold Rolling

To provide thinner tinplate with higher strength and yet adequate formability, secondary cold rolling is carried out. Strip that has been reduced, usually to thicknesses in the range 0.009 to 0.018-inch and annealed, is rolled on a 2-CR mill and given reductions in the range of 30 to 40 percent. The types of mills utilized for this purpose are described in Section 2-17 and the product rolled from these facilities is often the thinnest strip rolled in sheet and tin mills. Energy requirements for secondary cold reduction are illustrated in Figure 12-33. After reduction, the strip is processed on coating lines in the same manner as conventional black plate.

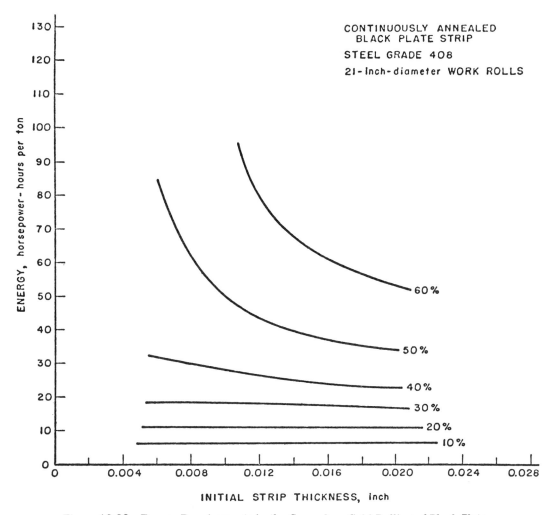

Figure 12-33: Energy Requirements in the Secondary Cold Rolling of Black Plate.

Secondary cold-rolled strip was first introduced commercially in 1960 and the percentage of tinplate or tin-free steel manufactured in these lighter gages has been steadily increasing.

12-19 The Production of Foil

In 1964, tin-coated steel foil 0.002-inch thick produced on secondary cold-rolling mills became commercially available.[1,φ] Because of the interest developed in this material, a specially designed foil mill (Figure 1-26) was installed in 1965 at the Gary Sheet and Tin Works of the U. S. Steel Corporation. This single-stand mill and its mode of operation have been described in greater detail in Section 2-18.

The typical chemical composition of the foil is 0.010% C maximum, 0.25-0.50% Mn, 0.040% P maximum and 0.050% S maximum. The mechanical properties of the foil are dependent on the reduction given to the strip in the rolling operation (or to its final gage). Table 12-6 presents these properties for strip thicknesses ranging from 0.001 to 0.0035-inch.

Table 12-6

Mechanical Properties of Steel Foil
(Tin-Coated or Uncoated As-Rolled)

Property	Thickness (in.)					
	0.0010	0.0015	0.0020	0.0025	0.0030	0.0035
Tensile Strength (ASTM D828) Longitudinal, psi Transverse, psi	105,000 115,000	105,000 105,000	105,000 105,000	105,000 105,000	105,000 105,000	105,000 115,000
Yield Strength (ASTM D828) Longitudinal, psi Transverse, psi	100,000 100,000	100,000 100,000	100,000 100,000	100,000 100,000	100,000 100,000	105,000 105,000
Elongation (ASTM D987) Longitudinal and Transverse, % in 2"	1.0	1.0	1.0	1.5	2.0	2.0
Drawing, Olsen Cup depth, in.	0.04	0.05	0.06	0.06	0.06	0.08
Impact Strength (ASTM D781) Beach puncture, in. − oz./in.	10	15	30	35	55	70
Burst Strength (ASTM D774) Mullen, psi	65	100	200	210	300	430

Both the tinned and uncoated foil may be annealed. In the former case, however, the tin is completely alloyed on the surface. Annealing produces considerable changes in the mechanical properties as is evidenced by a comparison of Tables 12-6 and 12-7.

12-20 Shearing, Side-Trimming and Slitting

Although much less shearing is being done in sheet and tin mills than formerly, because an increasing tonnage of product is being shipped in coil form, shearing is still one of the most

---

[1] R. C. Haab and M. B. Jacobs, "Steel Foil Production and Applications", Regional Technical Meetings of American Iron & Steel Institute, 1967, pp. 233-252.

[φ] H. Silman, "Steel and Iron Foil — Its Manufacture and Uses", Sheet Metal Industries, November, 1971, pp. 886-888.

common and most important operations. There are two general types of shears: (a) those for making cuts across the width of the strip and (b) those known as side-trimmers or slitters that make continuous cuts along the length of a moving strip.

Table 12-7

Mechanical Properties of Steel Foil
(Tin-Coated or Uncoated Annealed)

Property	Thickness (in.)					
	0.0010	0.0015	0.0020	0.0025	0.0030	0.0035
Tensile Strength (ASTM D828)						
Longitudinal, psi	45,000	45,000	45,000	45,000	45,000	50,000
Transverse, psi	45,000	45,000	45,000	45,000	45,000	50,000
Yield Strength (ASTM D828)						
Longitudinal, psi	35,000	35,000	35,000	35,000	35,000	40,000
Transverse, psi	35,000	35,000	35,000	35,000	35,000	40,000
Elongation (ASTM D987)						
Longitudinal and Transverse, % in 2"	5	12	20	20	30	30
Drawing, Olsen Cup Depth, in.	0.15	0.22	0.24	0.26	0.27	0.29
Impact Strength (ASTM D781)						
Beach puncture, in .02"	10	25	50	70	100	130
Burst Strength (ASTM D774)						
Mullen, psi	75	165	270	315	390	510

Much of the sheet product derived from shearing coils of cold-reduced material are cut to length on continuous-shearing lines (See Figure 12-34) employing flying shears that may be of either the guillotine type or the rotary type. As the name "flying shear" implies, both types perform the cutting operation on strip passing through the shearing line at some pre-set speed.

The guillotine type flying shear operates on the same principle as a stationary type guillotine shear, except that it is mounted in a movable housing that can move in the same direction and at the same speed as the moving strip while the knives perform the cutting operation. After each cut, the housing moves back to its original position in preparation for the next cut. Flying guillotine shears are used in lines capable of cutting strip moving at speeds up to 350 feet per minute.

As the strip passes through the shearing line, it drives a pair of idler pinch rolls, one of which is connected to a pulse generator which measures the length of strip passing through the rolls.[1] When a predetermined length of material has passed, the shear is accelerated to strip speed and the shearing operation initiated.

Rotary type flying shears consist of two horizontal cylinders, mounted one above the other in a housing, each carrying a knife that is parallel to the axes of the cylinders and that extends a suitable distance from the cylinder face. The cylinders can be operated so that the knife edges are brought together at intervals to achieve the proper length of cut. Lines employing rotary type flying shears can operate at strip speeds up to 1,000 feet per minute.

---

[1] "New Cut-to-Length Line Speeds Up Productivity", Sheet Metal Industries, June, 1969, pp. 539-543.

Figure 12-34: Cut-to-length Line with 17-Roll Leveller and Flying Shear Unit.

The speed advantage of the rotary shear is best realized on lines employed to cut large numbers of sheets of the same size. When small lots of sheets of different lengths are to be cut, it is possible to cut the same number of sheets per day on a line employing a guillotine type flying shear.

It should also be noted that hump-type lines, as illustrated in Figure 12-35, are also used for shearing light gage strip. In these lines, a guillotine operates at a rate close to 180 strokes per minute and while the cut is being made, the portion of the strip where the cut is to be made is held stationary and the incoming material forms a hump ahead of the shear.

Figure 12-35: Line with stationary shear and hump table.

PROPERTIES OF THE ROLLED STRIP

For thicker sheet products, stop-start cut-up lines are often used. To obtain good accuracy of cut length on such lines, it is absolutely essential that some form of slow-down device should be fitted to the leveller so that the strip approaches the final gaging position at crawl speed, thus allowing the leveller and the strip to be stopped in a fraction of a second when the final gaging position is reached.[∓]

A number of different types of strip processing lines (i.e., continuous annealing lines and continuous tinning lines) generally incorporate side trimmers in their design. These trimmers employ mating circular knives, mounted on arbors, to remove continuously the desired amount from both edges of the strip, thereby establishing an accurate and uniform width and producing parallel and reasonably smooth edges. Many sheet-shearing lines, however, do not have side trimmers because practically all sheet-mill strip is side trimmed at the continuous pickling line and may not need a second trimming.

The general arrangement of a line for slitting cold-rolled sheet and strip is illustrated in Figure 12-36 [8] with the machine operated on the pull-through principle, the cutters being driven only when feeding and inching.[Ψ]

A line of this design usually has a magazine feed for two or three coils, the strip lineup is generally controlled either by microswitches or photocells and the decoiler is automatically positioned by hydraulic means during the slitting operation.

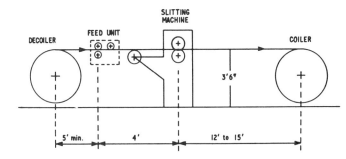

Figure 12-36: General arrangement of a line for slitting cold rolled sheet and strip.

Figure 12-37 shows the general arrangement of the cutters and spacers assembled on the arbors. The arbors, the diameters of which are dependent upon their lengths, normally run in precision roller bearings. The cutters are usually made from carbon/chrome steel and their thickness ranges from 0.25-inch to 0.75-inch, according to the gage of the material being slit. Clearances

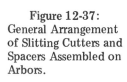

Figure 12-37: General Arrangement of Slitting Cutters and Spacers Assembled on Arbors.

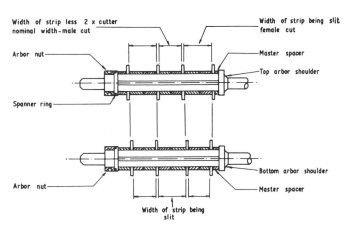

[∓] N. H. Josephy, "Processing Coil and Strip With Particular Reference to Coil Cut-Up Lines for the Steel Service Centre", Sheet Metal Industries, September, 1970, pp. 801-814.

[8] "Rotary Slitting of Sheet and Strip", Sheet Metal Industries, March 1971, pp. 213-226.

[Ψ] For power requirements in slitting lines, the reader is referred to the article "Horsepower Required for Slitting", by F. M. Butrick, American Machinist, May 17, 1971, pp. 89-90.

between the cutters of 10-15% of the strip thickness are recommended. Edge scrap is usually wound on a baler mounted in front of the slitting machine.

During multistrand slitting, the edge strips tend to coil loosely due to their slightly thinner gage. To overcome this difficulty, short lengths of paper may be fed into the edge strips during coiling. Even so, double slitting is sometimes necessary to reduce the number of strands or alternatively, the coils have to be cut halfway to reduce the coil diameter.

12-21 Coating Processes for Sheet and Strip

To provide a satisfactory degree of rust and corrosion resistance, steel sheet and strip products are usually coated with a protective material of some type. In the case of sheet products to be used in automobile and appliance manufacture, an oil is usually applied to their surfaces. This oil not only prevents the product from rusting during short-period storage and transit, but also provides lubricity during pressing and forming operations. In other cases, sheet products are painted, lacquered, or provided with some other nonmetallic coating. However, considerable tonnages of sheet and strip are coated with tin, chromium, zinc, and other metallic coatings such as aluminum and terne metal. Such coating operations are usually performed on continuously operating processing lines which are described in greater detail in subsequent sections of this chapter.

The sheet or strip is generally coated by passing it through a bath of a molten coating metal, by electrolysis or by vacuum deposition. Regardless of the coating method, however, the coating must be uniformly applied with an accurately controlled thickness and sometimes with a different thickness on each side of the sheet or strip. In some cases, the metallic coating is partially or wholly alloyed to the steel. For example, to obtain the desired luster in tinplate, the electrolytically deposited tin with a dull matte appearance is melted so as to flow and subsequently solidify into the well-known shiny coating characteristic of tinplate. In this reflowing process, however, the tin becomes partially alloyed to the steel base as illustrated in Figure 12-41. Also, in the galvannealing process, the hot-dipped galvanized strip is heated so that the entire coating becomes alloyed to the steel.

In many cases, the metallic coatings are subsequently covered with other materials. For example, tinplate is coated with a very thin film of dioctylsebacate or cottonseed oil and galvanized coatings may be chemically treated to prevent the formation of white rust.

To apply metallic coatings to strip, the more important processes are as follows:

<u>Hot Dip Processes</u> — Here the steel article to be coated is first thoroughly cleaned and then immersed in a bath of the molten coating metal. Zinc, tin, terne metal, aluminum and lead are applied commercially in this manner.

<u>Electroplating</u> — Essentially all electroplated coatings, with the exception of zinc, are cathodic to steel and provide protection of the steel surface by coverage. Metals applied electrolytically include cadmium, chromium, copper, gold, tin, lead, nickel, silver and zinc and alloys such as brass, bronze and lead-tin. Some of the electroplating baths and the appropriate operating data are shown in Table 12-8[±] and the use of this method with respect to tin, chromium, and zinc is discussed more fully in Sections 12-22 to 12-24.

<u>Vacuum or Vapor Deposition</u> — Light coatings of certain metals, such as aluminum, may be conveniently applied by deposition of the metallic vapor in vacuum chambers. On these processing lines, the strip enters and leaves the chamber with the highest vacuum through a series of chambers containing different levels of air pressure. In the chamber of highest vacuum, each side of the strip is exposed to the liquified coating metal so that the metallic vapor may condense on the strip in a uniform manner.

[±] The Making, Shaping and Treating of Steel, Edited by H. E. McGannon, Eighth Edition, U. S. Steel, Pittsburgh, 1964.

PROPERTIES OF THE ROLLED STRIP

Slurry Deposition — In this process, strip is coated with a slurry of a metallic powder, such as nickel, passed through a furnace with a protective atmosphere, and rolled while still under the same atmosphere.⊖

Table 12-8
Electroplating Baths

Kind of Coating	Type of Coat	Typical Composition of Baths Water to make one gallon	Operating Conditions			
			pH	°F	Amps/ft²	Volts
Nickel	Matte or Dull	40 oz. nickel sulphate, (1) 8 oz. nickel chloride, (2) 5 oz. boric acid	1.5 — 2.5	110 — 125	40	4 — 6
	Bright	40 oz. nickel sulphate, (1) 6 oz. nickel chloride, (2) 4½ oz. boric acid and 1 to 2 percent of addition agents	2 to 5	125 — 140	10	2 — 3
	Hard	24 oz. nickel sulphate, (1) 3.3 oz. ammonium chloride, 4.0 oz. boric acid	5.6 — 5.9	120 — 140	25/50	4 — 6
Cadmium		3 oz. cadmium oxide 14.5 oz. sodium cyanide, plus brighteners.	13.0	70 — 95	25	4 — 6
Copper	Cyanide	3 oz. copper cyanide, 4.5 oz. sodium cyanide, 2 oz. sodium carbonate (5)	11.8 — 12.2	75 — 12.2	15	1.5 — 2
	Acid	28 oz. copper sulphate, (5) 6.5 oz. sulphuric acid (3)		70 — 80	30	4 — 6
Brass		3.6 oz. copper cyanide, 1.2 oz. zinc cyanide, 7.5 oz. sodium cyanide, 4 oz. sodium carbonate (5)	11	75 — 100	3 — 5	2 — 3

(1) $NiSO_4 7H_2O$ (3) $H_2SO_4 - 100\%$ (5) Na_2CO_2 (anhydrous)
(2) $NiCl_2 6H_2O$ (4) $CuSO_4 - 5H_2O$

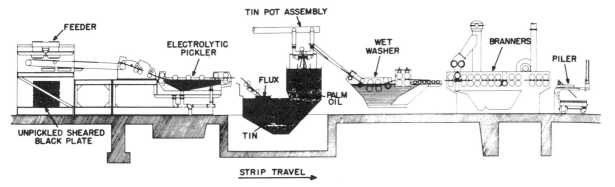

Figure 12-38: Schematic arrangement of equipment comprising a tinning stack.

12-22 Tin Coating Processes

A "tinning stack" for tin coating black plate sheets by the hot dip process is illustrated in Figure 12-38. The principal parts of this line consist of a feeder, electrolytic pickler, the tin pot and tinning machine, the wet-washing machine, the branners and the pilers. The tinning machine shown in Figure 12-39 consists of a series of guides and rolls to convey the sheets being coated from the feeding mechanism downward through the molten flux and tin baths, then upward out of the

⊖ U. S. Patent 2,989,944.

molten tin through a bath of hot palm oil on to the conveyor. The wet-washing machine is used to remove the excess palm oil from the coated sheet and it uses a hot water solution of soda ash or trisodium phosphate at a concentration of 0.10 to 0.15 per cent total alkalinity. A light uniform oil film is distributed across the sheet surfaces in the branner which consists principally of thousands of canton-flannel discs about 4 inches in diameter, lightly compressed on a long, square steel mandril. Bran or middlings (from the milling of wheat and rye) are fed to a number of pairs of pinch rolls and another group of pinch rolls are used to minimize the presence of bran on the sheets.

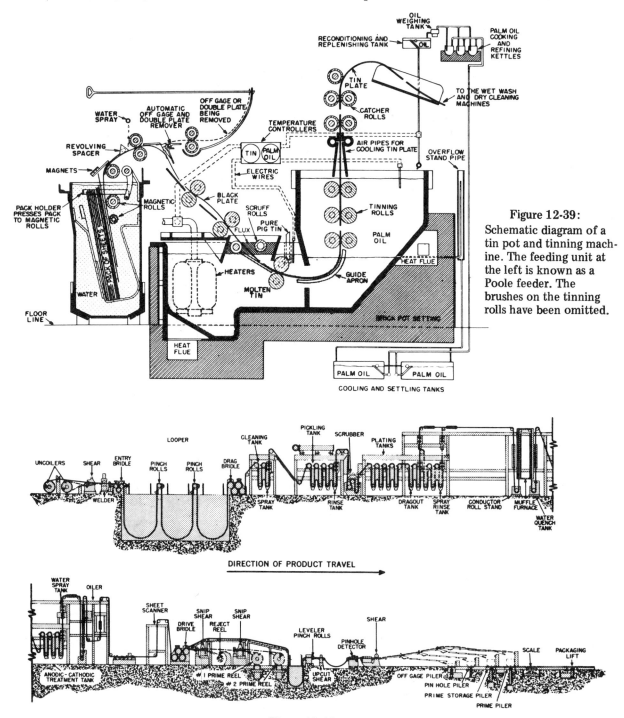

Figure 12-39: Schematic diagram of a tin pot and tinning machine. The feeding unit at the left is known as a Poole feeder. The brushes on the tinning rolls have been omitted.

Figure 12-40:
Schematic arrangement of the handling and processing units comprising a sulphonic acid electrolytic tinning line.

Electrolytic tinning lines have almost completely supplanted the tinning stacks. A line of this type is shown in schematic form in Figure 12-40. It successively cleans, pickles, plates, rinses, reflows the tin coating and provides the tinned surface with a chemical treatment and a very thin oil film.[*] (See Figure 12-41.)

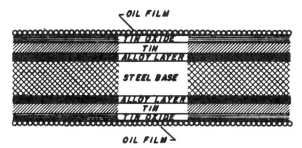

Figure 12-41:
Schematic enlarged cross section of a sheet of tinplate, showing approximate relative thicknesses of the various "layers". The approximate thickness in inches of each of the individual layers is as follows:

Layer	Thickness (In.)
Oil film	10^{-7}
Tin oxide	10^{-7}
Alloy layer	10^{-4}
Tin	10^{-5}
Steel base	10^{-2}

The main difference in the various electrolytic tinning lines lies in the type of electrolyte used (See Table 12-9). Phenolsulphonic-acid lines are designed for use with high current densities and banks of tin anodes. Alkaline type plating tanks are essentially of the same basic design but require appreciably more flow space. On the other hand, halogen-type electrolyte units consist of a series of small tanks with the strip barely immersed in the electrolyte and plated on the bottom side only. After passing through a number of these cells, the strip is deflected upward and backward so that the other side may be coated, sometimes with a differential thickness.[∞]

Table 12-9
Electroplating Baths for Tin

Type of Coat	Typical Composition of Baths Water to make one Gallon	Operating Conditions		
		°F	$\frac{Amps}{ft^2}$	Volts
Alkaline	16 oz. sodium stannate, 1 oz. sodium hydroxide, 2 oz. sodium acetate, 1/16 oz. hydrogen peroxide.	160 – 200	10 – 60	4 – 6
Acid	8 oz. stannous sulphate, 9 oz. sulphuric acid, 13 oz. phenol sulphonic acid, plus addition agent.	70	25 – 500	1 – 18

After plating, the tin has a matte appearance and the tinplate must be heated to a temperature high enough so that the coating melts and reflows to provide the usual lustrous tinplate surface. This heating is provided by electrical resistance heating of the strip in some cases, by high frequency induction heating in others, and by radiant heating in still other cases.

The thickness of tinplate is generally described in terms of its "base weight" or the weight of 1 base box (or 217.78 square feet) of the material. A listing of standard base weights is given in Table 12-10 together with the equivalent weight and the thickness.

---

[*] T. B. Bruce, R. L. Dowell and P. P. Gottschall, "Expansion of Electrolytic Tinning Facilities at Pittsburg Works", Iron and Steel Engineer Year Book, 1960, pp. 492-502.

[∞] S. S. Johnston, "Differential Thickness Coating on Electrolytic Tin Plate", Iron and Steel Engineer Year Book, 1952, pp. 280-283.

Table 12-10

Nominal Weights and Thicknesses of Tinplate

Weight (lb. per base box)	Equivalent Weight (lb. per sq. ft.)	Thickness (inches)	Weight (lb. per base box)	Equivalent Weight (lb. per sq. ft.)	Thickness (inches)
55	0.2526	0.00605	155	0.7117	0.01705
60	0.2755	0.00660	168	0.7714	0.01848
65	0.2985	0.00715	175	0.8036	0.01925
70	0.3214	0.00770	180	0.8265	0.01980
75	0.3444	0.00825	188	0.8633	0.02068
80	0.3673	0.00880	196	0.8954	0.02156
85	0.3903	0.00935	208	0.9551	0.02288
90	0.4133	0.00990	210	0.9643	0.02310
95	0.4362	0.01045	215	0.9872	0.02365
100	0.4592	0.01100	228	1.0469	0.02508
107	0.4913	0.01177	235	1.0791	0.02585
112	0.5143	0.01232	240	1.1020	0.02640
118	0.5418	0.01298	248	1.1388	0.02728
128	0.5878	0.01408	255	1.1709	0.02805
135	0.6197	0.01485	268	1.2306	0.02948
139	0.6383	0.01529	270	1.2398	0.02970
148	0.6796	0.01628	275	1.2628	0.03025

The coating thickness of the tin is also designated in terms of the weight of the tin (on both surfaces) per base box. In the case of hot-dipped tinplate, the coating thicknesses are relatively heavy and are classified as listed in Table 12-11.

In the case of electrolytic tinplate, the coating thickness is generally less than that obtained by the hot dip process and is usually given in 1/4 pound increments, viz. 0.25, 0.5, 0.75, and 1.0 pound per base box. It is of interest to note that a coating weight of 1 lb./bb corresponds to an average coating thickness of 60 microinches.

Table 12-11

Coating Weights of Standard Grades of Hot-Dipped Tinplate

Class Designation	Minimum Average Coating Weight Test Value (Pounds per base box)
Common Cokes	0.85
Standard Cokes	1.05
Best Cokes	1.19
Kanners Special Cokes	1.40
1A Charcoal	1.80
2A Charcoal	2.30
3A Charcoal	2.80
4A Charcoal	3.50
5A Charcoal	4.20
Premier Charcoal	4.90

12-23 The Production of Chromium-Coated Sheet and Strip

With the development of methods other than soldering for fabricating the side seams of food and beverage cans, it became possible to utilize for can manufacture thin steel strip with coatings other than tin. Accordingly, in the mid 1960's very thin coatings of chromium — chromium oxide were commercially produced, the product being named "tin-free steel" (TFS).

PROPERTIES OF THE ROLLED STRIP

A variety of manufacturing methods were developed for the production of this material.[Ω] In many instances, modifications were made to electrolytic tinning lines to permit the TFS coating to be applied. National Steel Corporation[ε,ʎ] first electrolytically cleans the strip using a proprietary alkaline cleaner, and then scrubs and pickles it with dilute sulphuric acid. The strip is then chromium plated, chemically treated in electrolytic cells, dried and oiled on the line shown schematically in Figure 12-42 to produce a strip with a cross section shown schematically in Figure 12-43.

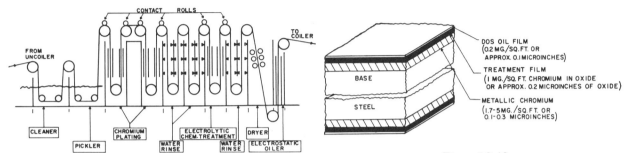

Figure 12-42:
Schematic chromium-plating line. (National Steel)

Figure 12-43:
Schematic representation of cross section of a typical chromium-plated steel strip.

To provide sheet steel with corrosion resistant properties similar to those of ferritic stainless steels, chromizing may be used. This method utilizes the high temperature diffusion of chromium metal into the steel surfaces to produce a material with a wide range of desirable properties.[d] In addition to good corrosion resistance, the chromized coating has sufficient ductility and adherence to withstand the most severe forming operations such as deep-drawing and spinning and it can be maintained during conventional welding procedures.

The essentials of the process, as practiced by Bethlehem Steel Corporation, may be summarized as follows:

1. Prior to the diffusion treatment, a precise amount of finely divided ferrochrome powder is applied to the surface of cold rolled low-carbon steel by roll compacting.

2. Conversion of this roll-compacted layer into a continuous iron-chromium alloy coating is then achieved by heating the open coil at 1650° to 1700° F in hydrogen.

In terms of starting materials, the process can coat decarburized or titanium-stabilized hot-rolled coils or annealed cold-rolled coils. Chromium powder (mixtures of chromium and iron powders) or ferrochrome powders with varying chromium contents can be used. However, a very effective combination of material economy and product quality is obtained by using titanium stabilized low-carbon sheet steel and a low-carbon ferrochrome powder containing 68 to 72 percent chromium.

A schematic diagram of the powder application and compacting mill is shown in Figure 12-44. The coil stock is fed from the compacting mill pay-off reel into a roller coater which applies a very thin film of tridecyl alcohol to both surfaces. (Tridecyl alcohol is a nonflammable viscous

---

[Ω] S. Yonezaki, "Research and Development on Tin-Free Steel Sheet", Sheet Metal Industries, January, 1971, pp. 25-36.

[ε] E. J. Smith, "Chromium-Coated Steel for Container Application", Iron and Steel Engineer Year Book, 1967, pp. 485-490.

[ʎ] J. E. Allen, "Tin Line Conversion to Chrome", Iron and Steel Engineer Year Book, 1970, pp. 285-289.

[d] E. H. Mayer and R. M. Willison, "Bethlehem's Sheet Chromizing Process", Iron and Steel Engineer Year Book, 1967, pp. 481-485.

fluid.) The strip is then passed over a specially designed fluidized bed containing the ferrochrome powder. A steady stream of powder is lifted upward from the fluidized bed by rotating brushes. As it contacts the strip surface, a portion of the powder adheres to the alcohol film. A uniform layer of powder measuring 10 to 18 g per sq. ft. can be applied in this manner. The thickness of the alcohol film as well as the particle shape and size distribution control the amount of powder that adheres to the strip.

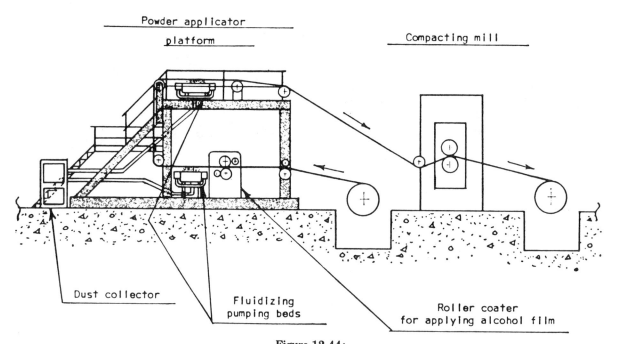

Figure 12-44:
Schematic Diagram of Line for the Application of Ferrochrome Powder to Cold-Rolled, Low-Carbon Steel Strip.

The direction of the moving strip is reversed by passing over guide rolls so that the opposite side can be coated by a second fluidized bed. The strip is then passed through a single-stand 2-high rolling mill where the powder is roll bonded to the surface. Strip speeds of 200 to 400 fpm are used in the powder application and roll compacting operation. The roll pressure required to give satisfactory compacting causes the strip to elongate one to two percent. The outer surface of the powder layer remains dry, i.e., free of alcohol during application and compacting. Thus, the compacting rolls remain dry and clean and there is no tendency for the powder particles to cling to them.

To achieve proper bonding, the steel base must be comparatively soft. Pickled hot rolled strip can be coated satisfactorily without additional heat treatment, but cold reduced strip must be annealed prior to compacting.

Bonding the powder on the strip surface prevents the powder from being removed by the open coil winding operation and by the circulating gases in the subsequent diffusion treatment. Compacting and bonding also provide the intimate contact necessary for the diffusion to occur.

Powder made up of angular particles gives the desired coating weight and also bonds well to the surface during roll compacting, whereas flat or flaky powders give coating weights that are too light and will not bond well.

Following the powder-compacting operation, the coil is open wound using a kinked spacer wire positioned at both the top and bottom edge. The wire spacers are left in place during the diffusion treatment to maintain uniform spacing between the coil laps. After open winding, the coil is placed on a high-temperature open-coil annealing base for the diffusion treatment at temperatures in the range 1650° to 1700°F for about 28 hours.

PROPERTIES OF THE ROLLED STRIP

12-24 Zinc Coating Processes

As in the case of tin coating, zinc coating or galvanizing may be accomplished either by the hot-dip process or by electroplating. The hot-dip process may be designed either for processing single sheets or for continuous strip, whereas the electroplating process is designed solely for continuous operation.

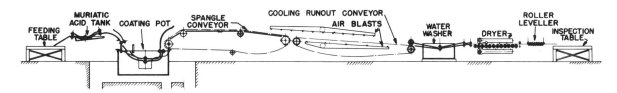

Figure 12-45: Schematic side elevation of a conventional sheet galvanizing line.

The arrangement of a conventional sheet galvanizing line is shown in Figure 12-45. A low-carbon steel pot is often used for the zinc bath, with a baffled section at the entry end of the pot containing a floating fused-chloride flux prepared from salammoniac and the zinc of the bath.

Continuous hot-dip galvanizing lines are designed in several different ways.[℧] In the simplest, strip that has been box annealed and temper rolled is passed through a long hydrochloric acid pickling tank into the galvanizing pot through a layer of flux. In a more elaborate design, the strip is passed through a cleaning unit (either acid or alkaline) treated with a thin film of "dry flux" in aqueous solution, dried and then conveyed into the zinc pot (containing some aluminum in the zinc).[⊥] Another design utilizes a continuous annealing furnace with a complex gaseous atmosphere containing hydrogen chloride, a cooling compartment, an acid bath, and the galvanizing pot with its flux box at the entry end. Asbestos wipes are normally used instead of exit rolls at the exit end of the pot and the coatings produced are thin but extremely adherent. Yet another type of line (the Sendzimir process)[+] features controlled oxidation and reduction prior to coating.

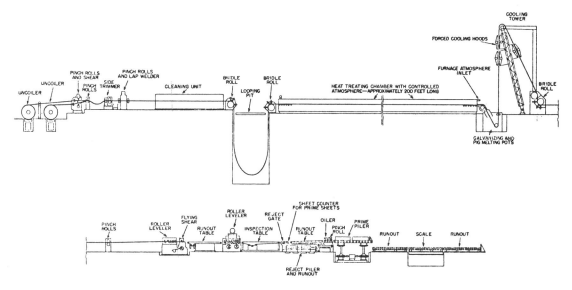

Figure 12-46:
Schematic arrangement (not to scale) of the units comprising one type of continuous galvanizing line for applying zinc coatings to cold-reduced light-gage strip steel.

[℧] G. C. Turner, "Modern Continuous Galvanizing Lines", Iron and Steel Engineer Year Book, 1963, pp. 245-252.
[⊥] N. E. Cook and M. D. Ayers, "Continuous Hot-Dipped Galvanizing", Iron and Steel Engineer Year Book, 1956, pp. 253-257.
[+] C. Cone, "Continuous Strip Galvanizing Developments", Iron and Steel Engineer Year Book, 1962, pp. 188-195.

A commonly used design is shown in Figure 12-46.⁹ Such a line can process strip 0.0157 to 0.050-inch thick at speeds up to 300 fpm. The cleaning unit utilizes a hot alkaline solution or electrolytic alkali cleaning. After rinsing and drying, the strip is annealed in an atmosphere of cracked ammonia and NX gas and the discharge end of the furnace lies below the level of the molten zinc (which contains 0.12 to 0.18 percent aluminum). From the exit rolls of the pot, the strip passes vertically upward to a large diameter roll mounted at the top of a cooling tower.

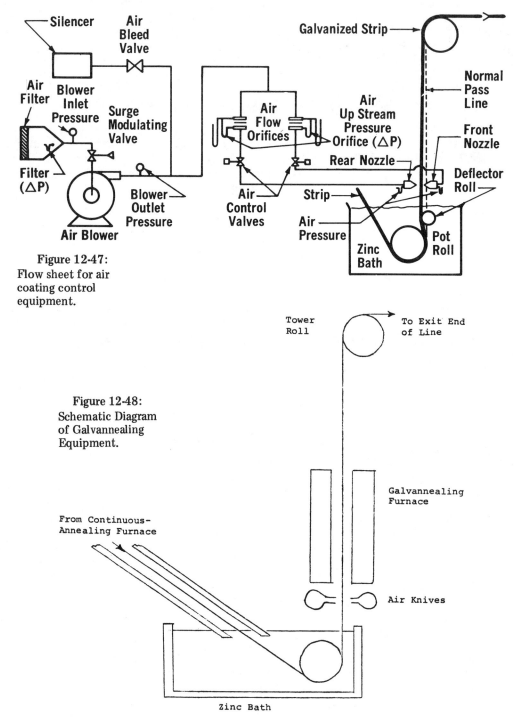

Figure 12-47: Flow sheet for air coating control equipment.

Figure 12-48: Schematic Diagram of Galvannealing Equipment.

---

⁹ M. D. Baughman, Jr., "High Speed Continuous Galvanizing Lines", Iron and Steel Engineer Year Book, 1970, pp. 455-462.

PROPERTIES OF THE ROLLED STRIP

Although coating rolls have long been used to regulate the thickness of the zinc coating, air or steam "knives" directing high velocity gaseous jets against the molten zinc on the strip surface now provide coatings of much improved uniformity.[α] A schematic diagram of an air-knife system is illustrated in Figure 12-47.[◁]

Regular zinc coatings applied by the above methods are composed of about 5 percent interfacial zinc-iron alloy and about 95 percent unalloyed zinc. However, by heating the zinc-coated strip (galvannealing) as illustrated in Figure 12-48, it is possible to produce coatings that are totally composed of zinc-iron intermetallic alloys. Compared with regular galvanized product, galvannealed strip has a better weldability and appearance after painting, but the coating, usually in the thickness range of 0.00025 to 0.00075-inch, has less ductility and adherence in forming operations.

Galvanized coatings may also be applied electrolytically,[T] and such coatings exhibit (1) no spangle, (2) easy workability and (3) lightweight and controllable thicknesses. This technique is particularly useful where a thin zinc coating is required on only one surface of the strip. Electrolytic cells may be of the conventional planar type where precautions must be taken to prevent plating on the rear side of the sheet, or of the new radial cell design[⇑] (See Figure 12-49) which, using a solution of zinc sulphate and sulphuric acid as the electrolyte, operates much more efficiently than the conventional units.

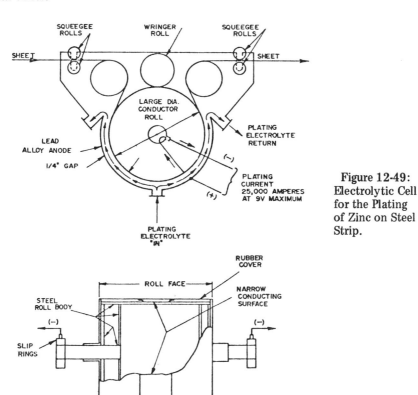

Figure 12-49: Electrolytic Cell for the Plating of Zinc on Steel Strip.

[α] J. M. Sheehan and J. A. Henrickson, "Effects of Processing Variables on the Formation Mechanisms and Properties of Galvannealed Coatings" — paper presented at the Annual Convention and Exposition of the AISE in Cleveland, Ohio, September 14-17, 1970.

[◁] J. J. Butler, D. J. Beam and J. C. Hawkins, "The Development of Air Coating Control for Continuous Strip Galvanizing", Iron and Steel Engineer Year Book, 1970, pp. 151-160.

[T] D. T. Carter, "A Practical Process for Production of Galvanized Sheets with Coating on One Side", Iron and Steel Engineer Year Book, 1971, pp. 526-530.

[⇑] Y. Uto, D. Yamasaki, T. Morihara, S. Shigemura, S. Suzuki and Y. Shimozato, "Study on Continuous Electrogalvanizing Line", Sheet Metal Industries, September, 1969, pp. 793-801.

The thickness of galvanized sheet product is normally specified in terms of gage, as presented in Table 12-12. It is to be noted, however, that these gage numbers differ from those for uncoated sheet product as given in Table 12-5.

Table 12-12
Galvanized Sheet Gage Numbers with Equivalent Unit Weights

Galvanized Sheet Gage Number	Ounces Per Square Foot	Pounds Per Square Foot	Pounds Per Square Inch	Thickness Equivalents (Inches)
8	112.5	7.03125	0.048828	0.1681
9	102.5	6.40625	0.044488	0.1532
10	92.5	5.78125	0.040148	0.1382
11	82.5	5.15625	0.035807	0.1233
12	72.5	4.53125	0.031467	0.1084
13	62.5	3.90625	0.027127	0.0934
14	52.5	3.28125	0.022786	0.0785
15	47.5	2.96875	0.020616	0.0710
16	42.5	2.65625	0.018446	0.0635
17	38.5	2.40625	0.016710	0.0575
18	34.5	2.15625	0.014974	0.0516
19	30.5	1.90625	0.013238	0.0456
20	26.5	1.65625	0.011502	0.0396
21	24.5	1.53125	0.010634	0.0366
22	22.5	1.40625	0.0097656	0.0336
23	20.5	1.28125	0.0088976	0.0306
24	18.5	1.15625	0.0080295	0.0276
25	16.5	1.03125	0.0071615	0.0247
26	14.5	0.90625	0.0062934	0.0217
27	13.5	0.84375	0.0058594	0.0202
28	12.5	0.78125	0.0054253	0.0187
29	11.5	0.71875	0.0049913	0.0172
30	10.5	0.65625	0.0045573	0.0157
31	9.5	0.59375	0.0041233	0.0142
32	9.0	0.56250	0.0039062	0.0134

The amount of zinc on a galvanized sheet is stated in terms of ounces per square foot of sheet; since the sheet is normally coated on both sides, the stated weight of coating is twice the average weight of coating per square foot on either side. Coatings applied to sheets by the hot-dip process range, in general, from 0.6 to 2.5 ounces per square foot of sheet.

12-25 Terne Coating

For certain applications, such as the fabrication of automobile gas tanks, it is desirable to coat strip with lead. Lead alone does not alloy with iron, however, so that it is necessary to incorporate another element in the lead, in this case, tin, which alloys with the steel base and so enhances the "wetting" properties of the lead.

Single steel sheets may be terne coated by the flux process similar to that used in hot dip tinning. The pickled sheets are carried through a hydrochloric acid wash tank, through a flux box and downward through the molten terne metal, where the coating is applied, then upward through a bath of oil floating on the top of the metal. Excess oil on the coated sheet is then removed by a branner similar to that used for hot dipped tinplate.

The continuous production of terne plate is carried out on a line similar to that shown in Figure 12-46. On such a line, the strip is subjected to a pickling operation, a dip in hydrochloric acid, and then submersion in the terne-coating pot. More modern lines have eliminated the use of oil as part of the coating operation by the use of steam or gas wipes with the flux being applied to the strip at its entry into the bath.

Long terne sheets are also specified in unique gage weights as shown in Table 12-13. Each gage number may have various weights depending on the coating weight (expressed in ounces/square foot).

Table 12-13
Gage Weights for Long Terne Sheets of Various Coating Weights

Long Terne Gage No.	Gage Weights in Ounces and Pounds per Square Foot, for the Gages and Coatings Given													
	Commercial		0.35 Ounce		0.45 Ounce		0.55 Ounce		0.75 Ounce		1.10 Ounce		1.45 Ounce	
	Oz. per Sq. Ft.	Lb. per Sq. Ft.	Oz. per Sq. Ft.	Lb. per Sq. Ft.	Oz. per Sq. Ft.	Lb. per Sq. Ft.	Oz. per Sq. Ft.	Lb. per Sq. Ft.	Oz. per Sq. Ft.	Lb. per Sq. Ft.	Oz. per Sq. Ft.	Lb. per Sq. Ft.	Oz. per Sq. Ft.	Lb. per Sq. Ft.
10	90.25	5.641												
11	80.25	5.016												
12	70.25	4.391												
13	60.25	3.766												
14	50.25	3.141												
15	45.25	2.828												
16	40.25	2.516	40.35	2.522										
17	36.25	2.266	36.35	2.272										
18	32.25	2.016	32.35	2.022	32.45	2.028								
19	28.25	1.766	28.35	1.772	28.45	1.778								
20	24.25	1.516	24.35	1.522	24.45	1.528	24.55	1.534	24.75	1.547				
21	22.25	1.391	22.35	1.397	22.45	1.403	22.55	1.409	22.75	1.422				
22	20.25	1.266	20.35	1.272	20.45	1.278	20.55	1.284	20.75	1.297	21.10	1.319	21.45	1.341
23	18.25	1.141	18.35	1.147	18.45	1.153	18.55	1.159	18.75	1.172	19.10	1.194	19.45	1.216
24	16.25	1.016	16.35	1.022	16.45	1.028	16.55	1.034	16.75	1.047	17.10	1.069	17.45	1.091
25	14.25	0.892	14.35	0.897	14.45	0.903	14.55	0.909	14.75	0.922	15.10	0.944	15.45	0.966
26	12.25	0.766	12.35	0.722	12.45	0.778	12.55	0.784	12.75	0.797	13.10	0.819	13.45	0.841
27	11.25	0.703	11.35	0.709	11.45	0.716	11.55	0.722	11.75	0.734	12.10	0.756	12.45	0.778
28	10.25	0.641	10.35	0.647	10.45	0.653	10.55	0.659	10.75	0.672	11.10	0.694	11.45	0.716
29	9.25	0.578	9.35	0.584	9.45	0.591	9.55	0.597	9.75	0.609	10.10	0.631	10.45	0.653
30	8.25	0.516	8.35	0.522	8.45	0.528	8.55	0.534	8.75	0.547	9.10	0.569	9.45	0.591
Nominal Coating Weights, pounds per double base box														
	6		9		12		15		20		30		40	

12-26 The Continuous Aluminum Coating of Strip

Aluminum–coated steel sheets are marketed in the United States and England for applications that require heat resistance (Type 1 product) and atmospheric-corrosion resistance (Type 2 product). For heat-resistance applications, an aluminum coating is applied that contains 7 to 9 percent silicon, which is required to provide the coating adhesion necessary in forming automotive mufflers and heating-furnace combustion chambers. In addition, the weight of the aluminum is controlled to two nominal levels, namely 0.5 oz./sq. ft. of sheet (about 1 mil thick per side) and 0.3 oz./sq. ft. of sheet (about 0.6 mil thick per side) with the lighter weight coating supplied for the most severe forming operations. For weather exposure applications, the Type 2 product is coated with one nominal weight, namely 1.0 oz./sq. ft. of sheet (about 2 mils thick per side). Good corrosion resistance is related to the thickness of the coating rather than the coating composition. Thus, Type 2 product is also produced with the aluminum – 7 to 9 percent silicon alloy used for Type 1 coatings.

To date, the only commercially successful processes for continuously aluminum coating steel sheets are those using an in-line continuous annealing operation prior to coating. That is, although the process in which the steel is pretreated before coating with chloride fluxes has been used in wire coating operations, the flux process does not produce coatings that are free from uncoated areas and entrapped flux. The preferred process used initially in 1954 and permanently since 1958 for the aluminum coating of sheet steel is essentially the same as that used in continuous galvanizing operations. In this process, coils of cold reduced or preannealed steel pass through an electrolytic alkaline cleaner, scrubbers and rinsers into a continuous annealing furnace that contains a hydrogen-nitrogen atmosphere. Without exposure to air and at a temperature close to 1217° F, the steel then passes into the aluminum-coating bath. To insure that the molten aluminum uniformly wets and alloys with the steel, adjustable baffles are used at the point where the steel enters the molten aluminum[⊖] and steam knives or coating rolls can be used to control the coating thickness. After coating, all product is then temper rolled to ensure flatness and a smooth surface finish.

12-27 Coating Strip with Paints and Organic Materials

A continuous line capable of applying two coats of paint to strip is illustrated in Figure 12-50.[∂] The entry terminal equipment of the line consists of two expanding mandrel-type uncoilers, dual laterally retracting floor-type coil cars, feed pinch-roll units with thread tables, entry squaring shear, die-type splicer press, entry bridle incorporating a strip-edge deburring device, entry accumulator and a 180-degree wrap, center-pivoted steering roll. The process section of the line contains the metal-preparation machine followed by a blow-off unit, the No. 2 tension bridle (commonly called the coater holdback, a double-roll, center-pivoted steering unit preceding the prime coater, the prime-paint coater, the prime-bake oven, an air-cooling unit and water quench, turn-around rolls for deflecting the strip toward the entry end of the finish coater, a single-roll center-pivoted steering assembly, the finish-paint coater, the finish-bake oven again followed by an air-cooling unit and water quench and, finally, the No. 3 tension or pull out bridle.

In the metal-preparation unit, the strip surface is cleaned by chemical sprays and subsequently brushed. Then the strip passes through a spray-rinse stage, is treated with zinc or iron phosphate and rinsed. Residual moisture is then removed from the strip by air jets.

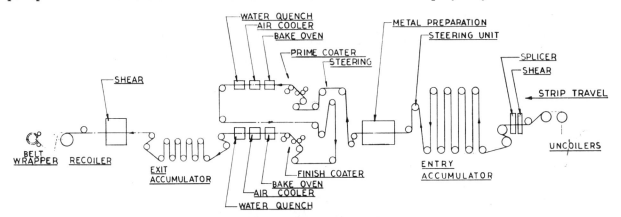

Figure 12-50: Basic components of a typical heavy-duty continuous strip paint line.

The prime and finish coaters used in the line are both of the type illustrated in Figure 12-51. Each coating head contains two rolls; namely, a steel metering and a urethane-covered applicator roll.

Natural gas, with about 1,000 Btu per cubic foot, used to heat the air stream, has been found to be the most satisfactory fuel for baking the strip. With this fuel, an inexpensive drier can be used which does not contaminate the painted strip and the heated air can be applied by convection to the moving strip in a very uniform manner.

A more recent paint coating line features the use of sprays (both air and electrostatic) instead of roll coaters.[Π] In general, the air sprays are used for the finished coating and the

---

[⊖] U. S. Patent 2,989,944.

[∂] W. M. Bevis and A. S. Dawe, "Organic Coatings for Steel Sheets — Continuous Painting Line", Iron and Steel Engineer Year Book, 1966, pp. 738-743.

[Π] "Strip-Coating at Kockums", Sheet Metal Industries, July, 1970, pp. 639-641.

PROPERTIES OF THE ROLLED STRIP

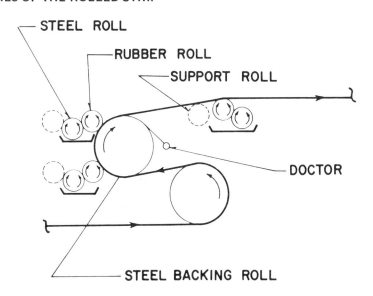

Figure 12-51: Two-roll heads for coating both sides of strip.

Table 12-14
Applications of Organic Coatings

(The following table lists typical applications for pre-paint along with the coating normally recommended for these uses.)

Coating	Alkyd	Acrylic appl	Acrylic arch	Solution vinyl	Epoxy	Organosol	Plastisol or vinyl laminate	'Timber-grain' woodgrain
Automotive								
underbody parts				x		x		
trim							x	x
fascia panels	x					x	x	x
Appliance								
wraparounds		x					x	x
heater cabinets		x					x	x
heat resistant parts		x						x
deep drawings						x	x	
TV cabinets	x						x	x
light fittings (fluorescent)		x				x		
Building								
ceiling tiles	x	x				x		
suspended ceiling sections				x		x		
cladding external			x				x	
curtain walling						x	x	
infill panels		x				x	x	
skirting boards	x	x				x	x	
partitioning	x	x				x	x	
cable trunking	x					x		
roofing			x				x	
roof decking			x				x	
Caravans			x				x	
Shelving	x					x	x	x
Steel furniture	x						x	x
Cold storage rooms								
inside skin		x				x	x	
outside cladding						x	x	
Garage doors								
primed	x				x			
finished			x			x	x	
Venetian blind boxes		x				x		
Skirting board heaters		x				x		

electrostatic sprays for the thinner undercoatings. Five separate recirculating paint systems are provided in order that color changes may be made quickly. In fact, color changeover takes only about one minute and may be accomplished while the line is running.

Subsequent to the successful development of continuous paint lines, the coil-coating industry now applies a wide range of organic coatings or plastics to steel strip on continuous coating lines.† Several types of organic resins are applied on high-speed lines in the United Kingdom. These include alkyds, acrylics, vinyls, phenolics, organosols and plastisols. (See Table 12-14.)

Alkyds are the least expensive but are limited by their lack of heat resistance, lack of resistance to chemical attack under adverse conditions and, except in the case of a few modified formulations, poor weatherability. Thermosetting acrylics, based on acrylamide copolymers, have extremely good heat resistance and resistance to breakdown in a corrosive environment, and under certain conditions have good weatherability at a reasonable price. Epoxy coatings are more suitable for thin lacquers requiring very good fabricating properties and chemical resistance.

The polyvinyl chloride (pvc) organosol type system has all the flexibility to be expected from a purely thermoplastic finish but lacks chemical resistance due to the plasticizer present, and the combination of the latter with high pigmentation makes good weathering formulations difficult to achieve. Pvc plastisols and films are similar in properties but are applied as much thicker coatings which may be embossed or otherwise decoratively finished. A lower pigmentation is required to hide the substrate metal because the coating is thicker and, with careful plasticization and stabilization, good weather resistance may be attained. The embossed pattern prevents any scratch marks incurred during site handling from becoming noticeable. Dirt pickup and retention may be reduced by the application of a hard mar-resistant lacquer over the top of the pvc.

For interior use, it may be appropriate to use an acrylic or polyester overlacquer to overcome nicotine staining and various kinds of discoloration. For a decorative finish, laminated films are available in a far greater variety of textures and multicolor prints than would be economically convenient to produce with pvc liquid coatings on a strip line.

Solution vinyl coating is useful in those applications where a mar-resistant surface is required, in addition to sufficient flexibility for the more severe deep-drawing fabricating operations.

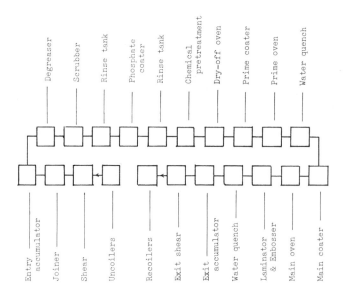

Figure 12-52: Schematic Diagram of Coil Coating Line.

---

† D. S. Newton and P. D. Winchcombe, "Recent Developments in Coil Coating — Part I", Sheet Metal Industries, September, 1973, p. 508 ff.

PROPERTIES OF THE ROLLED STRIP

A modern coating line which chemically cleans and treats the strip prior to coating is shown in Figure 12-52.[20] It utilizes a proprietary chemical pretreatment Accumet C, developed by Albright and Wilson, Ltd., which is capable of processing a variety of metals in a single stage (thereby eliminating bath changes when a different metal is coated). The line can handle cold-rolled and hot-dipped galvanized steel, aluminum, brass and phosphor bronze in thicknesses from 0.004-inch up to 0.080-inch.

Lamination can also be carried out on the line using coils of various laminates fed from pay-off units mounted above the embossing and laminating machine and thermosetting adhesives applied at the main coater unit.

Figure 12-53: Exit End of the New "Stelvetite" Plastic Coating Line at British Steel Corporation's Shotton Works.

The lamination of polyvinylchloride to zinc coated strip (to form a product designated "Stelvetite") is carried out on the line shown in Figure 12-53.[+] This new line, one of the largest in the world, has a design maximum speed of 65m (200 ft.) a minute and laminates zinc-coated strip between 0.67 and 1.35m (24 inches and 52 inches) wide and 0.35 and 1.6 mm (0.015-inch and 0.064-inch) thick, with polyvinyl chloride (pvc) in a wide range of colors, wood grains and embosses. The substrate material is either hot-dipped galvanized steel or uncoated cold-reduced steel, the latter being electrolytically zinc plated in the Stelvetite line prior to lamination.

12-28 Embossing Sheet Products

Rotary embossing is the reshaping of a flat surface into a three-dimensional pattern possessing depth and width. It stretches and bends the material to conform to a design imposed by pressure-loaded, matched rotary male and female rolls.[∞]

In the metalworking field, rotary embossing serves three purposes: It provides decoration; it strengthens the material, permitting use of lighter gage metal without sacrificing rigidity; and it permits salvaging of marred or scratched metal sheet.

---

[20] "Custom Coil Coaters Installs New Coating Line", Sheet Metal Industries, September, 1972, pp. 537-548.
[+] "Plastic-Coated Steel Makes Yet Another Step Forward", Sheet Metal Industries, August, 1973, pp. 437-440.
[∞] "Embossing Uncoated and Precoated Coil", Sheet Metal Industries, January, 1971, pp. 50-52.

Applications of embossed metals are found in appliances, TV sets, radios, automotive scuff plates, dashboard panels, and interior and exterior trim. In architecture, it is becoming an important design element in siding, roofing, and curtain wall panels.

Basically, the rotary embossing machine is an offspring of the conventional rolling mill. It was pioneered by Modern Engraving & Machine Company, USA, a few years after World War II, when the need arose for the high-speed embossing of metals on a mass production basis. The machine consists of a pair of side frames with integral hydraulic rams securely mounted to move the engraved bottom roll under controlled pressure against the engraved top roll. The frames have side rails that are precision machined to accept sliding ball-bearing boxes. These rotate the top and bottom rolls with a minimum of friction and maintain alignment between the bearing centers.

The embosser, coupled to an ac variable speed or dc drive, requires a power unit to supply hydraulic pressure to the rams. Dynamic braking, smooth acceleration and jogging are prerequisites for a smooth functioning unit.

The embossed design is imparted by matched, hardened forged-steel embossing rolls (Figure 12-54). Tools have been designed to produce 100,000 patterns and rolls can be produced in varying degrees of hardness, ranging up to the tool steel category.

Figure 12-54: Embossing Pre-painted Strip.

If the embosser is used in a high-speed line with pay-off and recoiling reels, pneumatically actuated bridle assemblies are commonly used to support and align the strip each side of the embossing rolls. Sometimes the embosser is used in a roller levelling or slitting line, in which case it is usually carriage mounted so that it may be conveniently moved in or out of the line as required.

Embossing produces elongation and bending, which creates a stress differential between the bond of the paint and the base metal. Because of this, the pattern must be rounded to produce a flowing effect in the embossing. Patterns should not exceed 0.015-inch in depth on painted strip.

The largest volume of embossing throughout the world is of aluminum and its alloys. In lesser degrees, stainless steel, copper-base alloys and, increasingly, mild steels are embossed. All these embossing operations benefit from the use of fully hardened engraved rolls.

PROPERTIES OF THE ROLLED STRIP

Bearing in mind the fact that the embossing of aluminum represents the largest volume application of the art, there are four standard universal patterns, which are commercially in demand. In order of importance they are stucco, diamond, square and woodgrain. Ribbed, hammered, leathergrain and pebble patterns are used in minor volumes.

Embossed stainless steel for purposes of decoration and permanency is being increasingly used. Doors, elevator interiors and surrounds, floor tiles and wall panels are all being produced and used with a variety of embossed patterns. Many of these patterns are one sided, produced by a coining process using one engraved roll and one plain roll.

In an attempt to extend the application of metal cladding with a compromise between the softness of aluminum, the greater wear resistance of stainless steel and their respective corrosion resistance in differing environments, without the high cost of stainless steel, embossed galvanized steel sheet is emerging as a new building material.

Embossing of plain or channeled panels of galvanized steel achieves some of the objective of stiffening, produces a more pleasing texture and surface and reduces reflectivity and glare. Thinner sheets can be used. Such embossing over wide steel coil or sheet has only been made possible by the use of fully hardened forged steel engraved embossing rolls in the mills. Only these rolls will resist the high pressures and stresses involved, while permitting operation at high speed with high rates of production.

12-29 Cold Roll-Forming

Cold-rolled steel strip is often roll-formed into a variety of products such as angles, channels, tubing, etc., with typical cross-sectional profiles as shown in Figure 12-55. Such profiles exhibit essentially uniform metal thickness at all parts in the cross section and are held to very close dimensional tolerances.[ω]

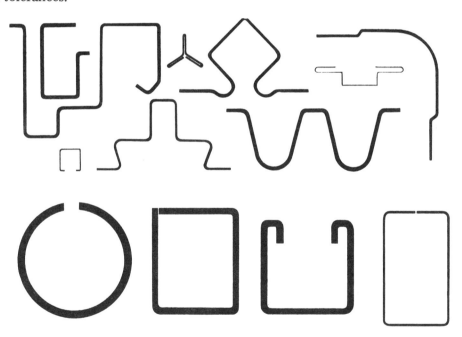

Figure 12-55: A typical range of cold-formed section profiles (Ayrshire Metal Products, Ltd.).

Materials used in the manufacture of cold-formed sections are either in coiled strip or in sheet form and may be of almost any metal with sufficient ductility to be bent cold. These include low-carbon steels, low-alloy steels, stainless steels, aluminum and its alloys, copper, brass,

---

[ω] "What is a Cold-Rolled Section?", Sheet Metal Industries, August, 1971, pp. 596-597.

phosphor-bronze and zinc. Although low-carbon steels are most commonly used, high-strength steels are also being utilized in ever increasing quantities.

For the cold roll-forming process, strip is supplied in coils up to 5 feet wide and weighing up to 15 tons. This strip has, of necessity, to be slit on multiple gang slitters to the exact strip width required for any particular section. Commonly used strip thicknesses range from 0.0625-inch (0.1587 mm) to 0.1563-inch (0.3987 mm) in widths close to 18 inches.

The cold roll forming process is such that strip mounted on a drum or in a coil box is fed into one end of a rolling mill where it passes through a series of rolls or forming cones of the required contour and arranged in tandem. These progressively shape the strip to its desired final profile.

The mill, essentially a series of roll stands, generally supports the pairs of forming cones on horizontal spindles. A simple profile, say a plain channel section, may be produced with as few as six pairs of rolls. A complex or more intricate section may require 15 sets or more of such rolls. The various states or settings in one method of rolling an inwardly lipped channel are shown in Figure 12-56.

Various methods are used to cut the finished section as it leaves the mill, some of which involve the use of a movable table with a mounted jig stopper travelling with the section being rolled. This allows a flying shear or similar device to cut the section to required length without interrupting the speed of the material passing through the mill, the table returning to position in time for cutting the next length. The speed of rolling may vary from 20 to over 200 ft./min. As the coil may be several hundred feet long, almost any length of section can be rolled. Ease of handling and transport facilities are usually the only limiting factors.

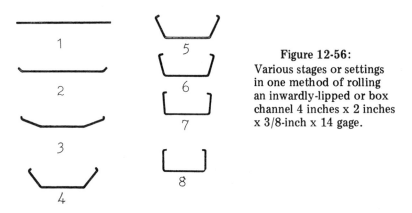

Figure 12-56: Various stages or settings in one method of rolling an inwardly-lipped or box channel 4 inches x 2 inches x 3/8-inch x 14 gage.

12-30 Silicon-Steel Electrical Sheets

Silicon-steel electrical sheets fall into two general classifications (1) grain-oriented sheets and (2) non-oriented steels. The former contain about 3-1/4% silicon and the latter are further classified on the basis of silicon content. Low-silicon steels contain 1/2 to 1-1/2 percent silicon, intermediate-silicon steels contain 2-1/2 to 3-1/2 percent silicon and high-silicon steels about 3-3/4 to 5% silicon. In each of these classifications, sheets are "graded" on the basis of core loss defined as the energy expended in magnetizing the material (expressed in terms of watts per pound) with a 60 cps alternating electric current providing a maximum flux density of 10 or 15 kilogauss). Such core losses usually range from about 0.5 to 2.5 watts per pound at the 10 kilogauss level and from 0.6 to 6 watts per pound at the 15 kilogauss level.

In the processing of electrical steels of this type, only those containing less than 3-1/2 percent silicon can be satisfactorily cold reduced because of the brittleness of the steels at room temperature. Hot-rolled coils that are to be cold rolled are processed either on single-stand reversing mills or tandem mills, followed by heat treatments to develop the desired electrical properties and levelling to achieve the desired flatness. In the case of grain-oriented steel, however, the cold

PROPERTIES OF THE ROLLED STRIP

reduction takes place in two stages, each followed by a suitable continuous anneal and a final, high-temperature box anneal.

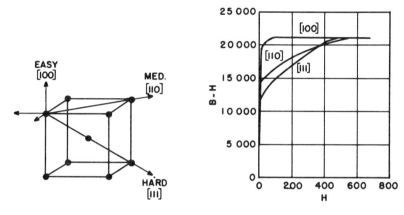

Figure 12-57:
Effect of orientation on the magnetic properties of a crystal, showing relative ease with which the cubes comprising the iron-silicon space lattice can be magnetized in different directions.

The ease with which a bcc crystal of iron may be magnetized depends on the orientation of the lattice with respect to the magnetic field. As shown in Figure 12-57, the "easy" direction is the {100}, and the "hard" direction is the {111}, with the {110} being of an intermediate nature. After the completion of the processing of grain-oriented steels, most of the grains are so arranged that edges of the unit cubes comprising each grain (the {100} direction) are parallel to the rolling direction and face diagonals are aligned in the transverse direction. This orientation represents the "cube-on-edge" texture as shown in Figure 12-58, with the best magnetic properties in the rolling direction, poorer properties in the transverse direction and the poorest at an angle of 55 degrees to the rolling direction. Fortunately, certain impurities favor the growth of the cube-on-edge texture, and sulfides and nitrides have both been used to impede the growth of primary grains other than those of the desired texture.

Cube-on-face orientation (as shown in Figure 12-59), such as is commonly found in the nickel-iron alloys, can also be developed in silicon iron. However, in this case, the material should be free of impurities and cold rolled enough to provide an adequate number of nuclei of the correct orientation.

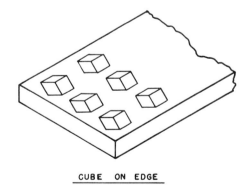

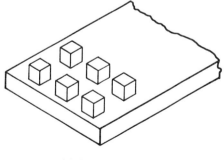

Figure 12-58: Cube-on-Edge Crystal Orientation.

Figure 12-59: Cube-on-Face Crystal Orientation.

R. H. Trapp, "Processing and Properties of Magnetic Materials", Met. Soc. Conf. Vol. 6 — Flat Rolled Products II, Semifinished and Finished — Interscience Publishers, New York, 1960, pp. 99-114.

J. E. May and D. Turnbull, "Effect of Impurities on the Temperature Dependence of the (110) [001] Texture in Silicon Iron", Journal of Applied Physics 30S, April, 1959, pp. 210S-211S.

12-31 The Processing of Stainless-Steel Strip

Stainless steels are grouped into three classes:

1. Martensitic — iron-chromium alloys hardenable by heat treatment and which include Types 403, 410, 414, 416, 420, 431, 400A, 440B, 440C, 501 and 502.
2. Ferritic — iron-chromium alloys that are largely ferritic and not hardenable by heat treatment (ignoring the $885°$ F embrittlement). They include Types 405, 430, $430°$ F and 446.
3. Austenitic — iron-chromium-nickel alloys not hardenable by heat treatments and predominantly austenitic as commercially heat treated. Such stainless steels include Types 301, 302, 302B, 303, 304, 304L, 305, 308, 309, 310, 314, 316, 316L, 317, 320 and 347.

With the exception of the high-carbon hardenable types, all the stainless steels can be cold worked, although certain precautions must be taken in some cases. Ferritic types, especially those containing over 20% chromium, are extremely notch sensitive at room temperature and care must be taken to avoid notching. However, between $400°$ F and $600°$ F, the sheets are tough and cold rolling may be satisfactorily carried out in this temperature range.

Cold work causes some austenitic steels to transform partially to a low-carbon martensite. This effect, in addition to the strain hardening caused by the cold work itself, causes such austenitic steels to have a high rate of work hardening and necessitates intermediate anneals.

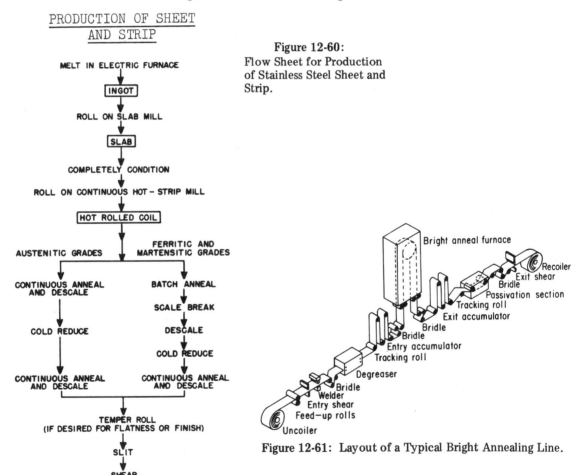

Figure 12-60: Flow Sheet for Production of Stainless Steel Sheet and Strip.

Figure 12-61: Layout of a Typical Bright Annealing Line.

PROPERTIES OF THE ROLLED STRIP

The processing of stainless-steel sheet and strip is illustrated in Figure 12-60. After rolling and annealing, all of the stainless steels are descaled usually by pickling in acids. For straight chromium grades containing up to about 21% chromium, the hot-rolled coils are batch annealed at subcritical temperatures, whereas the straight chromium grades containing in excess of 21% chromium and the austenitic grades are annealed and quenched on a continuous basis. For the austenitic steels, this annealing is carried out in an oxidizing atmosphere which produces a heavy scale and "burns off" the surface defects. The quenching practice used depends upon the thickness of the material. For thick material, high-pressure water sprays are used, but for thin strip, air cooling is adequate.

Pickling may be carried out using acids (hydrochloric followed by a mixture of hydrofluoric and nitric acids or sulphuric followed by nitric acid) or by molten-salt descaling processes. Cold reduction is often undertaken on a reversing mill although tandem mills may also be satisfactorily used. Intermediate anneals are usually carried out in an atmosphere of pure hydrogen with a dew point of $-100°$ F.

After annealing and pickling, the surface of the cold-rolled material exhibits what is called a No. 2D (dull, cold-rolled) finish for sheet or a No. 1 cold-rolled finish for strip. If a brighter finish is required, the material is rolled on a temper mill to provide a No. 2-B (bright-cold-rolled) finish for sheet or a No. 2 cold-rolled finish for strip. Strip processed on a bright annealing line (Figure 12-61) has a brighter luster than material conventionally annealed and pickled.[λ] Even shinier finishes are obtainable by mechanically polishing sheets or strip.[ω][∩]

Austenitic stainless steels (such as Type 301) may be produced to tensile strengths as high as 200,000 psi. Usually Type 301 is supplied to four standard minimum tensile strength levels of 125,000, 150,000, 175,000 and 185,000 psi (being respectively designated 1/4, 1/2, 3/4 and full-hard temper).

12-32 Surface Finish and Its Influence on Formability and Final Appearance

The final surface finish of sheet products is attained both by the cold-reduction and temper-rolling operations.[Γ] In the former, the surface finish is affected significantly only by the final rolling pass which customarily reduces the sheet about 10 percent. The surface pattern of the work rolls in this last stand is largely, but not entirely, impressed or coined into the sheet surfaces. In temper rolling, not only is the discontinuous yielding characteristics of the steel eliminated and the strip flattened, but the final surface texture is imparted to the surface.

To provide as-annealed product with consistent surface finish, it is most important to maintain cold-rolling conditions that produce a large number of peaks per inch on the strip surface. Moreover, such a surface condition is desirable for temper rolling since the effectiveness of temper rolling in eliminating the discontinuous yield point and associated Luders bands depends on the production of a fine-scale pattern of nonuniform microstrains in the temper-rolled sheet.

It is now generally accepted that there is a marked relationship between surface finish and press performance. The optimum surface finish for cold-rolled sheets intended for press forming is one with a roughness height of 30 to 60 microinches arithmetic average (AA) and a large number of peaks per inch (in excess of 150 as determined by the Proficorder or about 250 using the Surfacount instrument).[✕] The principal attribute of a surface with this finish is its ability to supply effective lubrication to reduce frictional forces between the die and the sheet throughout the

---

[λ] R. C. Bongartz and G. L. Steffel, "Bright Annealing of Stainless Steel", Iron and Steel Engineer Year Book, 1973, pp. 460-464.

[ω] E. Pole, "Surface Finishing of Stainless-Steel Sheet and Strip", Sheet Metal Industries, January, 1972, pp. 47-53.

[∩] D. V. Roland, "Modern Production Methods for Stainless Steels", Sheet Metal Industries, January, 1972, pp. 47-53.

[Γ] M. Kotyk and J. W. Stewart, "The Influence of Surfaces and Shape on Formability of Carbon-Steel Sheet and Strip", Paper presented at Material Forming Seminar sponsored by the ASTME, January 20, 1969.

[✕] For a description of these instruments, see Section 4-14.

forming operation, thereby preventing seizing and stick-slip conditions and the development of sharp stress gradients.

In the many applications where the sheet-steel surface is exposed to view, the final appearance after coating is of major importance. This is significantly dependent on the surface roughness of the metal surface and, at the same time, the best appearance is attained when the surface exhibits a high degree of uniformity of roughness. Experience has shown that medium surface roughnesses are to be preferred and too smooth surface finishes (<20 microinches) should be avoided.

Where sheet surfaces are to be painted, the influence of the sheet roughness is reduced as the paint thickness is increased. Butler & Pope[a] showed that prior sheet roughness had no effect on the appearance of a three-coat paint system over a phosphate primer having a total thickness about 16 times the peak-to-valley height of the surface roughness. However, when the thickness of the paint film was decreased to two coats with no initial primer, the effect of sheet-surface roughness on paint appearance was found to be considerable (see Figure 12-62).

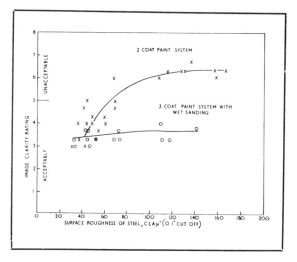

Figure 12-62: Influence of surface roughness of steel sheet on paint finish (after Butler and Pope).

12-33 The Application of Forming Lubricants to Steel Products

Originally lubricants were applied to sheets by the user just prior to his forming operations. Recently, however, purchasers of sheet product are requesting that such lubricants be applied at the sheet mills to both hot-rolled and cold-rolled product. Various proprietary type lubricants are available for this purpose, some of which can be applied simply by spraying or dipping, others requiring a two-stage application with heating of the strip being necessary.

Undoubtedly, considerable development in this area of forming lubricants and their application will be necessary to establish the composition and the film thickness of an "ideal" lubricant. At the present time, the desired lubricant should be

(1) dry (such as a soap)
(2) easily applied (in about 10 feet of strip travel)
(3) economical (not costing more than a good drawing compound)
(4) readily removed by alkaline cleaning
(5) provide enough lubricity to permit forming the steel without the application of more lubricant at the presses, and
(6) afford good rust protection on blanks and uncleaned stampings stored outdoors.

---

[a] R. D. Butler and R. J. Pope, "Surface Roughness and Lubrication in Press Working of Autobody Sheet Steel", Sheet Metal Ind., 44, September, 1967, pp. 579-592 and 597.

Name Index

Abersychen Iron Works, 6
Achey, F. A., 217
Adams, J. R., 1, 3, 14
Ainslie, N. G., 726
Alexander, J. M., 246, 565
Alford, M. H. T., 79, 192
Algoma Steel Corp., Ltd., 62
Allegheny Ludlum Steel Corp., 14
Allen, J. E., 759
Allison, F. H., Jr., 115, 116, 118, 158
Alters, M. F., 713
American Iron and Steel Co., 14
American Rolling Mill Co., 14
American Steel and Wire Co., 10
Anderson, E., 421-422
Anderson, J. T., 355
Angeid, E., 214
Anmura, T., 18
Arant, N. R., 747
Archard, J. F., 340
Arimura, T., 227, 643
Arimuza, M., 679, 707
Arnold, J., 727
Arnold, L. E., 140
Arrowsmith, J. M., 724
Arthur, D. F., 317
Atkinson, M., 736
August Schmitz Co., 15
Aul, F. W., 721, 724
Avner, S. H., 142
Ayers, M. D., 761

Baark, W. R., 209
Bach, H., 458
Backofen, W. A., 288, 736-737
Backstrom, B., 742
Badlam, S., 8
Bain, E. C., 733
Baird, J. D., 474, 724, 742, 743
Baker, J. L. L., 362
Baker, R., 661, 663
Baldwin, W. M., 661, 663

Ball, C. J., 474
Ball, C. S., 743
Baltz, W. E., 213
Barney, D. V., Jr., 133-134
Barnitz, R. W., 183
Barreto, J. P., 675
Bates, R. K., 511
Baumann, R., 458
Bauscher, J. A., 727, 735
Beam, D. J., 763
Becker, R., 459
Bedi, D. S., 244
Bedson, G., 7
Belevskii, L. S., 114
Belkin, M. Y., 170
Bell, R. J., 711-712
Belyayev, N., 149
Bennett, G. V., 140
Bentz, R. J., 249-250, 260, 264, 269, 300, 316, 321, 334, 471-472, 481, 533, 559, 587, 604, 704
Benzhega, A. S., 170
Berger, B., 683
Bergh, S., 742
Bernsmann, G. P., 657
Bessemer, H., 7
Bessemer Works, 10
Bessen, I. I., 741, 743
Bessent, T. A., 56, 82, 96
Bever, M. B., 476
Bevis, W. M., 766
Beynon, R. E., 110, 116
Biffin, F. M., 329
Bilby, B. A., 741
Billingmann, J., 255
Biot, J. B., 345
Birkbeck, G., 460
Birkenshaw, J., 6
Bland, D. R., 493, 511, 528, 557, 575, 576, 588, 595, 626, 645
Bland, R. A., 713
Bleckley, 8

Blickwede, D. J., 721, 726, 734, 735, 741, 743
Blockley, T., 3
Blok, H., 339, 342
Bodner, S. R., 463, 472
Bongartz, R. C., 775
Boos, R. T., 217
Bosenberg, K., 702
Brachet, N. A., 111, 114, 167-168
Bradd, A. A., 111, 114, 118, 149, 167-168
Bradford Iron Works, 7
Bradley, B. C., 48, 236, 239
Brandywine Rolling Mill, 10
Brant, R. K., 256, 315
Bray, D. B., 229
Bressanelli, J. P., 266
Briggs, P. R. A., 207
British Steel Corp., 55, 58
Bruce, T. B., 757
Brulier, 1
Brunner, O. G., 106, 230
Bryan, T. E., 222
Bryant, G. F., 236, 557
Buchinski, J. J., 715
Bugakow, W., 661
Bulmer, B., 1, 11
Burke, J. E., 734
Burns, R. S., 736, 748
Burton, M. S., 460
Butler, J. F., 743
Butler, J. J., 743, 763
Butler, R. D., 776
Butrick, F. M., 753
Butylkina, L. I., 170

Cahn, J. W., 723
Cambria Iron Works, 10
Cameron, A., 342
Campbell, J. E., 463
Capus, J. M., 246, 294
Carmi, A., 355
Carrier, G. F., 706
Carter, D. J., 763
Chakko, M. K., 158
Chamin, I. A., 269
Charlton, F., 452
Cheng, H. S., 243, 540
Chisholm, S. F., 257
Choi, H., 345
Christ, B. W., 460
Christof, G. R., 655, 658-659, 661
Christopherson, D. G., 476

Cichelli, A. E., 183
Climax Molybdenum Co., 109
Cockroft, M. G., 246, 294
Cockshutt, J., 5, 11
Cohen, J. B., 450
Coleman, V. S., 56, 228
Collins, D. W., 61, 266
Colombier, L., 110
Cone, C., 761
Conklin, W. F., 15
Conley, R. C., 99
Consett Iron Co., 6
Cook, G., 453
Cook, M., 511, 526
Cook, N. E., 761
Cooper, Hewitt and Co., 10
Cooper, W. L., 19, 183
Cort, H., 6
Cosse, P., 519
Cottrell, A. H., 448, 741
Courcoulas, J. H., 626
Cowan, C., 9
Cox, H. N., 48, 236-237, 239
Crane, F. A. A., 246
Crawshay, R., 5, 11
Cross, H. C., 304
Curry, A. M., 92
Cuyahoga Plant, 63

Daelen, R. M., 7
Dahl, W., 256, 284, 551, 585
Dahle, O., 674, 683
Dahm, J. R., 689
Danzig, Q. B., 130
Dauberman, W. H., 106, 108
Davenport, E. S., 733
Davidenkow, N., 661
Davies, J., 42
Davies, W. E., 674, 683
da Vinci, L., 1, 2
Davis, N., 683
Dawe, A. S., 766
de Bochs, G., 1, 11
de Buigne, J., 8
de Caus, S., 1, 2
Dedek, V., 284
Deering, J. M., 233
Derringer, B. W., Jr., 81
de Sars, M., 746
De Young, D. R., 106
DiGioia, A. M., 145
Dillamore, I. L., 721, 734

NAME INDEX

Dinnik, A. N., 482
Diolot, L., 698
Dion, P., 715
D'Isa, F. A., 113, 155
Dolphin, T. J., 106
Douglas, J., 262
Douthwaite, R. M., 460
Dow, T. A., 341
Dowlais Works, 8
Dowler, L. K., 389
Drake, H. J., 263-264
Draper, D. W., 227
Draxler, J. A., 103
Dropkin, D., 355
Drutowski, R. C., 338
Druzhinin, A. N., 226
Druzhinin, N. N., 226
Dugan, J. M., 122, 127, 158
Dunn, C. G., 734

Eckert, E. R. G., 353
Economopoulos, M., 519
Edgar Thomson Works, 13
Edington, J. W., 742
Edmundsen, M. R., 409
Edwards, C. A., 733
Eichinger, A., 453
Ekelund, S., 478, 526
Ellis, F., 557, 576, 626, 645
Emanuel, N. M., 275
Emicke, O., 551
Emori, T., 679, 707
Eppelsheimer, D., 1, 2, 15
Erzurum, H., 113, 155
Ess, T. J., 15
Evans, I. W., 127, 256

Fairfield Works, 58
Fairless Works, 47
Fanning, F. H., 3
Farrington, G. E., 36
Fedeli, E., 275
Fletcher, S. R., 734
Fontana Plant, 61
Fontley Iron Mills, 6
Foppl, 491
Ford, H., 5, 290, 317, 457, 471, 493, 511, 528, 552, 557, 565, 571-572, 575-576, 581, 588, 594-595, 626, 645
Forster, G. A., 75
Fourier, J. B. J., 345
Fox, T. A., 694

Frank, F. C., 448
Freeman, J. B., 205
French, H. J., 304
Friedel, J., 474
Fritz, G., 10
Fritz, J., 10
Frolov, E. J., 550
Fuchs, S., 482
Fujii, S., 18
Fukuda, T., 662-663, 666, 672
Fukuyama Works, 18
Funke, P., 256, 284

Gary Steel Works, 13, 20, 59
Geiringer, H., 454
Geleji, A., 26, 69, 113, 171, 172, 552
Gensamer, M., 721
Gibson, D. G., 172, 174, 184
Gifford, J. D., 229
Gilbertson, W. F., 658
Giles, M. E., 166
Godenow, R. H., 734
Golubev, T. M., 550
Gordon, G., 726
Gorecki, W., 287
Goss, B. L., 87
Gough, J. W., 87
Gray, R., 694-696, 706
Green, S. J., 462
Greenberger, J. I., 27, 75, 112
Gregory, W., 263-264
Griffin, J. F., 655, 658-659
Griffith, J. F., 210
Gronbeck, R. W., 75
Gross, J., 113, 155
Grubb, D. F., 95
Grunhofer, H. G., 746
Guetlbauer, F. G., 193
Gurin, S. M., 171
Guy, A. G., 723-724
Guy, V. H., 300, 321

Haab, R. C., 20, 59, 750
Hahn, G. T., 471, 519
Halliday, D., 398
Halter, A. C., 54, 592
Ham, J. M., 626
Hanbury, J., 2, 3
Hanci, G., 702
Harbaugh, W. Le C., 304
Hashimoto, Y., 733
Hawkins, J. C., 763

Hawley, R., 83, 85
Hazeldine, J., 5
Hedges, E. S., 24, 717
Heger, R. H., 736
Hencky, H., 454
Henrickson, J. A., 763
Henze, M., 674, 683
Herman, R., 353
Hertz, H., 148, 482
Hessenberg, W. C. F., 207, 519, 572, 574-575, 584, 610, 612-613, 619-620, 623
Hewitt, C., 10
Heys, J. D., 214, 227
Hill, R., 454, 520, 579, 583, 645
Hillier, M. J., 244, 675
Hinton, R., 450
Hirano, H., 256-258
Hirasawa, T., 714, 716
Hitchcock, J., 246, 478, 484, 491
Hoare, W. E., 24, 717
Hocket, J. E., 511
Hodge, J. M., 118
Hoffman, R. E., 726
Hogshead, T. H., 353, 358
Hollomon, J. H., 459, 466, 471
Homestead Works, 10
Horn, J. E., 222
Horst, R. J., 205
Horsford, W. F., 736-737
Hotta, M., 681, 707
Huber, M. T., 482
Hultgren, F. A., 734
Hundy, B. B., 741, 743, 748
Hunter Engineering Co., 31
Hutchison, M. M., 460

Irvin Works, 17
Iwao, Y., 256-258

Jackman, R., 75
Jackson, G., 300
Jackson, R. W., 42
Jacobs, M. B., 20, 59, 750
Jakimovich, A., 606, 609
Jamieson, R., 709
Jenkins, W. N., 610, 620, 623
Johannsen, 2
Johansson, B., 674
Johnson, F. G., 215, 222, 231-232
Johnson, T. E., 736
Johnson, W. A., 723
Johnson, W. R., 256

Johnston, S. S., 757
Jolley, W., 734
Jones and Laughlin, 14
Jones and Lauth, 10
Jones, F. W., 159
Jortner, D., 478, 485, 511, 515
Josephy, N. H., 753
Josepsson, A., 742
Ju, V., 748
Juvinall, R. C., 447

Kading, G., 525, 559
Kaiser Aluminum and Chemical Corp., 43
Kaiser Steel, 61
Kajiwara, T., 681, 707
Kamata, M., 227, 643
Kamijo, T., 451
Kannel, J. W., 341
Kaplanov, V. I., 258
Kawaguchi, S., 159
Kawamata, T., 681, 707
Kawasaki, B., 730
Kawasoko, Y., 18
Keh, A. S., 423, 473, 721, 742
Keller, A. T., 706
Kenyon, R. L., 92, 724
Kimes, T. H., 131
Kimmel, J. J., 81
King, W. D., 226
Kinoshita, M., 310
Kishida, A., 230, 239
Kiss, E., 372
Kitao, S., 230, 239
Kloman, A., 10
Kneschke, A., 548, 588
Knowlton, A. E., 87, 91
Kobasa, D., 492, 533
Kobayashi, 160
Koehler, H. B., 92
Kokubo, I., 256-258
Kolydich, V. M., 226
Koot, L. W., 379
Korolev, A. A., 550
Kotyk, M., 655, 775
Krashers, D., 140
Kubotera, H., 734
Kuntz, H., 664
Kusakabe, T., 714, 716
Kuz'min, V. I., 251, 258

Labee, C. J., 239
Langer, J., 551

NAME INDEX

Lankau, G., 166
Lankford, W. T., 735
Lapham, G. C., 230
Larke, E. C., 36, 150, 154, 288, 480, 511, 572
Larkman, E. C., 61, 266
Latorre, J. V., 158
Latour, U. H., 748
Lauth, B., 6
Lay, K., 450
Leach, R. V., 7
LeMay, I., 321
Leslie, W. C., 721, 724, 734, 742
Levinson, D. W., 718
Lianis, G., 528, 626
Lillie, C. R., 718
Lindholm, U. S., 465-466
Lindley, T. C. 742
Liputkin, Y. V., 343
Litz, D. C., 249, 269, 316, 321, 471, 481, 668
Lloyd, K. A., 262
Llwyd, E., 2
Lode, W., 453
Loewy-Robertson, 30
Lomax, D., 172
Long, R. F., 56, 82, 96
Lucas, K. H., 551
Ludwig, J. R., 273
Leug, W., 246, 256, 284, 550, 594
Lukens, C., 9
Lukens Steel Corp., 9-10

Mackenzie, C. R., 474
Mackintosh-Hemphill Co., 1-4, 10-11, 115-116
MacNaughton, L., 149
MacQueen, A. J. F., 62, 556
Madejski, J., 287
Maiden, C. J., 462
Manjoine, M. J., 461, 462
Mann, 14
Marks, L. S., 539
Marshall, J. G., 75
Mason, W., 453
Mathey, 36
Matsuda, K., 734
Matsudo, K., 733
Matsushima, T., 679, 707
Matsuura, Y., 258, 293
Mattke, G. H., 233
Mavis, F. T., 148, 157
Maxey, J. C., 227

May, J. E., 773
Mayer, E. H., 734, 739
McAdams, W. H., 352
McCallum, N., 718
McConnell, W. M., 34-35
McCoy, J. R., 174, 184, 393
McCoy, W. E., 172, 179
McGannon, H. E., 83, 120, 272, 332, 719, 731, 733, 739, 754
McLean, C. J., 746, 747
Meason, I., 9
Mehl, R. F., 431, 723
Melford, D. A., 166
Melloy, G. F., 110, 125
Menelaus, 8
Michalak, J. T., 721, 723-724
Micheev, M. A., 355
Mirer, A. G., 226
Misaka, Y., 626
Mizukoshi, T., 210
Montgomery, A. B., 34, 35
Montgomery, T. B., 608-609
Moon, D. P., 463
Morgan Construction Co., 338
Morgan, J. P., 747
Morihara, T., 763
Morris and Bailey Steel Co., 14
Morrison, J. L. M., 453
Morrison, W. B., 722
Mould, P. R., 736
Muller, H. E., Jr., 158, 169
Murata, K., 562
Murphy, H. L., 266

Nadai, A., 455, 461, 471, 480, 504, 507
Nagoya Works, 238
Nakada, Y., 473
Nakajima, K., 562
Nakaoka, N., 734
National Steel Corp., 43
Neckervis, R. J., 256
Newton, D. S., 768
Newton, I., 345
Newton, J. T., 7
Nichol, W. G., 130
Nikolaev, V. A., 273, 363
Nileshwar, V. B., 166
Nippon Kokan, 43, 52
Nisshin Steel Co., 17, 43, 50
Noble, R. E., 43, 46-47, 61
North Chicago Rolling Mill Co., 10
Norton, A. S., 237

Obernikhin, S. A., 251, 258
O'Brien, J. J., 233, 707
O'Connor, H. W., 657, 706
Ohama, T., 17, 50
Ohm, A., 265
Okado, M., 227
Okamoto, T., 18
Oliver, P., 9
Oppermann, K., 686
Orehoski, M. A., 118
Orellana, I. G., 196, 198, 205
Orowan, E., 228, 459, 462, 478, 490, 594
Osborn, R., 557
Osterle, J. F., 478, 485, 511, 515
Ostermann, F., 661
O'Sullivan, D., 58
Owen, L. J., 736

Paliwoda, E. J., 741, 743
Pallone, G. T., 349
Panknin, W., 716
Pannek, H., 264, 272
Parke, D. M., 362
Parker, R. J., 526
Parkgate Works, 7
Parkins, R. N., 453
Parsons, B., 476
Patula, E. J., 365
Pavlov, I. M., 321
Pawelski, O., 491, 525, 559, 683
Paxton, H. W., 723
Payne, J., 3, 4
Pearce, R., 735
Pearson, W. J. K., 673, 675
Peck, C. F., Jr., 145, 148, 157
Peebles, J. E., 333
Peeples, R. M., 54, 592
Pelloux, R., 140
Pennoch, I., 9
Perrault, G., Jr., 45
Petch, N. J., 460
Peterson, C. E., 115-116, 118
Petin, Gaudet et Cie., 8
Petraske, K. A., 214
Phillips, A., 140
Phillips, R. A., 626
Piehler, H., 737
Pimenov, A. F., 171
Pinnow, K. E., 266
Pittsburgh Steam Engine Co., 9
Playfield, W., 5
Pohlem, C., 3, 14

Polakowski, N. H., 655, 701, 715-716
Pole, E., 775
Polukhin, V. P., 112-113, 363
Pope, R. J., 776
Powell, N., 114
Prandtl, L., 507
Prentiss, F. L., 730
Prescott, J., 486
Puppe, J., 36
Purnell, J., 4
Putman, W. J., 453

Rabinowicz, E., 338
Radzimovsky, E. I., 149
Randich, E. A., 728
Rasselstein, AG, 42
Rea, D. E., 215, 222, 231-232
Read, W. T., 448
Reddy, D. M., 655, 701
Reebel, D., 27
Reis, G. D., 129
Renner, S. E., 689
Resnick, R., 398
Reynolds, E. S., 84
Rickett, R. L., 734
Ricksecker, R., 661, 663
Ride, J. S., 658
Riley, J. E., 61, 266
Ripperger, E. A., 463, 465
Roark, R. J., 530
Robb, G. C., 133-134
Roberts, J. M., 707, 711, 721
Roberts, W. L., 239, 246, 249-250, 255-257,
    264-266, 269, 273, 279, 296, 300, 302,
    305, 316, 321, 323, 328, 334, 366, 382,
    471-472, 481, 484-485, 529, 533, 537,
    559, 584-585, 587, 603-604, 668, 671, 704
Roberts, W. T., 733
Robertson, A., 453
Robertson's of Bedford, 36
Roden, R. B., 6
Rodman, S., 256-258, 261, 266, 282, 315
Rohn, W., 36-37
Rohsonow, W. M., 345
Roland, D. V., 775
Rollason, E. C., 450
Roos, G. W., 75
Ros, M., 453
Rosenberg, S. J., 304
Rosenfield, A. R., 471, 519
Roumanis, P. J., 92
Royce, R. E., 166

NAME INDEX

Rudbakh, V. N., 503
Russell, B., 742

Sabatini, B., 659, 690, 692
Sachs, G., 158, 565
Sadamura, Y., 734
Sadre, M., 158
Saeki, K., 252
Saiki, T., 730
Saito, M., 643
Sakabe, K., 159, 166
Sakaki, T., 17, 679, 707
Sankaran, S., 31
Sansome, D. H., 467
Santo, T., 679, 707
Sasaki, S., 50
Sass, S., 450
Saunders, O. A., 355
Savery, T., 12
Saxl, K., 700
Schey, J. A., 246, 252, 262, 264, 272, 284, 314, 321
Schiff, A. N., 92
Schloemann, Aktiengesellschaft, 25
Schmeissing, H. N., 655, 701
Schmidt, H., 239
Schoenberger and Co., 10
Schoenberger Works, 10
Scholefield, H. H., 61, 266
Schonholzer, E. T., 92
Schroeder, H., 491
Schuler, V., 683
Schultz, R. A., 492, 533
Schwartzbart, H., 256
Scoble, A. A., 453
Scott, F. H., 288
Seeger, A., 723
Seely, F. B., 453
Sellers, J. F., 54, 592, 608, 609
Sendzimir, M. G., 17, 50
Sendzimir, T., 37
Severdenko, V. P., 503
Seybolt, A. V., 726
Shannon, R. W., 5, 15, 24
Shayne, A., 619
Sheehan, J. M., 763
Sheehan, J. P., 256
Shelesnow, J. D., 655, 686
Shellenberger, D. J., 747
Sheppard, T., 707, 711
Sherwood, W., 658
Shibata, H., 679, 707

Shigemura, S., 763
Shikhanovich, B. A., 226
Shimizu, M., 210
Shimomura, T., 733
Shimozato, Y., 763
Shipley, T. G., Jr., 130
Shuman, J. R., 205
Shunan Works, 17
Shutt, A., 250
Shvartsman, Z. M., 171
Siadak, J. C., 45
Sibakin, J. G., 658
Sidey, M. P., 736
Siebel, E., 246, 248, 288, 594, 550
Silken, V., 748
Sills, R. M., 214, 226
Silman, H., 750
Simizu, M., 734
Sims, R. B., 205, 207, 317, 320, 519, 528, 552, 557, 582, 584, 596, 612-613, 619
Singer, A. R. E., 707
Sivilotti, O. G., 674, 683
Skefco Ball Bearing Co., Ltd., 503
Skovokhodov, N. E., 114
Sleeswyk, A. W., 467
Slyusarev, A. T., 261
Smallman, R. E., 732, 742
Smith, C. A. M., 453
Smith, C. L., 288
Smith, E. J., 759
Smith, E. P., 87
Smith, G. V., 460
Smith, M. O., 48, 236, 239
Smirnov, V. V., 74, 78-79, 103, 150, 172, 192, 710
Snell, F. D., 329
Snyder, S. C., 735
Soehnghen, 353
Somers, R. R., 255, 260, 265, 273, 302, 305
Somerville, R. A., 694
Sorokin, V. O., 214
Sparling, L. M. G., 552, 587
Spenceley, G. D., 272
Spivak, L. J., 158
Spreadborough, J., 421-422
Starchenko, D. I., 251, 258
Starling, C. W., 27, 66, 197, 201, 206
Stasko, S., 148, 158
Steffel, G. L., 775
Steinbrecher, O., 224, 226
Steiner, J. E., 118
Stepanek, R., 147

Stewart, J. W., 655, 775
Stickler, S. V., Jr., 106, 108
Stone, M. D., 8, 14, 61, 75, 148, 158, 226, 248, 479, 532, 537, 559, 585, 694-696, 698, 706, 712-713, 728
Stringer, L. F., 92
Struttman, H. S. 179
Sumitomo Metal Industries, Ltd., 640
Superior Steel Co., 14-15
Suzuki, S., 763
Swift, H. W., 471
Swisshelm, W. H., 729
Sykes, C., 159
Sylwestrowicz, W., 288
Symm, G. T., 342

Tabor, D., 338
Takahashi, K., 562
Takahaski, N., 648
Takemura, K., 230, 239
Tanaka, E., 662-663, 666, 672
Tardif, H. P., 743
Tarokh, M., 690, 692
Terekhov, V. N., 363
Tereshko, A. K., 363
Thiele, H., 716
Thieme, J. C., 109
Thomas, P., 58
Thomson, O. B., 56, 228
Thorpe, J. M., 262, 315, 321
Tichener, A. L., 476
Titarenko, I. A., 363
Toda, K., 730
Togashi, N., 230, 239
Tokar, I. K., 269
Tong, K. N., 258, 565
Torrance, J. H., 75
Townsend, N. A., 676
Tracy, J. A., 42
Trapp, R. H., 773
Trenton Iron Works, 10
Trentwood Plant, 43
Tresca, H., 452
Trinks, W., 498
Trostre Works, 55, 58
Tselikov, A. I., 74, 78-79, 103, 150, 172, 192, 499, 710
Tsivitse, P. J., 92
Tsunokawa, K., 662-663, 666, 672
Turley, J. W., 700
Turnbull, D., 773
Turner, G. C., 3, 97, 761

Ullmann, J., 95
Umeda, S., 210
Underwood, L. R., 452, 489, 495, 497, 503, 511, 529, 569
Ungerer, K. F., 700
Union Rolling Mill, 9
United States Steel Corp., 10, 20, 47-48, 59, 63
Usinor-Schloemann, 25
Uto, Y., 763

Van der Walt, C. M., 450
Van Rooyen, G. T., 288
Vassily, G. R., 711-712
Veck, D., 59
Verduzco, M., 716
Verheyden, V. E., 13
Vigneron, F. R., 321
Vojnovic, S. N., 302, 305
Volegov, V. P., 748
Von Karman, T., 478
Von Mises, R., 452

Wakayama Works, 640
Walker, J. B., 92
Wallace, J. F., 587
Wallace, J. W., 226, 552
Wallquist, G., 551
Wang, C. T., 440
Wantanabe, H., 648
Ward, H. W., 40, 266
Wardle, R. V., 131
Warrington, D. H., 474
Washburn and Moen Co., 14
Weber, K. H., 548, 551, 588
Weinstein, A. S., 657, 706
Weissman, S., 721
West, C. H., 179
Westinghouse Electric Corp., 15
West Leechburg Steel Co., 14
Westwood, J., 14
Wettach, J. P., 79, 387, 390, 395
Whetzel, J. C., 256-258, 261, 268, 282, 315
White, C., 7
Whiteley, R. L., 733, 735-737
Whitton, P. W., 290
Wiesner, F., 61
Winchcombe, P. D., 768
Wildschutz, E., 551
Wilkinson, J., 12
Wilks, P. E., 179
Williams, R. C., 256, 315

NAME INDEX

Willison, R. M., 759
Wilmot, 14
Wilson, D. V., 742-743
Wilson, R. G., 92, 95
Winch, J. C. E., 676
Winkler, J. J., 81
Winsper, C. E., 467
Wise, D. E., 733-735, 737
Wistreich, J. G., 252, 663, 676, 687
Witner, D. A., 724
Witt, F. A., Jr., 320-321
Wood, W. A., 453
Woodbury, F. A., 91
Woodthorpe, J., 735
Worcester Works, 14
Worthington, P. J., 460
Wusatowski, Z., 154
Wyle, C., 268

Yamanouchi, H., 258, 293
Yamasaki, D., 763
Yoemans, K. A., 659
Yonezaki, S., 759
Yorke, D., 158
Youngstown Sheet and Tube Co., 45

Zdanov, A. A., 748
Zeigler, W., 716
Zeitlin, A., 619
Zener, C. J., 459, 466, 471
Zhuchin, V. N., 321
Zimmer, J., 10
Zores, 6
Zorowski, C. F., 478, 485, 511, 515

Subject Index

Abrasion rail, 171
Acceleration of mill stands, 33, 43, 66
Accumulator, strip, 18, 43, 51
Additives for rolling lubricants, 255-258
A. G. C. (Automatic Gage Control) Systems, 18, 43, 52, 56, 63, 193, 225-231, 648-654, 692-693, 699
Aging, quench, 731-733
Aging, strain, 741-743
Air cooling, 352-358
Alkaline cleaning, 717-718
Allotropy of iron, 399-401
Alloying elements
    in iron rolls, 110-111
    in steel, 424-425, 429, 734
    in steel rolls, 110-111
Alloy iron rolls, 110
Alloy steels, 423-440
Alpha iron, 399-401
Aluminum
    alloys, rolling of, 290, 342
    coating of strip, 765-766
    coated strip, rolling of, 269-270
Amanton's law, 479
Ammeters, 189-191
Amplification factor, 97-98
Amplimeter, 138
Amsler machine, 302-303
Analog computer control system, 239-242
Analog-to-digital conversion, 106, 219
Angle of bite in rolling, 244-245
Anisotropy, 736-737
Annealed stained edges, 748
Annealing, 723-726
Annealing, box, 19, 717
    box bottom, 24
    continuous, 43
    facilities for, 726-731
    of cast steel rolls, 121, 125
    textures, 733-734
Antioxidants, 253

Application
    of rolling force models, 556-559
    of rolling lubricants, 258-262, 387-397
Aqueous emulsions, 258-262
Arc of contact, 288, 340, 490-494
Armature
    current, 87-88
    inertia of, 608
    voltage, 87-88
Armco iron, 460, 463
Arrhenius equation, 467
A.S.T.M. grain size, 401-402
"As-U-Roll" shape control system, 40
Audio energy, 467-468
Austenite, 142, 426-434
    decomposition of, 430-434
    formers, 427-430
    isothermal formation of, 427-430
Austenitizing temperature, 120, 426-427
Automatic controls
    for decelerating and stopping reversing mills, 224-225
    for extension in temper rolling, 222-223, 231
    for sequencing, 234-236
    for strip entry angle, 232
    for thermal roll crowns, 707-708
Automatic roll changing, 25, 75, 77
Average rolling index, 302

Backing shafts, 37-39
Backup rolls, 7, 26-27, 32, 41-42, 66, 362-363
    bearings for, 30, 41, 66-67, 174
    bending of, 697-700
    cast iron, 14
    chocks for, 28, 31, 171-172
    driven, 26, 33, 46, 66
    eccentricity of, 46, 332
    sleeves for, 118-121

SUBJECT INDEX

Backup rolls (continued)
    spalling of, 112, 158, 168-170
    square, 66, 112
    temperatures of, 365-366
Bacteriacides, 253
Bainite, 123-127
Balance
    energy, for torque calculations, 585-587
    thermal, for single stand mills, 372-378
Balancing
    of rolls, 33, 35, 72-73
    of spindles, 82
Barrel, roll, 64, 109
Bars, 2, 6, 9
Bar spacers, 26
Basebox, 757-758
Base weights, 757-758
Bases, annealing, 726-727
Batch annealing, 726-729
B.C.C. (body centered cubic) structure, 450-451, 720
Beam mill, 10
Beams, 6, 8, 70
Bearings
    antifriction, 109
    backup roll, 30, 41, 66-67, 172
    concentrically mounted, 37
    fabric, 67
    lubrication of, 39, 67-68, 338
    "Morgoil," 67, 109, 172, 181-184, 186
    oil film, 67
    outboard, 51, 101
    pinion, 83-84
    plain, 67, 83
    roller, 14, 67-68, 83, 95, 172-181
    roll neck, 14, 25
    tapered roller, 172, 174
    thrust, 67
    sleeve-type, 29
    white-metal, 67
    workroll, 32, 36, 41, 64, 66
Bedplate for mill stand, 70
Belt wrapper, 46, 51, 101-102
Bending
    of backup rolls, 30, 252
    of workrolls, 30, 63, 66, 153, 693-704
Beta-ray gages, 210-212
B-H curves, 438-773
Biot number, 347, 368
BISRA
    magnetic shape sensor, 687-690
    mechanical shape sensor, 676-679

Bite, roll, 18, 39
Black plate, 19, 56
Bland and Ford's approximate theory, 511-515
Bloch wall, 438
Blooming mill, 10
Bode diagram, 652
Body-centered-cubic structure, 399-401
Bore, roll, 166
Boundaries, grain, 449-451
Boundary lubrication, 243
Box, annealing, 19, 24, 717
Box, coil, 46, 99
Bridgeman correction factor, 469-470
Bridle rolls, 96-98
Bright annealing, 774-775
Build-up on strip, 660-661
Built-up rolls, 126-129
Burgers vector, 415-418
Burnishing of rolls, 170-171
Burn-off characteristics of lubrications, 329-331, 718

Camber, of strip, 655-657, 663, 665
Can-making, 20
Cans, drawn and ironed, 20
Carbide, iron, 424-432, 723, 731-733
Carbide rolls, 38, 64, 109, 342
Carbon chain length, 314
Carbon residues, 264, 718
Carburization, 415
Cast housings, 5
Castigliano's theorem, 69
Casting, 407
Cast-iron rolls, 109, 116
Cast-steel backup rolls, 109
C.C.T. (continuous cooling transformation) diagram, 125-127, 430-431
Cells, load, 30-31, 196-202
Cells, within grains, 422-423
Cementite, 424-432
C-hooks, 103-104
Changes in rolling solutions, 273-275
Changing rolls, 25, 33, 75-77
Charcoal forges, 9
Chill zone, 115-118
Chocks, roll, 28, 30-31, 66, 73, 171-172
Chrome-steel rolls, 110
Circulating oil systems, 389-397
Classification of products by dimensions, 16
Cleanability of lubricant, 329-331
Cleaning, electrolytic, 19, 717

Closed housings, 26
Cluster mills, 23, 35, 41, 61, 79, 266, 363
Clutches, 3, 15, 89, 101
Coated products
    manufacture of, 754-769
    rolling of, 269-270, 321, 326-328
Coatings, roll, 270-272
Coefficient of friction, 97-98, 244-247, 286-302
    dependence on mill speed, 316-322
    dependence on reduction, 316-322
    dependence on work-roll diameter, 316-322
    measurement on bench machines, 302-306
    measurement under rolling conditions, 286-302
Coefficient of heat transfer, 353, 358-362
Coercive force, 438-439
Coil
    box, 46, 99
    breaks, 748
    buggy, 103
    carriers, 103
    diameter measurement, 106, 215-217
    end detection, 62
    length measurement, 215-217
    set, 665-668
    skids, 51
    tracking, 53
    weight, 46
Coilers
    design of, 100-102
    hot strip, 15
    power requirements for, 103
Coiling speed, 48, 103, 370
Coiling temperature, 19, 368-382
Coils, hot-rolled, 409-410
Cold-rolled strip, 14, 15
Combination mills, 22
Compensating for roll deflection, 153-154
Compliances, 444
Components of rolling mills, 64-108
Composition of rolls, 109-111
Composition ranges of steels, 402-406
Compound-wound d.c. motors, 87
Computer control, 48-49, 53, 106, 187, 235-242
Concentration of rolling solution, 204-205
Concentrically mounted bearings, 37
Conduction, thermal, 340, 342, 345-352
Cone, speed, 43, 608
Consoles, 48, 187-188, 222
Constant, Lamé's, 445

Constrained yield strength, 454-455
Contact, arc of, 288, 490-494
Contact gages, 207-208
Continuous
    annealing, 43
    casting, 407
    mills, 8, 17, 43, 50-52
    shearing lines, 751-753
    stretcher levelling, 711-713
Control equations, 610-654
Controls
    computer, 48, 53, 222-223, 235-242
    crown, 39
    entry angle, 231-232
    location of, 187-188
    preprogrammed, 62
    strip guidance, 232-234
    strip shape, 40, 690-708
    tension, 38, 52
    Ward-Leonard, 36
Conversion
    a-c frequency, 95
    a-c to d-c, 93-94, 107
    analog-to-digital, 106
Coolants, 18, 391
Cooling
    curves, 426-431
    of rolls, 18, 46, 51, 79, 332-338, 352-358, 382-384
    of strip, 46, 332, 366-378
    sprays, 384-387
Cope neck of roll, 118
Core loss, 438-440, 772-773
Cores, 57, 117
Corrosion, 332
Corrugated products, 9
Cottonseed oil, 18, 256
Couplings, 5, 79-82, 85-86
Cracking, season, 118, 156, 159-160
Cracks, 166, 415
Criteria
    for evaluating rolling lubricants, 284, 300-302, 306, 313-316
    for yielding, 452-453, 455
Critical
    temperatures, 123, 426-428
    shear stress, 449
Crossbow, 663-644, 668-670
Cross breaks, 748
Cross rolling, 706-707
Cross ties, 70-71, 73
Crown of rolls, 33, 39, 131, 153, 332
Crystal lattice defects, 414-418

SUBJECT INDEX

Crystallographic
  directions, 410-413, 473-474
  planes, 410-413
  symmetry, 412
  texture, 736-738
Cube-on-corner texture, 736-737
Cube-on-edge texture, 773
Cube-on-face texture, 736-737, 773
Curie point, 438
Curl, 664-668
Current regulators, 38
Current transformers, 190-191

Data accumulation systems, 48-49, 106, 217-223
D.C.R. (double cold reduction) mills, 19, 23
Decarburization, 415
Deceleration of mills, 33, 43, 66, 224-225
Decorative coatings, 766-771
Defects in crystals, 414-418, 477
Defects, shape, 664-673
Defects, surface, 746-749
Deflection, roll, 31, 150-154, 296, 482-494
Deformation, 58, 141, 448-449, 451-453, 475-476, 481-482, 569, 570
Degassing, 119
Degreasing, 717
Delta iron, 426
Densities of steels, 340, 406
Design of headers, 384-387
Detergents, 264, 717
Diameters
  of mill rolls, 32, 41, 52, 111-113, 133, 296, 317, 332
  of roll necks, 113
Diatomaceous earth, 395
Dies, rotating, 27
Dimensions of mill rolls, 133
Dimensions of units, 348
Direct application lubricating systems, 46, 57, 57, 59, 79, 387-389
Directions, crystallographic, 410-413
Direct methods for computing rolling force, 526-540
Dislocations, 414-418, 448-449
Dissipation of energy in rolling, 332-333
Double cold reduction, 23, 56-57, 749-750
Double pouring of rolls, 117-118
Doublings, 746
Drafting, 49, 335
Drag generators, 100
Drag neck of roll, 118
Drawing, 28

Drive power requirements, 20, 42, 52, 57, 59, 332, 568, 603-610
Drives
  backup roll, 26, 33, 46
  work roll, 37
  electric motors for, 27, 332
Dry friction, 249-252, 340
Dry rolling, 249-252, 341, 479
Dynamic behavior of tandem mills, 642-654

Earth
  diatomaceous, 395
  Fuller's, 394
Eccentric bearing sleeves, 29, 332
Eccentricity, backup roll, 46
Edge dislocation, 414-418
Efficiency of the rolling process, 591-595
Elastic deformation of rolls, 150-154, 246, 296, 338, 482-494
Electrical
  discharge machining, 132
  drives, 27
  generators, 13
  motors, 13, 27
  resistance, 415, 440
  steels, 61, 772-773
Electrogalvanizing, 763
Electrolytic cleaning, 19, 717
Electromagnets, 99, 103-104
Electrons, 398-400
Electro-plating, 754, 757-758
Electro-tinning, 757-758
Elements, transition, 398
Elongation, 49, 241-242, 569-570
Embossing strip, 769-771
Emissivity, 345
Emulsifiers, 255, 216
Emulsions
  changes in, 261, 273-275
  concentration of, 204-205
  rolling oil, 204-205, 258-262, 361-362
  stability of, 261, 328-329
End scrap, 61
Energy
  balance in rolling, 585-587
  deformation, 58, 69, 475-476
  frictional, 79, 341-342
  grain boundary, 419
  requirements in rolling, 44, 49, 332-335, 568-610, 749-750
  stored, 476-477
  strain, 69

Engines, 9, 12
Entry angle factor, 324
Equilibrium diagrams, 402-403, 425-428, 434-437
Eutetics, 426-428
Eutectoid steels, 426-428
Evaluation of rolling lubricants, 283-331
Exciters, 88, 188
Exponent, strain hardening, 519-526
Extension, 231, 739-741
Extensometers, 212-214
Extreme pressure additives, 255-258

Face of roll, 111
Failures of rolls, 156-160, 166-167
Fatigue damage to rolls, 160-167
Fatty acids, 255-257
Fault, stacking, 418
F.C.C. (face centered cubic) structure, 142, 399-401, 450-451
Feathering, 748
Feet, housing, 70-71
Ferrite, 424-434, 719
Ferromagnetic domain, 438
Ferromagnetism, 438
Field rheostats, 49
Field weakening, 87
Figures, pole, 413
Filtration of lubricants, 388-397
Finish
  surface, 20, 134-138, 775-776
  of rolls, 66, 134-138
  of strip, 134, 275-280
Finishing stands, 8
Firecrack transfer marks, 747
First frictional characteristic, 323-328
Flaked graphite iron rolls, 109
Flaking of forged steel, 118
Flatness of rolled products, 21, 655-657, 700-704
Flattening, 148, 482-494
Flexible couplings, 85
Flinger rings, 185-186
Flow rate, 202-205
Flying micrometer, 207-208
Flywheels, 11-12, 66
Foil mills, 20, 59-61, 750
Foils, 59-61, 269, 750
Force, coercive, 439
Force rolling, 18, 30, 33, 58, 105, 196-200, 248-252, 296-302, 494-495

Ford and Bland, approximate theory of, 511-515
Forged rolls, 64, 109, 118-129
Forges, charcoal, 9
Forward slip, 249, 596-603
Foundations, mill, 71
Four-high mills, 8, 14, 23, 26, 41
Fourier number, 347
Frank-Read sources, 448-449
Free fatty acids, 306, 314, 329
Frenkel defect, 414
Friction
  coefficient of, 287-302, 480, 504-505
  hill, 244, 248, 490-491, 495-511, 533-537, 574-584
  in roll bite, 243-252, 332
  sticking, 479
Frictional characteristics, 323-328
Frictional energy, 79, 133, 332, 335, 341-342
Frohling low-expansion mill, 28-29
Full center, 664, 670
Full edges, 664, 670-671
Fuller's earth, 394
Furnaces
  annealing, 726-731
  arc, 118, 407
  car-bottom, 120
  normalizing, 728-729
  open-hearth, 116, 407
  oxygen steelmaking, 407
  reheat, 407
  reverbatory, 116

Gage
  changing, dynamic, 53
  control, automatic, 43, 46, 52, 63, 225-231, 648-654
  minimum, 248-249, 284-287
  Manufacturers' Standard, 744-745
  numbers, 743-745, 764
  tolerance, 49, 655-657
  United States Standard, 744
  variations, 30
  weights, 765
Gages
  continuous contact, 207-208
  radioisotope, 210-212
  strain, 197-198
  X-ray thickness, 57, 62-63, 209-210, 226
Gaging, Sims' method of, 205-207
Galvanized products, 43, 761-764

SUBJECT INDEX

Gamma iron, 399-401, 425-432
Gamma rays, 210-212
Gap, roll, 29-31, 33
Gears, 3, 41, 74, 79-82, 84-85
Gear-type spindle couplings, 81
Generators, 13, 100
Gold, 11
Grab, 103-104
Graetz number, 347
Grain
    boundaries, 418-423, 449, 474
    growth, 477
    oriented electrical sheets, 773
    size, 401-402, 421-423, 437, 460, 722-723
    structure, 418-423
Grammer oiler, 258-260
Grashof number, 347-348
Grinding of rolls, 19, 31, 61, 129-133
Grit blasting, 132-133
Grooved rolls, 3, 6
Guidance of strip, 232-234
Guides, 10
Guttering, 664, 668-670

Halogen tinning lines, 757
Hammer, tilt, 6
Hand-powered mills, 1, 11
Hardening of rolls, 122, 127, 144
Hardening, strain, 468-475
Hardness
    of rolls, 123, 138-140
    of strip, 54, 733
    testing, 138-140
H-beams, 8
Header design, 384-387
Heat
    balance, 367, 372-384
    exchangers, 391
    transfer, 345-384
    treatment, 121-129, 717
Herringbone, 664
High-carbon steels, 403
Hill's method of computing torque, 579-582
History of rolling, 1-22
Hitchcock's equation, 246, 484-494, 579
Holes, 746
Hoods, 46
Hooked edges, 668-670
Hooke's law, 444
Horizontal deflection of workrolls, 702-704
Hot-dip processes, 754-756
Hot-rolled coils, 15, 409, 657-661

Hot rolling, 2, 14, 26, 407-409
Hot-strip mill, 8, 408
Hot-strip scratches, 747
Housings, mill, 5, 26, 30-31, 37, 41, 49, 56, 68-70
Hydraulic components of mills, 25-26, 28-31, 39, 42, 72, 75, 196, 202-204, 220
Hydronamic rolling lubrication, 243-244, 480
Hydrogen, 118
Hydrostatic pressure, 411
Hypereutectoid steels, 403
Hypoeutectoid steels, 403
Hysteresis, 338, 438-439

I-beams, 6
Idler rolls, 26
Idling losses, 54
Indicators, roll opening, 42
Indices, Miller, 410-413
Induction heating of rolls, 124, 128
Inertia, of armatures, 608
Ingot iron, 400
Ingots, 3, 120
Instrumentation, 19, 56, 105-106, 187-242
Inter-atomic distance, 400, 411-412
Intermediate rolls, 37-38
Interstand tension, 15
Interstitialices, 414
Iodine value, 265, 313, 329
Iron, allotropy of, 339-401
    alloys, 423-440
    enamelling, 724
    gamma, 339-401
    lattice parameters of, 400
    properties of, 398-402
    quantum numbers of, 399
Iron-carbon equilibrium diagram, 403, 425-428
Iron-chromium-carbon equilibrium diagram, 435
Iron-chromium equilibrium diagram, 436
Iron-chromium-nickel equilibrium diagram, 436
Iron-nickel equilibrium diagram, 436
Isothermal formation of austenite, 427-430
Isothermal transformation (TTT) diagram, 430-431
Isotopes, 210-212
Isotropic materials, 477
I unit of shape, 674

Jacks, hydraulic, 28-29
Jaws, gripping, 33
Jortner's analysis, 515-519
Journals, 83

K (stress factor), 154-156
Keeper plates, 68
Knives, air, 762-763

Lamé's constant, 445
Latches, 68
Lattice
  crystal, 399-401, 410-423
  defects, 414-418
  parameters, 400-401, 412
Lauth mill, 3, 12
Lead
  ingots, 3
  sheets, 1-3
Ledeburite, 426-428
Levelling, 51, 709-715
Lever arm method of torque calculation, 570-574
Life, roll, 42, 167-171
Lifting tables, 6
Liners, 172
Load bearing capacity, 302-306
Load cells, 30-31, 196-202
Long ternes, 765
Losses, running, 54, 337-338
Low-carbon steels, 403
Lower yield point, 457
Lubricating systems, 46, 59, 79-80, 105-106, 387-397
Lubrication
  bearing, 67-68, 179-181
  boundary, 243
  cold rolling, 243-331, 704
  hot rolling, 409
  hydrodynamic, 243-244, 314-315
  "Morgoil" bearing, 67, 181-184, 186
  plastohydrodynamic, 540-548
  pinion, 83
  spindle-coupling, 81
Lubricants, rolling, 18, 35, 39, 45, 79-80, 243-331, 717
Luster of rolled strip, 247, 275-280

Magnetic
  alloys, 438-440
  clutches, 89
  strip shape monitors, 685-690

Magnetic (continued)
  susceptibility, 438
Magnetization curves, 438
Main drives, 332, 568
Mandrils, 99-102
Manganese, 118
Manufacturers' Stangard Gage, 744-745
Martensite, 123, 142, 430-434
Martensitic
  roll surfaces, 123
  transformation, 142, 432-434
  stainless steels, 434-435
M_D, M_F, and M_S temperatures, 433
Measuring
  coil dimensions, 215-217
  roll diameters, 133
  roll finish, 134-138
  roll temperatures, 363-366
  rolling force, 196-200
  slip, 215
  spindle torques, 200-202
  stresses in the mill housing, 197
  yield stress, 444-458
Medium carbon steels, 403
Melting points of lubricants, 252
Metallic coatings, 321, 326-328
Micrometers, 133, 226
Microstructures, 120, 123-127, 142, 170, 421-438, 460-461, 719-721
Mills
  beam, 10
  blooming, 10
  combination, 22
  continuous, 8, 17, 25, 43, 50-52
  cluster, 23, 41, 61, 79
  double cold reduction, 19, 23, 56-59
  experimental, 284
  finishing, 10
  five-stand, 17, 44
  foil, 20, 23, 59
  four-stand, 17
  four-high, 8, 14, 23, 26, 30, 41
  Frohling low-expansion, 28-29
  hand-powered, 1, 11
  hot-strip, 8, 408
  hydraulic, 75-76, 193, 652-654
  Lauth, 3, 12
  multistand, 23
  planishing, 14
  plate, 11, 13
  prestressed, 28, 30
  primary, cold, 23

Mills (continued)
  primary, hot, 407
  reversing, 3, 7, 14-15, 18, 27, 29, 33, 35, 40, 61, 75, 372-378, 390-391
  Rohn, 36
  roughing, 10
  Schloemann, 32-34
  secondary cold reduction, 19, 23, 56-59
  Sendzimir, 18, 23, 36-41, 50-52, 61-63, 79, 266, 363, 376-378, 668
  semicontinuous, 8
  sheet, 17, 23, 43, 45, 264-265
  sheet-temper, 739-741
  single-stand, 21-23, 25, 27, 59, 61, 65, 227-228
  six-roll cluster, 35-36
  six-stand, 17, 44-45, 48
  skin-pass, 21, 25, 53
  slitting, 1, 2, 12
  soft, 32
  Steckel, 23, 27-28
  stiff, 32
  structural, 10
  tandem, 5, 8, 14-15, 17, 25, 41-44, 46-48, 52, 56, 77, 228-231, 236-242, 379-382, 387-397, 623-654
  Taylor, 702-704
  temper, 21-24, 53-55, 231
  tin, 12, 23-24, 45-47, 265-266
  three-high, 6, 10, 26
  three-stand, 17, 43, 618-623
  twelve-high, 37
  twenty-high, 38-39
  two-high, 5, 10, 12, 14, 23-25
  two-stand, 21-22, 616-618
  unidirectional, 27
  Y-, 34-35
Miller indices, 410-413
Mineral oils, 39
Minimum
  coefficient of friction, 244-247
  gage, 248-249, 284-287, 559-567
  work-roll diameter, 382-384
Mist lubrication of bearings, 179-181
Models
  rolling force, 53, 243, 296-297, 494-567
  heat transfer, 349-352
Modulus
  bulk, 446
  of rigidity, 152
  of elasticity, 151, 296, 445, 447, 454-456
Molds for casting rolls, 115-118

Moments, 69-70, 74
Monitoring
  hydraulic systems, 202-204
  rolling solutions, 202-204
  strip shape, 675-690
Mons, 674
"Morgoil" bearings, 67, 181-184, 186
Mosaic structure, 474, 720
Motor generator sets, 91-92, 187
Motor rooms, 13, 187
Motors
  a-c, 88
  d-c, 13, 42, 87-88
  drive, 48-49, 52, 57, 79, 606-610
  efficiency of, 603-604
  screwdown, 42
  selection of, 606-610
Moto-trace unit, 138
Mottle, 276-278
Multiple passes, 27

Nail rods, 9
Necks, roll, 9, 64, 67, 76, 118, 154-156
Neumann bands, 449
Neutral point, 275, 294, 525, 546-548
Normalizing, 122, 723-731
Nozzles, spray, 384-387
Nucleation and growth of austenite, 428
Nusselt number, 347
Nuts
  screwdown, 41
  Wheeler, 75

Oilfilm
  bearing, 67
  residues on strip, 253, 280-283
Oil-mist lubrication, 179-181, 258
Oils
  cottonseed, 18, 256
  mineral, 39, 257
  palm, 18, 254-255, 300
  palm-oil substitutes, 255-258, 264
  petroleum, 257
  pickle, 262-264, 389
  rapeseed, 256, 287
  soluble, 39
  tramp, 391
Open-coil annealing, 727-729
Optical shape sensor, 679
Orange peel surface defect, 273, 665, 748

Organic coatings, 270
Outboard bearings, 51
Overaging
Oxide films, 332

Pack-rolling, 5
Paints, 766-769
Palm-oil, 18, 255-258
Palm-oil substitutes, 255-258
Paper interleaving, 51
Paraffins, 257
Paramagnetism, 438
Partial dislocations, 418
Partitioning of frictional energy, 341-342
Pass-line, 95
Pass schedules, 27, 61, 594
Pay-off reels, 18
Pearlite, 424-432, 719
Peclet number, 347
Pendants, control, 106
Peritectic reaction, 426
Permeability, magnetic, 438-439
Peroxide value, 313-314, 329
Petroleum oils, 257
Perturbation equations, 612-654
pH value, 314, 329
Pickup on rolls, 255
Pickle oils, 262-264, 389
Pickling, 274, 409-410
Pinch rolls, 46
Pinion
  gears, 5, 79, 82-84
  stands, 82-84
Pitting, 747
Plain bearings, 67
Plane strain compression test,
Planes
  crystallographic, 410-413
  slip,
Planishing, 14
Plastic
  deformation, 141, 442-444, 448-449, 481-482
  strain ratio, 735-736
Plastohydrodynamic lubrication, 540-548
Plate mills, 7, 11, 13
Plates, 3, 5, 9, 28, 68, 78
Plating
  of rolls, 96
  of strip,
"Plating-out" of oils, 253
Plyrolling, 5

Point
  Curie, 438
  defects, 414-418, 477
  neutral, 275, 294, 525, 546-548
Poisson's ratio, 152, 445, 447, 454-455
Pole figures, 413, 451-452, 736-738
Polished rolls, 25
Polishing of iron plates, 3
Polymerization, 281-283
Positioning of rolls, 73-75
Posts, mill, 41, 68, 70-71
Pour points, 252, 306-310
Pouring techniques in roll manufacture, 116-118
Power
  drive requirements, 20, 42, 49, 54, 59, 106, 241, 568-610
  steam, 9, 12-13
  systems, static, 92-95
Prandtl number, 347
Preheating work rolls, 342-345
Preprogrammed control systems, 234-236
"Pressductor" load cells, 196, 199-202
Pressure
  distribution along arc of contact, 495-511
  gages, 202-204
  hydrostatic, 441
  multiplication factor, 537
  plates, 28
Prestressed mill housings, 28, 30-31, 56
Primary cold reduction, 19, 23
Profile of hot-rolled strip, 657-661
Profilometer, 135-138
Projections, stereographic, 412-413
Protection devices, 78
Pulpits, control, 108
Pure iron, 460

Quality of strip, 18
Quantum numbers of iron, 399-400
Quarter buckles, 665
Quench aging, 731-733
Quenching
  of cast steel rolls, 122, 124, 141
  properties of rolling emulsions, 312-313

Races, bearing, 67
Radial stresses, 146, 166-167
Radiation-type gages, 205, 209-212, 226
Rail-rolling mill, 6-7
Rails, guide, 10
Rate, strain, 480-481

SUBJECT INDEX

Ratio
  plastic strain, 736
  Poisson's, 152, 445, 447, 454-455
  work-roll diameter to strip thickness, 56
Recirculation of lubricants, 46, 79-80
Reconditioning forged steel rolls, 126-129
Recorders, 217-219
Recovery, 477, 723
Recrystallization, 451, 477, 723-726
Reduction gears, 85
Reels
  cold mill, 15, 27, 35-36, 232-234
  drives for, 609-610
  pay-off, 18, 232-234
  turntables for, 33
Regulators
  current, 38
  speed, 88
Replication techniques, 140-141
Residual stresses, 661-664
Residues, 253, 280-283, 717
Resistance, electrical, 415, 440
Reverbatory air furnace, 126
Reversing mills, 3, 7, 10, 14-15, 18, 27, 29, 33, 35, 40, 61-63, 75, 96, 224-225, 234-235, 390-391
Reynolds number, 347-348
Rheostats, 49, 88
Rockwell hardness testers, 139-140
Rods, 2, 9, 28
Rohn mill, 36
Rokes, 746
Roll
  balancing, 33, 35, 72-73, 172
  bearing losses, 337-338
  bearings, 36, 41, 66-67
  bending, 30, 63, 153, 252, 680, 693-704
  bite, 18, 39
  body, 64
  burnishing, 170
  changing 25, 33, 45, 52, 75-77, 332
  chocks, 28, 171-172
  coatings, 170
  compositions, 109-111
  coolant, 18
  cooling, 27, 46, 51, 79, 105, 352-384
  crowns, 33, 64, 253, 332
  deflection, 31, 150-154
  diameter, 32, 106, 111, 296-297
  eccentricity, 332
  face, 64, 111
  failures, 156-160

Roll (continued)
  finishes, 134-138
  flattening, 148
  forming, 771-772
  gap, 29-31, 33, 193
  grinding, 19, 34, 61, 109, 129-133
  life, 42, 167-171
  marks, 747
  necks, 9, 64, 67, 76, 109, 117
  neck bearing, 14, 25
  opening indicators, 42, 192-195
  positioning, 73-75, 192-195, 226
  sleeves, 114, 332
  spalls, 158, 169-170
  surface hardness, 138-140
  temperatures, 49, 79, 332, 338-341, 336-366
  wear, 141, 167-171, 270-272, 332
Rolled-in scale, 746
Roller bearings, 14, 67-68, 83, 172-181
Roller levelling, 709-711
Rolling
  dry, 249-252
  efficiency of, 591-595
  energy, 44, 569-570
  force, 18, 30, 33, 49, 58, 196-197, 248-252, 296-297, 494-567, 579
  index, 302
  heavy sections, 6
  hot, 2, 14, 26, 407-409
  lubricants, 18, 35, 39, 45, 79, 204-205, 243-332, 479-480
  pack, 5
  rails, 6
  schedules, 59
  secondary cold, 19
  silicon steels, 18
  speeds, 18, 317-319, 325, 332, 340
  stainless steels, 17-18, 50
  temper, 41, 341, 537-540, 567, 589-592
  textures, 451-452
Rolls
  air cooling of, 352-358
  backup, 7, 14, 26-27, 30, 41-42, 66
  bending of, 693-704
  bridle, 95-98
  burnishing of, 170-171
  carbide, 64, 109, 342
  cast-iron, 116, 160
  cast-steel, 64, 109, 116, 121-122
  chilled-iron, 9
  chrome-steel, 110

Rolls (continued)
  cooling of, 352-384
  dimensions of, 66, 109
  double-poured, 117-118, 145, 160
  grinding of, 129-132
  grooved, 3, 6
  idler, 26, 95-96
  intermediate, 37-38
  pinch, 46
  plain, 65
  plating of, 170-171
  polished, 25
  preheating, 342-345
  rough surfaces of, 64
  sleeved, 109, 114
  spalling of, 27
  tungsten carbide, 38
  work, 7, 14, 27, 33, 37, 41-42
ROLMAX system, 197
Rooms, motor, 13, 187
Rotating dies, 27
Rough roll surfaces, 132-133
Roughness of roll and strip surfaces, 134-138
Running losses, 54
Rust, 749

Safety devices, 70
Saponification value, 265, 306, 313
Saturated fatty acids, 255
Saybolt viscosimeter, 307-310
Scale, 746
Scaling factors
  for metallic coatings, 326-328
  for surface roughness, 326
Schedules, rolling-pass, 59, 61, 371, 594
Schlerscope, 138-139
Schloemann mills, 25, 32-34
Schmidt number, 347
SCR (silicon-controlled rectifier), 92-95
Scrap, end, 61
Scratches, 747
Screw dislocation, 414-418
Screw down
  hydraulic, 25
  mechanisms, 41-42, 53, 198
  motors, 42
Screws, 5, 24, 28, 30, 36, 73-75, 105-106, 652, 654
Seals, 184-186
Seams, 746

Season cracking, 118, 156, 159-160
Secondary cold reduction, 19, 749-750
Secondary cold reduction mills, 23, 56-59
Second frictional characteristic, 323-328
Semicontinuous mills, 8
Sendzimir mills, 18, 23, 36-41, 43, 50-52, 61, 79, 110, 266, 363, 376-378, 668
Sequencing, automatic, 62
Set-up, mill, 48
Shadow lines, 748
Shafts, backing, 37-39
Shape, strip, 20, 40, 56, 247, 252, 278-280, 332, 655-716
Shear
  modulus of, 445
  plane theory of flat rolling, 552-556, 587-588
  stresses, 150-154, 441, 453, 473-474, 505
Shearing, 750-754
Shears, 18, 51, 751
Sheet
  lead, 1-3
  mills, 23, 43, 45, 264-265
  products, 2, 15, 743-745
  tin, 2, 23
Shell, 118, 746
Shell, 4-Ball E.P. Tester, 304-306
Shims, 34
Shoe, 71
Shunts, 190
Side-trimming, 750-754
Silicon
  as a deoxidizer, 118
  steels, 18, 33, 61, 266-269, 772-773
Silver, 11
Sims' method of computing torque, 582-584
Sims' method of gaging, 205-207
Single-stand mills, 18, 23, 25, 27, 59, 61, 65, 227-228, 372-378
Six-roll cluster mills, 35-36
Six-stand mills, 44-45, 48
Skids, for coils, 51
Skimmers, 396
Sled, roll changing, 45
Sleeves
  backup roll, 109, 114-155, 120, 147, 332
  eccentric bearing, 29
  safety, 78
  use on coilers, 57
Slip
  forward, 215, 249, 290-295, 596-603

Slip (continued)
   line field theory, 453-454
   planes, 415-418, 450-451
   regulators, 90
   systems, 417-418, 449-451, 473-474
   velocity of, 505-506
Slipper-type universal spindles, 81
Slitters, 750-754
Slitting mill, 1-2
Sludge, 273-274
Soaps, 274-275
Soft mills, 32
Solid rolling lubricants, 252-253
Soluble oils, 39
Solutions, rolling, 49, 204-205
Spacers, 26
Spalling of rolls, 27, 112, 157, 160-166
Speed
   Cone, 43
   mill, 18, 27, 33, 48, 105-106, 340, 370, 480
   reducer, 85
   of response of roll positioning systems, 75
   regulators, 62, 88, 106
Spherodite, 430
Spindle,
   couplings, 79-82
   torque, 66, 69, 83, 133, 292-293, 568-592
Spindles, 79-82
Split molds, 115-116
Spray nozzles, 384-388
Sprays, 57, 766
Spread in rolling, 246
Spring, mill, 31-32
Stacking fault, 418, 423
Staining of rolled strip, 276-278
Stainless steels, 17, 18, 36, 39, 50, 61, 63, 266-269, 342, 434-437, 774-755
Stands, pinion, 82-84
Stanton number, 347
Static power systems, 92-95
Steam engines, 9, 12-13
Steckel mills, 23, 27-28
Steels
   carbon, 390, 402-406
   chrome, 110
   compositions of, 402-406
   electrical, 37, 61, 402-405, 772-773
   euctectoid, 403
   hypereuctectoid, 403
   hypoeuctectoid, 403
   sheet, 15

Steels (continued)
   silicon, 18, 33, 37, 61, 266-269, 390, 402-405, 772-773
   stainless, 17-18, 36-37, 39, 61, 63-64, 266-269, 342, 390, 402-406, 410, 434-437, 774-775
   tin-free, 20, 758-760
Step-heating, 120
Stereographic projections, 412-413
Sticker breaks, 748
Sticking friction, 479
Stiffness
   coefficients, 444
   of drive system, 609, 625
   of mill stand, 32
Stored energy, 476-477
Strain
   aging, 741-743
   energy, 69
   engineering, 443
   gages, 197-199, 290
   hardening, 460-475
   hardening exponent, 469-471, 721-722, 741
   pure shear, 442
   rate, 296, 332-335, 461-468
   rate effect, 296, 332-335, 480-481
   stretcher, 21
   tensor, 442
Strains, 332-335, 442
Stress
   biaxial, 441-446
   concentration factor, 154-156
   flow, 340, 474
   pure shear, 441, 473-474, 505
   residual, 125, 142, 145-147, 166, 661-664
   -strain relationship, 444, 451-453, 457-458, 468-475, 721
   tensile, 20
   tensor, 441
   thermal, 147-148
   uniaxial, 441, 446
   yield, 30, 46, 322, 332, 455-474
Stresses, 69-70, 112, 141-150, 160-166, 197, 228-290, 440-442
Stressometer, 683-685
Stretcher leveling, 711-713
Stretcher strains, 21, 748
Stretch of housings, 30, 49
Strip
   accumulators, 18, 43, 51

Strip (continued)
  coatings, 321
  cold-rolled, 14-15
  cooling, 336-378
  embossed, 769-771
  guidance, 232-234
  hardness, 54
  hot-rolled, 15
  painted, 766-769
  quality, 18
  luster of, 247
  temperature of, 366-378
  shape, 20, 247, 252, 278-280, 332, 655-716
  temper-rolled, 456
  tension, 38, 69, 105, 214-215, 705-706
  tracking, 51
Structural mills, 10
Sub-boundaries, 422-423
Substitute palm oils, 255-258
Superficial hardness test, 140
Surface
  defects, 273, 409, 746-749
  finish, 20, 715-716, 775
  finish, effect of lubricant on, 266-269, 275-280, 320-321
  finish, influence on formability and appearance, 775-776
  temperatures, 351
  tension, 310-311
Surfacount unit, 135-137
Surfindicator, 135-137
Susceptibility, magnetic, 415
Sweep, 665
Synchronous motors, 88-90
Systems
  automatic gate control, 18, 43, 52, 56, 66, 692-693, 699
  clutch and gearing, 74
  data, 217-223
  slip, 417-418, 449-451, 473-474

Tables
  lifting, 6
  roller, 3, 26
Tachometers, 191-192
Tallows, 255
Tandem mills, 5, 8, 14-15, 17, 41-43, 45-48, 52, 56, 77, 236-242, 272, 335, 368-372 379-382, 387-389
Taylor mill, 702-704

Television, 53
Temperature
  austeniziting, 120, 426-427
  coiling, 19, 368-382
  critical, 123, 426-428
  film, 353
  forging, 120
  roll, 49, 79, 338, 342-345, 352, 363-384
  roll-bite, 339-342
  strip, 351, 366-384
Temper rollings, 21, 23, 41, 53-55, 114, 212-214, 222-223, 231, 249-252, 341, 537-540, 567, 589-592, 739-741
Tensile stresses, 20, 680
Tensile testing, 455-458
Tensiometers, 38, 57, 214-215
Tension
  control of, 38, 52, 106, 108
  interstand, 15
  levelling, 713-714
  reels, 27
  strip, 38, 69
  surface, 310-311
Tensors, second rank, 440, 442
Terne coatings, 764
Tetrakaidecahedron, 419-420
Textures, 451-452, 720-721, 733-734
TFS (tin-free steel) products, 20, 758-760
Thermal
  aspects of the cold-rolling process, 332-397
  balance of rolling process, 372-378
  conductivity, 340, 342, 345-352
  crowning of rolls, 332, 338
  models for tandem mills, 368-372, 379-382
  shock, 343
Thickness gaging, 105, 205-212
Three-high mills, 6, 10, 26
Three-stand mills, 43
Thrust bearings, 67
Thyristor control, 58, 92-95, 188
Tie rods, 28
Tiger stripes, 277-278
Tilt, angle of, 418-419
Tilt hammer, 6
Tin
  coatings, 269-270, 326-328, 755-758
  -free steel, 20, 758-759
  mills, 23-24, 45-47, 265-266
  plate, 14, 332
  sheets, 2
  temper mills, 23-24

SUBJECT INDEX

Tolerances
    camber, 655-657
    flatness, 655-657
    gage of thickness, 49, 655-657
Torque
    delivered by d-c motors, 87-88
    equations, 568-592
    monitoring devices, 200-202
    spindle, 66, 69, 200-202
Tracking of strip, 51, 53
Transducers, 106
Transfer heat, 345-352, 358-384
Transformation, martensitic, 432-434
Transition elements, 398-399
Tresca's yielding criterion, 452-453
TTT (time-temperature-transition) curves, 125, 142
Tungsten carbide work rolls, 38
Turntable, reel, 33
Three-high mills, 6
Twelve-high mills, 38-39
Twenty-high mills, 38-39
Twin boundaries, 422-423
Twist (shape defect), 663-664, 670
Twist boundaries, 418-419
Two-high mills, 5, 12, 14, 23-24

Ultrasonic-energy, 467-468
Uncoilers, 99-100
Undershot water wheels, 9, 11
Units, dimensions of, 348
Universal mills, 13
Unsaturated hydrocarbons, 255-262, 265, 314
Upper yield point, 457
U.S.S.G. (United States Standard Gage), 744

Vacancies, 414-418, 477
Vacuum
    degassing, 119
    deposition, 754
Vapor degreasing, 717
Vector, Burgers, 415-418
"Vidimon" shapemeter, 683
Viscometers, 307-310
Viscosity index, 306-310

Viscosity of rolling oils, 306-310
Viscous mass, workpiece considered as, 548-552, 588-589
Voltmeters, 181-191
Von Mises' yielding criterion, 452-453, 455

Ward-Leonard control system, 36, 90-92
Water
    as a rolling lubricant, 254, 300
    break test, 331
    cooling, 358-362
    wheels, 9, 11-12
Waxes, 287
Wear, roll, 141, 167-171, 270-272, 332
Wedge assemblies, 31, 73
Weld detectors, 53
Welders, 17, 51-52, 410
Welding to recondition rolls, 127-128
Welds, 43, 51-52, 410
Wet-temper rolling, 21, 53, 251, 258
Wheeler nut, 75
Windows, 68
Wings of backup rolls, 171
Wobblers, 65, 80, 117, 122
Workhardening, 332, 468-475, 482, 519-526, 721-723
Workrolls, 7, 14, 27, 33, 41-42, 52, 63-64, 66, 171-172, 317, 342-345, 362-365, 482-494
Workpiece considered as a viscous mass, 548-552, 588-589
Worm gears, 41, 74
Wound rotor, 88-89
Wrapper, belt, 46, 51, 101-102

X-ray thickness gages, 57, 62-63, 209-210

Y-mill, 34-35
Yielding criteria, 452-453, 455
Yield-point drop, 743
Yield strength, 30, 46, 322, 332, 455-474, 721-722
Yield strength, constrained, 332, 454, 481-482
Young's modulus, 445, 447, 454-455

Zinc coating processes, 761-764